Lecture Notes in Computer Science 1200
Edited by G. Goos, J. Hartmanis and J. van Leeuwen

Springer
Berlin
Heidelberg
New York
Barcelona
Budapest
Hong Kong
London
Milan
Paris
Santa Clara
Singapore
Tokyo

Rüdiger Reischuk   Michel Morvan (Eds.)

STACS 97

14th Annual Symposium
on Theoretical Aspects of Computer Science
Lübeck, Germany
February 27 — March 1, 1997
Proceedings

Volume Editors

Rüdiger Reischuk
Medizinische Universität zu Lübeck, Institut fürTheoretische Informatik
Wallstraße 40, D-23560 Lübeck, Germany
E-mail: reischuk@informatik.mu-luebeck.de

Michel Morvan
Université Denis Diderot Paris 7, LITP-IBP
2, place Jussieu, F-75251 Paris Cedex 05, France
E-mail: morvan@litp.ibp.fr

Cataloging-in-Publication data applied for

Die Deutsche Bibliothek - CIP-Einheitsaufnahme

STACS <14, 1997, Lübeck>:
Proceedings / STACS 97 / 14th Annual Symposium on Theoretical Aspects of Computer Science, Lübeck, Germany, February 27 - March 1, 1997. Rüdiger Reischuk ; Michel Morvan (ed.). - Berlin ; Heidelberg ; New York ; Barcelona ; Budapest ; Hong Kong ; London ; Milan ; Paris ; Santa Clara ; Singapore ; Tokyo : Springer, 1997
(Lecture notes in computer science ; Vol. 1200)
ISBN 3-540-62616-6

NE: Reischuk, Rüdiger [Hrsg.]; GT

CR Subject Classification (1991): F, D.1, D.4, G.1-2, I.3.5, E.3

ISSN 0302-9743
ISBN 3-540-62616-6 Springer-Verlag Berlin Heidelberg New York

Typesetting: Camera-ready by author
SPIN 10549519 06/3142 – 5 4 3 2 1 0 Printed on acid-free paper

Preface

STACS, the Symposium on Theoretical Aspects of Computer Science, is held annually, alternating between France and Germany. STACS is organized jointly by the Special Interest Group for Theoretical Computer Science of the Gesellschaft für Informatik (GI) in Germany and the Special Interest Group for Applied Mathematics of the Association Française des Sciences et Technologies de l'Information et des Systemes (AFCET) in France. Its format is a two-and-a-half day conference (Thursday morning to Saturday noon) at the end of February each year, presently with parallel sessions in two tracks and plenary talks given by invited speakers.

STACS'97, the 14th in this series, is scheduled to be held in Lübeck from February 27 to March 1, 1997. Previous STACS symposia took place in Paris (1984), Saarbrücken (1985), Orsay (1986), Passau (1987), Bordeaux (1988), Paderborn (1989), Rouen (1990), Hamburg (1991), Cachan (1992), Würzburg (1993), Caen (1994), München (1995), and Grenoble (1996).

STACS has become one of the most important annual meetings in Europe for the theoretical computer science community. It covers a wide range of topics in the area of foundations of computer science. For STACS'97, 139 submissions from all over the world and most European countries were received. This year, for the first time papers could be submitted to STACS electronically. About two-thirds of all authors chose this way, and the vast majority of such papers arrived by the submission deadline, August 30, 1996.

Thanks to *Jochen Bern* and *Christoph Meinel* in Trier and *Gerhard Buntrock* and *Christian Schindelhauer* in Lübeck, who made the electronic submission procedure possible by setting up a submission robot in Trier, then forwarding the files to Lübeck and finally handling the files here.

The submitted papers address basic problems from many areas of computer science, in particular algorithms and data structures, computational complexity, automata and formal languages, structural complexity, parallel and distributed systems, parallel algorithms, semantics, specification and verification, logic, computational geometry, cryptography, learning and inductive inference.

The members of the program committee are: *Jean-Paul Allouche* (Orsay), *Jürgen Dassow* (Magdeburg), *Afonso Ferreira* (Lyon), *José Luiz Fiadeiro* (Lisbon), *Rusins Freivalds* (Riga), *Martin Kummer* (Karlsruhe), *Mirek Kutylowski* (Wrocław/Paderborn), *Christian Lengauer* (Passau), *Michel Morvan* (Paris, co-chair), *Laurence Puel* (Orsay), *Prabhakar Ragde* (Waterloo), *Rüdiger Reischuk* (Lübeck, chair), *Miklos Santha* (Orsay), *Val Tannen* (Philadelphia), and *Dorothea Wagner* (Konstanz). Every submission was carefully evaluated by several members of the program committee (typically five), partially using the help of colleagues who served as subreferees. The committee was impressed by the scientific quality of the submissions as well as the broad spectrum they cover.

The program committee met in Lübeck November 1 and 2, 1996. By unanimous vote 46 papers were selected for presentation. Because of the constraints imposed by the format of the conference, a number of good papers could not be included in the program.

We would like to thank the program committee for its demanding work in evaluating the significance and scientific merits of all submissions within such a short period of time between the submission deadline and the meeting. It turned out that electronic correspondence before the real meeting helped a lot to accelerate and support the decision process. The program committee has tried to do its best to make a fair selection, but their is no doubt that the "optimal solution" (if it exists at all) can only be approximated given the bounded amount of time and space. Our gratitude extends to the numerous subreferees who have assisted in this process (listed separately).

The proceedings of this conference as of all previous ones will be published in the series Lecture Notes in Computer Science by *Springer-Verlag, Heidelberg*.

We also thank the three invited speakers *Bernhard Steffen* (Passau), *Stephane Gaubert* (Rocquencort), and *Oded Goldreich* (Rehovot) for accepting our invitation and sharing with us their insights on some new and exciting developments in their areas.

Thanks to all members of the *Institut für Theoretische Informatik der Med. Universität zu Lübeck,* who have helped us to organize the meeting: *Gerhard Buntrock, Gabriele Claasen, Karin Genther, Jens Heinrichs, Andreas Jakoby, Christian Schindelhauer,* and *Stephan Weis.*

Further information about STACS'97 can be found in the WorldWideWeb on its homepage `http://www.itheoi.mu-luebeck.de/stacs97` (professionally installed and maintained by *Karin Genther* and *Andreas Jakoby*).

Thanks also to the various sources who have supported STACS'97: *Wissenschaftsministerium des Landes Schleswig-Holstein, Deutsche Forschungsgemeinschaft, European Community Directorate for Science, Research and Development, Hewlett Packard, Delcom Vertriebsgesellschaft, Commerzbank Lübeck, Computerfachhandel JessenLenz, Buchhandlung Dreier, Weingut Jakoby-Blümling,* and other organizations.

Lübeck, January 1997

Rüdiger Reischuk

Michel Morvan

List of Subreferees

Ablayev, F.
Albers, S.
Allender, E.
Alt, H.
Alur, R.
Ambos-Spies, K.
Ameur, F.
Andersson, A.
Annexstein, F.
Arvind, V.
Atallah, M.
Bäumker, A.
Barrington, D.
Barth, D.
Barzdins, J.
Bazgan, C.
Beigel, R.
Bender, M.
Benevides, M.
Bertet, K.
Berthome, P.
Best, E.
Bezrukov, S.
Biehl, I.
Blondeel, S.
Boucheron, S.
Boudet, A.
Brandes, U.
Brandstädt, A.
Brlek, S.
Brzoska, C.
Buntrock, G.
Buss, J.
Camenisch, J.
Carmo, J.
Caromel, D.
Chakravarty, M.
Charatonik, W.
Chlebus, B.
Chretienne, P.
Ciancarini, P.
Condon, A.
Contejean, E.
Cori, R.
Costa, F.
Couveignes, J.
Creignou, N.
Cypher, R.
Czumaj, A.
Dahlhaus, E.
Dal Cin, M.
Dehne, F.
Delorme, C.
Denise, A.
Diekert, V.
Diettrich, W.
Dietzfelbinger, M.
Droste, M.
Durand, B.
Dutheillet, C.
Esparza, J.
Fachini, E.
Fagin, R.
Farach, M.
Favaron, O.
Fellows, M.
Felsner, S.
Fernandez de la Vega, W.
Ferreira, F.
Fischer, M.
Fleury, E.
Foessmeier, U.
Fortnow, L.
Fourré, J.-Y.
Fraigniaud, P.
Fuhrman, C.
Gąsieniec, L.
Gavoille, C.
Geisler, R.
Genther, K.
Geser, A.
Giavitto, J.-L.
Girard, J.-Y.
Gorlatch, S.
Gouyou-Beauchamps, D.
Grädel, E.
Gupta, A.
Gustedt, J.
Guérin, I.
Hähnle, R.
Hajnal, P.
Hambrusch, S.
Handke, D.
Hartlieb, S.
Hemaspaandra, L.
Hertrampf, U.
Hill, J.
Hivert, F.
Hölldobler, S.
Hofbauer, D.
Hoffmann, F.
Hofmeister, T.
Holzer, M.
Homer, S.
Hromkovič, J.
Hühne, M.
Hurd, L.
Ibel, M.
Jakoby, A.
Jantzen, M.
Jenner, B.
Jukna, S.
Jurdziński, T.
Juurlink, B.
Kanarek, P.
Kannan, S.
Kapron, B.
Karaiwasoglou, E.
Karpinski, M.
Kenyon, C.
Kesner, D.
Kirchner, C.
Klein, R.
Köbler, J.
Kozen, D.
Krause, M.

Kriegel, K.
Krob, D.
Kunde, M.
Kutrib, M.
Laforest, C.
Lauer, H.
Lautemann, C.
Lehmann, D.
Li, M.
Liebers, A.
Liśkiewicz, M.
Lopes, A.
Lopez-Ortiz, A.
Loryś, K.
Lubiw, A.
Lucks, S.
Lueling, R.
Lukovszki, T.
Lutz, J.
Maass, W.
Maggs, B.
Marché, C.
Marcinkowski, J.
Mattern, F.
Maurer, U.
Mayordomo, E.
Mayr, E.
Mazoyer, J.
McKenzie, P.
Meinel, C.
Mendler, M.
Meyer auf der Heide, F.
Michel, P.
Miltrup, M.
Mongenet, C.
Mostowski, A.
Muthukrishnan, S.
Mysliwietz, P.
Nestmann, U.
Niedermeier, R.
Nishimura, N.
Novelli, J.-C.
Oberschelp, W.
Oesterdiekhoff, B.
Ogihara, M.
Otto, M.
Pelc, A.
Petit, A.
Phan, H.
Poizat, B.
Porto, A.
Priebe, V.
Reagan, K.
Reichel, H.
Reinhardt, K.
Reisig, W.
Resende, P.
Richter, M.
Rieping, I.
Rolim, J.
Rosaz, L.
Rousset, M.-C.
Rozoy, B.
Sangiorgi, D.
Schäffter, M.
Scheideler, C.
Schindelhauer, C.
Schlink, S.
Schmeck, H.
Schmitt, P.
Schnitger, G.
Schöning, U.
Schröder, K.
Schulz, A.
Schwentick, T.
Seese, D.
Seidel, R.
Sharell, A.
Sibeyn, J.
Simon, H.
Sobrinho, L.
Spirakis, P.
Stachowiak, G.
Steffen, M.
Stephan, F.
Stern, J.
Strothmann, W.
Tan, S.
Tavangarian, D.
Thiemann, P.
Thierauf, T.
Thierry, E.
Thomas, W.
Tison, S.
Toran, J.
Treinen, R.
Turner, D.
Unger, W.
Upfal, E.
van Emde Boas, P.
Vasconcelos, V.
Veiga, P.
Verissimo, P.
Viennot, L.
Vöcking, B.
Voisin, F.
Vollmer, H.
Waack, S.
Wagener, H.
Wagner, F.
Wagner, K.
Wanka, R.
Watanabe, O.
Wedler, C.
Wegener, I.
Weihe, K.
Weiler, F.
Weinstein, S.
Weis, S.
Welzl, E.
Wiehagen, R.
Wierzbicki, T.
Wilke, T.
Wirsing, M.
Wolff, A.
Worsch, T.
Zielonka, W.
Zissimopoulos, V.

Table of Contents

Invited Talk

Algorithms I

Automata Theory I

Algorithms II

Structural Complexity I

Probabilism

Specification and Verification

Boolean Functions

Logic and Learning

Invited Talk

Automata Theory II

Structural Complexity II

Complexity Theory I

Parallel and Distributed Systems I

Complexity Theory II

Parallel and Distributed Systems II

Complexity Theory III

Parallel Algorithms

Algorithms III

Structural Complexity III

Algorithms IV

Automata Theory III

Invited Talk

Unifying Models

Bernhard Steffen

Lehrstuhl für Programmiersysteme
Universität Passau
Innstr.33, D-94032 Passau (Germany)
steffen@fmi.uni-passau.de

Abstract. In this paper we illustrate the unifying power and flexibility of an operational model-based approach by treating the problem dilemma of lack of consistency between the various description methods used in software systems design. The success of this approach strongly relies on the definition of adequate *unifying model structures*, which must be powerful enough to capture the interference potential between the different description methods, while remaining simple enough to support (automatic) verification, the key for formal methods to enter industrial practice.

Keywords: abstraction, architecture, behaviour, (in-)completeness, consistency, decidability, expressivity, (temporal) models, partial evaluation, intermediate language, operational semantics, process, refinement, temporal logic, tools, transition system, verification.

1 Motivation

The exploding size and conceptual complexity of programs or, even more strikingly, distributed systems have led to the development of numerous descriptions methods typically aiming at rigorous structured and/or focussed system design. Each of these description methods or paradigms has its particular application profile – severe strengths and weaknesses – which forces people to employ several methods or paradigms in the course of the overall system development in order to capture *all* vital aspects. However, as necessary as this development is, as unsatisfactory is its current state of the art: there is still no sufficient control guaranteeing the *consistency* of intentionally complementary descriptions of one and the same system. In fact, already single description methods usually lack a coherent semantic description, and the situation gets increasingly worse with the number of considered paradigms, even when they are all integrated in a single language, environment or tool. This situation is not too surprising when considering the heterogeneity of focus and aspects addressed in system descriptions:

– *control constraints*: these operation-oriented descriptions are often directly formulated in terms of automata or transition systems, and are therefore extremely close to the unifying models developed below.

- *architectural, network or process descriptions*, coming in two flavours:
  - *syntactic overhead*, i.e. structure that is meant to clarify the system's structure for the designers, like hierarchy and modularization, but which is irrelevant for the actual run time behaviour of the system as well as for its realization, and
  - *realization constraints*, like degree of parallelization, which each realization must satisfy, and which may well be observable, e.g. in terms of efficiency.
- *communication or synchronization structures*, typically used for controlling systems, where they are of primary importance and therefore observable.
- *abstract liveness and safety properties*, guaranteeing the progress of a reactive system and excluding the possibility of a malfunction of a system. Also these properties are often considered observable.

Unfortunately, in reality, the different aspects interfere in a highly non-trivial way, and even many single-paradigm description methods do not have a sound semantic basis, which often leads to severe mistakes and misunderstandings. Thus there is a strong need for a sound (abstract) semantic basis for description languages, including those that are usually only explained intuitively, and/or comprise multiple paradigms, which are of increasing practical interest [ZaJa96].

In this paper we illustrate the unifying power and flexibility of an operational model-based approach by treating the problem of lack of consistency between the various description methods used in software systems design. The success of this approach strongly relies on the definition of adequate *unifying model structures*:

> *The models must be powerful enough to capture (an adequate amount of) the interference potential between the different description methods, while remaining simple enough to support (automatic) verification.*

Clearly, there is a trade-off between expressive power and manageability, which we resolve here by giving priority to simplicity. However, already the simple modelling proposed here is rather powerful, and there are straightforward ways for extensions (Section 2.4).

Our notion of consistency is based on an operational intuition: two descriptions are consistent if there exists a concrete system whose operational behaviour satisfies the requirements of both of them. In the basic setting considered here this is formally realized in terms of temporal model structures, which can be regarded as combinations of Kripke structures and finite state transition systems [Stir95]. In fact, combined with adequate abstraction, refinement, and (non-standard) interpretation, this delivers an intuitive and flexible framework: varying the granularity or the concept of observation, i.e. of what is considered a visible 'action' of the system, these structures are able to cover a wide range of consistency aspects even for distributed systems.

Whereas the need for abstraction and refinement for adapting descriptions of different granularity is quite clear, *non-standard interpretations* of the model structures are the key for obtaining a satisfactory framework for consistency checking in the presence of true *heterogeneity*: establishing a match between the imposed constraints may required to interpret the same model structure differently in different contexts (Section 2.2). This goes beyond standard semantic approaches that merely abstract from the representations chosen for the various description methods.

Striking is the elaborate ***tool support*** for operational methods: there exist powerful decision tools based on model checking and model construction algorithms, which in the case of failure detection even provide intuitive diagnostic information for its cause. These tools can in particular be applied to decide consistency for the modelling considered here, which dramatically reduces the required expertise for a competent use of multi-paradigm description methods. This reduction is a necessary precondition for a wider dissemination of formal techniques, which leads us to stress the 'tool builder's perspective' in the remainder of the paper.

The proposed consistency modelling and checking is only a first step towards a general theory for a fully consistent and tool-controlled specification framework which flexibly covers various paradigms. Nevertheless, we are convinced that the current proposal is a major step towards reliable and consistent system development with a strong practical impact: similar to type checking, which dramatically increased reliability of programming while being (even in its intent) far from full verification, using a consistency model as proposed here will help to automatically detect misconceptions already in early design phases.

It is exactly this winning principle of type checking, ***fully automatic systematic fault detection, rather than full verification***, which we consider also as the key for model checking to become a widely used technique [StMa96,StMa96a]. We believe that future system design tools without model checking facilities will no longer be accepted as state of the art (for type checking the similar statement has been already true for a long time), and we are aiming at a similar success also for unifying model-based verification methods like consistency checking.

The rest of the paper is organized as follows: after introducing the notion of unifying models in section 2, where we focus on their possibles roles and in particular on their application to consistency checking, we address the corresponding tool support in Section 3. Finally, we discuss intentional and technical differences of some structurally related approaches in Section 4, and give our conclusions and perspectives in Section 5.

While reading, one should keep the intent behind this paper in mind: sketching a complex application scenario where various theories interact. Moreover, according to the addressed audience, the focus of the discussion is on the interplay between theoretical approaches, while the application side merely provides the 'playing field'. Thus software engineering aspects are only roughly sketched, and the corresponding bibliography is extremely selective.

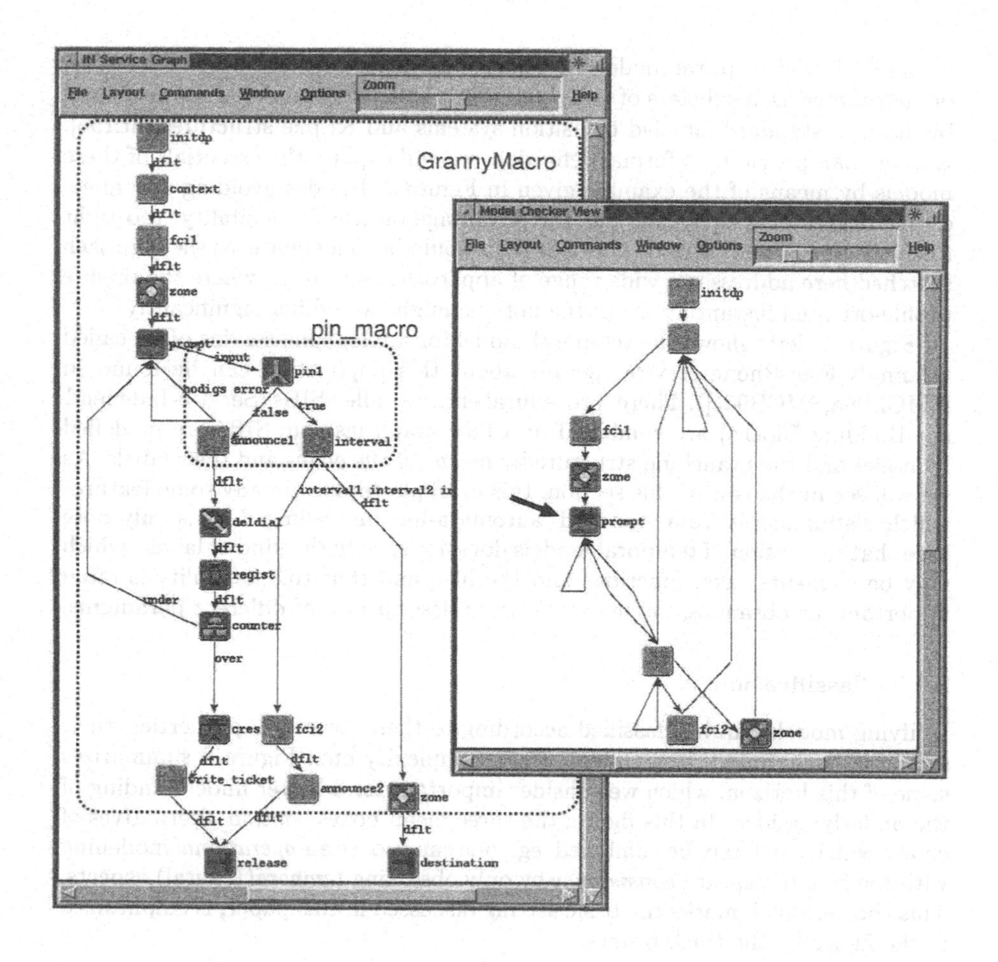

Fig. 1. The Granny's Free-Phone Service with Error View

2 Unifying Models

Necessary precondition for our approach is an agreement on a 'primary' semantic level, on which all description methods can be compared. In our case this means that there must be agreement about what the elementary visible steps (actions) are, and it must be formally defined when an operational behaviour in terms of these actions is in accordance with any of the underlying descriptions. Consistency of different descriptions then amounts to the existence of a system, whose 'operational behaviour' is in accordance with each of them.

As an initial step, we model the operational behaviour in terms of *temporal* or *causal models*, a subclass of unifying models which essentially arises as a combination of standard labelled transition systems and Kripke structures [Stir95]. Rather than providing a formal definition, we will explain the essentials of these models by means of the example given in Figure 1. Besides avoiding an unnecessary degree of details, this has the advantage of greater flexibility: too tight formally given constraints on the definition would be inadequate, as the approach sketched here addresses a wide range of application scenarios, where the precise profile-oriented instantiations of the notions might well differ significantly.

Figure 1 (left) shows the temporal model for a telephone service often called Granny's Free-Phone service (details about this application can be found in [SMCB96a,SMCB96c]). There, procedural entities called SIBs (Service-Independent Building Blocks) are combined in a flow graph fashion: SIBs are modelled by nodes and the branching structure by means of the edges and their labels. As we will see in the rest of this section, this example entails already some features which distinguish it from standard 'automata-like' modelling. Let us only note here that the notion of temporal models does not specify the kind of labels, which may be elements, sets, functions and the like, and that this flexibility is often important for obtaining the match between descriptions of different paradigms.

2.1 Classification

Unifying models can be classified according to their structural properties, their semantic background, their intent, their complexity etc.. Figure 2 summarizes some of this horizon, which we consider important for a better understanding of the underlying idea. In this figure, the three menu boxes contain alternatives of choice which can freely be combined, eg., one can choose an *operational* modelling with the goal to capture *consistency* by only observing *temporal* (causal) aspects. This choice, which marks the basic setting discussed in this paper, is emphasized in the figure by the thick frames.

The organization of choice shown in Figure 2 reflects the intuition that users will typically live in a certain semantic framework, eg. people modelling distributed systems will usually prefer an operational approach, and that they have a particular goal in mind, eg. consistency modelling in the context of this paper. It is usually only after these decisions that one considers the aspect of expressivity, represented by the selection of alternatives shown in the third line.

As indicated above, there are in general certainly more choices to be taken, like allowed complexity, requirements for hierarchical modelling etc., which are not displayed in Figure 2 for reasons of readability.

2.2 Variations and Non-Standard Interpretations

Initially it might be advantageous to focus on *regular* temporal models, which are characterized by having a finite number of nodes only. Their simplicity keeps the mind free for grabbing the essence of the new proposal. However, even considering decidability, this restriction is not vital as there are e.g. the much larger

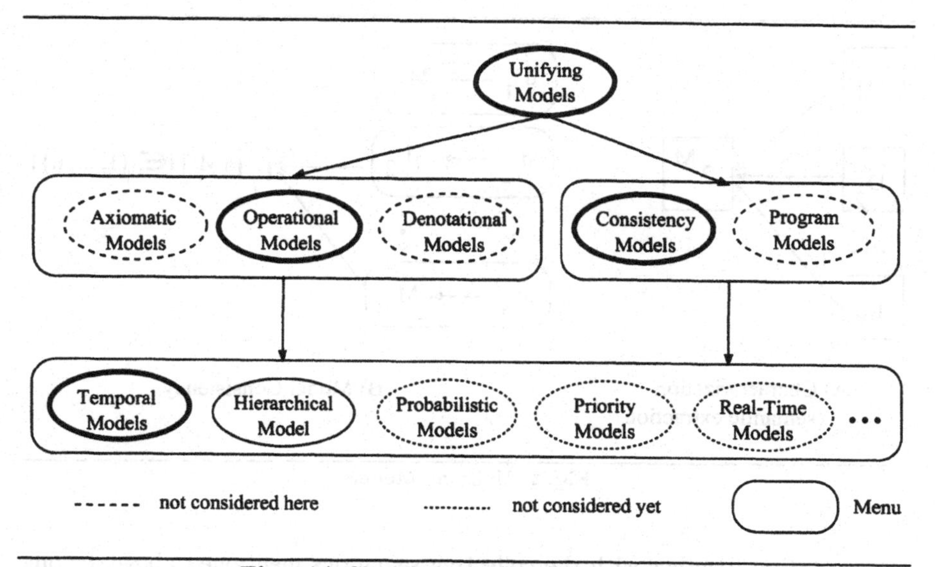

Fig. 2. A Classification of Unifying Models

classes of *context-free* or *macro* models, which still support decidability of many important properties [BuSt92,BuSt94,Hung94].

Orthogonal to the classification in terms of size is the classification in terms of structure. In future it may well be important to consider hierarchical models, in the flavour of Statecharts [Hare87], for consistency arguments. However, this still requires adequate semantic foundations of such models. The numerous proposals found in the literature for such foundations are rather unsatisfactory.

A third dimension of variation concerns the interpretation of node and edge labels. Eg. in the example of Figure 1 node labels may well be considered as transformations and edge labels as filters in order to allow a complete modelling in terms of a *collecting semantics* [JoNi95]. This interpretation is extremely powerful but fails to support decidability. In an industrial project with Siemens, we therefore went for a much simpler interpretation by considering both node and edge labels as *atomic syntactic objects*, which can only be compared up to syntactic identity. It turned out that already this extremely simple interpretation suffices to cover a wide range of consistency considerations.

The two interpretations sketched above illustrate the range of possibilities for an adequate application-dependent choice, which leaves tool-designers quite a lot of freedom.

It is worth noting that there is a trade-off between model size and complexity of interpretation: in order to achieve the same expressive power, one can often either choose a concise model with a rather complex interpretation or a large model with a simple interpretation. For example, in order to completely model a program over a finite data domain, one can either use a standard flow graph

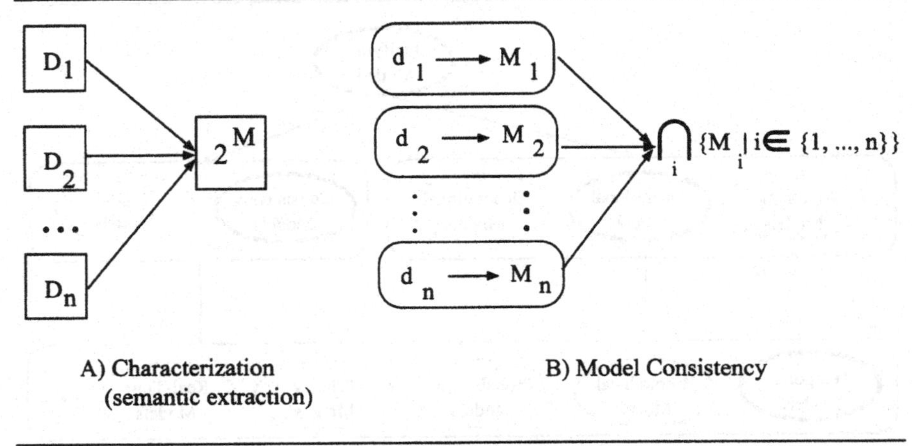

Fig. 3. Unifying Models

representation together with the collecting semantics mentioned above, or one could combine fully unrolled transition systems with a purely syntactic interpretation. This freedom of choice illustrates the flexibility and power of temporal modelling.

2.3 Consistency

The notion of consistency considered here is induced by the operational behaviour, which itself is expressed in terms of a temporal model structure: two descriptions are consistent if there exists a temporal model structure whose 'runs' are in accordance with the requirements and constraints imposed by both of them. Conflicts which may arise on a syntactic level, like different modularizations, are neglected. This point of view is directly reflected by viewing each description semantically as the set of all temporal models satisfying the described requirements, as sketched in Figure 3 (A), where D_i denotes description languages and M the set of all temporal model structures. Consequently, two descriptions are consistent, if the intersection of their semantics yields a non-empty set of temporal models, as shown in Figure 3 (B), where the d_i denote system descriptions and M_i their corresponding sets of temporal models.

In our basic setting, all the 'meta-structural' information in the descriptions, concerning e.g. architectural design decisions, is currently not taken into account. Rather it is considered as (at run time) unobservable 'syntactic overhead' which is 'flattened' during the model construction.

Also other language concepts like real-time constraints, priority, probability etc., even though they could be considered observable at run time, cannot be expressed in the basic 'automata-like' temporal models considered here. Rather, all these concepts are only modelled indirectly (and partially) in the unifying model as far as supported by the model structure, e.g. real-time constraints get reflected only in terms of imposed causality, and priority constraints may lead to the elimination of certain alternatives under certain circumstances.

2.4 Correctness and Completeness

We consider a consistency model as *correct* if detecting an inconsistency on the model level implies the inconsistency of the corresponding system descriptions, i.e. the *detection of inconsistency* is correct. As usual, the notion of completeness is complementary: a consistency model is *complete* if it allows to detect *all* inconsistencies between the considered system descriptions, i.e. each inconsistency in a heterogeneous system description is directly reflected in terms of an inconsistency on the consistency model level.

These notions of correctness and completeness, where correctness is considered a *must* while completeness is optional, reflect the fact that we are viewing the situation from the 'tool builder's perspective', who wants to construct a tool that detects as many faults (inconsistencies) as possible, while trying to avoid at any price that the designers' creativity be restricted by unnecessary constraints due to a too narrow consistency model. Given that correctness *and* completeness usually cannot be achieved together, it is our experience that the described viewpoint is extremely important for the acceptance of a new method, since users are typically not willing to change habits or style, unless there is no other choice. In fact, people being used to no control are much better prepared to accept a fast and non-restrictive but weak method than a more powerful but either slow or restrictive method: additional safety is fine, as long as one does not need to pay for it. Thus for a start we propose to go for simple consistency models with strong tool support.

Whereas most of the syntactic overhead should also in future be considered unobservable, except for e.g. parallelization constraints imposed by certain parallel operators, there will certainly be a demand for modelling conceptual extensions covering e.g. real time, probability, priority and the like. One should note however that, focusing simply on consistency between different description methods, the basic models considered here are sufficient, as long as the language extensions are dealt with in an orthogonal fashion, i.e. each through one description method in a non-interfering way (or such that the interference can be expressed within the basic model structure).

2.5 Model Class Representation

As mentioned, system descriptions are interpreted here as sets of models (for concrete systems), and, at least in their basic variant, these sets can be uniformly described in terms of temporal constraints (the corresponding language will grow with the requirements to be satisfied). Our basic constraint language is the modal μ-calculus [Koze83]. This choice is due to its great expressiveness and the fact that satisfiability is decidable [Emer90,Stir95]. – Interesting alternative choices are variants of monadic second order logic, which have similar qualities (one interpretation does in fact coincide with the μ-calculus in its expressive power) [JaWa96,Klar95,KMMG97,Thom90,Thom96].

The importance of having the μ-calculus as a common language for expressing the meaning of descriptions in terms of temporal model structures becomes clear in Figure 4: once each of the meanings of the single descriptions has been expressed in terms of the μ-calculus, their conjunction expresses the meaning of their combined impact. Since for the μ-calculus satisfiability is decidable, consistency can then be automatically checked by checking the satisfiability of the conjunction.

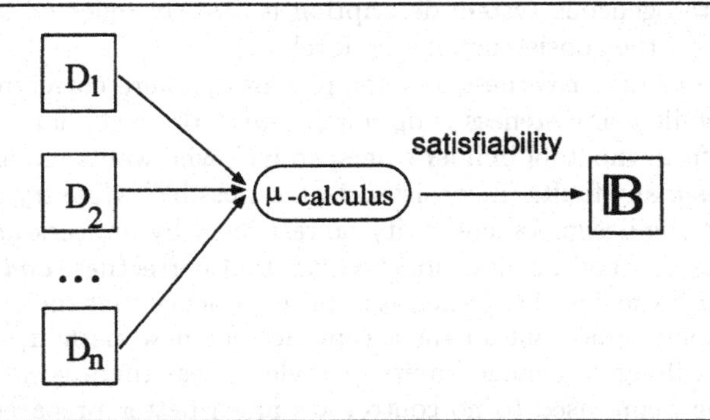

Fig. 4. Consistency as Satisfiability

3 Tool Support

Besides providing an intuitive kernel for a consistency model, the current focus on operational aspects is also motivated by the strong ***tool support*** available for the imposed consistency decisions, even in the presence of real time constraints [BoCa89,BoRS89,CeGL93,ClPS93,HeHW95,LaGZ89,Made92,MSGS93] [NOSY93,SCKK95,SMCB96c]. Various description mechanisms like process algebras, (temporal) logics and other constraint languages come in fact with effective decision tools solving eg. the consistency problem, and, in the case of failure detection, providing diagnostic information for its cause.

The power and applicability of these tools can be dramatically enhanced by means of partial evaluation and abstraction facilities, which are described in the following two subsections before we sketch a particular verification scenario in Section 3.3.

3.1 Refinement and Partial Evaluation

In order to bridge the gap between hierarchical descriptions it is often required to adapt the levels of granularity, as the (temporal) consistency modelling considered here only investigates the top-level transition structure: thus refinement,

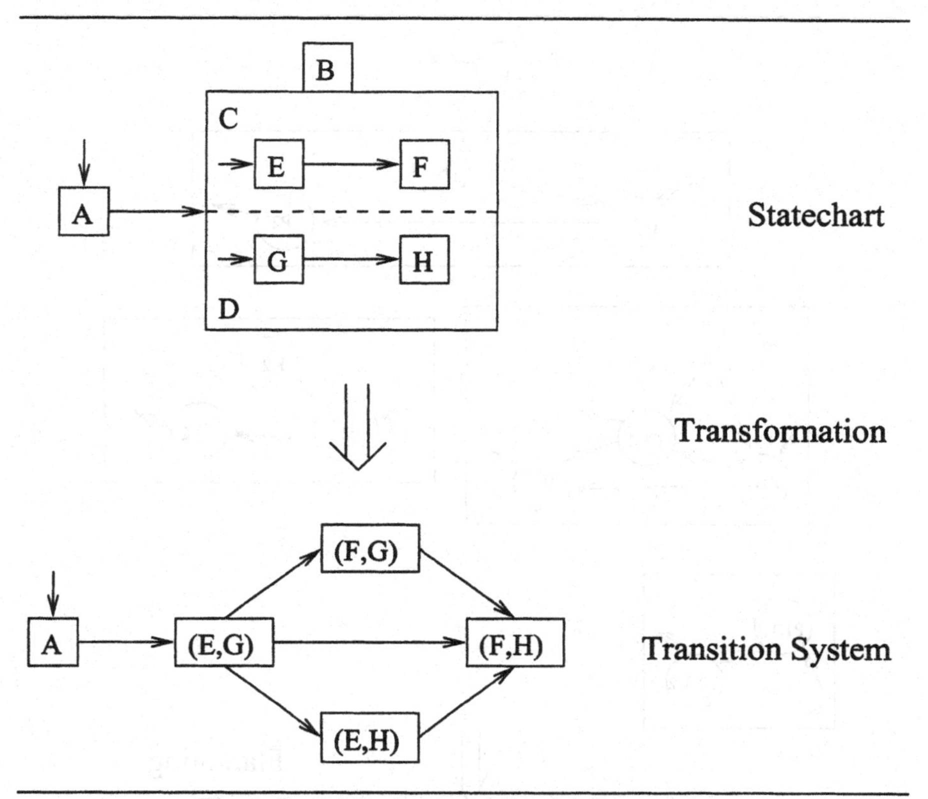

Fig. 5. Partial Evaluation of Parallel Compositions

eliminating some of the hierarchical overhead, needs to be employed in order to bring the essential details to the top level. Typical examples are the ***partial evaluation*** of a parallel operator according to the interleaving semantics [Miln89], as illustrated in Figure 5 on a Statechart, in order to control the consistency with purely sequential requirements, or the ***inline expansion*** of some module or procedure as illustrated in Figure 6, in order to present the requirements in sufficient detail. – In fact, as indicated by the dotted boxes in Figure 1 (left), also this model has been derived by inline expansion of the underlying macros structure.

We regard such refinement steps as ***partial evaluations*** of the control structure. Technically, they can be elegantly described and implemented using Plotkin-style structured operational semantics ([Plot81]): the required refinement steps are simply a matter of applying some syntactic rules.

3.2 Abstraction and Views

In addition to differences in granularity on the *syntactic level*, which preferably are resolved by means of refinement along the lines sketched above, descriptions

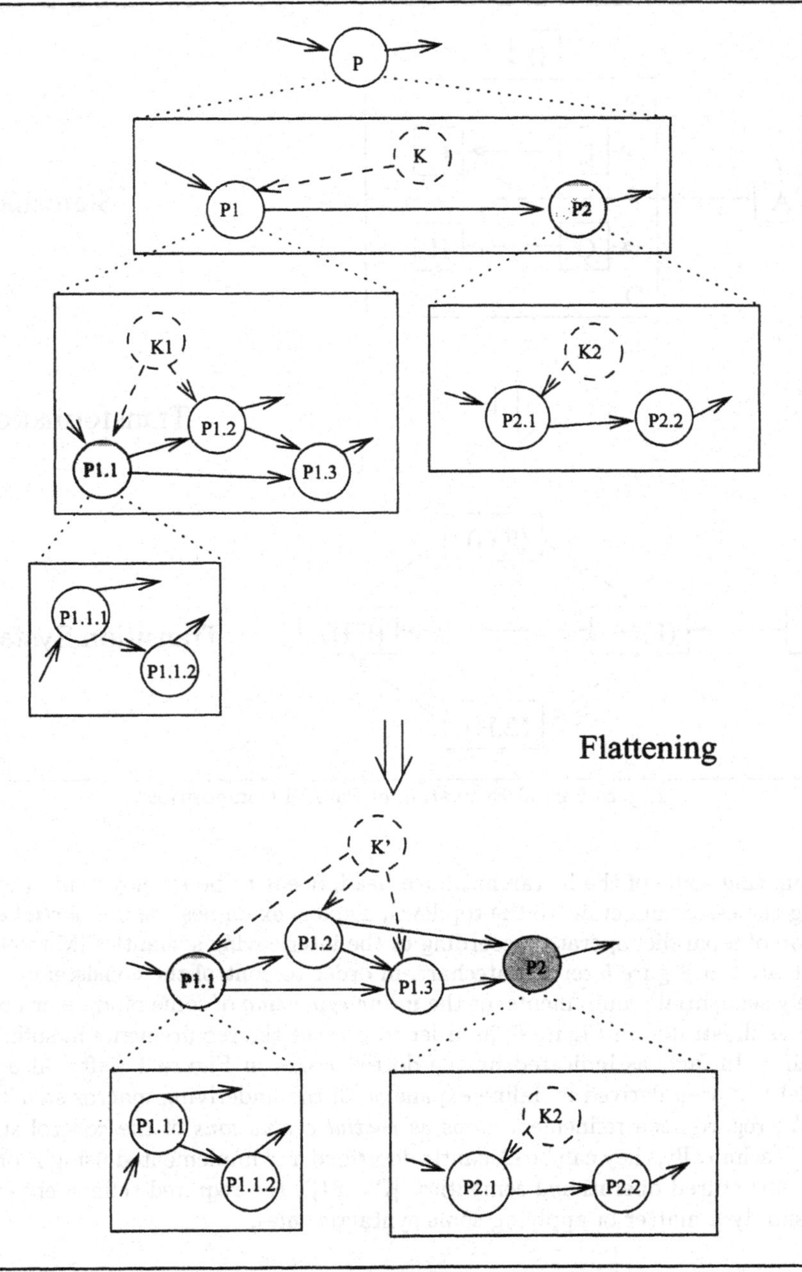

Fig. 6. Partial Evaluation of Hierarchical Overhead

may also differ in their *semantic* granularity, i.e. the notion of ***observation*** may be different. As conflicts between two descriptions of different notions of observation can only concern aspects expressible by means of both corresponding notions, adaptation for consistency checking is preferably done by abstraction. A simple example is the so-called ***model collapse***, which abstracts a (concrete) model according to the observation level of a constraint usually given in terms of a temporal (or modal) formula [Stir95]. Figure 1 (right) shows the result of such an abstraction applied to the model of Figure 1 (left), details of which can be found in [SMCB96a]: technically, nodes in the model collapse correspond to sets of nodes of the original (concrete) model, and node labels to subformulas of the modal formula. Semantically, a model collapse is characterized as the smallest 'collapse' of the model structure by means of node and edge identification of the concrete model, which still shows all the effects of the original model as far as they can be expressed in terms of the entities (subformulas) of the modal formula.

Note that this example is quite general: as our approach is based on the characterization of model classes in terms of μ-calculus formulas, the same principle can be applied to adequately adapt a model to any of the considered descriptions.

In fact, it turned out that the principle of model collapse is also of major importance for managing the conceptual complexity of consistency models: using this principle users can be provided with *views* of these models, which essentially look like the models obtained from the individual description methods, and which highlight portions being involved in consistency violations. This turned out to be an essential feature for getting our formal techniques accepted and applied in an industrial setting [SMCB96a,SMCB96c].

3.3 Decision Procedures

Important advantage of the operational (temporal) model-based uniform treatment of descriptions is the easy combination of synthesis-based consistency checks [MaWo84,SiCl85,StMC96] with other verification methods like model checking, which is much faster and covers more expressive specification languages. Whereas most of the ideas presented in this paper are only proposals, which have been only partially realized in our METAFrame environment [SMCB96b], the principle of unifying models has been successfully applied for the implementation of a uniform model checking-based verification scenario (Figure7): the use of a uniform and flexible model structure was the key for obtaining a uniform tool covering a wide range of applications, here summarized in the top line.

The standard models used for these applications can, according to their conceptual complexity, be classified in three categories, each of which could be represented in terms of adequate 'unifying models'. This reduces all the initial application problems to three versions of model checking (Figure7, 2nd line), which, in fact, could even be further unified in terms of a single model structure used in our fixpoint analysis machine. Admittedly, it is still sensible to maintain the three category distinction for performance reasons (and we do so by parameterization in our implementation), but in principle a single (unified) model structure is sufficient [SCKK95].

4 Associations

The structure of our approach reminds of 1) *common intermediate language* techniques of the compiler and programming language community used since the seventies and 2) attempts for the *formalization of informal specification languages*. In this section we will briefly sketch these approaches in order to highlight the intentional and technical distinctions.

4.1 Common Intermediate Languages

Common intermediate languages are the key to the uniform and efficient treatment of several programming languages in a compiler system. The compilation process is decomposed into two parts: a language-dependent part, the front end, which translates the source programs into the intermediate language, and a machine-dependent part, the back end, which generates code for the various host machines (Figure 8). This simplifies adaptations to new machines by saving coding: a front end written for a source language can be combined with (or reused for) all back ends, and vice versa.

This 'winning principle' has also been applied by Peter Mosses in order to obtain the common semantic framework of *action semantics* [Moss92]: it is based on a common intermediate language, the *action notation*, which itself has a sound semantic basis in terms of structured operational semantics or, if you wish, on a model basis. The framework is now characterized by giving semantics to programming languages in terms of a *translation* into this intermediate language. Remarkable from our point of view is the fact that the common intermediate language is here considered as a *semantic* entity: one does not need to go beyond action notation, because it is fully understood and designed to semantically model *all* programming language concepts.

4.2 Formalization of Informal Specification Languages

Most attempts towards the formalization of informal specification and description languages (eg. [Beec94a,Broy91,ElLP93,FGLB94,FrKV91,ShNi96]) are intentionally very similar to Peter Mosses' approach: semantics is given in terms of translations into well-studied languages with a formal semantics, e.g. Z [Spiv89], VDM-SL [BSI91], Petri-Nets [Reis85], CSP [Hoar85] or SCCS [Miln89].

However, the lack of expressive power of the underlying formally defined languages forced complicated adaptations or omissions in the semantic formalization, which altogether led to rather unsatisfactory solutions. – The use of actions semantics would have helped here, since it is carefully designed for 'universal expressivity'.

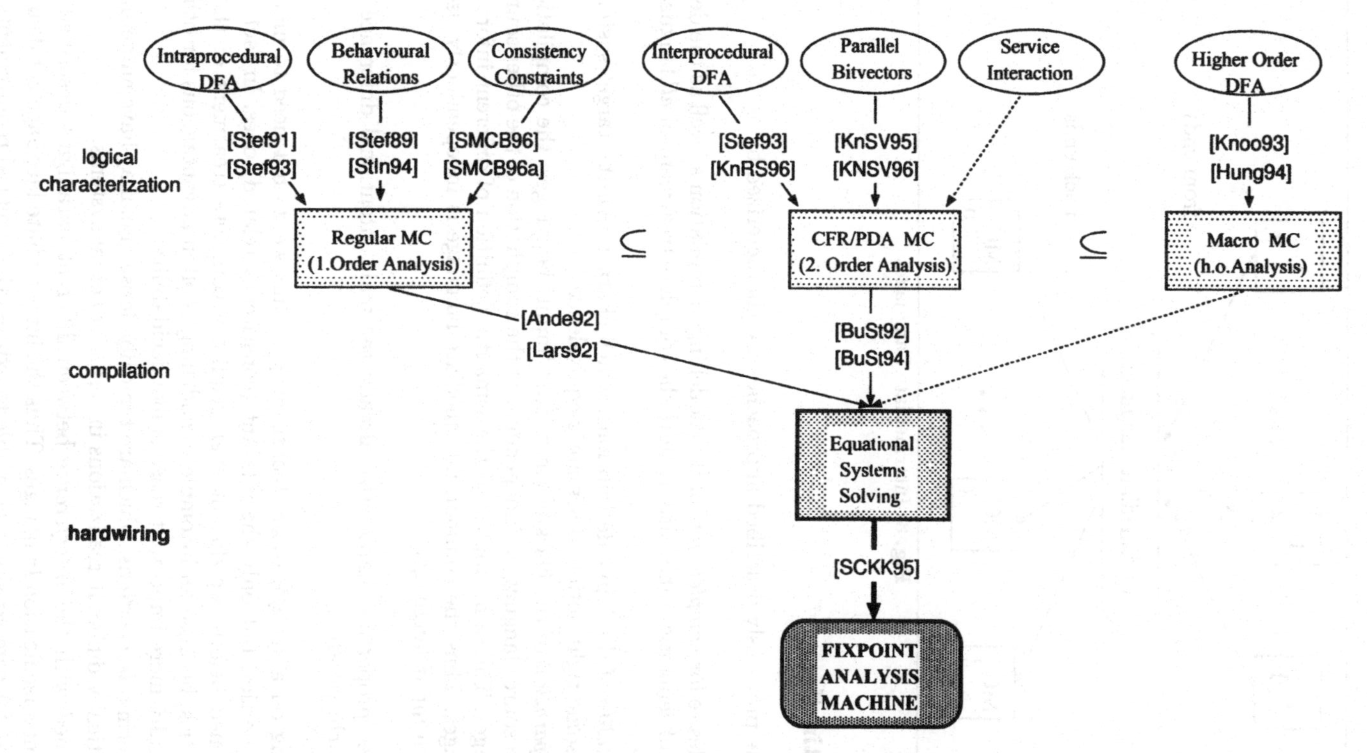

Fig. 7. Setup of the Analysis Environment: A Verification Scenario

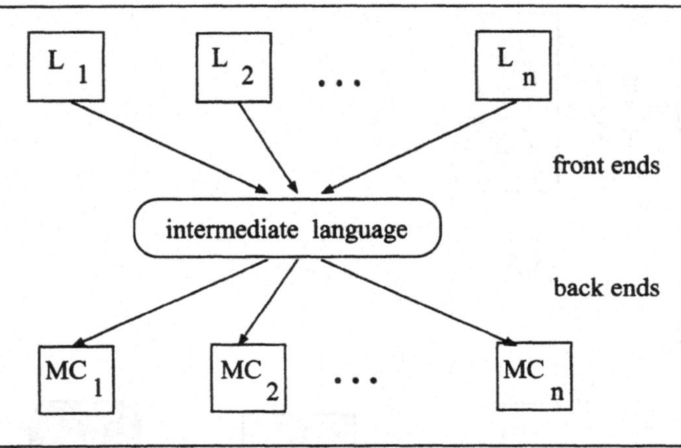

Fig. 8. Intermediate Languages

4.3 Distinctions

Both of the previously described approaches are characterized by

- their desire for *completeness* of the modelling: compilation as well as the definition of semantics inherently require the complete treatment of *all* language aspects.
- the *totality* of the input descriptions, which characterize the target systems up to some tight notion of semantic equivalence.
- the *separation* of the class of considered 'input' languages: the compilation of each source language is independent of the compilation of the other source languages. The same applies to the semantic definition of different informal languages. Thus the common intermediate language is independently used for each input language.

These three properties characterize already the main intentional difference to our approach: we are

- focusing on a *specific aspect* (consistency). Thus we do not need to model *all* properties, but only the relevant (consistency-related) ones. In fact, as mentioned before, we do not even require a complete treatment of these properties, because an incomplete modelling is often advantageous for other practically more important reasons like decidability.
- considering *descriptions*, which are typically 'loose', i.e. may have numerous, semantically different realizations in terms of actual systems.
- concerned with the *interference* between different paradigms described in different specification languages. Thus the intermediate language is taken as the level for comparision: being able to express the essential requirements in the μ-calculus allows us to automatically check for their consistency.

All these distinctions have a direct practical impact: whereas the first property provides flexibility, which is required in order to adapt the considered scenario for achieving a practical solution, the combination of the second and third property is necessary for the application of the paradigm of 'consistency modelling by conjunction'. This rather straightforward paradigm has also been applied with a similar intent in a first order logic framework by Zave and Jackson [ZaJa93]. Moreover, the idea of 'loose', application specific temporal modelling has also been considered in [JaJD96], but in a single paradigm scenario. Nevertheless this shows that model-based approaches are starting to reach the software engineering community.

5 Conclusions and Perspectives

An approach has been proposed to overcome the problem of lack of consistency between the various description methods used in software system design. Central idea is the definition of a common (temporal) model structure capturing the (operational) interference potential between the different description methods. This gets rid of 'syntactic chains' in the more flexible world of (operational) semantics, which allows consistency considerations independently of the representations chosen for the various descriptions, and without requiring a complete modelling of all the semantic aspects captured by the individual description methods. In fact, consistency checking may well be complete, even though the underlying consistency model does not address aspects which are only captured by one of the description methods in an 'orthogonal' fashion, i.e. without any interference with other descriptions. And even if consistency checking is not complete, our semantics-based approach still provides a new dimension for consistent multi-paradigm system development.

Nevertheless, in future we will consider how to widen the 'consistency kernel' by adding at least some of the vital meta-structures and new features like real time, probability and priority. We believe that here too the more flexible semantic treatment of these structures, which allows us to consider all the described requirements as constraints imposed on possible solutions, will be the key to a satisfactory solution.

Even though much work has still to be done before we have a complete, fully consistent, and tool-controlled specification environment covering various paradigms, we believe that the current proposal is a major step towards reliable and consistent system development with a strong practical impact: like type checking, which dramatically increased the reliability of programming by covering certain classes of errors, using a consistency model as proposed here will help strengthening the automatic control of *early* design phases. This is very important, since the earlier the fault detection, the better the cost reduction factor.

The experience in an industrial cooperation on Intelligent Networks in the world of telecommunications convinced us that (automatic) tool support is the key towards a wider dissemination of formal techniques in general [StMa96,StMa96a]. Tools supporting or encapsulating advanced theoretical concepts are becoming easier to handle: many of them are fully automatic, and tool designers care more and more about practitioners being able to use them while maintaining the style of specification they are accustomed to. In fact, we believe in the general power of broad, *tool-based* communication as the main channel for advanced technology transfer: complex theories can be communicated effectively in terms of illustrative prototype implementations. This concept starts to be accepted also by other research groups [ClWi96,StMa96], and it led to the foundation of **STTT** (Software Tools for Technology Transfer, [STTT]), a new international Springer journal, which is intended to actively support this kind of direct communication and to act as a bridge between various research communities as well as between industry and academia.

STTT's most challenging objective is to provide an electronic tool integration platform (ETI), which is intended to serve as an active repository with which users cannot only experiment with individual tools but also investigate their interplay by combining them into complex *heterogeneous* systems that exploit various methodologies and algorithms. ETI is based on the idea of unifying models too. Having such models as an internal structure at hand, Passau's METAFrame environment [SMCB96b], which supports also non-experts in their choice of adequate tools for tackling their problems, can easily be maintained and updated. Note, however, that similarly to the situation described in 4.1 and 4.2, this requires a much more complex model structure than for consistency checking, as these models must be able to capture *all* the aspects of the individual methods.

As METAFrame and in particular the ETI platform are intended to support multi-paradigm approaches, we are planning to integrate the consistency checking method presented here in near future.

Acknowledgment

I am very grateful to Gerald Lüttgen, Tiziana Margaria, Markus Müller-Olm, and Michael von der Beeck for numerous fruitful discussions, proof reading and help in the final writing phase. I am also indebted to the METAFrame team headed by Volker Braun and Andreas Claßen, whose feedback from the conceptual, development and implementation work strongly influenced the research direction of the whole group.

References

[Ande92] H. Andersen: *Model Checking and Boolean Graphs*, Proc. of ESOP '92, LNCS 582, Springer, 1992.

[Beec94a] M. von der Beeck: *Method Integration and Abstraction from Detailed Semantics to Improve Software Quality*, Proc. 1st Int. Worksh. on Requirements Engineering: Foundation of Software Quality REFSQ'94, Augustinus Buchh., Aachen, pp. 102-111, 1994

[BoCa89] T. Bolognesi, M. Caneve: *Squiggles: A tool for the analysis of LOTOS specifications*, in K. Turner Ed., Formal Description Techniques, pp. 201-216, North–Holland, 1989.

[BoRS89] G. Boudol, V. Roy, R. de Simone, D. Vergamini: *Process calculi, from theory to practice: Verification tools*, Rapport de Recherche RR1098, INRIA, 1989.

[Broy91] M. Broy: *Towards a Formal Foundation of the Specification and Description Language SDL*, Formal Aspects of Computing, vol. 3, 1991.

[BSI91] British Standards Institution: *VDM Specification Language – Proto-Standard*, Technical Report, BSI ist/5/50 N-231, 1991.

[BuSt92] O. Burkart, B. Steffen: *Model Checking for Context–Free Processes*, Proc. CONCUR '92, Stony Brook (NJ), Aug. 1992, LNCS 630, pp. 123-137, Springer.

[BuSt94] O. Burkart, B. Steffen: *Pushdown Processes: Parallel Composition and Model Checking*, Proc. CONCUR '94, Uppsala (Sweden), August 1994, LNCS 836, pp. 98-113, Springer.

[CeGL93] K. Cerans, J. Godskesen, K. Larsen: *Timed Modal Specification - Theory and Tools*, Proc. CAV, LNCS 697, Springer, pp. 253-267, 1993.

[ClPS93] R. Cleaveland, J. Parrow, B. Steffen: *The Concurrency Workbench: A Semantics-Based Verification Tool for Finite State Systems*, ACM TOPLAS, Vol. 15, No. 1, pp. 36-72, 1993.

[ClWi96] E. Clarke, J.M. Wing: *Position Statement of the Formal Methods Working Group*, ACM Worksh. on Strategic Directions in Computing Research, Boston (USA), June 14-15 1996. *ACM Computing Surveys*, 28(4), Dec. 1996.

[DoMS90] G. Doumenc, E. Madelaine, R. de Simone: *Proving process calculi translations in ECRINS: The PureLotos -> Meije example*, Rapport de recherche RR1192, INRIA, 1990.

[ElLP93] R. Elmstrom, R. Lintulampi, M. Pezzé: *Giving Semantics to SA/RT by Means of High-Level Timed Petri Nets*, Real-Time Systems, Vol. 5, pp. 249-271, Academic Publishing, 1993.

[Emer90] E.A. Emerson: *Temporal and Modal Logic*, In J. van Leeuwen, editor, *Handbook of Theoretical Computer Science*, vol. B, p. 995-1072, MIT Press/Elsevier, 1990.

[FGLB94] P. Fencott, A. Galloway, M. Lockyer, S. O'Brien, S. Pearson: *Formalising the Semantics of Ward/Mellor SA/RT Essential Models using a Process Algebra*, Proc. FME'94, LNCS 873, pp. 681-702, Springer-Verlag, 1994.

[FrKV91] M. Fraser, K. Kumar, V. Vaishnavi: *Informal and Formal Requirements Specification Languages: Bridging the Gap*, IEEE Transact. on Softw. Eng., vol. 17, no. 5, pp. 454-466, 1991.

[Hare87] D. Harel: *Statecharts: A visual formalism for complex systems*, Science of Computer Programming, Vol. 8, pp. 231-274, 1987.

[HeHW95] T. Henzinger, P. Ho, H. Wong-Toi: *A User Guide to HyTech*, Proc. TACAS'95, LNCS 1019, Springer, pp. 41-71, 1995.

[Hoar85] C. Hoare: *Communicating Sequential Processes*, Prentice-Hall Int., 1985.

[Hung94] H. Hungar: *Model Checking of Macro Processes*, Proc. of CAV'94, Palo Alto (CA), June 1994, LNCS 818, Springer, pp.169-181.

[JaJD96] D. Jackson, S. Jha, C.A. Damon: *Faster checking of software specifications by eliminating isomorphs*, Proc. ACM POPL'96, St. Petersburg Beach, FL (USA), Jan. 1996.

[JoNi95] N.D. Jones, F. Nielson. *Abstract Interpretation: A Semantics-Based Tool for Program Analysis*. In *Handbook of Logics in Computer Science*, Vol. 4, pp. 527 – 637, Oxford University Press, 1995.

[Klar95] J. Henriksen, J. Jensen, M. Jørgensen N. Klarlund, R. Paige, T. Rauhe, A. Sandholm: "*Mona: Monadic second-order logic in practice*," Proc. of TACAS'95, Århus (DK), May 1995, LNCS 1019, Springer, pp. 89-110.

[KMMG97] P. Kelb, T. Margaria, M. Mendler, C. Gsottberger: "MOSEL: *A Flexible Toolset for Monadic Second-Order Logic*," Appears at TACAS'97, Enschede (NL), April 1997, LNCS, Springer.

[Knoo93] J. Knoop: *Optimal Interprocedural Program Optimization: A new Framework and its Application*, PhD thesis, Dep. of Computer Science, Univ. of Kiel, Germany, 1993. To appear as LNCS monograph, Springer.

[KnRS94] J. Knoop, O. Rüthing, B. Steffen. *Partial Dead Code Elimination*, ACM SIGPLAN PLDI Conf.'94, *ACM SIGPLAN Notices 29*, Orlando, June 1994.

[KnRS96] J. Knoop, O. Rüthing, B. Steffen: *A Tool Kit for Constructing Optimal Interprocedural Data Flow Analyses*, appears in Journal of Programming Languages, Chapman & Hall

[KnSV95] J. Knoop. B. Steffen. J. Vollmer: *Parallelism for Free: bitvector analyses* $\Longrightarrow$ *no state explosion!* TACAS'95, Selected Papers, Aarhus (DK), LNCS 1019, pp. 264-290, Springer, 1995.

[KnSV96] J. Knoop. B. Steffen. J. Vollmer: *Parallelism for free: Efficient and optimal bitvector analyses for parallel programs*, ACM TOPLAS, Vol. 18, 3 (1996), pp.268-299.

[Koze83] D. Kozen. *Results on the Propositional mu-Calculus.* TCS 27, 333-354, 1983

[LaGZ89] K.G. Larsen, J.C. Godskesen, M. Zeeberg: *TAV, tools for automatic verification, user manual*, Technical Report R 89–19, Department of Mathematics and Computer Science, Ålborg University (DK), 1989.

[Lars92] K.G. Larsen: *Efficient Local Correctness Checking*, Proc. of CAV'92, Montreal (CAN), LNCS 663, pp. 410-422, Springer.

[Made92] E. Madelaine: *Verification tools from the Concur project*, EATCS Bulletin, Vol. 47, 1992.

[MaWo84] Z. Manna, P. Wolper. *Synthesis of Communicating Processes from Temporal Logic Specifications*, ACM TOPLAS Vol.6, N.1, Jan. 1984, pp.68-93.

[Miln89] R. Milner: *Communication and Concurrency*, Prentice-Hall, 1989.

[Moss92] P.D. Mosses: *Action Semantics*, Cambridge Tracts in Theoretical Computer Science, Vol. 26, Cambridge Univ. Press, 1992.

[MSGS93] J. Malhotra, S.A. Smolka, A. Giacalone, R. Shapiro: *Winston: A Tool for Hierarchical Design and Simulation of Concurrent Systems*, Work. on Specification and Verification of Concurrent Systems, Univ. of Stirling, Scotland, 1988.

[NOSY93] X. Nicollin, A. Olivero, J. Sifakis, S. Yovine: *An Approach to the Description and Analysis of Hybrid Systems*, Proc. Work. on Theory of Hybrid Systems, LNCS 736, Springer, pp. 149-178, 1993.

[Plot81] G. Plotkin: *A Structural Approach to Operational Semantics*, University of Aarhus (DK), DAIMI FN-19, 1981.

[Reis85] W. Reisig: *Petri Nets. An Introduction.*, Springer-Verlag, 1985.

[SCKK95] B. Steffen, A. Claßen, M. Klein, J. Knoop, T. Margaria: *The Fixpoint Analysis Machine*, (*invited paper*) to CONCUR'95, Pittsburgh (USA), August 1995, LNCS 962, Springer.

[ShNi96] L. Shi, P. Nixon: *An Improved Translation of SA/RT Specification Model to High-Level Timed Petri Nets*, Proc. FME'96, LNCS 1051, pp. 518-537, 1996.

[SiCl85] A.P. Sistla, E.M. Clarke. *The Complexity of the Propositional Linear Temporal Logics*, Journal of the ACM, Vol.32, 3, July 1985, pp.733-749.

[SMCB96a] B. Steffen, T. Margaria, A. Claßen, V. Braun, M. Reitenspieß: *A Constraint-Oriented Service Creation Environment* , Proc. PACT'96, 2nd Int. Conf. on Practical Application of Constraint Technology - April 1996, London (UK), Ed. by The Practical Application Company, pp. 283-298.

[SMCB96b] B. Steffen, T. Margaria, A. Claßen, V. Braun: *The* METAFrame *'95 Environment* , Proc. CAV'96, Juli-Aug. 1996, New Brunswick, NJ (USA), LNCS 1102, pp.450-453, Springer.

[SMCB96c] B. Steffen, T. Margaria, A. Claßen, V. Braun: "*Incremental Formalization: A Key to Industrial Success* ", In "SOFTWARE: Concepts and Tools", Vol. 17, No 2, pp. 78-91, Springer, July 1996.

[Spiv89] J. Spivey: *The Z Notation: A Reference Manual*, Prentice-Hall, 1989.

[Stef89] B. Steffen. *Characteristic Formulae.* Proc. ICALP'89, Stresa (Italy), LNCS 372, Springer, 1989.

[Stef91] B. Steffen. *Data Flow Analysis as Model Checking.* Proc. of TACS'91, Sendai (Japan), LNCS 526, pp. 346-364, Springer, 1991.

[Stef93] B. Steffen. *Generating Data Flow Analysis Algorithms from Modal Specifications*, Science of Computer Programming, N. 21, 1993, pp. 115-139.

[StIn94] B. Steffen, A. Ingólfsdóttir: *Characteristic Formulae for Finite State Processes*, Information and Computation, Vol. 110, No. 1, 1994.

[Stir95] C. Stirling: *Modal and Temporal Logics*, In *Handbook of Logics in Computer Science*, Vol. 2, pp. 478 – 551, Oxford Univ. Press, 1995.

[StMa96] B. Steffen, T. Margaria: *Method Engineering for Real-Life Concurrent Systems*, position statement, ACM Works. on *Strategic Directions in Computing Research*, Working Group on Concurrency (Chair S. Smolka). Appears in ACM Computing Surveys 28A(4), Dec. 1996, `http://www.acm.org/surveys/1996/SteffenMethod/`.

[StMa96a] B. Steffen, T. Margaria: *Tools Get Formal Methods into Practice*, position statement, ACM Works. on *Strategic Directions in Computing Research*, Working Group on Formal Methods (Co-Chairs E. Clarke, J. Wing). Appears in ACM Computing Surveys 28A(4), Dec. 1996, `http://www.acm.org/surveys/1996/SteffenTools/`.

[StMC96] B. Steffen, T. Margaria, A. Claßen. *Heterogeneous Analysis and Verification for Distributed Systems*, In "SOFTWARE: Concepts and Tools", vol. 17, N.1, pp. 13-25, Springer, 1996.

[STTT] International Journal on *Software Tools for Technology Transfer* (STTT), Springer Verlag, coming September 1997 `http://brahms.fmi.uni-passau.de/bs/sttt`.

[Thom90] W. Thomas: "*Automata on infinite objects*," In J. van Leeuwen, ed., *Handbook of Theoretical Computer Science*, vol.B, pp.133-191. MIT Press/Elsevier, 1990.

[Thom96] W. Thomas: "*Languages, automata, and objects*," to appear in the forthcoming new edition of the *Handbook of Theoretical Computer Science*, MIT Press/Elsevier.

[JaWa96] D. Janin, I. Walukiewicz: *On the expressive completeness of the propositional mu-calculus with respect to the Monadic Second Order logic*, Proc. CONCUR'96, Pisa (I), LNCS 1119, Springer, pp.263-277, Aug. 1996.

[ZaJa93] P. Zave, M. Jackson: *Conjunction as Composition*, ACM TOSEM 2(4), pp. 379-411, October'93.

[ZaJa96] P. Zave, M. Jackson: *Where do operations come from? A Multiparadigm specification technique*, To appear on Trans. on Softw. Eng.

Predecessor Queries in Dynamic Integer Sets

Gerth Stølting Brodal*

BRICS**, Department of Computer Science, University of Aarhus
Ny Munkegade, DK-8000 Århus C, Denmark
gerth@brics.dk

Abstract. We consider the problem of maintaining a set of n integers in the range $0..2^w - 1$ under the operations of insertion, deletion, predecessor queries, minimum queries and maximum queries on a unit cost RAM with word size w bits. Let $f(n)$ be an arbitrary nondecreasing smooth function satisfying $\log\log n \le f(n) \le \sqrt{\log n}$. A data structure is presented supporting insertions and deletions in worst case $O(f(n))$ time, predecessor queries in worst case $O((\log n)/f(n))$ time and minimum and maximum queries in worst case constant time. The required space is $O(n2^{\epsilon w})$ for an arbitrary constant $\epsilon > 0$. The RAM operations used are addition, arbitrary left and right bit shifts and bit-wise boolean operations. The data structure is the first supporting predecessor queries in worst case $O(\log n/\log\log n)$ time while having worst case $O(\log\log n)$ update time.

1 Introduction

We consider the problem of maintaining a set S of size n under the operations:

INSERT(e) inserts element e into S,
DELETE(e) deletes element e from S,
PRED(e) returns the largest element $\le e$ in S, and
FINDMIN/FINDMAX returns the minimum/maximum element in S.

In the comparison model INSERT, DELETE and PRED can be supported in worst case $O(\log n)$ time and FINDMIN and FINDMAX in worst case constant time by a balanced search tree, say an (a, b)-tree [8]. For the comparison model a tradeoff between the operations has been shown by Brodal *et al.* [6]. The tradeoff shown in [6] is that if INSERT and DELETE take worst case $O(t(n))$ time then FINDMIN (and FINDMAX) requires at least worst case $n/2^{O(t(n))}$ time. Because predecessor queries can be used to answer member queries, minimum queries and maximum queries, PRED requires worst case $\max\{\Omega(\log n), n/2^{O(t(n))}\}$ time. For the sake of completeness we mention that matching upper bounds can be

* Supported by the Danish Natural Science Research Council (Grant No. 9400044). Partially supported by the ESPRIT Long Term Research Program of the EU under contract #20244 (ALCOM-IT).
** Basic Research in Computer Science, a Centre of the Danish National Research Foundation.

Reischuk, Morvan (Eds.): STACS'97 Proceedings, LNCS 1200

achieved by a $(2,4)$-tree of depth at most $t(n)$ where each leaf stores $\Theta(n/2^{t(n)})$ elements, provided DELETE takes a pointer to the element to be deleted.

In the following we consider the problem on a unit cost RAM with word size w bits allowing addition, arbitrary left and right bit shifts and bit-wise boolean operations on words in constant time. Miltersen [10] refers to this model as a *Practical RAM*. We assume the elements are integers in the range $0..2^w - 1$. A tradeoff similar to the one for the comparison model [6] is not known for a Practical RAM.

A data structure of van Emde Boas *et al.* [15,16] supports the operations INSERT, DELETE, PRED, FINDMIN and FINDMAX on a Practical RAM in worst case $O(\log w)$ time. For word size $\log^{O(1)} n$ this implies an $O(\log\log n)$ time implementation.

Thorup [14] recently presented a priority queue supporting INSERT and EXTRACTMIN in worst case $O(\log\log n)$ time independently of the word size w. Thorup notes that by tabulating the multiplicity of each of the inserted elements the construction supports DELETE in amortized $O(\log\log n)$ time by skipping extracted integers of multiplicity zero. The data structure of Thorup does not support predecessor queries but Thorup mentions that an $\Omega(\log^{1/3-o(1)} n)$ lower bound for PRED can be extracted from [9,11]. The space requirement of Thorup's data structure is $O(n2^{\epsilon w})$ (if the time bounds are amortized the space requirement is $O(n + 2^{\epsilon w})$).

Andersson [2] has presented a Practical RAM implementation supporting insertions, deletions and predecessor queries in worst case $O(\sqrt{\log n})$ time and minimum and maximum queries in worst case constant time. The space requirement of Andersson's data structure is $O(n + 2^{\epsilon w})$. Several data structures can achieve the same time bounds as Andersson [2], but they all require constant time multiplication [3,7,13].

The main result of this paper is Theorem 1 stated below. The theorem requires the notion of *smooth* functions. Overmars [12] defines a nondecreasing function f to be smooth if and only if $f(O(n)) = O(f(n))$.

Theorem 1. *Let $f(n)$ be a nondecreasing smooth function satisfying $\log\log n \leq f(n) \leq \sqrt{\log n}$. On a Practical RAM a data structure exists supporting* INSERT *and* DELETE *in worst case $O(f(n))$ time,* PRED *in worst case $O((\log n)/f(n))$ time and* FINDMIN *and* FINDMAX *in worst case constant time, where n is the number of integers stored. The space required is $O(n2^{\epsilon w})$ for any constant $\epsilon > 0$.*

If $f(n) = \log\log n$ we achieve the result of Thorup but in the worst case sense, i.e. we can support INSERT, EXTRACTMIN and DELETE in worst case $O(\log\log n)$ time. We can support PRED queries in worst case $O(\log n/\log\log n)$ time. The data structure is the first allowing predecessor queries in $O(\log n/\log\log n)$ time while having $O(\log\log n)$ update time. If $f(n) = \sqrt{\log n}$, we achieve time bounds matching those of Andersson [2].

The basic idea of our construction is to apply the data structure of van Emde Boas *et al.* [15,16] for $O(f(n))$ levels and then switch to a packed search tree of height $O(\log n/f(n))$. This is very similar to the data structure of Andersson [2].

But where Andersson uses $O(\log n/f(n))$ time to update his packed B-tree, we only need $O(f(n))$ time. The idea we apply to achieve this speedup is to add *buffers* of delayed insertions and deletions to the search tree, such that we can work on several insertions concurrently by using the word parallelism of the Practical RAM. The idea of adding buffers to a search tree has in the context of designing I/O efficient data structures been applied by Arge [4].

Throughout this paper we w.l.o.g. assume DELETE only deletes integers actually contained in the set and INSERT never inserts an already inserted integer. This can be satisfied by tabulating the multiplicity of each inserted integer.

In the description of our data structure we in the following assume n is a constant such that the current number of integers in the set is $\Theta(n)$. This can be satisfied by using the general dynamization technique described by Overmars [12], which requires $f(n)$ to be smooth. In Sect. 2 if we write $\log^5 n \leq k$, we actually mean that k is a function of n, but because we assume n to be a constant k is also assumed to be a constant.

In Sect. 2 we describe our packed search trees with buffers. In Sect. 3 we describe how to perform queries in a packed search tree and in Sect. 4 how to update a packed search tree. In Sect. 5 we combine the packed search trees with a range reduction based on the data structure of van Emde Boas *et al.* [15,16] to achieve the result stated in Theorem 1. Section 6 contains some concluding remarks and lists some open problems.

2 Packed search trees with buffers

In this and the following two sections we describe how to maintain a set of integers of w/k bits each, for k satisfying $\log^5 n \leq k \leq w/\log n$. The bounds we achieve are:

Lemma 2. *Let k satisfy $\log^5 n \leq k \leq w/\log n$. If the integers to be stored are of w/k bits each then on a Practical RAM* INSERT *and* DELETE *can be supported in worst case $O(\log k)$ time,* PRED *in worst case $O(\log k + \log n/\log k)$ time and* FINDMIN *and* FINDMAX *in worst case constant time. The space required is $O(n)$.*

The basic idea is to store $O(k)$ integers in each word and to use the word parallelism of the Practical RAM to work on $O(k)$ integers in parallel in constant time. In the following we w.l.o.g. assume that we can apply Practical RAM operations to a list of $O(k)$ integers stored in $O(1)$ words in worst cast constant time. Together with each integer we store a *test bit*, as in [1,2,14]. An integer together with the associated test bit is denoted a *field*. Figure 1 illustrates the structure of a list of maximum capacity k containing $\ell \leq k$ integers $x_1, \ldots, x_\ell$. A field containing the integer x_i has a test bit equal to zero. The remaining $k-\ell$ empty fields store the integer zero and a test bit equal to one.

Essential to the data structure to be described is the following lemma due to Albers and Hagerup [1].

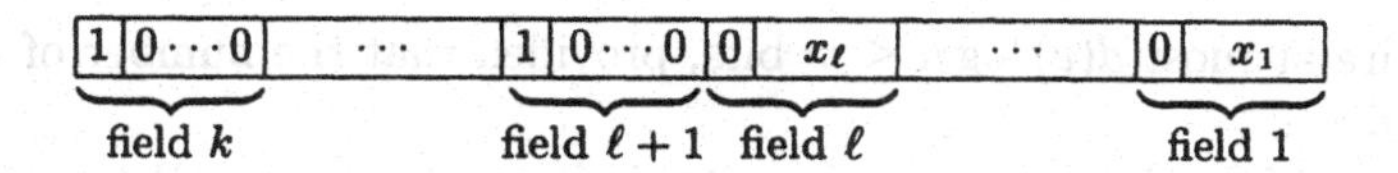

Fig. 1. The structure of a list of maximum capacity k, containing integers $x_1, \ldots, x_\ell$.

Lemma 3 (Albers and Hagerup). *On a Practical RAM two sorted lists each of at most $O(k)$ integers stored in $O(1)$ words can be merged into a single sorted list stored in $O(1)$ words in $O(\log k)$ time.*

Albers and Hagerup's proof of Lemma 3 is a description of how to implement the bitonic merging algorithm of Batcher [5] in a constant number of words on the Practical RAM. The algorithm of Albers and Hagerup does not handle partial full lists as defined (all test bits are assumed to be zero), but it is straightforward to modify their algorithm to do so, by considering an integer's test bit as the integer's most significant bit. A related lemma we need for our construction is the following:

Lemma 4. *Let k satisfy $k \leq w/\log n$. Let A and B be two sorted and repetition free lists each of at most $O(k)$ integers stored in $O(1)$ words on a Practical RAM. Then the sorted list $A \setminus B$ can be computed and stored in $O(1)$ words in $O(\log k)$ time.*

Proof. Let C be the list consisting of A merged with B twice. By Lemma 3 the merging can be done in worst case $O(\log k)$ time. By removing all integers appearing at least twice from C we get $A \setminus B$. In the following we outline how to eliminate these repetitions from C. Tedious implementation details are omitted.

First a mask is constructed corresponding to the integers only appearing once in C. This can be done in worst case constant time by performing the comparisons between neighbor integers in C by subtraction like the mask construction described in [1]. The integers appearing only once in C are compressed to form a single list as follows. First a prefix sum computation is performed to calculate how many fields each integer has to be shifted to the right. This can be done in $O(\log k)$ time by using the constructed mask. Notice that each of the calculated values is an integer in the range $0, \ldots, |A| + 2|B|$, implying that each field is required to contain at least $O(\log k)$ bits. Finally we perform $O(\log k)$ iterations where we in the i'th iteration move all integers x_j, 2^i fields to the right if the binary representation of the number of fields x_j has to be shifted has the i'th bit set. A similar approach has been applied in [1] to reverse a list of integers.

The main component of our data structure is a search tree T where all leaves have equal depth and all internal nodes have degree at least one and at most $\Delta \leq k/\log^4 n$. Each leaf v stores a sorted list I_v of between $k/2$ and k integers. With each internal node v of degree $d(v)$ we store $d(v)-1$ keys to guide searches. The $d(v)$ pointers to the children of v can be packed into a single word because they require at most $d(v) \log n \leq w$ bits, provided that the number of nodes is less than n.

This part of the data structure is quite similar to the packed B-tree described by Andersson [2]. To achieve faster update times for INSERT and DELETE than Andersson, we add buffers of delayed INSERT and DELETE operations to each internal node of the tree.

With each internal node v we maintain a buffer I_v containing a sorted list of integers to be inserted into the leaves of the subtree T_v rooted at v, and a buffer D_v containing a sorted list of integers to be deleted from T_v. We maintain the invariants that I_v and D_v are disjoint and repetition free, and that

$$\max\{|I_v|, |D_v|\} < \Delta \log n \ . \tag{1}$$

The set S_v of integers stored in a subtree T_v can recursively be defined as

$$S_v = \begin{cases} I_v & \text{if } v \text{ is a leaf,} \\ I_v \cup ((\bigcup_{w \text{ a child of } v} S_w) \setminus D_v) & \text{otherwise.} \end{cases} \tag{2}$$

Finally we maintain two nonempty global buffers of integers L and R each of size $O(k)$ to be able to answer minimum and maximum queries in constant time. The integers in L are less than all other integers stored, and the integers in R are greater than all other integers stored.

Let h denote the height of T. In Sect. 4 we show how to guarantee that $h = O(\log n / \log k)$, implying that the number of nodes is $O(hn/k) = O(n)$.

3 Queries in packed search trees

By explicitly remembering the minimum integer in L and the maximum integer in R it is trivial to implement FINDMIN and FINDMAX in worst case constant time. A PRED(e) query can be answered as follows. If $e \leq \max(L)$ then the predecessor of e is contained in L and can be found in worst case $O(\log k)$ time by standard techniques. If $\min(R) \leq e$ then the predecessor of e is contained in R. Otherwise we have to search for the predecessor of e in T.

We first perform a search for e in the search tree T. The implementation of the search for e in T is identical to how Andersson searches in a packed B-tree [2]. We refer to [2] for details. Let λ be the leaf reached and $w_1, \ldots, w_{h-1}$ be the internal nodes on the path from the root to λ. Define $w_h = \lambda$. Because we have introduced buffers at each internal node of T the predecessor of e does not necessarily have to be stored in I_λ but can also be contained in one of the insert buffers I_{w_i}. An integer $a \in I_{w_i}$ can only be a predecessor of e if it has not been deleted by a delayed delete operation, i.e. $a \notin D_{w_j}$ for $1 \leq j < i$. It seems necessary to *flush* all buffers I_{w_i} and D_{w_i} for integers which should be inserted in or deleted from I_λ to be able to find the predecessor of e. If dom_λ denotes the interval of integers spanned by the leaf λ, the buffers I_{w_i} and D_{w_i} can be flushed for elements in dom_λ by the following sequence of operations:

$$\begin{aligned}
I_{w_{i+1}} &\leftarrow I_{w_{i+1}} \setminus (D_{w_i} \cap \text{dom}_\lambda) \cup (I_{w_i} \cap \text{dom}_\lambda) \setminus D_{w_{i+1}} \ , \\
D_{w_{i+1}} &\leftarrow D_{w_{i+1}} \setminus (I_{w_i} \cap \text{dom}_\lambda) \cup (D_{w_i} \cap \text{dom}_\lambda) \setminus I_{w_{i+1}} \ , \\
I_{w_i} &\leftarrow I_{w_i} \setminus \text{dom}_\lambda \ , \\
D_{w_i} &\leftarrow D_{w_i} \setminus \text{dom}_\lambda \ .
\end{aligned}$$

Let $\hat{I}_\lambda$ denote the value of I_λ after flushing all buffers I_{w_i} and D_{w_i} for integers in the range dom_λ. From (2) it follows that $\hat{I}_\lambda$ can also be computed directly by the expression

$$\hat{I}_\lambda = \text{dom}_\lambda \cap (((\cdots((I_\lambda \setminus D_{w_{h-1}}) \cup I_{w_{h-1}})\cdots) \setminus D_{w_1}) \cup I_{w_1}) \ . \tag{3}$$

Based on Lemmas 3 and 4 we can compute this expression in $O(h \log k)$ time. This is unfortunately $O(\log n)$ for the tree height $h = \log n / \log k$. In the following we outline how to find the predecessor of e in $\hat{I}_\lambda$ without actually computing $\hat{I}_\lambda$ in $O(\log k + \log n / \log k)$ time.

Let I'_{w_i} be $I_{w_i} \cap \text{dom}_\lambda \cap]\infty, e]$ for $i = 1, \ldots, h$. An alternative expression to compute the predecessor of e in $\hat{I}_\lambda$ is

$$\max \bigcup_{i=1,\ldots,h} (I'_{w_i} \setminus \bigcup_{j=1,\ldots,i-1} D_{w_j}) \ . \tag{4}$$

Because $|\bigcup_{j=1,\ldots,h-1} D_{w_j}| < \Delta \log^2 n$ we can w.l.o.g. assume $|I'_{w_h}| \leq \Delta \log^2 n$ in (4) by restricting our attention to the $\Delta \log^2 n$ largest integers in I'_{w_h}, i.e. all sets involved in (4) have size at most $\Delta \log^2 n$. The steps we perform to compute (4) are the following. All implementation details are omitted.

- First all buffers I_{w_i} and D_{w_i} for $i < h$ are inserted into a single word $\mathcal{W}$ where the contents of $\mathcal{W}$ is considered as $2h - 2$ independent lists each of maximum capacity $\Delta \log^2 n$. This can be done in $O(h) = O(\log n / \log k)$ time.
- Using the word parallelism of the Practical RAM we now for all I_{w_i} compute I'_{w_i}. This can be done in $O(\log k)$ time if $\min(\text{dom}_\lambda)$ is known. The integer $\min(\text{dom}_\lambda)$ can be computed in the search phase determining the leaf λ. $\mathcal{W}$ now contains I'_{w_i} and D_{w_i} for $i < h$.
- The value of I'_{w_h} is computed (satisfying $|I'_{w_h}| \leq \Delta \log^2 n$) and appended to $\mathcal{W}$. This can be done in $O(\log k)$ time. The contents of $\mathcal{W}$ is now

$$I'_{w_h} D_{w_{h-1}} I'_{w_{h-1}} \cdots D_{w_1} I'_{w_1} \ .$$

- Let $\mathcal{W}_I = (I'_{w_h})^{h-1} \cdots (I'_{w_1})^{h-1}$ and $\mathcal{W}_D = (D_{w_{h-1}} \cdots D_{w_1})^h$. See Fig. 2. The number of fields required in each word is $h(h-1)\Delta \log^2 n \leq \Delta \log^4 n \leq k$. The two words can be constructed from $\mathcal{W}$ in $O(\log k)$ time.
- From $\mathcal{W}_I$ and $\mathcal{W}_D$ we now construct $h(h-1)$ masks $M_{i,j}$ such that $M_{i,j}$ is a mask for the fields of I'_{w_i} which are not contained in D_{w_j}. See Fig. 2. The

$\mathcal{W}_I$	I'_{w_h}	$\cdots$	I'_{w_h}	I'_{w_h}	$\cdots$	I'_{w_1}	$\cdots$	I'_{w_1}	I'_{w_1}
$\mathcal{W}_D$	$D_{w_{h-1}}$	$\cdots$	D_{w_2}	D_{w_1}	$\cdots$	$D_{w_{h-1}}$	$\cdots$	D_{w_2}	D_{w_1}
$\mathcal{W}_M$	$M_{h,h-1}$	$\cdots$	$M_{h,2}$	$M_{h,1}$	$\cdots$	$M_{1,h-1}$	$\cdots$	$M_{1,2}$	$M_{1,1}$

Fig. 2. The structure of the words $\mathcal{W}_I$, $\mathcal{W}_D$ and $\mathcal{W}_M$.

construction of a mask $M_{i,j}$ from the two list I'_{w_i} and D_{w_j} is very similar to the proof of Lemma 4 and can be done as follows in $O(\log k)$ time.
First I is merged with D twice (we omit the subscripts while outlining the mask construction). Let C be the resulting list. From C construct in constant time a mask C' that contains ones in the fields in which C stores an integer only appearing once in C and zero in all other fields. By removing all fields from C having *exactly* one identical neighbor we can recover I from C. By removing the corresponding fields from C' we get the required mask M. As an example assume $I = (7, 5, 4, 3, 1)$ and $D = (6, 5, 2)$. Then $C = (7, \underline{6}, \underline{6}, \underline{5}, 5, \underline{5}, 4, 3, \underline{2}, \underline{2}, 1)$, $C' = (1, \underline{0}, \underline{0}, \underline{0}, 0, \underline{0}, 1, 1, \underline{0}, \underline{0}, 1)$ and $M = (1, 0, 1, 1, 1)$ where underlined fields are the fields in C having exactly one identical neighbor.

- We now compute masks $M_i = \bigwedge_{j=1,\ldots,i-1} M_{i,j}$ for all i. By applying M_i to I'_{w_i} we get $I'_{w_i} \setminus \bigcup_{j=1,\ldots,i-1} D_{w_j}$. This can be done in $O(\log k)$ time from $\mathcal{W}_M$ and $\mathcal{W}_I$.
- Finally we in $O(\log k)$ time compute (4) as the maximum over all the integers in the sets computed in the previous step. Notice that it can easily be checked if e has a predecessor in $\hat{I}_\lambda$ by checking if all the sets computed in the previous step are empty.

We conclude that the predecessor of e in $\hat{I}_\lambda$ can be found in $O(\log k + h) = O(\log k + \log n / \log k)$ time.

If e does not have a predecessor in $\hat{I}_\lambda$ there are two cases to consider. The first is if there exists a leaf $\bar{\lambda}$ to the left of λ. Then the predecessor of e is the largest integer in $\hat{I}_{\bar{\lambda}}$. Notice that $\hat{I}_{\bar{\lambda}}$ is nonempty because $|\bigcup_{j=1,\ldots,h-1} D_{\bar{w}_j}| < |I_{\bar{\lambda}}|$. If λ is the leftmost leaf the predecessor of e is the largest integer in L. We conclude that PRED queries can be answered in worst case $O(\log k + \log n / \log k)$ time on a Practical RAM.

4 Updating packed search trees

In the following we describe how to perform INSERT and DELETE updates. We first give a solution achieving the claimed time bounds in the amortized sense. The amortized solution is then converted into a worst case solution by standard techniques.

We first consider INSERT(e). If $e < \max(L)$ we insert e into L in $\log k$ time, remove the maximum from L such that $|L|$ remains unchanged, and let e become the removed integer. If $\min(R) < e$ we insert e in R, remove the minimum from R, and let e become the removed integer.

Let r denote the root of T. If $e \in D_r$, remove e from D_r in worst case $O(\log k)$ time, i.e. INSERT(e) cancels a delayed DELETE(e) operation. Otherwise insert e into I_r.

If $|I_r| < \Delta \log n$ this concludes the INSERT operation. Otherwise there must exist a child w of r such that $\log n$ integers can be moved from I_r to the subtree rooted at w. The child w and the $\log n$ integers X to be moved can be found by a binary search using the search keys stored at r in worst case $O(\log k)$ time. We omit the details of the binary search in I_r. We first remove the set of integers X from I_r such that $|I_r| < \Delta \log n$. We next remove all integers in $X \cap D_w$ from X and from D_w in $O(\log k)$ time by Lemma 4, i.e. we let delayed deletions be cancel out by delayed insertions. The remaining integers in X are merged into I_w in $O(\log k)$ time. Notice that I_w and D_w are disjoint after the merging and that if w is an internal node then $|I_w| < (\Delta + 1) \log n$.

If $|I_w| \geq \Delta \log n$ and w is not a leaf we recursively apply the above to I_w. If w is a leaf and $|I_w| \leq k$ we are done. The only problem remaining is if w is a leaf and $k < |I_w| \leq k + \log n \leq 2k$. In this case we split the leaf w into two leaves each containing between $k/2$ and k integers, and update the search keys and child pointers stored at the parent of w. If the parent p of w now has $\Delta + 1$ children we split p into two nodes of degree $\geq \Delta/2$ while distributing the buffers I_p and D_p among the two nodes w.r.t. the new search key. The details of how to split a node is described in [2]. If the parent of p gets degree $\Delta + 1$ we recursively split the parent of p.

The implementation of inserting e in T takes worst case $O(h \log k)$ time. Because the number of leaves is $O(n)$ and that T is similar to a B-tree if we only consider insertions we get that the height of T is $h = O(\log n / \log \Delta) = O(\log n / \log(k / \log^4 n)) = O(\log n / \log k)$ because $k \geq \log^5 n$. It follows that the worst case insertion time in T is $O(\log n)$. But because we remove $\log n$ integers from I_r every time $|I_r| = \Delta \log n$ we spend at most worst case $O(\log n)$ time once for every $\log n$ insertion. All other insertions require worst case $O(\log k)$ time. We conclude that the amortized insertion time is $O(\log k)$.

We now describe how to implement DELETE(e) in amortized $O(\log k)$ time. If e is contained in L we remove e from L. If L is nonempty after having removed e we are done. If L becomes empty we proceed as follows. Let λ be the leftmost leaf of T. The basic idea is to let L become $\hat{I}_\lambda$. We do this as follows. First we flush all buffers along the leftmost path in the tree for integers contained in dom_λ. Based on (3) this can be done in $O(h \log k)$ time. We can now assume $(I_w \cup D_w) \cap \mathrm{dom}_\lambda = \emptyset$ for all nodes w on the leftmost path and that $I_\lambda = \hat{I}_\lambda$. We can now assign L the set I_λ and remove the leaf λ. If the parent p of λ gets degree zero we recursively remove p. Notice that if p gets degree zero then I_p and D_p are both empty. Because the total size of the of insertion and deletion buffers on the leftmost path is bounded by $h\Delta \log n \leq k / \log^2 n$ it follows that $\log n \leq k/2 - k/\log^2 n \leq |L| \leq k + k/\log^2 n$. It follows that L cannot become empty throughout the next $\log n$ DELETE operations. The case $e \in R$ is handled symmetrically by letting λ be the rightmost leaf.

If $e \notin L \cup R$ we insert e in D_r provided $e \notin I_r$. If $e \in I_r$ we remove e from I_r in $O(\log k)$ time and are done. If $|D_r| \geq \Delta \log n$ we can move $\log n$ integers X from D_r to a child w of r. If w is an internal node we first remove $X \cap I_w$ from X and I_w, i.e. delayed insertions cancels delayed insertions, and then inserts the remaining elements in X into D_w. If $|D_w| \geq \Delta \log n$ we recursively move $\log n$ integers from D_w to a child of w. If w is a leaf λ we just remove the integers X from I_λ. If $|I_\lambda| \geq k/2$ we are done. Otherwise let $\bar{\lambda}$ denote the leaf to the right or left of λ (If $\bar{\lambda}$ does not exist the set only contains $O(k)$ integers and the problem is easy to handle. In the following we w.l.o.g. assume $\bar{\lambda}$ exists). We first flush all buffers on the paths from the root r to λ and $\bar{\lambda}$ such that the buffers do not contain elements from $\mathrm{dom}_\lambda \cup \mathrm{dom}_{\bar{\lambda}}$. This can be done in $O(h \log n)$ time as previously described. From

$$k/2 + k/2 - \log n - 2h\Delta \log n \leq |I_\lambda \cup I_{\bar{\lambda}}| \leq k/2 + k - 1 + 2h\Delta \log n$$

it follows that $k/2 \leq |I_\lambda \cup I_{\bar{\lambda}}| \leq 2k$. There are two cases to consider. If $|\lambda + \bar{\lambda}| \geq k$ we redistribute I_λ and $I_{\bar{\lambda}}$ such that they both have size at least $k/2$ and at most k. Because all buffers on the path from λ ($\bar{\lambda}$) to the root intersect empty with $\mathrm{dom}_\lambda \cup \mathrm{dom}_{\bar{\lambda}}$ we in addition only need to update the search key stored at the nearest common ancestor of λ and $\bar{\lambda}$ in T which separates dom_λ and $\mathrm{dom}_{\bar{\lambda}}$. This can be done in $O(h + \log k)$ time. The second case is if $|\lambda + \bar{\lambda}| < k$. We then move the integers in I_λ to $I_{\bar{\lambda}}$ and remove the leaf λ as described previously. The total worst case time for a deletion becomes $O(h \log k) = O(\log n)$. But again the amortized time is $O(\log k)$ because L and R become empty for at most every $\log n$'th DELETE operation, and because D_r becomes full for at most every $\log n$'th DELETE operation.

In the previous description of DELETE we assumed the height of T is $h = O(\log n / \log k)$. We argued that this was true if only INSERT operations were performed because then our search tree is similar to a B-tree. It is easy to see that if only $O(n)$ leaves have been remove, then the height of T is still $h = O(\log n / \log k)$. One way to see this is by assuming that all removed nodes still resist in T. Then T has at most $O(n)$ leaves and each internal node has degree at least $\Delta/2$, which implies the claimed height. By rebuilding T completely such that all internal nodes have degree $\Theta(\Delta)$ for every n'th DELETE operation we can guarantee that at most n leaves have been removed since T was rebuild the last time. The rebuilding of T can easily be done in $O(n \log k)$ time implying that the amortized time for DELETE only increases by $O(\log k)$.

We conclude that INSERT and DELETE can be implemented in amortized $O(\log k)$ time. The space required is $O(n)$ because each node can be stored in $O(1)$ words.

To convert the amortized time bounds into worst case time bounds we apply the standard technique of incrementally performing a worst case expensive operation over the following sequence of operations by moving the expensive operation into a shadow process that is executed in a quasi-parallel fashion with the main algorithm. The rebuilding of T when $O(n)$ DELETE operations have been performed can be handled by the general dynamization technique of Overmars [12] in worst case $O(\log k)$ time per operation. For details refer to [12].

What remains to be described is how to handle the cases when L or R becomes empty and when I_r or D_r becomes full. The basic idea is to handle these cases by simply avoiding them. Below we outline the necessary changes to the amortized solution.

The idea is to allow I_r and D_r to have size $\Delta \log n + O(\log n)$ and to divide the sequence of INSERT and DELETE operations into phases of $\log n/4$ operations. In each phase we perform one of the transformations below to T incrementally over the $\log n/4$ operations of the phase by performing worst case $O(1)$ work per INSERT or DELETE operation. We cyclic choose which transformation to perform, such that for each $\log n$'th operation each transformation has been performed at least once. Each of the transformations can be implemented in worst case $O(\log n)$ time as described in the amortized solution.

- If $|L| < k$ at the start of the phase and λ denotes the leftmost leaf of T we incrementally merge L with $\hat{I}_\lambda$ and remove the leaf λ. It follows that L always has size at least $k - O(\log n) > 0$.
- The second transformation similarly guarantees that $|R| > 0$ by merging R with $\hat{I}_\lambda$ where λ is rightmost leaf of T if $|R| < k$.
- If $|I_r| \geq \Delta \log n$ at the start of the phase we incrementally remove $\log n$ integers from I_r. It follows that the size of I_r is bounded by $\Delta \log n + O(\log n) = O(k)$.
- The last transformation similarly guarantees that the size of D_r is bounded by $\Delta \log n + O(\log n)$ by removing $\log n$ integers from D_r if $|D_r| \geq \Delta \log n$.

This finishes our description of how to achieve the bounds stated in Lemma 2.

5 Range reduction

To prove Theorem 1 we combine Lemma 2 with a range reduction based on a data structure of van Emde Boas *et al.* [15,16]. This is similar to the data structure of Andersson [2], and for details we refer to [2]. We w.l.o.g. assume $w \geq 2^{f(n)} \log n$.

The idea is to use the topmost $f(n)$ levels of the data structure of van Emde Boas *et al.* and then switch to our packed search trees. If $f(n) \geq 5 \log\log n$ the integers we need to store are of $w/2^{f(n)} \leq w/\log^5 n$ bits each and Lemma 2 applies for $k = 2^{f(n)}$. By explicitly remembering the minimum and maximum integer stored FINDMIN and FINDMAX are trivial to support in worst case constant time. The remaining time bounds follow from Lemma 2. The space bound of $O(n2^{\epsilon w})$ follows from storing the arrays at each of the $O(n)$ nodes in the data structure of van Emde Boas *et al.* as a trie of degree $2^{\epsilon w}$.

6 Conclusion

We have presented the first data structure for a Practical RAM allowing the update operations INSERT and DELETE in worst case $O(\log\log n)$ time while answering PRED queries in worst case $O(\log n / \log\log n)$ time. An interesting open problem is if it is possible to support INSERT and DELETE in worst case $O(\log\log n)$ time and PRED in worst case $O(\sqrt{\log n})$ time. The general open problem is to find a tradeoff between the update time and the time for predecessor queries on a Practical RAM.

Acknowledgments

The author thanks Theis Rauhe, Thore Husfeldt and Peter Bro Miltersen for encouraging discussions, and the referees for comments.

References

1. Susanne Albers and Torben Hagerup. Improved parallel integer sorting without concurrent writing. In *Proc. 3rd ACM-SIAM Symposium on Discrete Algorithms (SODA)*, pages 463–472, 1992.
2. Arne Andersson. Sublogarithmic searching without multiplications. In *Proc. 36th Ann. Symp. on Foundations of Computer Science (FOCS)*, pages 655–663, 1995.
3. Arne Andersson. Faster deterministic sorting and searching in linear space. In *Proc. 37th Ann. Symp. on Foundations of Computer Science (FOCS)*, pages 135–141, 1996.
4. Lars Arge. The buffer tree: A new technique for optimal I/O-algorithms. In *Proc. 4th Workshop on Algorithms and Data Structures (WADS)*, volume 955 of *Lecture Notes in Computer Science*, pages 334–345. Springer Verlag, Berlin, 1995.
5. Kenneth E. Batcher. Sorting networks and their applications. In *Proc. AFIPS Spring Joint Computer Conference, 32*, pages 307–314, 1968.
6. Gerth Stølting Brodal, Shiva Chaudhuri, and Jaikumar Radhakrishnan. The randomized complexity of maintaining the minimum. In *Proc. 5th Scandinavian Workshop on Algorithm Theory (SWAT)*, volume 1097 of *Lecture Notes in Computer Science*, pages 4–15. Springer Verlag, Berlin, 1996.
7. Michael L. Fredman and Dan E. Willard. Surpassing the information theoretic bound with fusion trees. *Journal of Computer and System Sciences*, 47:424–436, 1993.
8. Scott Huddleston and Kurt Mehlhorn. A new data structure for representing sorted lists. *Acta Informatica*, 17:157–184, 1982.
9. Peter Bro Miltersen. Lower bounds for Union-Split-Find related problems on random access machines. In *Proc. 26th Ann. ACM Symp. on Theory of Computing (STOC)*, pages 625–634, 1994.
10. Peter Bro Miltersen. Lower bounds for static dictionaries on RAMs with bit operations but no multiplications. In *Proc. 23rd Int. Colloquium on Automata, Languages and Programming (ICALP)*, volume 1099 of *Lecture Notes in Computer Science*, pages 442–453. Springer Verlag, Berlin, 1996.
11. Peter Bro Miltersen, Noam Nisan, Shmuel Safra, and Avi Wigderson. On data structures and asymmetric communication complexity. In *Proc. 27th Ann. ACM Symp. on Theory of Computing (STOC)*, pages 103–111, 1995.
12. Mark H. Overmars. *The Design of Dynamic Data Structures*, volume 156 of *Lecture Notes in Computer Science*. Springer Verlag, Berlin, 1983.

13. Rajeev Raman. Priority queues: Small, monotone and trans-dichotomous. In *ESA '96, Algorithms*, volume 1136 of *Lecture Notes in Computer Science*, pages 121–137. Springer Verlag, Berlin, 1996.
14. Mikkel Thorup. On RAM priority queues. In *Proc. 7th ACM-SIAM Symposium on Discrete Algorithms (SODA)*, pages 59–67, 1996.
15. Peter van Emde Boas. Preserving order in a forest in less than logarithmic time and linear space. *Information Processing Letters*, 6:80–82, 1977.
16. Peter van Emde Boas, R. Kaas, and E. Zijlstra. Design and implementation of an efficient priority queue. *Mathematical Systems Theory*, 10:99–127, 1977.

Semi-Dynamic Shortest Paths and Breadth-First Search in Digraphs*

Paolo Giulio Franciosa[1], Daniele Frigioni[1,2], and Roberto Giaccio[1]

[1] Dipartimento di Informatica e Sistemistica, Università di Roma "La Sapienza", via Salaria 113, I–00198 Roma, Italy. E-mail: {pgf,giaccio}@dis.uniroma1.it
[2] Dipartimento di Matematica Pura ed Applicata, Università di L'Aquila, via Vetoio, I–67010 Coppito (AQ), Italy. E-mail: frigioni@univaq.it

Abstract. We show how to maintain a shortest path tree of a general directed graph G with unit edge weights and n vertices, during a sequence of edge deletions or a sequence of edge insertions, in $O(n)$ amortized time per operation using linear space. Distance queries can be answered in constant time, while shortest path queries can be answered in time linear in the length of the retrieved path. These results are extended to the case of integer edge weights in $[1, C]$, with a bound of $O(Cn)$ amortized time per operation.
We also show how to maintain a breadth-first search tree of a directed graph G in an incremental or a decremental setting in $O(n)$ amortized time per operation using linear space.

1 Introduction

A graph problem is dynamic if it is required to maintain some property on a graph during edge deletions and edge insertions. When both insertions and deletions of edges are allowed we refer to the *fully dynamic problem*; if only insertions or only deletions are allowed then we refer to the *semi-dynamic incremental* or *decremental* problem, respectively. Dynamic graph problems are very interesting in practice since, if the graph represents a transportation or a communication networks, then insertion and deletion of edges reflect the real network changes as links that go up and down during the lifetime of the network.

Various approaches have been considered in the literature to deal with dynamic shortest paths problems both for *single-source* and *all-pairs* versions [1,4,6,8,10,11,14]. In [10] a fully dynamic solution to maintain all-pairs shortest paths for planar graphs with unrestricted edge weights has been proposed, while in [11] the authors provide a sub-linear approximation scheme for the fully dynamic single-source shortest path problem on planar graphs. Both these solutions are complex and far from being practical. In [2] the authors consider efficient dynamic solutions for bounded tree-width graphs when the weight of

* Work partially supported by EC ESPRIT Long Term Research Project ALCOM-IT under contract no. 20244, and by *Progetto Finalizzato Trasporti 2 (PFT 2)* of the Italian National Research Council (*CNR*).

Reischuk, Morvan (Eds.): STACS'97 Proceedings, LNCS 1200

edges might change, but without considering insertion and deletion of edges. An efficient solution for the all-pairs incremental problem has been proposed in [1], assuming that edge weights are restricted in the range of integers in $[1, C]$. In [8] an efficient fully-dynamic output bounded solution is proposed for general graphs with positive edge weights.

From now on we restrict our attention to the single source case. To the best of our knowledge, if there is no restriction on the class of considered graphs then neither a fully dynamic solution nor a decremental solution for the single source shortest path problem is known in the standard models (***worst case*** and ***amortized*** [15]) that is asymptotically better than recomputing the new solution from scratch. This applies even to the case of unit edge weights.

A problem which is strongly related to that of determining a single source shortest path tree (sp-tree for short) is computing a ***breadth-first search tree*** (bfs-tree). A bfs-tree is a spanning tree consisting of the edges traversed by a breadth-first visit of a directed graph, as defined for example in [12], and can be considered as a sp-tree, assuming unit edge weights, with some additional topological constraints. These constraints make the problem of maintaining a bfs-tree possibly harder than maintaining a sp-tree in the case of unit edge weights. In fact, no result is known in the literature for maintaining a bfs-tree of a directed graph in any dynamic setting. In [5] an algorithm is given for maintaining the ***bfs-structure*** of an undirected graph with unit edge weights under edge deletions in $O(n)$ amortized time per deletion, where a bfs-structure is the union of all possible shortest paths from a fixed vertex. Note that the approach in [5] does not maintain a bfs-tree within the same time bounds.

Conversely, another fundamental graph traversal algorithm, the depth-first visit, has been studied in an incremental setting in [7], where an $O(n)$ amortized time algorithm is given for maintaining a depth-first search tree of a directed acyclic graph under edge insertions.

In this paper we study the problems of maintaining a sp-tree and a bfs-tree for a directed graph, in either an incremental or decremental setting. Given a directed graph, we represent a sp-tree or a bfs-tree by associating to each vertex x its parent $\mathrm{p}(x)$ on the current tree and its current distance $\mathrm{d}(x)$ from the source vertex. Thus, a shortest path can be returned in time linear in the number of edges in the path. We maintain this representation during a sequence of edge insertions or a sequence of edge deletions.

In more detail, we maintain a sp-tree of a general directed graph G with unit edge weights during a sequence of q edge deletions or a sequence of q edge insertions in $O(m \cdot \min\{q, n\})$ total time over the sequence, where n is the number of vertices of G and m is the final number of edges of G in the case of insertions or the initial number of edges of G in the case of deletions. This gives a $O(n)$ amortized time for each operation, if the sequence has length $\Omega(m)$. Our algorithm for deletions represents the first decremental solution for directed graphs with unit edge weights that is asymptotically better than recomputing the sp-tree from scratch after each deletion, which can be accomplished in $O(mq)$ total time over the whole sequence (thus giving $O(m)$ time per deletion).

The result for unit edge weights is extended to handle the case of integer edge weights in $[1, C]$, maintaining a sp-tree during a sequence of q edge deletions or a sequence of q edge insertions in total time $O(m \cdot \min\{q, Cn\})$, thus obtaining a $O(Cn)$ amortized time per operation.

We also maintain a bfs-tree of a graph G from a given source vertex, during a sequence of q edge insertions or a sequence of q edge deletions in total time $O(m \cdot \min\{q, n\})$. These are the first results concerning the dynamic maintenance of a bfs-tree for digraphs that are better than applying a breadth-first visit after each update, that can be accomplished in $O(mq)$ total time over a sequence of q updates.

All the proposed algorithms require $O(n + m)$ space ($O(n + m + C)$ in the case of integer weights in $[1, C]$).

2 Preliminaries and notations

We assume the reader familiar with the standard graph terminology as contained, for example, in [9]. Let $G = (V, E)$ be a directed graph with $|V| = n$ and $|E| = m$, and let $s \in V$ be a fixed *source vertex*. The *length* of a path P is the number of edges in P. The *distance function dist* : $V \to \{1, \ldots, n\}$ associates to each vertex x the length *dist*(x) of the shortest path from s to x in G. For the sake of clarity we assume that all vertices are reachable from s, the extension being straightforward.

A spanning tree $T(s) = (V, E_T)$ of G rooted in s is a *single source shortest path tree* from s if for any vertex x the path from s to x in $T(s)$ has length *dist*(x). For any $x \in V$, $S(x)$ denotes the set of vertices in the subtree of $T(s)$ rooted in x. Any vertex $x \in V$ has one *parent* (except for source vertex s), and a (possibly empty) set of *children* in $T(s)$.

In order to efficiently implement our algorithms we represent the graph and the current sp-tree tree by the following simple data structures. Each vertex x is associated the following information:

- $\mathtt{bs}(x)$ = the list of edges $\langle y, x\rangle \in E$, (the *backward star* of x);
- $\mathtt{fs}(x)$ = the list of edges $\langle x, y\rangle \in E$, (the *forward star* of x).
- $\mathtt{p}(x)$ = the parent of x (according to the current sp-tree);
- $\mathtt{ch}(x)$ = the list of children of x (according to the current sp-tree);
- $\mathtt{d}(x)$ = a suitable integer labeling.

Before and after applying our update procedures, the value of $\mathtt{d}(x)$ coincides with *dist*(x). Furthermore, the algorithms explicitly update parent pointers, but not children lists. In fact, children lists are needed only at the start of each update processing, before any modification of the sp-tree is performed.

Children lists can be rebuilt in $O(n)$ worst case time after processing each insertion/deletion as follows:

1. for each vertex x, set $\mathtt{ch}(x)$ to the empty set;
2. for each vertex y, append y to $\mathtt{ch}(\mathtt{p}(y))$.

3 Semi-dynamic sp-tree

In this section we study the problem of maintaining a sp-tree for a directed graph, in either an incremental or decremental setting. More specifically, we maintain the two functions $\mathtt{p}(\cdot)$ and $\mathtt{d}(\cdot)$, which completely represent the sp-tree.

After inserting or deleting an edge, this information is still correct for a subset of the vertices (a *rooted set*), while parents and/or distances of the other vertices have to be recomputed. The basic idea of our solution consists in finding a *linking set*, i.e., a suitable subset of edges connecting vertices in the *rooted set* to the other vertices, which allows us to propagate the correct distances from vertices in the rooted set.

Given a graph $G = (V, E)$, a source vertex $s \in V$, a parent function $\mathtt{p}(\cdot)$ and an integer function $\mathtt{d}(\cdot)$ on the vertex set, we define:

Definition 1. A set of vertices $R \subseteq V$ is a ***rooted set*** if there exists a sp-tree T of G rooted in s such that the following conditions hold:

- $\mathtt{d}(s) = 0$;
- $x \in R - \{s\} \Leftrightarrow \langle \mathtt{p}(x), x\rangle \in T \wedge \mathtt{d}(x) = \mathit{dist}(x)$.

According to this definition, functions $\mathtt{p}(\cdot)$ and $\mathtt{d}(\cdot)$ completely represent a sp-tree if and only if the whole V is a rooted set.

Definition 2. A set of edges L is a ***linking set*** for a rooted set R if there exists a sp-tree $T = (V, E_T)$ s.t. $E_T \cap (R \times \overline{R}) \subseteq L \subseteq E \cap (R \times \overline{R})$.

The above definition says that a set L of edges from R to $\overline{R}$ is a linking set if for at least a sp-tree T, all edges of T from R to $\overline{R}$ are in L. This ensures that if we want to expand a rooted set R by adding a vertex $x \in \overline{R}$ adjacent to a vertex in R, an edge connecting it to a suitable parent can be found in any corresponding linking set.

In order to correctly update distances of vertices in $\overline{R}$, we need the following:

Definition 3. A rooted set R is ***propagable*** if $\forall x \in \overline{R}$, $\mathtt{d}(x) > \mathit{dist}(x)$.

Finally, functions $\mathtt{p}(\cdot)$ and $\mathtt{d}(\cdot)$ are actually updated by Procedure **Propagate**, shown in Figure 1.

Lemma 4. *If Procedure* **Propagate** *is invoked on a linking set L for some propagable rooted set R, then when it terminates V is a rooted set.*

Proof. We first show that each iteration of the **while** loop adds a new vertex to R and updates $\mathtt{L}$ accordingly.

By Definition 2, there exists a sp-tree $T = (V, E_T)$ rooted in s such that all vertices in $\overline{R}$ can be reached from s without using edges in $(R \times \overline{R}) - \mathtt{L}$.

Since $\langle a, b\rangle$ is selected from $\mathtt{L}$ in line 1 as the edge having minimum $\mathit{dist}(a)$, then all vertices in $\overline{R}$ have distance at least $\mathit{dist}(a)+1$. This implies that $\mathit{dist}(b) = \mathit{dist}(a) + 1$. Thus b is correctly added to R.

```
procedure Propagate(L: set of edges)
begin
    while L ≠ ∅ do
1.      select edge ⟨a, b⟩ from L having minimum d(a) + 1
2.      remove all edges ⟨a', b⟩ from L
        p(b) := a
        d(b) := d(a) + 1
        for each edge ⟨b, c⟩ ∈ fs(b) s.t. d(c) > d(b) + 1 do
3.          insert ⟨b, c⟩ into L
        end do
    end do
end
```

Fig. 1. Procedure Propagate.

When vertex b is moved from $\overline{R}$ to R, we have to update L; in line 2 of the procedure we remove from L all edges from R to b (these edges are no longer in $R \times \overline{R}$); in line 3 we insert into L all edges $\langle b, c\rangle$ such that $\mathtt{d}(c) > \mathtt{d}(b) + 1$. Since $dist(c) \leq dist(b) + 1 = \mathtt{d}(b) + 1$, we have $\mathtt{d}(c) > dist(c)$, and Definition 3 ensures that $c \in \overline{R}$.

Furthermore, by Definition 2, a linking set is empty if and only if the corresponding rooted set consists of all the vertices. This implies that when Procedure Propagate stops, all the vertices have been moved to R. □

Lemma 5. *Procedure* Propagate *requires* $O\left(n + m_q + \sum_{x\in\overline{R}} |\mathtt{fs}(x)|\right)$ *time, where* m_q *is the number of edges initially in* L.

Proof. We represent L by means of an array of edge sets, one for each possible value of $\mathtt{d}(\cdot)$; $\mathtt{L}[k]$ contains all edges $\langle a, b\rangle$ such that $\mathtt{d}(a)+1 = k$. Each edge can be trivially inserted in L in constant time, while edges $\langle a, b\rangle$ are extracted from L by non decreasing $\mathtt{d}(a)$. Since any time we extract $\langle a, b\rangle$ in line 1, edges $\langle b, c\rangle$ inserted in L have $\mathtt{d}(b) > \mathtt{d}(a)$, then the total time needed for inserting and extracting edges in lines 1 and 3, respectively, is $O(m_q + \textit{number of edges inserted into } \mathtt{L})$ plus $O(n)$ for scanning the array L for all possible distance values.

Furthermore we maintain links from each vertex b to all the edge entries $\langle a', b\rangle$ currently in L. This allows us to remove those edges from L in line 2 in constant time per edge. Since the number of edges inserted into L is bounded by $\sum_{x\in\overline{R}} |\mathtt{fs}(x)|$, the lemma follows. □

We describe in the sequel a straightforward extension of Dijkstra's algorithm [3] for maintaining a sp-tree under edge insertions, which is shown in Figure 2. It represents a first application of Lemmas 4 and 5.

Theorem 6. *Procedure* Insert *correctly maintains a sp-tree.*

```
procedure Insert(⟨x,y⟩: edge)
begin
    insert ⟨x,y⟩ into fs(x)
    insert ⟨x,y⟩ into bs(y)
    if d(y) > d(x) + 1 then
        Propagate(⟨x,y⟩)
    end if
end
```

Fig. 2. Procedure `Insert`.

Proof. Let R be the set of vertices whose distance does not change after the insertion of edge $\langle x, y\rangle$. R is a propagable rooted set with respect to functions $\mathtt{d}(\cdot)$ and $\mathtt{p}(\cdot)$. In fact, vertices having uncorrect parents after the insertion also have uncorrect distances, and such distances can only be greater than the correct ones.

If $\mathtt{d}(y) > \mathtt{d}(x) + 1$ then $\{\langle x, y\rangle\}$ is a linking set for R, since any vertex whose distance changes must have all its new shortest paths passing through edge $\langle x, y\rangle$. The theorem now follows from Lemma 4. □

Theorem 7. *The total time required by Procedure* `Insert` *to handle a sequence of q edge insertions is $O(m \cdot \min\{q, n\})$.*

Proof. Due to an edge insertion vertices cannot increase their distance. Since the distance of a vertex can decrease at most $\min\{q, n\}$ times, each vertex can be in $\overline{R}$ at most $\min\{q, n\}$ times during any sequence of q edge insertions. Since in this case $m_q = 1$, by Lemma 5 the total time spent by Procedure `Propagate` is $O(nq + \min\{q, n\} \cdot \sum_{x \in \overline{R}} |\mathtt{fs}(x)|) = O(nq + m \cdot \min\{q, n\})$. Since $qn \leq qm$ and $qn \leq mn$, the theorem follows. □

3.1 Decremental sp-tree

We describe now Procedure `Delete`, which maintains a sp-tree of a directed graph during a sequence of edge deletions. Some considerations are worth noting: when an edge $\langle x, y\rangle$ is deleted, distances cannot decrease; furthermore, all vertices whose distance increases are contained in $S(y)$.

The algorithm works in two phases: first a parent is assigned to all vertices in $S(y)$ whose distance from s does not increase due to the deletion of edge $\langle x, y\rangle$; then a parent is assigned to vertices in $S(y)$ whose distance from the source increases; this is done by setting their distance to $+\infty$, thus fulfilling Definition 3, computing a linking set and calling Procedure `Propagate`. The algorithm is shown in detail in Figure 3. Note that, for the sake of clarity, we do not specify what to do when a vertex becomes disconnected from the source, the extension being straightforward.

A crucial consideration is that the relative position in which edges are stored in each backward star does not change under edge deletions. So, at the beginning we assume that edges are stored in each backward star according to an arbitrary given ordering. We also assume that at the beginning (before any edge deletion) the parent of each vertex is the first possible one according to that ordering, i.e. edge $\langle \mathtt{p}(x), x\rangle$ is stored in $\mathtt{bs}(x)$ in a position such that there is no edge $\langle a, x\rangle$ with $\mathtt{d}(a) = \mathtt{d}(\mathtt{p}(x))$ preceding $\langle \mathtt{p}(x), x\rangle$ in that order. We say vertex x *appears* in $\mathtt{bs}(y)$ if there exists edge $\langle x, y\rangle$ (i.e., edge $\langle x, y\rangle$ is stored in $\mathtt{bs}(y)$).

```
procedure Delete(⟨x,y⟩: edge)
begin
     remove ⟨x,y⟩ from fs(x)
     remove ⟨x,y⟩ from bs(y)
     if x = p(y) then
          L := ∅
          for each vertex b in S(y), scanned by increasing tree level, do
1.             search the first edge ⟨a,b⟩ in bs(b) with d(a) + 1 = d(b),
                    starting from edge ⟨p(b),b⟩
               if such an edge ⟨a,b⟩ exists then
                    p(b) := a
               else
                    d(b) := +∞
2.                  select the first edge ⟨a,b⟩ in bs(b) s.t. d(b) is minimum
                    if d(a) ≠ +∞ then
                         insert ⟨a,b⟩ into L
                    end if
               end if
          end do
          Propagate(L)
     end if
end
```

Fig. 3. Procedure Delete.

Let us prove the correctness of Procedure Delete. The following property is guaranteed by line 1.

Lemma 8. *Given a vertex b, for each edge $\langle z, b\rangle$ stored in* $\mathtt{bs}(b)$ *before* $\langle \mathtt{p}(b), b\rangle$, $\mathtt{d}(z) > \mathtt{d}(\mathtt{p}(b))$ *holds.*

Proof. We assume that the thesis is true before processing a deletion. While processing the deletion the following cases may occur:

- if $\mathtt{p}(b)$ preserves its distance from the source then the thesis still holds, since vertices appearing before $\mathtt{p}(b)$ in $\mathtt{bs}(b)$ cannot decrease their distance after the deletion;
- else, let a be the first vertex with $\mathtt{d}(a)+1 = \mathtt{d}(b)$ appearing after $\mathtt{p}(b)$ in $\mathtt{bs}(b)$. If such vertex exists then it is found by procedure `Delete` in line 1, and $\mathtt{p}(b)$ is set to a; otherwise procedure `Delete` scans all $\mathtt{bs}(b)$ from the beginning (line 2), setting $\mathtt{p}(b)$ to the first vertex found with minimum distance. In both cases the lemma still holds. □

Theorem 9. *Procedure* `Delete` *correctly maintains a sp-tree.*

Proof. After deleting edge $\langle x, y\rangle$, only the distance of vertices in $S(y)$ might increase. Procedure `Delete` finds the set U of all vertices in $S(y)$ whose distance remains unchanged (line 1), and assigns a parent to them, as described in Lemma 8.

Set U is built by scanning vertices in $S(y)$ by increasing tree level using children lists. When a vertex b is scanned, if its distance does not change then all vertices in the shortest path from s to b either are in $\overline{S(y)}$ or in the already scanned portion of $S(y)$, and their distances have not changed, too. Hence, $R = \overline{S(y)} \cup U$ is a propagable rooted set with respect to function $\mathtt{d}(\cdot)$, provided that $\mathtt{d}(b)$ has been set to $+\infty$ for each vertex in $S(y) - U$.

A linking set for R is determined by scanning the whole backward star of each vertex in $\overline{R}$, and selecting for each vertex b the first edge $\langle a, b\rangle$ such that $a \in R$ and $\mathtt{d}(a)$ is minimum (line 2). It is easy to see that any edge of a sp-tree from R to $\overline{R}$ is contained in this set; in fact for each vertex $b \in \overline{R}$, either $\mathtt{p}(b) \in \overline{R}$ or $\langle \mathtt{p}(b), b\rangle$ has been selected. The theorem now follows from Lemma 4. □

The following Lemma provides a bound for the time spent in lines 1 and 2 of Procedure `Delete` during the whole sequence of deletions.

Lemma 10. *Given a vertex b, the total time required by Procedure* `Delete` *for scanning* $\mathtt{bs}(b)$ *during a sequence of q edge deletions is* $O(\min\{q, n\} \cdot |\mathtt{bs}_i(b)| + q)$, *where* $\mathtt{bs}_i(b)$ *is the initial backward star of b.*

Proof. The backward star of vertex b is completely scanned at most once for each value assumed by $\mathtt{d}(b)$. In fact, when b's distance assumes a new value k, the first parent in $\mathtt{bs}(b)$ having distance $k - 1$ is found in line 2, while line 1 scans the remaining portion of $\mathtt{bs}(b)$ during a subsequence of deletions that do not affect b's distance.

Since on a sequence of edge deletions $\mathtt{d}(b)$ is non decreasing, and at any time $|\mathtt{bs}(b)| \le |\mathtt{bs}_i(b)|$, the lemma follows. □

Theorem 11. *The total time required by Procedure* `Delete` *to handle a sequence of q edge deletions is* $O(m \cdot \min\{q, n\})$.

Proof. Since the distance of a vertex can increase at most $\min\{n, q\}$ times, each vertex can be in $\overline{R}$ at most $\min\{n, q\}$ times during any sequence of q edge deletions. Moreover, for each vertex b at most one edge $\langle a, b\rangle$ is inserted into L, so term m_q in lemma 5 is bounded by n. Hence the total time spent by Procedure `Propagate` during a sequence of q edge deletions is $O(nq + \min\{q, n\} \cdot (\sum_{x \in \overline{R}} |\mathtt{fs}(x)|)) = O(nq + m \cdot \min\{q, n\})$.

Finally, the total time spent by Procedure `Delete` in lines 1 and 2, which can be derived by applying Lemma 10 to all vertices, sums up to $O(\min\{q, n\} \cdot (\sum_{x \in V} |\mathtt{bs}(x)|) + qn) = O(m \cdot \min\{q, n\} + qn)$.

Since $qn \leq qm$ and $qn \leq mn$, we have $qn \leq \min\{q, n\} \cdot m$, and the theorem follows. □

The $O(m \cdot \min\{q, n\})$ total time bound for the unit weight case mainly derives from the fact that the distance of each vertex changes at most n times during the whole sequence of q updates.

It is possible to adapt this algorithm in order to deal with integer edge weights in the range $[1, C]$. In this case, since distances between vertices are now bounded by Cn, we can prove what follows.

Theorem 12. *The total time required by Procedure* `Delete` *to handle a sequence of q edge deletions, in the case of edge weights in $[1, C]$, is $O(m \cdot \min\{q, Cn\})$.*

Theorem 13. *Procedure* `Delete`*, in the case of edge weights in $[1, C]$, requires $O(n + C + m)$ space.*

More formal arguments will be given in the full version of the paper.

4 Semi-dynamic bfs-tree

A *breadth-first search tree* (bfs-tree for short) is a spanning tree consisting of the edges traversed by a breadth-first visit of G. A breadth-first visit is a well known graph traversal technique, defined in several textbooks (see for example [12], p. 539), that scans all vertices of the graph one level at a time and, for each scanned vertex, traverses all outgoing edges leading to unvisited vertices, as described in Figure 4. The resulting tree T strongly depends on the order in which vertices in each level are considered in line 1. Clearly, the algorithm described above may be easily implemented by storing all newly visited vertices in a queue (line 2), extracting them before traversing outgoing edges (line 1).

Using a queue would lead to a proper subclass of bfs-trees, with respect to the definition of bfs-trees given in [12]. Actually, in some papers concerning breadth-first visits only this subclass is considered; for this subclass a structural characterization in the case of undirected graphs is provided in [13].

It is easy to see that, in the case of unit weight edges, a bfs-tree is always a sp-tree, and the distance of any vertex x from s is given by `level(x)`. However, note that the opposite is not true, since there are sp-trees that cannot be obtained by the algorithm in Figure 4.

Figure 5 shows three spanning trees for the same graph G: T_1 is a sp-tree but not a bfs-tree. T_2 is a bfs-tree but cannot be obtained by using a queue, while T_3 belongs to the more restricted class of bfs-trees characterized in [13].

```
T := ∅
for each x ∈ V do level(x) := +∞
level(s) := 0
k := 0
repeat
1.  for each x ∈ V with level(x) = k do
        for each ⟨x, y⟩ do
            if level(y) = +∞ then
2.              insert ⟨x, y⟩ into T
                level(y) := k + 1
            end if
        end do
    end do
    k := k + 1
until all vertices have level(x) ≠ +∞
```

Fig. 4. A breadth-first visit algorithm.

In the following we refer to bfs-trees as defined in [12]. We give here a novel characterization of such bfs-trees as follows:

Lemma 14. *Given a digraph $G = (V, E)$ with a source s, a spanning tree $T = (V, E_T)$ for G rooted in s is a bfs-tree if and only if the following conditions hold:*

1. *for each edge $\langle x, y\rangle \in E_T$, $dist(y) = dist(x) + 1$;*
2. *it is possible to order vertices of G in such a way that, if $\langle x, y\rangle \in E - E_T$, $\langle z, y\rangle \in E_T$, and $\texttt{level}(x) = \texttt{level}(z)$, then z precedes x in this order.*

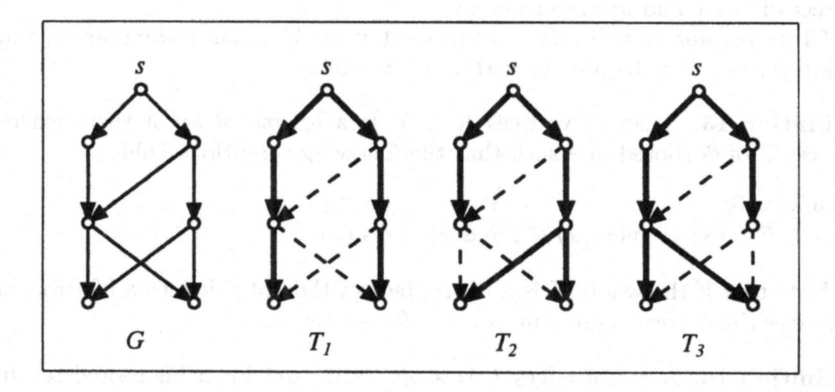

Fig. 5. Different spanning trees of a graph.

Proof. Only if: since algorithm in Figure 4 visits all vertices ordered by distance, Property 1 is a necessary condition for a spanning tree in order to be a bfs-tree. Property 2 is fulfilled by the ordering in which vertices are considered in line 1.

If: Spanning tree T can be generated by algorithm in Figure 4 if in line 1, vertices in the same level are considered according to the ordering provided by Property 2. □

Note that property 1 in Lemma 14 completely characterizes sp-trees in the case of unit edge weights.

Since bfs-trees are characterized by a suitable ordering of the vertices, we also need to represent such a total ordering; this is accomplished by a one to one ranking function $rank : V \to [1, n]$. The rank of a vertex is fixed before the first update takes place and never changes.

Our algorithms for semi-dynamic maintenance of bfs-trees use exactly the same simple data structures described in Section 2 for sp-trees, with the only difference that, in order to bound the time complexity, for each vertex y, edges $\langle x, y \rangle$ in $\mathtt{bs}(y)$ and vertices x in $\mathtt{ch}(y)$ are sorted by increasing value of $rank(x)$.

As in the case of sp-tree maintenance, children lists can be rebuilt in $O(n)$ worst case time after processing each insertion/deletion as follows:

1. for each vertex x, set $\mathtt{ch}(x)$ to the empty set;
2. scan vertices by increasing rank and, for each vertex y, append y to $\mathtt{ch}(\mathtt{p}(y))$.

While maintaining a bfs-tree, a vertex b whose distance does not change might change its parent; in the case of edge insertions, this may happen when the old parent a' can be replaced by a new parent a whose distance has decreased to that of a', and such that $rank(a) < rank(a')$; in the case of edge deletions, this may happen when the old parent a' increases its distance but a new parent a at the same distance as a' exists with $rank(a) < rank(a')$. Hence in the definition of (propagable) rooted set for a bfs-tree we have to consider also vertices with correct distance and incorrect parent.

Given a graph $G = (V, E)$, a source vertex $s \in V$, a parent function $\mathtt{p}(\cdot)$ and an integer function $\mathtt{d}(\cdot)$ on the vertex set, we define:

Definition 15. A set of vertices $R \subseteq V$ is a ***bfs-rooted set*** if there exists a bfs-tree T of G rooted in s such that the following conditions hold:

- $\mathtt{d}(s) = 0$;
- $x \in R - \{s\} \Leftrightarrow \langle \mathtt{p}(x), x \rangle \in T \wedge \mathtt{d}(x) = dist(x)$.

Note that if the whole V is a bfs-rooted set then $\mathtt{p}(\cdot)$ defines a bfs-tree, and $\mathtt{d}(\cdot)$ gives the correct minimum distance for all vertices.

Definition 16. A set of edges L is a ***bfs-linking set*** for a bfs-rooted set R if there exists a bfs-tree $T = (V, E_T)$ s.t. $E_T \cap \left(R \times \overline{R}\right) \subseteq L \subseteq E \cap \left(R \times \overline{R}\right)$.

Definition 17. A bfs-rooted set R is a ***propagable*** bfs-rooted set if $\forall x \in \overline{R}$, $\mathtt{d}(x) > dist(x) \vee (\mathtt{d}(x) = dist(x) \wedge \langle \mathtt{p}(x), x \rangle \notin T)$.

Procedure BFS_Propagate, shown in Figure 6, is used to propagate the distance values from vertices whose distance and parent is correct to vertices whose distance and/or parent has to be changed. It is similar to Procedure Propagate; in addition it behaves differently on vertices in L having correct distance but incorrect parent.

```
procedure BFS_Propagate(L: set of edges)
begin
    while L ≠ ∅ do
        let ⟨a', b⟩ be an edge in L having minimum d(a') + 1
1.      select ⟨a, b⟩ from L s.t. d(a) = d(a') having minimum rank
2.      remove all edges ⟨a'', b⟩ from L
        p(b) := a
        if d(b) > d(a) + 1 then
            d(b) := d(a) + 1
            for each edge ⟨b, c⟩ ∈ fs(b) s.t. d(c) > d(b) + 1
                    or (d(c) = d(b) + 1 and rank(b) < rank(p(c))) do
                insert ⟨b, c⟩ into L
            end do
        end if
    end do
end
```

Fig. 6. Procedure BFS_Propagate.

Lemma 18. *If Procedure* BFS_Propagate *is invoked on a bfs-linking set for some propagable bfs-rooted set R, then when it terminates V is a bfs-rooted set.*

Proof. Concerning vertices $b \in \overline{R}$ having $\mathtt{d}(b) > dist(b)$, considerations similar to those exploited in Lemma 4 can be used to prove that Procedure BFS_Propagate correctly assigns a parent and a distance to each of them. Note that the new parent of b is chosen as the vertex having minimum rank among all possible parents having distance $dist(b) - 1$.

Regarding vertices having $\mathtt{d}(b) = dist(b)$, Procedure BFS_Propagate assigns them a new parent having the same distance and a lower rank.

Since the chosen parent for each vertex added to R is the one having minimum rank over all possible parents, the theorem follows from Lemma 14. □

The following Lemma is the equivalent of Lemma 5.

Lemma 19. *Procedure* BFS_Propagate *requires* $O\left(n + m_q + \sum_{x \in W} |\mathtt{fs}(x)|\right)$ *time, where m_q is the number of edges initially in* L, *and $W \subseteq \overline{R}$ is the set of all vertices x having* $\mathtt{d}(x) > dist(x)$.

Proof. The proof is analogous to the proof of Lemma 5. The main difference derives from the fact that if $x \in \overline{R}$ and its distance does not change, then $\mathtt{fs}(x)$ is not examined. The selection step in line 1 can be performed within the same time bound of the removal in line 2, which is the same as the removal step in line 2 of Procedure `Propagate`. □

4.1 Incremental bfs-tree

In order to maintain a bfs-tree of a directed graph during a sequence of edge insertions, we can use an algorithm which is quite similar to Procedure `Insert` for sp-trees. The only change required is the following: when edge $\langle x, y\rangle$ is inserted, in addition to the case in which the distance of y decreases, $\mathtt{p}(y)$ has to be updated also when the distance of y remains unchanged but $rank(x) < rank(\mathtt{p}(y))$. In both these cases Procedure `BFS_Propagate` is called on the set consisting of the single edge $\langle x, y\rangle$. It is easy to see that $\{\langle x, y\rangle\}$ is a bfs-linking set for the set R of vertices whose distance and/or parent changes. In fact, if $\mathtt{d}(y) = \mathtt{d}(x) + 1$ and $rank(x) < rank(\mathtt{p}(y))$ then $\overline{R} = \{y\}$ and x is a better parent for y, hence $\{\langle x, y\rangle\}$ is a bfs-linking set. On the other hand, if $\mathtt{d}(y) > \mathtt{d}(x) + 1$, then $\{\langle x, y\rangle\}$ is a bfs-linking set for R, since all vertices whose distance or parent changes must have all their possible shortest paths passing through edge $\langle x, y\rangle$.

The following Theorem is the equivalent of Theorem 7, and is a direct consequence of Lemma 19.

Theorem 20. *The total time required for maintaining a bfs-tree of a digraph under a sequence of q edge insertions is $O(m \cdot \min\{q, n\})$.*

4.2 Decremental bfs-tree

We now briefly describe how to maintain a bfs-tree of a directed graph during a sequence of edge deletions. Similarly to the case of sp-tree, vertices whose distance increases and/or whose parent changes can only be in $S(y)$.

The only difference with respect to sp-tree maintenance is that, when we assign a new parent to a vertex x, it must be chosen as the vertex having minimum rank among those having minimum distance, thus fulfilling Property 2 in Lemma 14. Due to Lemma 8, this can be accomplished as in line 1 and line 2 of Procedure `Delete`, provided that vertices appearing in each backward star are stored by increasing rank.

Thus, by calling Procedure `BFS_Propagate` instead of `Propagate`, and by storing vertices in backward stars by increasing rank, Procedure `Delete` also maintains a bfs-tree. The correctness of the algorithm directly derives from Lemma 18, and from the fact that, by the above considerations, set `L` is a bfs-linking set.

The time complexity of the algorithm for decremental bfs-tree maintenance just derives from the analysis of Procedure `Delete` and Lemma 19.

Theorem 21. *The total time required for maintaining a bfs-tree of a digraph under a sequence of q edge deletions is $O(m \cdot \min\{q, n\})$.*

Acknowledgments

We like to thank Pino Italiano and Alberto Marchetti-Spaccamela for useful discussions and constructive comments.

References

1. G. Ausiello, G. F. Italiano, A. Marchetti-Spaccamela, and U. Nanni. Incremental algorithms for minimal length paths. *Journal of Algorithms*, 12(4):615–638, 1991.
2. S. Chaudhuri and C. D. Zaroliagis. Shortest path queries in digraphs of small treewidth. In *Proc. of 22nd Int. Colloq. on Automata, Languages and Programming (ICALP '95)*, volume 944 of *Lecture Notes in Computer Science*, pages 244–255. Springer-Verlag, 1995.
3. E. W. Dijkstra. A note on two problems in connexion with graphs. *Numerische Mathematik*, 1:269–271, 1959.
4. S. Even and H. Gazit. Updating distances in dynamic graphs. *Methods of Operations Research*, 49:371–387, 1985.
5. S. Even and Y. Shiloach. An on-line edge deletion problem. *Journal of ACM*, 28:1–4, 1981.
6. E. Feuerstein and A. Marchetti-Spaccamela. On-line algorithms for shortest paths in planar graphs. *Theoretical Computer Science*, 116:359–371, 1993.
7. P. G. Franciosa, G. Gambosi, and U. Nanni. On the structure of DFS-Forests on directed graphs and the dynamic maintenance of DFS on DAG's. In *Proc. of 2nd European Symposium on Algorithms (ESA '94)*, volume 855 of *Lecture Notes in Computer Science*, pages 343–353. Springer-Verlag, 1994.
8. D. Frigioni, A. Marchetti-Spaccamela, and U. Nanni. Fully dynamic output bounded single source shortest path problem. In *Proc. of 7th Annual ACM-SIAM Symposium on Discrete Algorithms (SODA '96)*, pages 212–221, Atlanta, Georgia, 1996.
9. F. Harary. *Graph Theory*. Addison-Wesley, Reading, Mass., 1969.
10. P. N. Klein, S. Rao, M. H. Rauch, and S. Subramanian. Faster shortest-path algorithms for planar graphs. In *Proc. of 23rd ACM Symposium on Theory of Computing (STOC '94)*, pages 27–37. Assoc. for Computing Machinery, 1994.
11. P. N. Klein and S. Subramanian. A fully dynamic approximation scheme for all-pairs shortest paths in planar graphs. In *Proc. of 3rd Workshop on Algorithms and Data Structures (WADS '93)*, volume 709 of *Lecture Notes in Computer Science*, pages 442–451. Springer-Verlag, 1993.
12. J. van Leeuwen. Graph algorithms. In J. van Leeuwen, editor, *Algorithms and Complexity*, volume A of *Handbook of Theoretical Computer Science*, pages 527–631. Elsevier, Amsterdam, 1990.
13. U. Manber. Recognizing breadth-first search trees in linear time. *Information Processing Letters*, 34:167–171, 1990.
14. H. Rohnert. A dynamization of the all-pairs least cost problem. In *Proc. of 2nd Sympos. on Theoretical Aspects of Computer Science (STACS '85)*, volume 182 of *Lecture Notes in Computer Science*, pages 279–286. Springer Verlag, 1985.
15. R. E. Tarjan. Amortized computational complexity. *SIAM Journal on Algebraic Discrete Methods*, 6(2):306–318, 1985.

Greibach Normal Form Transformation, Revisited

Robert Koch and Norbert Blum

Informatik IV, Universität Bonn
Römerstr. 164, D-53117 Bonn, Germany
email: blum@cs.uni-bonn.de

Abstract. We develop a direct method for placing a given context-free grammar into Greibach normal form with only polynomial increase of its size; i.e., we don't use any algebraic concept like formal power series. Starting with a cfg G in Chomsky normal form, we will use standard methods for the construction of an equivalent context-free grammar from a finite automaton and vice versa for transformation of G into an equivalent cfg G' in Greibach normal form. The size of G' will be $O(|G|^3)$, where $|G|$ is the size of G. Moreover, we show that it would be more efficient to apply the algorithm to a context-free grammar in canonical two form, obtaining a context-free grammar where, up to chain rules, the productions fulfill the Greibach normal form properties, and then to use the standard method for chain rule elimination for the transformation of this grammar into Greibach normal form. The size of the constructed grammar is $O(|G|^4)$ instead of $O(|G|^6)$, which we would obtain if we transform G into Chomsky normal form and then into Greibach normal form.

1 Introduction and Definitions

We assume that the reader is familiar with the elementary theory of finite automata and context-free grammars as written in standard text books, e.g. [1, 4, 6, 9]. First, we will review the notations used in the subsequence.

A *context-free grammar* (cfg) G is a 4-tuple (V, Σ, P, S) where V is a finite, nonempty set of symbols called the *total vocabulary*, $\Sigma \subset V$ a finite set of *terminal symbols*, $N = V \setminus \Sigma$ the set of *nonterminal symbols* (or *variables*), P a finite set of *rules* (or *productions*), and $S \in N$ is the *start symbol*. The productions are of the form $A \to \alpha$, where $A \in N$ and $\alpha \in V^*$. α is called *alternative* of A. $L(G)$ denotes the *context-free language* generated by G. The *size* $|G|$ of the cfg G is defined by

$$|G| = \sum_{A \to \alpha \in P} lg(A\alpha),$$

where $lg(A\alpha)$ is the length of the string $A\alpha$. Two context-free grammars G and G' are equivalent if both grammars generate the same language; i.e., $L(G) = L(G')$.

Reischuk, Morvan (Eds.): STACS'97 Proceedings, LNCS 1200

Let ε denote the empty word. A production $A \to \varepsilon$ is called *ε-rule*. A production $A \to B$ with $B \in N$ is called *chain rule*.

A *leftmost derivation* is a derivation where, at every step, the variable replaced has no variable to its left in the sentential form from which the replacement is made.

A cfg $G = (V, \Sigma, P, S)$ is in *canonical two form* if each production is of the form

i) $A \to BC$ with $B, C \in N \setminus \{S\}$,
ii) $A \to B$ with $B \in N \setminus \{S\}$,
iii) $A \to a$ with $a \in \Sigma$, or
iv) $S \to \varepsilon$.

A cfg G is in *Chomsky normal form* if G is in canonical two form and G contains no chain rule.

A cfg $G = (V, \Sigma, P, S)$ is in *extended Greibach normal form* if each production is of the form

i) $A \to aB_1B_2 \dots B_t$ with $B_1, B_2, \dots B_t \in N \setminus \{S\}, a \in \Sigma$,
ii) $A \to B$ with $B \in N \setminus S$,
iii) $A \to a$ with $a \in \Sigma$, or
iv) $S \to \varepsilon$.

A cfg G is in *extended 2-standard form* if G is in extended Greibach normal form and with respect to rules of kind (i), always $t \leq 2$.

A cfg $G = (V, \Sigma, P, S)$ is in *Greibach normal form* (Gnf) if G is in extended Greibach normal form and G contains no chain rule. G is in *2-standard form* (2-Gnf) if G is in extended 2-standard form and G contains no chain rule.

A *nondeterministic finite automaton* M is a 5-tuple $(Q, \Sigma, \delta, q_0, q_f)$, where Q is a finite set of states, Σ is a finite, nonempty set of input symbols, δ is a transition function mapping $Q \times \Sigma$ to 2^Q, $q_0 \in Q$ is the initial state, and $q_f \in Q$ is the finite state.

Given an arbitrary cfg $G = (V, \Sigma, P, S)$, it is well known that G can be transformed into an equivalent cfg G' which is in Gnf [3, 4, 6, 9]. But the usual algorithms possibly construct a cfg G', where the size of G' is exponential in the size of G (see [4], pp. 113–115 for an example). Given a cfg G without ε-rules and without chain rules, Rosenkrantz [8] has given an algorithm which produces an equivalent cfg G' in Gnf such that $|G'| = O(|G|^3)$. Rosenkrantz gave no analysis of the size of G'. For an analysis, see [4], pp. 129–130 or [7]. Rosenkrantz's algorithm uses formal power series.

We will develop a direct method for placing a given cfg into Gnf with only polynomial increase of its size; i.e., we don't use any algebraic concept like formal power series. Given any cfg G, in a first step the grammar will be transformed into an equivalent cfg G' in Cnf. Then, we will use standard methods for the construction of an equivalent cfg from an nfa and vice versa for transformation of

G' into an equivalent cfg G'' in 2-Gnf. The size of G'' will be $O(|G'|^3)$. Moreover, we show that it would be more efficient to apply the algorithm to a cfg in canonical two form, obtaining a cfg in extended 2-standard form and then to use the standard method for chain rule elimination for the transformation of the grammar into 2-Gnf. The size of the constructed grammar is $O(|G|^4)$ instead of $O(|G|^6)$, which we would obtain if we transform G into Cnf and then into Gnf.

2 The Method

Let $G = (V, \Sigma, P, S)$ be an arbitrary cfg in Cnf. Productions of type $A \to a$ with $a \in \Sigma$ already fulfill the Greibach normal form properties. Our goal is now to replace the productions of type $A \to BC$, $B, C \in N \setminus \{S\}$ by productions which fulfill the Greibach normal form properties. Let $B \in N \setminus \{S\}$. We have an interest in leftmost derivations of the form

$$B \Rightarrow a \text{ or } B \Rightarrow^*_{lm} B_0 B_1 \ldots B_t \Rightarrow a B_1 B_2 \ldots B_t,$$

where $a \in \Sigma$ and $B_0, B_1, \ldots, B_t \in N \setminus \{S\}$. Up to the last replacement, only alternatives from $(N \setminus \{S\})^2$ are chosen. The last replacement chooses for B_0 an alternative in Σ. Such a leftmost derivation is called *terminal leftmost derivation* and is denoted by

$$B \Rightarrow_{tlm} a \text{ and } B \Rightarrow^*_{tlm} a B_1 B_2 \ldots B_t, \text{ respectively.}$$

Let $L_B = \{a\alpha \in \Sigma(N \setminus \{S\})^* \mid B \Rightarrow^*_{tlm} a\alpha\}$. Our goal is to construct a cfg $G_B = (V_B, V, P_B, S_B)$ such that

a) $L(G_B) = L_B$, and
b) each alternative of a variable begins with a symbol in Σ.

N_B denotes the set of nonterminals of G_B; i.e. $N_B = V_B \setminus V$. Assume that for $B, C \in N \setminus \{S\}$, $B \neq C$, the grammars G_B and G_C are constructed such that

i) $N_B \cap N_C = \emptyset$, or
ii) each production with a variable from $N_B \cap N_C$ on the left side is contained in both set of rules P_B and P_C.

If we replace each production $A \to BC \in P$ by the set

$$\{A \to a\gamma C \mid a\gamma \text{ is an alternative of } S_B \text{ in } G_B\},$$

then we obtain a cfg $G' = (V', \Sigma, P', S)$ in Gnf, where

$$\begin{aligned} V' &= V \cup \{N_B \mid B \in N \setminus \{S\}\}, \text{ and} \\ P' &= \{A \to a \in P \mid a \in \Sigma\} \\ &\quad \cup \{A \to a\gamma C \mid A \to BC \in P \text{ and } a\gamma \text{ is an alternative of } S_B \text{ in } G_B\} \\ &\quad \cup \textstyle\bigcup_{B \in N \setminus \{S\}} P_B \setminus \{S_B \to \alpha \mid \alpha \in \Sigma N_B^*\} \end{aligned}$$

The conditions (i) and (ii) above ensure that in a derivation in G' no illegal changes between two grammars G_B and G_C are possible. It remains the construction of the cfg $G_B = (V_B, V, P_B, S_B)$, $B \in N \setminus \{S\}$. The central observation is that L_B is a regular set. Let

$$L_B^R = \{B_t B_{t-1} \dots B_1 a \mid a B_1 B_2 \dots B_t \in L_B\};$$

i.e., L_B^R is L_B reversed.

First, we will construct a nfa $M_B = (Q, V, \delta, B_B, \mathcal{S}_B)$ with $L(M_B) = L_B^R$. Using a standard method (see [4], pp. 55–56), it is easy to construct from M_B a nfa M'_B with $L(M'_B) = L_B$. For doing this, $\mathcal{S}_B$ will be the initial state, B_B will be the unique finite state, and we turn around the transitions of M_B. Then, we will use for each $B \in N \setminus \{S\}$ the nfa M'_B for the construction of a cfg $G'_B = (V_B, V, P'_B, \mathcal{S}_B)$ which fulfills

1. $L(G'_B) = L_B$,
2. G'_B is in Gnf. (Note that V is the terminal alphabet of G'_B.)
3. $\alpha \in \Sigma N_B^*$ for each production $\mathcal{S}_B \to \alpha \in P'_B$.
4. The right side of every other production begins with a variable of G; i.e., a symbol in $N \setminus \{S\}$.
5. $N_B \cap N_C = \emptyset$ for all $B, C \in N \setminus \{S\}$, $B \neq C$.

Now, G_B will be constructed from G'_B by the replacement of every rule of the form $D_B \to EC_B$ and $D_B \to E$, respectively in P'_B by a set of productions, analogously to the construction of P' from P. Altogether, we obtain the following method for the construction of the cfg $G_B = (V_B, V, P_B, \mathcal{S}_B)$.

(1) Define $M_B = (Q, V, \delta, B_B, \mathcal{S}_B)$ by

$$\begin{aligned} Q &= \{A_B \mid A \in N\} \cup \{\mathcal{S}_B\}, \text{ and} \\ \delta(C_B, E) &= \{D_B \mid C \to DE \in P\} \\ \delta(C_B, a) &= \begin{cases} \{\mathcal{S}_B\} & \text{if } C \to a \in P \\ \emptyset & \text{otherwise} \end{cases} \end{aligned}$$

for all $C_B \in Q$, $E \in N \setminus \{S\}$, $a \in \Sigma$.

$L(M_B) = L_B^R$ can easily be proven by induction on the length of the terminal left derivation and on the length of the computation of the nfa, respectively.

(2) Define $M'_B = (Q, V, \delta', \mathcal{S}_B, B_B)$ by

$$\begin{aligned} \delta'(\mathcal{S}_B, a) &= \{C_B \mid \{\mathcal{S}_B\} = \delta(C_B, a)\}, \text{ and} \\ \delta'(D_B, E) &= \{C_B \mid D_B \in \delta(C_B, E)\} \end{aligned}$$

for all $D_B \in Q$, $E \in N \setminus \{S\}$ and $a \in \Sigma$.

Also $L(M'_B) = L_B$ can easily be proven by induction. Using the standard method for the construction of an equivalent cfg from a given nfa (see [4], pp. 61–62), we obtain the cfg G'_B.

(3) Define $G'_B = (V_B, V, P'_B, \mathcal{S}_B)$ by

$$
\begin{array}{lll}
V_B = \{A_B & \mid A \in N \setminus \{S\}\} \cup \{\mathcal{S}_B\} \cup V, \text{ and} \\
P'_B = \{\mathcal{S}_B \to aC_B & \mid C_B \in \delta'(\mathcal{S}_B, a) \text{ and} \\
 & (C_B \neq B_B \text{ or } \delta'(\{B_B\} \times N \setminus \{S\} \neq \emptyset))\} \\
\quad \cup\{\mathcal{S}_B \to a & \mid B_B \in \delta'(\mathcal{S}_B, a)\} \\
\quad \cup\{D_B \to EC_B & \mid C_B \in \delta'(D_B, E) \text{ and} \\
 & (C_B \neq B_B \text{ or } \delta'(\{B_B\} \times N \setminus \{S\} \neq \emptyset))\} \\
\quad \cup\{D_B \to E & \mid B_B \in \delta'(D_B, E)\}
\end{array}
$$

for all $D_B \in V'_B$, $E \in N \setminus \{S\}$, $a \in \Sigma$.

By inspection, Properties 1 – 5 can easily be proven. It is easy to show that $L(G'_B) = L_B$. For obtaining only productions fulfilling the Greibach normal properties with respect to the terminal alphabet Σ, simultaneously to all G'_B, $B \in N \setminus \{S\}$, we apply the trick used for the construction of P' from P again. Note that for all grammars G'_E, $E \in N\setminus\{S\}$, the alternatives of the start symbol $\mathcal{S}_E$ fulfill the Greibach normal form properties with respect to the terminal alphabet Σ.

(4) For each $B \in N \setminus \{S\}$ we define $G_B = (V_B, \Sigma, P_B, \mathcal{S}_B)$ by

$$
\begin{array}{ll}
P_B = & \{\mathcal{S}_B \to aC_B \mid \mathcal{S}_B \to aC_B \in P'_B\} \\
 & \{\mathcal{S}_B \to a \mid \mathcal{S}_B \to a \in P'_B\} \\
 & \cup \bigcup_{D_B \to EC_B \in P'_B} \{D_B \to \alpha C_B \mid \alpha \text{ is an alternative of } \mathcal{S}_E \text{ in } G'_E\} \\
 & \cup \bigcup_{D_B \to E \in P'_B} \{D_B \to \alpha \mid \alpha \text{ is an alternative of } \mathcal{S}_E \text{ in } G'_E\}
\end{array}
$$

Note that $L(G_B) \neq L_B$, since the same trick was applied twice and hence, the terminal alphabet of G_B is Σ and not V. We have to add to P_B all productions in G_E with left side $\neq \mathcal{S}_E$. Since we use for each $E \in N \setminus \{S\}$ the grammar G_E for the construction of the cfg $G' = (V', \Sigma, P', S)$ and all these productions are added to P', we do not have to add these productions explicitely to P_B. For the same reasons, we have not to extend N_B explicitely.

By construction, the cfg G_B fulfills the Greibach normal form properties with respect to the terminal alphabet Σ for all $B \in N \setminus \{S\}$. Furthermore, for all $B, C \in N \setminus \{S\}, B \neq C$ the grammar G_B and G_C are constructed, such that

i) $N_B \cap N_C = \emptyset$, or
ii) each production with a variable from $N_B \cap N_C$ on the left side is contained implicitely in both set of rules P_B and P_C.

Since $L(G'_B) = L_B$ for all $B \in N \setminus \{S\}$, $L(G') = L(G)$ follows immediately. Furthermore, G' is in Gnf. Moreover, since every alternative α of the start symbol $\mathcal{S}_B$ of the grammars G'_B and G_B are of the form $\alpha = aE$, $a \in \Sigma, E \in N \setminus \{S\}$, the grammar G' is in 2-Gnf.

Let us analyze the size of G' in dependence of the size of G. Since each transition of M_B and hence, each transition of M'_B corresponds to a rule of P

and vice versa, it is easy to see that $|G'_B| \le 5|G|$ for all $B \in N\backslash\{S\}$. Constructing G_B from G'_B, every production $D_B \to EC_B$ and $D_B \to E$, respectively of P'_B is replaced by at most $|G'_E|$ rules. Hence, $|G_B| \le |G'_B||G| = O(|G|^2)$. Similary, each production of type $A \to BC$ of P is replaced by at most $|G|$ productions. Note that for all $B \in N \setminus \{S\}$, the number of alternatives of the start symbol S_B of G_B is bounded by $|G|$. Altogether, we have obtained

$$|G'| \le 5|G|^2 + \sum_{B \in N\backslash\{S\}} |G_B| = O(|G|^3).$$

Hence, we have proven the following theorem.

Theorem 1 *Let $G = (V, \Sigma, P, S)$ be a cfg in Cnf. Then there is a cfg $G' = (V', \Sigma, P', S)$ in 2-Gnf such that $L(G') = L(G)$ and $|G'| = O(|G|^3)$.*

Given an arbitrary cfg $G = (V, \Sigma, P, S)$, the usual algorithm for the transformation of G into Cnf can square the size of the grammar (see [4], pp. 102 for an example). No better algorithm is known. This observation leads directly to the following corollary.

Corollary 1 *Let $G = (V, \Sigma, P, S)$ be a cfg. Then there is a cfg $G' = (V', \Sigma, P', S)$ in 2-Gnf such that $L(G') = L(G)$ and $|G'| = O(|G|^6)$.*

In [2], it is shown that for all $\epsilon > 0$ and sufficiently large n there is a context-free language CL_n with the following properties:

a) CL_n has a cfg of size $O(n)$.
b) Each chain rule free cfg for CL_n has size $O(\epsilon^3 n^{3/2-\epsilon})$.

3 Improving the Size of Gnf Grammars

Harrison and Yehudai [5] have developed an algorithm which eliminates for a given cfg $G = (V, \Sigma, P, S)$ the ε-rules in linear time. Moreover, the constructed cfg $G' = (V', \Sigma, P', S)$ is in canonical two form and $|G'| \le 12|G|$. For obtaining an equivalent cfg G'' in Cnf from G', it would suffice to eliminate the chain rules from G'. Hence, the expensive part of the transformation of an arbitrary cfg $G = (V, \Sigma, P, S)$ into Cnf is the elimination of the chain rules. The standard method for chain rule elimination computes for all $A \in N$ the set

$$W(A) = \{B \in N \mid A \Rightarrow^* B\}$$

and deletes the chain rules from P and adds the set

$$P(A) = \{A \to \alpha \mid \alpha \in V^+ \setminus N \text{ and } \exists B \in W(A) : B \to \alpha \in P\}$$

of productions to P. Note that $|P(A)| \le |G|$. Hence, $|G''| = O(n^2)$.

Given an arbitrary cfg $G = (V, \Sigma, P, S)$, our goal is now to construct an equivalent cfg G' in Gnf such that $|G'| = O(|G|^4)$. Since the transformation of G into canonical two form enlarges the size of G at most by the factor 12, we can assume that G is already in canonical two form. First, we will show that an appropriate extension of our algorithm will produce an equivalent cfg $\bar{G}$ in extended 2-standard form. Then we will show that applying the standard method for chain rule elimination produces a cfg G' in 2-standard form with $L(G') = L(G)$ and $|G'| = O(|G|^4)$.

Let $G = (V, \Sigma, P, S)$ be a cfg in canonical two form. Then the chain rules already fulfill the properties for extended 2-standard form. Hence, the grammar $\bar{G} = (V', \Sigma, \bar{P}, S)$, where

$$\begin{aligned} V' &= V \cup \{N_B \mid B \in N \setminus \{S\}\}, \text{ and} \\ \bar{P} &= \{A \to a \in P \mid a \in \Sigma\} \\ &\quad \{A \to B \in P \mid B \in N\} \\ &\quad \cup \{A \to a\gamma C \mid A \to BC \in P \text{ and } a\gamma \text{ is an alternative of } \mathcal{S}_B \text{ in } G_B\} \\ &\quad \cup \textstyle\bigcup_{B \in N\setminus\{S\}} P_B \setminus \{\mathcal{S}_B \to \alpha \mid \alpha \in \Sigma N_B^*\} \end{aligned}$$

would be in extended 2-standard form, assumed that the grammars G_B, $B \in N \setminus \{S\}$ are constructed correctly. It remains to discuss, how to treat the chain rules during the construction of G_B. For doing this, we will extend the four steps of the algorithm in an appropriate manner.

(1) We add the following transitions to δ.

$$\delta(C_B, \varepsilon) = \{D_B \mid C \to D \in P\}$$

for all $C_B \in Q$.

(2) For the correct definition of M'_B, we have to add the following transitions to δ'.

$$\delta'(D_B, \varepsilon) = \{C_B \mid D_B \in \delta(C_B, \varepsilon)\}$$

for all $D_B \in Q$.

(3) The standard algorithm for the construction of a cfg from a given nfa add the following extra rules to P'_B.

$$\{D_B \to C_B \mid C_B \in \delta'(D_B, \varepsilon) \text{ and } (C_B \neq B_B \text{ or } \delta'(\{B_B\} \times N \setminus \{S\}) \neq \emptyset)\}$$

for all $D_B \in V'_B$.

Step 4 has not to be extended.

For the elimination of the chain rules $A \to B$ from $\bar{P}$, we add the set

$$\{A \to a\gamma \mid \exists B \in W(A) \text{ and } a\gamma \text{ is an alternative of } \mathcal{S}_B \text{ in } G_B\}$$

of productions to P'. Since $|W(A)| \leq |N|$ for all $A \in N$ and since every start symbol $\mathcal{S}_B$ of a grammar G_B has at most $|G|$ alternatives, this enlarges the size of $\bar{G}$ at most by an amount of $O(|G|^2)$. Now, we apply the standard method for chain rule elimination to G_B for all $B \in N \setminus \{S\}$. Note that for all $C_B \in N_B$, $|W(C_B)| \leq |N_B|$ and hence, $|W(C_B)| \leq |G|$. Furthermore, the number of distinct alternatives of variables in V_B is bounded by $O(|G|^2)$. Hence, the standard method for chain rule elimination enlarges the size of G_B at most by a factor $|G|$. Hence, $|G_B| = O(|G|^3)$. Altogether, we have obtained the following theorem.

Theorem 2 *Let $G = (V, \Sigma, P, S)$ be a cfg. Then there is a cfg $G' = (V', \Sigma, P', S)$ in 2-Gnf such that $L(G') = L(G)$ and $|G'| = O(|G|^4)$.*

Remark: Given an ε-rule free cfg $G = (V, \Sigma, P, S)$, the transformation of G into canonical two form can enlarge the number of production considerably. If one does not wish to enlarge the number of productions in such a way, one can transform G directly into extended Greibach normal form. Let $A \to B\alpha$ be any production of G. We consider α as one symbol and built in Steps 1 and 2 nfa's $M_B = (Q, V_{ext}, \delta, B_B, S_B)$ and $M'_B = (Q, V_{ext}, \delta', S_B, B_B)$, respectively, where $V_{ext} = V \cup \{\alpha \mid \exists E \in V$ with $E\alpha$ is an alternative of a variable$\}$. Then we construct the cfg G'_B from M'_B. During Step 4, α will be considered as an element of V^*.

Acknowledgment: We thank Claus Rick for helpful comments.

References

1. A. V. Aho, and J. D. Ullman, *The Theory of Parsing, Translation, and Compiling*, Vol. I: Parsing, Prentice-Hall (1972).
2. N. Blum, More on the power of chain rules in context-free grammars, *TCS* **27** (1983), 287–295.
3. S. A. Greibach, A new normal-form theorem for context-free, phrase-structure grammars, *JACM* **12** (1965), 42–52.
4. M. A. Harrison, *Introduction to Formal Language Theory*, Addison-Wesley (1978).
5. M. A. Harrison, and A. Yehudai, Eliminating null rules in linear time, *The Computer Journal* **24** (1981), 156–161.
6. J. E. Hopcroft, and J. D. Ullman, *Introduction to Autmata Theory, Languages, and Computation*, Addison-Wesley (1979).
7. A. Kelemenová, Complexity of normal form grammars, *TCS* **28** (1984), 299–314.
8. D. J. Rosenkrantz, Matrix equations and normal forms for context-free grammers, *JACM* **14** (1967), 501–507.
9. D. Wood, *Theory of Computation*, Harper & Row (1987).

Translating Regular Expressions into Small ε-Free Nondeterministic Finite Automata

Juraj Hromkovič*, Sebastian Seibert*, and Thomas Wilke

Institut für Informatik und Praktische Mathematik
Christian-Albrechts-Universität zu Kiel, 24098 Kiel, Germany
email:{jhr,ss,tw}@informatik.uni-kiel.de

Abstract. It is proved that every regular expression of size n can be converted into an equivalent nondeterministic finite automaton (NFA) of size $\mathcal{O}(n(\log n)^2)$ in polynomial time. The best previous conversions result in NFAs of worst case size $\Theta(n^2)$. Moreover, the nonexistence of any linear conversion is proved: we give a language L_n described by a regular expression of size $\mathcal{O}(n)$ such that every NFA accepting L_n is of size $\Omega(n \log n)$.

Classification: theory of formal languages, descriptive complexity

1 Introduction

One of the central tasks of formal language theory is to describe infinite objects (languages) by finite formalisms (automata, grammars, expressions, etc.), and to investigate the descriptional power and complexity of these formalisms. In this paper, we consider two standard models for the description of regular languages: nondeterministic finite automata without ε-transitions (NFAs) and regular expressions. The size (descriptional complexity) of an NFA is considered to be the number of its transitions (the size of the memory to store it); the size of a regular expression is the number of symbols occurring in it.

For these two fundamental descriptional complexity measures of regular languages it is known [EZ76] that the conversion of NFAs into equivalent regular expressions may lead to a considerable increase of the descriptional complexity, i. e., there are regular languages requiring regular expressions of size exponential in the size of their minimal NFAs. On the other hand, previously described conversions from regular expressions into NFAs [RS59,BEGO71,HU79] produce automata whose size is in the worst case quadratic in the size of the input. In [SS88][1] it is even claimed that for each n the regular language defined by $(a_1+\varepsilon)(a_2+\varepsilon)\ldots(a_n+\varepsilon)$ requires NFAs of size $\Omega(n^2)$, which would imply that the above conversions are optimal.

In this paper, we devise a polynomial time conversion from regular expressions to NFAs that produces automata of size $\mathcal{O}(n(\log n)^2)$ where n denotes the size of the input. This is an essential improvement over the previously known

* Supported by the Deutsche Forschungsgemeinschaft under project no. HR 14/3-1.

[1] See page 77 and Exercise 3.7 on page 106.

Reischuk, Morvan (Eds.): STACS'97 Proceedings, LNCS 1200

conversions and disproves the lower bound claimed in [SS88]. We show that our construction is almost optimal by proving a lower bound of $\Omega(n \log n)$ for the above-mentioned example from [SS88]. This also implies the nonexistence of linear size conversions from regular expressions to NFAs.

The starting point of our construction is what we call the "position automaton" for a regular expression; this automaton, first described in [BEGO71] and also known as "nondeterministic Glushkov automaton" [Brü93], is based on ideas already explained in [MY60] and [Glu61]. The basic idea of our construction is as follows. Each state of the "position automaton" for a given regular expression is split up into a small number of new states in a way such that several of the new states (coming from different old states) that account for a lot of transitions are "equivalent" and can thus be "merged", leading to an overall small number of transitions. The crucial task is splitting the states of the "position automaton" in the right way; this is done by a recursive procedure.

This paper is organized as follows. Section 2 describes our terminology and states the main results; Section 3 recalls the "position automaton construction"; Section 4 describes our construction: First we explain the general idea of "splitting" and "merging" states in a "position automaton" and show that it results in an equivalent automaton. Afterwards we describe the recursive procedure that yields a particularly economical "splitting"; and finally the actual upper bound on the size of the output of this procedure is established. Section 5 gives the lower bound for the example from [SS88]; and we conclude this paper with open questions and an outlook.

2 Terminology and results

When we speak of a *regular expression* over an alphabet $\mathbb{A}$, we mean a finite expression built from the symbols in $\mathbb{A}$ and the special symbols "$\emptyset$" and "ε" using the binary operation symbols "+" and "·" and the unary operation symbol "*". Parentheses are used to indicate grouping, the operators "+" and "·" are written in infix notation, "*" is written in postfix notation, and "·" is often omitted. Given a regular expression E over an alphabet $\mathbb{A}$, we write $\mathcal{L}(E)$ for the subset of $\mathbb{A}^*$ that is denoted by E. The *size* of a regular expression E, denoted $\text{size}(E)$, is the number of occurrences of elements from $\mathbb{A}$ in E.

When we speak of a *nondeterministic finite automaton (NFA)* over an alphabet $\mathbb{A}$, we mean a tuple (Q, q_I, Δ, Q_F) where Q is a finite set of states, q_I is the initial state, $\Delta \subseteq Q \times \mathbb{A} \times Q$ is the finite transition relation, and Q_F is the set of final states. We thus don't allow ε-transitions in NFAs. Given an NFA $\mathcal{A}$, we denote by $\mathcal{L}(\mathcal{A})$ the subset of $\mathbb{A}^*$ recognized by $\mathcal{A}$.

In our examples, we use the following alphabets: for each $n > 0$, we assume an alphabet $\mathbb{A}_n$ of cardinality n whose letters are denoted by [1], [2], ..., [n].

The main results of this paper are the following two theorems.

Theorem 1. *There exists a polynomial time algorithm for converting every regular expression of size n into an equivalent NFA with at most $\mathcal{O}(n)$ states and $\mathcal{O}(n(\log n)^2)$ transitions.*

Theorem 2. *Let* $E_n = ([1]+\varepsilon)([2]+\varepsilon)\cdots([n]+\varepsilon)$. *Every NFA that recognizes* $\mathcal{L}(E_n)$ *has at least* $n+1$ *states and* $\lfloor \frac{n \log n - n}{4} \rfloor$ *transitions.*

In the rest of this section, we use E_5 as an example to illustrate the basic idea of our conversion from regular expressions to NFAs.

Depicted in Figure 1 is the NFA—the so-called "position automaton"—that is produced by the conversion described in [BEGO71] when applied to E_5. It has 15 transitions and 6 states.[2]

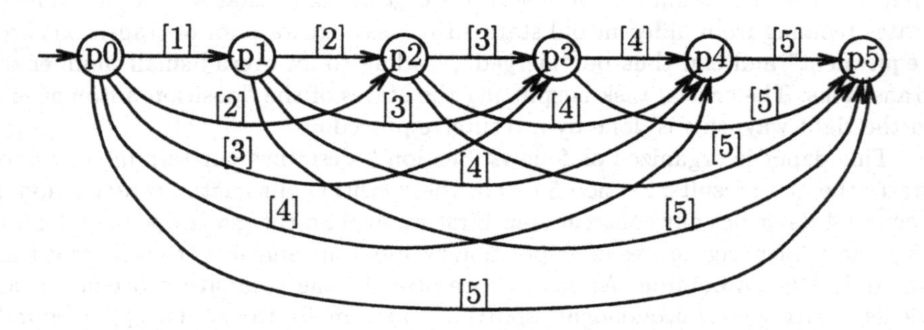

Fig. 1. The position automaton for E_5 (all states are final)

Our construction results in automata which we call "common follow sets automata". Such an automaton can be seen as a derivative of a position automaton, obtained by dividing each state into some pieces and identifying equivalent pieces from different states. For E_5, our construction produces the NFA depicted in Figure 2, which can be obtained from the position automaton in Figure 1 as follows.

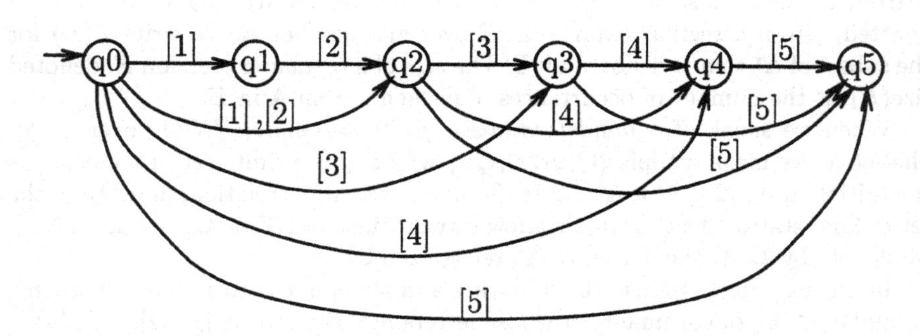

Fig. 2. A common follow sets automaton of E_5 (all states are final)

[2] In general, the conversion from [BEGO71], when applied to E_n, produces NFAs with $\frac{n(n-1)}{2}$ transitions and $n+1$ states.

State p1 of the position automaton is split into two pieces q1 and q2, each transition into p1 is redirected to both q1 and q2, and the outgoing transitions of p1 are distributed among q1 and q2 such that p2 and q2 have the same outgoing transitions. The states p2 and q2 can then be identified, leading to an overall smaller number of transitions. The depicted common follow sets automaton needs only 13 transitions (as opposed to 15 in the position automaton).

3 Positions, first, last, and follow sets

In this section, we recall the definitions of some sets containing positions of a regular expression which are the basis of the "position automaton" as well as of the new construction. We follow the exposition in [Brü93].[3]

From now on, we assume that every regular expression E over an alphabet $\mathbb{A}$ comes with a set of ***positions***, denoted $\text{pos}(E)$, whose elements are in one-to-one correspondence with the occurrences of letters from $\mathbb{A}$ in E. A position can be best thought of as pointing to a particular occurrence of a letter in E. Given a regular expression E and a position $x \in \text{pos}(E)$, we write $\langle E \rangle_x$ for the letter in E that x points to (occurs in position x).

This assumption has a consequence for inductive definitions and proofs: when we consider a regular expression of the form EF or $E + F$, the sets $\text{pos}(E)$ and $\text{pos}(F)$ are disjoint.

Let E be a regular expression over an alphabet $\mathbb{A}$. When scanning a word from $\mathcal{L}(E)$, each letter scanned matches a particular occurrence of this letter in E, or—in our terminology—corresponds to a particular position of E. Of course, there is often more than just one way to scan a word from $\mathcal{L}(E)$ (due to ambiguities of E). The sets $\text{first}(E)$ and $\text{last}(E)$ are defined in a way such that a position x belongs to $\text{first}(E)$ (respectively $\text{last}(E)$) if and only if it corresponds to the first (respectively last) letter scanned in some scanning process.

Formally, $\text{first}(E)$ is defined by induction according to the following rules:

$$\text{first}(\emptyset) = \emptyset\,, \tag{1}$$

$$\text{first}(\varepsilon) = \emptyset\,, \tag{2}$$

$$\text{first}(a) = \text{pos}(a) \qquad \text{for } a \in \mathbb{A}\,, \tag{3}$$

$$\text{first}(F + G) = \text{first}(F) \cup \text{first}(G)\,, \tag{4}$$

$$\text{first}(FG) = \begin{cases} \text{first}(F) & \text{if } \varepsilon \notin \mathcal{L}(F)\,, \\ \text{first}(F) \cup \text{first}(G) & \text{if } \varepsilon \in \mathcal{L}(F)\,, \end{cases} \tag{5}$$

$$\text{first}(F^*) = \text{first}(F)\,. \tag{6}$$

In order to obtain rules for $\text{last}(E)$ substitute "last" for "first" and replace the last but one rule by:

$$\text{last}(FG) = \begin{cases} \text{last}(G) & \text{if } \varepsilon \notin \mathcal{L}(G)\,, \\ \text{last}(F) \cup \text{last}(G) & \text{if } \varepsilon \in \mathcal{L}(G)\,. \end{cases} \tag{7}$$

[3] In [BEGO71], slightly different definitions for the "first" and "last" set of a regular expression are given, which are less useful for our purposes than the definitions from [Brü93].

Given a position $x \in \text{pos}(E)$, the set $\text{follow}(E,x)$ is defined in a way such that $y \in \text{follow}(E,x)$ if and only if x is immediately followed by y in some scanning process. It is defined inductively according to the following rules:

$$\text{follow}(a,x) = \emptyset \qquad \text{for } a \in \mathbb{A}\,, \tag{8}$$

$$\text{follow}(F+G,x) = \begin{cases} \text{follow}(F,x) & \text{if } x \in \text{pos}(F)\,, \\ \text{follow}(G,x) & \text{if } x \in \text{pos}(G)\,. \end{cases} \tag{9}$$

$$\text{follow}(FG,x) = \begin{cases} \text{follow}(F,x) & \text{if } x \in \text{pos}(F) \setminus \text{last}(F)\,, \\ \text{follow}(F,x) \cup \text{first}(G) & \text{if } x \in \text{last}(F)\,, \\ \text{follow}(G,x) & \text{if } x \in \text{pos}(G)\,, \end{cases} \tag{10}$$

$$\text{follow}(F^*,x) = \begin{cases} \text{follow}(F,x) & \text{if } x \in \text{pos}(F) \setminus \text{last}(F)\,, \\ \text{follow}(F,x) \cup \text{first}(F) & \text{if } x \in \text{last}(F)\,. \end{cases} \tag{11}$$

The first, last, and follow sets of a regular expression E can be used to define an NFA that recognizes $\mathcal{L}(E)$ in the following way, as was already described in [BEGO71]. Define the *position automaton* for E, denoted $\mathcal{A}_E$, as the tuple (Q, q_I, Δ, Q_F) where, for some $q_I \notin \text{pos}(E)$,

$$Q = \{q_I\} \cup \text{pos}(E)\,,$$

$$Q_F = \begin{cases} \{q_I\} \cup \text{last}(E) & \text{if } \varepsilon \in \mathcal{L}(E)\,, \\ \text{last}(E) & \text{otherwise,} \end{cases}$$

$$\Delta = \{(q_I, \langle E\rangle_x, x) \mid x \in \text{first}(E)\} \cup \{(x, \langle E\rangle_y, y) \mid y \in \text{follow}(E,x)\}\,.$$

The correctness of this construction is stated in the following lemma:

Lemma 1. *For every regular expression E, $\mathcal{L}(E) = \mathcal{L}(\mathcal{A}_E)$.* □

4 Converting regular expressions into NFA

In this section we prove Theorem 1 by giving a polynomial-time construction converting every regular expression of size n into an equivalent NFA with at most $\mathcal{O}(n)$ states and $\mathcal{O}(n(\log n)^2)$ transitions.

In Subsection 4.1, we define for each regular expression E a variety of NFAs (the common follow sets automata for E) all of which recognize E. In Subsection 4.3, the main part of this section, we describe a recursive procedure that constructs for every regular expression E a particularly small common follow sets automaton. Subsection 4.2 provides technical groundwork, and Subsection 4.4 establishes the upper bound for the output of our construction.

4.1 The common follow sets automaton

The idea behind our construction is to decompose the follow set of each position, i.e. the set of successors of a state of the position automaton, into some subsets. These subsets become the states of our automaton, each subset C being responsible for the transitions from the original state to the elements of C. This means, when the position automaton is in a state x (a position of E), our automaton will be in one of the chosen subsets of follow(E, x) instead, say in C. It has thus nondeterministically restricted the set of possible next states to the elements of C. Each transition from x to every $x' \in C$ is replaced by transitions from C to every C' belonging to the decomposition of follow(E, x').

The common use of subsets in the decomposition of different follow sets will give the desired complexity bounds.

So far, we have not considered the distinction between final and non final states. To cope with this, we will add a flag to the state set.

Definition 1. Let E be a regular expression, given with its set of positions pos(E). A *system of common follow sets* for E is given by a *decomposition* $\text{dec}(x) \subseteq \mathcal{P}(\text{pos}(E))$ for each $x \in \text{pos}(X)$ such that

$$\text{follow}(E, x) = \bigcup_{C \in \text{dec}(x)} C .$$

The *family of common follow sets* $\mathcal{C}$ associated with this system is defined by

$$\mathcal{C} = \{\text{first}(E)\} \cup \bigcup_{x \in \text{pos}(E)} \text{dec}(x).$$

The *common follow sets NFA* (Q, q_I, Δ, Q_F), associated with this system is given by

$$Q = \mathcal{C} \times \{0, 1\};$$

$$q_I = \begin{cases} (\text{first}(E), 1) & \text{if } \varepsilon \in \mathcal{L}(E), \\ (\text{first}(E), 0) & \text{otherwise}; \end{cases}$$

$$\Delta = \{((C, f), \langle E \rangle_x, (C', f')) \mid x \in C,\ C' \in \text{dec}(x), \text{ and } f' = 1 \text{ iff } x \in \text{last}(E)\};$$

$$Q_F = \mathcal{C} \times \{1\} .$$

Note that if follow$(E, x) = \emptyset$, then necessarily dec$(x) = \{\emptyset\}$.

Lemma 2. *Let E be a regular expression and $\mathcal{A}$ the common follow sets NFA associated with any system of common follow sets for E. Then $\mathcal{L}(E) = \mathcal{L}(\mathcal{A})$.*

Proof (sketch). By a straightforward induction, we show that the following conditions are equivalent for any nonempty word w:

- After reading w, the automaton $\mathcal{A}$ can be in state (C, f).
- After reading w, the position automaton $\mathcal{A}_E$ can be in some state x where $C \in \text{dec}(x)$, and x is a final state in $\mathcal{A}_E$ iff $f = 1$.

The rest follows from Lemma 1. □

The complexity of the common follow sets automaton obviously depends on the choice of the system of common follow sets. The rest of this section deals with the problem of finding appropriate common follow sets.

4.2 Follow sets and the tree representation of expressions

From now on, we fix a regular expression E and a decomposition of E into subexpressions (i. e., we resolve ambiguities that arise from iterated products and sums). Our aim is to give an alternative definition of the follow sets of E.

We represent E as a tree, which is denoted by t_E. In this tree, each node corresponds to exactly one occurrence of a subexpression of E. We identify the node and the subexpression. So when we speak of a subexpression of E we really mean an occurrence of a subexpression.

If a subexpression F of E is an ancestor of G, we write $F \prec G$. The notion of a *subtree* will be understood in the graph theoretic manner, i. e., as a subgraph being a tree. As a special case, we consider *the subtree of t below F* as the tree consisting of a node F and all its descendants $F' \succeq F$ in t. (This is what is sometimes called a subtree, whereas our notion of subtree allows for instance a single path of the original tree to be a subtree.)

For any subtree t of t_E, $\text{pos}(t)$ denotes the set of positions of expressions $a \in \mathbb{A}$ being leaves of t. If the root of t is the expression F, we have $\text{pos}(t) \subseteq \text{pos}(F)$. The inclusion may be strict since t need not be the full subtree of t_E below F. As a measure of t we use the cardinality of $\text{pos}(t)$; we set $|t| = |\text{pos}(t)|$.

Let's analyze the rules (8)–(11) which describe how a set $\text{follow}(E, x)$ is obtained. We observe that on certain occasions, when $x \in \text{last}(F)$ for some subexpression F, the set $\text{first}(F)$ or $\text{first}(G)$ for a related subexpression G is added to the follow set of E. In order to be able to formalize this observation, we define an injective function "next" between subexpressions of E as follows:

$$\text{next}(F) = \begin{cases} F & \text{if } F \text{ is a son of } F^* \text{ in } t_E, \\ G & \text{if } F \text{ is a son of } FG \text{ in } t_E, \\ \bullet & \text{otherwise.} \end{cases}$$

We set $\text{first}(\bullet) = \emptyset$ and obtain as a reformulation of (8)–(11) item (a) of the following lemma.

Lemma 3. *For all positions $x \in \text{pos}(E)$ and for all subexpressions F, G of E, the following holds:*

(a) $\text{follow}(E, x) = \bigcup_{\substack{H \succ E \\ x \in \text{last}(H)}} \text{first}(\text{next}(H))$;

(b) if $F \prec G$, then $\text{first}(\text{next}(G)) \subseteq \text{pos}(F)$; *and*

(c) if $F \prec G$, then $\text{last}(F) \supseteq \text{last}(G)$ *or* $\text{last}(F) \cap \text{pos}(G) = \emptyset$. □

Note that in the union in item (a) the root E can be excluded since always $\text{next}(E) = \bullet$. The other items result rather directly from the fact that the first and last sets are obtained by computing unions bottom up.

4.3 The construction of a system of common follow sets

We now describe a procedure that, for the given regular expression E, computes a system of common follow sets that yields a small common follow sets automaton.

Using the observations of the last subsection, we recursively compute on certain subtrees t of t_E (starting with $t = t_E$) families of common follow sets $\mathcal{C}(t) \subseteq \mathcal{P}(\text{pos}(t))$ and decompositions $\text{dec}(x,t) \subseteq \mathcal{C}(t)$ for all $x \in \text{pos}(t)$. When F is the root of a tree t treated in the recursion, these decompositions will be constructed in such a way that the following equation holds:[4]

$$\bigcup_{C \in \text{dec}(x,t)} C = \text{pos}(t) \cap \bigcup_{\substack{G \succ F \\ x \in \text{last}(G)}} \text{first}(\text{next}(G)) \quad \text{for } x \in \text{pos}(t)\,. \tag{12}$$

Applied to the entire tree, (12) gives by Lemma 3:

$$\bigcup_{C \in \text{dec}(x,t_E)} C = \bigcup_{\substack{G \succ E \\ x \in \text{last}(G)}} \text{first}(\text{next}(G)) = \text{follow}(E,x) \quad \text{for } x \in \text{pos}(E).$$

Hence we may set

$$\begin{aligned} \text{dec}(x) &= \text{dec}(x,t_E) \qquad\qquad \text{for } x \in \text{pos}(E),\\ \mathcal{C} &= \{\text{first}(E)\} \cup \mathcal{C}(t_E) \end{aligned}$$

in order to obtain a system of follow sets for E. That this yields a small follow sets automaton will be discussed in the next subsection.

The recursive procedure. Assume t is a subtree of t_E such that $|t| > 1$. (The end of the recursion, the case $|t| = 1$, is discussed below.) Let F be the root of t. Divide t into two subtrees t_1, t_2 according to the following rule: starting from the root of t, search downwards for some node F_1 such that $\frac{|t|}{3} \le |t_1| \le \frac{2|t|}{3}$, where t_1 is the subtree of t below F_1; let t_2 be the rest of t after removing t_1. Finally, set

$$C_1 = \text{pos}(t) \cap \bigcup_{\substack{F \prec G \preceq F_1 \\ \text{last}(G) \supseteq \text{last}(F_1)}} \text{first}(\text{next}(G)), \tag{13}$$

$$C_2 = \text{pos}(t) \cap \text{first}(F_1), \tag{14}$$

$$\text{dec}(x,t) = \begin{cases} \text{dec}(x,t_1) \cup \{C_1\} & \text{if } x \in \text{pos}(t_1), x \in \text{last}(F_1),\\ \text{dec}(x,t_1) & \text{if } x \in \text{pos}(t_1), x \notin \text{last}(F_1),\\ \text{dec}(x,t_2) \cup \{C_2\} & \text{if } x \in \text{pos}(t_2), \text{first}(F_1) \subseteq \text{follow}(E,x),\\ \text{dec}(x,t_2) & \text{if } x \in \text{pos}(t_2), \text{first}(F_1) \not\subseteq \text{follow}(E,x), \end{cases} \tag{15}$$

$$\mathcal{C}(t) = \{C_1, C_2\} \cup \mathcal{C}(t_1) \cup \mathcal{C}(t_2)\,. \tag{16}$$

The correctness argument. We show that (12) holds, provided it holds for t_1 and t_2 as constructed above. We consider only the case $x \in \text{pos}(t_1) \cap \text{last}(F_1)$. The other cases can be dealt with similarly.

[4] As Volker Diekert has pointed out to the authors, one could use instead $\text{pos}(t) \cap \text{follow}(E,x)$ on the right hand side of (12). We use the above version since it is the starting point of some optimizations to be described in the full version of the paper.

The left hand side of (12) evaluates to $C_1 \cup \bigcup_{C \in \mathrm{dec}(x,t_1)} C$. From the assumption that (12) holds for t_1, we can conclude that this term is identical with

$$C_1 \cup (\mathrm{pos}(t_1) \cap \bigcup_{\substack{H \succ F_1 \\ x \in \mathrm{last}(H)}} \mathrm{first}(\mathrm{next}(H)))\,. \tag{17}$$

By Lemma 3 (b), the term in (17) coincides with

$$C_1 \cup (\mathrm{pos}(t) \cap \bigcup_{\substack{H \succ F_1 \\ x \in \mathrm{last}(H)}} \mathrm{first}(\mathrm{next}(H)))\,. \tag{18}$$

But the right hand side of (12) is the union of the term $(\mathrm{pos}(t) \cap \bigcup \ldots)$ from (18) with

$$\mathrm{pos}(t) \cap \bigcup_{\substack{F \prec G \preceq F_1 \\ x \in \mathrm{last}(G)}} \mathrm{first}(\mathrm{next}(G))\,. \tag{19}$$

It remains to show that (19) equals C_1. Consider a node G such that $F \prec G \preceq F_1$. From Lemma 3 (c), we conclude $x \in \mathrm{last}(G)$ iff ($x \in \mathrm{last}(F_1)$ and $\mathrm{last}(F_1) \subseteq \mathrm{last}(G)$). Thus, if $x \in \mathrm{last}(F_1)$, then $x \in \mathrm{last}(G)$ iff $\mathrm{last}(F_1) \subseteq \mathrm{last}(G)$, hence C_1 is identical with (19). □

To conclude the description of our recursive procedure, we finally explain what happens to a tree t with $|t| = 1$, i.e., when $\mathrm{pos}(t) = \{x\}$ for some x. In this case, we set

$$\mathrm{dec}(x,t) = \begin{cases} \{\{x\}\} & \text{if there exists a node } F \text{ in } t \text{ other than the root} \\ & \qquad \text{such that } x \in \mathrm{last}(F) \cap \mathrm{first}(\mathrm{next}(F)), \\ \{\emptyset\} & \text{otherwise,} \end{cases}$$

$$\mathcal{C}(t) = \mathrm{dec}(x,t)\,.$$

In this case, (12) holds immediately.

The resulting systems of common follow sets allow some optimization, which will be given in the full version of the paper. For example it is possible to reduce the number of states of the common follow sets automaton to $2 \cdot |\mathrm{size}(E)|$, whereas the rough estimation in Lemma 4 (i) gives only $6 \cdot |\mathrm{size}(E)|$.

A trivial optimization, the leaving-out of the empty set in the decomposition of a nonempty follow set, is already done in the construction of the example automaton for E_5 (see Figure 2). With this hint, the reader may re-enact the construction, when starting from the expression $([1]+\varepsilon)\cdot(([2]+\varepsilon)\cdot(([3]+\varepsilon)\cdot(([4]+\varepsilon)\cdot([5]+\varepsilon))))$.

4.4 The descriptive complexity

In order to establish the upper bound of Theorem 1 we show that the above construction guarantees the following properties, where $n = \mathrm{size}(E)$.

Lemma 4. *For* $\mathcal{C} = \{\text{first}(E)\} \cup \mathcal{C}(t_E)$ *and* $\text{dec}(x) = \text{dec}(x, t_E)$ *the following holds:*

(i) $|\mathcal{C}| \leq 3n - 1$,

(ii) $\sum_{C \in \mathcal{C}} |C| \leq \frac{3}{\log 3/2} n \log n + n + 1$, *and*

(iii) $|\text{dec}(x)| \leq \frac{1}{\log 3/2} \log n + 1$ *for all* $x \in \text{pos}(E)$.

Since the common follow sets automaton has, for every $C \in \mathcal{C}$, at most two transitions for each $x \in C$ and each $C' \in \text{dec}(x)$, namely from $(C, 0)$ and $(C, 1)$ to either $(C', 0)$ or $(C', 1)$, items (ii) and (iii) imply the claimed bound on the number of transitions of $\mathcal{A}$, whereas (i) gives immediately the bound on the number of states.

Proof (of Lemma 4, sketch).

(i) By straightforward induction we get $|\mathcal{C}(t)| \leq 3|t| - 2$ for all subtrees t used in the construction.

(ii) Likewise, we get inductively $\sum_{C \in \mathcal{C}(t)} |C| \leq \frac{3}{\log 3/2} |t| \log |t| + 1$. For $t = t_E$ this gives the claimed result when adding $|\text{first}(E)| \leq n$.

(iii) Each decomposition $\text{dec}(x)$ contains at most one element for each subtree t used in the construction with $x \in \text{pos}(t)$. In this sequence of subtrees containing x each tree contains at most 2/3 of the positions of its predecessor. Therefore, we can have at most $\frac{\log n}{\log 3/2} + 1$ subtrees containing x. □

5 An $n \log n$ lower bound

For proving the lower bound of Theorem 2, we consider the regular expression $E_n = ([1]+\varepsilon)([2]+\varepsilon) \cdot \ldots \cdot ([n]+\varepsilon)$, respectively the language $L_n = \mathcal{L}(E_n)$. Note that for any factor $[i][j]$ of any word in L_n we have $i < j$. We want to show that any NFA $\mathcal{A}$ recognizing L_n has at least $n+1$ states and $\lfloor \frac{n \log n - n}{4} \rfloor$ transitions.

Proof (of Theorem 2). We may trivially restrict to automata having only productive and reachable states.

The minimal number of states of any automaton recognizing L_n can be obtained as a consequence of the more general fact that the state set Q of any automaton $\mathcal{A} = (Q, q_I, \Delta, Q_F)$ recognizing L_n can be divided into $n+1$ disjoint *nonempty* subsets, i.e.,

$$Q = Q_1 \cup Q_2 \cup \ldots \cup Q_{n+1},$$

where each $Q_i, i = 1, \ldots, n$, is defined by

$$q \in Q_i \text{ iff } i = \min\{j \mid \exists q' \in Q : (q, [j], q') \in \Delta\},$$

and the states in Q_{n+1} are those having no outgoing transitions.

For showing the nonemptiness of these sets, we will first prove the following property of the transitions of $\mathcal{A}$.

$$\text{For all } (p, [i], q) \in \Delta \text{ we have } p \in Q_{\leq i} \text{ and } q \in Q_{\geq i+1}, \tag{20}$$

where we use $Q_{\leq i}$ and $Q_{\geq i+1}$ as abbreviations of $Q_1 \cup Q_2 \cup \ldots \cup Q_i$ and $Q_{i+1} \cup Q_{i+2} \cup \ldots \cup Q_{n+1}$, respectively.

The claim $p \in Q_{\leq i}$ is implied by the definition of the partition of Q; it remains to show $q \in Q_{\geq i+1}$. If the state q were in Q_j for some $j \leq i$, we would have, by definition of Q_j, a transition $(q, [j], q')$ for some $q' \in Q$. Since any state was supposed to be reachable and productive, we would obtain $[i][j]$ as a factor of some word in $\mathcal{L}(\mathcal{A})$, in contradiction to the assumption of $\mathcal{A}$ recognizing L_n.

From condition (20) we obtain immediately the nonemptiness of each set Q_i for $i = 2, \ldots, n$, since when reading (and accepting) the word $[1][2]\ldots[i-1][i]\ldots[n] \in L_n$, $\mathcal{A}$ must by (20) be in some state from $Q_{\geq i}$ after reading $[i-1]$ and in $Q_{\leq i}$ before reading $[i]$, hence it has to be in some state from Q_i. The nonemptiness of Q_1 and Q_{n+1} is implied by (20) directly when looking at the source, respectively target, of any transition labeled $[1]$, respectively $[n]$.

For showing the lower bound on the number of transitions, we define, for say $p \in Q_i$, $q \in Q_k$, $[j] \in \mathbb{A}$, the *length of the transition* $(p, [j], q)$ as the difference $k - i$.

We now look at successful runs of $\mathcal{A}$ on the following words. For $k = 0, \ldots, \lfloor \log n \rfloor$ and $j = 1, \ldots, 2^k$ we set

$$w_{k,j} = [j][j+2^k][j+2\cdot 2^k][j+3\cdot 2^k]\ldots[j+l\cdot 2^k],$$

where l is maximal such that $j + l \cdot 2^k \leq n$.

We conclude the proof by showing that

(a) at least one of every two consecutive transitions of any successful run on $w_{k,j}$ is of length greater than 2^{k-1};

(b) in any successful run on $w_{k,j}$, all transitions are of length less than 2^{k+1}.

Note that for each $k = 0, \ldots, \lfloor \log n \rfloor$ every letter $[i] \in \mathbb{A}_n$ occurs exactly in one word $w_{k,j}$, namely for $j = i \mod 2^k$. In other words, there are $\sum_{j=1}^{2^k} |w_{k,j}| = n$ different transitions used, when $\mathcal{A}$ is reading all words $w_{k,j}$ for a fixed $k = 0, \ldots, \lfloor \log n \rfloor$.

Assuming (a) and (b), we obtain for each word $w_{k,j}$ at least (odd word length being the worse case) $\frac{|w_{j,k}|-1}{2}$ transitions of length greater than 2^{k-1} which cannot occur in a run on any $w_{k',j'}$ for $k' \leq k-2$. Summing up over all words $w_{k,j}$ for a fixed k we will obtain $\lfloor \frac{n-2^k}{2} \rfloor$ new transitions for at least every second k from $\{0, \ldots, \lfloor \log n \rfloor\}$. This gives the desired lower bound of $\lfloor \frac{n \log n - n}{4} \rfloor$ transitions.

It remains to prove (a) and (b). Claim (a) holds since before reading the factor $[j+i2^k][j+(i+1)2^k]$ of $w_{k,j}$, $\mathcal{A}$ has to be in some state $p \in Q_{\leq j+i2^k}$, according to (20). After reading this factor, $\mathcal{A}$ is in $q \in Q_{\geq j+(i+1)2^k+1}$. Consequently the sum of the lengths of the two consecutive transitions is at least $2^k + 1$, which implies that at least one of these has to be of a length greater than 2^{k-1}.

For showing the upper bound on the transition length (b), we look at a transition in the context of previous and following transition, i.e. at a factor $[j+(i-1)2^k][j+i2^k][j+(i+1)2^k]$. (The first and last transition of the run on $w_{k,j}$ are treated similarly.) Now we apply (20) to the transitions labeled by the "surrounding" letters $[j+(i-1)2^k]$ and $[j+(i+1)2^k]$. This shows that $\mathcal{A}$ has to be in $Q_{\geq j+(i-1)2^k+1}$ before reading $[j+i2^k]$, and in $Q_{\leq j+(i+1)2^k}$ thereafter. So the length of the transitions used for reading $[j+i2^k]$ is at most $2^{k+1}-1$. □

6 Conclusion

In this paper we have shown how one can efficiently convert regular expressions of size n into NFAs of size $\mathcal{O}(n(\log n)^2)$. Moreover, the nonexistence of any linear conversion has been proved. Note that if one considers nondeterministic finite automata with ε-transitions (ε-NFAs) instead of NFAs the standard conversion [HU79] results in ε-NFAs of size $\mathcal{O}(n)$. The best conversion of ε-NFAs into NFAs known so far results in automata of size $\Theta(n^2)$ in the worst case [HU79]. We do not see any straightforward application of our method of common follow sets to improve this conversion. On the other hand, the nonexistence of any linear conversions of ε-NFAs into NFAs is a direct consequence of our $\Omega(n \log n)$ lower bound on the size of NFAs accepting $\mathcal{L}(E_n)$. Thus the following problems remain open.

Problem 1. Can the quadratic conversion of ε-NFAs into NFAs be improved or do there exist regular languages where the difference between the sizes of ε-NFAs and NFAs is of a factor greater than $\log n$?

Problem 2. Comparing the upper bound of $\mathcal{O}(n(\log n)^2)$ and the lower bound of $\Omega(n \log n)$ on the conversion of regular expressions into NFAs, naturally the question arises, which of these two can be improved? Which is the exact size of a minimal NFA for the language $\mathcal{L}(E_n)$?

References

[BEGO71] Ronald Book, Shimon Even, Sheila Greibach, and Gene Ott. Ambiguity in graphs and expressions. *IEEE Trans. Comput.*, C-20(2):149–153, 1971.

[Brü93] Anne Brüggemann-Klein. Regular expressions into finite automata. *Theoret. Comput. Sci.*, 120:197–213, 1993.

[EZ76] Andrzej Ehrenfeucht and Paul Zeiger. Complexity measures for regular expressions. *J. Comput. System Sci.*, 12:134–146, 1976.

[Glu61] V. M. Glushkov. The abstract theory of automata. *Russian Math. Surveys*, 16:1–53, 1961. Translation from Usp. Mat. Naut. 16:3–41, 1961 by J. M. Jackson.

[HU79] John E. Hopcroft and Jeffrey D. Ullman. *Introduction to Automata Theory, Languages and Computation.* Addison-Wesley, Reading, Mass., 1979.

[MY60] Robert F. McNaughton and H. Yamada. Regular expressions and state graphs for automata. *IRE Trans. Electron. Comput.*, EC-9(1):39–47, 1960.

[RS59] Michael O. Rabin and Dana Scott. Finite automata and their decision problems. *IBM J. Res. Develop.*, 3:114–125, 1959.

[SS88] Seppo Sippu and Eljas Soisalon-Soininen. *Parsing Theory, Vol. I: Languages and Parsing*, volume 15 of *EATCS Monographs on Theoret. Comput. Sci.* Springer-Verlag, Berlin, 1988.

Memory Management for Union-Find Algorithms

Christophe Fiorio[1] and Jens Gustedt[2]

Technische Universität Berlin, Sekr. MA 6-1, D-10623 Berlin, Germany

Abstract. We provide a general tool to improve the real time performance of a broad class of *Union-Find* algorithms. This is done by minimizing the random access memory that is used and thus to avoid the well-known von Neumann bottleneck of synchronizing CPU and memory. A main application to image segmentation algorithms is demonstrated where the real time performance is drastically improved.

1 Introduction

A main obstacle for really efficient implementations of random access data structures on todays computers is still the so-called von Neumann bottleneck. It addresses the fact that background memory of such a computer usually has an access time that is much larger than the cycle rate of the CPU. In particular it states that a computation that randomly accesses data elements over and over again (eg by pointer jumping) mainly has to **wait** and so the CPU will be idle most of the time – `nop` is certainly the assembler instruction that is among the most executed ones.

Modern architectures try to circumvent this problem by introducing a hierarchical memory model, consisting of registers, cache, RAM and disk. In particular cache that has a cycle rate comparable with the CPU is used to hold those data elements that are suspected to be accessed in the near future. Whereas most programmers seem to be sensible to the problem that occurs when data located in virtual memory must be fetched from disk they seem to be less aware of the problem that arises when the data transfer between RAM, cache and CPU must be handled.

Clearly such a hierarchical architecture is useless if it is not supported by appropriate software that allows easy estimates on what elements to load into cache and/or registers. Our goal here is to provide such a scheme for *Union-Find* (UF) data structures and algorithms.

The best way to control which data elements are going to be used next by an algorithm is when data is accessed sequentially, i.e when data is consecutively read or written into an array or (pseudo) file. We model this in distinguishing to different kinds of memory access, *random* and *sequential*, and give a tool to reduce the memory that is accessed randomly to a neglectable portion.

For a UF algorithm $\mathcal{A}$ we say that an element v of the groundset is **open** from the moment in time it is accessed for the first time until $\mathcal{A}$ *knows* that it is accessed for the last time. For an instance I, by $Width_{\mathcal{A}}(I)$ we denote the maximum cardinality of the

[1] Supported by DIMANET. Current address: LIRMM, F-34392 Montpellier Cedex 5, France
[2] Supported by the IFP "Digitale Filter"

Reischuk, Morvan (Eds.): STACS'97 Proceedings, LNCS 1200

set of open elements during the run of $\mathcal{A}$ on I. For an integer n let $Width_{\mathcal{A}}(n)$ denote the maximum of $Width_{\mathcal{A}}(I)$ taken over all instances I of size n. Our main result is given by the following theorem.

Main Theorem. *Any UF algorithm $\mathcal{A}$ may be implemented in such a way that the amount of random access memory needed during the execution of $\mathcal{A}$ is bounded by $O(Width_{\mathcal{A}}(n))$.*

The main idea is an improvement of the techniques found in [2], namely to delete elements that are not further needed for a UF algorithm and to reuse the memory that was occupied by them. For practical purposes any access of arrays using only *pointer increments* (or decrements) and no other pointer arithmetics can be subsumed under being a sequential access. We use that fact to save those deleted elements into a (pseudo) *file*, that allows us to reconstruct the final partition of the ground set after having done all necessary *Union* operations. This technique applies to a broad class of algorithms for which we show how to improve them in a straight forward manner.

We tackle the UF problems by modeling their access to individual elements by a *graph*; the elements of the set are identified with vertices of the graph and edges represent permissible *Union* operations. We assume that an algorithm that uses a UF data structure has an estimation of a time interval for each element in which this element might be accessed. This is equivalent of saying that an interval supergraph (or **path decomposition**) of such graph is given as well. This is introduced and developed in Section 3.

Path decompositions are then used in Section 4 to prove the Main Theorem. In addition we also give a technique that after all *Unions* being performed allows a reconstruction of the final partition of the groundset.

Our main application, see Sections 2.3 and 5, are UF algorithms for image segmentation. Here we are able to combine and extend results of [2,3,5] to achieve linear time algorithms with a very low demand for random access memory that already have proven to perform very well in practice, [4].

2 Basic Definitions and Facts

2.1 Basics of Union-Find

Union-Find algorithms solve the disjoint set union problem. It can be stated as follows: let S be a set of elements that form one-element subsets at the beginning, perform a sequence of ***Union*** operations on these subsets; ***Find*** operation identifies for one element the set it belongs to (for a more general presentation see e.g [7,14]). This must be done in a more efficiently way, but algorithms that solve the *Union-Find* problem in the general case are not known to have a linear time solution. The best complexity known has been obtained by an algorithm of McIllroy and Morris that has been shown to perform in $O(\alpha(m,n)m)$ by Tarjan, [11], where α is a very slowly growing function and $n < m$ are the amount of calls to an *Union* and *Find* operation respectively. This bound has been proven to be sharp for some classes of pointer machines [12,1,13] and extended to general pointer machines by La Poutré in [6].

Algorithm 1: *FindComp(p)*

```
1  if p is the root of the tree then
   |  return p
   else
2  |  p.par ←FindComp(p.par )
3  |_ return p.par
```

Efficient implementations represent sets by trees, the root of a tree being the unique representative of the set. This can, e.g, easily be done by giving each element e a pointer to another element in the same set, its parent in the tree, denoted e.`par`. The *Union* of two sets is realized by linking the root of one tree to the root of the other one. *Find* identifies the root of the tree, i.e the unique representative of the set, by an iterative pointer search. In the following we will always assume that any algorithm doing UF will work with such a representation, we will refer to such an algorithm as tree *Union-Find* algorithm, **TUF** for short. Clearly the cost of a TUF algorithm is dominated by the number of pointer jumps of *Find* operations. In *Union* operations, the choice of which root to link and which to remain a root has an influence on the number of pointer jumps to be done for future *Find* operations and so on the overall complexity of the algorithm. Usually one links the smaller tree under the bigger one, the so-called "*weighted union rule*".

There is also an commonly used refinement of the *Find* operation, see Algorithm 1. Clearly after *FindComp(p)*, all elements on the path from p to the root have direct access to the root, i.e are linked directly to the root. Let S_0 be an arbitrary subset and let *Impl*(S_0) be S_0 together with those elements that lye on the path from any element $s \in S_0$ to its root. We denote by *Flatten*(S_0) the operation that consists of applying *FindComp* to each element of S_0. We get the following statements.

Remark 1. *After Flatten(S_0) all elements of S_0 have direct access to their root and* $|Impl(S_0)| \leqslant 2 \cdot |S_0|$.

Proposition 2. *Flatten(S_0) performs with at most* $2 \cdot |Impl(S_0)|$ *pointer jumps.*

Proof. Any element $s \in Impl(S_0)$ is used for pointer jumps in two different roles. First, *FindComp* may be called directly on s. This occurs at most once. Second, s is found on the path of some element to its root. This can occur as often as s is parent pointer for other elements *at the beginning*. Clearly that each of these two possibilities sums up in total to at most $|Impl(S_0)|$. □

In this paper, we will show that, given an *Union-Find* algorithm, we can improve its practical efficiency by a better memory management thanks to *Flatten*. This will be verified with an application to image processing. In fact, we consider a particular type of *Union-Find* problem: the so-called *Union-Find* with graphical restrictions.

2.2 Union-Find with Graphical Restrictions

Not only that UF appears as a subproblem of many algorithmic graph problems, see [8], graphs can also be quite useful to *model* algorithmic features of UF problems, as we intend to prove in this paper.

Union-Find problem with graphical restriction can be defined as follows: given is a graph G, vertices are element of the set S and *Union*'s can only be done according to the edges, i.e at each step of the algorithm, subsets are obtained form connected subgraphs of G. For such a problem, the **maximal sets** are the connected subgraphs obtained after several *Union* and *Find* operations when the process of UF is considered to be terminated.

In addition we assume that each demand for an *Union* operation is explicitly given by an edge joining the corresponding sets. The problem of eventually finding such an edge is not part of the UF problem itself.

This problem has been studied in [5] and shown to be linear for several classes of graphs, trees and partial k-trees, for any fixed parameter k, d-dimensional grids for fixed d and 8-neighborhood graphs of a 2-dimensional grid, and planar graphs.

2.3 Image Segmentation and Union-Find

Image segmentation can be seen as an attempt to capture the essential features of a scene, i.e an image. One way to do that is to extract significant regions from an image. One possible technique of extraction is **region growing**, first described in [9]. It consists of starting with the smallest possible regions, i.e pixels, and merging them until they are considered to be optimal. The merging criterion is some oracle that should guarantee the significance of the newly created region. Clearly the specification of such oracles is a matter of its own rights and can not be the subject of this paper.

As has already been observed by Dillencourt et al. in [2], region growing as defined above *leads naturally* to the *Union-Find* problem. In addition to usual UF the problem of image segmentation requires also that each set is connected. UF with graphical restriction easily models that situation: choose an appropriate grid as underlying graph, where vertices of the grid are pixels of the image and edges denote the adjacency relation. The maximal sets represent the regions of the image. Moreover, due to the *Find* operation, we are able to tell for each pixel the region it belongs to. So in terms of image processing a "connected component labeling" has been realized in the same time as the segmentation by the UF algorithm.

In [3] segmentation algorithms using *Union-Find*, and as well an extension to a restricted version of the *Union-Find* problem on planar graphs, have been shown to perform in linear time. In Algorithm 2 we give an implementation of *Union-Find* applied to segmentation of 2-dimensional image.

As already said above, *Oracle* is a criterion to make the decision whether or not to merge the two sets (regions in this case). This algorithm uses a special rule to decide which tree must be linked to the other one when an *Union* is realized: the sets seen the first on a given line is always linked under the one seen later on the line. For a proof of the linearity see [3].

Algorithm 2: Image Segmentation with Union-Find: Line by Line

Data : A grid *bm* of size $w \times l$
Result : A segmented image with connected components labeling

```
special treatment of the first line
for i = 2 to l do
    Flatten(line i − 1 of bm)
    special treatment of the first pixel
    for j = 2 to w do
        left = FindComp(bm[j, i − 1])
        up = FindComp(bm[j − 1, i])
        current = FindComp(bm[j, i])
        if Oracle(left,current ) then
            Union(left,current )
        if Oracle(up,current ) then
            Union(up,current )
```

The image processed can be very large, so there is a lot of data (about 1 million elements for a 1024×1024 image), and an efficient memory management has a direct influence on the practical efficiency of the algorithm. In Algorithm 2 *Flatten* is necessary to achieve the linear complexity, but as we will see in Section 3 it is also useful for a better memory management. In the general case we will show that applying *Flatten* on some particular elements and at some particular moments allows a better memory management without increasing the complexity of *Union-Find* algorithm and so gives a better practical efficiency. These results are emphasized by some experimental results in image processing given in Section 5.

3 Memory Management by Path Decompositions

3.1 Random versus Sequential Memory Access

As already addressed in the introduction our main issue is to avoid the von Neumann bottleneck for UF algorithms. There we must provide a tool to ease an estimation for compilers and operating systems which data elements are going to be used next.

The best way to control which data elements are going to be used next by an algorithm is when data is accessed sequentially, i.e when data is consecutively read or written into an array or (pseudo) file. If done so, modern operation systems and compilers perform quite well in optimizing the performance since they are capable to shuffle entire blocks of memory between cache, RAM and disk.

For the context of this paper we propose thus a distinction between two types of access to data and thus to the memory that is used to store it:

RAM random access memory, a part of the memory that may be accessed in a unpredictable way, and

SAM sequential access memory, a part of the memory that we only access sequentially in a predictable way, i.e e.g as a *stack* or a *file*.

For practical purposes any access of arrays using only *pointer increments* (or decrements) and no other pointer arithmetics can be subsumed under the SAM model.

Our goal in this paper is to optimize the use of memory under these aspects and thus improve the real time performance of TUF algorithms.

3.2 UF in Phases

To allow minimization of RAM we assume that a virtual algorithm $\mathcal{A}$ requiring TUF operations proceeds in phases, $1,\ldots,\ell$ say. In fact the main idea is, for a given phase, to recycle as much memory as possible that was used previously.

We will assume that the amount of memory that is potentially accessed randomly in each phase is the resource that we want to minimize. In fact such an approach is not a restriction to the TUF algorithms that are to be used since we may introduce a phase for each *Union* and *Find* operation. On the other hand it allows to combine consecutive *groups of such operations* in order to improve the behavior of certain algorithms.

We make the assumption that for each phase i a set V_i' of **active** elements is known, i.e a set of elements for which phase i will possibly do *Find* operations[3]. In our example, Algorithm 2, the set of active elements are e.g. pairs of consecutive lines in the image. We also assume that each individual element must only be created once and freed later on. To cover that we call an element $v \in V$ **open** in phase i if there are $j \leqslant i \leqslant k$ such that v is active for both phase j and k. Note that every active element is open as well. By V_i we denote the set of open elements for phase i.

V_i is the least set of elements that any TUF algorithm $\mathcal{A}$ with the same sets of active elements has to administrate at phase i if in addition it is only allowed to create and free elements once. So $\max_{1\leq i\leq \ell}|V_i|$ measures the minimal amount of RAM needed by each such algorithm.

3.3 Path Decompositions

Since our overall goal is to solve UF problems with graphical restrictions we now develop a tool that turns out to combine the underlying graph with the idea of doing TUF in phases: *path decompositions*. These originally have been developed in the framework of the Graph Minor Project of Robertson and Seymour, see [10] for the original definition.

Two key observations lead us there; the first is that for TUF running in phases a set V_i forms a separator of the graph between those elements that already have been processed and those that are not yet touched. The second is that for every element $v \in V$ the indices i such that $v \in V_i$ are consecutive numbers.

Formally a path decomposition $V_1,\ldots, V_\ell$ of a graph $G = (V,E)$ fulfills

(1) $\bigcup_{1\leq i\leq \ell} V_i = V$.
(2) For all $e \in E$ there is i such that $e \subseteq V_i$.

[3] Including those *Find*'s that are needed to perform a *Union*.

(3) For all $v \in V$ and all $j \leq i \leq k$ with $v \in V_j$ and $v \in V_k$ we also have that $v \in V_i$.

In our context Requirement (2) covers the fact that the endpoints of an edge that might be used for a *Union* operation must be active simultaneously in some phase.

The notion of path decomposition is closely related to interval extensions of the graph G: blowing up each set V_i to a clique defines such an extension. On the other hand a consecutive clique arrangement of an interval extension is easily checked to verify the necessary conditions of a path decomposition. For the graphical TUF problem in phases such an interval extension thus adds those edges to the graph that still would give rise to exactly the same sequence of sets of open elements as G.

The ***Width*** [4] of a path decomposition is $\max_{1 \leq i \leq \ell} |V_i| - 1$. For a TUF algorithm $\mathcal{A}$ we denote with *Width*($\mathcal{A}$) the width of the path decomposition corresponding to the sequence of open sets. As we have seen above this parameter is of particular interest in our context — it measures the least amount of elements that our algorithm $\mathcal{A}$ must keep in RAM. Below, Corollary 4, we will see that in fact it always can be realized up to a (small) constant factor.

A lot of efforts are made for several theoretical and practical applications to keep the width of path decompositions as small as possible, i.e to chose a particular path decomposition of a graph that minimizes the width. Often one aims to bound this parameter by a constant for graph classes of particular interest. But our situation is much better: we are not seeking to minimize it but only to keep it inside certain bounds in terms of the size of the graph.

3.4 Bounding the Memory Requirement

Our algorithms use tree data structures to represent the current subsets of the UF process. So if we don't want to exceed *Width*($\mathcal{A}$) by more than a linear factor we have to be careful how many elements are used as internal nodes of some UF tree. For $\mathcal{A}$ call $Impl_{\mathcal{A}}(V_i)$ the total set of elements that are either open for i or accessed during a possible *Find* operation of one of the open elements. Clearly, for a given phase i, we only need to keep the elements of $Impl_{\mathcal{A}}(V_i)$ in RAM. So all the elements that don't belong to $Impl_{\mathcal{A}}(V_i)$ can be deleted and the memory space they used can be recycled.

In fact an easy estimation for $\mathcal{A}$ is given by the following proposition.

Proposition 3. *Let $\mathcal{A}$ be a TUF strategy that runs in ℓ phases with open sets $V_1, \ldots, V_\ell$ such that before each phase $i > 1$ Flatten(V_{i-1}) is invoked.*

Then $|Impl_{\mathcal{A}}(V_i)| \leq 2 \cdot |V_i|$ and the total amount of additional work introduced by the calls to Flatten is linear in $\sum_{1 \leq i \leq \ell} |V_i|$.

Proof. For the bound on $|Impl_{\mathcal{A}}(V_i)|$ just observe that when starting phase i after the Flatten every active element, i.e element of V_i, has direct access to the root of its tree[5] and so the elements that are accessed during phase i itself are at most the active ones and these roots. The estimation of the work now follows easily with Proposition 2. □

[4] The "−1" is included historical reasons only.

[5] All new elements of $V_i \setminus V_{i-1}$ introduced after *Flatten*(V_{i-1}) are one-element sets and so have direct access to their root, i.e themselves

Algorithm 3: TUF with optimizing memory

let $\mathcal{A}_i$ be the part of $\mathcal{A}$ done during in phase i

```
1 A_1
2 for i = 2···ℓ do
3   | Flatten(V_{i-1})
4   | foreach e ∈ V_{i-1} \ V_i do
5   |   | if e is not the root of the tree then
6   |   |   └ Decr(e.par,1)
7   |   └ Decr(e,1)
8   └ A_i
```

If we assume that for each phase i we are able to delete all elements not in $Impl_{\mathcal{A}}(V_i)$ Proposition 3 leads us to the following corollary:

Corollary 4. *Let $\mathcal{A}$ be as in Proposition 3. Then the amount of RAM used by $\mathcal{A}$ for the UF data structure is linear in* Width($\mathcal{A}$).

Observe that estimating the additional work in terms of the *open* elements instead of the *active* ones may already overshoot the budget given by the complexity of $\mathcal{A}$. So for a particular algorithm that produces much more open elements than active ones it might be interesting to refine the ideas given so far to call *Flatten* on active elements only. We will see an example where this is possible in Section 5.

4 Union-Find Algorithm With Memory Management

From Corollary 4 we know that the memory needed by a TUF in phase algorithm $\mathcal{A}$ can be linear in *Width($\mathcal{A}$)*. This corollary holds only if, for a given phase i, we are able to delete all elements not in *Impl(V_i)*. In the following we present an implementation of such an improved algorithm, see Algorithm 3. Then in Section 4.2 we show that the data what is freed can in fact be mapped into a SAM in order to be able to reconstruct the maximal sets. Then in Section 4.3 we prove a linear time bound for the additional work introduced by the memory management. Thus the global complexity of final Algorithm 3 remains the same.

4.1 Implementation of a Low Memory Consuming TUF

To achieve our goal we must be able to delete all elements not in *Impl(V_i)*, for a given phase i. This problem can be decomposed into two sub-problems. First we have to know that a given element *does not belong* to *Impl(V_i)*. Then we have to ensure that *all* such elements are deleted.

To solve the first point we provide each element e with a counter, denoted `e.impl`.

Invariant 1. *For a given phase i and for a given element e,* `e.impl` *records the number of elements, below it in the tree , including itself, that belong to Impl(V_i).*

Algorithm 4: *Decr(e,n)*

Input: an element e and an integer n.

$e.\mathtt{impl} \leftarrow e.\mathtt{impl} - n$
if $e.\mathit{impl} = 0$ **then**
  delete e

Algorithm 5: *FindComp(p)* with update of `impl`

1 **if** *p is the root of the tree* **then**
  return p
else
2.1 `origin` ← 2.2 `par` p.`par` ← *FindComp(p.par)*
2.3 **if** $p.par \neq origin$ **then**
  Decr(origin, p.impl)
3 **return** $p.par$

Now with Invariant 1 we know that, for a given phase i, a given element e does not belong to *Impl(V_i)* if e.`impl` is equal to 0.

Invariant 2. *An element e with* $e.\mathtt{impl} = 0$ *is immediately deleted.*

Provided Invariant 1 2 are guaranteed we get easily the following invariant:

Invariant 3. *For a given phase i, all elements that do not belong to Impl(V_i) are deleted.*

Note that Invariant 2 is verified if each time a value is subtracted from `impl` we check its value and delete the corresponding element if $\mathtt{impl} = 0$. So we add the new function ***Decr*** , see Algorithm 4 and by that easily verify Invariant 2.

In order to verify Invariant 1, we must check that (1) every time trees are modified, and (2) every time an element will not be active anymore, `impl` is well maintained.

The first point occurs only during *Union-Find* operations, i.e *Union*, *FindComp* or when a new element is created. For *Union* it is sufficient to add to the new root the `impl` value of the root linked to it. When a new element is created we just have to initialize `impl` to the value of 1. The only difficulty comes from *FindComp* which messes up the tree. Algorithm 5 shows an appropriate update of *FindComp*.

It is easy to see that if Invariant 1 is verified before a *FindComp*, it remains after. Indeed the only place where the tree can be modified is at line 2.2. At line 2.3 we check if this really happens. In fact if the parent has changed, the old parent has "lost" all the elements of *Impl(V_i)* below p. This number of elements is recorded by p.`impl` since Invariant 1 is assumed to be verified before doing the *FindComp*. So we just have to subtract this value to the value of `impl` of the old parent. This is realized by the instruction *Decr(origin, p.impl)*.

Observe that by now we already have shown that *Flatten* also maintains `impl` properly if this new version of *FindComp* is used.

So we have proven that every time trees are modified, `impl` is well maintained. Now we must update `impl` of elements that will no more be active. This happens only during the transition between two phases. Clearly, all such elements are in $V_{i-1} \setminus V_i$. So at the beginning of a given phase i we must decrement the value of `impl` of all these elements by one. Of course this operation will be done by a call to *Decr*. But changing the value of `impl` of a given element implies to update the value of `impl` of all the elements above in the tree. After *Flatten*(V_{i-1}) we know that all elements of $V_{i-1} \setminus V_i$ have direct access to its root. Thus for a given element $e \in V_{i-1} \setminus V_i$, in the same time we decrement e.`impl`, we just have to decrement the value of `impl` of e.`par`. This is realized by lines 4, 5, 6 and 7 of Algorithm 3.

We just have proven that Invariant 1 remains true along the algorithm, Invariant 2 remains true too, since we always call *Decr* to reduced the value of `impl` of a given element. Algorithm 3 implements these solutions and thus respects Invariant 3. So it is an implementation of $\mathcal{A}$ according to Corollary 4.

Since the amount of necessary RAM is reduced, one can expect that all the data is allocated consecutively, or at least on the same page of memory. So this reduces the number of page faults of large scale applications drastically. But minimizing the amount of necessary RAM is not only useful for such large applications: for medium sized applications a great part of the data now fits into the memory cache of the CPU. So the practical efficiency will then be greatly improved for such applications as well.

4.2 Mapping Data into Sequential Memory

Algorithm 3 presented above allows to minimize the amount of RAM necessary to a TUF algorithm by deleting elements that are not necessary anymore, and by reusing the memory that was freed. But UF is generally a part of a more global process and after running an *Union-Find* algorithm, one may need to access the maximal sets. We now present a way to reconstruct the maximal sets while keeping the amount of necessary RAM linear in *Width*($\mathcal{A}$).

The principle of the method is to save informations about the deleted elements on a stack. Clearly a stack can be implemented on a SAM and so it doesn't use RAM. At the end of the process, unstacking the information saved should allow to reconstruct the maximal sets. The only information we need is the element itself and the current root of its tree. We must also ensure that this root is always saved after all elements below it. So it will be unstacked first and the tree will be easily reconstruct when unstacking the elements.

This can easily realized in Algorithm 3. Note that elements are deleted, either in a *FindComp*, or when updating elements of $V_{i-1} \setminus V_i$. In a *FindComp* the root of the tree is known, and updating elements of $V_{i-1} \setminus V_i$ occurs just after *Flatten*(V_{i-1}), so either the element itself or its parent is a root.

So when deleting an element we know the root of the tree. We just have to check if roots are always deleted after elements below them and so delete operation can be replaced by a stack-in (push) operation. In *FindComp* of Algorithm 5 we never call *Decr* on a root element, so a root will never be deleted during a *FindComp*. But when updating elements of $V_{i-1} \setminus V_i$, no assumption is done on the order the elements are processed. In particular a root can be deleted before elements below it. So to guarantee that roots will be stacked after leaves we just have to check the elements two times: the first time we update an element only if it is a leaf, the second time we update the roots. The delete operation can thus be replaced by a push operation which saves the element and the name of its root on a stack (SAM).

The efficiency of the future memory accesses to the maximal sets can be improved if elements of the same set are consecutive in RAM. This can be easily achieved. Indeed the number of elements of a set can be recorded in the root. So when saving a root, one can stack also this number. Thus the necessary amount of memory can be allocated in a whole, when reconstructing the maximal sets.

4.3 Complexity of the Additional Work

First of all, let us recall that the total amount of additional work introduced by the calls to *Flatten* is linear in *Width*($\mathcal{A}$), ie in $\sum_{1\leq i\leq \ell}|V_i|$, see Proposition 3. It is easy to see that operation *Decr* is done in constant time, even after replacing delete by push. Additional work added in *Union-Find* functions are also constant time operations since we only change value of `impl` and make call to *Decr*. The process of updating elements of $V_{i-1}\setminus V_i$ is not realized in constant time, but clearly is done in linear time of $|V_{i-1}\setminus V_i|$ so it doesn't cost more than the additional work introduced by *Flatten*.

Thus the complexity of the additional work introduced for the memory management is bounded by $O(\mathit{Width}(\mathcal{A}))$.

5 Application to Image Segmentation

As already mentioned in Section 2.3, *Union-Find* is well suited to implement region growing segmentation in image processing. But a major drawback is the space consuming data structure. Indeed we need at least one pointer for each element, and for such an application as region segmentation, some additional data are needed in order to help the *Oracle* function to take its decision. For example in our application, an element requires a structure of at least 16 bytes. So for an image of about 1024×1024, i.e about 1 million of element, we need 16 MBytes to keep the entire *Union-Find* data structure in RAM. In such a situation our approach of using the main part of this memory only sequentially pays off in a qualitative improvement of the running times. If we go even further and try to treat 3-dimensional images we reach – by analogous computations – a memory demand of 16GB, something nobody is currently capable to install as RAM for a reasonable price. So it is clear that a part of the structure must be kept on a SAM to improve efficiency of the memory accesses, and Algorithm 3 is well-suited for such a purpose.

In fact Algorithm 2 follows already the scheme of Algorithm 3. Indeed the algorithm proceeds line by line, and only two are involved in the same time. So we have a phase decomposition where each V_i is the set of pixels of two consecutive lines. Moreover *Flatten* is done at the end of the phase and this is the same as doing it at the beginning of the following phase. So we just have to add the process of updating elements of $V_{i-1}\setminus V_i$ after Flatten and to modify *Union-Find* operations accordingly to Section 4.1. Note that $V_{i-1}\setminus V_i$ is in fact the line $j-1$.

Algorithm 6: Image Segmentation with Union-Find: Merging Squares

```
Input: An integer k and a bitmap bm of size 2^k × 2^k
if k = 0 then return
hm ← 2^(k-1) − 1; hp ← 2^(k-1)
for dir=NM to SE do
  └ MergeSquare(bm[dir], k − 1)
for i = 0 to 2^k do
  | left ←Find(bm[i,hm])
  | right ←Find(bm[i,hp])
  | if Oracle(left,right) then
  |   └ Union(left,right)
  └
for i = 0 to 2^k do
  | up ←Find (bm[hm,i])
  | down ←Find (bm[hp,i])
  | if Oracle(up,down) then
  |   └ Union(up,down)
  └
```

In [3] a second segmentation algorithm based on UF is presented, see Algorithm 6. Assuming the image is a square of size $n \times n$, this algorithm proceeds recursively by dividing the image into 4 sub-squares of size $n/2 \times n/2$. After coming up the recursion, the regions in the 4 sub-squares are merged together along the common boundary. For a proof of the linearity of this algorithm we again refer to [3].

For the formulation of Algorithm 6, *bm*[*NW*] denotes the northwestern sub-matrix of *bm*, *bm*[*NE*] the northeastern, etc... Here again, the decomposition in phases is trivial, and to assure a better memory management we just need to add the necessary stuffs before the first **for** loop of Algorithm 6. But in the case of this algorithm the sets of open elements can be very large, e.g all the image at the lower level of the recursion. So the cost of the additional works will completely overshoot the linear time complexity of this algorithm. But here, the process can be refined by doing *Flatten* only on active elements. Indeed, due to the recursion and to the particular decomposition of the bitmap, we are sure that open elements which do not belong to the current bitmap (i.e *bm* in Algorithm 6) cannot be part of a tree of an active element. Then we only need to do *Flatten* on active elements of *bm* (medians of *bm*), i.e $bm[0\cdots 2^k, hm]$, $bm[0\cdots 2^k, hp]$, $bm[hm, 0\cdots 2^k]$ and $bm[hp, 0\cdots 2^k]$. So additional work introduced by *Flatten* is not too much costly.

Note that these two algorithms have been implemented and gives short running times: about 2*s* for a 512×512 image.

In addition in [4] we have presented a method to segment 3-dimensional images using this principle. Using such a strategy, we have reduced the need of RAM from 710MB to only 10MB. So the algorithm can be executed on a common workstation.

6 Conclusion

We have presented a general tool to minimize the random access memory that is used of a broad class of *Union-Find* algorithms. Not only this allows to use *Union-Find* strategy on large application, but also this improves the real time performance. Indeed the well-known von Neumann bottleneck of synchronizing CPU and memory is thus avoided. This method has been successfully implemented for the image segmentation problem.

Acknowledgment

We like to thank Thomas Rehm for successfully implementing parts of the strategies presented in this paper.

References

1. L. BANACHOWSKI, *A complement to Tarjan's result about the lower bound on the complexity of the set union problem*, Inform. Process. Lett., 11 (1980), pp. 59–65.
2. M. B. DILLENCOURT, H. SAMET, AND M. TAMMINEN, *A general approach to connected-component labeling for arbitrary image representations*, J. Assoc. Comput. Mach., 39 (1992), pp. 253–280. Corr. p. 985-986.
3. C. FIORIO AND J. GUSTEDT, *Two linear time Union-Find strategies for image processing*, Theoret. Comput. Sci., 154 (1996), pp. 165–181.
4. ——, *Volume segmentation of 3-dimensional images*, Tech. Rep. 515/1996, Technische Universität Berlin, 1996.
5. J. GUSTEDT, *Efficient union-find for planar graphs and other sparse graph classes*, in Graph-Theoretic Concepts in Computer Science, 22nd International Workshop WG '96, Ausiello et al., eds., Lecture Notes in Computer Science, Springer-Verlag, 1996, pp. 181–195. *to appear.*
6. J. A. LA POUTRÉ, *New techniques for the union-find problem*, in Proceedings of the first annual ACM-SIAM Symposium on Discrete Algorithms, A. Aggarwal et al., eds., Society of Industrial and Applied Mathematics (SIAM), 1990, pp. 54–63.
7. K. MEHLHORN, *Data Structures and Algorithms 1: Sorting and Searching*, Springer, 1984.
8. K. MEHLHORN AND A. TSAKALIDIS, *Handbook of Theoretical Computer Science*, vol. A, Algorithms and Complexity, Elsevier Science Publishers B.V., Amsterdam, 1990, ch. 6, Data Structures, pp. 301–314.
9. J. MUERLE AND D. ALLEN, *Experimental evaluation of techniques for automatic segmentation of objects in a complex scene*, in Pictorial Pattern Recognition, G. C. Cheng et al., eds., Thompson, Washington, 1968, pp. 3–13.
10. N. ROBERTSON AND P. SEYMOUR, *Graph minors I, excluding a forest*, J. Combin. Theory Ser. B, 35 (1983), pp. 39–61.
11. R. E. TARJAN, *Efficiency of a good but not linear set union algorithm*, J. Assoc. Comput. Mach., 22 (1975), pp. 215–225.
12. ——, *A class of algorithms which require non-linear time to maintain disjoint sets*, J. Comput. System Sci., 18 (1979), pp. 110–127.
13. R. E. TARJAN AND J. VAN LEEUWEN, *Worst-case analysis of set union algorithms*, J. Assoc. Comput. Mach., 31 (1984), pp. 245–281.
14. M. J. VAN KREVELD AND M. H. OVERMARS, *Union-copy structures and dynamic segment trees*, J. of the Association for Computing Machinery, 40 (1993), pp. 635–652.

Fast Online Multiplication of Real Numbers*

Matthias Schröder

Theoretische Informatik I, FernUniversität Hagen,
D-58084 Hagen, Germany
e-mail: Matthias.Schroeder@fernuni-hagen.de

Abstract. We develop an online-algorithm for multiplication of real numbers which runs in time $O(M(n)\log(n))$, where M denotes the Schönhage-Strassen-bound for integer multiplication which is defined by $M(n) = n\log(n)\log\log(n)$, and n refers to the output precision $(\frac{1}{2})^n$. Our computational model is based on Type-2-machines. The real numbers are given by infinite sequences of symbols which approximate the reals with increasing precision. While reading more and more digits of the input reals, an algorithm for a real function produces more and more precise approximations of the desired result. An algorithm M is called online, if for every $n \in \mathbb{N}$ the input-precision, which M requires for producing the result with precision $(\frac{1}{2})^n$, is approximately the same as the topologically necessary precision.

Topics: Computable real analysis, computational complexity, online computations.

1 Introduction

The main idea of the Type-2-Theory ([Wei87, Wei95, Wei96]) is to represent the objects of a given topological space (X, τ) with cardinality of the continuum by sequences of symbols of a finite alphabet Σ, on which the actual computation is performed by Type-2-machines. The corresponding partial surjection $\delta :\subseteq \Sigma^\omega \to X$ [1,2] which maps every name to the represented element of X is called a *representation* of X, and should fulfil some effectivity properties like "admissibility", in order to provide a reasonable computability theory on X.

An often used admissible representation of the real numbers $\mathbb{R}$ is the "modified binary representation" ρ_{sd} (cf. [Wei95]), which differs from the ordinary binary or decimal representation in using an additional negative digit $\bar{1}$ with value -1. Briefly, each ρ_{sd}-name has the form $a_k \ldots a_0.a_{-1}a_{-2}\ldots$ with $a_i \in \{\bar{1}, 0, 1\}$ representing the real number $\sum_{i=-\infty}^{k} a_i \cdot 2^i$. The use of the negative digit $\bar{1}$ guarantees effectivity, whereas for topological reasons the ordinary binary and the ordinary

---

* This work was supported by the Deutsche Forschungsgemeinschaft.

[1] $\Sigma^\omega = \{p \mid p : \mathbb{N} \to \Sigma\}$ is the set of sequences over Σ.

[2] $f :\subseteq X \to Y$ denotes that f is a partial function with a domain $dom(f) \subseteq X$ which is not necessarily equal to X.

R. Reischuk, M. Morvan (Eds.): STACS 97 Proceedings, LNCS 1200

Fast Online Multiplication of Real Numbers

Matthias Schröder *

Theoretische Informatik I, FernUniversität Hagen
D-58084 Hagen, Germany
e-mail: Matthias.Schroeder@fernuni-hagen.de

Abstract. We develop an online–algorithm for multiplication of real numbers which runs in time $\mathcal{O}(\mathcal{M}(n)\log(n))$, where $\mathcal{M}$ denotes the Schönhage–Strassen–bound for integer multiplication which is defined by $\mathcal{M}(m) = m\log(m)\log\log(m)$, and n refers to the output precision $(\frac{1}{2})^n$. Our computational model is based on Type–2–machines: The real numbers are given by infinite sequences of symbols which approximate the reals with increasing precision. While reading more and more digits of the input reals, an algorithm for a real function produces more and more precise approximations of the desired result. An algorithm M is called online, if for every $n \in \mathbb{N}$ the input–precision, which M requires for producing the result with precision $(\frac{1}{2})^n$, is approximately the same as the topologically necessary precision.

Topics: Computable real analysis, computational complexity, online computations.

1 Introduction

The main idea of the Type–2–Theory ([We87, We95a, We95b]) is to represent the objects of a given topological space (X, τ) with cardinality of the continuum by sequences of symbols of a finite alphabet Σ, on which the actual computation is executed by Type–2–machines. The corresponding partial surjection $\delta :\subseteq \Sigma^\omega \to X$ [1], which maps every name to the represented element of X, is called *a representation* of X, and should fulfil some effectivity properties like "admissibility" in order to provide a reasonable computability theory on X.

An often used admissible representation of the real numbers $\mathbb{R}$ is the "modified binary representation" ϱ (cf. Def. 1), which differs from the ordinary binary or decimal representation in using an additional negative digit $\bar{1}$ with value -1. Briefly, each ϱ–name has the form $a_{-m} \dots a_0 : a_1 a_2 \dots$ with $a_i \in \{\bar{1}, 0, 1\}$ representing the real number $\sum_{i=-m}^{\infty} a_i \cdot (\frac{1}{2})^i$. The use of the negative digit $\bar{1}$ guarantees effectivity, whereas for topological reasons the ordinary binary and the ordinary

* This work was supported by the Deutsche Forschungsgemeinschaft.

[1] $\Sigma^\omega := \{p : \mathbb{N} \to \Sigma\}$ is the set of sequences over Σ.
$f :\subseteq X \to Y$ denotes that f is a partial function with a domain $dom(f) \subseteq X$ which not necessarily equals X.

Reischuk, Morvan (Eds.): STACS'97 Proceedings, LNCS 1200

decimal representation lead to the uncomputability of some basic real functions like multiplication by 3 (cf. [We95b]) [2].

Type–2–machines induce a notion of computability for functions $\Gamma :\subseteq (\Sigma^\omega)^k \to \Sigma^\omega$. Briefly, a Type–2–machine M is an usual Turing machine with changed semantics. It has k one–way input tapes, finite many work tapes, a single *one–way* output tape and is controlled by a finite flowchart with the usual commands of Turing machines. The function $\Gamma_M :\subseteq (\Sigma^\omega)^k \to \Sigma^\omega$ computed by M is defined by:

$$\Gamma_M(p_1,\ldots,p_k) := \begin{cases} q & \text{if } M \text{ with input } p_1,\ldots,p_k \text{ writes step by step} \\ & \text{the sequence } q \in \Sigma^\omega \text{ onto the output tape} \\ \bot & \text{if } M \text{ with input } p_1,\ldots,p_k \text{ does not write in-} \\ & \text{finitely many symbols on the output tape} \end{cases}$$

Since a Type–2–machine can never change a symbol already written on the output tape (because it is an one–way tape), each prefix of the output only depends on a prefix of each input sequence (finiteness property).

Based on Type–2–machines, computability of real functions is defined as follows. A real function $f :\subseteq \mathbb{R}^k \to \mathbb{R}$ is called computable iff there is a Type–2–machine M with

$$\varrho\Big(\Gamma_M(p_1,\ldots,p_k)\Big) = f\Big(\varrho(p_1),\ldots,\varrho(p_k)\Big) \tag{1}$$

for all $p_1,\ldots,p_k \in dom(\varrho)$ with $\big(\varrho(p_1),\ldots,\varrho(p_k)\big) \in dom(f)$, i.e. M transforms every name of a real tuple $\bar{x} \in dom(f)$ to a sequence representing the result $f(\bar{x})$. This notion of computability for real functions corresponds to the ones considered by M. Pour–El/J. Richards ([PR89]), by K. Ko ([Ko91]) and by H.J. Hoover ([Ho90]). The finiteness property guarantees that Type–2–machines, although they formally handle infinite objects, yield a realistic computational model for $\mathbb{R}$, because they can be simulated by physical machines. In contrast to this model, the BSS model of real computability (cf. [BSS89]) seems to be unrealistic, because it expects registers to be able to store real numbers and therefore to be able to store an *infinite* amount of information at the same time.

The computation of a real function f by a Type–2–machine M can be regarded as a non-ending process which produces more and more precise approximations of the result $f(\bar{x})$ while reading and considering more and more precise approximations of the input $\bar{x}$. For $n \in \mathbb{N}$, the *time complexity* of M is the time

[2] The two names of the real number 1 under the decimal representation are of course the sequences $q_1 = 1{:}000\ldots$ and $q_2 = 0{:}999\ldots$. The approximation given by the first two symbols of q_1 states that the represented number is contained in $[1;2]$, whereas the first two symbols of q_2 claim that the represented number is in $[0;1]$. But from no approximation given by the name $p = 0{:}333\ldots$ the corresponding informations over p stating that the number represented by p is in $[\frac{1}{3};\frac{2}{3}]$ resp. in $[0;\frac{1}{3}]$ can be derived, because every prefix of p can be lengthened to a name representing a number $< \frac{1}{3}$ as well as to one representing a number $> \frac{1}{3}$. Therefore, it is impossible to compute $x \mapsto 3x$ under the ordinary decimal representation.

M needs for computing an approximation with precision $(\frac{1}{2})^n$, more precisely the number of steps M executes before M outputs the n–th digit to the right of the binary point "$:$". The *input–lookahead* (Ila_M) of M is the maximal number of digits to the right of "$:$" which M has read from one of its inputs until this moment. The number of digits to the right of "$:$" which must be known for producing an approximation with precision $(\frac{1}{2})^n$ is called the *dependency* (Dep_f) of the real function f.

We say that M works *strictly online* iff the dependency of the computed real function f and the input–lookahead of M coincide. K. Weihrauch showed that there are computable real functions which cannot be computed by a strictly online Type–2–machine (cf. [We91]). Nevertheless, for every computable real function f with a reasonable domain there is a Type–2–machine M for f which for computing an approximation with precision $(\frac{1}{2})^n$ reads at most as much digits as necessary to produce the $(n+2)$–th digit[3] to the right of "$:$". Unfortunately, there are functions such that the time complexity rises rapidly, if "online"–machines are used (cf. [We91]). Nevertheless, if the costs for producing the input digits are high (for example if the input is the output of another Type–2–machine), it might be faster to use a slow "online"–machine than a fast machine which reads its input wastefully.

In the following, we will show that there is an online algorithm for the multiplication of real numbers, which runs in time $\mathcal{O}(\mathcal{M}(n)\log_2(n))$ on every bounded interval of $\mathbb{R}$ and reads at most 2 digits more than necessary for producing the n–th digit to the right of "$:$". $\mathcal{M}$ denotes the Schönhage–Strassen–bound ($\mathcal{M}(m) = m \cdot \log_2(m) \cdot \log_2 \log_2(m)$) for multiplication of m–bit integers (cf. [SS71]). $\mathcal{O}(\mathcal{M}(n)\log_2(n))$ is exactly the time complexity of the algorithm developed by M. Fischer and L. Stockmeyer for integer online multiplication (cf. [FS74]) which starts to read and to write the *least* significant digit of the input resp. the output instead of the most significant digit as in the case of real multiplication. The fastest known algorithm for real multiplication requires time $\mathcal{O}(\mathcal{M}(n))$ uniformly on every bounded interval and was presented by N. Müller (cf. [Mü86]). For the most of the inputs, the difference between the input–lookahead of this algorithm and the dependency can be only bounded by $n + \mathcal{O}(1)$. An algorithm running in time $\mathcal{O}(\mathcal{M}(n)\log_2(n))$, such that this difference is bounded by $\log_2 \log_2(n) + \mathcal{O}(1)$, has been presented by B. Landgraf (cf. [La95]). Strictly online multiplication, however, seems to require quadratic time.

For the sake of simplicity, we will restrict us to construct a *weakly online* Type–2–machine M_{w} for multiplication on the subset $[-1;1]$ with time complexity $\mathcal{O}(\mathcal{M}(n)\log_2(n))$. "Weakly online" means that the input–lookahead for $n \in \mathbb{N}$ is bounded by $n + c_{\bar{x}}$, where $c_{\bar{x}}$ is a constant that only depends on the input $\bar{x} \in \mathbb{R}^k$, not on n, and is called *the delay* at $\bar{x}$ of the algorithm (cf. [DHM89]). Thus, for every input p, q (except for $\varrho(p) = \varrho(q) = 0$) the difference between the input–lookahead of M_{w} and the dependency is bounded by a constant that only depends on $(\varrho(p), \varrho(q))$. In Section 4, we will briefly describe how to transform

[3] Even "$n+1$" instead of "$n+2$" does not suffice.

M_w into a machine satisfying the above online condition.

2 Basic Definitions

In the following, let Σ be the alphabet $\{\bar{1}, 0, 1, :\}$. ":" serves as the binary point. For a word resp. a sequence $p = a_{-m} \ldots a_0 : a_1 \ldots a_n$ with $m \in \mathbb{N} \cup \{-1\}$, $n \in \mathbb{N} \cup \{\infty\}$ and $a_{-m}, a_{-m+1} \ldots \in \{\bar{1}, 0, 1\}$ and for $z \in \mathbb{Z}$ and $w \in \Sigma^*$ we denote by

$p^{\leq z}$	the word $\begin{cases} \varepsilon & \text{if } z < 0 \\ a_{-m} \ldots a_0 : a_1 \ldots a_z & \text{if } 0 \leq z \leq n \\ a_{-m} \ldots a_0 : a_1 \ldots a_n 0^{z-n} & \text{if } z > n; \end{cases}$
$p^{=z}$	the integer $\begin{cases} a_z \text{ if } -m \leq z \leq n \\ 0 \text{ else;} \end{cases}$
$\mathrm{lk}(p)$	the length of the integral part of p, i.e. $m + 1$;
$\mathrm{prec}(p)$	the length of the fractional part of p, i.e. n;
$w\Sigma^\omega$	the set of all $q \in \Sigma^\omega$ with prefix w.

Now we give the definition of the representation ϱ mentioned in Section 1.

Definition 1 (The Representation ϱ of $\mathbb{R}$).
Define $\varrho :\subseteq \Sigma^\omega \to \mathbb{R}$ by

$$dom(\varrho) := \Big\{ p \in \{\bar{1}, 0, 1\}^* \{:\} \{\bar{1}, 0, 1\}^\omega \mid p \text{ does not begin with "0" or "}\bar{1}1\text{" or "}1\bar{1}\text{"} \Big\}$$

and $\varrho(p) := \sum_{i=-\infty}^{\infty} (\frac{1}{2})^i \cdot p^{=i}$ for all $p \in dom(\varrho)$.

Obviously, the mapping ϱ is surjective. Furthermore, ϱ is a continuous[4], "admissible" and "effective" representation of ϱ (cf. [We95a, We95b]). Due to the restriction in the definition of its domain, ϱ can be shown to be a perfect representation, thus $\varrho^{-1}(K)$ is compact for every compact set $K \subseteq \mathbb{R}$ (cf. [We95a] or [Mü86]). The perfectness of ϱ guarantees that ϱ provides a reasonable complexity theory (cf. [Sch95]). The following lemma summarizes some important properties of the representation ϱ.

Lemma 2 (Properties of ϱ).
(1) $-2^{\mathrm{lk}(p)} \leq \varrho(p) \leq 2^{\mathrm{lk}(p)}$ *holds for all* $p \in dom(\varrho)$.
(2) For all words $w \in \{\bar{1}, 0, 1\}^* \{:\} \{\bar{1}, 0, 1\}^*$ *with* $\varrho(w\Sigma^\omega) \neq \emptyset$

$$\varrho(w\Sigma^\omega) = \left[d - (\tfrac{1}{2})^{\mathrm{prec}(w)};\, d + (\tfrac{1}{2})^{\mathrm{prec}(w)} \right]$$

holds, where d is the dyadic rational $d = \varrho(w0^\omega) = \sum_{z=-\mathrm{lk}(w)}^{\mathrm{prec}(w)} (\frac{1}{2})^z \cdot w^{=z}$.

[4] We use the Cantor topology on Σ^ω which is defined by the base $\{w\Sigma^\omega \mid w \in \Sigma^*\}$.

(3) *For all* $x \in \mathbb{R}$ *and* $n \in \mathbb{N}$ *there is a word* $w \in \Sigma^*$ *with* $\mathrm{prec}(w) = n$ *and*

$$\left[x - (\tfrac{1}{2})^{n+1}; x + (\tfrac{1}{2})^{n+1}\right] \subseteq \varrho(w\Sigma^\omega) .$$

(4) *For all words* $u, v \in \{\bar{1}, 0, 1\}^*\{:\}\{\bar{1}, 0, 1\}^*$ *with* $\mathrm{prec}(u) \leq \mathrm{prec}(v)$ *there is a word* $w \in \{\bar{1}, 0, 1\}^{\mathrm{prec}(v)-\mathrm{prec}(u)}$ *with* $\mathrm{prec}(uw) = \mathrm{prec}(v)$ *and*

$$\varrho(u\Sigma^\omega) \cap \varrho(v\Sigma^\omega) \subseteq \varrho(uw\Sigma^\omega) .$$

Lemma 2(4) applies in the following situation: the present output of a Type–2–machine is u and a better approximation has been computed which can be represented by some ϱ–prefix v. Lemma 2(4) states that u can be lengthened to a ϱ–prefix u' which approximates the desired result with the same precision as v.

Definition 3 (Time complexity, Input–lookahead, Dependency). Let $f :\subseteq \mathbb{R}^k \to \mathbb{R}$ be a continuous real function and let M be a k–ary Type–2–machine.

(1) Define the dependency $\mathrm{Dep}_f : (dom(\varrho))^k \to (\subseteq \mathbb{N} \to \mathbb{Z})$ of f, the time complexity $\mathrm{Time}_M : (dom(\varrho))^k \to (\subseteq \mathbb{N} \to \mathbb{N})$ of M and the input–lookahead $\mathrm{Ila}_M : (dom(\varrho))^k \to (\subseteq \mathbb{N} \to \mathbb{Z})$ of M by

$$\begin{aligned}\mathrm{Dep}_f(p_1,\ldots,p_k)(n) := \min\Big\{r \geq -1 \,\big|\, &(\exists q \in dom(\varrho))\\ &f\Big(\varrho(p_1^{\leq r}\Sigma^\omega) \times \ldots \times \varrho(p_k^{\leq r}\Sigma^\omega)\Big) \subseteq \varrho(q^{\leq n}\Sigma^\omega)\Big\},\end{aligned}$$

$$\mathrm{Time}_M(\bar{p})(n) := \begin{cases} \bot & \text{if } M \text{ with input } \bar{p} \text{ never outputs ``:'' or never outputs the } n\text{-th digit to the right of ``:''} \\ \text{the number of steps which } M \text{ with input } \bar{p} \text{ executes until the } n\text{–th digit to the right of ``:'' has been written on the output tape,} & \text{otherwise} \end{cases}$$

and

$$\mathrm{Ila}_M(\bar{p})(n) := \max\Big\{\mathrm{Ila}_{M,1}(\bar{p})(n), \ldots, \mathrm{Ila}_{M,k}(\bar{p})(n)\Big\},$$

where

$$\mathrm{Ila}_{M,i}(\bar{p})(n) := \begin{cases} \bot & \text{if } \mathrm{Time}_M(\bar{p})(n) = \bot \\ -1 & \text{if } M \text{ with input } \bar{p} \text{ has not yet read some ``:'' from the tape } i \text{ during the first } \mathrm{Time}_M(\bar{p})(n) \text{ steps of execution} \\ r_n^i & \text{if } r_n^i \text{ is the number of digits to the right of ``:'' which } M \text{ with input } \bar{p} \text{ has read from the tape } i \text{ during the first } \mathrm{Time}_M(\bar{p})(n) \text{ steps of execution} \end{cases}$$

(2) For $\bar{x} \in \mathbb{R}^k$, $X \subseteq \mathbb{R}^k$ and $n \in \mathbb{N}$ define additionally[5]:

$$\mathrm{Time}_M(\bar{x})(n) := \sup\{\mathrm{Time}_M(\bar{p})(n) \mid \varrho^k(\bar{p}) = \bar{x}\}$$
$$\mathrm{Time}_M(X)(n) := \sup\{\mathrm{Time}_M(\bar{p})(n) \mid \varrho^k(\bar{p}) \in X\}$$

[5] $\varrho^k :\subseteq (\Sigma^\omega)^k \to \mathbb{R}^k$ denotes the k–fold juxtaposition of ϱ.

Let $f :\subseteq \mathbb{R}^k \to \mathbb{R}$ be continuous and M be a Type–2–machine computing f. By Lemma 2(4), for all $\bar{p} = (p_1, \dots, p_k) \in dom(f\varrho^k)$ there is a name q of $f(\varrho^k(\bar{p}))$ which satifies

$$f\Big(\varrho(p_1^{\leq \mathrm{Dep}_f(\bar{p})(n)}\Sigma^\omega) \times \dots \times \varrho(p_k^{\leq \mathrm{Dep}_f(\bar{p})(n)}\Sigma^\omega)\Big) \subseteq \varrho(q^{\leq n}\Sigma^\omega)$$

simultaneously for all $n \in \mathbb{N}$. For topological reasons, $\mathrm{Dep}_f(\bar{p})$ is total and

$$(\forall n \in \mathbb{N})\ \mathrm{Dep}_f(\bar{p})(n) \leq \mathrm{Ila}_M(\bar{p})(n) \leq \mathrm{Time}_M(\bar{p})(n)$$

holds for all $\bar{p} \in dom(f\varrho^k)$. The perfectness of ϱ guarantees that $\mathrm{Time}_M(\bar{x})(n)$ and $\mathrm{Time}_M(K)(n)$ exist for all $\bar{x} \in dom(f)$ and for all compact sets $K \subseteq dom(f)$ (cf. [Sch95]). Hence, Definition 3(2) establishes a well–defined complexity theory for real functions. It corresponds to the one considered in [Ko91].

3 Fast Weakly Online Multiplication on [-1;1]

In this section, we will develop a fast algorithm M_w for weakly online multiplication on $[-1;1]$ running in time $\mathcal{O}(\mathcal{M}(n)\log_2(n))$.

Theorem 4.
Weakly online multiplication on the subset $[-1;1]$ *can be performed uniformly in time* $\mathcal{O}(\mathcal{M}(n)\log_2(n))$, *i.e. there are a Type–2–machine* M_w *and some* $c \in \mathbb{N}$ *such that*

(a) $\varrho\Big(\Gamma_{M_w}(p,q)\Big) = \varrho(p) \cdot \varrho(q)$
(b) $\mathrm{Time}_{M_w}(x,y)(n) \leq c \cdot \mathcal{M}(n) \cdot \log_2(n) + c$
(c) $\mathrm{Ila}_{M_w}(p,q)(n) \leq n+3$

hold for all $x, y \in [-1;1]$, $p, q \in \{:\}\{\bar{1},0,1\}^\omega$ *and* $n \in \mathbb{N}$.

For the proof, we will only describe the basic ideas of the algorithm M_w. In the following, let $p, q \in \{:\}\{\bar{1},0,1\}^\omega$ and define $x_r := \varrho(p^{\leq r}0^\omega)$, $y_r := \varrho(q^{\leq r}0^\omega)$ for $r \in \mathbb{N}$. It is easy to prove

$$\varrho(p^{\leq r}\Sigma^\omega) \cdot \varrho(q^{\leq r}\Sigma^\omega) \subseteq [x_r \cdot y_r - (\tfrac{1}{2})^{r-1}; x_r \cdot y_r + (\tfrac{1}{2})^{r-1}] \quad (2)$$

By Lemma 2(3), we obtain $\mathrm{Dep}_*(p,q)(n) \leq n+2$. Furthermore, every word w_n with $prec(w_n) = n$ and $|\varrho(w_n0^\omega) - x_{n+3} \cdot y_{n+3}| \leq (\frac{1}{2})^{n+1}$ satifies $\varrho(p^{\leq r}\Sigma^\omega) \cdot \varrho(q^{\leq r}\Sigma^\omega) \subseteq \varrho(w_n\Sigma^\omega)$. However, an algorithm for weakly online multiplication which determines the n–th digit to the right of ":" by computing $x_{n+3} \cdot y_{n+3}$, finding w_n and applying Lemma 2(4) seems to require quadratic time, because computing $x_{n+3} \cdot y_{n+3}$ simply by adding $(\frac{1}{2})^{n+3} \cdot (p^{=(n+3)} \cdot y_{n+3} + q^{=(n+3)} \cdot x_{n+2})$ to $x_{n+2} \cdot y_{n+2}$ needs time $\mathcal{O}(n)$ at each step n. Therefore, the algorithm M_w will only compute an appropriate approximation α_{n+3} of $x_{n+3} \cdot y_{n+3}$ for determining the n–th output digit.

Since $\varrho(p) \cdot \varrho(q)$ equals $\sum_{i,j\geq 1} (\frac{1}{2})^{i+j} \cdot p^{=i} \cdot q^{=j}$, each digit of p should be multiplied by each digit of q at some step. For getting a fast algorithm, as many pairs of input digits as possible should be multiplied simultaneously by one call of a fast integer multiplication algorithm. However, the length of possible blocks of input digits multiplied in one step are restricted, because due to the online–condition $p^{=i} \cdot q^{=j}$ has to be considered before the digits $i + j + 4$ to the right of "**:**" are known. Therefore, an intelligent schedule of the multiplications is necessary.

The schedule implicitly used by the "almost online" algorithm in [La95] (and in similar version as well by the integer online multiplication algorithm in [FS74]) can be *approximately* illustrated by Fig. 1. Each rectangle $P =$

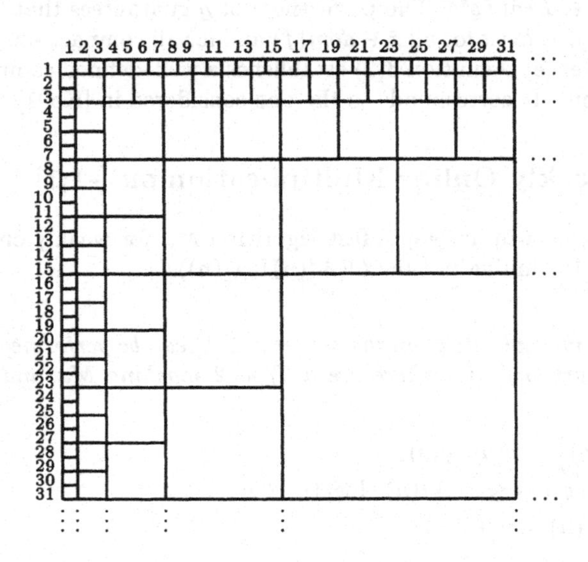

Fig. 1. Multiplication–schedule of the almost online algorithm

$\{i_1, \ldots, i_2\} \times \{j_1, \ldots, j_2\}$ in the diagram denotes that the algorithm computes $\sum_{i=i_1}^{i_2} \sum_{j=j_1}^{j_2} (\frac{1}{2})^{i+j} p^{=i} q^{=j}$ by a single call of the Schönhage–Strassen integer multiplication algorithm (cf. [SS71]), as soon as the i_2–th digit of p and the j_2–th digit of q are known. Thus, the approximation α_r of $x_r \cdot y_r$ is defined by $\sum_{(i,j)\in A_r} (\frac{1}{2})^{i+j} p^{=i} q^{=j}$, where A_r is the union of those rectangles in Fig. 1 which contain only pairs of indices ι with $\iota \le r$. Unfortunately, $|\alpha_r - x_r \cdot y_r|$ cannot be bounded by $(\frac{1}{2})^{r+\mathcal{O}(1)}$, hence this schedule does not lead to a weakly online algorithm.

The algorithm M_{w} uses a slightly modified schedule of multiplication. Roughly speaking, we increase the number of digits, by which every input digit $p^{=r}$ resp. $q^{=r}$ is multiplied at least before the next digit is read, from 1 to $\approx \log_2(r)$ (even $\approx \log_2 \log_2(r)$ suffices). Figure 2 illustrates the schedule of M_{w}. The formal definition of the rectangles and the approximations α_r of $x_r \cdot y_r$ used by M_{w} is given in Definition 5.

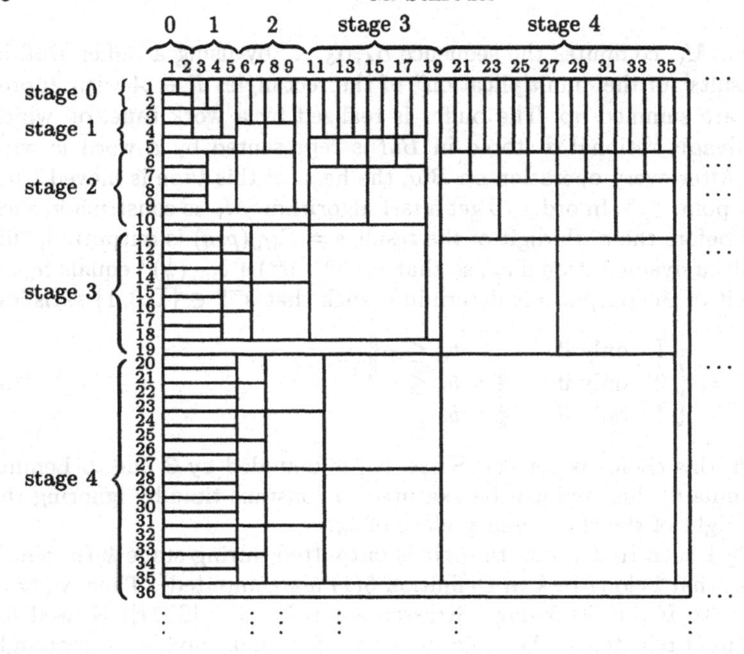

Fig. 2. Multiplication–schedule of the weakly online algorithm

Definition 5 (Multiplication–schedule $\mathcal{A}$ of M_w).
(1) Define $\mathcal{A} \subseteq 2^{\mathbb{N}\times\mathbb{N}}$ by the set that contains the sets
(a) $\{(i+k, j+k) \mid 2^k + m\cdot 2^l \le i < 2^k + (m+1)\cdot 2^l,\ 2^l \le j < 2^{l+1}\}$,
$\{(i+k, j+k) \mid 2^k + m\cdot 2^l \le j < 2^k + (m+1)\cdot 2^l,\ 2^l \le i < 2^{l+1}\}$,
(b) $\{2^k + m' + k\} \times \{1, \ldots, 1+k\}$, $\{1, \ldots, 1+k\} \times \{2^k + m' + k\}$,
(c) $\{2^{k+1} + k\} \times \{1, \ldots, 2^{k+1} + k\}$, $\{1, \ldots, 2^{k+1} - 1 + k\} \times \{2^{k+1} + k\}$
for all "stages" $k \in \mathbb{N}$ and all $1 \le l \le k$, $0 \le m < 2^{k-l}$ and $0 \le m' < 2^k$.
(2) For $r \ge 1$, define

$$\alpha_r := \sum \{(\tfrac{1}{2})^{i+j} \cdot p^{=i} \cdot q^{=j} \mid (\exists P \in \mathcal{A}_r)\,(i,j) \in P\},$$

where $\mathcal{A}_r := \{P \in \mathcal{A} \mid (\forall (i,j) \in P)\,(i \le r \text{ and } j \le r)\}$.

Lemma 6 states that the schedule $\mathcal{A}$ leads to a weakly online algorithm.

Lemma 6.
(a) For all $r \ge 1$, $|\alpha_r - x_r \cdot y_r| \le (\frac{1}{2})^r$ holds.
(b) For all $n \in \mathbb{N}$, $|\alpha_{n+3} - \varrho(p)\cdot\varrho(q)| \le \frac{3}{4}\cdot(\frac{1}{2})^{n+1}$ holds.
(c) If for $\tau \in \mathbb{N}$ "$+k$" in Definition 5 is replaced by "$+k+\tau$" and $\{1, \ldots, \tau\}^2$ is added to $\mathcal{A}$, we obtain $|\alpha_r - x_r \cdot y_r| \le (\frac{1}{2})^{r+\tau}$ for $r \ge \tau$.

The algorithm M_w computes the sequence $\alpha_4, \alpha_5, \ldots$ by using a buffer *Buf*, in which the results[6] of the "multiplication" of the rectangles $P \in \mathcal{A}$ with appropriate shifts are summed up. The buffer is realized by a work tape, on which the current dyadic rational d stored in *Buf* is represented by a word w with $\varrho(w0^\omega) = d$. After every operation on *Buf*, the head of this tape is moved back to the binary point ":". In order to get a fast algorithm, M_w is constructed such that directly before the n–th digit of the result $s = \Gamma_{M_w}(p,q)$ is outputted, this buffer contains a dyadic rational b_n, so that $\varrho(s^{\leq n-1}0^\omega) + b_n \cdot (\frac{1}{2})^n$ equals α_{n+3}. The n–th digit of the output s is determined such that $s^{=n} \in \{\bar{1}, 0, 1\}$ satisfies:

$$s^{=n} = \begin{cases} \bar{1} & \text{only if} \quad b_n \leq -\frac{3}{8} \\ 0 & \text{only if} \quad -\frac{5}{8} \leq b_n \leq \frac{5}{8} \\ 1 & \text{only if} \quad \frac{3}{8} \leq b_n \end{cases} \tag{3}$$

By Lemma 6, this choice is correct. Since $|b_n|$ is bounded by 3 due to Lemma 2(2) and Lemma 6, this step can be executed in constant time by ignoring the digits to the right of the third binary place of b_n.

Let $k, n \geq 1$ such that the n–th digit is outputted during stage k (in which the rectangles that belong to k in Definition 5(1) are computed). Then we have $k \leq \log_2(n+3)$. If the Schönhage–Strassen algorithm (cf. [SS71]) is used for the internal multiplications, the time necessary for "multiplying" a rectangle $P \in \mathcal{A}_{n+3} \setminus \mathcal{A}_{n+2}$ of type (a) (cf. Definition 5) with length 2^l and for adding the result to the buffer is bounded by $c_1 \cdot (\mathcal{M}(2^l) + k)$, because only the first $k + 2 \cdot 2^l + 3$ digits to right of ":" in *Buf* are affected by the addition. Since the number of rectangles $P \in \mathcal{A}_{n+3}$ of type (a) with length 2^l is bounded by $2^{k-l+2} - 3$, the costs caused by these rectangles can be estimated by

$$\sum_{l=1}^{k} (2^{k-l+2} - 3) \cdot c_1 \cdot (\mathcal{M}(2^l) + k) \leq c_2 \cdot \mathcal{M}(n) \cdot \log_2(n) + c_2 \ .$$

The time necessary for handling the rectangles in $\mathcal{A}_{n+3}$ of type (b) and (c) can be bounded by $c_3 \cdot n \cdot k + c_3 \leq c_4 \cdot n \cdot \log_2(n) + c_4$. We conclude that M_w runs in time $\mathcal{O}(\mathcal{M}(n)\log_2(n))$.

4 The Algorithm for the Fast Online Multiplication

In this section, we briefly decribe how to transform M_w into a fast algorithm M_{on} for multiplication on $\mathbb{R}$, which for producing the first n digits to the right of ":" reads at most 2 digits more from each input than topologically necessary.

Theorem 7.
There are a Type–2–machine M_{on} which computes real multiplication and some $c \in \mathbb{N}$ such that for all $x, y \in \mathbb{R}$, $p, q \in dom(\varrho)$ and $n \in \mathbb{N}$

[6] i.e. $\left(\sum_{(i,j)\in P} (\frac{1}{2})^{i+j} p^{=i} q^{=j} \right) \cdot 2^\mu$ with a suitable $\mu \in \mathbb{N}$ according to the following definition of the b_n's

(a) $\mathrm{Time}_{M_{on}}(x,y)(n) \leq c \cdot \mathcal{M}(n+\mu_{x,y}) \cdot \log_2(n+\mu_{x,y}) + c$
(b) $\mathrm{Ila}_{M_{on}}(p,q)(n) \leq \mathrm{Dep}_*(p,q)(n) + 2$
(c) $\mathrm{Ila}_{M_{on}}(p,q)(n) \leq \mathrm{Dep}_*(p,q)(n+2)$
hold, where $\mu_{x,y}$ *denotes the number* $\lceil \log_2(\max\{|x|,|y|,1\}) \rceil$.

In the following, let $p,q \in \{:\}\{\bar{1},0,1\}^\omega$ and define $x_r := \varrho(p^{\leq r}0^\omega)$, $y_r = \varrho(q^{\leq r}0^\omega)$ and $radius(r) := \frac{1}{2} \cdot \mathrm{diam}(\varrho(p^{\leq r}\Sigma^\omega) \cdot \varrho(q^{\leq r}\Sigma^\omega))$.[7] It is easy to see that

$$\varrho(p^{\leq r}\Sigma^\omega) \cdot \varrho(q^{\leq r}\Sigma^\omega) \tag{4}$$
$$= \begin{cases} [\, x_r \cdot y_r + (\frac{1}{2})^{2r} \cdot \mathrm{sgn}(x_r y_r) \pm (\frac{1}{2})^r \cdot (|x_r| + |y_r|) \,] & \text{if } |x_r|, |y_r| \geq (\frac{1}{2})^r \\ [\, 0 \pm (\frac{1}{2})^r \cdot (|x_r| + (\frac{1}{2})^r) \,] & \text{if } y_r = 0 \\ [\, 0 \pm (\frac{1}{2})^r \cdot (|y_r| + (\frac{1}{2})^r) \,] & \text{if } x_r = 0 \end{cases}$$

holds, where $[z \pm \varepsilon]$ denotes the interval $[z-\varepsilon; z+\varepsilon]$.
The idea for constructing M_{on} is the fact that for all $n \in \mathbb{N}$

$$\mathrm{Ila}_{M_{on}}(p,q)(n) \leq \mathrm{Dep}_*(p,q)(n) + 2 \quad \text{and} \quad \mathrm{Ila}_{M_{on}}(p,q)(n) \leq \mathrm{Dep}_*(p,q)(n+2)$$

are guaranteed, if for all $r \in \mathbb{N}$

$$radius(r) \leq \tfrac{7}{8} \cdot (\tfrac{1}{2})^{n+1} \quad \text{implies} \quad \mathrm{Ila}_{M_{on}}(p,q)(n) \leq r. \tag{5}$$

This can be shown by (4) and Lemma 2. Therefore, M_{on} with input p,q implicitly computes two suitable sequences α_r and δ_r of dyadic rationals with

$$\varrho(p^{\leq r}\Sigma^\omega) \cdot \varrho(q^{\leq r}\Sigma^\omega) \subseteq [\alpha_r - \delta_r; \alpha_r + \delta_r] \quad \text{and} \quad \delta_r \leq \tfrac{8}{7} \cdot radius(r) \tag{6}$$

by a multiplication–schedule like the one presented in Definition 5. When α_r and δ_r have been computed, M_{on} adds to the present output as many digits as possible such that the new output prefix w satisfies

$$[\alpha_r - \delta_r; \alpha_r + \delta_r] \subseteq \varrho(w\Sigma^\omega) .$$

From (6) and Lemma 2 it follows that the property (5) holds for every $n \in \mathbb{N}$. This implies that the desired online–condition is fulfilled by M_{on}. Though the necessary precision of the approximations α_r now depends on $radius(r)$, the sequence α_r can be computed uniformly on the set $(\{:\}\{\bar{1},0,1\})^2$ in time $\mathcal{O}(\mathcal{M}(r)\log_2(r))$ by using Lemma 6(c) with an appropriate τ that depends on $z = \min\{r \in \mathbb{N} \mid |x_r| \geq 2(\frac{1}{2})^r \text{ or } |y_r| \geq 2(\frac{1}{2})^r\}$.

For inputs $(p,q) \in (dom(\varrho))^2 \setminus (\{:\}\{\bar{1},0,1\}^\omega)^2$, M_{on} first moves the binary point ":" $\max\{\mathrm{lk}(p),\mathrm{lk}(q)\}$ digits to the left. Then M_{on} works on the resulting sequences $p',q' \in \{:\}\{\bar{1},0,1\}^\omega$ essentially as decribed above, except that the binary point of the output has to be placed differently and the restriction on the domain of ϱ must be obeyed. Since for determining the n–th digit of $\varrho(p) \cdot \varrho(q)$ to the right of ":" it suffices to compute $\varrho(p') \cdot \varrho(q')$ with precision $(\frac{1}{2})^{n+2\cdot\max\{\mathrm{lk}(p),\mathrm{lk}(q)\}}$, M_{on} has the desired time complexity.

[7] For a bounded subset $J \subset \mathbb{R}$, $\mathrm{diam}(J)$ denotes the diameter of J, i.e. $\sup(J) - \inf(J)$.

5 Final Remarks

By using a representation which has two additional digits with value $\frac{1}{2}$ resp. $-\frac{1}{2}$ one can realize fast real multiplication by a machine which reads at most one digit more than topologically necessary.

Theorem 4 can be extended to functions $f : ([-1;1])^k \to \mathbb{R}$ that have a Lipschitz–continuous derivative: if f is computable on $([-1;1])^k$ uniformly in time $\mathcal{O}(T(n))$, where $T : \mathbb{N} \to \mathbb{N}$ is "regular", then f can be computed by some weakly online machine in time $\mathcal{O}(T(n) + \mathcal{M}(n)\log_2(n))$. $T : \mathbb{N} \to \mathbb{N}$ is called regular, iff T is monotonic increasing almost everywhere and

$$\{n \mapsto n, \quad n \mapsto \sum_{i=0}^{\lceil \log_2(n) \rceil} T(2^i + 1)\} \subseteq \mathcal{O}(T)$$

holds (cf. [Mü86]). Thus, the most of the polynomially bounded functions usually appearing as time bounds are regular functions.

An open problem is to find out non–trivial lower bounds for the three types of online multiplication presented here. For integer online multiplication starting with the least significant digit, M. Paterson, M. Fischer and A. Meyer proved as a lower bound $\Omega(n \log_2(n))$ (cf. [PFM74]). Thus, it is probable that a lower bound of the time complexity of online multiplication on $[-1;1]$ is $\Omega(n \log_2(n))$.

References

[BSS89] L. Blum, M. Shub, S. Smale: *On a theory of computation and complexity over the real numbers*, Bull. Amer. Math. Soc. 21, pp. 1–46 (1989)

[DHM89] J. Duprat, Y. Herreros, J. M. Muller: *Some results about on-line computation of functions*, Proc. 9th IEEE Symposium on Computer Arithmetic, IEEE Computer Society Press, Los Alamitos, pp. 112–118 (1989)

[FS74] M. Fischer, L. Stockmeyer: *Fast On-line Integer Multiplication*, Journal of Computer and System Sciences 9 (1974)

[Ho90] H.J. Hoover: *Feasible real functions and arithmetic circuits*, SIAM Journal of Computing 19, pp. 182-204 (1990)

[Ko91] Ker–I Ko: *Complexity Theory of Real Functions*, Birkhäuser, Boston (1991)

[La95] B. Landgraf: *Schnelle beinahe Online–Multiplikation reeller Zahlen*, Diplomarbeit, Fernuniversität Hagen (1995)

[Mü86] N. Müller: *Computational Complexity of Real Functions and Real Numbers*, Informatik Berichte Nr. 59, Fernuniversität Hagen (1986)

[PFM74] M. Paterson, M. Fischer, A. Meyer: *An Improved Overlap Argument For On-Line Multiplication*, SIAM–AMS Proceedings Volume 7 (1974)

[PR89] M. Pour–El, J. Richards: *Computability in Analysis and Physics*, Springer–Verlag, Berlin, Heidelberg (1989)

[SS71] A. Schönhage, V. Strassen: *Schnelle Multiplikation großer Zahlen*, Computing 7, (1971)

[Sch95] M. Schröder: *Topological Spaces Allowing Type 2 Complexity Theory*, in: Workshop on Computability and Complexity in Analysis, Informatik Berichte Nr. 190, Fernuniversität Hagen (1995)

[We91] K. Weihrauch: *On the Complexity of Online Computations of Real Functions*, Journal of Complexity 7, pp. 340-394 (1991)

[We87] K. Weihrauch: *Computability*, Springer–Verlag, Berlin, Heidelberg (1987)

[We95a] K. Weihrauch: *A Simple Introduction to Computable Analysis*, Informatik Berichte Nr. 171, Fernuniversität Hagen (1995)

[We95b] K. Weihrauch: *A Foundation of Computable Analysis*, in: EATCS Bulletin Nr. 57, pp. 167-182 (October 1995)

The Operators min and max on the Polynomial Hierarchy*

Harald Hempel** and Gerd Wechsung***

Institut für Informatik, Friedrich-Schiller-Universität Jena, 07743 Jena

Abstract. Starting from Krentel's class OptP [Kre88] we define a general maximization operator max and a general minimization operator min for complexity classes and show that there are other interesting optimization classes beside OptP. We investigate the behavior of these operators on the polynomial hierarchy, in particular we study the inclusion structure of the classes max · P, max · NP, max · coNP, min · P, min · NP and min · coNP. Furthermore we prove some very powerful relations regarding the interaction of the operators max, min, U, Sig, C, $\oplus$, $\exists$ and $\forall$. This gives us a tool to show that the considered min and max classes are distinct under reasonable structural assumptions. Besides that, we are able to characterize the polynomial hierarchy uniformly by three operators.

1 Introduction

Maximization functions and the complexity of maximization problems play a central role in structural complexity. Krentel [Kre88] defined the first function class based on the concept of optimization. His class OptP, was subsequently studied in a series of papers [Kre88, Kre92, GKR95, Köb89, KST89].

Recall that OptP = max-P ∪ min-P, where max-P is the set of all functions f such that there exists a nondeterministic polynomial-time Turing machine N (with output device), and for all inputs x, the lexicographically maximal output of $N(x)$ equals $f(x)$. Similarly one defines min-P.

In [Kre88] it is shown that the function MAXIMUM SATISFYING ASSIGNMENT, which given a boolean formula outputs the lexicographically largest satisfying assignment, is, though in max-P, metric complete for $F\Delta_2^p$ which is a seemingly larger class then max-P. But note that computing MAXIMUM SATISFYING ASSIGNMENT does not need the full computational power of the class max-P, namely it can be computed by a nondeterministic polynomial-time Turing machine in the sense that the largest *accepting path* (not output) is the value of the function (we guess all assignments and accept if the guessed

* A full version of this paper, including proofs of all claims, is available as Friedrich-Schiller-Universität Jena, Fakultät für Mathematik und Informatik, Technical Report Math/Inf/96/8.

** Email: `hempel@pdec01.mipool.uni-jena.de`.

*** Email: `wechsung@minet.uni-jena.de`.

Reischuk, Morvan (Eds.): STACS'97 Proceedings, LNCS 1200

assignment satisfies the input formula). Does this observation lead to a smaller class of optimization functions?

Besides that in each of the above mentioned papers dealing with the class OptP it is pointed out that OptP and its subclasses max-P and min-P are constructed from NP in a natural way. What happens, if this construction is applied to P, coNP and other complexity classes?

There have been various attempts to relate function and complexity classes by defining operators in order to map one kind of class to the other. Two major examples of operators which map complexity classes to function classes are # [Tod91] as a generalization of Valiants class #P [Val79a, Val79b] and Mēd [VW93]. Both operators have turned out to be of considerable interest in structural complexity as can be seen by a long series of papers, for instance [HV95, OH93, VW93] to name only a few of them. In other words if we apply those operators to central complexity classes we obtain central function classes. Note that the # and the Mēd operator capture the essence of counting and finding the middle element, respectively.

So it is quite natural to define "pure" optimization operators. Let $\mathcal{K}$ be a complexity class. We define the function classes max·$\mathcal{K}$ and min·$\mathcal{K}$:

$$f \in \max\cdot\mathcal{K} \iff \bigvee_{A\in\mathcal{K}} \bigvee_{p\in Pol} \bigwedge_{x\in\Sigma^*} f(x) = max\{y : 0 \leq y < 2^{p(|x|)} \wedge \langle x,y\rangle \in A\}$$

and if this set is empty let $f(x) = 0$,

$$g \in \min\cdot\mathcal{K} \iff \bigvee_{A\in\mathcal{K}} \bigvee_{p\in Pol} \bigwedge_{x\in\Sigma^*} g(x) = min\{y : 0 \leq y < 2^{p(|x|)} \wedge \langle x,y\rangle \in A\}$$

and if this set is empty let $g(x) = 2^{p(|x|)}$.

Obviously we have max · NP = max-P. But we do not know similar equivalences for our classes max · P and max · coNP offhand. However. it is not hard to see that MAXIMUM SATISFYING ASSIGNMENT $\in$ max · P and clearly, max · P $\subseteq$ max · NP. We will give strong evidence for this inclusion to be strict.

We study the inclusion relations between the min and max classes and also their inclusion relations with the other well known function classes such as # · P, span-P, # · coNP and the classes FP and FΔ_2^p. It turns out that none of the interesting new classes such as max · P and max · coNP is equal to some known class. And under reasonable structural assumptions all min and max classes are distinct.

Another maximization operator can be found in [VW95, Vol94]. Note that their operator when applied to classes from the polynomial hierarchy yields only our classes max·Σ_i^p and min·Σ_i^p but not max·Δ_i^p, min·Δ_i^p, max·Π_i^p or min·Π_i^p.

A different model of maximization was investigated by Chen and Toda in [CT95].

In this paper we prove a number of powerful relations regarding the interaction of the operators max, min, U, Sig, C, $\oplus$, $\exists$ and $\forall$. In particular we show for complexity classes $\mathcal{K}$ having some reasonable closure properties, such as closure under $\leq_m^p$:

$$\begin{array}{ll} \mathrm{U}\cdot\max\cdot\mathcal{K}=\mathcal{K}, & \mathrm{Sig}\cdot\min\cdot\mathcal{K}=\mathrm{co}\mathcal{K},\\ \mathrm{U}\cdot\min\cdot\mathcal{K}=\mathrm{co}\mathcal{K}, & \mathrm{C}\cdot\max\cdot\mathcal{K}=\exists\cdot\mathcal{K},\\ \mathrm{Sig}\cdot\max\cdot\mathcal{K}=\exists\cdot\mathcal{K}, & \mathrm{C}\cdot\min\cdot\mathcal{K}=\forall\cdot\mathrm{co}\mathcal{K},\end{array}$$
$$\oplus\cdot\max\cdot\mathcal{K}=\oplus\cdot\min\cdot\mathcal{K}=\mathrm{P}^{\exists\cdot\mathcal{K}}.$$

Note that we define the operators C and $\oplus$ on function classes and not on complexity classes as originally done in the literature [Wag86, PZ83]. Our operator C appeared already in [VW95].

The above relations turn out to be a very useful tool in order to show that various inclusions between function classes are unlikely. Our study yields a huge number of new characterizations of central complexity classes such as P, NP, coNP and PP and allows a characterization of the polynomial hierarchy based on only three operators.

Classes of optimization functions are very related to approximability problems. The authors note that including results along this line of research would certainly exceed the tight space limitations and decrease the papers readability. However, a self-contained paper clarifying the relationship between the concept of approximation and the min-max classes is in preparation.

2 Preliminaries

We adopt the notations commonly used in structural complexity. For details we refer the reader to any of the following standard books [BDG95, Pap94]. Denote the characteristic function of a set A by c_A. Denote the set of all polynomials by *Pol*. Let FP be the set of all functions computable in polynomial time.

As already noted we want to study classes of "pure" optimization functions and thus define for a complexity class $\mathcal{K}$ the function classes $\max\cdot\mathcal{K}$ and $\min\cdot\mathcal{K}$ as already done in the previous section. Throughout the paper we will omit any statement concerning the empty set. Since all complexity classes $\mathcal{K}$ in the context of this paper are at least closed under $\leq_m^p$, we can always without loss of generality assume that for all $A \in \mathcal{K}$ and all x, in the case of maximization $\langle x, 0\rangle \in A$ and in the case of minimization $\langle x, 2^{p(|x|)}\rangle \in A$. In this paper we want to concentrate on classes of optimization functions where the underlying complexity class $\mathcal{K}$ is a class Δ_i^p, Σ_i^p or Π_i^p from the polynomial hierarchy. For short we will call these classes the min-max classes.

We define the following four operators:

Definition 1. For any function class $\mathcal{F}$ let

$$A \in \mathrm{U}\cdot\mathcal{F} \iff c_A \in \mathcal{F},$$
$$A \in \mathrm{Sig}\cdot\mathcal{F} \iff \bigvee_{f\in\mathcal{F}} \bigwedge_{x\in\Sigma^*} (x \in A \iff f(x) > 0),$$
$$A \in \mathrm{C}\cdot\mathcal{F} \iff \bigvee_{f\in\mathcal{F}} \bigvee_{g\in\mathrm{FP}} \bigwedge_{x\in\Sigma^*} (x \in A \iff f(x) \geq g(x)),$$
$$A \in \oplus\cdot\mathcal{F} \iff \bigvee_{f\in\mathcal{F}} \bigwedge_{x\in\Sigma^*} (x \in A \iff f(x) \equiv 1 \bmod 2).$$

The above operators C and $\oplus$ are usually defined on complexity classes and not on function classes (see [Wag86, PZ83]). But note that we have for every operator $\mathrm{OP} \in \{\mathrm{C}, \oplus\}$ (let op be its counterpart defined on complexity classes), $\mathrm{op} \cdot \mathcal{K} = \mathrm{OP} \cdot \# \cdot \mathcal{K}$. So it is quite natural to denote the above defined operators in that manner, since they capture the essential properties of their original definitions. Note that all of the above operators are monotone.

For every set of operators OP defined on a (complexity or function) class $\mathcal{C}$ we denote the algebraic closure of $\mathcal{C}$ under the operators of OP by $\Gamma_{\mathrm{OP}}(\mathcal{C})$.

3 Inclusion Relations

As already mentioned in Section 1 we have $\max \cdot \mathrm{NP} = \text{max-P}$ and $\min \cdot \mathrm{NP} = \text{min-P}$. The operators max and min are obviously monotone and the following claims are either well known or easy to prove.

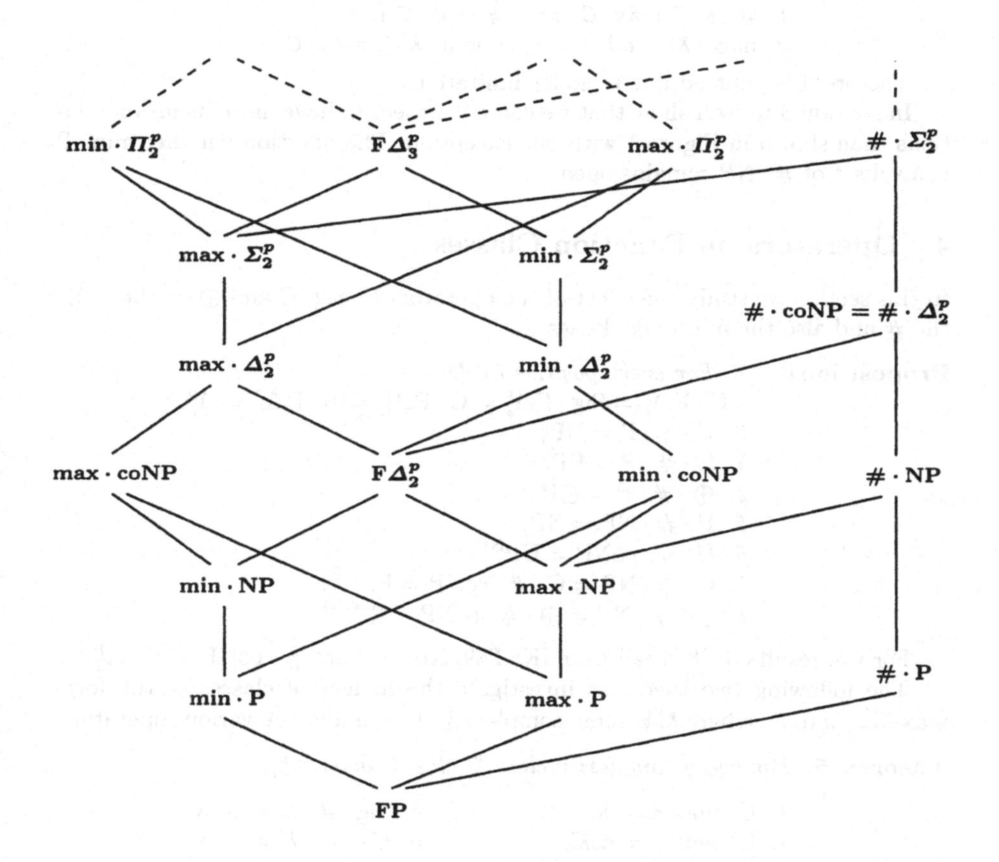

Fig. 1. Inclusion structure of the min-max classes

Proposition 2. *For all positive integers i we have,*

1. $\mathrm{F}\Delta_i^p \subseteq \max\cdot\Delta_i^p \subseteq \max\cdot\Sigma_i^p \subseteq \mathrm{F}\Delta_{i+1}^p$,
2. $\max\cdot\Delta_i^p \subseteq \max\cdot\Pi_i^p \subseteq \max\cdot\Delta_{i+1}^p \subseteq \mathrm{F}\Delta_{i+2}^p$,
3. $\max\cdot\Sigma_i^p \subseteq \min\cdot\Pi_i^p$,
4. $\mathrm{F}\Delta_i^p \subseteq \min\cdot\Delta_i^p \subseteq \min\cdot\Sigma_i^p \subseteq \mathrm{F}\Delta_{i+1}^p$,
5. $\min\cdot\Delta_i^p \subseteq \min\cdot\Pi_i^p \subseteq \min\cdot\Delta_{i+1}^p \subseteq \mathrm{F}\Delta_{i+2}^p$,
6. $\min\cdot\Sigma_i^p \subseteq \max\cdot\Pi_i^p$.

Note that we have $\max\cdot\mathrm{NP} \subseteq \mathrm{F}\Delta_2^p$ according to Proposition 2. One would expect that also $\max\cdot\mathrm{coNP} \subseteq \mathrm{F}\Delta_2^p$ holds. We will see later, that this is not the case under reasonable structural assumptions.

Recall from [KST89, Köb89] that $\max\cdot\mathrm{NP} \subseteq \#\cdot\mathrm{NP}$, $\mathrm{F}\Delta_2^p \subseteq \#\cdot\mathrm{coNP}$ and $\#\cdot\mathrm{coNP} = \#\cdot\Delta_2^p$. The known inclusion relations between the considered function classes are presented in Figure 1.

Theorem 3. *For every pair of complexity classes $\mathcal{K}$ and $\mathcal{C}$ closed under $\leq_m^p$,*

1. $\#\cdot\mathcal{K} \subseteq \max\cdot\mathcal{C} \iff \#\cdot\mathrm{co}\mathcal{K} \subseteq \min\cdot\mathcal{C}$,
2. $\max\cdot\mathcal{K} \subseteq \min\cdot\mathcal{C} \iff \min\cdot\mathcal{K} \subseteq \max\cdot\mathcal{C}$.

The proof is omitted due to space limitations.

In Section 5 we will show that we can not expect to have more inclusion relations than shown in Figure 1 with one exception. The question whether $\min\cdot\mathrm{P}$ is a subset of $\#\cdot\mathrm{NP}$ remains open.

4 Operators on Function Classes

In this section we study the effect of the operators U, Sig, C and $\oplus$ on the $\mathrm{F}\Delta_i^p$, the # and also the min-max classes.

Proposition 4. 1. *For every positive $i \in \mathbb{N}$,*
$\mathrm{U}\cdot\mathrm{F}\Delta_i^p = \mathrm{Sig}\cdot\mathrm{F}\Delta_i^p = \mathrm{C}\cdot\mathrm{F}\Delta_i^p = \oplus\cdot\mathrm{F}\Delta_i^p = \Delta_i^p$.
2. $\mathrm{U}\cdot\#\cdot\mathrm{P} = \mathrm{UP}$,
3. $\mathrm{C}\cdot\#\cdot\mathrm{P} = \mathrm{PP}$,
4. $\oplus\cdot\#\cdot\mathrm{P} = \oplus\mathrm{P}$,
5. $\mathrm{U}\cdot\#\cdot\mathrm{NP} = \mathrm{NP}$,
6. $\mathrm{U}\cdot\#\cdot\mathrm{coNP} = \mathrm{UP}^{\mathrm{NP}}$,
7. $\mathrm{C}\cdot\#\cdot\mathrm{NP} = \mathrm{C}\cdot\#\cdot\mathrm{coNP} = \mathrm{PP}^{\mathrm{NP}}$,
8. $\oplus\cdot\#\cdot\mathrm{NP} = \oplus\cdot\#\cdot\mathrm{coNP} = \oplus\mathrm{P}^{\mathrm{NP}}$.

For the results 6 - 8 recall from [KST89, Köb89] that $\#\cdot\mathrm{coNP} = \#\cdot\Delta_2^p$.

The following two theorems investigate the images of classes of the form $\max\cdot\mathcal{K}$ or $\min\cdot\mathcal{K}$, where $\mathcal{K}$ is some complexity class, under the various operators.

Theorem 5. *For every complexity class $\mathcal{K}$ closed under $\leq_m^p$,*

1. $\mathrm{U}\cdot\max\cdot\mathcal{K} = \mathcal{K}$,
2. $\mathrm{U}\cdot\min\cdot\mathcal{K} = \mathrm{co}\mathcal{K}$,
3. $\mathrm{Sig}\cdot\max\cdot\mathcal{K} = \exists\cdot\mathcal{K}$,
4. $\mathrm{Sig}\cdot\min\cdot\mathcal{K} = \mathrm{co}\mathcal{K}$,
5. $\mathrm{Sig}\cdot\#\cdot\mathcal{K} = \exists\cdot\mathcal{K}$,
6. $\mathrm{C}\cdot\max\cdot\mathcal{K} = \exists\cdot\mathcal{K}$,
7. $\mathrm{C}\cdot\min\cdot\mathcal{K} = \forall\cdot\mathrm{co}\mathcal{K}$.

The proofs are omitted due to space restrictions.

One general result regarding the image of function classes under the operator $\oplus$ can be found in [VW95], namely if $\mathcal{K}$ has certain closure properties then $\oplus \cdot \mathrm{F}\cdot\mathcal{K} = \mathrm{P}^{\mathcal{K}}$, where F· is the operator already spoken of in Section 1. Note that our result holds for a wider range of classes since we only require closure under $\leq^{p}_{ctt}$. Furthermore we are able to show that $\mathrm{F}\cdot\mathcal{K} = \max\cdot\mathcal{K} \iff \mathcal{K} = \exists\cdot\mathcal{K}$ and $\mathrm{F}\cdot\mathcal{K} = \min\cdot \mathrm{co}\mathcal{K} \iff \mathcal{K} = \forall\cdot\mathcal{K}$.

Theorem 6. *For every complexity class $\mathcal{K}$ closed under $\leq^{p}_{ctt}$,*

1. $\oplus\cdot\max\cdot\mathcal{K} = \mathrm{P}^{\exists\cdot\mathcal{K}}$, *2.* $\oplus\cdot\min\cdot\mathcal{K} = \mathrm{P}^{\exists\cdot\mathcal{K}}$.

Proof. Let $\mathcal{K}$ be a complexity class closed under $\leq^{p}_{ctt}$.

(1.) Let $A \in \oplus\cdot\max\cdot\mathcal{K}$. Hence there exists a function $f \in \max\cdot\mathcal{K}$ such that for all $x \in \Sigma^*$,

$$x \in A \iff f(x) \equiv 1 \bmod 2.$$

Let B be the set $B = \{\langle x,z\rangle : f(x) \geq z\}$. Obviously $B \in \mathrm{C}\cdot\max\cdot\mathcal{K}$ and thus by Theorem 5 $B \in \exists\cdot\mathcal{K}$.

For a given $x \in \Sigma^*$ the value $f(x)$ can be determined deterministically in polynomial-time by binary search using queries to B. And given the value $f(x)$, checking whether $f(x)$ is odd or even can also be done deterministically in polynomial-time. So we have $A \in \mathrm{P}^B$ and thus $A \in \mathrm{P}^{\exists\cdot\mathcal{K}}$.

Now let $A \in \mathrm{P}^{\exists\cdot\mathcal{K}}$. So by definition there exist a deterministic oracle machine M, a set $B \in \exists\cdot\mathcal{K}$ and a nondecreasing polynomial p such that M^B accepts A in time p. Since $B \in \exists\cdot\mathcal{K}$ there exist a set $C \in \mathcal{K}$ and a polynomial q such that for all y,

$$y \in B \iff \bigvee_{|w|=2^{q(|y|)}} (\langle y,w\rangle \in C).$$

Let us without loss of generality assume that the machine M on input x runs exactly $p(|x|)$ steps, making one query of length $p(|x|)$ to B in every step. Define a set D to be

$D = \{\langle x, a_1a_2\cdots a_{p(|x|)}y_1y_2\cdots y_{p(|x|)}w_1w_2\cdots w_{p(|x|)}c\rangle$:

(1) $x \in \Sigma^*$, and for every $1 \leq i \leq p(|x|)$, $a_i \in \{0,1\}$, $y_i \in \Sigma^*$, $|y_i| = p(|x|)$, $w_i \in \Sigma^*$ and $|w_i| = q(p(|x|))$, and $c \in \{0,1\}$, **and**

(2) $M^{(\cdot)}(x)$ yields the result c given that the answers to the queries asked are $a_1, a_2, \cdots a_{p(|x|)}$ (in this order) and the queries asked are $y_1, y_2, \cdots, y_{p(|x|)}$ (in this order), **and**

(3) $\bigwedge_{1\leq i\leq p(|x|)} (a_i = 1 \Rightarrow \langle y_i, w_i\rangle \in C)\}$.

Note that for every $\langle x,z\rangle \in D$, where z is of the form

$$z = a_1a_2\cdots a_{p(|x|)}y_1y_2\cdots y_{p(|x|)}w_1w_2\cdots w_{p(|x|)}c,$$

we have, w_i is a witness for y_i (being an element of B) for all i with $a_i = 1$. With other words there is no element $\langle x,z\rangle$ in D such that for some i, $a_i = 1$

and $y_i \notin B$. Hence the largest z, call it $z_0(x)$, such that $\langle x, z\rangle \in D$ describes the computation process of $M^B(x)$ in the sense that $z_0(x)$ contains all queries asked by $M^B(x)$, all their answers and also the overall answer of $M^B(x)$. Thus the least significant bit of $z_0(x)$ is equal to $c_A(x)$ and we have,

$$x \in A \iff max\{z : \langle x, z\rangle \in D\} \equiv 1 \bmod 2.$$

Note furthermore that condition (1) and (2) of D can be checked in deterministic polynomial-time and condition (3) is clearly a $\mathcal{K}$ predicate due to the closure of $\mathcal{K}$ under $\leq^p_{ctt}$. Hence $D \in \mathcal{K}$. This proves $A \in \oplus \cdot \max\cdot\mathcal{K}$.

(2.) The proof is similar. Note that for any complexity class $\mathcal{K}$ closed under $\leq^p_{ctt}$ we have, $\forall \cdot \text{co}\mathcal{K} = \exists \cdot \mathcal{K}$. Furthermore one can easily modify the definition of the set D in such a way that the smallest z, such that $\langle x, z\rangle \in D$, describes the computation of $M^B(x)$. □

The last two theorems, Theorem 5 and 6, give us new characterizations of central complexity classes, for example:

Corollary 7.
1. $\mathrm{U} \cdot \max \cdot \mathrm{P} = \mathrm{P}$,
2. $\mathrm{Sig} \cdot \max \cdot \mathrm{P} = \mathrm{NP}$,
3. $\mathrm{Sig} \cdot \min \cdot \mathrm{P} = \mathrm{P}$,
4. $\oplus \cdot \min \cdot \mathrm{P} = \Delta^p_2$,
5. $\mathrm{C} \cdot \max \cdot \mathrm{NP} = \mathrm{NP}$,
6. $\oplus \cdot \min \cdot \mathrm{NP} = \Delta^p_2$,
7. $\mathrm{Sig} \cdot \max \cdot \mathrm{coNP} = \Sigma^p_2$,
8. $\mathrm{C} \cdot \max \cdot \mathrm{coNP} = \Sigma^p_2$,
9. $\mathrm{Sig} \cdot \min \cdot \mathrm{coNP} = \mathrm{NP}$,
10. $\oplus \cdot \min \cdot \mathrm{coNP} = \Delta^p_3$.

Some of the results related to max · NP and min · NP were previously known and mentioned in a series of papers [Kre88, Köb89, Wag87, GKR95, VW95].

Note that according to Theorem 5 and 6 we can verify the results obtained by Ogiwara [Ogi91], namely: $\mathrm{P}^{\mathrm{C_=P}} = \oplus \cdot \min \cdot \mathrm{coC_=P}$, $\mathrm{NP}^{\mathrm{C_=P}} = \mathrm{C} \cdot \max \cdot \mathrm{C_=P}$ and $\mathrm{P}^{\mathrm{NP}^{\mathrm{C_=P}}} = \oplus \cdot \max \cdot \mathrm{C_=P}$.

Similarly we can now characterize all levels of the polynomial hierarchy relative to every class (having the closure properties mentioned in the theorems). Especially we are now able to characterize the polynomial hierarchy itself by three combined operators, namely C · max, C · min and ⊕ · max.

Corollary 8.

$$\Sigma^p_0 = \Pi^p_0 = \Delta^p_0 = \mathrm{P},$$
$$\Sigma^p_{i+1} = \mathrm{C} \cdot \max \cdot \Delta^p_i,$$
$$\Pi^p_{i+1} = \mathrm{C} \cdot \min \cdot \Delta^p_i,$$
$$\Delta^p_{i+1} = \oplus \cdot \max \cdot \Delta^p_i,$$
$$\mathrm{PH} = \Gamma_{\{\mathrm{C}\cdot\max\cdot, \mathrm{C}\cdot\min\cdot, \oplus\cdot\max\cdot\}}(\mathrm{P}).$$

5 Structural Consequences

As already noted in [Vol94] the operator theoretical approach in order to provide evidence that two function (or complexity) classes are incomparable is very powerful and elegant. In this section we will completely analyze the inclusion relations of the main function classes which are the min-max classes, the # classes and the $\mathrm{F}\Delta^p_i$ classes.

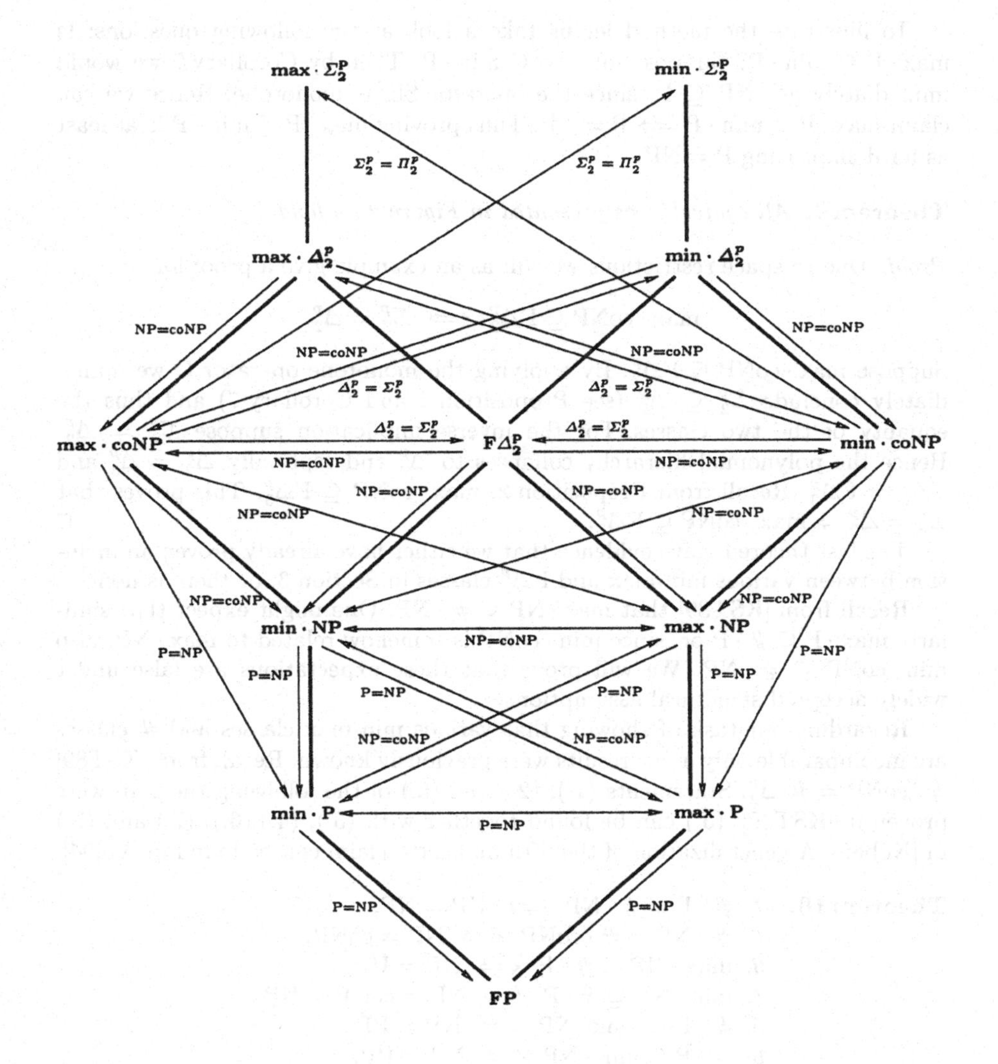

Key: a bold line indicates an inclusion of the lower in the upper class

$\mathcal{F}_1 \xrightarrow{\alpha} \mathcal{F}_2$ means : $\mathcal{F}_1 \subseteq \mathcal{F}_2 \iff \alpha$

$\mathcal{F}_1 \xleftrightarrow{\alpha} \mathcal{F}_2$ means : $(\mathcal{F}_1 \subseteq \mathcal{F}_2 \iff \alpha) \wedge (\mathcal{F}_1 \supseteq \mathcal{F}_2 \iff \alpha)$

Fig. 2. Structural consequences I

To illustrate the method let us take a look at the following questions: Is $\max \cdot \mathrm{P} \subseteq \min \cdot \mathrm{P}$? Suppose $\max \cdot \mathrm{P} \subseteq \min \cdot \mathrm{P}$. Then by Corollary 7 we would immediately get $\mathrm{NP} \subseteq \mathrm{P}$, since the operator Sig is monotone. Hence we can claim $\max \cdot \mathrm{P} \subseteq \min \cdot \mathrm{P} \Longrightarrow \mathrm{P} = \mathrm{NP}$. Thus proving $\max \cdot \mathrm{P} \subseteq \min \cdot \mathrm{P}$ is at least as hard as proving $\mathrm{P} = \mathrm{NP}$.

Theorem 9. *All equivalences presented in Figure 2 do hold.*

Proof. Due to space restrictions we will as an example give a proof for

$$\max \cdot \mathrm{coNP} \subseteq \mathrm{F}\Delta_2^p \iff \Sigma_2^p = \Delta_2^p.$$

Suppose $\max \cdot \mathrm{coNP} \subseteq \mathrm{F}\Delta_2^p$. By applying the monotone operator C we immediately conclude $\Sigma_2^p \subseteq \Delta_2^p$ (see Proposition 4 and Corollary 7) and thus the equality of the two classes. For the inverse implication suppose $\Sigma_2^p = \Delta_2^p$. Hence the polynomial hierarchy collapses to Δ_2^p and especially $\Delta_2^p = \Delta_3^p$ and $\mathrm{F}\Delta_2^p = \mathrm{F}\Delta_3^p$. Recall from Proposition 2, $\max \cdot \mathrm{coNP} \subseteq \mathrm{F}\Delta_3^p$. This proves that $\Sigma_2^p = \Delta_2^p \Rightarrow \max \cdot \mathrm{coNP} \subseteq \mathrm{F}\Delta_2^p$. □

The last theorem gave evidence that we either have already proven an inclusion between various min-max and $\mathrm{F}\Delta_i^p$ classes in Section 3, or there is none.

Recall from [KST89] that $\max \cdot \mathrm{NP} \subseteq \# \cdot \mathrm{NP}$. One might expect that similarly $\max \cdot \mathrm{P} \subseteq \# \cdot \mathrm{P}$ or, since $\min \cdot \mathrm{coNP}$ is somehow related to $\max \cdot \mathrm{NP}$, also $\min \cdot \mathrm{coNP} \subseteq \# \cdot \mathrm{NP}$. We will prove that these expectations are false under widely accepted structural assumptions.

Regarding our task of showing that various min-max classes and # classes are incomparable only a few results were previously known. Recall from [KST89] $\# \cdot \mathrm{coNP} = \# \cdot \Delta_2^p$. Statements (1.), (2.) and (5.) of the following theorem were proven in [KST89]. (5.) can be found together with (3.), (4.),(6.), (7.) and (8.) in [Köb89]. A generalization of the fifth and sixth claim can be found in [Vol94].

Theorem 10.
1. $\# \cdot \mathrm{P} = \# \cdot \mathrm{NP} \iff \mathrm{UP} = \mathrm{NP}$,
2. $\# \cdot \mathrm{NP} = \# \cdot \mathrm{coNP} \iff \mathrm{NP} = \mathrm{coNP}$,
3. $\max \cdot \mathrm{NP} \subseteq \# \cdot \mathrm{P} \iff \mathrm{NP} = \mathrm{UP}$,
4. $\min \cdot \mathrm{NP} \subseteq \# \cdot \mathrm{P} \iff \mathrm{NP} = \mathrm{coNP} = \mathrm{UP}$,
5. $\# \cdot \mathrm{P} \subseteq \max \cdot \mathrm{NP} \iff \mathrm{NP} = \mathrm{PP}$,
6. $\# \cdot \mathrm{P} \subseteq \min \cdot \mathrm{NP} \iff \mathrm{NP} = \mathrm{PP}$,
7. $\min \cdot \mathrm{NP} \subseteq \# \cdot \mathrm{NP} \iff \mathrm{NP} = \mathrm{coNP}$,
8. $\# \cdot \mathrm{NP} \subseteq \min \cdot \mathrm{NP} \vee \# \cdot \mathrm{NP} \subseteq \max \cdot \mathrm{NP} \Longrightarrow \mathrm{NP} = \mathrm{coNP}$.

We will now turn to the second task of this section, namely show that various min-max classes and the # classes are incomparable under reasonable structural assumptions.

Theorem 11. *All equivalences and implications presented in Figure 3 do hold.*

Proof. As an example we will show that

$$\# \cdot \mathrm{NP} \subseteq \max \cdot \mathrm{NP} \iff \mathrm{NP} = \mathrm{PP}.$$

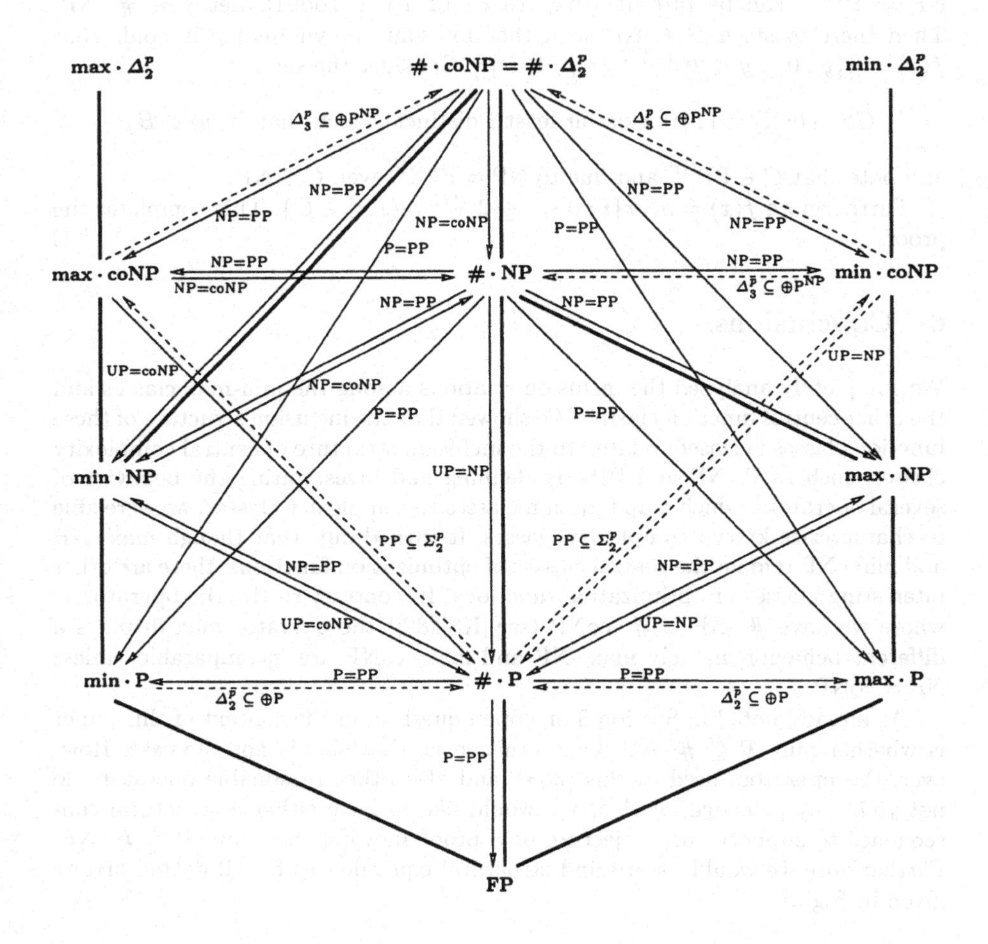

Key: a bold line indicates an inclusion of the lower in the upper class

$\mathcal{F}_1 \overset{\alpha}{- - \rightarrow} \mathcal{F}_2$ means : $\mathcal{F}_1 \subseteq \mathcal{F}_2 \Longrightarrow \alpha$

$\mathcal{F}_1 \overset{\alpha}{\longrightarrow} \mathcal{F}_2$ means : $\mathcal{F}_1 \subseteq \mathcal{F}_2 \Longleftrightarrow \alpha$

Fig. 3. Structural consequences II

Suppose $\# \cdot \mathrm{NP} \subseteq \max \cdot \mathrm{NP}$. By applying the monotone operator C we immediately conclude $\mathrm{PP}^{\mathrm{NP}} \subseteq \mathrm{NP}$ and thus NP = PP.

For the inverse implication suppose NP = PP. Note that $\mathrm{NP} = \mathrm{PP} \iff \mathrm{NP} = \mathrm{PP}^{\mathrm{NP}}$ can be proven using results of Toda [Tod91]. Let $f \in \# \cdot \mathrm{NP}$. Then there exists a $B \in \mathrm{NP}$ such that for some polynomial p it holds that $f(x) = ||\{y : 0 \leq y < 2^{p(|x|)} \wedge \langle x, y\rangle \in B\}||$. Consider the set

$$C = \{\langle x, i\rangle : \text{ There exist at least } i \text{ distinct } y \text{ such that } \langle x, y\rangle \in B\}$$

and note that $C \in \mathrm{PP}^{\mathrm{NP}}$ and due to $\mathrm{NP} = \mathrm{PP}^{\mathrm{NP}}$ even $C \in \mathrm{NP}$.

Furthermore $f(x) = max\{i : 0 \leq i \leq 2^{p(|x|)} \wedge \langle x, i\rangle \in C\}$. This completes the proof. □

6 Conclusions

We completely analyzed the inclusion relations among the min-max classes and the other central function classes. We showed that the inclusion structure of these function classes is closely related to the inclusion structure of central complexity classes, such as P, NP and PP. By defining and investigating the behavior of several operators, which map function classes to complexity classes, we were able to characterize known complexity classes. It turned out, that though $\max \cdot \mathrm{NP}$ and $\min \cdot \mathrm{NP}$ remain the central classes of optimization functions, there are other interesting classes of optimization functions. In contrast to the the operator # where we have $\# \cdot \mathrm{NP} \subseteq \# \cdot \mathrm{coNP}$ (see [KST89]) the operator max displays a different behavior, namely $\max \cdot \mathrm{NP}$ and $\max \cdot \mathrm{coNP}$ are incomparable, unless NP = coNP.

As already noted in Section 3 one open question in the context of this paper is whether $\min \cdot \mathrm{P} \subseteq \# \cdot \mathrm{NP}$. Our intuition is, that this is not the case. However, the operators used in this paper and also other reasonable operators do not yield any evidence for this. We would like to have either a structural consequence to support our conjecture or a proof showing that $\min \cdot \mathrm{P} \subseteq \# \cdot \mathrm{NP}$. Furthermore we would like to find structural equivalences for all dotted arrows given in Figure 3.

References

[BDG95] J.L. Balcázar, J. Díaz and J. Gabarró. Structural Complexity I. *Springer Berlin-Heidelberg-New York, 2nd ed.,1995.*

[BDG90] J.L. Balcázar, J. Díaz and J. Gabarró. Structural Complexity II. *Springer Berlin-Heidelberg-New York, 1990.*

[CT95] Z.-Z. Chen and S. Toda. The Complexity of Selecting Maximal Solutions. *Information and Computation, 119(1):231-239, 1995.*

[GKR95] W.I. Gasarch, M.W. Krentel and K.J. Rappoport. OptP as the Normal Behavior of NP-Complete Problems. *Mathematical Systems Theory, 28:487-514, 1995.*

[HV95] L. Hemaspaandra and H. Vollmer. The Satanic notations: Counting classes beyond $\# \cdot$ P and other definitional adventures. *SIGACT News, 26(1):2-13, 1995.*

[Köb89] J. Köbler. Strukturelle Komplexität von Anzahlproblemen. *PhD thesis, Universität Stuttgart, Fakultät für Informatik, Stuttgart, Germany, 1989.*

[KST89] J. Köbler, U. Schöning and J. Torán. On counting and approximation. *Acta Informatica, 26:363-379, 1989.*

[Kre88] M.W. Krentel. The complexity of optimization problems. *Journal of Computer and Systems Sciences, 36:490-509, 1988.*

[Kre92] M.W. Krentel. Generalizations of OptP to the polynomial hierarchy. *Theoretical Computer Science, 97:183-198. 1992.*

[Ogi91] M. Ogiwara. Characterizing low levels of the polynomial-time hierarchy relative to $C_{=}P$ via metric turing machines. *Technical Report C-101, Tokyo Institute of Technology, Department of Information Science, Tokyo, Japan, 1991.*

[OH93] M. Ogiwara and L. Hemachandra. A complexity theory for feasible closure properties. *Journal of Computer and Systems Sciences, 46:295-325, 1993.*

[Pap94] C. Papadimitriou. Computational Complexity. *Addison-Wesley, 1994.*

[PZ83] C. Papadimitriou and S. Zachos. Two remarks on the power of counting. *Proceedings 6th GI Conference on Theoretical Computer Science, 269-276. Springer Verlag, Lecture Notes in Computer Science # 145, 1983.*

[Tod91] S. Toda. Computational Complexity of Counting Complexity Classes. *PhD thesis, Tokyo Institute of Technology, Department of Computer Science, Tokyo, Japan, 1991.*

[Val79a] L.G. Valiant. The complexity of computing the permanent. *Theoretical Computer Science, 8:189-201, 1979.*

[Val79b] L.G. Valiant. The complexity of enumeration and reliability problems. *SIAM Journal on Computing, 8(3):410-421, 1979.*

[Vol94] H. Vollmer. Komplexitätsklassen von Funktionen. *PhD thesis, Universität Würzburg, Institut für Informatik, Würzburg, Germany, 1994.*

[VW93] H. Vollmer and K.W. Wagner. The complexity of finding middle elements. *International Journal of Foundations of Computer Science, 4:293-307. 1993.*

[VW95] H. Vollmer and K.W. Wagner. Complexity Classes of Optimization Functions. *Information and Computation, 120(2): 198-219, 1995*

[Wag86] K.W. Wagner. Some observations on the connection between counting and recursion. *Theoretical Computer Science, 47:131-147, 1986.*

[Wag87] K.W. Wagner. More complicated questions about maxima and minima, and some closures of NP. *Theoretical Computer Science, 51:53-80, 1987.*

Resource-Bounded Kolmogorov Complexity Revisited*

Harry Buhrman**[1] and Lance Fortnow***[2]

[1] CWI, PO Box 94079, 1090 GB Amsterdam, The Netherlands
[2] CWI & University of Chicago, Department of Computer Science, 1100 E. 58th St., Chicago, IL 60637

Abstract. We take a fresh look at **CD** complexity, where $\mathbf{CD}^t(x)$ is the smallest program that distinguishes x from all other strings in time $t(|x|)$. We also look at a **CND** complexity, a new nondeterministic variant of **CD** complexity.

We show several results relating time-bounded **C**, **CD** and **CND** complexity and their applications to a variety of questions in computational complexity theory including:

- Showing how to approximate the size of a set using **CD** complexity avoiding the random string needed by Sipser. Also we give a new simpler proof of Sipser's lemma.
- A proof of the Valiant-Vazirani lemma directly from Sipser's earlier **CD** lemma.
- A relativized lower bound for **CND** complexity.
- Exact characterizations of equivalences between **C**, **CD** and **CND** complexity.
- Showing that a satisfying assignment can be found in output polynomial time if and only if a unique assignment can be found quickly. This answers an open question of Papadimitriou.
- New Kolmogorov-based constructions of the following relativized worlds:
  - There exists an infinite set in **P** with no sparse infinite subsets in **NP**.
  - **EXP** = **NEXP** but there exists a nondeterministic exponential time Turing machine whose accepting paths cannot be found in exponential time.
  - Satisfying assignment cannot be found with nonadaptive queries to **SAT**.

---

* At http://www.cs.uchicago.edu/~fortnow/papers/ a full version of this paper including complete proofs can be found.

** E-mail: buhrman@cwi.nl. Part of this research was done while visiting The University of Chicago. Partially supported by the Dutch foundation for scientific research (NWO) through NFI Project ALADDIN, under contract number NF 62-376 and SION project 612-34-002, and by the European Union through NeuroCOLT ESPRIT Working Group Nr. 8556, and HC&M grant nr. ERB4050PL93-0516.

*** Email: fortnow@cs.uchicago.edu. Supported in part by NSF grant CCR 92-53582 and the Fulbright scholar program and NWO.

Reischuk, Morvan (Eds.): STACS'97 Proceedings, LNCS 1200

1 Introduction

Originally designed to measure the randomness of strings, Kolmogorov complexity has become an important tool in computability and complexity theory. A simple lower bound showing that there exist random strings of every length has had several important applications (see [LV93, Chapter 6]).

Early in the history of computational complexity theory, many people naturally looked at resource-bounded versions of Kolmogorov complexity. This line of research was initially fruitful and led to some interesting results. In particular, Sipser [Sip83] invented a new variation of resource-bounded complexity, **CD** complexity, where one considers the size of the smallest program that accepts the given string and no others. Sipser showed that one can approximate the size of sets using **CD** complexity with random advice.

Complexity theory has marched on for the past two decades, but resource-bounded Kolmogorov complexity has seen little interest. Now that computational complexity theory has matured a bit, we ought to look back at resource-bounded Kolmogorov complexity and see what new results and applications we can draw from it.

First, we use algebraic techniques to give a new upper bound lemma for **CD** complexity without the random advice required of Sipser's Lemma [Sip83].

We also give a new simpler proof of Sipser's Lemma and show how it implies the important Valiant-Vazirani lemma [VV86] that randomly isolates satisfying assignments. Surprisingly, Sipser's paper predates the result of Valiant and Vazirani.

We define **CND** complexity, a variation of **CD** complexity where we allow nondeterministic computation. We prove a lower bound for **CND** complexity where we show that there exists an infinite set A such that every string in A has high **CND** complexity even if we allow access to A as an oracle. We use this lemma to prove some negative result on nondeterministic search vs. deterministic decision.

Once we have these tools in place, we use them to unify several important theorems in complexity theory. We answer an open question of Papadimitriou [Pap96] characterizing exactly when the set of satisfying assignments of a formula can be enumerated in output polynomial-time. We create relativized worlds where assignments to **SAT** cannot be found with non adaptive queries to **SAT** (first proven by Buhrman and Thierauf [BT96]), and where **EXP** = **NEXP** but there exists a nondeterministic exponential time Turing machine whose accepting paths cannot be found in polynomial time (first proven by Impagliazzo and Tardos [IT89]).

These results in their original form require a great deal of time to fully understand the proof because either the ideas and/or technical details are quite complex. We show that by understanding resource-bounded Kolmogorov complexity, one can see full and complete proofs of these results without much additional effort. We also look at when polynomial-time **C**, **CD** and **CND** complexity collide. We give a precise characterization of when we have equality of these classes, and some interesting consequences thereof.

2 Preliminaries

We use basic concepts and notation from computational complexity theory texts like Balcázar, Díaz, and Gabarró [BDG88] and Kolmogorov complexity from the excellent book by Li and Vitányi [LV93]. We use $|x|$ to represent the length of a string x and $\|A\|$ to represent the number of elements in the set A. All of the logarithms are base 2. **EXP** is defined as DTIME(2^{poly}) and **NEXP** is defined as NTIME(2^{poly}).

Formally, we define the Kolmogorov complexity function $\mathbf{C}(x|y)$ by $\mathbf{C}(x|y) = \min_p\{|p| : U(p,y) = x\}$ where U is some fixed universal deterministic Turing machine. We define unconditional Kolmogorov complexity by $\mathbf{C}(x) = \mathbf{C}(x|\epsilon)$.

A few basic facts about Kolmogorov complexity:

- The choice of U affects the Kolmogorov complexity by at most an additive constant.
- For some constant c, $\mathbf{C}(x) \leq |x| + c$ for every x.
- For every n and every y, there is an x such that $|x| = n$ and $\mathbf{C}(x|y) \geq n$.

We will also use time-bounded Kolmogorov complexity. Fix a fully time-computable function $t(n) \geq n$. We define the $\mathbf{C}^t(x|y)$ complexity function as

$$\mathbf{C}^t(x|y) = \min_p\{|p| : U(p,y) = x \text{ and } U(p) \text{ runs in at most } t(|x|+|y|) \text{ steps}\}.$$

As before we let $\mathbf{C}^t(x) = \mathbf{C}^t(x|\epsilon)$. A different universal U may affect the complexity by at most a constant additive factor and the time by a $\log t$ factor.

While the usual Kolmogorov complexity asks about the smallest program to *produce* a given string, we may also want to know about the smallest program to *distinguish* a string. While this difference affects the unbounded Kolmogorov complexity by only a constant it can make a difference for the time-bounded case. Sipser [Sip83] defined the distinguishing complexity $\mathbf{CD}^t$ by

$$\mathbf{CD}^t(x|y) = \min_p \left\{ |p| : \begin{array}{l} (1)\ U(p,x,y) \text{ accepts.} \\ (2)\ U(p,z,y) \text{ rejects for all } z \neq x. \\ (3)\ U(p,z,y) \text{ runs in at most } t(|z|+|y|) \text{ steps} \\ \qquad \text{for all } z \in \Sigma^*. \end{array} \right\}$$

Fix a universal nondeterministic Turing machine U_n. We define the nondeterministic distinguishing complexity $\mathbf{CND}^t$ by

$$\mathbf{CND}^t(x|y) = \min_p \left\{ |p| : \begin{array}{l} (1)\ U_n(p,x,y) \text{ accepts.} \\ (2)\ U_n(p,z,y) \text{ rejects for all } z \neq x. \\ (3)\ U_n(p,z,y) \text{ runs in at most } t(|z|+|y|) \text{ steps} \\ \qquad \text{for all } z \in \Sigma^*. \end{array} \right\}$$

Once again we let $\mathbf{CND}^t(x) = \mathbf{CND}^t(x|\epsilon)$.

We can also allow for relativized Kolmogorov complexity. For example for some set A, $\mathbf{CD}^{t,A}(x|y)$ is defined as above except that the universal machine U has access to A as an oracle.

Since one can distinguish a string by generating it we have

Lemma 1. $\forall t\ \exists c\ \forall x, y : \mathbf{CD}^{ct}(x \mid y) \leq \mathbf{C}^{t}(x \mid y) + c$

where c is a constant. Likewise, since every deterministic computation is also a nondeterministic computation we get

Lemma 2. $\forall t\ \exists c\ \forall x, y : \mathbf{CND}^{ct}(x \mid y) \leq \mathbf{CD}^{t}(x \mid y) + c.$

In Section 6 we examine the consequences of the converses of these lemmas.

3 Approximating Sets with Distinguishing Complexity

In this section we derive a lemma that enables one to deterministically approximate the density of a set, using polynomial-time distinguishing complexity.

Lemma 3. *Let $S = \{x_1, \ldots, x_d\} \subseteq \{0, \ldots, 2^n - 1\}$. For all $x_i \in S$ and at least half of the primes $p \leq 4dn^2$, $x_i \not\equiv x_j \bmod p$ for all $j \neq i$.*

Proof: For each $x_i, x_j \in S$, $i \neq j$, it holds that for at most n different prime numbers p, $x_i \equiv x_j \bmod p$ by the Chinese Remainder Theorem. For x_i there are at most dn primes p such that $x_i \equiv x_j \bmod p$ for some $x_j \in S$. The prime number Theorem (see for example [Ing32]) states that for any m there are approximately $m/\ln(m) > m/\log(m)$ primes less than m. There are at least $4dn^2/\log(4dn^2) > 2dn$ primes less than $4dn^2$. So at least half of these primes p must have $x_i \not\equiv x_j \bmod p$ for all $j \neq i$. □

Lemma 4. *Let A be any set. For all strings $x \in A^{=n}$ it holds that $\mathbf{CD}^{p,A^{=n}}(x) \leq 2\log(\|A\|) + O(\log n)$ for some polynomial p.*

Proof: Fix n and let $S = A^{=n}$. Fix $x \in S$ and a prime p_x fulfilling the conditions of Lemma 3 for x.

The $\mathbf{CD}^{poly}$ program for x works as follows:

> input y
> If $y \notin A^{=n}$ then REJECT
> else if $y \bmod p_x = x \bmod p_x$ then ACCEPT
> else REJECT

The size of the above program is $|p_x| + |x \bmod p_x| + O(1)$. This is $2\log(\|A\|) + O(\log n)$. It is clear that the program runs in polynomial time, and only accepts x. □

We note that the above Lemma also works for $\mathbf{CND}^{p}$ complexity for p some polynomial.

Corollary 5. *Let A be a set in $\mathbf{P}$. For each string $x \in A$ it holds that: $\mathbf{CD}^{p}(x) \leq 2\log(\|A^{=n}\|) + O(\log(n))$ for some polynomial p.*

Proof: We will use the same scheme as in Lemma 4, now using that $A \in \mathbf{P}$ and specifying the length of x, yielding an extra $\log(n)$ term for $|x|$ plus an additional $2\log\log(n)$ penalty for concatenating the strings. □

Corollary 6. *1. A set S is sparse if and only if for all $x \in S$, $\mathbf{CD}^{p,S}(x) \leq O(\log(|x|))$, for some polynomial p.*
2. A set $S \in \mathbf{P}$ is sparse if and only if for all $x \in S$, $\mathbf{CD}^{p}(x) \leq O(\log(|x|))$, for some polynomial p.
3. A set $S \in \mathbf{NP}$ is sparse if and only if for all $x \in S$, $\mathbf{CND}^{p}(x) \leq O(\log(|x|))$, for some polynomial p.

Proof: Lemma 4 yields that all strings in a sparse set have $O(\log(n))$ $\mathbf{CD}^p$ complexity. On the other hand simple counting shows that for any set A there must be a string $x \in A$ such that $\mathbf{CND}^A(x) \geq \log(\|A\|)$. □

3.1 Sipser's Lemma

We can also use Lemma 3 to give a simple proof of the following important result due to Sipser [Sip83].

Lemma 7 Sipser. *For every polynomial-time computable set A there exists a polynomial p and constant c such that for every n, for most r in $\Sigma^{p(n)}$ and every $x \in A^{=n}$,*

$$\mathbf{CD}^{p,A^{=n}}(x|r) \leq \log \|A^{=n}\| + c \log n$$

Proof: For each k, $1 \leq k \leq n$, let r_k be a list of $4k(n+1)$ randomly chosen numbers less than 2^k. Let r be the concatenation of all of the r_k.

Fix $x \in A^{=n}$. Let $d = \|A^{=n}\|$. Fix k such that $2^{k-1} < 4dn^2 \leq 2^k$. Consider one of the numbers y listed in r_k. By the Prime Number Theorem [Ing32], the probability that y is prime and less than $4dn^2$ is at least $\frac{1}{2(\log 4dn^2)}$. The probability that y fulfills the conditions of Lemma 3 for x is at least $\frac{1}{4 \log 4dn^2} > \frac{1}{4k}$. With probability about $(1 - 1/e^{n+1}) > (1 - 1/2^{n+1})$ we have that some y in r_k fulfills the condition of Lemma 3.

With probability at least 1/2, for every $x \in A$ there is some y listed in r_k fulfilling the conditions of Lemma 3 for x.

We can now describe x by $x \bmod y$ and the pointer to y in r. □

Note: Sipser can get a tighter bound than $c \log n$ but for most applications the additional $O(\log n)$ additive factor makes no substantial difference.

Comparing our Lemma 4 with Sipser's lemma 7, we are able to eliminate the random string required by Sipser at the cost of an additional $\log |A^{=n}|$ bits.

4 Lower Bounds

In this section we show that there exists an infinite set A such that every string in A has high **CND** complexity, even relative to A.

Fortnow and Kummer [FK96] prove the following result about relativized **CD** complexity:

Theorem 8. *There exists an infinite set A such that for every polynomial p, $\mathbf{CD}^{p,A}(x) \geq |x|/5$ for almost all $x \in A$.*

We extend and strengthen their result for **CND** complexity:

Theorem 9. *There exists an infinite set A such that $\mathbf{CND}^{2^{\sqrt{|x|}},A}(x) \geq |x|/4$ for all $x \in A$.*

The proof of Fortnow and Kummer of Theorem 8 uses the fact that one can start with a large set A of strings of the same length such that any polynomial-time algorithm on an input x in A cannot query any other y in A. However, a nondeterministic machine may query every string of a given length. Thus we need a more careful proof.

This proof is based on the proof of Corollary 10 of Goldsmith, Hemachandra and Kunen [GHK92]. In Section 5, we will also describe a rough equivalence between this result and an "X-search" theorem of Impagliazzo and Tardos [IT89].

Using Theorem 9 we get the following corollary first proved by Goldsmith, Hemachandra and Kunen [GHK92].

Corollary 10 Goldsmith-Hemachandra-Kunen. *Relative to some oracle, there exists an infinite set in* **P** *with no infinite sparse subsets in* **NP**.

Proof: Let A from Theorem 9 be both the oracle and the set in P^A. Suppose A has an infinite sparse subset S in NP^A. Pick a large x such that $x \in S$. Applying Corollary 6(3) it follows that $\mathbf{CND}^{A,p}(x) \leq O(\log(n))$. This contradicts the fact that $x \in S \subseteq A$ and Theorem 9. □

The above argument shows actually something stronger:

Corollary 11. *Relative to some oracle, there exists an infinite polynomial-time computable set with no infinite subset in* **NP** *of density less than $2^{n/9}$.*

5 Search vs. Decision in Exponential-Time

If **P** = **NP** then given a satisfiable formula, one can use binary search to find the assignment.

One might expect a similar result for exponential-time computation, i.e., if **EXP** = **NEXP** then one should find a witness of a nondeterministic exponential-time computation in exponential time. However, the proof for polynomial-time breaks down because as one does the binary search the input questions get too long. Impagliazzo and Tardos [IT89] give relativized evidence that this problem is indeed hard.

Theorem 12 [IT89]. *There exists a relativized world where* **EXP** = **NEXP** *but there exists a nondeterministic exponential-time Turing machine whose accepting paths cannot be found in exponential time.*

We can give a short proof of this theorem using Theorem 9.

Proof of Theorem 12: Let A be from Theorem 9.

We will encode a tally set T such that $\mathbf{EXP}^{A\oplus T} = \mathbf{NEXP}^{A\oplus T}$. Let M be a nondeterministic oracle machine such that M runs in time 2^n and for all B, M^B is $\mathbf{NEXP}^B$-complete.

Initially let $T = \emptyset$. For every string w in lexicographic order, put 1^{2w} into T if $M^{A\oplus T}(w)$ accepts.

Let $B = A \oplus T$ at the end of the construction. Since $M(w)$ could only query strings with length at most $2^{|w|} \leq w$, this construction will give us $\mathbf{EXP}^B = \mathbf{NEXP}^B$.

We will show that there exists a nondeterministic exponential time Turing machine with access to B whose accepting paths cannot be found in time exponential relative to B.

Consider the nondeterministic machine M that on input n guesses a string y of length n and accepts if y is in A. Note that M runs in time $2^{|n|} \leq n$.

Suppose accepting computations of M^B can be found in time $2^{|n|^k} = 2^{\log^k n}$ relative to B. By Theorem 9, we can fix some large n such that $A^{=n} \neq \emptyset$ and for all $x \in A^{=n}$,

$$\mathbf{CND}^{2^{\log^k n},A}(x) \geq n/4. \tag{1}$$

Let $w_i = \|\{1^m \mid 1^m \in T \text{ and } 2^i < m \leq 2^{i+1}\}\|$. We will show the following lemma.

Lemma 13. $\mathbf{CND}^{2^{\log^k n},A}(x|w_1, \ldots, w_{\log^k n}) \leq \log n + O(1)$.

Assuming Lemma 13, Theorem 12 follows since for each i, $|w_i| \leq i + 1$. We thus have our contradiction with Equation (1).

Proof of Lemma 13: We will construct a program p^A to nondeterministically distinguish x. We use $\log n$ bits to encode n. First p will reconstruct T using the w_i's.

Suppose we have reconstructed T up to length 2^i. By our construction of T, strings of T of length at most 2^{i+1} can only depend on oracle strings of length at most $2^{i+1}/2 = 2^i$. We guess w_i strings of the form 1^m for $2^i < m \leq 2^{i+1}$ and nondeterministically verify that these are the strings in T. Once we have T, we also have $B = A \oplus T$ so in time $2^{\log^k n}$ we can find x. □

Impagliazzo and Tardos [IT89] prove Theorem 12 using an "X-search" problem. We can also relate this problem to **CND** complexity and Theorem 9.

Definition 14. The X-search problem has a player who given N input variables not all zero, wants to find a one. The player can ask r rounds of l parallel queries of a certain type each and wins if the player discovers a one.

Impagliazzo and Tardos use the following result about the X-search problem to prove Theorem 12.

Theorem 15 [IT89]. *If the type of the queries is restricted to k-DNFs and $N > 2(klr)^2(l+1)^r$ then the player will lose on some non-zero setting of the variables.*

One can use a proof similar to that of Theorem 12 to prove a similar bound for Theorem 15. One needs just to apply Theorem 9 relative to the strategy of the player.

One can also use Theorem 15 to prove a variant of Theorem 9. Suppose Theorem 9 fails. For any A and for every x in A there exists a small program that nondeterministically distinguishes x. For some x suppose we know p. We can find x by asking a **DNF** question based on p about the ith bit of x.

We do not in general know p but there are not too many possibilities. We can use an additional round of queries to try all programs and test all the answers in parallel. This will give us a general strategy for the X-search problem contradicting Theorem 15.

6 CD vs. C and CND

This section deals with the consequences of the assumption that one of the complexity measures **C**, **CD**, and **CND** coincide for polynomial time. We will see that these assumptions are equivalent to well studied complexity theoretic assumptions. This allows us to apply the machinery developed in the previous sections. We will use the following function classes:

Definition 16. 1. The class $\mathbf{FP}^{\mathbf{NP}[\log(n)]}$ is the class of functions computable in polynomial time that can adaptively access an oracle in **NP** at most $c\log(n)$ times, for some c.
2. The class $\mathbf{FP}^{\mathbf{NP}}_{tt}$ is the class of functions computable in polynomial time that can non-adaptively access an oracle in **NP**.

Theorem 17. *The following are equivalent:*

1. $\forall p_2\, \exists p_1, c\, \forall x, y : \mathbf{C}^{p_1}(x \mid y) \leq \mathbf{CND}^{p_2}(x \mid y) + c\log(|x|)$.
2. $\forall p_2\, \exists p_1, c\, \forall x, y : \mathbf{CD}^{p_1}(x \mid y) \leq \mathbf{CND}^{p_2}(x \mid y) + c\log(|x|)$.
3. $\mathbf{FP}^{\mathbf{NP}[\log(n)]} = \mathbf{FP}^{\mathbf{NP}}_{tt}$.

For the next corollary we will use some results from [JT95]. We will use the following class of limited nondeterminism defined in [DT90].

Definition 18. Let $f(n)$ be a function from $\mathbb{N} \mapsto \mathbb{N}$. The class $\mathbf{NP}[f(n)]$ denotes that class of languages that are accepted by polynomial-time bounded nondeterministic machines that on inputs of length n make at most $f(n)$ nondeterministic moves.

Corollary 19. *If* $\forall p_2\, \exists p_1, c\, \forall x, y : \mathbf{CD}^{p_1}(x \mid y) \leq \mathbf{CND}^{p_2}(x \mid y) + c\log(|x|)$ *then for any* k*:*

1. $\mathbf{NP}[\log^k(n)]$ *is included in* $\mathbf{P}$.
2. $\mathbf{SAT} \in \mathbf{NP}[\frac{n}{\log^k(n)}]$.
3. $\mathbf{SAT} \in \mathbf{DTIME}(2^{n^{O(1/\log\log n)}})$.

4. *There exists a polynomial q such that for every m formulae $\phi_1, \ldots, \phi_m$ of n variables each such that at least one is satisfiable, there exists a i such that ϕ_i is satisfiable and*

$$\mathbf{CND}^q(\phi_i | \langle \phi_1, \ldots, \phi_m \rangle) \leq O(\log\log(n+m))$$

Proof: The consequences in the corollary follow from the assumption that $\mathbf{FP}^{\mathbf{NP}[\log(n)]} = \mathbf{FP}^{\mathbf{NP}}_{tt}$ [JT95]. $\mathbf{FP}^{\mathbf{NP}[\log(n)]} = \mathbf{FP}^{\mathbf{NP}}_{tt}$ follows from Theorem 17. □

We can use Corollary 19 to get a complete collapse if there is only a constant difference between **CD** and **CND** complexity.

Theorem 20. *The following are equivalent:*

1. $\forall p_2\, \exists p_1, c\, \forall x, y : \mathbf{C}^{p_1}(x \mid y) \leq \mathbf{CND}^{p_2}(x \mid y) + c.$
2. $\forall p_2\, \exists p_1, c\, \forall x, y : \mathbf{CD}^{p_1}(x \mid y) \leq \mathbf{CND}^{p_2}(x \mid y) + c.$
3. $\mathbf{P} = \mathbf{NP}.$

In fact Theorem 20 holds if we replace the constant c with $a \log n$ for any $a < 1$.

For the next corollary we will need the following definition (see [ESY84]).

Definition 21. A promise problem is a pair of sets (Q, R). A set L is called a solution to the promise problem (Q, R) if $\forall x (x \in Q \Rightarrow (x \in L \Leftrightarrow x \in R))$. For any function f, fSAT denotes the set of boolean formulas with at most $f(n)$ satisfying assignments for formulae of length n.

The next theorem states that nondeterministic computations that have few accepting computations can be "compressed" to nondeterministic computations that have few nondeterministic moves if and only if $\mathbf{C}^{poly} \leq \mathbf{CD}^{poly}$.

Theorem 22. *The following are equivalent:*

1. $\forall p_2\, \exists p_1, c\, \forall x, y : \mathbf{C}^{p_1}(x \mid y) \leq \mathbf{CD}^{p_2}(x \mid y) + c.$
2. *(1SAT,SAT) has a solution in* $\mathbf{P}$.
3. *For all time constructible f, (fSAT,SAT) has a solution in* $\mathbf{NP}[2\log(f(n)) + O(\log(n))]$.

Corollary 23. $\mathbf{FP}^{\mathbf{NP}[\log(n)]} = \mathbf{FP}^{\mathbf{NP}}_{tt}$ *implies the following:*

1. *For any k the promise problem ($2^{\log^k(n)}SAT$,SAT) has a solution in* $\mathbf{P}$.
2. *For any k, the class of languages that is accepted by nondeterministic machines that have at most $2^{\log^k(n)}$ accepting paths on inputs of length n is included in* $\mathbf{P}$

Proof: This follows from Theorem 17, Theorem 22, and Corollary 19. □

7 Satisfying Assignments

We show several connections between **CD** complexity and finding satisfying assignments of boolean formulae. By Cook's Theorem [Coo71], finding satisfying assignments is equivalent to finding accepting computation paths of any nondeterministic polynomial-time computation.

7.1 Enumerating Satisfying Assignments

Papadimitriou [Pap96] mentioned the following proposition:

Proposition 24. *There exists a Turing machine that given a formula ϕ will output the set A of satisfying assignments of ϕ in time polynomial in $|\phi|$ and $\|A\|$.*

We can use **CD** complexity to show the following.

Theorem 25. *Proposition 24 is equivalent to $(1SAT, SAT)$ has a solution in* **P**.

In Proposition 24, we do not require the machine to halt after printing out the assignments. If the machine is required to halt in time polynomial in ϕ and $\|A\|$ we have that Proposition 24 is equivalent to $\mathbf{P} = \mathbf{NP}$.

Proof of Theorem 25: The implication of $(1SAT, SAT)$ having a solution in **P** is straightforward. We concentrate on the other direction.

Let $d = \|A\|$. By Lemma 4 and Theorem 22 we have that for every element x of A, $\mathbf{C}^q(x|\phi) \leq 2\log d + c\log n$ for some polynomial q and constant c. We simply now try every program p in length increasing order and enumerate $p(\phi)$ if it is a satisfying assignment of ϕ. □

7.2 Computing Satisfying Assignments

In this section we turn our attention to the question of the complexity of generating a satisfying assignment for a satisfiable formula [WT93, HNOS96, Ogi96, BKT94]. It is well known [Kre88] that one can generate (the leftmost) satisfying assignment in $\mathbf{FP}^{\mathbf{NP}}$. A tantalizing open question is whether one can compute some (not necessary the leftmost) satisfying assignment in $\mathbf{FP}^{\mathbf{NP}}_{tt}$. Formalizing this question, define the function class $\mathbf{F}_{sat}$ by $f \in \mathbf{F}_{sat}$ if when $\varphi \in \mathbf{SAT}$ then $f(\varphi)$ is a satisfying assignment of φ.

The question now becomes $\mathbf{F}_{sat} \bigcap \mathbf{FP}^{\mathbf{NP}}_{tt} = \emptyset$? Translating this to a **CND** setting we have the following.

Lemma 26. $\mathbf{F}_{sat} \bigcap \mathbf{FP}^{\mathbf{NP}}_{tt} \neq \emptyset$ *if and only if for all $\phi \in$* **SAT** *there exists a satisfying assignment a of ϕ such that $\mathbf{CND}^p(a \mid \phi) \leq c\log(|\phi|)$ for some polynomial p and constant c.*

Toda and Watanabe [WT93] showed that $\mathbf{F}_{sat} \bigcap \mathbf{FP}^{\mathbf{NP}}_{tt} \neq \emptyset$ relative to a random oracle. On the other hand Buhrman and Thierauf [BT96] showed that there exists an oracle where $\mathbf{F}_{sat} \bigcap \mathbf{FP}^{\mathbf{NP}}_{tt} = \emptyset$. Their result also holds relative to the set constructed in Theorem 9.

Theorem 27. *Relative to the set A constructed in Theorem 9, $\mathbf{F}_{sat} \bigcap \mathbf{FP}_{tt}^{\mathbf{NP}} = \emptyset$.*

Proof: For some n, let ϕ be the formula on n variables such that $\phi(x) = \top$ if and only if $x \in A$. Suppose $\mathbf{F}_{sat} \bigcap \mathbf{FP}_{tt}^{\mathbf{NP}} \neq \emptyset$. It now follows by Lemma 26 that there exists an $x \in A$ such that $\mathbf{CND}^{p,A}(x) \leq O(\log(|x|))$ for some polynomial p, contradicting the fact that for all $x \in A$, $\mathbf{CND}^{2^{\sqrt{|x|}},A}(x) \geq |x|/4$. □

7.3 Isolating Satisfying Assignments

In this section we take a Kolmogorov complexity view of the statement and proof of the famous Valiant-Vazirani lemma [VV86]. The Valiant-Vazirani lemma gives a randomized reduction from a satisfiable formula to another formula that with a non negligible probability has exactly one satisfying assignment.

We state the lemma in terms of Kolmogorov complexity.

Lemma 28. *There is some polynomial p such that for all ϕ in* **SAT** *and all r such that $|r| = p(|\phi|)$ and $\mathbf{C}(r) \geq |r|$, there is some satisfying assignment a of ϕ such that $\mathbf{CD}^p(a|\langle\phi, r\rangle) \leq O(\log|\phi|)$.*

The usual Valiant-Vazirani lemma follows from the statement of Lemma 28 by choosing r and the $O(\log|\phi|)$ program randomly.

We show how to derive the Valiant-Vazirani Lemma from Sipser's Lemma (Lemma 7). Note Sipser's result predates Valiant-Vazirani by a couple of years.

Proof of Lemma 28: Let $n = |\phi|$.

Consider the set A of satisfying assignments of ϕ. We can apply Lemma 7 conditioned on ϕ using part of r as the random strings. Let $d = \lfloor \log \|A\| \rfloor$. We get that every element of A has a **CD** program of length bounded by $d + c\log n$ for some constant c. Since two different elements from A must have different programs, we have at least $1/n^c$ of the strings of length $d+c\log n$ must distinguish some assignment in A.

We use the rest of r to list n^{2c} different strings of length $d + c\log n$. Since r is random, one of these strings w must be a program that distinguishes some assignment a in A. We can give a **CD** program for a in $O(\log n)$ bits by giving d and a pointer to w in r. □

Acknowledgments

We would like to thank José Balcázar and Leen Torenvliet for their comments on this subject. We thank John Tromp for the current presentation of the proof of Lemma 4. We also thank Sophie Laplante for her important contributions to Section 6. We thank Richard Beigel, Bill Gasarch and Leen Torenvliet for comments on earlier drafts. We thank the anonymous reviewers for helpful comments.

References

[BDG88] J. Balcázar, J. Díaz, and J. Gabarró. *Structural Complexity I*. Springer-Verlag, 1988.

[BKT94] H. Buhrman, J. Kadin, and T. Thierauf. On functions computable with nonadaptive queries to NP. In *Proc. Structure in Complexity Theory 9th Annual Conference*, pages 43–52. IEEE computer society press, 1994.

[BT96] H. Buhrman and T. Thierauf. The complexity of generating and checking proofs of membership. In C. Pueach and R. Reischuk, editors, *13th Annual Symposium on Theoretical Aspects of Computer Science*, number 1046 in Lecture Notes in Computer Science, pages 75–86. Springer, 1996.

[Coo71] S. Cook. The complexity of theorem-proving procedures. In *Proc. 3rd ACM Symposium Theory of Computing*, pages 151–158, Shaker Heights, Ohio, 1971.

[DT90] J. Díaz and J. Torán. Classes of bounded nondeterminism. *Math. Systems Theory*, 23:21–32, 1990.

[ESY84] S. Even, A. L. Selman, and Y. Yacobi. The complexity of promise problems with applications to public-key cryptography. *Information and Control*, 61(2):159–173, May 1984.

[FK96] L. Fortnow and M. Kummer. Resource-bounded instance complexity. *Theoretical Computer Science A*, 161:123–140, 1996.

[GHK92] J. Goldsmith, L. Hemachandra, and K. Kunen. Polynomial-time compression. *Computational Complexity*, 2(1):18–39, 1992.

[HNOS96] L. Hemaspaandra, A. Naik, M. Ogihara, and A. Selman. Computing solutions uniquely collapses the polynomial hierarchy. *SIAM J. Comput.*, 25(4):697–708, 1996.

[Ing32] A.E. Ingham. *The Distribution of Prime Numbers*. Cambridge Tracts in Mathematics and Mathematical Physics. Cambridge University Press, 1932.

[IT89] R. Impagliazzo and G. Tardos. Decision versus search problems in super-polynomial time. In *Proc. 30th IEEE Symposium on Foundations of Computer Science*, pages 222–227, 1989.

[JT95] Jenner and Toran. Computing functions with parallel queries to NP. *Theoretical Computer Science*, 141, 1995.

[Kre88] M. Krentel. The complexity of optimization problem. *J. Computer and System Sciences*, 36:490–509, 1988.

[LV93] Ming Li and P.M.B. Vitányi. *An Introduction to Kolmogorov Complexity and Its Applications*. Springer-Verlag, 1993.

[Ogi96] M. Ogihara. Functions computable with limited access to NP. *Information Processing Letters*, 58:35–38, 1996.

[Pap96] C. Papadimitriou. The complexity of knowledge representation. Invited Presentation at the Eleventh Annual IEEE Conference on Computational Complexity, May 1996.

[Sip83] M. Sipser. A complexity theoretic approach to randomness. In *Proc. 15th ACM Symposium on Theory of Computing*, pages 330–335, 1983.

[VV86] L. Valiant and V. Vazirani. NP is as easy as detecting unique solutions. *Theoretical Computer Science*, 47:85–93, 1986.

[WT93] O. Watanabe and S. Toda. Structural analysis on the complexity of inverse functions. *Mathematical Systems Theory*, 26:203–214, 1993.

Las Vegas Versus Determinism for One-way Communication Complexity, Finite Automata, and Polynomial-time Computations

Pavol Ďuriš[1], Juraj Hromkovič[2]*, José D. P. Rolim[3], and Georg Schnitger[4]

[1] Department of Computer Science, Comenius University, Mlynská dolina, 842 15 Bratislava, Slovakia
[2] Institut für Informatik, Universität zu Kiel, 24098 Kiel, Germany
[3] Centre Universitaire d'Informatique, Université de Genève, 1211 Genéve 4, Switzerland
[4] Fachbereich Informatik, Johann Wolfgang Goethe–Universität Frankfurt, Robert Mayer Strasse 11–15, 60054 Frankfurt am Main, Germany

Abstract. The study of the computational power of randomized computations is one of the central tasks of complexity theory. The main aim of this paper is the comparison of the power of Las Vegas computation and deterministic respectively nondeterministic computation. An at most polynomial gap has been established for the combinational complexity of circuits and for the communication complexity of two-party protocols. We investigate the power of Las Vegas computation for the complexity measures of one-way communication, finite automata and polynomial-time relativized Turing machine computation.

(i) For the one-way communication complexity of two-party protocols we show that Las Vegas communication can save at most one half of the deterministic one-way communication complexity.
We also present a language for which this gap is tight.

(ii) For the size (i.e., the number of states) of finite automata we show that the size of Las Vegas finite automata recognizing a language L is at least the root of the size of the minimal deterministic finite automaton recognizing L. Using a specific language we verify the optimality of this lower bound.
Note, that this result establishes for the first time an at most polynomial gap between Las Vegas and determinism for a uniform computing model.

(iii) For relativized polynomial computations we show that Las Vegas can be even more powerful than nondeterminism with a polynomial restriction on the number of nondeterministic guesses.
On the other hand superlogarithmic many advice bits in nondeterministic computations can be more powerful than Las Vegas (even Monte Carlo) computations in a relativized word.

Keywords: computational and structural complexity, Las Vegas, determinism, communication complexity, automata

* The work of this author has been supported by DFG Project HR 14/3-1.

Reischuk, Morvan (Eds.): STACS'97 Proceedings, LNCS 1200

1 Introduction and Definitions

The comparative study of the computational power of nondeterministic, deterministic, and randomized computations is one of the central tasks of complexity theory. In this paper we focus on the relationship between Las Vegas and determinism and between Las Vegas and nondeterminism.

The relationship between the complexity classes P and ZPP, the class of languages accepted by polynomial-time Las Vegas Turing machines, is unresolved. The two classes coincide for two-party communication [7] and for the combinational complexity of non-uniform circuits.

We consider the P *versus* ZPP problem, respectively the "ZPP *versus* NP" problem for the following three complexity measures:

(i) one-way communication complexity,
(ii) the size (i.e., the number of states) of finite automata, and
(iii) the time complexity of relativized Turing machine computations.

To define Las Vegas computations we follow [2]. In particular we consider *self-verifying nondeterminism* (introduced for communication complexity in [5]). A self-verifying nondeterministic machine M is allowed to give three possible answers *yes, no, I do not know.* M is not allowed to make mistakes: If the answer is *yes*, then the input must be in $L(M)$. If the answer is *no*, then the input *cannot* be in $L(M)$.

We say that M is a Las Vegas machine recognizing a language L if and only if M is a self-verifying nondeterministic machine recognizing L *and* for each input the answer "I do not know" is given with probability at most $\frac{1}{2}$.

The main results of this paper are as follows:

(i) **One-way communication complexity**
We consider the one-way version of two-party protocols as introduced by Yao [8] for a fixed partition of the input. Computer C_I receives the first half of the input and computer C_{II} receives the rest. Informally, a deterministic one-way protocol P determines first the message sent from computer C_I to computer C_{II} and then decides, whether C_{II} accepts or rejects.
The one-way communication complexity, $\mathrm{cc}_1(P)$, of P is the length of the longest message sent by C_I. Finally for a language L, $\mathrm{cc}_1(L)$ denotes the one-way communication complexity of the best protocol computing L. Let $\mathrm{ncc}_1(L)$ [resp. $\mathrm{svncc}_1(L), \mathrm{lvcc}_1(L)$] denote the one-way nondeterministic [resp. self-verifying, Las Vegas] communication complexity of L.
In our main result of this part we show that

$$\mathrm{lvcc}_1(L) \geq \mathrm{cc}_1(L)/2$$

for every language L. This result is quite surprising, because there is a quadratic gap between Las Vegas and determinism for the general (two-way) model of communication complexity [7]. Moreover, for a specific language $L \subseteq \{0,1\}^*$, we show that $\mathrm{cc}_1(h_n(L)) = 2 \cdot \mathrm{lvcc}_1(h_n(L))$, where $h_n(L)$ is the

Boolean function of n variables defined by $h_n(L)(\alpha) = 1$ iff $\alpha \in L \cap \{0,1\}^n$ and hence our relationship between Las Vegas and deterministic one-way communication is best possible.
It is well known that there exist languages A with an exponential gap between $\mathrm{cc}_1(h_n(L))$ and $\mathrm{ncc}_1(h_n(L))$. Here, we show that there is a language L with an exponential gap between $\mathrm{cc}_1(h_n(L))$ and $\mathrm{svncc}_1(h_n(L))$. Thus, self-verifying nondeterminism may be much more powerful than determinism. As a consequence we have found another substantial difference between one-way and two-way communication complexity, where self-verifying nondeterminism is polynomially related to determinism [1, 5].

(ii) **Finite automata**
We consider the model of one-way finite automata. In the following $L(A)$ denotes the language accepted by the computing device A. We introduce self-verifying nondeterministic finite automata (*SNFA*) as nondeterministic automata whose states are partitioned into three disjoint groups: accepting states, rejecting states, and neutral states. An input word is accepted (rejected) by an $SNFA$ if there exists a computation finishing in an accepting (rejecting) state. Moreover, for no input there exist computations finishing in both accepting and rejecting states and for each input at least one computation is accepting or rejecting.
We introduce Las Vegas finite automaton (*LVFA*) as a *SNFA* A which for any $x \in L(A)$ reaches an accepting state with probability at least $\frac{1}{2}$ and which for any $x \notin L(A)$ reaches a rejecting state with probability at least $\frac{1}{2}$. (The probability of a computation of a $LVFA$ is defined through the transition probabilities of the automaton.)
For any regular language L we define $\mathrm{s}(L)$, $\mathrm{ns}(L)$, $\mathrm{svns}(L)$ and $\mathrm{lvs}(L)$ respectively as the size of a minimal deterministic, nondeterministic, self-verifying nondeterministic and Las Vegas finite automaton for L.
The main result of this part shows that

$$\mathrm{lvs}(L) \geq \sqrt{\mathrm{s}(L)}$$

for every regular language L. The optimality of this lower bound on $\mathrm{lvs}(L)$ is verified by constructing a language L' with $\mathrm{s}(L') = \Omega((\mathrm{lvs}(L'))^2)$. Note, that this is the first result showing a polynomial relation between Las Vegas and determinism for a uniform computing model.
It is well known that there are regular languages with $\mathrm{s}(L) \sim 2^{\mathrm{ns}(L)}$. Here, we show that for some regular languages A,B there are exponential gaps between $\mathrm{s}(A)$ and $\mathrm{svns}(A)$, and between $\mathrm{svns}(B)$ and $\mathrm{ns}(B)$.

(iii) **Relativized polynomial-time computations**
It is well known that Las Vegas may be more powerful than determinism in a relativized world. We strengthen this result by showing that Las Vegas computations may be even more powerful than nondeterministic computations with at most $f(n)$ advice bits (nondeterministic decisions) for any f bounded by a polynomial. This shows the existence of a relativization (oracle) for which any polynomial restriction on the number of nondeterministic

guesses (advice bits) essentially decreases the power of polynomial-time nondeterministic computations.
On the other hand, for any polynomial-time constructible function h growing faster than $\log_2 n$, there exists an oracle B such that polynomial-time nondeterministic computations with oracle B and at most $h(n)$ advice bits are more powerful than polynomial-time two-sided error Monte Carlo computations with oracle B. Thus, there exists a relativization for which a small number of nondeterministic guesses (for instance, $\log_2 n \cdot \log\log n$) gives more power than Monte Carlo (Las Vegas) computations with any number of random bits. The last result solves an open question stated in [3].

This extended abstract is organized as follows. Section 2 is devoted to the formal presentation of the results and some proofs are presented in section 3. Because of space restrictions we present only the proofs (respectively proof outlines) of three of the main results ($\text{lvcc}_1(f) \geq \text{cc}_1(f)/2$ for every f; $\text{lvs}(L) \geq \sqrt{\text{s}(L)}$ for every L, and that Las Vegas can be much more powerful than bounded nondeterminism in a relativized word).

2 Results

2.1 One-way Communication Complexity

First, we present our main result.

Theorem 1. *For every language L*

$$\text{lvcc}_1(L) \geq \text{cc}_1(L)/2.$$

To show that the lower bound of Theorem 1 cannot be improved we consider the language $L = \{xy \in \{0,1\}^* \mid |x| = |y| \text{ and if } y = 0^i 1z, \text{ then } x_{i+1} = 1\}$. We obtain a Las Vegas protocol for $h_{4n}(L)$ exchanging n bits as follows. The first computer flips an unbiased coin and sends accordingly the first respectively the second half of its input. Obviously the second computer is now able to determine the result with probability $\frac{1}{2}$.

Theorem 2. *For every positive integer n,*

(i) $\text{cc}_1(h_{4n}(L)) = 2n$,
(ii) $\text{lvcc}_1(h_{4n}(L)) = n$, *and*
(iii) $\text{svncc}_1(h_{4n}(L)) = \lceil \log_2 2n \rceil$.

Thus, Theorem 2 shows not only the optimality of the lower bound of Theorem 1, but also a surprising exponential gap between determinism and self-verifying nondeterminism. Note, that this exponential gap cannot be improved because $\text{ncc}_1(L) \leq 2^{\text{cc}_1(L)}$ for every language L. To show also an exponential gap between nondeterminism and self-verifying nondeterminism, it suffices to consider the identity language $ID^C = \{xy \in \{0,1\}^* \mid x \neq y\}$. It is quite easy to show that $\text{ncc}_1(h_{2n}(ID^C)) \leq \lceil \log n \rceil + 1$, whereas $\text{svncc}_1(h_{2n}(ID^C)) = n$.

2.2 Finite Automata

First we give our main result showing a quadratic gap between Las Vegas and determinism.

Theorem 3. *For every regular language L*

$$\mathrm{lvs}(L) \geq \sqrt{\mathrm{s}(L)}.$$

The language $L_k = \{w \in \{0,1\}^* \mid w = u1v \text{ and } |v| = k-1\}$ is a well known example of a language producing an exponential gap between $\mathrm{s}(L)$ and $\mathrm{ns}(L)$ [6]. We use L_k to show that Theorem 3 cannot be improved.

Theorem 4. *For every positive integer k,*

(i) $\mathrm{s}(L_k) = 2^{k+1} - 1$,
(ii) $\mathrm{lvs}(L_k) \leq 4 \cdot 2^{k/2} = O(\sqrt{\mathrm{s}(L_k)})$,
(iii) $\mathrm{svns}(L_k) \leq 2k + 3$, *and*
(iv) $\mathrm{ns}(L_k) = k + 1$.

Again we see that self-verifying nondeterminism may be much more powerful than determinism (resp. Las Vegas). The following result is typical for self-verifying nondeterminism.

Observation 2.2.1 *For any regular language L:*

$$\max\{\mathrm{ns}(L), \mathrm{ns}(L^C)\} \leq \mathrm{svns}(L) \leq 1 + \mathrm{ns}(L) + \mathrm{ns}(L^C).$$

Thus, if one wishes to demonstrate a large difference between self-verifying nondeterminism and nondeterminism, one has to choose a regular language with a large difference between $\mathrm{sn}(L)$ and $\mathrm{sn}(L^C)$:

Observation 2.2.2 *For* $m \in N$ *we set* $U_m = \{v2u \mid u, v \in \{0,1\}^m,\ u \neq v\}$. *Then,*

(i) $\mathrm{ns}(U_m) \leq 3m$,
(ii) $\mathrm{ns}({U_m}^C) \geq 2^m$, *and*
(iii) $\mathrm{svns}((U_m)) \geq 2^m$.

2.3 Polynomial-time Turing Machines

Here we consider nondeterministic and probabilistic polynomial-time bounded complexity classes as defined in [2].

- $ZPP = \{L \mid L = L(M)$ for a polynomial-time Las Vegas Turing machine $M\}$
- $R = \{L \mid L = L(M)$ for a polynomial-time one-sided error Monte Carlo Turing machine $M\}$ and

– $BPP = \{L \mid L = L(M)$ for a polynomial-time two-sided error Monte Carlo Turing machine $M\}$.

For any function $f : N \to N$, we define the class $\beta_f = \{L \mid L = L(M)$ for a polynomial-time nondeterministic Turing machine using at most $O(f(n))$ nondeterministic guesses of inputs of length $n\}$.

For any language A and any complexity class X, X^A is the complexity class X relativized by the oracle A. For any language class X, $co\text{–}X = \{L \mid \bar{L} \in X\}$. The following inclusion between complexity classes are well-known:

$$P \subseteq ZPP = R \cap co - R \subseteq R \subseteq NP$$

Diaz and Torán [3] have asked for the relation between β_f and the classes between P and NP in the relativized world, i.e., whether limited nondeterminism can be more powerful than Las Vegas (resp. Monte Carlo) or whether Las Vegas (resp. Monte Carlo) can be more powerful than limited nondeterminism. The answer for this question is given in the following two theorems.

Theorem 5. *For every function $f : N \to N$ bounded by a polynomial, there exists a language L and an oracle B such that*

$$L \in ZPP^B - {\beta_f}^B.$$

Theorem 6. *Let $f : N \to N$ be a polynomial-time constructible function bounded by a polynomial such that the function $2^{f(n)}$ majorizes each polynomial for almost all n. Then there exist a language L and an oracle B such that*

$$L \in {\beta_f}^B - BPP^B.$$

3 Proofs

3.1 The Proof of Theorem 1

First, we give an informal idea of the proof. Let $L \subseteq \{0,1\}^{2n}$ be a language and let $f : \{0,1\}^{2n} \to \{0,1\}$ be the corresponding Boolean function. One can represent f as a $2^n \times 2^n$ Boolean matrix $M(f) = [a_{u,v}]_{u,v \in \{0,1\}^n}$ with $a_{u,v} = f(uv)$. Then the number of different messages of an optimal one-way protocol P computing f is exactly the same as the number $r(M(f))$ of different rows of $M(f)$, i.e. $\mathrm{cc}_1(f) = \lceil \log_2(r(M(f))) \rceil$ [1].

Any one-way Las Vegas protocol P' may be considered as a collection of (say) m deterministic one-way protocols $P_1, \ldots, P_m$ with probabilities $p_1, \ldots p_m$. For any input α P_i may compute the results 0, 1 or 2 (i.e., I do not know). Since P' is a Las Vegas protocol, no protocol P_i ever errs and for every $(u,v) \in \{0,1\}^n \times \{0,1\}^n$, the protocols $P_1, \ldots, P_m$ produce the output 2 with probability at most $\frac{1}{2}$. To any protocol P_i $(i = 1, 2, \ldots, m)$, one can assign its 0/1/2 communication matrix $M(P_i) = [b^i_{uv}]_{u,v \in \{0,1\}^n}$, where $b^i_{uv} = a_{uv}$ if P_i does not give output 2 and $b^i_{uv} = 2$ otherwise.

Our goal is to find one protocol P_i such that $M(P_i)$ has at least $\sqrt{r(M(f))}$ different rows. In order to reduce the number of different rows a deterministic protocol will smartly replace certain entries of $M(f)$ by a 2. Obviously, "twoing" certain entries of $M(f)$ will help reduce the number of different rows far more than twoing other entries: For the identity matrix the diagonal entries play this helper role. For instance we can reduce the number of different rows to two by setting the upper left and the lower right quarter to 2. Observe that this radical reduction in the number of different rows is obtained after twoing only one half of the entries! On the other hand, any significant reduction in the number of different rows has to involve the diagonal entries and any such entry has to stay untouched with probability at least one half. Hence one deterministic protocol exists with at least $N/2$ different rows (if we consider the $N \times N$ identity matrix).

In the above example the diagonal entries form a fooling set and any Las Vegas computation has to send at least $\frac{|F|}{2}$ messages for a fooling set F. However we cannot expect to find large fooling sets in general. In particular, the $n \times \log_2 n$ communication matrix M^*, whose ith row contains the binary representation of i, possesses only fooling sets of logarithmic size, but it can be shown that any Las Vegas one-way protocol has to exchange $\sqrt{n}$ mesages.

Our proof will introduce a new notion of fooling sets. Set $M(f)=M$ and assume that M has r pairwise different rows and c columns. Our new notion of fooling sets is based on a real-valued weight assignment

$$\text{weight} : \{1, \ldots, r\} \times \{1, \ldots, c\} \rightarrow R$$

for M. Let $I = \{1, \ldots, r\}$. We define weight iteratively, processing column after column. We begin with column 1.

Case 1: Column 1 is monochromatic for all rows in I. Then set

$$\text{weight}(i, 1) = 0$$

for all rows i.

Case 2: Column 1 is not monochromatic for the rows in I. In particular assume that $c \cdot |I|$ rows have a 0 in column 1 (and $(1-c) \cdot |I|$ rows have a 1 in column 1). We set

$$\text{weight}(i, 1) = \begin{cases} \log_2(\frac{1}{c}) & \text{if } M[i, 1] = 0 \\ \log_2(\frac{1}{1-c}) & \text{otherwise.} \end{cases}$$

We repeat this procedure for column 2, but now with the subsets $I_0 = \{i \in I \mid M[i, 1] = 0\}$ and $I_1 = \{i \in I \mid M[i, 1] = 1\}$ replacing the set I. The procedure stops if the row sets are singletons (since then all columns will be monochromatic).

We begin our analysis with a technical fact, whose proof is omitted in this extended abstract.

Fact 3.1.1 *For any $x, y \geq 0$ and $c \in (0, 1)$,*

$$x \cdot \log_2 \frac{x}{c} + y \cdot \log_2 \frac{y}{1-c} \geq (x + y) \cdot \log_2(x + y).$$

For a subset $R \subseteq \{1, \ldots, r\}$ set

$$\text{differ}(R) = \{j \mid \exists i_1, i_2 \in R : M[i_1, j] \neq M[i_2, j]\}.$$

Now, we are ready to analyse the properties of our weight assignment.

Lemma 7. **(a)** *For each* $(i, j) \in \{1, \ldots, r\} \times \{1, \ldots, c\}$,

$$\text{weight}(i, j) \geq 0.$$

(b) *For each* i $(1 \leq i \leq r)$

$$\sum_{j=1}^{c} \text{weight}(i, j) = \log_2 r.$$

(c) For any $R \subseteq \{1, \ldots, r\}$,

$$\sum_{j \in \text{differ}(R)} \sum_{i \in R} \text{weight}(i, j) \geq |R| \cdot \log_2 |R|.$$

Proof. **(a)** is immediate by construction. We verify **(b)** by induction on r. The basis for $r = 1$ is trivial. For the inductive step we can assume without loss of generality that column 1 is not a monochromatic column. Let I_0 (resp. I_1) be the set of those rows with a zero (resp. one) in column 1 and assume that $|I_0| = c \cdot r$.

We apply the induction hypothesis to the rows in I_0 and I_1. For a row $i \in I_0$ we obtain

$$\sum_{j=2}^{c} \text{weight}(i, j) = \log_2(c \cdot r).$$

But $\text{weight}(i, 1) = \log_2(\frac{1}{c})$ and

$$\sum_{j=1}^{c} \text{weight}(i, j) = \log_2(\frac{1}{c}) + \log_2(c \cdot r) = \log_2 r.$$

The claim follows with a symmetric argument for the rows in I_1.

We apply induction on the size of R to verify part **(c)**. The basis for $|R| = 1$ is again trivial. We assume for the inductive step that column 1 is not monochromatic for the rows in R. Hence R splits into the subsets R_0 and R_1 of those rows in R with value zero (resp. one) in column 1. For I the set of *all* rows and I_0 the set of *all* rows with a zero in column 1 we assume that $|I_0| = c \cdot |I|$. Since we can apply the induction hypothesis on R_0 and R_1, we obtain

$$\begin{aligned}
&\sum_{j \in \text{differ}(R)} \sum_{i \in R} \text{weight}(i, j) \\
&= \sum_{i \in R} \text{weight}(i, 1) + \sum_{j \in \text{differ}(R), j \neq 1} \sum_{i \in R} \text{weight}(i, j) \\
&\geq |R_0| \cdot \log_2(\frac{1}{c}) + |R_1| \cdot \log_2(\frac{1}{1-c}) + |R_0| \cdot \log_2(|R_0|) + |R_1| \cdot \log_2(|R_1|)
\end{aligned}$$

Thus the claim follows from Fact 3.1.1.

Assume we have a one-way Las Vegas protocol P' for a Boolean function f represented by the matrix $M(f)$ with r pairwise different rows. Let the function weight be defined for $M(f)$ with the above three properties. Then we obtain a deterministic one-way protocol $P \in \{P_1, P_2, \ldots, P_m\}$ such that

(*) the sum of all weights of entries of $M(P)$ with value 2 is at most one half of the sum of all weights (i.e., at most $\frac{1}{2} \cdot \sum_{i=1}^{r} \sum_{j=1}^{c} \text{weight}(i,j) = \frac{r}{2} \cdot \log_2 r$ with property (b) of the weight assignment).

This follows, since for every input the output of P' is equal to 2 with probability at most one half. The deterministic protocol partitions the set of all rows of $M(f)$ into classes $R_1, \ldots, R_k$ of identical rows (after twoing). By property (c) of the function weight we obtain for any class R_s

$$\sum_{j \in \text{differ}(R_s)} \sum_{i \in R_s} \text{weight}(i,j) \geq |R_s| \cdot \log_2 |R_s|.$$

The quantity on the left hand side is a lower bound for the weight of all entries of $M(R_s)$ with value 2. Moreover observe that, for $x = \sum_{i=1}^{k} x_i$,

$$\sum_{i=1}^{k} x_i \log_2 x_i \geq x \log_2 x - x \log_2 k.$$

(The convex function $y \log_2 y$ is minimized for $x_1 = \cdots = x_k = \frac{x}{k}$, assuming $x = \sum_{i=1}^{k} x_i$.) And hence the sum of weights of entries of $M(P)$ with value 2 is at least

$$\sum_{i=1}^{k} |R_i| \log_2 |R_i| \geq r \log_2 r - r \cdot log_2 k.$$

But with (*)

$$\frac{r \log_2 r}{2} \geq \sum_{i=1}^{k} |R_i| \log_2 |R_i| \geq r \log_2 r - r \cdot log_2 k$$

and hence $k \geq \sqrt{r}$. In other words, $M(P)$ has at least $\sqrt{r}$ different rows (resp. the deterministic protocol P has to consist of at least $\sqrt{r}$ messages).

3.2 Sketch of the Proof of Theorem 3

Let $L \subseteq \Sigma^*$ be a regular language. We define an infinite Boolean matrix M_L whose rows and columns are labeled by words from Σ^*. We set $M_L(x,y) = 1$ iff $xy \in L$. Obviously we can again consider one-way communication protocols for M_L . The complexity of such a protocol D is the *number* $c(D)$ of different messages of the protocol D. The **counting communication complexity of L** is

$$\mathbf{cc}_1^*(\boldsymbol{L}) = \min\{c(D) \mid D \text{ is a one–way uniform protocol for } M_L\}.$$

Similarly, one can define Las Vegas one–way communication protocols for M_L and the **Las Vegas counting communication complexity of L – lvcc_1^* (L).** Our proof utilizes the two following crucial facts.
Claim 1 For every regular language L

$$\mathrm{cc}_1^*(L) = \mathrm{s}(L).$$

Claim 2 For every regular language L

$$\mathrm{lvs}(L) \geq \mathrm{lvcc}_1^*(L).$$

Now, it is sufficient to prove that $\mathrm{lvcc}_1^*(L) \geq \sqrt{\mathrm{cc}_1^*(L)}$.

A first problem is the infinite size of $M(L)$. But it suffices to only consider the finitely many different rows of $M(L)$. By picking one column from the finitely many classes of identical columns we then arrive at a finite submatrix M'_L of $M(L)$. This submatrix is equivalent to $M(L)$ from the perspective of one-way communication complexity.

Finally, we apply Theorem 1 to M'_L and obtain that any Las Vegas one-way protocol computing M'_L uses at least $\sqrt{\mathrm{cc}_1^*(L)}$ different messages.

□

3.3 Proof of Theorem 5

Let $M_1, M_2, M_3, \ldots$ be a sequence of all oracle Turing machines such that the running time of machine M_i on each input of length n is bounded by $p_i(n) = n^i + i$, and M_i on each input of length n can make at most $if(n)$ binary nondeterministic choices. Let $f(n)$ be bounded by the polynomial $n^m + m$ for some $m > 0$. Choose a sufficiently large $k > 0$ so that

$$2^{ik} > \log i + i(2^{im} + m) + i^2 + 3 \qquad (1)$$

for each $i \geq 1$. Let $p(n) = n^k$.

To construct the desired language L and the oracle B, first we set a natural number n_i and define B_i for each $i \geq 0$. Let $B_0 = \emptyset$ and let $n_i = 2^i$ for each $i \geq 0$. Having defined B_{i-1}, we introduce B_i as follows. Let

$$U_i = \{0^{n_i}u | u \in \{0,1\}^{p(n_i)}, 0^{n_i}u \text{ is not queried by } M_j^{\bigcup_{l<j} B_l} \text{ on } 0^{n_j} \text{ for each } j \leq i\},$$

and, let

$$V_i = \{0^{n_i}v | v \in \{0,1\}^{p(n_i)+1}, 0^{n_i}v \text{ is not queried by } M_j^{\bigcup_{l<j} B_l} \text{ on } 0^{n_j} \text{ for each } j \leq i\}.$$

If 0^{n_i} is accepted by $M_i^{\bigcup_{j<i} B_j}$ then set $B_i = V_i$ else set $B_i = U_i$.
Let $B = \bigcup_i B_i$ and let $L = \{0^{n_i} | i \geq 1, 0^{n_i} \text{ is not accepted by } M_i^B\}$. Hence, $L \notin \beta_f^B$.
Now, let us show that $L \in R^B \cap \text{co-}R^B$. To do so, we need the following claim and we also need to bound the cardinality of the U_i's and the V_i's.

Claim 1. *0^{n_i} is accepted by $M_i^{\bigcup_{j<i} B_j}$ if and only if 0^{n_i} is accepted by M_i^B for each $i \geq 1$.*

Proof: Since each B_l is either U_l or V_l, there is no computational path of $M_i^{\bigcup_{j<i} B_j}$ on 0^{n_i} with a query in B_l for $l \geq i$ (see the definition of U_l and V_l above). □

Let X be any oracle. One can easily observe that the number of queries of M_j^X on input 0^n is at most $2^{jf(n)}p_j(n)$. Thus, we have by (1) for each $i \geq 1$ (recall $p(n) = n^k$, $n_j = 2^j$, $p_j(n) = n^j + j$ and $f(n) \leq n^m + m$),

$$\begin{aligned}
|U_i| &\geq 2^{p(n_i)} - \sum_{j=1}^{i} 2^{jf(n_j)} p_j(n_j) \\
&\geq 2^{2^{ik}} - \sum_{j=1}^{i} 2^{j(2^{jm}+m)}(2^{j^2}+j) \\
&\geq 2^{2^{ik}} - i2^{i(2^{im}+m)}(2^{i^2}+i) \\
&\geq 2^{2^{ik}} - 2^{\log i} 2^{i(2^{im}+m)} 2^{i^2+1} \\
&\geq 3.2^{2^{ik}-2} \\
&= (3/4)2^{p(n_i)}.
\end{aligned}$$

Similarly, one can show that $|V_i| \geq (3/4)2^{p(n_i)+1}$ for each $i \geq 1$.

Now we are ready to prove that $L \in R^B$. Let us consider a probabilistic oracle Turing machine M operating on input x as follows. If x is not of the form 0^{n_i}, then M rejects x. If $x = 0^{n_i}$ for some $i \geq 1$, then M guesses a string $y \in \{0,1\}^{p(n_i)}$ and asks the oracle whether $0^{n_i}y \in B$. M accepts x if and only if $0^{n_i}y \in B$. Now our goal is to show that M^B accepts inputs in L with probability at least $3/4$, and that there is no accepting path of M^B on inputs not in L. Choose an arbitrary input x. There are three cases to be considered.

Case 1. $x = 0^{n_i} \in L$. By the definition of L above, 0^{n_i} is not accepted by M_i^B. By Claim 1, 0^{n_i} is not accepted by $M_i^{\bigcup_{j<i} B_j}$. By the definition of B_i above, $B_i = U_i$. Each string $0^{n_i}y$ queried by M^B on input 0^{n_i} belongs to $U_i = B_i \subseteq B$ with probability at least $3/4$, since $|U_i| \geq (3/4)2^{p(n_i)}$ and y is arbitrarily chosen string in $\{0,1\}^{p(n_i)}$ (see above). Thus, each input 0^{n_i} is accepted by M^B with probability at least $3/4$.

Case 2. $x = 0^{n_i} \notin L$. One can prove as in Case 1, that $B_i = V_i$. Any string $0^{n_i}y$ queried by M^B on input x cannot belong to B_i $(= V_i)$, since the length of $0^{n_i}y$ is $n_i + p(n_i)$, but V_i contains only strings of length $n_i + p(n_i) + 1$ (see the definition of V_i above). Similarly, $0^{n_i}y$ cannot belong to any B_j with $j \neq i$, since the length of $0^{n_i}y$ is $n_i + p(n_i) = 2^i + 2^{ik}$, but each B_j with $j \neq i$ is either U_j or V_j, and hence it contains only strings of length either $n_j + p(n_j) = 2^j + 2^{jk}$ or $n_j + p(n_j) + 1 = 2^j + 2^{jk} + 1$. Thus, $0^{n_i}y$ cannot belong to $B = \bigcup_j B_j$. Hence, there is no accepting computational path of M^B on x.

Case 3. x is not of the form 0^{n_i}. In such a case, M^B rejects x (see above). This completes the proof that $L \in R^B$.

The proof that $L \in$ co-R^B is similar and we omit details here. Note that instead of M one can use a machine M' that accepts any input not of the form 0^{n_i}, and M' guesses a string $y \in \{0,1\}^{p(n_i)+1}$ on input 0^{n_i} and accepts it if and only if $0^{n_i}y \in V_i$.

This completes the proof of the theorem.

References

1. Aho, A.V., Hopcroft, J.E., Yannakakis, M.: On notions of information transfer in VLSI circuits.*In: Proc. 15th Annual ACM STOC, ACM* 1983, 133–139.
2. Bovet,D.P., Crescenzi,P.: Introduction to the Theory of Complexity. Prentice Hall 1994.
3. Diaz, J. Torán, J.: Classes of bounded nondeterminism. *Mathematical Systems Theory,* **23** (1990), 21-32.
4. Freivalds, R.: Probabilistic two–way machines. *Lecture Notes in Computer Science* **118,** Springer–Verlag, Berlin 1981, 33–45.
5. Hromkovič, J., Schnitger, G.: On the power of the number of advice bits in nondeterministic computations. *Proc. ACM STOC'96,*ACM 1996, pp. 551-560.
6. Meyer, A.R., Fischer, M.J.: Economies of description by automata, grammars and formal systems. *In: Proceedings 12th SWAT Symp.* 1971, 188–191
7. Mehlhorn,K., Schmidt,E.: Las Vegas is better than determinism in VLSI and distributed computing. *Proc. 14th ACM STOC'82,* ACM 1982, pp. 330-337.
8. Yao, A.C.: Some complexity questions related to distributed computing. *In: Proc. 11th Annual ACM STOC, ACM* 1981, 308–311.
9. Csiszar, I., Körner, J.: Information theory: coding theorems for discrete memeoryless systems, *Academic Press,* 1986.

Interactive Proof Systems with Public Coin: Lower Space Bounds and Hierarchies of Complexity Classes

Maciej Liśkiewicz*

International Computer Science Institute
1947 Center Street, Berkeley, California 94704, USA
E-mail: liskiewi@icsi.berkeley.edu

Abstract. This paper studies small space-bounded interactive proof systems (IPSs) using public coin tosses, respectively Turing machines with both nondeterministic and probabilistic states, that works with bounded number of rounds of interactions. For this model of computations new impossibility results are shown. As a consequence we prove that for sublogarithmic space bounds, IPSs working in k rounds are less powerful than systems of $2k^{k-1}$ rounds of interactions. It is well known that such a property does not hold for polynomial time bounds. Babai showed that in this case any constant number of rounds can be reduced to 2 rounds.

1 Introduction

Interactive proof systems (IPSs) working in constant space seems to be very powerful devices. Dwork and Stockmeyer ([DwSt92]) showed that any language recognized in deterministic exponential time has an IPS, where the verifier is a probabilistic finite automaton. The power of computation of constant-space-bounded IPSs becomes more realistic when the random moves of the verifier are known to the prover, as it is the case for Arthur-Merlin games, resp. for Turing machines with both nondeterministic and probabilistic states. Condon ([Co89]) proved that any language recognized by such system is in P even when the verifier works in logarithmic space. Hence the difference between the privet and the public coin tossing is significant. This property does not hold for the corresponding polynomial time classes (see [GoSi86]).

In their paper [DwSt92], Dwork and Stockmeyer showed some further separating results proving a number of strong lower bounds for probabilistic finite automata and for small space-bounded IPSs tossing public coin. They proved for example that the language

$$\texttt{CENTER} := \{w0x : w, x \in \{0,1\}^* \text{ and } |w| = |x|\},$$

* On leave of Institute of Computer Science, University of Wrocław. The research was supported by KBN Grant 2 P301 034 07.

Reischuk, Morvan (Eds.): STACS'97 Proceedings, LNCS 1200

cannot be recognized by a sublogarithmic space-bounded probabilistic Turing machine with any error probability $\epsilon < 1/2$. On the other hand, there exists a constant-space-bounded interactive proof system (with public coin flips) for CENTER: in [DwSt92] a protocol for this language is given. Thus at least in case of sublogarithmic space bounds interactive proof systems enable the verifier to accept something it cannot accept on its own. It seems, however, that to accept CENTER in constant space, a huge number of rounds of interactions is crucial. The protocol of [DwSt92], for example, works in exponential number of rounds and it is unknown if the language can be recognized in smaller number of interactions. Hence, the lower and upper bound results of [DwSt92] prove that 1 probabilistic round is less powerful than exponential number of rounds and it is open whether there is also a difference in power of computations between IPSs of R and resp. R' interactions for any function R and R', with $1 \leq R < R' \leq \exp$. In the paper we will prove that for space bounds in $\mathcal{SUBLOG} := \Omega(\text{llog}) \cap o(\log)$ the problem has a positive answer for any function $R(n) \leq k$ and $R'(n) \geq 2k^{k-1}$, where k is an arbitrary constant. Here llog denotes the logarithmic function log iterated twice.

We refer to TMs which have both probabilistic and nondeterministic states as stochastic Turing machines (STMs). Denote by $AMSpace(S)$ the set of languages that can be accepted by STMs, equivalently by public coin IPSs, with space S. Let $MA_kSpace(S)$, resp. $AM_kSpace(S)$, denote the set of languages that can be accepted by such machines making at most $k-1$ alternations between nondeterministic and probabilistic configurations, equivalently making at most $k-1$ interaction, and starting in nondeterministic, resp. probabilistic, mode (for such a machine we also say that it works in k rounds). $BPSpace(S)$ denotes the class of languages accepted by S-space-bounded probabilistic TMs with bounded error. Then language CENTER yields the separations $BPSpace(S) = AM_1Space(S) \subset AMSpace(S)$ for any $S \in o(\log)$. In this paper we strengthen these result for $S \in \mathcal{SUBLOG}$ showing an infinite hierarchy of $AM_kSpace(S)$ complexity classes between $AM_1Space(S)$ and $AMSpace(S)$.

Let $\texttt{BIN}(m) := \text{bin}(0)\#\,\text{bin}(1)\#\,\text{bin}(2)\#\ldots\#\,\text{bin}(m)$, where $bin(i)$ is the binary representation of the number i. We define for any positive integer k

$$\texttt{PATTERN}_k := \{W_1\#\ldots\#W_k\#u\#\texttt{BIN}(2^d) : \; W_1,\ldots,W_k,u \in \{0,1,*\}^+, \; |u| = d \text{ for some } d \in \mathbb{N}, \text{ and } u \text{ is a substring of } W_i, \text{ for } i = 1,\ldots,k \} \; .$$

One of the main results of this paper says that for any integer $k \geq 2$

$$\texttt{PATTERN}_{k^{k-1}} \notin AM_kSpace(o(\log)) \cup MA_kSpace(o(\log)) \; . \qquad (1)$$

To prove these lower bounds we show that any STM M accepting an input $W_1\#\ldots\#W_{k^{k-1}}\#u\#\texttt{BIN}(2^d) \in \texttt{PATTERN}_{k^{k-1}}$ in k rounds, alternates between a nondeterministic and probabilistic move on some block W_i with very small probability. Using methods of [DwSt92] we choose the string W_i in such a way that M cannot distinguish in one round this specific string from some "wrong" string. Hence, if one replaces W_i by the wrong string, M to detect this difference has to alternate on this string; otherwise it behaves as previously. But because

M alternates on the i-th block with very small probability hence it can detect the difference with small probability, too.

From (1) it follows that $\mathtt{PATTERN}_2$ cannot be accepted by 2-round STMs. In [LiRe96b] we showed a little bit more for such machines. Namely, we proved that any $o(\log)$-space-bounded STM cannot recognize even $\mathtt{PATTERN}_1$ in 2 rounds when starting in probabilistic mode. On the other hand this language is in $MA_2Space(\mathrm{llog})$. Hence an optimal lower bound on the number of rounds has been founded in this case. Using a method of Freivalds ([Fr79]), we generalize in this paper the upper bound for recognizing $\mathtt{PATTERN}_1$ as follows:

$$\mathtt{PATTERN}_k \in MA_{2k}Space(\mathrm{llog}) , \tag{2}$$

for any $k \geq 2$. Note that the above upper bound does not match our lower bound (1). In fact the gap is rather large.

Therefore, we obtain the following separations:

Theorem 1. *For any integer $k \geq 2$ it holds that*

$$MA_{2k^{k-1}}Space(\mathrm{llog}) \not\subseteq AM_kSpace(o(\log)) \cup MA_kSpace(o(\log) .$$

This implies that round/alternation hierarchy for sublogarithmic space-bounded AM_k machines is infinite, similar as for standard alternating TMs (see e.g. [LiRe96a]):

Corollary 2. *For any function $S \in \mathcal{SUBLOG}$ and any integer $k \geq 2$*

$$AM_kSpace(S) \cup MA_kSpace(S) \subset MA_{2k^{k-1}}Space(S) .$$

This property does not hold for the corresponding polynomial time classes $AM_kTime(\mathcal{POL})$. Here, $\mathcal{POL}$ denotes the set of all polynomials. In [Ba85] Babai has shown that for polynomial time any constant number of rounds can be reduced to two rounds, that is

$$AM_2Time(\mathcal{POL}) = AM_kTime(\mathcal{POL}) ,$$

for any integer $k \geq 2$.

In the paper it is also shown an astonishing lower bound on space and on the number of rounds for recognizing the complement of $\mathtt{PATTERN}_1$. Though the language is in $MA_2Space(\mathrm{llog})$, we prove that that for any integer function $R \in O(\log / \mathrm{llog})$, its complement

$$\overline{\mathtt{PATTERN}_1} \notin AM_RSpace(o(\log)) . \tag{3}$$

Obviously, this result implies the following

Theorem 3. *For any $S \in \mathcal{SUBLOG}$, and for any integer function R, with $3 \leq R \in O(\log / \mathrm{llog})$ the classes $AM_RSpace(S)$ and $MA_RSpace(S)$ are not closed under complement.*

The remainder of this paper is organized as follows. In Section 2 some definitions and notions are introduced. Section 3 contains the proofs of our lower bound results (1) and (3). In Section 4 a machine for $\mathtt{PATTERN}_k$ is described what proves (2).

2 Definitions

A computation of a stochastic machine M on an input X can be described by a computation tree. In a probabilistic state a stochastic machine M chooses among the successor configurations with equal probability. To define acceptance of X, for each nondeterministic configuration one chooses a successor that maximizes the probability of reaching an accepting leaf. The acceptance probability of X is then given by the acceptance probability of the starting configuration in this truncated tree. M accepts a language L in space S if for all $X \in L$, the probability that M accepts X is more than $3/4$, every $X \notin L$ is accepted with probability less than $1/4$, and M never uses more than $S(|X|)$ space. We say that M accepts the language L in R rounds if with any input X M makes at most $R(|X|) - 1$ alternations between nondeterministic and probabilistic configurations. The above machines are equivalent the to S-space-bounded IPSs using public coin tosses and working with R rounds of interactions between verifier and prover (see [DwSt92] for a formal definition of S-space-bounded IPSs).

Let M be an STM. We will assume that M is equipped with a two-way read-only input tape and a single read-write work tape. A *memory state* of M is an ordered triple $\alpha = (q, u, i)$, where q is a state of M, u a string over the work tape alphabet, and i a position in u (the location of the work tape head). By $|\alpha|$ we denote the length of the string u of the memory state α. A *configuration* of M on an input X is a pair (α, j) consisting of a memory state α and a position j with $0 \leq j \leq |X| + 1$ of the input head. $j = 0$ or $j = |X| + 1$ means that this head scans the left, resp. the right end-marker. Let $h(\alpha, j) \stackrel{\text{def}}{=} j$ be the function describing the input head position for an configuration (α, j). We say that a configuration (α, j), with $\alpha = (q, u, i)$, is nondeterministic, probabilistic (or random), accepting or rejecting, according to q.

We call a phase of computation of M a *probabilistic* (or *random*) *round* if M starts the phase in a probabilistic configuration and makes only probabilistic steps during the phase and finally reaches a non-probabilistic configuration. Analogously we call a phase of computation a *nondeterministic round* if M starts in a nondeterministic configuration and performing only nondeterministic steps during the phase reaches probabilistic, accepting or rejecting state. Let for a probabilistic configuration c and a nondeterministic, accepting or rejecting configuration configuration c'

$$\mathcal{R}[c, c', X]$$

denote the probability that M with X on its input tape and starting in c reaches the configuration c' in a probabilistic round. Let for a nondeterministic configuration c

$$\mathcal{N}(c, c', X)$$

be 1 if M starting in c on the input X reaches the configuration c' in a nondeterministic round; otherwise $\mathcal{N}(c, c', X) = 0$. Denote by

$$\mathcal{A}_k[c, X]$$

the probability that M accepts the input X in k or less rounds starting in configuration c. Formally let for the accepting configuration c, $\mathcal{A}_0[c,X] := 1$ and for the rejecting c, $\mathcal{A}_0[c,X] := 0$. Then for any $k \geq 1$ if c is probabilistic, then

$$\mathcal{A}_k[c,X] \ := \ \sum_{\substack{c'-\text{nondet.}\\ \text{or accept}}} \mathcal{R}[c,c',X] \cdot \mathcal{A}_{k-1}[c',X] \,,$$

and if c is nondeterministic, then

$$\mathcal{A}_k[c,X] \ := \ \max_{\substack{c'-\text{random}\\ \text{or accept}}} \ \{\mathcal{A}_{k-1}[c',X] \ : \ \mathcal{N}(c,c',X)\,\}.$$

Let $\mathcal{A}_k[X] := \mathcal{A}_k[c_0,X]$, where c_0 be the initial configuration of M.

3 Lower Bounds

In this section we give proofs of our impossibility results:

Theorem 4. *For any $S \in o(\log)$, an S-space-bounded STM cannot recognize*

(1) the language $\mathtt{PATTERN}_{k^{k-1}}$ in k rounds, for any integer $k \geq 2$, and
(2) the complement of $\mathtt{PATTERN}_1$ in R rounds, for any $R \in O(\log / \text{llog})$.

We start with definitions and technical preliminaries that were originally be showed in [DwSt92] for probabilistic Turing machines. Here we extend them for STMs.

The *word probabilities* of M on a word Z over the input alphabet of M is defined as follows. A starting condition for the word probability is a pair $\langle \alpha, h \rangle$ where α is a probabilistic memory state of M and $h \in \{\text{Left}, \text{Right}\}$ what means that M starts according to the value of h on the leftmost or on the rightmost symbol of Z in memory state α. A stopping condition for the word probability is either:

- a pair $\langle \alpha, h \rangle$ as above meaning that in a probabilistic round the input head falls off according to h the leftmost, resp. the right most symbol of Z with M in memory state α,
- "Accept" meaning that M in a probabilistic round halts in the accepting state before the input head falls off either end of Z,
- "Reject" meaning that M in a probabilistic round halts in the rejecting state before the input head falls off either end of Z,
- "Alter" meaning that within Z M alternates from a probabilistic to a nondeterministic round, or
- "Loop" meaning that within Z the probabilistic computation of M loops forever.

For each starting condition σ and each stopping condition τ, let $p(Z,\sigma,\tau)$ be the probability that stopping condition occurs given that M started in starting condition σ on Z.

Computations of a probabilistic round of M are modeled by Markov chains with finite state space, say $1,2,\ldots,s$ for some s. A particular Markov chain is completely defined by its matrix $R = \{r_{ij}\}_{1\leq i,j\leq s}$ of transition probabilities. If the Markov chain is in state i, then it next moves to state j with probability r_{ij}. The chains we consider have the designated starting state, say, state 1, and some set T_R of trapping states, so $r_{tt} = 1$ for all $t \in T_R$. For $t \in T_R$, let $p^*[t,R]$ denote the probability that Markov chain R is trapped in state t when started in state 1.

Let $\beta \geq 1$. Say that two numbers r and r' are β-close if either $r = r' = 0$, or $r > 0$, $r' > 0$, and $\beta^{-1} \leq r/r' \leq \beta$. Two Markov chains $R = \{r_{ij}\}_{1\leq i,j\leq s}$ and $R' = \{r'_{ij}\}_{1\leq i,j\leq s}$ are β-close if r_{ij} and r'_{ij} are β-close for all pairs i,j.

Lemma 5 ([DwSt92]). *Let R and R' be two s-state Markov chains which are β-close, and let t be a trapping state of both R and R'. Then $p^*[t,R]$ and $p^*[t,R']$ are β^z-close where $z = 2s$.*

We characterize a word Z according to a nondeterministic round of M on Z by *word transitions*. As previously, a starting condition for the word transition is a pair $\langle\alpha,h\rangle$ where α is a nondeterministic memory state of M and $h \in \{\text{Left},\text{Right}\}$. A stopping condition for the word transition is either a pair $\langle\alpha,h\rangle$ as above, "Accept" meaning that M in a nondeterministic round halts in the accepting state before the input head falls off either end of Z, or "Reject" meaning that M in a nondeterministic round halts in the rejecting state before the input head falls off either end of Z. For each starting condition σ and each stopping condition τ, the word transition $t(Z,\sigma,\tau)$ equals to 1 if M starting in σ can reach τ on Z during a nondeterministic round; otherwise it is 0.

3.1 Constant Number of Rounds

In this section a proof for Theorem 4(1) will be given. Let $k \geq 2$ be arbitrary integer and assume that M is a STM, of space complexity $S \in o(\log)$ that works in k rounds. Moreover, let n be sufficiently large integer of the form 2^d. In the section we will consider the input words of the form

$$w_{1,1} * w_{1,2} * \ldots * w_{1,n} \# \ldots \# w_{k^{k-1},1} * \ldots * w_{k^{k-1},n} \# u \# \mathtt{BIN}(n)\ , \qquad \text{(i)}$$

with $w_{i,j}, u \in \{0,1\}^d$. It will be proved that if M accepts with high probability an input of this form that belongs to $\mathtt{PATTERN}_{k^{k-1}}$ than M has to accept with probability exceeding $1/4$ an input which does not belong to the language.

Denote by N the length of considered inputs, i.e. let

$$N := k^{k-1}(nd+n) + d + 1 + |\mathtt{BIN}(n)|\ ,$$

and let $\text{Vol}(N)$ be the number of possible memory states of the machine M on input words of length N. Note that $\text{Vol}(N) \leq 2^{O(S(N))}$. We will consider the word probabilities and the word transitions restricting the starting and the stopping conditions generated by memory states α to states with $|\alpha| \leq S(N)$. Let us fix some order of the pairs (σ, τ) of starting and stopping conditions for word probabilities as well as some order of the pairs (σ, τ) for word transitions. Let $\mathbf{p}(Z)$ be the vector of the word probabilities and let $\mathbf{t}(Z)$ be the vector of the word transitions according to these orderings. Define

$$\mu := 2^{-\sqrt{n}} . \tag{ii}$$

Lemma 6. *There exist two words* $W := w_1 * w_2 * \ldots * w_n \#$ *and* $\overline{W} := \overline{w}_1 * \overline{w}_2 * \ldots * \overline{w}_n \#$, *with* $w_j, \overline{w}_j \in \{0,1\}^d$, *for* $j = 1, \ldots, n$, *and* $\{w_1, w_2, \ldots, w_n\} \setminus \{\overline{w}_1, \overline{w}_2, \ldots, \overline{w}_n\} \neq \emptyset$ *such that* $\mathbf{t}(W) = \mathbf{t}(\overline{W})$ *and* $\mathbf{p}(W)$ *and* $\mathbf{p}(\overline{W})$ *are componentwise* 2^{μ}*-close.*

To prove the lemma one can adapt a counting argument of [DwSt92]. Now let us fix two words W and $\overline{W}$ as in the lemma above. Because word transitions of W and $\overline{W}$ are the same it means that for any string X and Y, with $|XWY| = N$, M cannot distinguish W from $\overline{W}$ in one nondeterministic round when starting on the prefix X or on the suffix Y. For a probabilistic round an analogous property holds for γ defined as follows

$$\gamma := 2^{2\mu(10\mathrm{Vol}(N)+20)} . \tag{iii}$$

Lemma 7. *Let* XWY *be an input string of the length* N, *and let* c *and* c' *be configurations such that* $h(c), h(c') \in [0..|X|] \cup [|XW| + 1..N + 1]$. *Then it holds that*

(a) $\mathcal{N}(c, c', XWY) = \mathcal{N}(c, c', X\overline{W}Y)$ *and* $\mathcal{A}_1[c, XWY] = \mathcal{A}_1[c, X\overline{W}Y]$, *if* c *is nondeterministic, and*
(b) the probabilities $\mathcal{R}[c, c', XWY]$ *and* $\mathcal{R}[c, c', X\overline{W}Y]$, *respectively,* $\mathcal{A}_1[c, XWY]$ *and* $\mathcal{A}_1[c, X\overline{W}Y]$, *are* γ*-close if* c *is probabilistic.*

Proof. A proof of (*a*) is straightforward and we will omit it here at all. For (*b*) we will sketch only a proof for configurations with $h(c) \leq |X|$ and $h(c') \geq |XW|+1$. The other cases can be showed in a very similar way.

We will describe Markov chains R_W and $R_{\overline{W}}$ which model the probabilistic round of the machine M on inputs XWY and $X\overline{W}Y$, respectively, when M starts in configuration c. This configuration will correspond to the starting state of this Markov chains and configuration c' will correspond to their trapping state.

Let us denote the prefix of X of the length $h(c)$ by X_1 and let the remaining part of X will be denoted by X_2. Similarly, let Y_1 be the prefix of Y such that M in configuration c' reads the last symbol of Y_1 and let Y_2 be a suffix such that $Y = Y_1\ Y_2$. Each Markov chain we describe has $s = 10 \cdot \mathrm{Vol}(N) + 20$ states. The first $10 \cdot \mathrm{Vol}(N)$ states have the form $\langle \alpha, h \rangle$, where α is a memory state of the length bounded by $S(N)$ and h is a position of the first or the last symbol of words $\$X_1$, X_2, W, Y_1, $Y_2\$$ on the input tape containing the string

$X_1X_2WY_1Y_2$ (remember that the left end-marker has position 0, and the last one – the position $N+1$). An intuitive meaning of a state $\langle\alpha, h\rangle$ of the chain R_W is: start M in configuration (α, h) on the input $X_1X_2WY_1Y_2$. The meaning of state $\langle\alpha, h\rangle$ for the chain $R_{\overline{W}}$ is analogous. The next twenty states are the following: $\text{Accept}_j, \text{Reject}_j, \text{Alter}_j$ and Loop_j, for $j = 1, \ldots, 5$. For chain R_W they mean that the probabilistic computation of M accepts, rejects, alternate or loops forever within $\$X_1$, X_2, W, Y_1, $Y_2\$$, respectively. For $R_{\overline{W}}$ the meaning is analogous.

The transition probabilities r_{ij} of R_W for non-trapping states i are obtained from word probabilities of M on the substrings: $\$X_1$, X_2, W, Y_1, $Y_2\$$. More precisely, the transitions r_{ij} such that states i, j are applied both to the same substring are equal to an appropriate word probabilities of M on this substring. E.g. if $i = \langle\alpha, |X_1X_2W|\rangle$ and $j = \langle\beta, |X_1X_2|+1\rangle$ then $r_{ij} = p(W, \langle\alpha, \text{Right}\rangle, \langle\beta, \text{Left}\rangle)$ since the position $|X_1X_2W|$ and $|X_1X_2|+1$ means the rightmost resp., the leftmost symbol of the substring W. Remaining values r_{ij}, i.e. transitions for states i, j connecting with two different substrings are defined as follows. If $i = \langle\alpha_1, h_1\rangle$ and $j = \langle\alpha_2, h_2\rangle$ then the transition equals to the probability that M reaches in one step the configuration (α_2, h_2) starting in (α_1, h_1). Otherwise $r_{ij} = 0$. The transition probabilities of $R_{\overline{W}}$ are obtained analogously.

The states $\text{Accept}_j, \text{Reject}_j, \text{Alter}_j, \text{Loop}_j$, for $j = 1, \ldots, 5$ as well as all states $\langle\alpha, h\rangle$, with non-probabilistic α, are defined to be trapping states for both R_W and $R_{\overline{W}}$. For any trapping state t the transitions $r_{t,t}$ are defined to be 1.

Let the memory state of c (c') be α_c ($\alpha_{c'}$, resp.). Then the initial state of each chain is $\langle\alpha_c, h(c)\rangle$. Note that according to the definition of the trapping states, $\langle\alpha_{c'}, h(c')\rangle$ is a trapping state of both Markov chains.

W.l.o.g. let us assume that the first symbols of W and $\overline{W}$ are the same. Remember that the last symbols of the both words are equal, too. Hence a transition for any pair of states i, j connecting with two different substrings, is the same in considered chains. From this and from the fact that $\mathbf{p}(W)$ and $\mathbf{p}(\overline{W})$ are componentwise 2^μ-close we have that chains R_W and $R_{\overline{W}}$ are 2^μ-close. Now, using Lemma 5 we obtain that for the configuration c' the probabilities $p^*[\langle\alpha_{c'}, h(c')\rangle, R_W]$ and $p^*[\langle\alpha_{c'}, h(c')\rangle, R_{\overline{W}}]$ are $2^{2s\mu}$-close, what proves (b). □

Let for the strings $W = w_1 * w_2 * \ldots * w_n\ \#$ and $\overline{W} = \overline{w}_1 * \overline{w}_2 * \ldots * \overline{w}_n\ \#$, w be a word such that $w \in \{w_1, w_2, \ldots, w_n\} \setminus \{\overline{w}_1, \overline{w}_2, \ldots, \overline{w}_n\}$. Define, for short, $W^j := \underbrace{W\ W\ \ldots\ W\ W}_{j \text{ times}}$, for any integer $j \geq 0$ and $\hat{w} := \#w\#\mathtt{BIN}(n)$.

Obviously $W^j\hat{w} \in \mathtt{PATTERN}_j$ but for any i, with $1 \leq i \leq j$, if in $W^j\hat{w}$ one replaces the i-th substring W by $\overline{W}$ then the new input string does not belong to $\mathtt{PATTERN}_j$ any more. Below we show that if M accepts in k rounds then there is an integer i such that the probability that M accepts the input $W^{i-1}\overline{W}W^{k^{k-1}-i}\hat{w}$ does not decrease drastically according to the probability that M accepts $W^{k^{k-1}}\hat{w}$.

Lemma 8 (Key). *There exists integer i, with $1 \leq i \leq k^{k-1}$, such that*

$$\mathcal{A}_k[\ W^{i-1}\overline{W}W^{k^{k-1}-i}\hat{w}\ ] \quad \geq \quad \gamma^{-\lceil k/2\rceil} \cdot (1 - 1/k)^{k-1} \cdot \mathcal{A}_k[\ W^{k^{k-1}}\hat{w}\ ]\ .$$

Proof. Proceeding by induction on the number of rounds r we will show a more general fact. Let U and V be words over the input alphabet of M such that the string $UW^{k^{r-1}}V$ is of the form (i) and let A be a set of configurations such that

(♠) all non-accepting configuration of A are either probabilistic or nondeterministic and for any $c \in A$, machine M on the input string $UW^{k^{r-1}}V$ scans, with the input head position $h(c)$, the prefix U or the suffix V, i.e. $h(c) \leq |U|$ or $h(c) \geq |UW^{k^{r-1}}| + 1$.

Additionally, let us assign to each configuration $c \in A$ a non-negative real p_c. The only condition we will assume is that $\sum_{c \in A} p_c \leq 1$. The intuitive meaning of the number p_c is: start the machine M in configuration c with probability p_c.

Fact. *For any r, with $1 \leq r \leq k$, there exists integer i_r, with $1 \leq i_r \leq k^{r-1}$, such that for $Z := UW^{k^{r-1}}V$ and $Z_r := UW^{i_r-1}\overline{W}W^{k^{r-1}-i_r}V$*

$$\sum_{c \in A} p_c \cdot \mathcal{A}_r[c, Z_r] \quad \geq \quad \gamma^{-\tilde{r}} \cdot (1 - 1/k)^{r-1} \cdot \sum_{c \in A} p_c \cdot \mathcal{A}_r[c, Z] \ ,$$

where $\tilde{r} = \lfloor r/2 \rfloor$ if A has no probabilistic states and $\tilde{r} = \lceil r/2 \rceil$, otherwise.

For $r := k$, $A := \{c_0\}$, where c_0 is the initial configuration, $p_{c_0} := 1$ and for the empty word U and $V := \hat{w}$ this fact proves the lemma.

Proof of Fact. We proceed by induction on r. Assume first that $r = 1$ and A does not contain probabilistic states. Then by Lemma 7(a) we have that $\mathcal{A}_1[c, U\overline{W}V] = \mathcal{A}_1[c, UWV]$ for any $c \in A$. For $r = 1$ and the set A that does not contain nondeterministic states, by Lemma 7(b) we conclude that the probabilities $\mathcal{A}_1[c, U\overline{W}V]$ and $\mathcal{A}_1[c, UWV]$ are γ-close. This means that $\mathcal{A}_1[c, U\overline{W}V] \geq \gamma^{-1}\mathcal{A}_1[c, UWV]$ for any $c \in A$. Hence for $r = 1$ the fact holds. Below we will prove that it holds for any $r \geq 1$.

Assume that A does not contain nondeterministic states. The symmetric case, i.e. that A does not have probabilistic states is similar and it will be omitted here. Let for the input Z, B be the set of all nondeterministic and accepting configurations. Define for any $c' \in B$ the real $p'_{c'} := \sum_{c \in A} p_c \cdot \mathcal{R}[c, c', Z]$. According to the definition of $\mathcal{A}$ it holds that

$$\sum_{c \in A} p_c \cdot \mathcal{A}_r[c, Z] = \sum_{c \in B} p'_c \cdot \mathcal{A}_{r-1}[c, Z] \ . \tag{iv}$$

Partition next the set B into $k + 1$ subsets $B_0, B_1, \ldots, B_k$ as follows: let for $j = 1, \ldots, k$

$$B_j := \{c \in B : \ |UW^{(j-1)k^{r-2}}| \ < \ h(c) \ \leq \ |UW^{jk^{r-2}}| \ \} \ ,$$

and let $B_0 := \{c \in B : \ h(c) \ \leq \ |U| \ \text{ or } \ |UW^{k^{r-1}}| \ < \ h(c) \ \}$. Clearly, by the pigeon-hole-principle, there exists integer i, with $1 \leq i \leq k$ such that

$$\sum_{c \in B_i} p'_c \cdot \mathcal{A}_{r-1}[c, Z] \ \leq \ (1/k) \sum_{c \in B} p'_c \cdot \mathcal{A}_{r-1}[c, Z] \ . \tag{v}$$

Apply now the inductive hypothesis for: $r-1$, $U' := UW^{(i-1)k^{r-2}}$, $V' := W^{k^{r-1}-ik^{r-2}}$, and for the set of configurations $A' := B \setminus B_i$. Note that for A' and the input $U'W^{k^{r-2}}V'$ the assumption (♠) is fulfilled. Hence, by the hypothesis there exists integer i_{r-1} such that for $Z_{r-1} := U'W^{i_{r-1}-1}\overline{W}W^{k^{r-2}-i_{r-1}}V'$ it holds that

$$\sum_{c \in B\setminus B_i} p'_c \cdot \mathcal{A}_{r-1}[c, Z_{r-1}] \geq \gamma^{-\tilde{r}'} \cdot (1-1/k)^{r-2} \cdot \sum_{c \in B\setminus B_i} p'_c \cdot \mathcal{A}_{r-1}[c, Z], \quad \text{(vi)}$$

where $\tilde{r}' = \lfloor (r-1)/2 \rfloor$ because $B \setminus B_i$ does not contain probabilistic states. Define $i_r := (i-1)\cdot k^{r-2} + i_{r-1}$. Obviously for this value i_r, words Z_{r-1} and Z_r are equal.

Now putting all of this together we conclude

$$\begin{aligned}
\sum_{c \in A} p_c \cdot \mathcal{A}_r[c, Z_r] &= \sum_{c\in A}\sum_{c' \in B} p_c \cdot \mathcal{R}[c, c', Z_r] \cdot \mathcal{A}_{r-1}[c', Z_r] && \text{by def. of } \mathcal{A} \\
&\geq \sum_{c' \in B\setminus B_i}\sum_{c\in A} p_c \cdot \mathcal{R}[c, c', Z_r] \cdot \mathcal{A}_{r-1}[c', Z_r] \\
&\geq \gamma^{-1} \sum_{c' \in B\setminus B_i}\sum_{c\in A} p_c \cdot \mathcal{R}[c, c', Z] \cdot \mathcal{A}_{r-1}[c', Z_r] && \text{by Lemma 7} \\
&= \gamma^{-1} \sum_{c' \in B\setminus B_i} p'_{c'} \cdot \mathcal{A}_{r-1}[c', Z_r] && \text{by def. of } p' \\
&\geq \gamma^{-\tilde{r}'-1}(1-1/k)^{r-2} \sum_{c' \in B\setminus B_i} p'_{c'} \cdot \mathcal{A}_{r-1}[c', Z] && \text{by (vi)} \\
&\geq \gamma^{-\tilde{r}'-1}(1-1/k)^{r-1} \sum_{c' \in B} p'_{c'} \cdot \mathcal{A}_{r-1}[c', Z] && \text{by (v)} \\
&= \gamma^{-\tilde{r}'-1}(1-1/k)^{r-1} \sum_{c \in A} p_c \cdot \mathcal{A}_r[c, Z] && \text{by (iv)}
\end{aligned}$$

what proves the fact. □

Now we are ready to prove Theorem 4(1). Let us assume that M is a STM accepting $\texttt{PATTERN}_{k^{k-1}}$ in $k \geq 2$ rounds and in sublogarithmic space S. Since $W^{k^{k-1}}\hat{w} \in \texttt{PATTERN}_{k^{k-1}}$, hence M has to accept $W^{k^{k-1}}\hat{w}$ with probability greater or equal to 3/4, which means that $\mathcal{A}_k[\, W^{k^{k-1}}\hat{w}\,] \geq 3/4$. From the Key Lemma we conclude, however, that there exists integer i, with $1 \leq i \leq k^{k-1}$, such that

$$\mathcal{A}_k[\, W^{i-1}\overline{W}W^{k^{k-1}-i}\hat{w}\,] \geq \gamma^{-\lceil k/2 \rceil}\cdot(1-1/k)^{k-1}\cdot 3/4 \geq \gamma^{-\lceil k/2 \rceil}\cdot 3/8 > 1/4,$$

since $\gamma^{-\lceil k/2 \rceil}$ tends to 1. But string $W^{i-1}\overline{W}W^{k^{k-1}-i}\hat{w}$ does not belong to $\texttt{PATTERN}_{k^{k-1}}$ – a contradiction.

3.2 Proof of Theorem 4(2)

Let W and $\overline{W}$ be the words as defined in the previous subsection. Let Z and $\bar{Z}$ be the strings W and $\overline{W}$, resp., where the # symbols are removed, e.g. let $Z := w_1 * w_2 * \ldots * w_n$ and $\bar{Z} := \overline{w}_1 * \overline{w}_2 * \ldots * \overline{w}_n$.

Lemma 9. *Let $r(n)$ be an integer function, with $r(n) \in O(\log n / \operatorname{llog} n)$. Then for any sufficiently large integer n there exists integer i, with $1 \le i \le n$, such that*

$$\mathcal{A}_{r(n)}[\ \bar{Z}^{i-1} Z \bar{Z}^{n-i} \hat{w}\ ] \quad \ge \quad \gamma^{-\lceil r(n)/2 \rceil} \cdot e^{-1} \cdot \mathcal{A}_{r(n)}[\ \bar{Z}^{n} \hat{w}\ ] .$$

Theorem 4(2) follows now straightforward from the above lemma.

4 Space Efficient Algorithm for PATTERN Languages

In this section we show for any integer $k \ge 1$ and for arbitrary small $\epsilon > 0$, a llog-space-bounded STM M that recognizes $\mathtt{PATTERN}_k$ with error probability ϵ. M works in $2k$ rounds starting in nondeterministic mode and in time bounded by a polynomial.

The machine M performs the following algorithm: Check deterministically at the beginning whether the input is of the form $W_1 \# W_2 \# \ldots \# W_k \# u \# \mathtt{BIN}(2^d)$, for some words $W_1, \ldots, W_k, u \in \{0, 1, *\}^+$, with $|u| = d$. Reject and stop if this condition does not hold. Otherwise let $i := 1$ and go to step 1 below.

1. Nondeterministically guess a substring w_i of the length d in W_i.
2. Randomly choose a prime q_i with $2 \le q_i \le d^2$ and then compute $r_i := n_{w_i} \bmod q_i$, where n_{w_i} denotes an integer with the binary representation w_i.
3. Reject and stop if $r_i \ne n_u \bmod q_i$; otherwise if $i = k$ accept and stop else increase i by 1 and go to step 1.

If for any i, with $1 \le i \le k$, the strings w_i and u are equal then of course $n_{w_i} = n_u \bmod q_i$ for any value q_i and machine M accepts correctly in step 3. If $w_i \ne u$ for some i, with $1 \le i \le k$, than it could happen that $n_{w_i} = n_u \bmod q_i$ and M reaches in step 3 the accepting state that is wrong. This event happens, however, with probability that tends to 0. Indeed. Since $|n_{w_i} - n_u| \le 2^d$ hence $n_{w_i} - n_u$ has at most d different prime divisors. On the other hand, M chooses from about $d^2/2 \ln d$ different primes at the beginning of step 2. So the probability that a wrong value q_i is chosen is at most $(2 \ln d)/d$.

Obviously, M uses $O(\operatorname{llog})$ space and works in $2k$ rounds.

5 Conclusions and Open Problems

In this paper separations were obtained for sublogarithmic $AM_k Space$ complexity classes. An interesting open problem is if our separations can be refined. Is it true that

$$AM_k Space(S) \subset AM_{k+1} Space(S) ,$$

for any integer k and sublogarithmic S?

How looks the hierarchy for at least logarithmic space bounds? Using a simple simulation of space-bounded NTMs by one-sided-error probabilistic TMs (see e.g. [Gil77] or the survey paper [Ma95]) one can show that $AM_2Space(\log) = AM_1Space(\log)$, that means the AM_2-class is quite weak in case of space bounds – contrary to time bounded classes. Is it also true that $AM_2Space(\mathcal{SUBLOG})$ is equal to $AM_1Space(\mathcal{SUBLOG})$?

What is the situation for space bounds S smaller than llog? The most interesting case seems to be space bounds restricted to constant functions. It is well known that $MA_1Space(\mathcal{CON})$ coincides with the class of regular languages. This result, however, does not extend to the class $AM_1Space(\mathcal{CON})$. Freivalds has shown the surprising result [Fr81] that COUNT $:= \{1^n01^m : n = m\}$, can be accepted by a probabilistic TM in constant space with an arbitrarily small constant for the error probability. Is there a language that separates $AM_1Space(\mathcal{CON})$ from $AM_kSpace(\mathcal{CON})$ classes, for some $k > 1$? Dwork and Stockmeyer showed that CENTER does not belong to $AM_1Space(\mathcal{CON})$ and that there exists a constant space interactive proof system for this language. Can their protocol be improved to make only constant number of rounds?

References

[Ba85] L. Babai, *Trading group theory for randomness,* in Proceedings of the 17th ACM Symposium on Theory of Computing, ACM Press, 1985, 421-429.

[Co89] A. Condon, *Computational model of games,* MIT Press, 1989.

[DwSt92] S. Dwork and L. Stockmeyer, *Finite state verifiers I: the power of interaction,* Journal of the ACM, 39, 1992, 800-828.

[Fr79] R. Freivalds, *Fast probabilistic algorithms,* in Proceedings of the 8th International Symposium on Mathematical Foundations of Computer Science, Springer-Verlag, Heidelberg, 1979, 57-69.

[Fr81] R. Freivalds, *Probabilistic 2-way machines,* in Proceedings of the 10th International Symposium on Mathematical Foundations of Computer Science, Springer-Verlag, Heidelberg, 1981, 33-45.

[Gil77] J. Gill, *Computational complexity of probabilistic Turing machines,* SIAM Journal on Computing, 7, 1977, 675-695.

[GoSi86] S. Goldwasser and M. Sipser, *Private coins versus public coins in interactive prove systems,* in Proceedings of the 18th ACM Symposium on Theory of Computing, ACM Press, 1986, 59-68.

[LiRe96a] M. Liśkiewicz and R. Reischuk, *The sublogarithmic alternating space world,* SIAM Journal on Computing, 24, 1996, 828-861.

[LiRe96b] M. Liśkiewicz and R. Reischuk, *Space Bounds for Interactive Proof Systems with Public Coins and Bounded Number of Rounds,* ICSI Technical Report No. TR-96-025, Berkeley, July 1996.

[Ma95] I. Macarie, *Space-bounded probabilistic computation: old and new stories,* SIGACT News, vol. 26(3), 1995, 2-12.

MOD$_p$-tests, Almost Independence and Small Probability Spaces
(Extended Abstract)

Claudia Bertram-Kretzberg and Hanno Lefmann

Lehrstuhl Informatik II, Universität Dortmund, D-44221 Dortmund, Germany
bertram,lefmann@ls2.informatik.uni-dortmund.de

Abstract. We consider approximations of probability distributions over $\mathbb{Z}_p^n$. We present an approach to estimate the quality of approximations towards the construction of small probability spaces which are used to derandomize algorithms. In contrast to results by Even et al. [13], our methods are simple, and for reasonably small p, we get smaller sample spaces. Our considerations are motivated by a problem which was mentioned in recent work of Azar et al. [5], namely, how to construct in time polynomial in n a good approximation to the joint probability distribution of i.i.d. random variables $X_1, \ldots, X_n$ where each X_i has values in $\{0, 1\}$. Our considerations improve on results in [5].

1 Introduction

During the last years, techniques have been developed to minimize the number of random bits which are used by randomized algorithms. These methods are based on the replacement of independent random variables by some weakly dependent random variables which can be generated using fewer bits, therefore, dropping the running times of several algorithms. Alon, Babai and Itai [1] observed that it suffices for certain algorithms to use only pairwise independent bits instead of mutually independent ones. In general, to generate k-wise independent bits sample spaces of size only $O(n^k)$ can be used, cf. Karloff and Mansour [15]. However, for some algorithms a large amount of independence is desirable. In view of this, Berger and Rompel [8] showed that sometimes it suffices to consider only $(\log n)^c$-wise independence of random variables. Small probability spaces are very desirable for derandomizing randomized algorithms. The resulting sample space, which should reflect the behaviour of the considered random variables, can be investigated by exhaustive search or the method of conditional probabilities, cf. Alon and Spencer [4], and Motwani, Naor and Naor [18].
Instead of looking for small probability spaces, Naor and Naor [19] considered approximations to probability distributions. They used the notion of the bias of a distribution, cf. Vazirani [22]. Let $X_1, \ldots, X_n$ be random variables with values in $\{0, 1\}$. The *bias* of a subset $S \subseteq \{X_1, \ldots, X_n\}$ w.r.t. linear tests is defined by $|Pr[\sum_{i \in S} X_i \equiv 0 \bmod 2] - Pr[\sum_{i \in S} X_i \equiv 1 \bmod 2]|$. In an ***$\epsilon$-biased distribution,*** each subset S of the random variables has bias at most ϵ. Clearly, for uniform random variables, the bias is zero. Naor and Naor gave in [19] constructions of ϵ-biased distributions where the sample space has size $poly(n, 1/\epsilon)$. A different construction based on Weil's theorem on quadratic residues was given by Peralta

Reischuk, Morvan (Eds.): STACS'97 Proceedings, LNCS 1200

[21]. Alon, Goldreich, Håstad and Peralta [3] gave three constructions, including Peralta's, for ϵ-biased sample spaces $S \subseteq \mathbb{Z}_2^n$ w.r.t. linear tests in $\mathbb{Z}_2$ of size $O(n^2/(\varepsilon^2(\log(n/\varepsilon))^\delta))$ where $\delta = 1, 0$ and 2, respectively. Azar, Motwani and Naor [5] generalized the work of [3] to random variables with values from arbitrary groups, in particular for $\mathbb{Z}_p = \{0, 1, \ldots, p-1\}$, the set of residues modulo p, cf. [14] for applications. There, they used Weil's theorem on character sums and Fourier transforms to measure the quality of approximations of the uniform distribution over $\mathbb{Z}_p^n$. Here, we use a more elementary way to achieve this, and we obtain sharper bounds.

Besides Weil's theorem on quadratic residues, a similar behaviour of the underlying structures is given by Lindsey's inequality [6] or by the corresponding inequalities for Expander- respectively Ramanujan graphs [17]. These phenommena can be summarized under the term *Quasirandomness*, see [11], namely, the structures behave approximately like random, that is, show small discrepancies. From that point of view, it is natural that the combinatorial notion of *discrepancy* was taken into account with the work of Even, Goldreich, Luby, Nisan and Veličković [13]. Indeed, Alon, Bruck, Naor, Naor and Roth [2] used Ramanujan graphs to construct good error-correcting codes which also yield small sample spaces for approximating the joint distribution of random variables.

Azar, Motwani and Naor stated in [5] the problem of finding good approximations for the joint distribution of i.i.d. random variables $X_1, \ldots, X_n$ with values in $\{0, 1\}$, where $Pr[X_1 = 0] = 1 - Pr[X_1 = 1] = q \neq \frac{1}{2}$. Even, Goldreich, Luby, Nisan and Veličković [13] considered this problem in a general setting, namely, for independent random variables $X_1, \ldots, X_n$ with values in $\{1, \ldots, m\}$ where $Pr[X_i = j] = p_{i,j}$, $1 \leq i \leq n$ and $1 \leq j \leq m$. In [13], constructions of small sample spaces were given which approximate the joint distribution of $X_1, \ldots, X_n$. To do so, they used the combinatorial notion of *discrepancy*, cf. [7]. Let R_n be the set of all axis aligned rectangles of the n-dimensional cube $[0, 1)^n$. For any finite set $S \subset [0, 1)^n$ and any rectangle $R \in R_n$ with volume $vol(R)$, the discrepancy of S on R_n is defined by $disc_S(R_n) = \sup_{R \in R_n} |vol(R) - |S \cap R|/|S||$.

A sample space $S \subseteq \{1, \ldots, m\}^n$ is *(ε, k)-independent* w.r.t. the joint distribution of the independent random variables $X_1, \ldots, X_n$ with values in $\{1, \ldots, m\}$ if for any sequence $(\alpha_{i_1}, \ldots \alpha_{i_k}) \in \{1, \ldots, m\}^k$ it holds $|Pr[(X_{i_1}, \ldots, X_{i_k}) = (\alpha_{i_1}, \ldots \alpha_{i_k})] - \prod_{j=1}^k p_{i_j, \alpha_{i_j}}| \leq \varepsilon$. In [13], Even et al. showed that sets S with small discrepancy, i.e. $disc_S(R_n) \leq \varepsilon$, yield sample spaces which are (ε, k)-independent w.r.t. the joint distribution of random variables. Their construction has the advantage to be universal. One construction in [13] yields an (ε, k)-independent sample space $S \subseteq \{1, \ldots, m\}^n$ of size $poly(\log n, 2^k, 1/\varepsilon)$, while the other two constructions yield (ε, k)-independent spaces $S \subseteq \{1, \ldots, m\}^n$ of size $poly\left((n/\varepsilon)^{\log(1/\varepsilon)}\right)$ and $poly\left((n/\varepsilon)^{\log n}\right)$, respectively. The results of [13] were extended and applied by Chari, Rohatgi and Srinivasan [10]. Again using the notion of discrepancy and projections, they constructed an (ε, k)-independent sample space S of size $poly(\log n, 1/\varepsilon, min\{2^k, k^{\log(1/\varepsilon)}\})$.

Our considerations here are motivated by the problem from Azar, Motwani and Naor [5]. In contrast to the work of [13] and [10] where the discrepancy of axis aligned rectangles is used, we offer a different approach for investigating approximations of probability spaces by using basic Linear Algebra. The intention behind our considerations is to give more insight towards the understanding of the underlying concepts for approximating random variables as asked for in [13]. Using our results on approximations to the uniform distribution over $\mathbb{Z}_p^n$, we show, by collapsing nonzero entries to 1, how good this strategy measures the deviation distance between these distributions. For uniformly distributed random variables and reasonably small values of p, the quality of our approximation is better than the one in [10], i.e., the sample spaces have size $O(p^2n^2/\varepsilon^2)$. Otherwise, the quality of our approximations is comparable to that of [10], i.e., for binary i.i.d. random variables $X_1, \ldots, X_n$ with $Pr[X_1 = 0] = 1-Pr[X_1 = 1] = 1/p$ and p a prime, the size of an (ε, k)-independent sample space S is $O(2^{2k}p^2n^2/\varepsilon^2)$. It should be mentioned that by using parity check matrices of BCH-codes as in [1] and [19], the n can be replaced by $O(k \cdot \log n)$ in all the upper bounds for $|S|$. However, for some applications our concepts seem to be more appropriate. Especially, if one wants to apply the results in circuit theory. Namely, Krause and Pudlák [16] show by a probabilistic argument that depth-2 $\{AND, OR, NOT\}$-circuits of quasipolynomial size can be realized by a threshold-MOD_n-circuit of quasipolynomial size for all integers n. Similarly, one can show that, small weights threshold-AND-circuits can be simulated by small weights threshold-MOD_p-circuits. By choosing MOD_p-gates with ε-biased weight vectors, one can construct such threshold-MOD_p-circuits approximatively, cf. [9].

2 (ϵ, k)-Independence

For fixed prime p, let $\mathbb{Z}_p = \{0, 1, \ldots, p-1\}$ be the set of residues modulo p. The set $\mathbb{Z}_p^n = \{0, 1, \ldots, p-1\}^n$ is the n-fold cartesian product of $\mathbb{Z}_p$. For sequences $\alpha = (\alpha_1, \ldots, \alpha_n), \beta = (\beta_1, \ldots, \beta_n) \in \mathbb{Z}_p^n$ let $<\alpha, \beta>_p \equiv \sum_{i=1}^n \alpha_i\beta_i \bmod p$ be the inner product of α and β modulo p. Let $0^n = (0, \ldots, 0)$ be the sequence of length n which has only zero entries. A subset $S \subseteq \mathbb{Z}_p^n$ is called *sample space*.

Definition 1. a) Let p be a prime. For a random variable X with values in $\mathbb{Z}_p$, let the *bias* of X be defined by $bias(X) = (p-1) \cdot \Pr[X = 0] - \Pr[X \neq 0]$. A random variable X with values in $\mathbb{Z}_p$ is called *ϵ-biased* if $|bias(X)| \leq \epsilon$.
b) A sample space $S \subseteq \mathbb{Z}_p^n$ is *ϵ-biased w.r.t.* MOD$_p$*-tests* if for each $c \in \mathbb{Z}_p$ and each sequence $\beta = (\beta_1, \ldots, \beta_n) \in \mathbb{Z}_p^n \setminus \{0^n\}$ the following is valid: if a sequence $X = (x_1, \ldots, x_n) \in S$ is chosen uniformly at random from S, then the random variable $(<\beta, X>_p + c \bmod p)$ is ϵ-biased.
c) For a fixed positive integer k, a sample space $S \subseteq \mathbb{Z}_p^n$ is *ϵ-biased w.r.t.* MOD$_p$*-tests of size at most k*, if for each $c \in \mathbb{Z}_p$ and each sequence $\beta = (\beta_1, \ldots, \beta_n) \in \mathbb{Z}_p^n \setminus \{0^n\}$ with at most k nonzero entries, the following is valid: if $X = (x_1, \ldots, x_n)$ is chosen uniformly at random from S, then the random variable $(<\beta, X>_p + c \bmod p)$ is ϵ-biased.
d) A sample space $S \subseteq \mathbb{Z}_p^n$ is called *(ϵ, k)-independent* if for each k positions $1 \leq i_1 < \ldots < i_k \leq n$ and any sequences $\alpha = (\alpha_1, \ldots, \alpha_k) \in \mathbb{Z}_p^k$ and $X = (x_1, \ldots, x_n) \in S$ where X is chosen uniformly at random from S, we have $|\Pr[(x_{i_1}, \ldots, x_{i_k}) = \alpha] - 1/p^k| \leq \epsilon$.

The following functions are convenient for our purposes. For fixed $c \in \mathbb{Z}_p$ and $\alpha = (\alpha_1, \ldots, \alpha_n), \beta = (\beta_1, \ldots, \beta_n) \in \mathbb{Z}_p^n$, let $\Phi_\beta^c : \mathbb{Z}_p^n \to \mathbb{R}$ be defined by

$$\Phi_\beta^c(\alpha) = \begin{cases} -\sqrt{p-1} & \text{if } \sum_{i=1}^n \alpha_i \beta_i + c \equiv 0 \bmod p \\ 1/\sqrt{p-1} & \text{else.} \end{cases}$$

The function Φ_β^c is a 'normalized' indicator function for the event $<\beta, \alpha>_p + c \equiv 0 \bmod p$ since $\sum_{c \in \mathbb{Z}_p} \Phi_\beta^c(\alpha) = -\sqrt{p-1} + (p-1)/\sqrt{p-1} = 0$.
Central in our argumentation is the following lemma which generalizes a result of Vazirani [22] who considered the case $p = 2$, cf. [3].

Lemma 2. *Let k be a fixed positive integer. Let $S \subseteq \mathbb{Z}_p^n$ be a sample space which is ϵ-biased w.r.t.* MOD$_p$*-tests of size at most k. Then, the space S is $(2 \cdot \epsilon/p \cdot (1 - p^{-k}), k)$-independent.*

An elementary proof of Lemma 2 using Linear Algebra is given in the Appendix. By Lemma 2, MOD_p-tests, i.e. linear tests, are appropiate to test (ε, k)-independence of sample spaces. As an immediate consequence, we obtain:

Corollary 3. *Let $S \subseteq \mathbb{Z}_p^n$ be a sample space which is ϵ-biased w.r.t.* MOD$_p$*-tests. Then, for every positive integer k, the space S is $(2 \cdot \epsilon/p \cdot (1 - p^{-k}), k)$-independent.*

Next, we consider the distance of two probability distributions.

Definition 4. For any sequence $\alpha = (\alpha_1, \ldots, \alpha_k) \in \mathbb{Z}_p^k$, let $||\alpha||_1 = \sum_{i=1}^k |\alpha_i|$ denote the L_1-norm of α (where addition is in the integers). The *distance* $d(\alpha, \beta)$ between two sequences $\alpha = (\alpha_1, \ldots, \alpha_k) \in \mathbb{Z}_p^k$ and $\beta = (\beta_1, \ldots, \beta_k) \in \mathbb{Z}_p^k$ is defined by $d(\alpha, \beta) = ||\alpha - \beta||_1$. For two probability distributions D_1, D_2 on $\mathbb{Z}_p^k$ the *variation distance* of D_1 and D_2 is $|| \left(D_1(x)\right)_{x \in \mathbb{Z}_p^k} - \left(D_2(x)\right)_{x \in \mathbb{Z}_p^k} ||_1$.
Let $X_1, \ldots, X_n$ be random variables with values in some set Y. The *joint distribution* $D(X_1, \ldots, X_n)$ is the distribution on Y^n, i.e., for any $(\alpha_1, \ldots, \alpha_n) \in Y^n$ one is interested in $Pr[(X_1, \ldots, X_n) = (\alpha_1, \ldots, \alpha_n)]$. For a subset $S \subseteq \{X_1, \ldots, X_n\}$ let $U(S)$ denote the uniform distribution on this subset S of random variables. For $Y = \mathbb{Z}_p$ the random variables $X_1, \ldots, X_n$ are called *k-wise δ-dependent* if for all subsets S with $|S| \leq k$, we have $||D(S) - U(S)||_1 \leq \delta$.

Theorem 5. *If the random variables $X_1, \ldots, X_n$, with values in $\mathbb{Z}_p$ are ϵ-biased w.r.t. to* MOD$_p$*-tests of size at most k, then they are also k-wise δ-dependent for $\delta = \varepsilon \cdot p^{k/2}/\sqrt{p-1}$.*

Thus, using a sample space of polynomial size, one can approximate well a $\log_p n$-wise independent uniform distribution, cf. [19]. Theorem 5 strengthens a result of Azar, Motwani and Naor [5] where $\delta = p^k \cdot \epsilon$ was shown. The case $p = 2$ was proven by Alon, Goldreich, Håstad and Peralta [3].

Lemma 6.

$$\sum_{\alpha \in \mathbb{Z}_p^k} p_\alpha^2 = p^{-(k+1)} \cdot \sum_{c \in \mathbb{Z}_p} \sum_{\beta \in \mathbb{Z}_p^k} (d_\beta^c)^2 . \quad (1)$$

Proof. We evaluate the right hand side of (1). Using (11) and (12) we infer

$$\sum\nolimits_{c\in\mathbb{Z}_p}\sum\nolimits_{\beta\in\mathbb{Z}_p^k}(d_\beta^c)^2=\sum\nolimits_{c\in\mathbb{Z}_p}\sum\nolimits_{\beta\in\mathbb{Z}_p^k}\left(\sum\nolimits_{\alpha\in\mathbb{Z}_p^k}\Phi_\beta^c(\alpha)\cdot p_\alpha\right)^2$$
$$=\sum_{\alpha\in\mathbb{Z}_p^k}p_\alpha^2\sum_{c\in\mathbb{Z}_p}\sum_{\beta\in\mathbb{Z}_p^k}\Phi_\beta^c(\alpha)^2+\sum_{\gamma\in\mathbb{Z}_p^k}p_\gamma\sum_{c\in\mathbb{Z}_p}\sum_{\beta\in\mathbb{Z}_p^k}\sum_{\alpha\in\mathbb{Z}_p^k;\alpha\neq\gamma}\Phi_\beta^c(\alpha)\cdot\Phi_\beta^c(\gamma)\cdot p_\alpha$$
$$=p^{k+1}\cdot\sum\nolimits_{\alpha\in\mathbb{Z}_p^k}p_\alpha^2\,.\qquad\square$$

Now we will prove Theorem 5 using Lemma 6.

Proof. By assumption and (9), we have $|d_\beta^c|\leq\epsilon/\sqrt{p-1}$. As $\sum_{\alpha\in\mathbb{Z}_p^k}p_\alpha=1$, we have $d_{0^k}^0=\sum_{\delta\in\mathbb{Z}_p^k}\Phi_{0^k}^0(\delta)\cdot p_\delta=-\sqrt{p-1}$ and for $c\neq 0$, it holds $d_{0^k}^c=\sum_{\delta\in\mathbb{Z}_p^k}\Phi_{0^k}^c(\delta)\cdot p_\delta=1/\sqrt{p-1}$. Thus,

$$p^{-(k+1)}\cdot\sum\nolimits_{c\in\mathbb{Z}_p}(d_{0^k}^c)^2=p^{-k}\,.\tag{2}$$

Let w.l.o.g. $S=\{X_1,\ldots,X_k\}$. Then, $||D(S)-U(S)||_1=\sum_{\alpha\in\mathbb{Z}_p^k}|p_\alpha-p^{-k}|$. We use the fact that $\sum_{\alpha\in\mathbb{Z}_p^k}p_\alpha=1$, the Cauchy-Schwarz inequality, (1) and (2), and

we obtain $\displaystyle\sum_{\alpha\in\mathbb{Z}_p^k}|p_\alpha-p^{-k}|\leq p^{\frac{k}{2}}\left(\sum_{\alpha\in\mathbb{Z}_p^k}(p_\alpha-p^{-k})^2\right)^{\frac{1}{2}}=p^{\frac{k}{2}}\left(\sum_{\alpha\in\mathbb{Z}_p^k}p_\alpha^2-p^{-k}\right)^{\frac{1}{2}}$

$$=p^{\frac{k}{2}}\cdot\left(p^{-(k+1)}\cdot\sum_{c\in\mathbb{Z}_p}\sum_{\beta\in\mathbb{Z}_p^k}(d_\beta^c)^2-p^{-k}\right)^{\frac{1}{2}}=p^{\frac{k}{2}}\cdot\left(p^{-(k+1)}\cdot\sum_{c\in\mathbb{Z}_p}\sum_{\beta\in\mathbb{Z}_p^k\setminus\{0^k\}}(d_\beta^c)^2\right)^{\frac{1}{2}}$$
$$\leq p^{\frac{k}{2}}\cdot\left(p^{-k}\cdot(p^k-1)\cdot\varepsilon^2/(p-1)\right)^{\frac{1}{2}}<p^{\frac{k}{2}}/\sqrt{p-1}\cdot\epsilon\,.\qquad\square$$

3 Approximating Nonuniform Distributions

In [5], Azar, Motwani and Naor stated the problem to construct in time polynomial in n a good approximation to the joint distribution of i.i.d. random variables $X_1,\ldots,X_n$ where X_i takes value 0 with probability $q\neq 1/2$ and value 1 with probability $1-q$. We approach this problem as follows. Let $q=1/p$ where p is a prime. The general case for arbitrary rational numbers q, $0\leq q\leq 1$, can be done similarly, cf. Section 5. We consider random variables $Z_1,\ldots,Z_n$ which take values in $\mathbb{Z}_p$ uniformly and independently, i.e., $Pr[Z_i=j]=Pr[Z_i=k]=1/p$ for all $j,k\in\mathbb{Z}_p$. Applying our results on ϵ-biased distributions to the joint distribution on $Z_1,\ldots,Z_n$, we consider what happens for the new distribution when all nonzero entries are collapsed to 1. In the unbiased case, i.e., $\epsilon=0$, the entry 0 occurs with probability $q=1/p$ and the entry 1 with probability $1-1/p$.

Definition 7. Let $\bar{\gamma}=(\gamma_1,\ldots,\gamma_n)\in\{0,*\}^n$ be a sequence where $*$ stands for any element from $\mathbb{Z}_p\setminus\{0\}$. A sequence $X=(x_1,\ldots,x_n)\in\mathbb{Z}_p^n$ is of *type* $\bar{\gamma}$, i.e., $\text{type}(X)=\bar{\gamma}$ if and only if it holds that $x_i=0$ iff $\gamma_i=0$ for $i=1,\ldots,n$.

Collapsing all nonzero entries of a sequence $\alpha \in \mathbb{Z}_p^n$ to 1, yields a new sequence $\alpha^* \in \{0,1\}^n$, the *reduced sequence of* α. For a sample space $S \subseteq \mathbb{Z}_p^n$, let the *reduced space* $\bar{S} \subseteq \mathbb{Z}_2^n$ be obtained from S by identifying in any sequence $X \in S$ every nonzero entry by 1.

Theorem 8. *Let $S \subseteq \mathbb{Z}_p^n$ be a sample space which is ϵ-biased w.r.t. MOD_p-tests of size at most k. Then, the reduced space $\bar{S} \subseteq \mathbb{Z}_2^n$ is $(\epsilon \cdot 2^k, k)$-independent.*

Proof. Let $X = (x_1, \ldots, x_n)$ be chosen uniformly at random from S. We consider w.l.o.g. the first k positions of X, i.e., $x_1, \ldots, x_k$. For $\bar{\gamma} \in \{0,*\}^k$ let $Pr[\bar{\gamma}]$ be the probability that $\bar{\gamma} = \text{type}(x_1, \ldots, x_k)$ and let $z(\bar{\gamma})$ be the number of entries of $\bar{\gamma}$ being zero. For $\alpha \in \mathbb{Z}_p^k$, let $p_\alpha = Pr[(x_1, \ldots, x_k) = \alpha]$. By Claim 1 in the appendix, we have

$$Pr[\bar{\gamma}] = \sum_{\alpha \in \mathbb{Z}_p^k; \text{type}(\alpha) = \bar{\gamma}} p_\alpha = p^{-(k+1)} \cdot \sum_{\alpha \in \mathbb{Z}_p^k; \text{type}(\alpha) = \bar{\gamma}} \sum_{\beta \in \mathbb{Z}_p^k} \sum_{c \in \mathbb{Z}_p} \Phi_\beta^c(\alpha) \cdot d_\beta^c \,.$$

Consider the sum for $\beta = 0^k$. With $p^{-k-1} \sum_{\alpha \in \mathbb{Z}_p^k; \text{type}(\alpha) = \bar{\gamma}} \sum_{c \in \mathbb{Z}_p} \Phi_{0^k}^c(\alpha) d_{0^k}^c = (p-1)^{k-z(\bar{\gamma})} p^{-k}$ we infer

$$\left| Pr[\bar{\gamma}] - (p-1)^{k-z(\bar{\gamma})}/p^k \right| = p^{-k-1} \left| \sum_{\beta \in \mathbb{Z}_p^k \setminus \{0^k\}} \sum_{c \in \mathbb{Z}_p} d_\beta^c \left(\sum_{\alpha \in \mathbb{Z}_p^k; \text{type}(\alpha) = \bar{\gamma}} \Phi_\beta^c(\alpha) \right) \right| \quad (3)$$

Let w.l.o.g. the first $g = z(\bar{\gamma})$ components of $\bar{\gamma}$ be zero. We partition the set $\mathbb{Z}_p^k \setminus \{0^k\}$ into subsets, i.e., $\mathbb{Z}_p^k \setminus \{0^k\} = B_0 \cup \ldots \cup B_{k-g}$, with $B_j = \{(\beta_1, \ldots, \beta_k) \in \mathbb{Z}_p^k \setminus \{0^k\} \mid |\{i \mid g+1 \leq i \leq k \text{ and } \beta_i \not\equiv 0 \bmod p\}| = j\}$. Then, $|B_0| = p^g - 1$ and $|B_j| = p^g \cdot \binom{k-g}{j} \cdot (p-1)^j$, $j = 1, \ldots, k-g$, and (3) becomes

$$\left| Pr[\bar{\gamma}] - (p-1)^{k-g}/p^k \right| \leq p^{-(k+1)} \cdot \sum_{c \in \mathbb{Z}_p} \sum_{j=0}^{k-g} \sum_{\beta \in B_j} |d_\beta^c| \cdot \left| \sum_{\alpha \in \mathbb{Z}_p^k; \text{type}(\alpha) = \bar{\gamma}} \Phi_\beta^c(\alpha) \right| \,.$$

Fix some $\beta \in B_0$. Then, $<\alpha, \beta>_p \equiv 0 \bmod p$ for each $\alpha \in \mathbb{Z}_p^k$ with $\text{type}(\alpha) = \bar{\gamma}$, thus, $\left| \sum_{\alpha \in \mathbb{Z}_p^k; \text{type}(\alpha) = \bar{\gamma}} \Phi_\beta^0(\alpha) \right| = (p-1)^{k-g} \cdot \sqrt{p-1}$, and, for $c \neq 0$, we obtain $\left| \sum_{\alpha \in \mathbb{Z}_p^k; \text{type}(\alpha) = \bar{\gamma}} \Phi_\beta^c(\alpha) \right| = (p-1)^{k-g} / \sqrt{p-1}$. With $|d_\beta^c| \leq \epsilon / \sqrt{p-1}$, we infer

$$\begin{aligned} & p^{-(k+1)} \cdot \sum\nolimits_{c \in \mathbb{Z}_p} \sum\nolimits_{\beta \in B_0} |d_\beta^c| \cdot \left| \sum\nolimits_{\alpha \in \mathbb{Z}_p^k; \text{type}(\alpha) = \bar{\gamma}} \Phi_\beta^c(\alpha) \right| \\ & \leq p^{-(k+1)} \cdot \frac{\epsilon}{\sqrt{p-1}} \cdot (p^g - 1) \cdot \left((p-1)^{k-g} \cdot \sqrt{p-1} + (p-1)^{k-g} \cdot \frac{p-1}{\sqrt{p-1}} \right) \\ & = 2 \cdot p^{-(k+1)} \cdot \epsilon \cdot (p^g - 1) \cdot (p-1)^{k-g} \,. \end{aligned} \quad (4)$$

For given $c \in \mathbb{Z}_p$ and $\beta = (\beta_1, \ldots, \beta_l) \in \{1, \ldots, p-1\}^l$, let N_l^c denote the number of solutions $X = (x_1, \ldots, x_l) \in (\mathbb{Z}_p \setminus \{0\})^l$ of $<\beta, X>_p + c \equiv 0 \bmod p$. It is easy to see that $N_l^c = (p-1)^{l-1} - N_{l-1}^c$. With $N_1^0 = 0$ and $N_1^c = 1$ for $c \neq 0$, we obtain

$N_l^0 = (p-1)^{l-1} - \sum_{i=1}^{l-2}(p-1)^i \cdot (-1)^{l+i}$ and $N_l^c = (p-1)^{l-1} - \sum_{i=0}^{l-2}(p-1)^i \cdot (-1)^{l+i}$, hence, for $\beta \in B_j, j \geq 1$, $\sum_{c \in \mathbb{Z}_p} \left| \sum_{\alpha \in \mathbb{Z}_p^k; \text{type}(\alpha)=\bar{\gamma}} \Phi_\beta^c(\alpha) \right| =$

$$= \sum_{c \in \mathbb{Z}_p} (p-1)^{k-g-j} \sqrt{p-1} \left| -N_j^c + \frac{(p-1)^j - N_j^c}{p-1} \right| = 2(p-1)^{k-g-j} \sqrt{p-1}. \quad (5)$$

With $|d_\beta^c| \leq \epsilon / \sqrt{p-1}$, it is $p^{-(k+1)} \cdot \sum_{j=1}^{k-g} \sum_{\beta \in B_j} \sum_{c \in \mathbb{Z}_p} |d_\beta^c| \cdot \left| \sum_{\alpha \in \mathbb{Z}_p^k; \text{type}(\alpha)=\bar{\gamma}} \Phi_\beta^c(\alpha) \right| \leq$

$$\leq p^{-(k+1)} \cdot \epsilon / \sqrt{p-1} \cdot \sum_{j=1}^{k-g} p^g \cdot \binom{k-g}{j} \cdot (p-1)^j \cdot 2(p-1)^{k-g-j} \cdot \sqrt{p-1}$$
$$= 2p^{-(k+1)} \cdot \epsilon \cdot p^g \cdot (p-1)^{k-g} \cdot (2^{k-g} - 1). \quad (6)$$

Altogether, with (4) and (6), we conclude $\left| Pr(\bar{\gamma}) - \frac{(p-1)^{k-z(\gamma)}}{p^k} \right| \leq$
$\leq p^{-(k+1)} \cdot \left[2\varepsilon(p^g - 1)(p-1)^{k-g} + 2\varepsilon p^g (p-1)^{k-g}(2^{k-g} - 1) \right] \leq \varepsilon \cdot 2^k$. □

Theorem 9. *Let $S \subseteq \mathbb{Z}_p^k$ be a sample space which is ϵ-biased w.r.t. MOD_p-tests of size at most k. Then, the reduced space $\bar{S} \subseteq \mathbb{Z}_2^k$ obtained from S is k-wise $(\epsilon \cdot 3^k)$-dependent.*

Thus, using a sample space of polynomial size, one can approximate well a $\log_p n$-wise independent nonuniform distribution, cf. [19].

Proof. Set $Pr(\bar{\gamma}) = \sum_{\alpha \in \mathbb{Z}_p^k; \text{type}(\alpha)=\bar{\gamma}} p_\alpha$. By assumption and (9), we have $|d_\beta^c| \leq \epsilon / \sqrt{p-1}$. The result follows by estimating $\sum_{\gamma \in \mathbb{Z}_2^k} | \sum_{\alpha \in \mathbb{Z}_p^k; \text{type}(\alpha)=\bar{\gamma}} p_\alpha - (p-1)^{k-z(\bar{\gamma})}/p^k |$, using (1), (4) and (6). Details are in the full version. □

4 Applications

Lemma 2 links the ability to pass MOD_p-tests with almost independence. We start with an ϵ-biased sample space $S \subseteq \mathbb{Z}_p^n$ hence, S is $(2 \cdot \epsilon/p \cdot (1 - p^{-k}), k)$-independent. If we replace in every $\alpha \in S$ each nonzero entry by one, the reduced space, which might be a multiset, will become a reasonable approximation, cf. Theorem 9, to the joint distribution of n binary i.i.d. random variables, where each variable takes value 0 resp. 1 with probability $1/p$ resp. $1 - 1/p$.
Alon, Goldreich, Håstad and Peralta gave in [3] three constructions for sample spaces which are ϵ-biased w.r.t. linear tests. Two of them were modified to the p-ary case such that they also yield sample spaces which are ϵ-biased w.r.t. MOD_p tests, cf. Azar, Motwani and Naor [5] and Even [12]. The generalization of the third construction is straightforward. To have a typical example, we give

it below. Another construction using Ramanujan graphs and Justesen codes is shown in [2] where an (ε, k)-independent sample space of size $O(n/\varepsilon^3)$ is given. For a fixed prime p consider the finite field $GF(p^m)$. Let $f\colon GF(p^m) \longrightarrow \mathbb{Z}_p^m$ be the standard representation of $GF(p^m)$ as a vector space over $GF(p)$ with $f(0) = 0^m$ and $f(u+v) \equiv (f(u) \oplus f(v)) \bmod p$ where addition $+$ is in $GF(p^m)$ and addition $\oplus$ in $\mathbb{Z}_p^m$ is componentwise modulo p. The elements of the sample space S_m^n are determined by pairs $x, y \in GF(p^m)$, where the ith entry of the sequence $s_{x,y} \in S_m^n$ equals $<f(x^i), f(y)>_p$, $i = 0, \ldots, n-1$.

Proposition 10. *The sample space S_m^n has size $|S_m^n| = p^{2m}$ and is $(p-1)(n-1)/p^m$-biased w.r.t. MOD_p-tests.*

Proof. Clearly, we have $|S_m^n| = p^{2m}$. Let $s(x,y) = (s_0(x,y), \ldots, s_{n-1}(x,y)) \in S_m^n$, where $s_i(x,y) \equiv <f(x^i), f(y)>_p$ and $x, y \in GF(p^m)$. For any $\alpha \in \mathbb{Z}_p^n$, we have $<\alpha, s(x,y)>_p \equiv \sum_{i=0}^{n-1} \alpha_i \cdot <f(x^i), f(y)>_p \equiv <f(\sum_{i=0}^{n-1} \alpha_i \cdot x^i), f(y)>_p$. Let $p_\alpha(t) = \sum_{i=0}^{n-1} \alpha_i \cdot t^i$ be a polynomial over $\mathbb{Z}_p$ with $u \leq n-1$ zeros. To determine the distribution of $<f(p_\alpha(x)), f(y)>_p$ where $x, y \in GF(p^m)$ are chosen uniformly at random, we fix $x \in GF(p^m)$. We distinguish two cases. If $p_\alpha(x) \neq 0$, i.e., x is not a zero of $p_\alpha(t)$, then, $f(p_\alpha(x)) \neq 0^m$ and for uniformly chosen $y \in GF(p^m)$ the values $<f(p_\alpha(x)), f(y)>_p$ are uniformly distributed in $\mathbb{Z}_p$, hence $<f(p_\alpha(x)), f(y)>_p$ is unbiased. Otherwise, if $p_\alpha(x) = 0$, then $<f(p_\alpha(x)), f(y)>_p \equiv 0 \bmod p$ for all $y \in GF(p^m)$. Hence, for each $\alpha \in \mathbb{Z}_p^n$ we have $|(p-1) \cdot Pr[<\alpha, s(x,y)>_p + c \equiv 0 \bmod p] - Pr[<\alpha, s(x,y)>_p + c \not\equiv 0 \bmod p]| \leq$

$$= \left|(p-1) \cdot \frac{u \cdot p^m + (p^m - u) \cdot p^{m-1}}{p^{2m}} - \frac{(p^m - u) \cdot (p^m - p^{m-1})}{p^{2m}}\right|$$

$= (p-1) \cdot u/p^m \leq (p-1) \cdot (n-1)/p^m$. □

Corollary 11. *Let $\varepsilon > 0$ and let p be a prime. One can explicitly construct a sample space $S \subseteq \mathbb{Z}_p^n$ of size $|S| < p^4 \cdot n^2/\varepsilon^2$ which is ε-biased w.r.t. MOD_p-tests.*

Proof. Let m be the largest positive integer such that $(p-1) \cdot (n-1)/p^m \leq \varepsilon$. By Proposition 10 the sample space $S := S_m^n$ is ε-biased and satisfies $|S| = p^{2m}$, i.e., $|S| < p^4 \cdot n^2/\varepsilon^2$. □

Indeed, if $n/p^{m-1} \sim \varepsilon$, then $|S| \leq c \cdot p^2 n^2/\varepsilon^2$ for some small constant $c > 0$.

Corollary 12. *Let $\varepsilon > 0$ and let p be a prime. One can explicitly construct a sample space $S \subseteq \mathbb{Z}_p^n$ of size $|S| < 4 \cdot (1-p^{-k})^2 \cdot p^2 \cdot n^2/\varepsilon^2$ which is (ε, k)-independent.*

Proof. Lemma 2 with $\varepsilon' := \varepsilon \cdot p/(2 \cdot (1-p^{-k}))$ and Corollary 11. □

Corollary 13. *Let $\varepsilon > 0$ and let p be a prime. Then, one can explicitly construct a sample space $S \subseteq \mathbb{Z}_2^n$ of size $|S| < 2^{2k+2} \cdot (1-p^{-k})^2 \cdot p^2 \cdot n^2/\varepsilon^2$ which is (ε, k)-independent (w.r.t. the probability $1/p$), i.e., if $X = (x_1, \ldots, x_n)$ is chosen uniformly at random from S, then for any k positions $1 \leq i_1 < \ldots < i_k \leq n$ and any sequence $\gamma = (\gamma_1, \ldots, \gamma_k) \in \{0, *\}^k$ with z entries being 0, we have*

$$\left|Pr[\text{type}(x_{i_1}, \ldots, x_{i_k}) = (\gamma_1, \ldots, \gamma_k)] - (p-1)^{k-z}/p^k\right| \leq \varepsilon .$$

Proof. By Corollary 12 we can construct a sample space $S \subseteq \mathbb{Z}_p^n$ of size $|S| < 2^{2k+2} \cdot (1-p^{-k})^2 \cdot p^2 \cdot n^2/\varepsilon^2$ which is $(\varepsilon/2^k, k)$-independent. Then, by Theorem 8 the reduced space $\overline{S} \subseteq \mathbb{Z}_2^n$ is (ε, k)-independent (w.r.t. the probability $1/p$). □

The sample spaces from above generate (ε, k)-independent random variables. Using as an additional tool parity check matrices of BCH-codes as in [1] and [19], the size of S can be further reduced by replacing in the upper bounds for S the n by $O(k \cdot \log_p n)$, i.e., for example in Corollary 11 we obtain $|S| = O(p^4 k^2 (\log n)^2/\varepsilon^2)$.

A simple application is the *heavy codeword problem* for linear codes over $\mathbb{Z}_p$. A matrix $M \in \mathbb{Z}_p^{m\times n}$ with no row containing only zeros, is given. One wants to find a vector $x \in \mathbb{Z}_p^n$ such that Mx has at least $(p-1)m/p$ nonzero entries. For a sample space $S \subseteq \mathbb{Z}_p^n$ which is ε-biased w.r.t. MOD_p-tests with $\varepsilon < 1/m$, let $x \in S$ be chosen uniformly at random. For $i = 1, \ldots, m$, let m_i be the ith row of matrix M. The weight $wt(x)$ of a vector is the number of nonzero entries of x. The expected weight $E(wt(Mx))$ fulfills $E(wt(Mx)) = \sum_{i=1}^{m} Pr[<m_i, x>_p \not\equiv 0] =$
$= \sum_{i=1}^{m} [(p-1) - bias(<m_i, x>_p)]/p \geq (p-1-\varepsilon)m/p$.

If $\varepsilon < 1/m$, then $E(wt(Mx)) > (p-1)m/p - 1/p$. As $wt(Mx)$ is an integer, there must be a codeword x with weight of Mx at least $\lceil (p-1)m/p \rceil$. Thus, the heavy codeword problem over $\mathbb{Z}_p$ is in NC.

Another example comes from testing circuits, namely, in order to test circuits in which each gate depends on at most k inputs, one uses (n, k, p)-universal sequences, cf. [19], [20]. Those vectors are taken from $\mathbb{Z}_p^n$, and for any k-subset of coordinates the projection on these contains all possible p^k sequences. If we have a k-wise δ-dependent probability space for $\delta < p^{-k}$, then it is also a (k, n)-universal set. The reason is simple. If for a k-subset $i_1, \ldots, i_k$, there is a sequence in the chosen sample space over $\mathbb{Z}_p^k$ which has probability 0, then the distance from the uniform distribution of $x_{i_1}, \ldots, x_{i_k}$ is at least $p^{-k} > \delta$. Using Theorem 5 and Corollary 11 together with the above mentioned BCH-codes one can construct (n, k, p)-universal sets of size $O(\log n \cdot p^{3k+O(\log k)})$.

5 Final Remarks

Our considerations can be extended to an arbitrary finite group G instead of the group $\mathbb{Z}_p$, but G should have no divisors of zero. If G has divisors of zero, this can be handled by taking only the multiples of an element under consideration. For approximating nonuniform distributions we considered the probabilities $q = 1/p$ where p is a prime. The general case, $0 \leq q \leq 1$, can be handled by choosing a prime p and an integer l such that $q \sim l/p$. Then, one distinguishes whether a MOD_p-test gives a result contained in $L = \{0, 1, \ldots, l-1\}$ or in $\{l, l+1, \ldots, p-1\}$. The calculations are along the lines we discussed in this paper, i.e., instead of the functions Φ_β^c we use $\Phi_\beta^{c,l} : \mathbb{Z}_p^n \to \mathbb{R}$ with

$$\Phi_\beta^{c,l}(\alpha) = \begin{cases} -(p-l)/\sqrt{p-1} & \text{if } \sum_{i=1}^n \alpha_i\beta_i + c \equiv j \bmod p \text{ for some } j \in L \\ l/\sqrt{p-1} & \text{else.} \end{cases}$$

References

1. N. Alon, L. Babai and A. Itai, A Fast and Simple Randomized Parallel Algorithm for the Maximal Independent Set Problem, J. Alg. 7, 1985, 567-583.
2. N. Alon, J. Bruck, J. Naor, M. Naor and R. Roth, Construction of Asymptotically Good Low-Rate Error-Correcting Codes Through Pseudo-Random Graphs, IEEE Trans. Inf. Th. 38, 1992, 509-516.
3. N. Alon, O. Goldreich, J. Håstad and R. Peralta, Simple Constructions of Almost k-wise Independent Random Variables, Random Structures & Algorithms 3, 1992, 289-304. Addendum: Random Structures & Algorithms 4, 1993, 119-120.
4. N. Alon and J. Spencer, The Probabilistic Method, Wiley & Sons, New York, 1992.
5. Y. Azar, R. Motwani and J. Naor, Approximating Probability Distributions using Small Sample Spaces, preprint, 1995.
6. L. Babai, P. Frankl and J. Simon, Complexity Classes in Communication Complexity Theory, Proc. 27th FOCS, 1986, 337-347.
7. J. Beck and W. Chen, Irregularities of Distribution, Cambridge Univ. Press, 1987.
8. B. Berger and J. Rompel, Simulating $(\log n)^c$-wise Independence in NC, J. ACM 38, 1991, 1026-1046.
9. C. Bertram-Kretzberg, TH-MOD_p-circuits, preprint, 1996.
10. S. Chari, P. Rohatgi and A. Srinivasan, Improved Algorithms via Approximations of Probability Distributions, Proc. 26th STOC, 1996, 584-592.
11. F. R. K. Chung, R. L. Graham and R. M. Wilson, Quasi-Random Graphs, Combinatorica 9, 1989, 345-362.
12. G. Even, Construction of Small Probability Spaces for Deterministic Simulation, M. Sc. thesis, Technion, Haifa, Israel, 1991.
13. G. Even, O. Goldreich, M. Luby, N. Nisan and B. Veličković, Approximations of General Independent Distributions, Proc. 24th STOC, 1992, 10-16.
14. J. Håstad, S. Phillips and S. Safra, A Well Characterized Approximation Problem, Inf. Proc. Lett. 47, 1993, 301-305.
15. H. Karloff and Y. Mansour, On Construction of k-wise Independent Random Variables, Proc. 24th STOC, 1994, 564-573.
16. M. Krause and P. Pudlák, On the Computational Power of Depth-2 Circuits with Threshold Modulo Gates, Proc. 26th STOC, 1994, 48-57.
17. A. Lubotzky, R. Phillips and P. Sarnak, Ramanujan Graphs, Combinatorica 8, 1988, 261-277.
18. R. Motwani, J. Naor and M. Naor, The Probabilistic Method Yields Deterministic Parallel Algorithms, J. Comp. Sys. Sci. 49, 1994, 478-516.
19. J. Naor and M. Naor, Small Bias Probability Spaces: Efficient Constructions and Applications, SIAM J. Comp. 22, 1993, 838-856.
20. M. Naor, L. J. Schulman and A. Srinivasan, Splitters and Near-Optimal Derandomization, Proc. 36th FOCS, 1995, 182-191.
21. R. Peralta, On the Randomness Complexity of Algorithms, CS Research Report TR 90-1, University of Wiskonsin, Milwaukee, 1990.
22. U. Vazirani, Randomness, Adversaries and Computation, PhD thesis, University of California, Berkeley, 1986.

6 Appendix

Proof of Lemma 2: Let $X = (x_1, \ldots, x_n)$ be chosen uniformly at random from the sample space S. By assumption, for each $c \in \mathbb{Z}_p$ and each $\beta = (\beta_1, \ldots, \beta_n) \in$

$\mathbb{Z}_p^n \setminus \{0^n\}$ with at most k nonzero entries, we have

$$|(p-1)\cdot \Pr[<\beta, X>_p + c \equiv 0 \bmod p] - \Pr[<\beta, X>_p + c \not\equiv 0 \bmod p]| \le \epsilon\,. \quad (7)$$

We consider w.l.o.g. the first k positions of X, i.e., $x_1, \ldots, x_k$. For each $\alpha = (\alpha_1, \ldots, \alpha_k) \in \mathbb{Z}_p^k$ let p_α be the probability that $x_i = \alpha_i$ for $i = 1, \ldots, k$. For $c \in \mathbb{Z}_p$ and $\beta \in \mathbb{Z}_p^k$, define

$$d_\beta^c = \sum\nolimits_{\alpha\in\mathbb{Z}_p^k} \Phi_\beta^c(\alpha)\cdot p_\alpha\,. \quad (8)$$

By definition of the functions Φ_β^c and by (7), we have

$$|d_\beta^c| = \left| - \sum_{\alpha\in\mathbb{Z}_p^k;<\alpha,\beta>_p+c\equiv 0 \bmod p} p_\alpha\sqrt{p-1} + \sum_{\alpha\in\mathbb{Z}_p^k;<\alpha,\beta>_p+c\not\equiv 0 \bmod p} p_\alpha/\sqrt{p-1}\right|$$
$$= 1/\sqrt{p-1}\cdot|(p-1)\cdot\Pr[<\beta,X>_p + c\equiv 0 \bmod p] - \Pr[<\beta,X>_p + c \not\equiv 0 \bmod p]|$$
$$\le \epsilon/\sqrt{p-1}\,. \quad (9)$$

Claim 1 *For every sequence* $\gamma \in \mathbb{Z}_p^k$, $p_\gamma = p^{-(k+1)}\cdot\sum_{c\in\mathbb{Z}_p}\sum_{\beta\in\mathbb{Z}_p^k} d_\beta^c\cdot\Phi_\beta^c(\gamma)\,.$

Proof. Let the sequence $\gamma \in \mathbb{Z}_p^k$ be given. By multiplying (8) by $\Phi_\beta^c(\gamma)$ and summing over all possible values of $c \in \mathbb{Z}_p$ and $\beta \in \mathbb{Z}_p^k$, we obtain

$$\sum\nolimits_{c\in\mathbb{Z}_p}\sum\nolimits_{\beta\in\mathbb{Z}_p^k} d_\beta^c\cdot\Phi_\beta^c(\gamma) = \sum\nolimits_{c\in\mathbb{Z}_p}\sum\nolimits_{\beta\in\mathbb{Z}_p^k}\sum\nolimits_{\alpha\in\mathbb{Z}_p^k}\Phi_\beta^c(\alpha)\cdot\Phi_\beta^c(\gamma)\cdot p_\alpha\,. \quad (10)$$

In the following we will show that $\sum_{c\in\mathbb{Z}_p}\sum_{\beta\in\mathbb{Z}_p^k}\Phi_\beta^c(\gamma)^2 p_\gamma = p_\gamma p^{k+1}$ (11)

as well as $\sum_{c\in\mathbb{Z}_p}\sum_{\beta\in\mathbb{Z}_p^k}\sum_{\alpha\in\mathbb{Z}_p^k;\alpha\neq\gamma}\Phi_\beta^c(\alpha)\cdot\Phi_\beta^c(\gamma)\cdot p_\alpha = 0\,.$ (12)

To evaluate (10), fix $\alpha \in \mathbb{Z}_p^k$, and consider $\displaystyle\sum_{c\in\mathbb{Z}_p}\sum_{\beta\in\mathbb{Z}_p^k}\Phi_\beta^c(\alpha)\cdot\Phi_\beta^c(\gamma)\cdot p_\alpha\,.$ (13)

We distinguish three cases according to the dependence of α and γ.
Case 1: Let $\alpha = \gamma$. Then, (13) becomes $\sum_{c\in\mathbb{Z}_p}\sum_{\beta\in\mathbb{Z}_p^k}\Phi_\beta^c(\gamma)^2\cdot p_\gamma =$
$= p_\gamma\cdot\sum_{\beta\in\mathbb{Z}_p^k}\left((-\sqrt{p-1})^2 + (p-1)\cdot\left(1/\sqrt{p-1}\right)^2\right) = p_\gamma\cdot p^{k+1}\,.$
Case 2: Let $\alpha = (\alpha_1, \ldots, \alpha_k), \gamma = (\gamma_1, \ldots, \gamma_k) \in \mathbb{Z}_p^k \setminus \{0^k\}$ be linearly independent. Then, there are i, j, $1 \le i < j \le k$ such that (α_i, α_j) and (γ_i, γ_j) are linearly independent in $\mathbb{Z}_p^2$. We want to count the number of terms $\Phi_\beta^c(\alpha)\cdot\Phi_\beta^c(\gamma)$

with the same value. To do so, for fixed $c \in \mathbb{Z}_p$, we partition $\mathbb{Z}_p^k$ into four sets, $\mathbb{Z}_p^k = A^c_{\equiv,\equiv} \cup A^c_{\equiv,\not\equiv} \cup A^c_{\not\equiv,\equiv} \cup A^c_{\not\equiv,\not\equiv}$ where $op_i, op_j \in \{\equiv, \not\equiv\}$

$$A^c_{op_i,op_j} = \{\beta \in \mathbb{Z}_p^k \mid <\alpha,\beta>_p +c \; op_i \; 0 \bmod p \text{ and } <\beta,\gamma>_p +c \; op_j \; 0 \bmod p\} .$$

As (α_i, α_j) and (γ_i, γ_j) are linearly independent, for any $\beta_1, \ldots, \beta_{i-1}, \beta_{i+1}, \ldots, \beta_{j-1}, \beta_{j+1}, \ldots, \beta_k \in \mathbb{Z}_p$ and any $r_1, r_2 \in \mathbb{Z}_p$ there exist unique $\beta', \beta^* \in \mathbb{Z}_p$ such that $<\alpha,\beta>_p +c \equiv r_1 \bmod p$ and $<\beta,\gamma>_p +c \equiv r_2 \bmod p$ for $\beta = (\beta, \ldots, \beta_{i-1}, \beta', \beta_{i+1}, \ldots, \beta_{j-1}, \beta^*, \beta_{j+1}, \ldots, \beta_k)$. Hence, the number of $\beta \in \mathbb{Z}_p^k$ with $<\alpha,\beta>_p +c \equiv r_1 \bmod p$ and $<\beta,\gamma>_p +c \equiv r_2 \bmod p$ is equal to p^{k-2}. Thus, $|A^c_{\equiv,\equiv}| = p^{k-2}$, and, similarly, $\left|A^c_{\equiv,\not\equiv}\right| = \left|A^c_{\not\equiv,\equiv}\right| = (p-1) \cdot p^{k-2}$, and $\left|A^c_{\not\equiv,\not\equiv}\right| = (p-1)^2 \cdot p^{k-2}$. Then, (13) becomes $\sum_{c\in\mathbb{Z}_p} \sum_{\beta\in\mathbb{Z}_p^k} \Phi^c_\beta(\alpha) \cdot \Phi^c_\beta(\gamma) \cdot p_\alpha$ $= p_\alpha \cdot \sum_{c\in\mathbb{Z}_p} \left[\left|A^c_{\equiv,\equiv}\right| \cdot (p-1) + 2 \cdot \left|A^c_{\equiv,\not\equiv}\right| \cdot (-1) + \left|A^c_{\not\equiv,\not\equiv}\right| \cdot 1/(p-1)\right] = 0$.

Case 3: Finally, let α and γ be linearly dependent, but $\alpha \neq \gamma$, i.e., $\alpha = l \cdot \gamma$ for some $l \in \mathbb{Z}_p \backslash \{1\}$. First let $l \neq 0$ and $\gamma \neq 0^k$. For fixed $c \in \mathbb{Z}_p$, as in Case 2, we consider the partition $\mathbb{Z}_p^k = A^c_{\equiv,\equiv} \cup A^c_{\equiv,\not\equiv} \cup A^c_{\not\equiv,\equiv} \cup A^c_{\not\equiv,\not\equiv}$. It is easy to see that the following holds: $|A^c_{\equiv,\equiv}| = \begin{cases} p^{k-1} & \text{if } c = 0 \\ 0 & \text{if } c \neq 0, \end{cases}$

$$|A^c_{\equiv,\not\equiv}| = |A^c_{\not\equiv,\equiv}| = \begin{cases} 0 & \text{if } c = 0 \\ p^{k-1} & \text{if } c \neq 0, \end{cases} \text{ and } |A^c_{\not\equiv,\not\equiv}| = \begin{cases} p^k - p^{k-1} & \text{if } c = 0 \\ p^k - 2p^{k-1} & \text{if } c \neq 0. \end{cases}$$

As in Case 2, (13) becomes $\sum_{c\in\mathbb{Z}_p} \sum_{\beta\in\mathbb{Z}_p^k} \Phi^c_\beta(\alpha) \cdot \Phi^c_\beta(\gamma) \cdot p_\alpha = 0$

For $\gamma = 0^k$ and $\alpha \neq 0^k$, the same holds. Details are in the full version. Hence, summarizing (11), (12) and (10), we proved Claim 1. □

We continue with the proof of Lemma 2. By Claim 1, we have for $\gamma \in \mathbb{Z}_p^k$ that

$$p_\gamma - p^{-k-1} \cdot \sum_{c\in\mathbb{Z}_p} d^c_{0^k} \cdot \Phi^c_{0^k}(\gamma) = p^{-k-1} \cdot \sum_{c\in\mathbb{Z}_p} \sum_{\beta\in\mathbb{Z}_p^k\backslash\{0^k\}} d^c_\beta \cdot \Phi^c_\beta(\gamma). \quad (14)$$

Since $\sum_{\alpha\in\mathbb{Z}_p^k} p_\alpha = 1$, we obtain $p^{-k-1} \cdot \sum_{c\in\mathbb{Z}_p} d^c_{0^k} \cdot \Phi^c_{0^k}(\gamma) =$
$$= p^{-k-1}\left((p-1) \cdot \sum_{\alpha\in\mathbb{Z}_p^k} p_\alpha + 1/(p-1) \cdot \sum_{c\in\mathbb{Z}_p\backslash\{0\}} \sum_{\alpha\in\mathbb{Z}_p^k} p_\alpha\right) = p^{-k} .$$

With (14), we infer $|p_\gamma - p^{-k}| = p^{-k-1} \cdot \left|\sum_{c\in\mathbb{Z}_p} \sum_{\beta\in\mathbb{Z}_p^k\backslash\{0^k\}} d^c_\beta \cdot \Phi^c_\beta(\gamma)\right|$.

By (9), i.e., $|d^c_\beta| \leq \epsilon/\sqrt{p-1}$, $|p_\gamma - p^{-k}| = p^{-k-1} \cdot \left|\sum_{c\in\mathbb{Z}_p} \sum_{\beta\in\mathbb{Z}_p^k\backslash\{0^k\}} d^c_\beta \cdot \Phi^c_\beta(\gamma)\right| \leq$

$$\leq \frac{p^{-k-1} \cdot \epsilon}{\sqrt{p-1}} \cdot \sum_{\beta\in\mathbb{Z}_p^k\backslash\{0^k\}} \sum_{c\in\mathbb{Z}_p} |\Phi^c_\beta(\gamma)| = \frac{p^{-k-1} \cdot \epsilon}{\sqrt{p-1}} \cdot \sum_{\beta\in\mathbb{Z}_p^k\backslash\{0^k\}} \left(\sqrt{p-1} + \frac{p-1}{\sqrt{p-1}}\right)$$
$= 2 \cdot \epsilon/p \cdot (1 - p^{-k})$. □

Hybrid Diagrams: A Deductive-Algorithmic Approach to Hybrid System Verification*

Luca de Alfaro   Arjun Kapur   Zohar Manna

Department of Computer Science
Stanford University

Abstract. We present a methodology for the verification of temporal properties of hybrid systems. The methodology is based on the deductive transformation of *hybrid diagrams*, which represent the system and its properties, and which can be algorithmically checked against the specification. This check either gives a positive answer to the verification problem, or provides guidance for the further transformation of the diagrams. The resulting methodology is complete for quantifier-free linear-time temporal logic.

1 Introduction

Specification and verification methodologies for hybrid systems range from algorithmic methods for the verification of linear-time temporal logic properties [2, 1], to deductive approaches for proving linear-time temporal logic properties [11, 7] and interval-based and duration properties [5, 3]. In this paper we present an approach that combines deductive and algorithmic methods into a methodology that is complete (relative to first-order reasoning) for proving linear-time temporal logic properties of hybrid systems, provided no temporal operator appears in the scope of a quantifier. The advantages of the proposed methodology over the rule-based approach of [11, 6] include the visual representation of the proof process, the provision of proof guidance, and the ability to prove specifications expressed by temporal formulas not in canonical form [10].

Hybrid diagrams are related to the *fairness diagrams* of [4] and to the *hybrid automata* of [2, 1]. They consist of a graph whose vertices are labeled by assertions and whose edges are labeled by transition relations; associated with each diagram are *fairness constraints*, that encode acceptance conditions similar to those of ω-automata. The diagrams represent the system behavior and the safety and progress properties that have been proved about it: the vertex and edge labels represent the safety properties, the fairness constraints represent the progress properties. Hybrid diagrams are sufficiently expressive to encode *phase transition systems* (PTSs) [9, 6], which will be the system model adopted in this paper.

The construction of the proof of a temporal specification begins by representing the system as a one-vertex diagram, whose single edge encodes the possible

* The research was supported in part by the National Science Foundation under grant CCR-9527927, by the Defense Advanced Research Projects Agency under contract NAG2-892, by ARO under grant DAAH04-95-1-0317, and by ARO under the MURI grant DAAH04-96-1-0341.

Reischuk, Morvan (Eds.): STACS'97 Proceedings, LNCS 1200

state transitions of the system. This initial diagram can be transformed using a set of rules that preserve the inclusion of system behaviors, producing a chain of diagram transformations. The aim of this process is to obtain a diagram that can be shown to satisfy the specification by purely algorithmic means.

After any number of transformations, an algorithmic procedure can applied to the last diagram, to either establish that the final diagram (and, by behavior inclusion, the original PTS) satisfies the specification, or it returns a set of *candidate counterexample paths* (CCP) in the diagram. The CCPs provide guidance for the extension of the chain of transformations, following the insights of [13]. Additionally, the CCPs can be used to guide the search for counterexamples, by directing the simulation of the original system along the CCPs.

There are four rules to transform diagrams. The *simulation rule* modifies the graph structure of the diagram, enabling the study of safety properties [4]. The *justice* and *compassion rules* prove progress properties of the diagrams, and represent them as additional fairness constraints. The *pruning rule* eliminates portions of the diagram that are never traversed by any computation along which time diverges. These rules generate first-order verification conditions that must be proved to justify the transformation. The justice and compassion rules are one of the main contributions of this paper, and are at the basis of the completeness results of the methodology. By relying on *ranking* and *delay* functions to measure progress towards given goals, the rules enable the proof of justice and compassion properties of the systems; these properties are then represented as fairness constraints which are added to the diagrams.

2 Phase Transition Systems

The hybrid system model we adopt in this paper is that of *phase transition systems* (PTS) [9, 6]. A PTS is a transition system that allows continuous state changes over time periods of positive duration, as well as discrete state changes in zero time. A PTS $\mathcal{S} = (\mathcal{V}, \theta, \mathcal{T}, \Pi, \mathcal{A})$ consists of the following components.

1. A set $\mathcal{V}$ of typed state variables, partitioned into the set $\mathcal{V}_d$ of *discrete variables*, the set $\mathcal{V}_c$ of *clock variables*, and the set $\mathcal{V}_h$ of *hybrid variables*. Clock variables have type $\mathbb{R}^+$ (i.e. the set of non-negative real numbers) and hybrid variables have type $\mathbb{R}$. We distinguish a special clock variable $T \in \mathcal{V}_c$, representing a *master clock* that measures the amount of time elapsed during the system behavior. The state space S consists of all type-consistent interpretations of the variables in $\mathcal{V}$; we denote by $s[\![x]\!]$ the value at state $s \in S$ of variable $x \in \mathcal{V}$.
2. An assertion θ over $\mathcal{V}$, which defines the set $\{s \in S \mid s \models \theta\}$ of initial states.
3. A finite set $\mathcal{T}$ of transition assertions over $\mathcal{V}$, $\mathcal{V}'$ representing the discrete state changes. Each assertion $\pi \in \mathcal{T}$ represents the transition relation $\{(s,t) \mid (s,t) \models \pi\}$, where (s,t) interprets $x \in \mathcal{V}$ as $s[\![x]\!]$ and $x' \in \mathcal{V}'$ as $t[\![x]\!]$. For all $\pi \in \mathcal{T}$, we require that the implication $\pi \rightarrow T = T'$ holds.
4. A *time-progress* assertion Π over $\mathcal{V}$, used to specify a restriction on the progress of time (see [6] for a discussion of its use).
5. A finite set $\mathcal{A}$ of *activities* representing the continuous state changes. Each activity $a \in \mathcal{A}$ consists of an *enabling assertion* C_a over $\mathcal{V}_d$ and of an evolution function $F_a : S \times \mathbb{R} \mapsto S$. At every $s \in S$ there must be exactly one

$a \in \mathcal{A}$ such that $s \models C_a$. If at time t the system is at a state $s \models C_a$, at time $t + \Delta$ the system will be at state $F_a(s, \Delta)$. For every $a \in \mathcal{A}$, the function F_a must satisfy the equations

$$\forall x \in \mathcal{V}_d \,.\, F_a(s,t)[\![x]\!] = s[\![x]\!] \qquad F_a(s,0) = s$$

$$\forall x \in \mathcal{V}_c \,.\, F_a(s,t)[\![x]\!] = s[\![x]\!] + t \qquad F_a(s,t) = F_a(F_a(s,t'), t - t')$$

for every $s \models C_a$, $t \geq 0$ and $0 \leq t' \leq t$. The function F_a is represented by the set of terms $\{F_a^x\}_{x \in \mathcal{V}}$ over $\mathcal{V} \cup \{\Delta\}$, where the term F_a^x gives the temporal evolution of the value of x as a function of the elapsed time Δ.

To define the set of computations of a PTS, we introduce the assertions $\{tick_a[\Delta]\}_{a \in \mathcal{A}}$, where each $tick_a[\Delta]$ is an assertion over $\mathcal{V} \cup \mathcal{V}'$ and over the parameter Δ, whose domain is the set $\mathbb{R}^+$ of non-negative real numbers. Assertion $tick_a[\Delta]$ describes a state change of the system due to activity a when an amount of time $\Delta \geq 0$ elapses, and is given by:

$$C_a \wedge \Big(\bigwedge_{x \in \mathcal{V}} (x' = F_a^x[\Delta])\Big) \wedge \forall t \,.\, \big(0 \leq t < \Delta \rightarrow \Pi\big[F_a^x[t]/x\big]_{x \in \mathcal{V}}\big) \,.$$

In the above formula, $\Pi[F_a^x[t]/x]_{x \in \mathcal{V}}$ denotes the result of simultaneously replacing for all $x \in \mathcal{V}$ each occurrence of x in Π with $F_a^x[t]$. The form of the assertion $tick_a[\Delta]$ insures that the progress constraint Π holds at every moment of a time-step, except possibly for the final one. As discussed in [6], if Π is used only to encode upper bounds on the transition waiting times, assertion $tick_a[\Delta]$ can be rewritten without quantifiers.

Definition 1 (PTS computations). A *computation* of a PTS $\mathcal{S} = (\mathcal{V}, \theta, \mathcal{T}, \Pi, \mathcal{A})$ is an infinite sequence $\sigma : s_0, s_1, s_2, \ldots$ of states of S that satisfies the following conditions.

1. *Initiality:* $s_0 \models \theta$.
2. *Consecution:* for each $i \geq 0$, either there is a transition $\pi \in \mathcal{T}$ such that $(s_i, s_{i+1}) \models \pi$, or there is an activity $a \in \mathcal{A}$ such that $(s_i, s_{i+1}) \models \exists \Delta \geq 0 \,.\, tick_a[\Delta]$.
3. *Time progress:* for each $t \in \mathbb{R}$ there is $i \in \mathbb{N}$ such that $s_i(T) \geq t$.

We denote by $\mathcal{L}(\mathcal{S})$ the set of computations of a PTS $\mathcal{S}$. □

A Room-Heater Example

As our running example throughout the paper, we consider a variant of the temperature control system presented in [2]. The system, which we call RH, consists of a room with a window and a heater. The window, controlled by some independent agent, may be opened or closed at will. The heater turns on when the temperature is below the threshold temperature of 68°F and turns off when the temperature is above the threshold temperature of 72°F. To prevent mechanical stress, the heater has an embedded clock that prevents it from changing state within 60 seconds of the last change. Initially, the room temperature is below 60°F and the environment temperature (i.e. the temperature outside the room) is 60°F. For simplicity, we assume that the temperature of the environment remains constant at 60°F. Our phase transition system $\mathcal{S} = (\mathcal{V}, \theta, \mathcal{T}, \Pi, \mathcal{A})$, is defined as follows.

1. $\mathcal{V}_d = \{H, W\}$, where H denotes the state of the heater and ranges over domain $\{On, Off\}$, and W denotes the state of the window and ranges over domain $\{Open, Closed\}$. $\mathcal{V}_c = \{T, y\}$, where T is the global clock, and y measures the time elapsed since the last switching *On*/*Off* of the heater. $\mathcal{V}_h = \{x\}$, where x is the temperature of the room.
2. $\theta : H = Off \wedge W = Closed \wedge x < 60 \wedge y = 0 \wedge T = 0$.
3. $\mathcal{T} = \{\tau_1, \tau_2, \tau_3\}$, where $\tau_i : E_i \wedge \rho_i$ for $i \in \{1, 2, 3\}$, and

$$\begin{array}{ll} E_1 : H = Off \wedge x \leq 68 \wedge y \geq 60 & \rho_1 : H' = On \wedge y' = 0 \\ E_2 : H = On \wedge x \geq 72 \wedge y \geq 60 & \rho_2 : H' = Off \wedge y' = 0 \\ E_3 : true & \rho_3 : W' = \neg W \end{array}$$

   where $\neg Open = Closed$ and $\neg Closed = Open$. Variables not mentioned in ρ_1, ρ_2, and ρ_3, respectively, are left unchanged by the transitions.
4. $\Pi = \neg E_1 \wedge \neg E_2$. This insures that τ_1 and τ_2 are taken as soon as they become enabled.
5. $\mathcal{A} = \{a_1, a_2, a_3, a_4\}$, where $F^T_{a_i} = T + \Delta$, $F^y_{a_i} = y + \Delta$, for every $i \in \{1, 2, 3, 4\}$, and C_{a_i} and $F^x_{a_i}$ are defined as follows:

$$\begin{array}{ll} C_{a_1} : H = Off \wedge W = Closed & F^x_{a_1} = 60 + e^{-\Delta/105}(x - 60) \\ C_{a_2} : H = Off \wedge W = Open & F^x_{a_2} = 60 + e^{-\Delta/70}(x - 60) \\ C_{a_3} : H = On \wedge W = Closed & F^x_{a_3} = 75 + e^{-\Delta/105}(x - 75) \\ C_{a_4} : H = On \wedge W = Open & F^x_{a_4} = 70 + e^{-\Delta/70}(x - 70) \, . \end{array}$$

The properties we wish to prove about *RH* state that the room temperature eventually reaches the range from 65°F to 75°F, and that once the temperature is in this range, it will remain in this range forever.

3 Hybrid Diagrams

To study the temporal behavior of a PTS, we introduce *hybrid diagrams*, derived from the *fairness diagrams* of [4]. A *hybrid diagram* (diagram, for short) $A = (\mathcal{V}, V, \rho, \theta, \tau, \mathcal{J}, \mathcal{C})$ consists of the following components.

1. A set $\mathcal{V}$ of typed state variables that includes the master clock T.
2. A set V of *vertices*.
3. A labeling ρ that assigns to each vertex $v \in V$ an assertion $\rho(v)$ over $\mathcal{V}$. A *location* of a diagram is a pair $(v, s) : v \in V, s \models \rho(v)$ composed of a vertex and of a corresponding state, and represents an instantaneous configuration of the diagram.
4. A labeling θ that assigns to each vertex $v \in V$ an initial assertion $\theta(v)$ over $\mathcal{V}$. This labeling defines the set of initial locations $\{(v, s) \mid v \in V, s \models \theta(v)\}$. For all $v \in V$, we require that $\theta(v) \rightarrow (\rho(v) \wedge T = 0)$.
5. A labeling τ that assigns to each edge $(u, v) \in V \times V$ a transition assertion $\tau(u, v)$ over $\mathcal{V} \cup \mathcal{V}'$ and Δ. For $u, v \in V$, assertion $\tau(u, v)$ represents the possible state changes of the system when going from vertex u to vertex v by a time-step of duration $\Delta \in \mathbb{R}^+$. We require that the assertion $\tau(u, v) \rightarrow T' = T + \Delta$ holds for all $u, v \in V$.
6. A set $\mathcal{J}$ of *justice constraints*, and a set $\mathcal{C}$ of *compassion constraints*. The elements of $\mathcal{J}$ and $\mathcal{C}$ are pairs $(R, G) : R \subseteq V, G \subseteq V^2$.

Fig. 1. Hybrid Diagram A_0.

Given an assertion ϕ over $\mathcal{V}$, we denote by ϕ' the formula obtained by replacing each free $x \in \mathcal{V}$ by $x' \in \mathcal{V}'$. Using this notation, we require that a diagram satisfies the requirement $\rho(u) \wedge \tau(u,v) \rightarrow \rho'(v)$, for all $u, v \in V$.

The justice and compassion constraints, collectively called fairness constraints, represent fairness properties that have been proved about the system. For a constraint (R, G), the set $R \subseteq V$ specifies a *request* region; the request is *gratified* when a transition from a vertex u to a vertex v is taken, with $(u, v) \in G$. A just constraint indicates that a request that is performed without interruptions will eventually lead to gratification; a compassionate constraint indicates that a request performed infinitely often will be gratified infinitely often [11, 4].

Definition 2 (diagram computations). A *run* of a diagram is an infinite sequence of locations $(v_0, s_0), (v_1, s_1), (v_2, s_2), \ldots$, satisfying the following conditions.

1. *Initiality:* $s_0 \models \theta(v_0)$.
2. *Consecution:* for all $i \geq 0$, $(s_i, s_{i+1}) \models \exists \Delta \,.\, \tau(v_i, v_{i+1})$.
3. *Time progress:* for each $t \in \mathbb{R}$ there is $i \in \mathbb{N}$ such that $s_i(T) \geq t$.
4. *Justice:* for each constraint $(R, G) \in \mathcal{J}$, if there is $k \in \mathbb{N}$ such that $v_i \in R$ for all $i \geq k$, then there is $j \geq k$ such that $(v_j, v_{j+1}) \in G$.
5. *Compassion:* for each constraint $(R, G) \in \mathcal{C}$, if $v_i \in R$ for infinitely many $i \in \mathbb{N}$, then there are infinitely many $j \in \mathbb{N}$ such that $(v_j, v_{j+1}) \in G$.

If $\sigma : (v_0, s_0), (v_1, s_1), (v_2, s_2), \ldots$ is a run of A, the sequence of states $s_0, s_1, s_2, \ldots$ is a *computation* of A. We denote by *Runs*(A) and $\mathcal{L}(A)$ the sets of runs and computations of A, respectively. □

Every PTS can be represented by a one-vertex diagram, as the following construction shows.

Construction 3. Given a PTS $\mathcal{S} = (\mathcal{V}, \theta, \mathcal{T}, \Pi, \mathcal{A})$, we define the diagram $hd(\mathcal{S}) = (\mathcal{V}, V, \rho, \theta', \tau', \mathcal{J}, \mathcal{C})$ by $V = \{v_0\}$, $\rho(v_0) = true$, $\theta'(v_0) = \theta$, $\mathcal{J} = \emptyset$, $\mathcal{C} = \emptyset$, and

$$\tau'(v_0, v_0) = \Big(\bigvee_{\pi \in \mathcal{T}} (\pi \wedge \Delta = 0)\Big) \vee \Big(\bigvee_{a \in \mathcal{A}} tick_a[\Delta]\Big) . \qquad \square$$

Theorem 4. *For a PTS $\mathcal{S}$, $\mathcal{L}(\mathcal{S}) = \mathcal{L}(hd(\mathcal{S}))$.*

Example 5. In Figure 1, we present the initial diagram $A_0 = hd(RH)$ corresponding to system RH. The transitions τ_1, τ_2, and τ_3 are as in RH (with the added conjunct $\Delta = 0$), and transitions τ_4, τ_5, τ_6, and τ_7 are $tick_{a_1}[\Delta]$, $tick_{a_2}[\Delta]$, $tick_{a_3}[\Delta]$, and $tick_{a_4}[\Delta]$, respectively. The single node u_0 is marked *Init* as a reminder that its initial label $\theta(u_0)$ is equal to the initial condition of the PTS. □

Hybrid diagrams vs. hybrid automata. Hybrid diagrams are related to *hybrid automata*, a formalism widely adopted for the modeling of hybrid systems and for the study of their temporal properties [2, 1]. While sharing a similar labeled-graph structure, the two formalisms differ in some respects.

In a hybrid automaton, the dynamical behavior of the system and the discrete state-transitions are described by different components: the first by differential equations labeling the vertices, the second by transition relations labeling the edges. A hybrid diagram describes both types of evolution using the edge labels: the assertions labeling the vertices represent instead inductive invariants. Moreover, hybrid automata use vertex labels to limit the amount of time for which the system can stay continuously at a vertex. In a hybrid diagram, this role is carried out by the edge labels, which can limit the duration Δ of a time-step.

These differences are motivated by the purposes hybrid automata and hybrid diagrams serve. Hybrid automata were proposed as a formal model of hybrid systems, to which various formal verification methods could be applied. Hybrid diagrams, on the other hand, are meant to provide a deductive representation of a hybrid system and of the safety and progress properties that have been proved about it, and are suited to the application of the diagram transformation rules that will be presented next.

4 Diagram Transformation Rules

The temporal properties of a PTS are studied by means of *transformation rules* [4]. There are four rules: the *simulation rule*, used to study safety properties; the *justice* and *compassion rules*, used to study progress properties; and the *pruning rule*, used to prune portions of a diagram that are never traversed by runs along which time diverges. If a diagram A can be transformed into a diagram B by one of these rules, we write $A \Rightarrow B$, and we indicate by $\stackrel{*}{\Rightarrow}$ the reflexive transitive closure of $\Rightarrow$. The rules preserve language containment: $A \Rightarrow B$ implies $\mathcal{L}(A) \subseteq \mathcal{L}(B)$. Given a PTS $\mathcal{S}$, the rules are used to construct a chain of transformations $hd(\mathcal{S}) \equiv A_0 \Rightarrow A_1 \Rightarrow \cdots \Rightarrow A_n$. At any time, it is possible to check algorithmically whether the last diagram of the chain comply with the specification. This test, discussed in the next section, provides a sufficient condition for the diagram to satisfy the specification, and returns either a positive answer to the verification problem, or guidance for the extension of the chain of transformations.

4.1 Simulation Rule

The *simulation rule*, derived from [4], enables the transformation of a diagram into a new one, such that the second diagram is capable of simulating the first one. A simulation relation between two diagrams A_1 and A_2 is induced by a function $\mu : V_1 \mapsto 2^{V_2}$ from the vertices of A_1 to those of A_2.

Rule 6 (simulation). Let $A_1 = (\mathcal{V}, V_1, \rho_1, \theta_1, \tau_1, \mathcal{J}_1, \mathcal{C}_1)$, $A_2 = (\mathcal{V}, V_2, \rho_2, \theta_2, \tau_2, \mathcal{J}_2, \mathcal{C}_2)$ be two diagrams sharing the same variables. If there is a function $\mu : V_1 \mapsto 2^{V_2}$ that satisfies the conditions below, then $A_1 \Rightarrow A_2$.

1. For all $u \in V_1$, $\theta_1(u) \rightarrow \bigvee_{v \in \mu(u)} \theta_2(v)$.

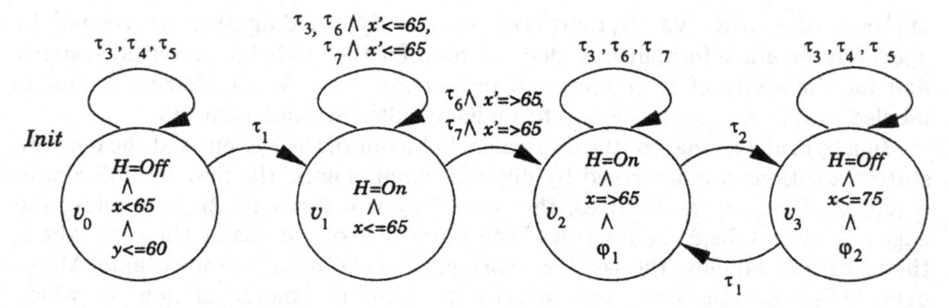

Fig. 2. Hybrid Diagram A_1, where $\varphi_1 : x \leq 75 - 7 \cdot e^{-y/105}$ and $\varphi_2 : x \geq 60 + 12 \cdot e^{-y/70}$. Edges labeled with *false* are not shown.

2. For all $u, u' \in V_1$ and $v \in \mu(u)$,
$$(\rho_1(u) \wedge \rho_2(v) \wedge \tau_1(u, u')) \rightarrow \bigvee_{v' \in \mu(u')} \tau_2(v, v') .$$
3. For each $(R_2, G_2) \in \mathcal{J}_2$ (resp. $\in \mathcal{C}_2$) there is $(R_1, G_1) \in \mathcal{J}_1$ (resp. $\in \mathcal{C}_1$) such that:
   (a) for all $u \in V_1$, if $\mu(u) \cap R_2 \neq \emptyset$ then $u \in R_1$;
   (b) for all $(u, u') \in G_1$ and $v \in \mu(u)$,
   $$(\rho_1(u) \wedge \rho_2(v) \wedge \tau_1(u, u')) \rightarrow \bigvee_{v' \in H(u', v)} \tau_2(v, v') ,$$
   where $H(u', v) = \{v' \mid v' \in \mu(u') \wedge (v, v') \in G_2\}$. □

Theorem 7 (soundness of Rule 6). *If $A_1 \Rightarrow A_2$ by Rule 6, then $\mathcal{L}(A_1) \subseteq \mathcal{L}(A_2)$.*

Example 8. By applying the simulation rule to the diagram A_0 of Figure 1, we obtain the diagram A_1 presented in Figure 2. The application of the rule is based on the function μ defined by $\mu(u_0) = \{v_0, v_1, v_2, v_3\}$. In Figure 2, v_0 is the only vertex satisfying the initial condition specified by θ. □

4.2 Progress Rules

The *justice* and *compassion rules* add new constraints to the justice or compassion sets of a diagram, respectively. Since the rules must preserve language containment, it is possible to add a constraint only if all runs of the diagram already obey it, implying that the constraint represents a progress property of the runs of the diagram. To prove that all runs obey the constraint, the rules rely on *ranking* and *delay* functions to measure progress towards its gratification. The delay functions are similar to the mappings of [8]; our results indicate that to achieve completeness they need to be used in conjunction with ranking functions.

Definition 9 (ranking and delay functions). Recall that *well-founded domain* is a set D together with a relation $>$, such that there is no infinite descending chain $d_0 > d_1 > d_2 > \cdots$ of elements in D.

Given a diagram $A = (\mathcal{V}, V, \rho, \theta, \tau, \mathcal{J}, \mathcal{C})$, let $loc(A) = \{(v, s) \in V \times S \mid s \models \rho(v)\}$ denote the set of locations of A. A *ranking function* $\delta : loc(A) \mapsto D$ for a diagram A is a function mapping locations of A into elements of a well-founded domain D. A *delay function* $\gamma : loc(A) \mapsto \mathbb{R}^+$ is a function mapping locations of A into non-negative real numbers. The ranking and delay functions δ, γ are represented by the families $\{\delta(u)\}_{u \in V}$, $\{\gamma(u)\}_{u \in V}$ of terms on $\mathcal{V}$. □

To add a constraint (R, G), the justice rule relies on ranking and delay functions δ, γ. While in R, δ cannot increase unless an edge in G is taken, and γ gives an upper bound to the amount of time before either an edge in G is taken or R is left.

Rule 10 (justice). Consider a diagram $A = (\mathcal{V}, V, \rho, \theta, \tau, \mathcal{J}, \mathcal{C})$ and a constraint $(R, G) : R \subseteq V, G \subseteq V^2$. Assume that there are ranking and delay functions δ, γ such that, for all $u, v \in R$ with $(u, v) \notin G$, the assertion

$$\rho(u) \wedge \tau(u, v) \quad \rightarrow \quad \delta(u) > \delta'(v) \vee (\delta(u) = \delta'(v) \wedge \gamma(u) \geq \gamma'(v) + \Delta)$$

holds. Then, $A \Rightarrow A'$, where $A' = (\mathcal{V}, V, \rho, \theta, \tau, \mathcal{J} \cup \{(R, G)\}, \mathcal{C})$. □

The rule to add compassion constraints is more involved, and requires the use of a family of assertions $\{\phi(v)\}_{v \in V}$, used to represent a set of locations $\{(v, s) \in loc(A) \mid s \models \phi(v)\}$ that plays the same role of R in leading to the goal.

Rule 11 (compassion). Given a diagram $A = (\mathcal{V}, V, \rho, \theta, \tau, \mathcal{J}, \mathcal{C})$ and a constraint $(R, G) : R \subseteq V, G \subseteq V^2$, assume that there are

1. a family of assertions $\{\phi(v)\}_{v \in V}$ over $\mathcal{V}$, such that $\phi(v) = \mathit{true}$ for all $v \in R$;
2. a ranking function δ and a delay function γ,

such that, for every $u, v \in V$ with $(u, v) \notin G$, the assertions

$$\begin{aligned} \rho(u) \wedge \tau(u, v) \quad &\rightarrow \quad \delta(u) \geq \delta'(v) \\ \rho(u) \wedge \phi(u) \wedge \tau(u, v) \quad &\rightarrow \quad \delta(u) > \delta'(v) \vee \neg\phi'(v) \vee \gamma(u) \geq \gamma'(v) + \Delta \\ \rho(u) \wedge \neg\phi(u) \wedge \tau(u, v) \quad &\rightarrow \quad \delta(u) > \delta'(v) \vee \neg\phi'(v) \end{aligned}$$

hold. Then, $A \Rightarrow A'$, where $A' = (\mathcal{V}, V, \rho, \theta, \tau, \mathcal{J}, \mathcal{C} \cup \{(R, G)\})$. □

Theorem 12 (soundness of Rules 10 and 11). *If a constraint (R, G) is added to a diagram A using Rules 10 or 11, obtaining diagram A', then $Runs(A) = Runs(A')$, and therefore $\mathcal{L}(A) = \mathcal{L}(A')$.*

Example 13. To show that the temperature eventually reaches the desired range, we apply Rule 10 to the diagram A_1 of Figure 2, adding the justice constraint $(\{v_0, v_1\}, \{(v_1, v_2)\})$; we denote by A_2 the resulting diagram. This constraint shows that a run of A_1 cannot stay forever in v_0 or v_1, and must eventually proceed to v_2. The rule uses a ranking function defined by $\delta(v_0) = 1$, $\delta(v_1) = \delta(v_2) = \delta(v_3) = 0$, and a delay function given by $\gamma(v_0) = 60 - y$, $\gamma(v_1) = \mathit{if}\ x \leq 60\ \mathit{then}\ 175 + 105 \ln((75 - x)/49)\ \mathit{else}\ 150 + 70 \ln((70 - x)/10)$, and $\gamma(v_2) = \gamma(v_3) = 0$. □

4.3 Pruning Rule

The *pruning rule* prunes from a diagram a subset of vertices that, because of the presence of a justice constraint, cannot appear in any run of the system.

Rule 14 (pruning). Let $A_1 = (\mathcal{V}, V_1, \rho_1, \theta_1, \tau_1, \mathcal{J}_1, \mathcal{C}_1)$ be a diagram, and let $U_1 \subseteq V_1$ be a subset of its vertices such that the following two conditions hold:

1. there is $(R_1, G_1) \in \mathcal{J}_1$ such that $U_1 \subseteq R_1$, $(U_1 \times V_1) \cap G_1 = \emptyset$;
2. for all $u \in U_1$ and $v \in V_1 - U_1$, $\tau_1(u, v) = \mathit{false}$.

Then, $A_1 \Rightarrow A_2$, where $A_2 = (\mathcal{V}, V_2, \rho_2, \theta_2, \tau_2, \mathcal{J}_2, \mathcal{P}_2)$ is obtained as follows:

1. $V_2 = V_1 - U_1$;
2. ρ_2, θ_2, τ_2 are obtained by restricting the domain of ρ_1, θ_1, and τ_1 to V_2, V_2, and $V_2 \times V_2$, respectively;
3. for each constraint $(R, G) \in \mathcal{J}_1$ (resp. $\in \mathcal{C}_1$), we insert the constraint $(R \cap V_2, G \cap (V_2 \times V_2))$ into $\mathcal{J}_2$ (resp. into $\mathcal{C}_2$). □

This rule can be used in conjunction with Rule 10 to prune from the diagram vertices reached only by invalid runs along which time does not diverge. The soundness of the rule follows from the observation that, if the conditions of the rule are satisfied, no run of the diagram can contain vertices in U. In fact, if a run entered U, it would not be able to leave it, and by staying forever in U it would violate at least one justice constraint of the diagram.

4.4 Completeness Results

Given a diagram $A = (\mathcal{V}, V, \rho, \theta, \tau, \mathcal{J}, \mathcal{C})$, we say that A is *deterministic* if $\theta(v) \wedge \theta(w) \leftrightarrow \mathit{false}$ and $\tau(u, v) \wedge \tau(u, w) \leftrightarrow \mathit{false}$ for all $u, v, w \in V$ with $v \neq w$. The following theorem holds.

Theorem 15. *If the set of computations of a PTS $\mathcal{S}$ is a subset of the set of computations of a deterministic diagram A, we can construct a chain of diagram transformations $hd(\mathcal{S}) \stackrel{*}{\Rightarrow} A$ using Rules 6, 10, 11, and 14.*

This completeness result is relative to first-order reasoning, and is proved by giving the construction of the chain of transformations $hd(\mathcal{S}) \stackrel{*}{\Rightarrow} A$ under the assumption $\mathcal{L}(\mathcal{S}) \subseteq \mathcal{L}(A)$. The proof, which uses ideas from [10, 4], has been omitted due to space constraints.

5 Proving Temporal Properties

In this section we present an algorithm to check whether a diagram satisfies a specification written in the linear-time temporal logic TL_s. The formulas of TL_s are obtained by combining first-order logic formulas by means of the future temporal operators $\bigcirc$ (next), $\Box$ (always), $\Diamond$ (eventually), $\mathcal{U}$ (until), and the corresponding past ones $\ominus$, $\boxminus$, $\diamondminus$ and $\mathcal{S}$ [11]. Given a diagram A and a formula $\phi \in TL_s$, the algorithm provides either a positive answer to $A \models \phi$, or information about the region of the diagram that can contain a counterexample to ϕ. This information can be used as guidance for the extension of the chain of transformations. The first step of the algorithm consists in constructing a Streett automaton $N_{\neg\phi}$ that accepts all the computations that do not satisfy ϕ. The automaton is a first-order version of a classical Streett automaton [12].

Definition 16 (Streett automaton). A (first-order) *Streett automaton* N consists of the components $(\mathcal{V}, (V, E), \rho, Q, \mathcal{B})$, where $\mathcal{V}$, ρ are as in hybrid diagrams; (V, E) is a directed graph with set of vertices V and set of edges $E \subseteq V^2$; $Q \subseteq V$ is the set of *initial vertices*, and $\mathcal{B}$, called the *acceptance list*, is a set of pairs $(P, R) : P, R \subseteq V$. A *run* σ of N is an infinite sequence of locations $(v_0, s_0), (v_1, s_1), (v_2, s_2), \ldots$ such that $v_0 \in Q$, and:

1. for all $i \geq 0$, $s_i \models \rho(v_i)$ and $(v_i, v_{i+1}) \in E$;
2. for each pair $(P, R) \in \mathcal{B}$, either $v_i \in R$ for infinitely many $i \in \mathbb{N}$, or there is $k \in \mathbb{N}$ such that $v_i \in P$ for all $i \geq k$.

If $\sigma : (v_0, s_0), (v_1, s_1), (v_2, s_2), \ldots$ is a run of N, the sequence of states $s_0, s_1, s_2, \ldots$ is a *computation* of N. The set of runs (resp. computations) of a Streett automaton N is denoted by *Runs*(N) (resp. $\mathcal{L}(N)$). □

To show that no behavior of A satisfies $\neg\phi$, the algorithm constructs the *graph product* $A \otimes N_{\neg\phi}$ and checks that no infinite path in it corresponds to a computation of both A and $N_{\neg\phi}$. The construction of the graph product relies on a terminating proof procedure $\vdash$ for the first-order language used in the specification and in the labels of the diagram. The procedure $\vdash$ should be able to prove a subset of the valid sentences that includes all substitution instances of propositional tautologies. Given a first-order formula ψ, we write $\vdash \psi$, $\not\vdash \psi$ depending on whether $\vdash$ terminates with or without a proof of ψ, respectively.

Construction 17 (graph product). Given diagram $A = (\mathcal{V}, U, \rho_A, \theta, \tau, \mathcal{J}, \mathcal{C})$ and Streett automaton $N_{\neg\phi} = (\mathcal{V}, (V, E), \rho_N, Q, \mathcal{B})$, the *graph product* $A \otimes N_{\neg\phi} = (W, Z, H)$ consists of a graph (W, H) and of a set of initial vertices $Z \subseteq W$, and is defined by:

1. $W = \{(u, v) \in U \times V \mid \not\vdash \neg(\rho_A(u) \wedge \rho_N(v))\}$;
2. $Z = \{(u, v) \in W \mid v \in Q \text{ and } \not\vdash \neg(\theta(u) \wedge \rho_N(v))\}$;
3. $H = \big\{((u_1, v_1), (u_2, v_2)) \in W^2 \;\big|\; (v_1, v_2) \in E \text{ and } \not\vdash \neg(\tau(u_1, u_2) \wedge \rho_N(v_1) \wedge \rho'_N(v_2))\big\}$. □

To show that there is no infinite path in the product that corresponds to a computation of both A and $N_{\neg\phi}$, we check that every infinite path in (W, H) starting from Z violates either a constraint of A or a pair in the acceptance list of $N_{\neg\phi}$. To this end, consider a *strongly connected subgraph* (SCS) $X \subseteq W$ of the graph (W, H). We say that X is *admissible* if the following conditions hold:

1. for all $(R, G) \in \mathcal{J}$ (resp. $\in \mathcal{C}$), if $X \subseteq R \times V$ (resp. if $X \cap (R \times V) \neq \emptyset$), then there are $(u_1, v_1), (u_2, v_2) \in X$ such that $(u_1, u_2) \in G$ and $((u_1, v_1), (u_2, v_2)) \in H$;
2. for all $(P, R) \in \mathcal{B}$, if $X \not\subseteq (U \times P)$ then $X \cap (U \times R) \neq \emptyset$.

The following theorem states that if there are no reachable admissible SCSs in the products, then $A \models \phi$. This check can be done in time polynomial in $|W|$ using efficient graph algorithms.

Theorem 18 (diagram checking). *Given a diagram A and a specification $\phi \in TL_s$, let $A \otimes N_{\neg\phi} = (W, Z, H)$. If all SCSs of (W, H) that are reachable in (W, H) from Z are not admissible, then $A \models \phi$.*

The following theorem states that the verification methodology presented in this paper is complete.

Theorem 19 (completeness for TL_s). *Given a PTS $\mathcal{S}$ and a specification $\phi \in TL_s$, if $\mathcal{S} \models \phi$ then there is a chain of transformations $hd(\mathcal{S}) \stackrel{*}{\Rightarrow} A$ such that $A \models \phi$ can be proved using Theorem 18.*

Obtaining Guidance

The presence of admissible and reachable SCSs in the product graph can be used to guide the further analysis of the system, following the insights of [13]. Given an admissible and reachable SCS X of $(W, Z, H) = A \otimes N_{\neg\phi}$, let $X_r \subseteq W$ be the set of vertices that can appear along a path from Z to X in (W, H). Consider the projections $Y = \{u \mid (u, v) \in X\}$, $Y_r = \{u \mid (u, v) \in X_r\}$ of X and X_r onto the diagram A: we say that Y_r and Y constitute a *candidate counterexample path* (CCP) in A. The CCPs correspond to regions of the diagram that can contain counterexamples: if a run $\sigma \in Runs(A)$ violates ϕ, there must be a CCP Y_r, Y such that σ first follows Y_r until it reaches Y, and then remains in Y forever while visiting all vertices of Y infinitely often.

The information provided by the CCPs can be used either to guide the search for a counterexample, or to extend the chain of transformations to show that no counterexample is contained in the CCPs.

Search for counterexample. Given a CCP Y_r, Y, it may be possible to prove that there is a behavior shared by the diagram A and the original PTS $\mathcal{S}$ that follows Y_r and then remains in Y forever, visiting all vertices of Y infinitely often. The existence of such a behavior would establish $\mathcal{S} \not\models \phi$.

Alternatively, the CCPs can be used to guide the simulation of the behavior of $\mathcal{S}$ by simulating $\mathcal{S}$ along the CCPs.

Search for proof. The CCPs provide guidance for the extension of the chain of transformations. The aim of the additional transformations is to show that, for every CCP Y_r, Y:

- either there is no path in Y_r from Z to Y;
- or, after following Y_r, a computation cannot remain in Y forever and visit all the vertices of Y infinitely often.

To show that there is no path in Y_r from Z to Y, it is possible to use the simulation rule to strengthen the assertions of the edges and vertices along Y_r, until the path is interrupted by labeling some edge or vertex with *false*. To show that a computation cannot stay in Y forever and visit all vertices of Y infinitely often, the simulation rule can be used to strengthen the labels of vertices and split vertices into new vertices, thus analyzing in more detail the structure of the SCS Y and possibly splitting it into several SCSs. The justice and compassion rules can be used to show that the system cannot stay forever in Y, or infinitely often in some subsets of Y.

Example 20. Using the algorithm presented in this section, it is possible to check that diagram A_1 of Figure 2 satisfies the specification $(65 \le x \le 75) \rightarrow \Box(65 \le x \le 75)$.

On the other hand, if we check A_1 against the specification $\diamond(65 \leq x \leq 75)$ we obtain two CCPs, corresponding to the SCS $\{v_0\}$, $\{v_1\}$. To prove the specification, we must thus show that either v_0 and v_1 are not reachable (which evidently is not possible), or that a run cannot be forever confined to v_0 or v_1. This is shown by adding the justice constraint $(\{v_0, v_1\}, \{(v_1, v_2)\})$ as in Example 13. The diagram-checking algorithm shows that the resulting diagram A_2 satisfies $\diamond(65 \leq x \leq 75)$. □

Acknowledgments. We thank Todd Neller and Henny Sipma for many useful comments.

References

1. R. Alur, C. Courcoubebetis, N. Halbwachs, T.A. Henzinger, P.-H. Ho, X. Nicollin, A. Olivero, J. Sifakis, and S. Yovine. The algorithmic analysis of hybrid systems. *Theor. Comp. Sci.*, 138(1):3–34, 1995.
2. R. Alur, C. Courcoubetis, T. Henzinger, and P. Ho. Hybrid automata: An algorithmic approach to the specification and analysis of hybrid systems. In *Workshop on Hybrid Systems*, volume 736 of *Lect. Notes in Comp. Sci.*, pages 209–229. Springer-Verlag, 1993.
3. Z. Chaochen, A.P. Ravn, and M.R. Hansen. An extended duration calculus for hybrid real-time systems. In *Hybrid Systems*, volume 736 of *Lect. Notes in Comp. Sci.*, pages 36–59. Springer-Verlag, 1993.
4. L. de Alfaro and Z. Manna. Temporal verification by diagram transformations. In *Computer Aided Verification*, volume 1102 of *Lect. Notes in Comp. Sci.*, pages 288–299. Springer-Verlag, 1996.
5. A. Kapur, T.A. Henzinger, Z. Manna, and A. Pnueli. Proving safety properties of hybrid systems. In *FTRTFT'94*, volume 863 of *Lect. Notes in Comp. Sci.*, pages 431–454. Springer-Verlag, 1994.
6. Y. Kesten, Z. Manna, and A. Pnueli. Verifying clocked transition systems. In *Hybrid Systems III*, volume 1066 of *Lect. Notes in Comp. Sci.*, pages 13–40. Springer-Verlag, 1996.
7. L. Lamport. Hybrid systems in TLA+. In *Hybrid Systems*, volume 736 of *Lect. Notes in Comp. Sci.*, pages 77–102. Springer-Verlag, 1993.
8. N.A. Lynch and H. Attiya. Using mappings to prove timing properties. *Distributed Computing*, 6:121–139, 1992.
9. O. Maler, Z. Manna, and A. Pnueli. From timed to hybrid systems. In *Proc. of the REX Workshop "Real-Time: Theory in Practice"*, volume 600 of *Lect. Notes in Comp. Sci.*, pages 447–484. Springer-Verlag, 1992.
10. Z. Manna and A. Pnueli. Completing the temporal picture. *Theor. Comp. Sci.*, 83(1):97–130, 1991.
11. Z. Manna and A. Pnueli. Models for reactivity. *Acta Informatica*, 30:609–678, 1993.
12. S. Safra. On the complexity of ω-automata. In *Proc. 29th IEEE Symp. Found. of Comp. Sci.*, 1988.
13. H.B. Sipma, T.E. Uribe, and Z. Manna. Deductive model checking. In *Computer Aided Verification*, volume 1102, pages 208–219. Springer-Verlag, 1996.

Temporal Logics for the Specification of Performance and Reliability*

Luca de Alfaro

Department of Computer Science
Stanford University

Abstract. In this paper we present a methodology for the verification of performance and reliability properties of discrete real-time systems. The methodology relies on a temporal logic that can express bounds on the probability of events and on the average time between them. The semantics of the logics is defined with respect to timed systems that exhibit both probabilistic and nondeterministic behavior. We present model-checking algorithms for the algorithmic verification of the specifications, and we discuss their complexity.

1 Introduction

Probabilistic temporal logics have been used for the formal study of correctness and reliability properties of both untimed systems and real-time systems. In this paper, we extend the range of its applications to include performance properties of real-time systems. By introducing an operator that expresses bounds on the average time between events, we propose a unified methodology for the specification and verification of performance, reliability and correctness properties of discrete-time probabilistic systems. The methodology is based on a probabilistic model for the systems, on a specification language derived from temporal logic, and on model-checking algorithms for the verification of system specifications.

We model probabilistic real-time systems as *Markov decision processes* with finite state space [9, 19]. To each state of the Markov decision process is associated a set of actions that can be chosen nondeterministically; the successor of the state is then determined according to the probability distribution arising from the action chosen. Thus, this model can describe both the nondeterministic and the probabilistic components of the system behavior. To each choice of action corresponds a cost, which is interpreted as the amount of time elapsed during the action. In our model, the cost of an action must be either 0 (immediate actions) or 1 (unitary time steps). This system model is closely related to the models proposed in [12, 22, 3].

The specification of system properties is based on the logics pTL and pTL*, and on the use of *instrumentation clocks* to measure the length of intervals of time. The logics pTL and pTL* are obtained by adding the probabilistic

* The research was supported in part by the National Science Foundation under grant CCR-9527927, by the Defense Advanced Research Projects Agency under contract NAG2-892, by ARO under grant DAAH04-95-1-0317, and by ARO under the MURI grant DAAH04-96-1-0341.

Reischuk, Morvan (Eds.): STACS'97 Proceedings, LNCS 1200

operators D and P to the branching-time temporal logics CTL and CTL*. The operator D, introduced in this paper, is used to express bounds on the average time between events. The operator P, already present in the probabilistic logics pCTL and pCTL* [13, 23, 2, 5, 17], is used to express bounds on the probability of system behaviors. The instrumentation clocks, related to the clocks used in timed automata [1], are reset depending on the transitions taken by the system, and their values can be used in the logic to reason about the timing behavior of the system.

To verify whether a system satisfies a specification written in pTL or pTL*, we present model-checking algorithms based on the properties of *stable* subsets of states. A subset of states is stable if there is a choice of nondeterministic actions such that every system behavior that enters the subset will not leave it. The characterization of the properties of stable sets is one of the contributions of this paper, and it leads to uniform algorithms that reduce the model checking problem to the solution of optimization problems on Markov decision processes. The algorithm for the operator P combines the ideas presented in [5] with automata-theoretic constructions, achieving the optimal complexity bound of the earlier algorithm of [6] while exhibiting a relatively simple structure. As discussed in [9, 8, 4], the optimization problems can then be solved by reducing them to linear programming problems. For the logic pTL*, this approach yields model-checking algorithms with time-complexity doubly exponential in the size of the specification, and polynomial in the size of the state space and in the number of bits used to encode the probabilities.

We conclude the paper by discussing an extension of pTL and pTL* that increases the expressive power of the logics by allowing the operator D to refer to arbitrary past formulas.

2 Timed Probabilistic Nondeterministic Systems

A *timed probabilistic nondeterministic system* (TPNS) is a Markov decision process in which the cost of the actions is either 0 or 1 [9].

Definition 1 (TPNS). A TPNS $\Pi = (\mathcal{P}, S, Acts, \kappa, p, c, s_{in})$ consists of the following components.

1. A set $\mathcal{P}$ of propositional symbols.
2. A finite state space S. Every state $s \in S$ assigns truth value $s[\![x]\!]$ to every symbol $x \in \mathcal{P}$.
3. A finite set of actions *Acts*.
4. A function κ, which associates with each $s \in S$ the non-empty set $\kappa(s) \subseteq Acts$ of actions that can be taken at state s.
5. A probability distribution p, that for all $s, t \in S$ and $a \in \kappa(s)$ gives the probability $p(t \mid s, a)$ of a transition from s to t under action a. For all $s \in S$ and $a \in \kappa(s)$, we require $\sum_{t \in S} p(t \mid s, a) = 1$.
6. a function c that associates with each $s \in S$ and $a \in \kappa(s)$ the cost (equal to the elapsed time) $c(s, a) \in \{0, 1\}$ of performing a at s.
7. an initial state $s_{in} \in S$. □

Given a state $s \in S$, the successor of s is chosen in two steps: first, an action $a \in \kappa(s)$ is selected nondeterministically; second, a successor state $t \in S$ is chosen according to the probability distribution $p(t \mid s, a)$. This process, iterated, gives rise to the *behaviors* of a TPNS.

Definition 2 (behaviors of TPNS). A *behavior* of a TPNS Π is an infinite sequence $\omega : s_0 a_0 s_1 a_1 \cdots$ such that $s_i \in S$, $a_i \in \kappa(s_i)$ and $p(s_{i+1} \mid s_i, a_i) > 0$ for all $i \geq 0$. Given a behavior $\omega : s_0 a_0 s_1 a_1 \cdots$, we denote by ω_i the state s_i, by ω_i^a the action a_i, and by $\omega_{\geq i}$ the behavior $s_i a_i s_{i+1} a_{i+1} \cdots$. □

Policies and probability of behaviors. For every state $s \in S$, we denote by Ω_s the set of behaviors starting from s, and we let $\mathcal{B}_s \subseteq 2^{\Omega_s}$ be the σ-algebra of *measurable* subsets of Ω_s, following the classical definition of [14]. To be able to talk about the probability of system behaviors, we would like to associate to each $\Delta \in \mathcal{B}_s$ its probability measure $\mu(\Delta)$. However, this measure is not well-defined, since the probability that a behavior $\omega \in \mathcal{B}_s$ belongs to Δ may depend on the criterion by which the actions are chosen.

To represent these choice criteria, we use the concept of *policy* [9, 19]. A policy η is a set of conditional probabilities $Q_\eta(a \mid s_0 a_0 s_1 \cdots s_n)$, where $a \in \kappa(s_n)$. A policy dictates the probabilities with which the actions are chosen: according to policy η, after the finite prefix $s_0 a_0 s_1 \cdots s_n$ starting at the root $s = s_0$ of Ω_s, action $a \in \kappa(s_n)$ is chosen with probability $Q_\eta(a \mid s_0 a_0 s_1 \cdots s_n)$. Thus, the probability of a direct transition to $t \in S$ after $s_0 \cdots s_n$ is given by

$$\Pr_s^\eta(t \mid s_0 a_0 s_1 \cdots s_n) = \sum_{a \in \kappa(s_n)} p(t \mid s_n, a)\, Q_\eta(a \mid s_0 a_0 s_1 \cdots s_n) .$$

These transition probabilities give rise to a unique probability measure μ_s^η on $\mathcal{B}_s$. We write $\Pr_s^\eta(A)$ to denote the probability of event A in Ω_s under policy η and probability measure μ_s^η, and we adopt the usual conventions to denote conditional probabilities and expectations.

Non-Zeno systems. We say that a TPNS $\Pi = (\mathcal{P}, S, Acts, \kappa, p, c, s_{in})$ is *non-Zeno* if a behavior from s_{in} follows with probability 1 infinitely many time steps, under any policy. Formally, we require that for any policy η, $\Pr_{s_{in}}^\eta\left(\sum_{i=0}^\infty c(\omega_1, \omega_i^a) = \infty\right) = 1$. In general, we are only interested in non-Zeno systems, and the proof of the above property should precede the proof of any other system specification. We will present algorithms to check whether a TPNS is non-Zeno.

Modeling a real-time system with TPNS. While TPNS provide a general model for real-time probabilistic systems, they model systems at a fairly low level. The report [7] introduces *stochastic real-time systems* (SRTS), which provide a more usable modeling language, and it describes how to translate SRTS into TPNS. Translation of stochastic process calculi into models related to TPNS have also been presented in [12]. Since the main focus of this paper are the specification language and the model-checking algorithms, we have omitted the description of these higher-level languages.

3 Specification Language: pTL and pTL*

The specification of performance and reliability properties of TPNS is based on the use of *instrumentation clocks* to measure the length of intervals of time, and on the probabilistic temporal logics pTL and pTL*, that extend pCTL and pCTL* by introducing an operator D to express bounds on the average time between events [13, 2, 5].

3.1 Instrumentation Clocks

An *instrumentation clock* ξ is defined by a propositional formula ξ_τ over $\mathcal{P} \cup \mathcal{P}'$, where $\mathcal{P}' = \{x' \mid x \in \mathcal{P}\}$. The formula ξ_τ specifies a transition relation $\{(s,s') \in S^2 \mid (s,s') \models \xi_\tau\}$, where (s,s') interprets $x \in \mathcal{P}$ as $s[\![x]\!]$ and $x' \in \mathcal{P}'$ as $s'[\![x]\!]$. When a transition from s to s' under action $a \in \kappa(s)$ occurs, clock ξ is incremented by $c(s,a)$ if $(s,s') \not\models \xi_\tau$, and is reset if $(s,s') \models \xi_\tau$. Thus, clock ξ measures the time elapsed since the last state transition that satisfies ξ_τ.

In previous approaches, the specification of timing properties of probabilistic systems relied on temporal operators augmented by time bounds [13, 12, 3]. The instrumentation clocks, derived from the clocks used in timed automata [1], and clocked transition systems [15], lead to a simpler definition of the logic and to a more compact presentation of the model-checking algorithms.

3.2 Syntax of pTL and pTL*

We distinguish two classes of pTL and pTL* formulas: the class *Stat* of *state formulas* (whose truth value is evaluated on the states), and the class *Seq* of *sequence formulas* (whose truth value is evaluated on infinite sequences of states). Given a set $\mathcal{P}$ of predicate symbols and a set C of instrumentation clocks, the classes *Stat* and *Seq* for pTL* are defined inductively as follows.

State formulas:

$$p \in \mathcal{P} \implies p \in \mathit{Stat} \qquad \xi \in C \implies \xi > k, \xi = k \in \mathit{Stat} \tag{1}$$

$$\phi, \psi \in \mathit{Stat} \implies \phi \wedge \psi, \neg\phi \in \mathit{Stat} \qquad \phi \in \mathit{Seq} \implies \mathrm{A}\phi, \mathrm{E}\phi \in \mathit{Stat} \tag{2}$$

$$\phi \in \mathit{Seq} \implies \mathrm{P}_{\bowtie b}\phi \in \mathit{Stat} \qquad \phi \in \mathit{Stat} \implies \mathrm{D}_{\bowtie d}\phi \in \mathit{Stat}\,. \tag{3}$$

Sequence formulas:

$$\phi \in \mathit{Stat} \implies \phi \in \mathit{Seq} \qquad \phi,\ \psi \in \mathit{Seq} \implies \phi \wedge \psi, \neg\phi \in \mathit{Seq} \tag{4}$$

$$\phi \in \mathit{Seq} \implies \Box\phi,\ \Diamond\phi \in \mathit{Seq} \qquad \phi,\ \psi \in \mathit{Seq} \implies \phi\, \mathcal{U}\, \psi \in \mathit{Seq}\,. \tag{5}$$

In the above definition, $\bowtie$ stands for one of $\{<, \leq, \geq, >\}$, $k \in \mathbb{N}$, $b \in [0,1]$ and $d \geq 0$. The temporal operators $\Box$, $\Diamond$, $\mathcal{U}$, and the path quantifiers A, E are taken from CTL* [10], the probabilistic operator P is taken from pCTL* [2, 5], and the operator D originates here. As usual, the other propositional connectives are defined in terms of $\neg$, $\wedge$. The logic pTL is a restricted version of pTL*; its definition is obtained by replacing the clauses (4), (5) with the single clause

$$\phi,\ \psi \in \mathit{Stat} \implies \Box\phi,\ \Diamond\phi,\ \phi\, \mathcal{U}\, \psi \in \mathit{Seq}\,. \tag{6}$$

3.3 Semantics

Given a TPNS Π and a set C of instrumentation clocks, the semantics of a formula ϕ of pTL or pTL* with respect to Π and C is defined in two steps. First, we construct an *instrumented* TPNS Π_C^ϕ, whose states keep track of the value of the clocks; second, we define the satisfaction of ϕ on Π_C^ϕ.

Instrumenting the TPNS. Given a TPNS $\Pi = (\mathcal{P}, S, Acts, \kappa, p, c, s_{in})$ and a set $C = \{\xi_1, \ldots, \xi_n\}$ of instrumentation clocks over S, we construct an *instrumented* TPNS Π_C^ϕ, whose states keep track of the value of the clocks. Note that if M_ξ is the largest constant with which the clock $\xi \in C$ is compared to in ϕ, we need to keep track of the value of ξ only up to $M_\xi + 1$, since no inequality of ϕ changes truth value when the value of ξ increases beyond $M_\xi + 1$. In light of this observation, we define the TPNS $\Pi_C^\phi = (\mathcal{P}, S^*, Acts, \kappa^*, p^*, c^*, s_{in}^*)$ as follows.

1. $S^* = S \times \prod_{\xi \in C}\{0, \ldots, M_\xi + 1\}$. A state $(s, c_1, \ldots, c_n) \in S^*$ assigns value $(s, c_1, \ldots, c_n)[\![x]\!] = s[\![x]\!]$ to $x \in \mathcal{P}$ and $(s, c_1, \ldots, c_n)[\![\xi_i]\!] = c_i$ to $\xi_i \in C$.
2. For $(s, c_1, \ldots, c_n) \in S^*$, $\kappa^*(s, c_1, \ldots, c_n) = \kappa(s)$.
3. For $(s, c_1, \ldots, c_n), (s', c_1', \ldots, c_n') \in S^*$ and $a \in \kappa^*(s, c_1, \ldots, c_n)$, $p^*\big((s', c_1', \ldots, c_n') \mid (s, c_1, \ldots, c_n), a\big)$ is equal to $p(s' \mid s, a)$ if for all $\xi_i \in C$:
$$c_i' = \begin{cases} 0 & \text{if } (s, s') \models \xi_{i_T}; \\ \max\{c(s,a) + c_i, M_{\xi_i} + 1\} & \text{otherwise;} \end{cases}$$
and is equal to 0 otherwise.
4. For $(s, c_1, \ldots, c_n) \in S^*$ and $a \in \kappa(s)$, $c^*\big((s, c_1, \ldots, c_n), a\big) = c(s, a)$.
5. $s_{in}^* = (s_{in}, 0, \ldots, 0)$.

Semantics over the instrumented TPNS. The truth value of pTL and pTL* formulas is then defined with respect to the instrumented TPNS $\Pi_C^\phi = (\mathcal{P}^*, S^*, Acts, \kappa^*, p^*, c^*, s_{in}^*)$. For $\phi \in Stat$, $\psi \in Seq$, we indicate with $s \models \phi$, $\omega \models \psi$ their satisfaction on $s \in S^*$, $\omega \in \bigcup_{s \in S^*} \Omega_s$ respectively. The base cases (1) and the cases for logical connectives are immediate.

Temporal operators. The truth value of $\omega \models \phi$ for a behavior ω and $\phi \in Seq$ is defined in the usual way (see for example [18]).

Path and probabilistic quantifiers. The semantics of the path and probabilistic quantifiers is defined as in pCTL and pCTL* [5]: for $\phi \in Seq$, $0 \le b \le 1$ and $s \in S^*$,

$$s \models \mathrm{A}\phi \;\; \textit{iff} \;\; \forall \omega \in \Omega_s \,.\, \omega \models \phi \qquad\qquad s \models \mathrm{E}\phi \;\; \textit{iff} \;\; \exists \omega \in \Omega_s \,.\, \omega \models \phi \tag{7}$$

$$s \models \mathrm{P}_{\bowtie b}\phi \;\; \textit{iff} \;\; \forall \eta \,.\, \Pr_s^\eta(\omega \models \phi) \bowtie b \,. \tag{8}$$

The intuitive meaning of (8) is that $\mathrm{P}_{\bowtie b}\phi$ holds at $s \in S$ if a behavior has probability $\bowtie b$ of satisfying ϕ, regardless of the policy.

Average-time operator. Given a behavior $\omega : s_0 a_0 s_1 a_1 \cdots$ and $\phi \in Stat$, let $T_{\omega,\phi} = \min\{i \mid \omega_i \models \phi\}$ be the first position at which ϕ holds along ω, with

$T_{\omega,\phi} = \infty$ if $\forall i . \omega_i \not\models \phi$. The *first-passage* cost from s to T under policy η is then defined by

$$C^T_{s,\eta} = \mathrm{E}_{s,\eta}\Big\{ \sum_{i=0}^{T_{\omega,\phi}-1} c(\omega_i, \omega^a_i) \Big\} ,$$

where $\mathrm{E}_{s,\eta}\{\cdot\}$ denotes as usual the expectation with respect to the measure μ^η_s. For $s \in S^*$ and $d \geq 0$ we define

$$s \models \mathrm{D}_{\bowtie d}\phi \quad \textit{iff} \quad \forall \eta . C^T_{s,\eta} \bowtie d . \tag{9}$$

The intuitive meaning of (9) is that $\mathrm{D}_{\bowtie d}\phi$ holds at $s \in S$ if the TPNS reaches a ϕ-state in average time $\bowtie d$, regardless of the policy. Note that this definition relies on the fact that the TPNS in non-Zeno: otherwise, the behaviors on which time never advances would affect the value of the first-passage cost.

Definition 3. Given a TPNS Π, a set C of instrumentation clocks and a specification $\phi \in \mathit{Stat}$, let Π^ϕ_C be the instrumented TPNS, and let s^*_{in} be its initial state. We say that Π instrumented with C satisfies ϕ, written $\Pi \models_C \phi$, iff $s^*_{in} \models \phi$. □

4 Model Checking

In this section, we present algorithms to decide whether an instrumented TPNS Π^ϕ_C is non-Zeno, and whether it satisfies a specification ϕ written in pTL or pTL*. Since the logics pTL and pTL* are obtained by adding the operators P and D to CTL and CTL*, we need to examine only the cases corresponding to these additional operators. The algorithms we introduce are based on the properties of certain subsets of states of a TPNS, called *stable sets.*

4.1 Stable Sets

Intuitively, a subset of the state space of a TPNS is *stable* if there is a policy such that all behaviors that enter the subset will never leave it [5]. Let $\mathrm{Supp}(s, a) = \{t \in S \mid p(t \mid s, a) > 0\}$. Stable sets are defined as follows.

Definition 4 (stable sets). Consider a TPNS $\Pi = (\mathcal{P}, S, \mathit{Acts}, \kappa, p, c, s_{in})$ and a subset $B \subseteq S$. We say that B is *stable* if, for all $s \in B$, there is $a \in \kappa(s)$ such that $\mathrm{Supp}(s, a) \subseteq B$. If B is stable, we define the relation

$$\rho_B = \Big\{ (s,t) \in B \times B \;\Big|\; \exists a \in \kappa(s) . \big(t \in \mathrm{Supp}(s,a) \wedge \mathrm{Supp}(s,a) \subseteq B\big) \Big\} .$$

Intuitively, if $(s, t) \in \rho_B$, then there is an action $a \in \kappa(s)$ that leads from s to t with non-zero probability while leading outside of B with probability 0. If the graph (B, ρ_B) is strongly connected, B is said to be a *strongly connected stable set* (SCSS). Given a subset $C \subseteq S$, we say that B is a *maximal* stable set in C (resp. a maximal SCSS in C) if B is stable (resp. a SCSS) and if there is no other $B' \subseteq C$ such that B' is stable (resp. a SCSS) and $B \subset B'$. □

An equivalent definition of SCSS is provided by the following remark, that we state without proof.

Remark (SCSS and closed recurrent classes). *A set B of states is an SCSS iff there is a Markovian policy η such that B is a closed recurrent class of the Markov chain arising from η.*

The following lemmas summarize the relevant properties of stable sets and SCSS.

Lemma 5. *The following assertions hold.*

1. *Let B be a stable set. Then, there is a policy such that any behavior that enters B will stay in B forever with probability 1.*
2. *Let B be an SCSS. Then, there is a policy such that any behavior that enters B with probability 1 will stay in B forever and will visit all states of B infinitely often.*

Given an infinite behavior ω, we denote by $\mathit{inft}(\omega)$ the set of states that appear infinitely often along ω. The next lemma states that this set is stable with probability 1, under any policy.

Lemma 6. *For all $s \in S$ and all policies η, $\Pr_s^\eta(\mathit{inft}(\omega)$ is an SCSS$) = 1$.*

Proof. Assume, towards the contradiction, that $\Pr_s^\eta(\mathit{inft}(\omega)$ *is an SCSS*$) < 1$. Then, there is a subset $B \subseteq S$ which is not at SCSS, and such that $\Pr_s^\eta(\mathit{inft}(\omega) = B) > 0$. Define $\rho_B = \big\{(t,t') \mid \exists a \in \kappa(t) \,.\, (t' \in \mathrm{Supp}(t,a) \wedge \mathrm{Supp}(t,a) \subseteq B)\big\}$, as for SCSS. Since B is not an SCSS, there are $t_1, t_2 \in B$ such that there is no path from t_1 to t_2 in (B, ρ_B). Define $\Omega_s^B = \{\omega \in \Omega_s \mid \mathit{inft}(\omega) = B\}$, and let

$$q = \min\{\sum_{t' \notin B} p(t' \mid t, a) \mid t \in B \wedge a \in \kappa(t) \wedge \mathrm{Supp}(t,a) \not\subseteq B\} \ .$$

The lack of a path from t_1 to t_2 in (B, ρ_B) implies that at most a fraction $1-q$ of the behaviors that pass from t_1 can then reach t_2 without leaving B. Since every behavior $\omega \in \Omega_s^B$ contains infinitely many disjoint subsequences from t_1 to t_2, we have that $\Pr_s^\eta(\omega \in \Omega_s^B) \leq (1-q)^k$ for all $k > 0$. As $q > 0$, this implies $\Pr_s^\eta(\mathit{inft}(\omega) = B) = 0$, leading to the required contradiction. □

The following corollary states that if a behavior is eventually confined in a set C, with probability 1 it is confined in the union of the SCSS maximal in C.

Corollary 7. *Consider any subset C of states of a TPNS. Let $D_1, \ldots, D_n$ be the SCSS that are maximal in C, and let $D = \bigcup_{i=1}^n D_i$. For any $s \in S$ and any policy η, $\Pr_s^\eta(\mathit{inft}(\omega) \subseteq C \wedge \mathit{inft}(\omega) \not\subseteq D) = 0$.*

Proof. Assume, towards the contradiction, that $\Pr_s^\eta(\mathit{inft}(\omega) \subseteq C \wedge \mathit{inft}(\omega) \not\subseteq D) > 0$. Then, by Lemma 6 there is an SCSS $B \subseteq C$ such that $B \not\subseteq D$ and $\Pr_s^\eta(\mathit{inft}(\omega) = B) > 0$. This contradicts the fact that D is the union of all SCSS maximal in C. □

Define the size $|\Pi|$ of a TPNS $\Pi = (\mathcal{P}, S, Acts, \kappa, p, c, s_{in})$ to be the length of its encoding, where we assume that p and c are represented by listing, for all $s, t \in S$ and $a \in \kappa(s)$, the values of $p(t \mid s, a)$ and $c(s, a)$ as fixed-precision binary numbers. Given a subset $C \subseteq S$, the following algorithm computes the maximal stable set $B \subseteq C$ and the maximal SCSS in C in time polynomial in $|\Pi|$.

Algorithm 8 (stable sets and SCSS).

Input: TPNS $\Pi = (\mathcal{P}, S, Acts, \kappa, p, c, s_{in})$, and a subset $C \subseteq S$.

Output: The stable set B maximal in C, and the list$E_1, \ldots, E_n$ of SCSS maximal in C.

Procedure: Define the functional $\Lambda : 2^S \mapsto 2^S$ by $\Lambda(D) = \{s \in D \mid \exists a \in \kappa(s) \,.\, \mathrm{Supp}(s, a) \subseteq D\}$. Then $B = \lim_{k \to \infty} \Lambda^k(C) = \Lambda^\infty(C)$, and the computation of the limit requires at most $|C|$ iterations, since the functional is monotonic. The SCSS $E_1, \ldots, E_n$ can be computed by computing the maximal strongly connected components of the graph (B, ρ_B). □

4.2 Checking Non-Zenoness

To check whether a TPNS is non-Zeno, we introduce *Zeno* stable sets.

Definition 9 (Zeno stable sets). Given a TPNS $\Pi_c^\phi = (\mathcal{P}, S, Acts, \kappa, p, c, s_{in})$, a *Zeno stable set* (ZSS) is a stable subset $B \subseteq S$ such that for every $s \in B$, there is an action $a \in \kappa(s)$ with $\mathrm{Supp}(s, a) \subseteq B$ and $c(s, a) = 0$. A Zeno stable set is *maximal* if it is not the proper subset of any other Zeno stable set. □

Note that since the union of two ZSS is still a ZSS, every TPNS has a single (possibly empty) maximal ZSS. The maximal ZSS can be computed in time polynomial in the size of the TPNS by the following algorithm.

Algorithm 10 (computation of maximal ZSS).

Input: A TPNS $(\mathcal{P}, S, s_{in}, \kappa, p, c)$.

Output: The maximal ZSS B of the TPNS.

Procedure: Define the functional $\Lambda : 2^S \mapsto 2^S$ by $\Lambda(D) = \{s \in D \mid \exists a \in \kappa(s).(\mathrm{Supp}(s, a) \subseteq D \wedge c(s, a) = 0)\}$. Then $B = \Lambda^\infty(S)$, and the computation requires at most $|S|$ iterations. □

The following theorem states that to check that the TPNS Π_C^ϕ is non-Zeno it suffices to check that there are no reachable ZSS, which by the above results can be done in time polynomial in $|\Pi_C^\phi|$.

Theorem 11 (checking non-Zenoness). *A TPNS is non-Zeno iff it does not contain any non-empty ZSS reachable in (S, ρ_S) from the initial state.*

Proof. If there is a ZSS reachable from the initial state, it is easy to see that there is also a policy under which the system is non-Zeno. In the other direction, assume that the TPNS does not contain any ZSS reachable from the initial state, and consider any policy η. By Lemma 6, $\Pr^\eta_{s_{in}}(\mathit{inft}(\omega)$ *is a SCSS and not a ZSS*$) = 1$. From this it can be shown that every path must take with probability 1 infinitely many actions with cost bounded away from 0, leading to the desired conclusion. □

4.3 Model Checking of pTL* Formulas

The model checking algorithms we present share the same basic structure of those proposed in [11] for CTL and CTL*. Given a TPNS Π_C^ϕ and a formula $\phi \in Stat$, the algorithms recursively evaluate the truth values of the state subformulas of ϕ at all states $s \in S$, following the recursive definitions (1)–(3), until the truth value of ϕ itself can be computed at all $s \in S$. For brevity, we will present algorithms only the logic pTL*, since pTL model checking can be done by combining the results of [5] with the methods presented for the D operator.

Model Checking for the Operator P

From definition (8), to compute whether $s \models \mathrm{P}_{\bowtie b}\psi$ for a state s and a formula $\psi \in Seq$ it suffices to consider the minimum and maximum probabilities with which a computation from s satisfies ψ. Even though these probabilities can be computed using the algorithm presented in [6], we will follow here a different approach. The algorithm we present relies on the properties of stable sets, and shares the insights of the one presented in [5]. However, by relying on the determinization of ω-automata instead of on canonical forms for temporal formulas, the algorithm achieves the optimal complexity bound of the one presented in [6] while exhibiting a relatively simple structure. This algorithm has been recently extended by [17] to logics with fairness assumptions on the policies.

The algorithm. By the results of [6, 5] there are optimal policies η^- and η^+ that minimize and maximize, respectively, the probability $\Pr_s^\eta(\omega \models \psi)$. From (8), to compute whether $s \models \mathrm{P}_{\bowtie b}\psi$ it suffices to compute either the minimum probability $\Pr_s^-(\omega \models \psi) = \Pr_s^{\eta^-}(\omega \models \psi)$ or the maximum one $\Pr_s^+(\omega \models \psi) = \Pr_s^{\eta^+}(\omega \models \psi)$, depending on the direction of the inequality $\bowtie$. Since $\Pr_s^-(\omega \models \psi) = 1 - \Pr_s^+(\omega \models \neg\psi)$, it suffices to give an algorithm for the computation of $\Pr_s^+(\omega \models \psi)$.

Let $\alpha_1, \ldots, \alpha_n \in Stat$ be the maximal state subformulas of ψ, i.e. the state subformulas of ψ that are not proper subformulas of any other state subformula of ψ. Define $\psi' = \psi[r_1/\alpha_1]\ldots[r_n/\alpha_n]$ to be the result of replacing each α_i with a new propositional symbol r_i. The formula ψ' is therefore a linear-time temporal formula constructed from $r_1, \ldots, r_n$ using the temporal operators $\Box$, $\Diamond$, $\mathcal{U}$. As the truth values of $\alpha_1, \ldots, \alpha_n$ have already been computed at all states, we define the *label* $l(t)$ of $t \in S$ by $l(t) = \{r_i \mid 1 \le i \le n \wedge t \models \alpha_i\}$.

It is known from automata theory that ψ' can be translated into a deterministic Rabin automaton $DR_{\psi'} = (Q, q_{in}, \Sigma, \gamma, U)$ with state space Q, initial state $q_{in} \in Q$, alphabet $\Sigma = 2^{\{r_1,\ldots,r_n\}}$, transition relation $\gamma : Q \times \Sigma \mapsto Q$, and acceptance condition U [24, 20, 21]. The acceptance condition is a list $U = \{(H_1, L_1), \ldots, (H_m, L_m)\}$ of pairs of subsets of Q. An infinite sequence $\sigma : b_0b_1b_2\cdots$ of symbols of Σ is accepted by $DR_{\psi'}$ if it induces a sequence $\omega_\sigma : q_0q_1q_2\cdots$ of states of Q s.t. $q_0 = q_{in}$, $\gamma(q_i, b_i) = q_{i+1}$ for all $i \ge 0$ and, for some $1 \le j \le m$, it is $\mathit{inft}(\omega_\sigma) \subseteq H_j$ and $\mathit{inft}(\omega_\sigma) \cap L_j \ne \emptyset$.

Given $\Pi_C^\phi = (\mathcal{P}, S, Acts, \kappa, p, c, s_{in})$, $DR_{\psi'} = (Q, q_{in}, \Sigma, \gamma, U)$ and $s \in S$ we construct the *product TPNS* $\Pi' = (\mathcal{P}, S', Acts, \kappa', p', c', s'_{in})$, where:

1. $S' = S \times Q$, where $(t,q)[\![p]\!] = t[\![p]\!]$ for all $(t,q) \in S'$ and $p \in \mathcal{P}$.
2. For $(t,q) \in S'$, $\kappa'(t,q) = \kappa(t)$.
3. For each $t \in S$ and $a \in \kappa(t)$, the probability $p'((t',q') \mid (t,q),a)$ of a transition to $(t',q') \in S'$ is equal to $p(t' \mid t,a)$ if $\gamma(q,l(t')) = q'$, and is equal to 0 otherwise.
4. For $(t,q) \in S'$ and $a \in \kappa(t)$, $c((t,q),a) = c(t,a)$.
5. $s'_{in} = (s, \gamma(q_{in}, l(s)))$.

Each pair (H_i, L_i), $1 \leq i \leq m$, induces a related pair (H'_i, L'_i) defined by $H'_i = S \times H_i$, $L'_i = S \times L_i$. For each pair (H'_i, L'_i), $1 \leq i \leq m$, we let $B_1^{(i)}, \ldots, B_{n_i}^{(i)}$ be the SCSS maximal in H'_i and having non-empty intersection with L'_i, and we let $T = \bigcup_{i=1}^m \bigcup_{j=1}^{n_i} B_j^{(i)}$. By the previous results, the set T can be computed in time polynomial in $|\Pi'|$. The following theorem states that to compute $\Pr_s^+(\omega \models \psi)$ it suffices to compute the maximum probability of reaching T from s'_{in} in Π'. As discussed in [9, 6], maximum reachability probabilities can be computed by solving a linear programming problem in time polynomial in $|\Pi'|$.

Theorem 12. $\Pr_s^+(\omega \models \psi) = \sup_\eta \Pr_{s'_{in}}^\eta (\exists k \,.\, \omega_k \in T)$.

Proof. Let $R_{\omega,T} \equiv \exists k \,.\, \omega_k \in T$. By construction of Π', it is $\Pr_s^+(\psi) = \sup_\eta \Pr_{s'_{in}}^\eta (\omega \models \psi')$. We can write

$$\Pr_{s'_{in}}^\eta (\omega \models \psi') = \Pr_{s'_{in}}^\eta (R_{\omega,T} \wedge \omega \models \psi') + \Pr_{s'_{in}}^\eta (\neg R_{\omega,T} \wedge \omega \models \psi') \,. \qquad (10)$$

A behavior $\omega \in \Omega_{s'_{in}}$ that satisfies ψ' must, for some $1 \leq i \leq m$, (a) be eventually confined to H'_i, (b) visit infinitely often L'_i. From (a), by Corollary 7, with probability 1 the behavior is eventually confined to the union of the SCSS in H'_i. From (b), the behavior can be eventually confined only to the SCSS in H'_i that have non-empty intersection with L'_i, that is, to $B_1^{(i)}, \ldots, B_{n_i}^{(i)}$. Since $\bigcup_{j=1}^{n_i} B_j^{(i)} \subseteq T$, a behavior that satisfies ψ' will enter T with probability 1, so that the second term on the right side of (10) is 0, and (10) reduces to $\Pr_{s'_{in}}^\eta (\omega \models \psi') = \Pr_{s'_{in}}^\eta (R_{\omega,T} \wedge \omega \models \psi')$. Taking $\sup_\eta$ of both sides and using Lemma 5, we have

$$\sup_\eta \Pr_{s'_{in}}^\eta (\omega \models \psi') = \sup_\eta \Pr_{s'_{in}}^\eta (R_{\omega,T} \wedge \omega \models \psi') = \sup_\eta \Pr_{s'_{in}}^\eta (R_{\omega,T}) \,,$$

and the result follows. □

Model Checking for the Operator D

Given $\psi \in \mathit{Stat}$ and $b \geq 0$, consider the problem of computing the truth value of $\mathrm{D}_{\bowtie b}\psi$ at state $s \in S$ of a TPNS Π_C^ϕ. We assume that Π_C^ϕ is non-Zeno, and that the truth value of ψ has already been computed at all states of S. Let $T = \{s \in S \mid s \models \psi\}$ be the set of states satisfying ψ. From (9), to decide whether $s \models \mathrm{D}_{\bowtie b}\psi$ we need to compute $\inf_\eta C_{s,\eta}^T$, $\sup_\eta C_{s,\eta}^T$. This corresponds to the computation of the minimum and maximum first-passage costs of a Markov decision process. As discussed in [9, 8, 4], these costs can be computed by solving linear-programming problems, which require time polynomial in $|\Pi_C^\phi|$.

Complexity of Model Checking

Combining the results of the previous sections with the results of [6, 5], we get the following theorem.

Theorem 13. *Given a TPNS Π, the following assertions hold:*

1. *Checking whether Π is non-Zeno can be done in polynomial time in $|\Pi|$.*
2. *Model checking a pTL formula ϕ with set of instrumentation clocks C has time-complexity linear in $|\phi|$ and polynomial in $|\Pi|$ and $\prod_{\xi \in C}(M_\xi + 1)$.*
3. *Model checking a pTL* formula ϕ with set of instrumentation clocks C has time-complexity doubly exponential in $|\phi|$, and polynomial in $|\Pi|$ and $\prod_{\xi \in C}(M_\xi + 1)$.*

5 Extending the D Operator to Past Formulas

To conclude, we discuss an extension of the logics pTL and pTL* that increases the expressive power of the logics by allowing the formula ϕ in $D_{\bowtie b}\phi$ to be a *past* temporal formula, instead of a state formula. A past temporal formula is a formula constructed from state formulas in *Stat* using the temporal operators $\boxminus$, $\ominus$ and $\mathcal{S}$ [18].[1]

The semantics of this extension can be defined as follows. Given a behavior ω and a past formula ϕ, let $T_{\omega,\phi} = \min\{i \mid \omega_0 \cdots \omega_i \models_i \phi\}$, where $\omega_0 \cdots \omega_i \models_i \phi$ indicates that ϕ holds at the last position i of the finite sequence of states $\omega_0 \cdots \omega_i$. The truth value of $D_{\bowtie b}\phi$ can be defined as in (9).

This extended version of the D operator can be model checked by combining the techniques of [5] with the algorithms presented in the previous section. Specifically, given a TPNS Π_C^ϕ and a subformula $D_{\bowtie b}\psi$ of ϕ, it is possible to construct a TPNS $\Pi_C^{\psi,\phi}$ in which the states keep track of the truth values of the past subformulas of ψ (ψ itself included). The truth value of $D_{\bowtie b}\psi$ can be computed by applying the algorithms of Section 4.3 to the TPNS $\Pi_C^{\psi,\phi}$. Since the complexity of the model checking is dominated by the doubly-exponential dependency arising from the operator P, the bounds expressed by Theorem 13 apply also to this extended version of the logic.

Acknowledgments. We wish to thank Andrea Bianco for many inspiring discussions and suggestions.

References

1. R. Alur and D. Dill. The theory of timed automata. In *Real-Time: Theory in Practice*, volume 600 of *Lect. Notes in Comp. Sci.*, pages 45–73. Springer-Verlag, 1991.
2. A. Aziz, V. Singhal, F. Balarin, R.K. Brayton, and A.L. Sangiovanni-Vincentelli. It usually works: The temporal logic of stochastic systems. In *Computer Aided Verification*, volume 939 of *Lect. Notes in Comp. Sci.* Springer-Verlag, 1995.

[1] The use of past temporal operators in non-probabilistic branching-time logics has been discussed in depth in [16].

3. D. Beauquier and A. Slissenko. Polytime model checking for timed probabilistic computation tree logic. Technical Report TR-96-08, Dept. of Informatics, Univ. Paris-12, April 1996.
4. D.P. Bertsekas. *Dynamic Programming.* Prentice-Hall, 1987.
5. A. Bianco and L. de Alfaro. Model checking of probabilistic and nondeterministic systems. In *Found. of Software Tech. and Theor. Comp. Sci.*, volume 1026 of *Lect. Notes in Comp. Sci.*, pages 499–513. Springer-Verlag, 1995.
6. C. Courcoubetis and M. Yannakakis. Markov decision processes and regular events. In *ICALP'90*, volume 443 of *Lect. Notes in Comp. Sci.*, pages 336–349. Springer-Verlag, 1990.
7. L. de Alfaro. Formal verification of performance and reliability of real-time systems. Technical Report STAN-CS-TR-96-1571, Stanford University, June 1996.
8. E.V. Denardo. Computing a bias-optimal policy in a discrete-time markov decision problem. *Operations Research*, 18:279–289, 1970.
9. C. Derman. *Finite State Markovian Decision Processes.* Acedemic Press, 1970.
10. E.A. Emerson. Temporal and modal logic. In J. van Leeuwen, editor, *Handbook of Theoretical Computer Science*, volume B, chapter 16, pages 995–1072. Elsevier Science Publishers (North-Holland), Amsterdam, 1990.
11. E.A. Emerson and C.L. Lei. Modalities for model checking: Branching time strikes back. In *Proc. 12th ACM Symp. Princ. of Prog. Lang.*, pages 84–96, 1985.
12. H. Hansson. *Time and Probability in Formal Design of Distributed Systems.* Elsevier, 1994.
13. H. Hansson and B. Jonsson. A framework for reasoning about time and reliability. In *Proc. of Real Time Systems Symposium*, pages 102–111. IEEE, 1989.
14. J.G. Kemeny, J.L. Snell, and A.W. Knapp. *Denumerable Markov Chains.* D. Van Nostrand Company, 1966.
15. Y. Kesten, Z. Manna, and A. Pnueli. Verifying clocked transition systems. In *Hybrid Systems III*, volume 1066 of *Lect. Notes in Comp. Sci.*, pages 13–40. Springer-Verlag, 1996.
16. O. Kupferman and A. Pnueli. Once and for all. In *Proc. 10th IEEE Symp. Logic in Comp. Sci.*, pages 25–35, 1995.
17. M. Kwiatkowska and C. Baier. Model checking for a probabilistic branching time logic with fairness. Technical Report CSR-96-12, University of Birmingham, June 1996.
18. Z. Manna and A. Pnueli. Models for reactivity. *Acta Informatica*, 30:609–678, 1993.
19. M.L. Puterman. *Markov Decision Processes.* John Wiley and Sons, 1994.
20. S. Safra. On the complexity of ω-automata. In *Proc. 29th IEEE Symp. Found. of Comp. Sci.*, 1988.
21. S. Safra. Exponential determinization for ω-automata with strong-fairness acceptance condition. In *Proc. ACM Symp. Theory of Comp.*, pages 275–282, 1992.
22. R. Segala. *Modeling and Verification of Randomized Distributed Real-Time Systems.* PhD thesis, MIT, June 1995. Technical Report MIT/LCS/TR-676.
23. R. Segala and N.A. Lynch. Probabilistic simulations for probabilistic processes. In *CONCUR'94*, volume 836, pages 481–496. Springer-Verlag, 1994.
24. M.Y. Vardi and P. Wolper. An automata-theoretic approach to automatic program verification. In *Proc. First IEEE Symp. Logic in Comp. Sci.*, pages 332–344, 1986.

Efficient Scaling-Invariant Checking of Timed Bisimulation

Carsten Weise and Dirk Lenzkes

Lehrstuhl für Informatik I,
Aachen University of Technology, Germany
{carsten|dlen}informatik.rwth-aachen.de

Keywords: algorithms and data structures, automata and formal languages, program specification and verification, decidability, real-time systems.

Abstract. Bisimulation is an important notion for the verification of distributed systems. Timed bisimulation is its natural extension to real time systems. Timed bisimulation is known to be decidable for timed automata using the so-called region technique. We present a new, top down approach to timed bisimulation which applies the zone technique from the theory of hybrid systems. In contrast to the original decision algorithm, our method has a better space complexity and is scaling invariant: altering the time scale does not effect the space complexity.

1 Introduction

Strong and weak bisimulation ([Mil89]) are useful notions of equivalence for the verification and analysis of distributed systems. Timed bisimulation is the suitable notion of bisimulation for real-time systems. We use timed graphs (or timed automata) ([NSY93,HNSY92,AD94,AC+95]) as the specification formalism for real time systems. Timed graph use the positive reals as time domain. While trace-based equivalences for timed automata are in general undecidable ([AD94]), timed bisimulation is known to be decidable ([Čer92]). The region-technique ([ACD93]) used in the algorithm leads to a high space-complexity, demanding for more efficient approaches. More recent publications on timed automata (e.g. [AC+95]) use a technique we shall refer to as *zone technique*. A zone is a union of regions. The space used by representations of convex zones (i.e. polyhedra) depends on the number of clocks of the automaton, but not on the size of the zone. Therefore keeping zones as large as possible will generally reduce the space complexity of an algorithm.

We will present a decision algorithm for timed bisimulation using the zone technique, which in practice will be more space efficient than the original algorithm. Our algorithm turns out to be *scaling invariant*: the number of zones computed by the algorithm is invariant against re-scaling the involved timed graphs. Re-scaling a timed graph is multiplication of all constants with a fixed factor, as e.g. used to change the resolution of the time scale.

Reischuk, Morvan (Eds.): STACS'97 Proceedings, LNCS 1200

For simplicity, we restrict our presentation to timed simulation (which is "half a bisimulation"). All challenging algorithmic problems are already present within this framework. Generalization to bisimulation is straightforward – though it has to be done with a certain care – and is discussed at the end of the paper together with possible applications and extensions of our algorithm.

We proceed as follows: after recalling the definitions of timed graphs and timed simulation, we give a new characterization of timed simulation. Then we introduce zones and zone graphs, which are the basis of our algorithm. We explain a general algorithm for strong and weak simulation, and discuss correctness and termination. The paper ends with a discussion of the implementation, the complexity, related work, applications and future work.

Due to lack of space we can only hint at how to prove our propositions. The reader is referred to [WL96] for a detailed treatment. In the formalism, impl. denotes implication, as we already use $\Rightarrow$ for the weak transition relation.

2 Timed Graphs

This section recalls the definition of timed graphs, the framework for the presentation of our algorithm. *Timed graphs* model the timing behavior of a real time system using a finite set of clocks, ranging over the positive reals. Timing constraints are expressed by formulae over these clocks. Legal formulae are `tt` (the value true) and finite conjunctions of inequalities comparing a natural number to a clock or a clock difference. All usual comparisons ($<, \leq, =, \geq, >$) are admitted. For a given clock set $\mathcal{C}$, the set of these *simple linear formulae* (SLF) is denoted by $\Phi(\mathcal{C})$, with typical elements ϕ, ψ, etc. For a subset $R \subseteq \mathcal{C}$, $\phi_0(R)$ is the formula which requires all clocks in R to be zero, i.e. $\phi_0(R) = \bigwedge_{C \in R} C = 0$.

A timed graph is a directed, labeled graph, whose nodes (called *locations*) have an associated *invariant* and whose edges (called *transitions*) are labeled with an action, a *guard* and a *reset set*. Invariants and guards are SLF, a reset set is a subset of the clocks.

Definition 1 (Timed Graphs). A *Timed Graph* is a tuple $P = (N, n_0, \mathcal{C}, A, \rightarrow, \mathsf{Inv})$, where N is a finite set of *locations*, $n_0 \in N$ is the *initial location* of P, $\mathcal{C}$ is a finite set of *clocks*, A is a finite set of (synchronization) actions, $\rightarrow \subseteq N \times (A \times \Phi(\mathcal{C}) \times 2^{\mathcal{C}}) \times N$ is the transition relation, and $\mathsf{Inv} : N \rightarrow \Phi(\mathcal{C})$ are the *invariants* of the locations.

We will refer to a timed graph as a *process*. The transition relation of a process P is usually denoted by $\rightarrow_P$. A timed graph is an abstract description of a real time system. The system starts in its initial location with all clocks set to zero. All clocks increase at the same speed, measuring exactly the elapse of time. As long as the clocks' values meet the invariant of a location, the process may choose to stay within the location. A transition can be taken if its guard is valid for the current values of clocks. If a transition is taken, the annotated action occurs with duration zero. Afterwards all clocks in the reset set are set to zero, while the other clocks retain their values. This intuition is made precise using *valuation graphs*, the semantic model of timed graphs.

A valuation is a mapping of the clocks into the positive reals. The set of all valuations for a given clock set $\mathcal{C}$ is $\mathcal{V}(\mathcal{C}) := \{v \mid v : \mathcal{C} \to \mathbb{R}^{\geq 0}\}$, with typical elements v, w, etc.. The following are useful operations on valuations:

time step: $\forall C \in \mathcal{C}.\ (v + d)(C) := v(C) + d, d \in \mathbb{R}^{\geq 0}$

future: $v^{\uparrow} := \{v + d \mid d \in \mathbb{R}^{\geq 0}\}$

past: $v^{\downarrow} := \{v' \mid \exists d \in \mathbb{R}^{\geq 0}.\ v = v' + d\}$

reset: $[R \to 0]v := \begin{cases} \forall C \in R.\ [R \to 0]v(C) := 0 \\ \forall C \notin R.\ [R \to 0]v(C) := v(C) \end{cases}$

restriction: $v|_R \in \mathcal{V}(R)$ where $\forall C \in R.\ v|_R(C) = v(C)$

embedding: $v|^{\mathcal{C}'} := \{w \in \mathcal{V}(\mathcal{C}') \mid w|_{\mathcal{C}} = v|_{\mathcal{C}}\}, \mathcal{C} \subseteq \mathcal{C}'$

The full valuation graph represents the semantics of a timed graph:

Definition 2 (Full Valuation Graph). A Timed Graph $P = (N, n_0, \mathcal{C}, A, \to_P, \mathsf{Inv})$ defines a valuation graph $G = (S, s_0, L, \to)$ where $S := N \times \mathcal{V}(\mathcal{C})$ is the set of states, $s_0 := (n_0, \phi_0(\mathcal{C}))$ is the start state, $L := A \cup \mathbb{R}^{\geq 0}$ is the set of transition labels, and $\to$ is the transition relation defined by

$$(A)\quad \frac{n \xrightarrow{a,\phi,R}_P n', \phi(v) = \mathtt{tt}, v \in \mathsf{Inv}(n), [R \to 0]v \in \mathsf{Inv}(n')}{(n, v) \xrightarrow{a} (n', [R \to 0]v)}$$

$$(T)\quad \frac{d \in \mathbb{R}^{\geq 0}, \forall 0 \leq d' \leq d.\ v + d' \in \mathsf{Inv}(n)}{(n, v) \xrightarrow{d} (n, v + d)}$$

Let $S_0 \subseteq S$ be the set of states reachable from the start state s_0 in G. Then the *full valuation graph of* P is the graph $G_f = (S_0, s_0, L, \to \cap\, S_0 \times L \times S_0)$.

We usually will denote the transition relation of the full valuation graph of a process P by $\to_{VP}$. Often the states of valuation graphs will be called points.

The full valuation graph describes the complete behavior of a process. In the sequel, we often will concentrate on a partial description of the process' behavior. For this we use *valuation graphs*, which are subgraphs of the full valuation graph. As usually, a labeled transition system $T' = (S_0, L, \to_0)$ is a *subgraph* of $T = (S, L, \to)$ (written $T' \leq T$), if $S_0 \subseteq S$ and $\to_0 \subseteq S_0 \times L \times S_0 \cap \to$.

Valuation graphs are *two phase transition systems*: continuous phases, where time passes, alternate with discrete steps, where actions are observable while no time passes.

3 Timed Simulation

This section recalls the definition of *timed (forward) simulation* ([LV91]). Intuitively, a process Q simulates a process P if Q can match every step of P by a step with the same label. Formally this is defined for valuation graphs by:

Definition 3 (Strong Simulation). Given two valuation graphs $V_i = (S_i, s_i, L, \rightarrow_i)(i \in \{1,2\})$, a relation $\mathcal{R} \subseteq S_1 \times S_2$ is a *strong simulation* if for all pairs $(p,v)\,\mathcal{R}\,(q,w)$ and all $\ell \in L$. $(p,v) \xrightarrow{\ell}_1 (p',v')$ impl. $\exists (q',w').\, (q,w) \xrightarrow{\ell}_2 (q',w')$ and $(p',v')\,\mathcal{R}\,(q',w')$.
A state (q,w) *strongly simulates* (p,v) (written $(p,v) \lesssim (q,w)$) if there is a simulation $\mathcal{R}$ such that $(p,v)\,\mathcal{R}\,(q,w)$.
The process Q *strongly simulates* the process P if there is a strong simulation of their valuation graphs which includes their respective start states.

In practice, (weak) simulation is the more important notion. Weak simulation abstracts from internals of the individual processes. To define (weak) simulation, we need the notion of *weak timed transition relation*, which differs from the classical notion ([Mil89]) of the weak transition relation by the additonal laws $(W2)$ and $(W3)$. These laws guarantee *time additivity.*

Definition 4 (Weak Transition Relation). A *timed transition relation* is a transition relation with labels L where $\mathbb{R}^{\geq 0} \cap L \neq \emptyset$.
Given a timed transition relation $\rightarrow$ and a special silent action $\tau \in L$, the *weak timed transition relation* is the least relation $\Rightarrow$ satisfying:

$$(W1)\ \frac{s \xrightarrow{\tau^n}\xrightarrow{\ell}\xrightarrow{\tau^m} s', \ell \in L, n, m \in \mathbb{N}}{s \stackrel{\ell}{\Rightarrow} s'}$$

$$(W2)\ \frac{s \xrightarrow{\tau^n} s', n \in \mathbb{N}}{s \stackrel{0}{\Rightarrow} s'} \qquad (W3)\ \frac{s \stackrel{d}{\Rightarrow} s', s' \stackrel{d'}{\Rightarrow} s'', d, d' \in \mathbb{R}^{\geq 0}}{s \stackrel{d+d'}{\Rightarrow} s''}$$

Then *(weak) timed simulation* is defined by replacing $\rightarrow_\lambda$ by $\Rightarrow_\lambda$ $(\lambda \in \{P, Q\})$ in the definition of strong simulation. If Q simulates P, we write $P \lessapprox Q$. Note that weak ε-transitions of the classical definition are replaced by 0-transitions. We will abuse notation and write $n \xrightarrow{a(d)} n'$ if there is a n'' with $n \xrightarrow{d} n'' \xrightarrow{a} n'$, and analogously for $n \stackrel{a(d)}{\Rightarrow} n'$.

Our approach follows the idea of Čerāns to decide bisimulation of timed graphs by examining the product graph of the involved processes:

Definition 5 (Product of Timed Graphs). Given two Timed Graphs $P_i = (N_i, n_0^i, C_i, A, \rightarrow_i, \mathsf{Inv}_i)(i \in \{1,2\})$ with $C_1 \cap C_2 = \emptyset$, their *strong product* is the Timed Graph $G = P_1 \times P_2 = (N_1 \times N_2, n_0, C_1 \cup C_2, A, \rightarrow, \mathsf{Inv})$ where $n_0 = (n_0^1, n_0^2)$, and $\mathsf{Inv}(n_1, n_2) = \mathsf{Inv}(n_1) \wedge \mathsf{Inv}(n_2)$ for all (n_1, n_2), and $\rightarrow$ is defined by

$$(S)\ \frac{n_1 \xrightarrow{a,\phi_1,R_1}_1 n_1', n_2 \xrightarrow{a,\phi_2,R_2}_2 n_2'}{(n_1, n_2) \xrightarrow{a,\phi_1\wedge\phi_2,R_1\cup R_2} (n_1', n_2')}$$

The *weak product* is the graph $G = P_1 \times_w P_2 = (N_1 \times N_2, n_0, \mathcal{C}_1 \cup \mathcal{C}_2 \cup \{T\}, A \cup \{\tau_Q\}, \rightarrow, \mathsf{Inv})$ where $\rightarrow$ is defined by

$$(W)\ \frac{n_1 \xrightarrow{a,\phi_1,R_1}_1 n_1', n_2 \xrightarrow{a,\phi_2,R_2}_2 n_2', a \neq \tau}{(n_1,n_2) \xrightarrow{a,\phi_1\wedge\phi_2,R_1\cup R_2\cup\{T\}} (n_1',n_2')}$$

$$(P)\ \frac{n_1 \xrightarrow{\tau,\phi_1,R_1}_1 n_1'}{(n_1,n_2) \xrightarrow{\tau,\phi_1,R_1\cup\{T\}} (n_1',n_2)} \qquad (Q)\ \frac{n_2 \xrightarrow{\tau,\phi_2,R_2}_2 n_2'}{(n_1,n_2) \xrightarrow{\tau_Q,\phi_2,R_2} (n_1,n_2')}$$

The product graph requires actions to happen synchronously in both processes. The weak product will be used in the decision of weak simulation. It allows τ-transition to occur independently in each process. Note that the autonomous τ-moves of Q are marked by a special label τ_Q. Both τ and τ_Q are interpreted as internal actions of the product, but must be distinguishable in the algorithm. The fresh clock T will be explained in the Subsect. 5.1. Typically, points of the valuation graph of the product are written $(p \times q, v_P \times v_Q)$, the point which is composed from (p, v_P) and (q, v_Q). We introduce the notion of P-closedness in order to give a new characterization of timed simulation:

Definition 6 (P-closed). Let $\rightarrow_P, \rightarrow_V$ be the transition relations of the full valuation graphs of P and $P \times Q$. A point $s = (p \times q, v_P \times v_Q)$ is *strongly $P(0)$-closed* by definition, and *strongly $P(n+1)$-closed* if for all $\alpha \in A \cup \mathbb{R}^{\geq 0}$, $(p, v_P) \xrightarrow{\alpha}_P (p', v_P')$ implies there is q', v_Q' such that $s \xrightarrow{\alpha}_V s' = (p' \times q', v_P' \times v_Q')$ and s' is strongly $P(n)$-closed.
A point is *strongly P-closed*, if it is strongly $P(n)$-closed for every $n \in \mathbb{N}$. Replacing $\rightarrow_V$ by $\Rightarrow_V$ yields the definition of *(weakly) $P(n)$-closed* and *(weakly) P-closed.*

Intuitively, a point $s = (p \times q, v_P \times v_Q)$ is $P(n+1)$-closed if it can match each step required by (p, v_P), reaching a $P(n)$-closed point. Thus if s is (strongly) P-closed, (q, v_Q) can (strongly) simulate (p, v_P).

Let n be a $P(n+1)$-closed point, then a set M of points is called *matching closed* for n if it contains all the endpoints s' as in Def. 6 necessary to establish $P(n+1)$-closedness. M is matching closed for a set M' if it is matching closed for every point in M'. Using this notion, existence of a simulation relation can be reduced to $P(1)$-closedness:

Theorem 7. *$P \lesssim Q$ iff there is a valuation graph of $P \times Q$ where all points are strongly $P(1)$-closed. $P \lessapprox Q$ iff there is a valuation graph of $P \times_w Q$ and a subset M of its nodes (including the start node) such that all points in M are $P(1)$-closed and M is matching closed for M itself.*

The proof is straightforward, using Čerāns' decidability result. We will use this theorem to establish the correctness of our algorithm.

4 Zone Graphs

The *zone technique* represents valuation graphs by *zone graphs*. This section defines zone graphs and especially ***backward stable zone graphs*** which are the basis of our decision procedure.

Definition 8 (Zone). The *characteristic set* of a simple linear formula ϕ is the set of all valuations for which ϕ holds.
A *zone* is a finite union of characteristic sets.

All operations on valuations (see page 179) can be extended to zones by taking the union of the pointwise application. In the sequel, we will identify characteristic sets with their generating formulae. We will also write $\phi_1 \vee \ldots \vee \phi_n$ for the finite union of the characteristic sets of the ϕ_i. The set of all zones is written $\Phi_\vee(C)$.

A *zone graph* is a graph where each node consists of a location and a zone:

Definition 9 (Zone Graph). For a process $P = (N, p_0, C, A, \rightarrow_P, \mathsf{Inv})$, a *zone-graph* is a transition system $(S, s_0, A, \rightarrow)$, where $S \subseteq N \times \Phi_\vee(C)$, $s_0 = (p_0, \phi_0(C))$ and $\rightarrow \subseteq S \times A \times S$ is connected.

Zone graphs represent valuation graphs. If a node $n = (p, \psi)$ is present in a zone graph, then for every point s in n, all admissible time steps $s \xrightarrow{d} s'$ are in the valuation graph. If additionally an edge $n \xrightarrow{a} n'$ is present, then all transitions $s' \xrightarrow{a} s''$ with s'' in n' are in the valuation graph:

Definition 10 (Valuation Graph of a Zone Graph). Given a process P, its full valuation graph with transition relation $\rightarrow_{VP}$, and a zone graph $Z = (S, s_0, A, \rightarrow)$, the valuation graph of Z is defined by:

$$\frac{(p,\psi) \xrightarrow{a} (p',\psi'), (p,v) \xrightarrow{a}_{VP} (p',v'), v \in \psi^{\uparrow} \wedge \mathsf{Inv}(p), v' \in \psi'}{(p,v) \xrightarrow{a}_V (p',v')}$$

$$\frac{(p,\psi) \in S, v, v+d \in \psi^{\uparrow} \wedge \mathsf{Inv}(p)}{(p,v) \xrightarrow{d}_V (p, v+d)}$$

Note that the valuation graph of a zone graph Z cannot be constructed from Z alone, but knowledge of the underlying timed graph is necessary. We will abuse notation and write $G \leq G'$ for two zone graphs if the valuation graph of G is a subgraph of the valuation graph of G'. For two nodes $(n, \psi), (n', \psi')$ of a zone graph, we will write $(n, \psi) \subseteq (n', \psi')$ iff $n = n'$ and $\psi \subseteq \psi'$.

A zone graph is ***backward stable*** (short: BS-graph) if along a transition $n \xrightarrow{a} n'$ every point in n' is reachable from some point in n:

Definition 11 (BS-Graph). A zone graph $Z = (N, n_0, A, \rightarrow)$ with valuation graph $V = (S, s_0, A \cup \mathbb{R}^{\geq 0}, \rightarrow)$ is called *backward stable* if $(p, \psi) \xrightarrow{a} (p', \psi')$ implies $\forall v' \in \psi'.\ \exists v \in \psi^{\uparrow}.\ (p, v) \xrightarrow{a}_V (p', v')$.

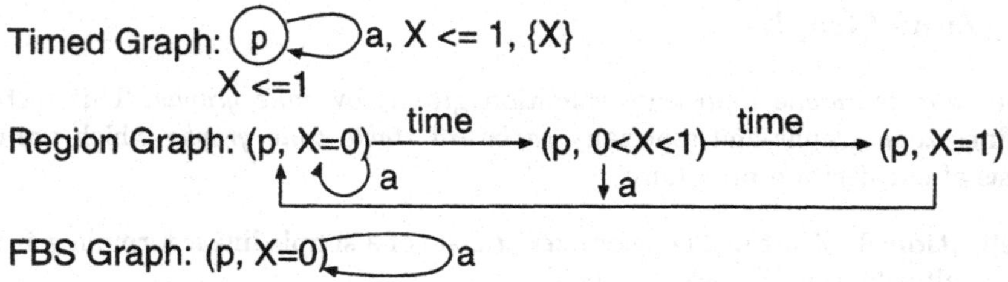

Timed Graph: (p) a, X <= 1, {X}
X <=1
Region Graph: (p, X=0) time → (p, 0<X<1) time → (p, X=1)
a a
FBS Graph: (p, X=0) a

Fig. 1. Timed Graph with Region and FBS-graph

A zone graph is a *full BS-graph* (short: FBS-graph) if it represents the full valuation graph of a process. As FBS-graphs are used in the decision algorithm, we present a construction method for FBS-graphs. For a given zone ψ and an edge $e = p \xrightarrow{a,\phi,R} p'$ of P, the e-successor of ψ is the set of all valuations reachable from (p, ψ) via e: $\mathsf{succ}(\psi, e) := \{v' \mid \exists d \in \mathbb{R}^{\geq 0}.\ (p, v) \xrightarrow{d}_{VP} (p, v+d) \xrightarrow{a}_{VP} (p', v'), \phi(v+d) = \mathtt{tt}, v' = [R \rightarrow 0]v\}$.
The e-successor can be computed using the operations on zones: $\mathsf{succ}(\psi, e) = [R \rightarrow 0](\phi \wedge \psi^{\uparrow} \wedge \mathsf{Inv}(p)) \wedge \mathsf{Inv}(p')$. An FBS-graph is constructed by starting from the initial node n_0. For every node $n = (p, \psi)$ already in the graph and every outgoing edge e of p, the transition $n \xrightarrow{a} (p', \mathsf{succ}(\psi, e))$ is added, until no more new nodes are generated.

In order to compare BS-graphs to the region graphs used by the original algorithm, we recall the definition of regions:

Definition 12 (Region). For a constant $k \in \mathbb{N}$, $\Phi_k(\mathcal{C})$ is the set of all simple linear formula over $\mathcal{C}$ with constants less than k. A *region* is then the smallest zone describable by a formula from $\Phi_k(\mathcal{C})$.
By $\Gamma_k(\mathcal{C})$ we denote the set of all regions w.r.t. $\Phi_k(\mathcal{C})$.

For a given timed graph, the constant c is chosen to be the largest constant appearing in the guards and the invariants of the timed graph. Fig. 1 gives an example of a process and its region- and FBS-graph. In many cases, the FBS-graph constructed by the method given above will be much smaller than the corresponding region graph. The size of the FBS-graph is scaling invariant. Multiplying all constants by a factor c in Fig. 1 will result in an FBS-graph of the same size, while the region graph will have $2*c+1$ nodes. As any finite union of regions is a zone, the number of zones is greater than the number of regions. Thus in principal a zone graph can be larger than a region graph. However this can be avoided by modifying the construction so that all nodes are disjoint, i.e. $\psi \cap \psi' = \emptyset$ for any pair (p, ψ) and (p, ψ'). This is achieved by splitting a node (p, ψ) already in the graph if a node (p, ψ') with $\psi \cap \psi' \neq \emptyset$ is added.

The result of this construction will no longer be backward stable, but has always a size less than or equal to the size of the region graph. Note that there is no way to prevent the worst case, where the zone and the region graph are identical.

5 The Decision Algorithm

Our algorithm uses a top down approach: starting from an FBS-graph G of $P \times Q$ (resp. $P \times_w Q$), all points which are not $P(1)$-closed in G are deleted. If a point is deleted, all its predecessors must be re-examined. The algorithm stops if either there is a set M of $P(1)$-closed points which is matching w.r.t. itself, or if the the start point is removed from G. In the latter case, Q cannot simulate P, while in the former the resulting graph represents a simulation relation.

For a point s, let $\Delta(s) := \{d \in \mathbb{R}^{\geq 0} \mid s \xrightarrow{d}\}$ be its admissible time steps. A point $s = (p \times q, v_P \times v_Q)$ is $P(1)$-closed if matching steps from s can be found for all $d \in \Delta(p, v_P)$ and for all action transitions leaving (p, v_P). For the matches of time steps it is sufficient to find matches which are on the same path in G (see Lemma 15 below). Two points are on the same path in a valuation graph if one of them is reachable from the other by a time step. To formalize this intuition we introduce the notion of *time sequences*. The points of a time sequence are the matches for the time steps of (p, v_P), and thus must stay within the control location p while the control location of Q may change. Time sequences are defined over a *grounded interval*. An interval of $\mathbb{R}^{\geq 0}$ is grounded if it is left-closed starting at zero, i.e. either an interval $[0, t]$ or $[0, t)$ or $\mathbb{R}^{\geq 0}$ itself:

Definition 13 (Time Sequence). Let $V = (S, s_0, L, \rightarrow)$ be a valuation graph of a (weak) product graph, and $s = (p \times q, v_P \times v_Q)$ a point in V. For a grounded interval I, a mapping $\delta : I \rightarrow S$ which fulfills $\delta(0) = s$ and $\forall t \in I.\, \exists q', v'_Q.\, \delta(t) = (p \times q', (v_P + t) \times v'_Q)$ and $\forall t < t' \in I.\, \delta(t) \xrightarrow{t'-t}_V \delta(t')$ is called a *strong time sequence* of s. The definition of a *(weak) time sequence* is yielded by replacing $\rightarrow_V$ by $\Rightarrow_V$ in the last requirement.

Assume a path $(p \times q, v) \xrightarrow{1} (p \times q, v + 1) \xrightarrow{\tau} (p \times q', w) \xrightarrow{2} (p \times q', w + 2)$ in V, then δ with $\delta(t) = (p \times q, v + t)(t \in [0, 1])$ is a strong time sequence, and δ with $\delta(t) = (p \times q, v + t)(t \in [0, 1))$ and $\delta(1 + t) = (p \times q', w + t)(t \in [0, 2])$ is a (weak) time sequence.

The straightforward approach to testing $P(1)$-closedness would be to find a time sequence of matching points for s, and additionally to find matches for the actions required in s. However we will require more: we call a point *good*, if we can find matches for the actions required by all the points in the time sequence:

Definition 14 (Good Points). Let V be a valuation graph of a product graph, $s = (p \times q, v_P \times v_Q)$ a point in V, and $I := \Delta(p, v_P)$. The point s is *strongly good* if there is a strong time sequence $\delta : I \rightarrow N$ such that $\forall d \in I, a \in A.(p, v_P + d) \xrightarrow{a}_P (p', v'_P)$ impl. $\exists q', v'_Q.\, \delta(d) \xrightarrow{a}_V (p' \times q', v'_P \times v'_Q)$.
The property *(weakly) good* is defined by replacing $\rightarrow_V$ by $\Rightarrow_V$.

INPUT: processes P and Q

1. compute the (weak) product of P and Q
2. compute an FBS-graph G of the (weak) product
3. find a subgraph of G where all relevant nodes are good by
   (a) mark the start node as relevant
   (b) initialize the list ex with the start node of G, and the list pcl with the empty set
   (c) repeat
      i. remove node n from ex
      ii. if n is not marked relevant, put all predecessors into ex (removing them from pcl if necessary) and start with the next iteration, else
      iii. compute n', the node containing the maximal subset of good points in n, marking all nodes in $\mathsf{rel}(n)$ as relevant,
      iv. if $(n' \neq n)$ replace n by n' in G, and put all predecessors of n into ex (removing them from pcl if necessary)
      v. add n' to pcl

      until ex empty or the start node is no longer in G

Fig. 2. Generic Algorithm For Strong and Weak Simulation

A *good sequence* is a time sequence used to establish that a point is good. All points of the time sequence of a good point are $P(1)$-closed:

Lemma 15. *If a point s of a product graph is (strongly) good, then all points $\delta(t)$ of the good sequence of s are (strongly) $P(1)$-closed. If a point s and all points reachable by time steps from s are (strongly) $P(1)$-closed, then s is (strongly) good.*

The proof of the first implication is straightforward. In the proof of the second implication, the existence of matching points for every time step of s follows from the definition of $P(1)$-closedness. It it not completely obvious that these points form a time sequence. The proof in [WL96] uses the region graph of the product to give a finite construction method for the time sequence.

Instead of testing all points of the valuation graph of G for $P(1)$-closedness, it is sufficient to test if the points of the nodes of G are good. In the case of weak simulation, we further need to keep track of the matching points. The matching points are the end points $(p' \times q', v'_P \times v'_Q)$ of the a-transitions in Def. 14. Our algorithm can now be described as in Fig. 2. The only problems left are how to compute the maximal set of good points and the set $\mathsf{rel}(n)$ in step (iii). We will explain this briefly in the next subsection.

5.1 Computing Good Points

Details of the computation of good points can be found in [WL96]. All computations use the operations on zones as defined on page 179. The principal idea is to compute two sets of points: in the set ur are all the points which are required to be reachable by time steps by process P but which are not reachable in the product graph, while the set op has all points which are reachable by time steps, but which cannot match all required actions. A point is good if it is not in the past of these two sets.

In the strong case, these sets are easily determined as due to time-determinism all points of a good time sequence are uniquely determined. In the weak case, all maximal paths consisting of τ_Q-steps only may contain good time sequences. The good points are computed relative to these paths. This done by stepping backwards from the end points of the paths to their start points, iterating the computation used for the strong case along the nodes. The additional clock T is needed as timing information in Q may get lost due to clock resets along τ_Q-transitions. Note that the necessity of an additional clock for checking timed bisimulation was already shown in [ACH94].

In the strong case, all nodes can be marked as relevant. In the weak case, only the end nodes of paths needed to find weak action steps are marked relevant.

5.2 Correctness and Termination

The correctness of the algorithm follows from Theorem 7 and Lemma 15. With the definitions we have given so far termination of our algorithm is however not guaranteed, as the number of zones is in fact infinite. Our implementation uses a closure operation on zones which is also present in the region technique. Let c be the largest constant occurring in guards and invariants of the processes. Then there is no need to distinguish between the values of a clock which are greater than this constant c. Formally, the closure operation is defined by:

Definition 16 (Closure). Given a zone $\psi \in \Phi_\vee(\mathcal{C})$ and the set of regions $\Gamma_k(\mathcal{C})$, the *closure of ψ w.r.t. $\Gamma_k(\mathcal{C})$* is defined by $\mathsf{cl}(\psi) := \bigcup\{\gamma \in \Gamma_k(\mathcal{C}) \mid \gamma \cap \psi \neq \emptyset\}$. Let $\mathsf{cl}_k(\mathcal{C})$ be the set of the closures of the zones from $\Phi_\vee(\mathcal{C})$ w.r.t. $\Gamma_k(\mathcal{C})$.

As the number of regions is finite, so is $\mathsf{cl}_k(\mathcal{C})$. Thus using $\mathsf{cl}(\psi)$ instead of ψ for all zones in our algorithm ensures termination, while it can be shown that this does not disturb the correctness of the algorithm. Due to space limitations we cannot present the details here.

6 Conclusion

We presented a new algorithm for deciding timed simulation, improving the original result of Čerāns.. Timed bisimulation and its generalization, timed weak refinement ([CGL93]), can be decided as well. For bisimulation, nodes must be tested for P/Q-closedness, which is closedness for P and Q at the same time. As the operations on zones we use are scaling invariant, so is our algorithm. From the remarks on page 183 it follows that our algorithm always has a better space complexity than the region technique method. The example in Fig. 1 gives a

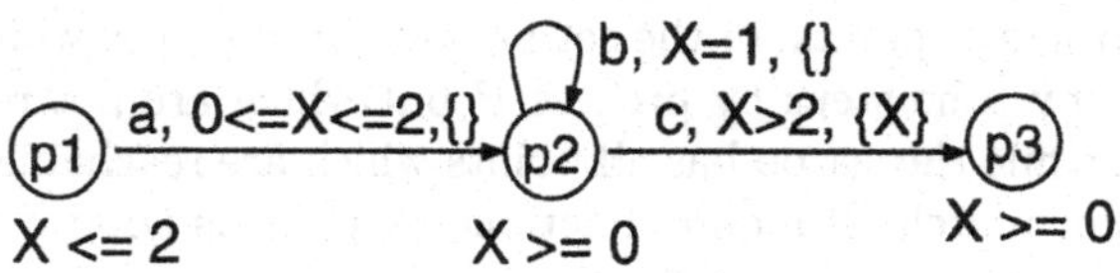

b, X=1, {}
p1 — a, 0<=X<=2,{} → p2 — c, X>2, {X} → p3
X <= 2 X >= 0 X >= 0

Fig. 3. Timed Graph with a small "simple" FBS-graph

hint on how to construct examples where it has a much better space complexity. We have implemented the algorithm in C++. Our implementation is on-the-fly, which even improves performance if Q cannot simulate P. It outperforms EPSILON ([CGL93]) – the only existing implementation of cerans's algorithm – by a factor of 80 even for simple examples, where the region graph is only double the size of the FBS graph.

In general, our implementation represents zones as union of polyhedra (i.e. convex zones), and polyhedra are implemented as difference bounds matrices (see [Dil89]). The zone ψ of a node n is instead stored as a list of zones $\psi_1, \ldots \psi_n$ such that $\psi = \psi_1 \setminus \psi_2 \setminus \ldots \psi_{n-1} \setminus \psi_n$. This representation comes natural, as the set of good points is in fact computed as a set difference between the original zone and the zone containing the "bad points". Whenever a node is diminished, a new ψ_i is added to the list. The representation has two advantages: first, the set difference of two polyhedra ψ_1, ψ_2 is in general a union of more than two polyhedra, so storing ψ_1, ψ_2 instead reduces the space used by the algorithm. Second, in the on-the-fly implementation to determine if a newly computed node is already in the graph we compare the new node's zone to ψ_1, the original zone of an old node. By this we avoid an infinite loop in which a node (p, ψ) is added, diminished to (p, ψ'), and then (p, ψ) is added again, and so forth. This again demonstrates that the main strenght of the algorithm lies in fact that we try to avoid splitting zones whenever possible.

The dual notion to backward stable is forward stable. Region graph are always forward stable. A node is stable if it is backward and forward stable. The algorithms given in [LY93,AC+95] for minimizing timed automata up to bisimulation use stable nodes. This indicates that our algorithm differs substantially from the known approaches to bisimulation, as we prefer to avoid splitting of zones instead of keeping (backward) stability of nodes. Note that the "improved" construction of an FBS-graph on page 183 can have a space complexity which is worse than the "simple" construction on top of page 183. Fig. 3 gives an example of a timed graph for which this is true. We plan to investigate this problem in more detail. While our algorithm reduces the space complexity in many practical cases, it cannot however be better than Čerāns' algorithm in the worst case. Note that our algorithm also tries to reduce space complexity in favor of time complexity, as advised in [HKV96]. We will use our implementation as a back end for the constraint oriented methodology ([LSW95]).

Acknowledgements We would like to thank Karlis Čerāns for pointing out that an efficient algorithm for timed bisimulation need to be scaling invariant.

References

[ACD93] R. Alur, C. Courcoubetis, D. Dill. Modelchecking in dense real-time. Information and Computation, 104(1):2-34, May 1993.

[AC+95] R. Alur,C. Courcoubetis, N. Halbwachs, T.A. Henzinger, et al. The algorithmic analysis of hybrid systems. TCS, February 1995.

[ACH94] R. Alur, C. Courcoubetis, T.A. Henzinger. The observational power of clocks. CONCUR 94, LNCS 836, Springer 1994, pp. 162-177.

[AD94] R. Alur, D.L. Dill. A Theory of Timed Automata. in: Theoretical Computer Science Vol. 126, No. 2, April 1994, pp. 183-236.

[AHV93] R. Alur, T.A. Henzinger, M.Y. Vardi. Parametric real-time reasoning. Proc. 25th STOC, ACM Press 1993, pp. 592–601.

[Čer92] K. Čerāns. Decidability of Bisimulation Equivalences for Parallel Timer Processes. in: Proc. CAV '92, LNCS 663, pp. 302 – 315.

[CGL93] K. Čerāns, J.C. Godsken, K.G. Larsen. Timed Modal Specification - Theory and Tools. in: Proc. CAV '93, LNCS 697, pp. 253–267.

[Dil89] D.L. Dill. Timing Assumptions and Verification of Finite-State Concurrent Systems. in: LNCS 407, Springer Berlin 1989, pp. 197-212.

[HKV96] T.A. Henzinger, O. Kupferman, M.Y. Vardi. A Space-Efficient On-the-fly Algorithm for Real-Time Model Checking. in: Proc. CONCUR 96.

[HNSY92] T.A. Henzinger, X. Nicollin, J. Sifakis, S. Yovine. Symbolic Model Checking for Real-time Systems. LICS '92, pp. 1-13.

[LY93] D. Lee, M. Yannakakis. An efficient algorithm or minimizing real-time transition systems. CAV '93, LNCS 697, Springer Berlin 1993, pp. 210–223.

[LSW95] K.G. Larsen, B. Steffen, C. Weise. Fischer's Protocol Revisited: A Simple Proof Using Modal Constraints. in: LNCS 1066, Springer 1996.

[LV91] N. Lynch, F. Vaandrager. Forward and Backward Simulations for Timing-Based Systems. in: REX Workshop, LNCS 600, pp. 397–446, 1991.

[Mil89] R. Milner. Communication and Concurrency. Prentice-Hall, 1989.

[NSY93] X. Nicollin, J. Sifakis, S. Yovine. From ATP to Timed Graphs and Hybrid Systems. Acta Informatica 30, 1993, S. 181-202.

[WL96] C. Weise, D. Lenzkes. A Fast Decision Algorithm for Timed Refinement. AIB 96-11, Technical Report University of Tech. Aachen, 1996.

Gossiping and Broadcasting versus Computing Functions in Networks

Martin Dietzfelbinger

Fachbereich Informatik, Universität Dortmund, 44221 Dortmund, Germany
email: dietzf@ls2.informatik.uni-dortmund.de

Abstract. The fundamental assumption in the classical theory of gossiping, broadcasting, and accumulation in networks is that atomic pieces of information are communicated in messages that consist of a set of such pieces. The communication mode in synchronous multiprocessor networks that are to compute a function does not fit this model. We show that, under certain assumptions about the way processors may communicate ("*predictable reception*"), computing an arbitrary n-ary function that has a "critical input" and distributing the result to all processors on an n-processor network takes at least as long as performing gossiping in the network graph. A similar relation exists between computing functions with the output appearing at only one processor and the complexity of broadcasting. Our methods can also be applied to extend known lower bounds for broadcasting a bit on EREW PRAMs to the much more general (randomized) distributed memory machines (DMMs).

1 Introduction

The purpose of this paper is to demonstrate that the well-established theory of dissemination of information in networks (gossiping, broadcasting, and accumulation) can directly be applied to characterize the complexity of computing functions in synchronous networks consisting of processors connected by bidirectional links.

1.1 Gossiping and broadcasting. A basic situation considered in *gossiping* theory is the following: each node of a network, which is described as a graph $G = (V, E)$, initially has an atomic piece of information, to be distributed to all other nodes. For this, in rounds, the nodes send each other messages consisting of an arbitrary number of such pieces. A standard restriction is that in one round a node can communicate with only one of its neighbors. One distinguishes 1-*way* (or *half-duplex*) mode, where in a round a link can be used in only one direction, and 2-*way* (or *full-duplex*) mode, where in a round two nodes may exchange all their information through a link that connects them. The most intensively studied efficiency criterion in this theory is the minimum number of rounds needed for disseminating all pieces of information to every node. The *broadcast* problem is similar, excepting that only one piece of information, initially located at one node, is to be spread to all others. The *accumulation* problem is the converse of the broadcast problem: the aim is to collect all pieces of information initially

Reischuk, Morvan (Eds.): STACS'97 Proceedings, LNCS 1200

located at the single nodes in one distinguished node. For an account of the history of the problem area and the intensive research devoted to it see the surveys [10, 13, 16].

1.2 Computing functions in processor networks. The computational model we consider is a network of n processors, $P_1, \ldots, P_n$, that are connected by bidirectional links, according to a network graph $G = (V, E)$. The processor network is to compute an n-ary function $f : A_1 \times \cdots \times A_n \to A$, for arbitrary sets $A_1, \ldots, A_n$, and A. Initially, processor P_i knows the ith component a_i of the input $a = (a_1, \ldots, a_n)$; at the end, all processors know the result $f(a)$ ("global output"). A central example for this output mode arises when a *synchronization barrier* is to be realized in the network, which requires that the Boolean function OR is computed with global output. We will also consider the situation where only one processor has to know the result ("local output"). The processors work synchronously in lock-step, i. e., in global steps $t = 1, \ldots, T$. In one step, a processor may communicate with at most one of its neighbors. Different models are obtained by allowing only 1-way traffic on a link in a step (half-duplex mode) or 2-way-traffic (full-duplex-mode).

It is clear that if a network has a T-round gossiping protocol, it can solve the synchronization problem in T steps with 1-bit messages, and that, under the assumption that messages may be arbitrarily long and computation is for free, it can compute any function with global output in T steps. The central theme of this paper is the question if this is optimal. The answer will depend on the way in which the processor network operates. We will study processor networks with a communication mode that includes the following restriction:

(*) "*Predictable reception*": A processor must know at the beginning of a step whether it is to receive a message in this step or not.

Intuitively, this restriction makes it impossible that information is transferred by *not* sending a message in a certain round. (The relevance of such a possibility in the context of computations on the parallel random-access machine (PRAM) was already observed in [3], and explored in more detail in [1, 5].) Apart from this restriction, the model is quite general. Algorithms that observe (*) are suitable to be executed on asynchronous networks as well (cf. remark after Def. 5).

1.3 Results. We show the following for processor networks with restriction (*): If an n-ary function f that has a "*critical input*" (see Def. 6(a)) can be computed on a processor network G with global output in T steps, then G has a T-round gossiping protocol. Moreover, if an n-ary function f that has a critical input can be computed on a processor network G in T steps with local output at processor P_{i_0}, or a function f that depends on component a_{i_0} can be computed on G in T steps with global output, then G has a T-round broadcast protocol.

These results hold if both in the processor network and in the gossiping network 1-way resp. 2-way communication is assumed. For networks without restriction (*), in §4 it is noted that the OR with global output can not be computed more than four times faster than by a gossiping protocol.

1.4 Applications. As corollaries of the main results we obtain a host of lower bounds for computing functions in networks of different topologies with algorithms that obey (*), since lower bounds for gossiping and broadcasting carry over. As is common in gossiping theory, many of these bounds are tight. E. g., we obtain that

• computing the OR in 1-way mode with global output on a complete network of n processors takes $1.44\ldots\log_2 n \pm O(1)$ steps [8, 18, 19, 22];

• computing the OR in 1-way mode with global output in a ring of n processors takes time $n/2 + \sqrt{2n} \pm O(1)$ [15, 18];

• computing the OR in a complete k-ary tree of depth d with local output at the root takes exactly kd steps; with global output, $2kd$ steps are necessary and sufficient [16].

Finally, in §4 we fully characterize the complexity of broadcasting one bit in some other models, e. g., distributed memory machines (DMMs) with the ARBITRARY access conflict resolution rule (see [7]).

1.5 Related work. In [18], the relevance of the gossiping and broadcasting model for real multiprocessor systems was discussed in depth, but informally, i. e., without making the model for the multiprocessor system explicit. There it is stated, but not proved formally, that lower bounds for gossiping carry over directly to the synchronization problem or, even more generally, to computing any multiple-output function in which all output components depend on all input components, like matrix inversion or computing the discrete Fourier transform of a vector, on arbitrary processor networks. To the best of the knowledge of the author, the problem has not been studied before on a comparable technical level, with the exception of [2], where methods from [8, 18, 19, 22] were combined with lower bound methods for CREW PRAMs from [6] to show that computing the OR function in complete networks that obey restriction (*) and the EXCLUSIVE-WRITE property takes exactly as long as gossiping. The methods used in the present paper are different. They are, however, related to the method introduced in [24] for analyzing computations on CRCW PRAMs with bounded communication width. The method used in §4 is a new and more general formalization of the idea of analyzing PRAM computations by keeping track of those cells and processors that are "affected" by some input bit [1, 3]. In the context of *asynchronous* communication in networks, Tel [23] considers "wave algorithms", which correspond to the problem of computing functions that have a critical input with "local output". Because of the absence of a notion of time in these models, Tel's results do not have direct applications in our setting.

2 Preliminaries

2.1 A formal view of gossiping and broadcasting. In the rigorous definitions given here, we partly follow [16]. Throughout this paper, by a graph $G = (V, E)$ we mean an undirected graph without loops or multiple edges, where $V = \{1, \ldots, n\}$ for some $n \in \mathbf{N}$. An undirected edge between i and j is denoted by $\{i, j\}$, a directed edge from i to j by (i, j). As usual, an *undirected matching* of G is a set $M \subseteq E$ consisting of node-disjoint edges; a *directed matching* in G is a set $M \subseteq V \times V$ of node-disjoint directed edges, such that each $(i, j) \in M$

corresponds to some edge $\{i, j\} \in E$.

Definition 1. A *communication protocol* for a graph $G = (V, E)$ in 1-way [2-way] mode is a sequence $\mathcal{M} = (M_1, \ldots, M_T)$ of directed [undirected] matchings of G. T is the number of rounds of $\mathcal{M}$. We define a sequence $\mathcal{K}(\mathcal{M}) = (K_0, K_1, \ldots, K_T)$ of mappings $K_t: V \to \mathcal{P}(V)$, where $\mathcal{P}(V)$ denotes the power set of V, as follows: $K_0(i) = \{i\}$, for $i \in V$, and, inductively, for $1 \le t \le T$, $i \in V$:

$$K_t(i) = \begin{cases} K_{t-1}(l) \cup K_{t-1}(i), & \text{for the (unique) } l \in V \text{ such that } (l,i) \in M_t \\ & [\{l,i\} \in M_t], \text{ if such an } l \text{ exists.} \\ K_{t-1}(i), & \text{otherwise.} \end{cases}$$

($K_t(i)$ models the set of processors P_j such that by the end of round t processor P_i has received the piece of information initially located at P_j.)

Definition 2. Let G be a graph, $\mathcal{M}$ a communication protocol for G in 1-way [2-way] mode, and let $\mathcal{K}(\mathcal{M})$ be as in the previous definition.
(a) $\mathcal{M}$ is a *gossip protocol* for G in 1-way [2-way] mode if $K_T(i) = V$ for all $i \in V$;
(b) $\mathcal{M}$ is a *broadcast protocol* for G in 1-way mode with source node $i_0 \in V$ if $i_0 \in K_T(i)$ for all $i \in V$;
(c) $\mathcal{M}$ is an *accumulation protocol* for G in 1-way mode with target node $i_0 \in V$ if $K_T(i_0) = V$.

Definition 3. Let $G = (V, E)$ be a graph. (a) The 1-way [2-way] *gossip complexity* $r(G)$ [$r_2(G)$] of G is the minimum T such that there is a gossip protocol for G in 1-way [2-way] mode with T rounds.[1] (b) The *broadcast complexity* $b(G, i_0)$ and the accumulation complexity $a(G, i_0)$ are defined analogously.

One can also define broadcast and accumulation complexity in 2-way mode. However, it is easily seen that these do not differ from their 1-way counterparts, and that we even have the following. (For the proofs see, e. g., [16].)

Fact 4. $a(G, i_0) = b(G, i_0)$ *for all* $G = (V, E)$ *and* $i_0 \in V$. □

2.2 The network model. The description of the machine model we will use will be slightly informal. A fully rigorous definition can quite easily be constructed, e. g., along the lines of the formal description of a CREW PRAM in [3]. We consider networks consisting of n processors, $P_1, \ldots, P_n$, which are connected by a set of bidirectional links. The topology of the network is described by a graph $G = (V, E)$ with $V = \{1, \ldots, n\}$, the edges $\{i, j\} \in E$ representing the links. If the network is to compute a function $f : A_1 \times \cdots \times A_n \to A$, input a_i is given to processor P_i, for $i \in V$, that means, the initial state of P_i depends on a_i. The computation proceeds synchronously in steps $t = 1, \ldots, T$. First, consider the 1-way case. In step t, processor P_i, $i \in V$, on the basis of its state

[1] Originally, the "r" was meant to abbreviate "round complexity".

after step $t-1$, chooses one of the following two possibilities.
(S): Choose a message $m_{i,t}$ and a set $V_{i,t} \subseteq \{j \in V \mid \{i,j\} \in E\}$ representing possible recipients of the message. We say that P_i SENDS a message in this step. (The choice $V_{i,t} = \emptyset$ means that P_i does nothing in this step.)
(R): Choose a nonempty set $W_{i,t} \subseteq \{l \in V \mid \{l,i\} \in E\}$ of neighbors, representing possible senders from which P_i wishes to **RECEIVE** a message.
For deciding which messages to deliver, the set $E_t = \{(i,j) \mid j \in V_{i,t}$ and $i \in W_{j,t}\}$ of possible sender-recipient pairs is considered. It will be important that edges from E_t to be used for communication may be chosen in a greedy manner, without preventing that processors that want to receive a message actually get one. For this, restriction $(*)$ is formulated technically as follows:

$(*)_1$ "*Predictable reception* (1-way)": For any directed matching $M' \subseteq E_t$ there is a directed matching M with $M' \subseteq M \subseteq E_t$ such that if P_j wants to receive a message in step t then $(i,j) \in M$ for some $i \in V$.

Some matching $M \subseteq E_t$ that "covers all recipients" is chosen, and for each pair $(i,j) \in M$ message $m_{i,t}$ is delivered to P_j. Messages that are not delivered are discarded. We require that no matter which decision is made here, the output produced is always correct. (This is analogous to the ARBITRARY write-conflict resolution rule for PRAMs, cf. [17].) Processors P_i that send a message in step t change their state only by noting that step t is finished; those processors P_j that have received a message $m_{i,t}$ assume a new state that also depends on this message. After step T, the result $f(a)$ is known to all processors (in "global output" mode) or to one processor P_{i_0} (in "local output" mode).

Remark. The model described here is quite general: If all senders P_i specify a set $V_{i,t}$ with $|V_{i,t}| = 1$ or $V_{i,t} = \emptyset$, and all recipients P_j specify $W_{j,t} = \{i \mid \{i,j\} \in E\}$, we obtain the natural model in which senders send a message to a specific recipient; $(*)_1$ simplifies to the condition that for each recipient there must be at least one message actually addressed to it. On the other hand, if all recipients P_j specify a set $W_{j,t}$ with $|W_{j,t}| = 1$, and all senders P_i specify $V_{i,t} = \{j \mid \{i,j\} \in E\}$, we obtain the situation in which senders offer their information via a "read window" to all their neighbors; $(*)_1$ turns into the EXCLUSIVE-READ rule known from PRAMs.

In the 2-way (full-duplex) variant of the model, the basic structure of the computation is similar. However, here at the beginning of step t processor P_i fixes a message $m_{i,t}$ and a set $V_{i,t} \subseteq \{j \in V \mid \{i,j\} \in E\}$ of neighbors that are possible partners. Consider the graph $G_t = (V_t, E_t)$, where $V_t = \{i \mid V_{i,t} \neq \emptyset\}$ and $E_t = \{\{i,j\} \mid i,j \in V_t$ and $i \in V_{j,t}$ and $j \in V_{i,t}\}$. The condition "predictable reception" here takes on the following form:

$(*)_2$ "*Predictable reception* (2-way)": Any partial matching $M' \subseteq E_t$ can be extended to a perfect matching M for G_t.

One such perfect matching M is chosen arbitrarily, and the processors communicate according to this matching, i. e., for $\{i,j\} \in M$ message $m_{i,t}$ is delivered to P_j and vice versa.

Definition 5. Let $G = (V, E)$ be a network, and let f be a function.
(a) A network algorithm (in 1-way or 2-way mode) is said to compute f with

"*global output*" ("g") if for all a, after T steps all processors know $f(a)$.
(b) The 1-way network complexity $T_G^{1,g}(f)$ is the minimum number T of steps of a 1-way algorithm with global output that computes f. The 2-way complexity $T_G^{2,g}(f)$ is defined analogously.
(c) A network algorithm in 1-way or 2-way mode is said to compute f with "*local output*" ("l") at node P_{i_0} if for all a, after T steps processor P_{i_0} knows $f(a)$.
(d) The network complexities $T_{G,i_0}^{1,l}(f)$ and $T_{G,i_0}^{2,l}(f)$ are defined for local output at processor P_{i_0} in analogy to the case of global output.

Remark. Restriction $(*)_1$ arises quite naturally in connection with the following problem: Assume that in an asynchronous network internal computations of processors and delivery of messages may be delayed for indefinite but finite periods of time. Still, we want to perform an algorithm written for a synchronous network in which processors can either receive or send one message in one step. The obvious idea is to have each processor keep an internal (virtual) step counter, and to keep the exchange of messages synchronized by the use of time stamps attached to the messages. Synchronous algorithms that obey restriction $(*)_1$ are suited for being run in this way on an asynchronous network; no matter what delays occur and what arbitrary decisions are made in the case of conflicting requests, it never happens that a processor waits indefinitely for a message with some time stamp t that will not arrive. (In the terminology of asynchronous distributed systems, this is a "liveness" property; see [23].)

2.3 Complexity measures for functions.

Definition 6. (Cf. [9].) Fix a function $f : A_1 \times \cdots \times A_n \to A$.
(a) For each input a, the *critical complexity* $c(f,a)$ is the maximal k such that for k different indices $i \in \{1,\ldots,n\}$ there is an input b that differs from a only in the ith component and satisfies $f(a) \neq f(b)$. An input a is *critical* if $c(f,a) = n$. The critical complexity $c(f)$ of f is $\max\{c(f,a) \mid a \text{ is an input}\}$.
(b) For each input a, the *sensitive complexity* (or *certificate complexity*) $s(f,a)$ of f at a is the minimal k such that there is a set $I_a \subseteq \{1,\ldots,n\}$ of cardinality k with the property that all inputs $b = (b_1,\ldots,b_n)$ with $\forall i \in I_a : a_i = b_i$ satisfy $f(a) = f(b)$. The sensitive complexity $s(f)$ of f is $\max\{s(f,a) \mid a \text{ is an input}\}$.
(c) The function f *depends* on input bit i if there are inputs a and b that differ only in the ith component and satisfy $f(a) \neq f(b)$.

3 Computing functions versus gossiping

3.1 The main result. Throughout this section, let $G = (V,E)$ be a graph and let $i_0 \in V$. We assume alternately that G represents a gossiping network and a processor network that operates in accordance with $(*)_1$ resp. $(*)_2$. The following is well known and an almost immediate consequence of the definitions.

Observation 7. *If* $f : A_1 \times \cdots \times A_n \to A$ *is an arbitrary n-ary function, then* (a) $T_G^{1,g}(f) \leq r(G)$, (b) $T_G^{2,g}(f) \leq r_2(G)$, *and* (c) *both* $T_{G,i_0}^{1,l}(f)$ *and* $T_{G,i_0}^{2,l}(f)$ *do* *not exceed* $a(G, i_0) = b(G, i_0)$, *for* $i_0 \in V$. *If* f *is the* OR *of* n *bits, it can be computed in the respective mode* "g" *or* "l" *within these time bounds by using 1-bit messages only.* □

The main result of this paper is essentially that, under the restriction "predictable reception", these upper bounds are optimal if f has a critical input.

Theorem 8. *Let* $G = (V, E)$ *be a graph, and let* f *be an* n*-ary function. If* f *has a critical input, then*
(a) $T_G^{1,g}(f) = r(G)$;
(b) $T_G^{2,g}(f) = r_2(G)$;
(c) $T_{G,i_0}^{1,l}(f) = T_{G,i_0}^{2,l}(f) = a(G, i_0) = b(G, i_0)$, *for* $i_0 \in V$.
Further, if f *depends on the* i_0*th input component for some* $i_0 \in V$, *then*
(d) $T_G^{1,g}(f), T_G^{2,g}(f) \geq b(G, i_0)$.

Corollary 9. *The complexity of computing the* OR *of* n *bits on* G *is* $r(G)$ $[r_2(G)]$ *for 1-way [2-way] communication and global output, and* $b(G, i_0)$ *for 1-way or 2-way communication and local output at processor* P_{i_0}. *In the upper bounds, 1-bit messages are sufficient.* □

Proof of Thm. 8. We deal with part (a) in detail; the proofs of the other parts, being similar, will only be sketched. In view of Obs. 7, only the inequality $T_G^{1,g}(f) \geq r(G)$ has to be proved. Assume that a 1-way algorithm for computing f on G in T steps is given, and that $a^* = (a_1^*, \ldots, a_n^*)$ is a critical input for f. We must construct a 1-way gossip protocol for G that has T rounds. The construction splits into two parts. First, we eliminate ambiguities from computations according to the algorithm, i. e., for each input a we fix a computation C_a, which essentially corresponds to a communication pattern. In the second part, we show that C_{a^*} induces a gossip protocol for G.

First, we arbitrarily fix a computation C_{a^*} for input a^*, by induction on steps $t = 1, \ldots, T$. Let $E_t(a^*)$ be the set of all possible pairs of senders and recipients determined on the basis of step $t - 1$ (cf. §2). Then an arbitrary matching $M_t^* \subseteq E_t(a^*)$ is chosen that covers all recipients, and messages are delivered according to M_t^*. Next, again by induction, we determine C_a, for all inputs $a \neq a^*$. (Intuitively, we try to make C_a act in the same way as C_{a^*} whenever possible.) Assume C_a has been fixed up to step $t - 1$, and consider the graph $(V, E_t(a))$ induced by the communication requests of the processors for step t. Let $M_t'(a) = M_t^* \cap E_t(a)$ be the set of those edges in $E_t(a)$ that are used in step t of C_{a^*}. By restriction $(*)_1$, we may extend $M_t'(a)$ to some directed matching $M_t(a) \subseteq E_t(a)$ that covers all recipients. Deliver messages according to $M_t(a)$. — For proving (a), it is sufficient to show the following.

CLAIM. $\mathcal{M}^* = (M_1^*, \ldots, M_T^*)$ is a 1-way gossip protocol for G.

Proof of Claim. Let $\mathcal{K}(\mathcal{M}^*) = (K_0, K_1, \ldots, K_T)$ be as in Def. 1. We must show that $K_T(i) = V$ for all $i \in V$. For this, it is sufficient to establish the following assertion $(\mathcal{A}_t)$ for $t = T$:

$(\mathcal{A}_t)$ For all inputs $a = (a_1, \ldots, a_n)$ and all $i \in V$ we have that if $a_j = a_j^*$ for all $j \in K_t(i)$, then P_i is in the same state in C_{a^*} and C_a after step t.

Indeed, if $K_T(i) \neq V$, e.g., $j \notin K_t(i)$, then by $(\mathcal{A}_T)$ processor P_i is in the same state after step T on input a^* and on each input that differs from a^* only in the jth component, which contradicts the assumption that a^* is critical for f and that the network computes the function f with global output.

We prove $(\mathcal{A}_t)$ by induction on t. $(\mathcal{A}_0)$ follows immediately from the definitions. Thus, assume $t > 0$, and that $(\mathcal{A}_{t-1})$ is true. Let $i \in V$ and a be an input so that $a_j = a_j^*$ for all $j \in K_t(i)$. There are two cases.

Case 1: There is no l such that $(l,i) \in M_t^*$. Then, by definition, $K_t(i) = K_{t-1}(i)$, in particular $a_j = a_j^*$ for all $j \in K_{t-1}(i)$. By $(\mathcal{A}_{t-1})$, P_i is in the same state after step $t-1$ in both C_{a^*} and C_a. Since no l satisfies $(l,i) \in M_t^*$, no processor P_l delivers a message to P_i in step t of C_{a^*}, i.e., by $(*)_1$, in step t of C_{a^*} processor P_i is a sender. Since this decision is based on the state at the end of step $t-1$, P_i does not receive a message in step t of C_a either, and enters the same state in C_a as in C_{a^*}.

Case 2: There is some (unique) l such that $(l,i) \in M_t^*$. In this case, by definition, $K_t(i) = K_{t-1}(l) \cup K_{t-1}(i)$. Since $(l,i) \in M_t^*$, in step t of computation C_{a^*} the message $m_{l,t}$ sent by processor P_l is delivered to P_i; in particular, P_l is a sender with $i \in V_{l,t}$ and P_i is a recipient with $l \in W_{i,t}$. We apply $(\mathcal{A}_{t-1})$ to a with respect to both $K_{t-1}(l)$ and $K_{t-1}(i)$ to conclude that P_l is in the same state after step $t-1$ of C_a and of C_{a^*}, and that the same is true for P_i. Thus, edge (l,i) is in $E_t(a)$ and in M_t^*, and P_l sends message $m_{l,t}$ in step t of C_a as well. By the construction of C_a, edge (l,i) will be chosen to be in $M_t(a)$, and message $m_{l,t}$ will be delivered to P_i in C_a. This implies that P_i receives identical messages in step t of C_{a^*} and of C_a; hence P_i will enter the same state at the end of step t in these two computations. This proves the claim, and (a).

The proof of (b) is essentially the same as that of (a), substituting undirected matchings for directed matchings and $(*)_2$ for $(*)_1$. In (c), the construction of the communication protocol is exactly the same as in (a) respectively (b). Since P_{i_0} knows the result at the end, we get $K_T(i_0) = V$, which means that $\mathcal{M}$ is an accumulation protocol for G. Fact 4 and the remark preceding it yield (c). Finally, part (d) is proved easily in a similar way by considering only two inputs a and a^* that differ in component i_0 and satisfy $f(a) \neq f(a^*)$.

3.2 Functions without critical input. In order to deal with arbitrary functions, we generalize the gossip complexity of a graph as follows: in the setting of Def. 1, for $1 \leq s \leq n$, let $r(G,s)$ denote the minimum number of rounds of a communication protocol for G that satisfies $|K_T(j)| \geq s$ for all $j \in V$. Further, recall the definition of the sensitive complexity $s(f)$ resp. $s(f,a)$ for inputs a from Def. 6. Using the techniques of the proof of Thm. 8, we obtain:

Proposition 10. *If G computes f in T steps with global output, then $T \geq r(G, s(f))$.* □

While the message complexity of the problem to obtain at least s different pieces of information was defined and investigated already in [21], its round complexity $r(G,s)$ does not seem to have received much attention, excepting that for the case of the complete network, in [2] it was established that the technique of [8,

18, 19, 22] can be adapted to prove a lower bound of about $\log_\rho(s)$, where $\rho = \frac{1}{2}(1+\sqrt{5})$, which is about $1.44...\log_2(s)$. — Alternatively, one can consider the critical complexity of f (Def. 6(a)), and show that an algorithm that computes f with global output induces a communication protocol for broadcasting from $k = c(f)$ fixed sources. Lower bound results for this problem for specific networks and specific placement of the sources have been obtained in [14].

3.3 Functions with multiple outputs. In applications, often functions $f : A_1 \times \cdots \times A_n \to B_1 \times \cdots \times B_n$, $a \mapsto (f_1(a), \ldots, f_n(a))$, must be computed in such a way that the jth component $f_j(a)$ appears at processor P_j at the end. In [18], matrix inversion, discrete Fourier transform, and sorting are listed as examples. We note the following. (In the proof we combine the methods from §3 with an adaptation of the method of [8, 18, 19, 22]).

Proposition 11. *If the multiple-output function $f : a \mapsto (f_1(a), \ldots, f_n(a))$ is computed on the complete network in 1-way mode in T steps, under restriction $(*)_1$, and a is any input, then, for $\rho = (1+\sqrt{5})/2$, we have:*
$T \geq \frac{1}{2}\log_\rho(\frac{1}{n}\sum_{1\leq j\leq n} s(f_j,a)^2) \geq 0.72\log_2(\frac{1}{n}\sum_{1\leq j\leq n} s(f_j,a)^2))$. □

Corollary 12. *Sorting n bits in 1-way mode on a complete network with restriction $(*)_1$ takes at least $1.44...\log_2 n - O(1)$ steps.* □

4 Broadcasting a bit in a synchronous network

In this section, we study the complexity of broadcasting one bit in a synchronous network of processors in the absence of restriction $(*)$.

4.1 Broadcasting a bit in 1-way mode. First, we focus on 1-way communication. Initially, one processor P_{i_0} is in either one of two different states, representing inputs 0 and 1, the other processors are in a state that is independent of the input. In step t, a processor may send one message and receive an arbitrary number of messages; it changes its state on the basis of *all* messages received. After step T, all processors must know the input bit. The *bit broadcast complexity* $\hat{T}_{G,i_0}$ of a processor network G in 1-way mode with source P_{i_0} is the smallest number of steps of such an algorithm.

It is not hard to show that with such a communication mode the lower bounds from the previous section fail to hold. As examples, we note the following. (The proof involves a simple trick, described, e. g., in [1, 5], that makes it possible that one processor informs two of its neighbors of the value of a bit in one step.)

Proposition 13. (a) *If G is a full binary tree of depth $d \geq 1$ with root P_1, then by synchronous algorithms in 1-way mode (which do not satisfy restriction $(*)_1$), the* OR *of n bits with local output at P_1 can be computed in d steps; a bit b located initially at P_1 can be broadcasted to all nodes in d steps; and the* OR *of n bits with global output can be computed in $2d$ steps.*
(b) *If G is the complete network of size n, then by synchronous algorithms in 1-way mode (which do not satisfy restriction $(*)_1$), the* OR *of n bits with local*

output can be computed in one step; a bit c located initially at some node can be broadcasted to all nodes in $\lceil \log_3 n \rceil \approx 0.63 \log_2 n$ *steps; and the* OR *of n bits with global output can be computed in* $1 + \lceil \log_3 n \rceil$ *steps.* □

The running times just given should be compared with the values $a(G, i_0) = b(G, i_0) = \lceil \log_2 n \rceil$ and $r(G) = 1.44... \log_2 n \pm O(1)$ for G the complete network.

It is straightforward to generalize broadcast protocols (see §2) to 2-*broadcast protocols*, in which one node may pass the information it has to two of its neighbors in one step. (This modification is discussed as "DMA-bound model" $H2$, an abbreviation for "half-duplex with outdegree 2", in [10, 18].) The 2-*broadcast complexity* $b_2(G, i_0)$ of G with source node i_0 is the minimum T such that there is a 2-broadcast protocol with T rounds. Clearly, we have $b(G, i_0) \leq 2b_2(G, i_0)$ for arbitrary networks G and nodes i_0 in G.

By using the trick mentioned before Prop. 13, it is easy to transform any 2-broadcast protocol for G into a bit broadcast algorithm in 1-way mode; thus we have $\hat{T}_{G,i_0} \leq b_2(G, i_0)$. Next, we show that this is optimal.

Theorem 14. $\hat{T}_{G,i_0} = b_2(G, i_0)$.

Proof. (Sketch.) Let a T-step bit broadcast algorithm for G with source i_0 be given. To construct a 2-broadcast protocol with at most T steps, we proceed similarly as in the proof of Thm. 8. Inductively, computations C_0 and C_1 on inputs 0 and 1 are fixed that take equal actions whenever possible. Next, we try to "peel off" inessential parts of the two computations, i. e., to identify "meaningless" messages and eliminate them. For this, we define a set $E_{0,1}$ of labeled, directed edges that run along some of the edges of G, as follows: Edge (i, j), with label t, for $1 \leq t \leq T$, is in $E_{0,1}$ if and only if there is some $b \in \{0, 1\}$ such that in computation C_b the message $m_{i,t}$ sent by processor P_i is delivered to P_j, but in the other computation $C_{\bar{b}}$ no message or a different one is sent from P_i to P_j in step t.

The following observations are crucial. (The formal proofs are omitted.)
Claim 1: Let $i \neq i_0$ and $j \neq i_0$. If (i, j) with label t is in $E_{0,1}$, and no edge (i', j) with a label $t' < t$ is in $E_{0,1}$, then there must be some $l \in V$ such that edge (l, i) with label t'' is in $E_{0,1}$ for some $t'' < t$.
Claim 2: If $i \neq i_0$ then there is at least one labeled edge that enters node i.

Finally, all edges that enter i_0 are removed, and for each $i \neq i_0$, all edges that enter i excepting one with a minimal label t are removed. The remaining edge set is called $E^*_{0,1}$. Using the two claims, one can check that $G^* = (V, E^*_{0,1})$ is a directed spanning tree for V with root i_0, and that the labels along directed paths in G^* are strictly increasing with respect to their t-parts. Moreover, for each node and each t there can be at most two edges leaving i that are labeled with t. This means that G^* describes a 2-broadcast protocol. □

By using this theorem, we may show that there is a limit to the speedup that can be attained by exploiting the absence of restriction $(*)_1$, i. e., the possibility of transmitting information by not writing. The following bound is tight, as the example of broadcasting a bit from the root in a binary tree shows.

Proposition 15. *If a nonconstant function f can be computed in T steps with global output in the network G (without restriction $(*)_1$), then $b(G, i_0) \leq 2T$ for some i_0, and $r(G) \leq 4T$.*

4.2 Broadcasting a bit in 2-way mode and in other parallel models. The complexity of broadcasting a bit on EREW PRAMs has been determined in [1]. Algorithms for EREW PRAMs with p processors and r memory cells can be regarded as algorithms for processor networks with $n = p + r$ processors, by representing each PRAM processor and each cell as a network processor. In a write phase all PRAM processors are senders and all cells are recipients; in a read phase these roles are switched. We want to generalize the lower bound from [1] to a PRAM in which concurrent read accesses or write accesses are not forbidden but rather are resolved by the ARBITRARY READ/WRITE rule, which says that if in a step several processors try to read from or write to the same cell, an arbitrary one of them succeeds; the the others receive a negative acknowledgement ("access failed"). This access rule is unusual for PRAMs; however, it has been proposed in the context of distributed memory machines (DMMs) as a relaxation of the COLLISION access rule, see [7, 20]. DMMs with the COLLISION access rule have been unter intensive investigation recently. It has turned out that they are very powerful in a randomized setting, in particular, they are able to perform routing and to simulate PRAMs in sublogarithmic time [11, 12, 20]. An interesting weakness of this model may be identified, by combining the technique from [1] with an adaptation of the proof of Thm. 14 to 2-way communication (proof omitted):

Theorem 16. *Broadcasting a bit from one cell to all p processors in a PRAM with the* ARBITRARY READ/WRITE *rule takes $\log_{2+\sqrt{3}} p \pm O(1)$ steps. The same bound holds for broadcasting a bit on a DMM with p processors and p memory modules, even if randomization is used.* □

Acknowledgement. The author thanks Danny Krizanc for a careful reading of the manuscript and many helpful suggestions, as well as for pointing out that the lower bound for broadcasting even holds for randomized DMMs.

References

1. P. Beame, M. Kutyłowski, and M. Kik, Information Broadcasting by exclusive-read PRAMs, *Parallel Processing Letters* **1 & 2** (1994) 159–169.
2. G. Belting, *Untere Schranken für die Berechnung von Booleschen Funktionen in vollständigen Prozessornetzwerken im Telefon- und Telegraf-Modus*, Diplomarbeit, Universität–Gesamthochschule–Paderborn, Paderborn, 1994.
3. S. Cook, C. Dwork, and R. Reischuk, Upper and lower time bounds for parallel random access machines without simultaneous writes, *SIAM J. Comput.* **15** (1986) 87–97.
4. M. Dietzfelbinger, Gossiping and broadcasting versus computing functions in networks, ECCC (`http://www.eccc.uni-trier.de/eccc/`), Report TR96-052, 1996.
5. M. Dietzfelbinger, M. Kutyłowski, and R. Reischuk, Exact time bounds for computing Boolean functions on PRAMs without simultaneous writes, in: Proc. 2nd Annual ACM Symp. on Parallel Algorithms and Architectures, 1990, pp. 125–135.

6. M. Dietzfelbinger, M. Kutyłowski, and R. Reischuk, Exact lower bounds for computing Boolean functions on CREW PRAMs, *J. Comput. Syst. Sci.* **48** (1994) 1231–254.
7. M. Dietzfelbinger and F. Meyer auf der Heide, Simple, efficient shared memory simulations, in: Proc. 5th ACM Symp. on Parallel Algorithms and Architectures, 1993, pp. 110–118.
8. S. Even and B. Monien, On the number of rounds necessary to disseminate information, in: Proc. ACM Symp. on Parallel Algorithms and Architectures, 1989, pp. 318–327.
9. F. Fich, The complexity of computation on the parallel random access machine, in J. H. Reif (ed.), *Synthesis of Parallel Computation*, Morgan Kaufmann, San Mateo, 1994, pp. 843–899.
10. P. Fraigniaud and E. Lazard, Methods and problems of communication in usual networks, *Discrete Applied Math.* **53** (1994) 79–134.
11. L. A. Goldberg, M. Jerrum, T. Leighton, and S. Rao, A doubly logarithmic communication algorithm for the completely connected optical communication parallel computer, in: Proc. 5th Annual ACM Symp. on Parallel Algorithms and Architectures, 1993, pp. 300–309.
12. L. A. Goldberg, Y. Matias, and S. Rao, An optical simulation of shared memory, in: Proc. 6th Annual ACM Symp. on Parallel Algorithms and Architectures, 1994, pp. 257–267.
13. S. M. Hedetniemi, S. T. Hedetniemi, and A. L. Liestman, A survey of gossiping and broadcasting in communication networks, *Networks* **18** (1986) 319–349.
14. I. Höltring, *Broadcast und Gossip in parallelen Netzwerken*, Diplomarbeit, Universität–Gesamthochschule–Paderborn, Paderborn, 1994.
15. J. Hromkovič, C.-D. Jeschke, and B. Monien, Optimal algorithms for dissemination of information in some interconnection networks, *Algorithmica* **10** (1993) 24–40.
16. J. Hromkovič, R. Klasing, B. Monien, and R. Peine, Dissemination of information in interconnection networks (broadcasting & gossiping), in: D.-Z. Du and D. F. Hsu (eds.), *Combinatorial Network Theory*, Kluwer Academic Publishers, Amsterdam, 1996, pp. 125–212.
17. R. M. Karp and V. Ramachandran, Parallel algorithms for shared-memory machines, in J. van Leeuwen (ed.), *Handbook of Theoretical Computer Science*, Vol. A, Algorithms and Complexity, Elsevier, Amsterdam, 1990, pp. 869–941.
18. D. W. Krumme, G. Cybenko, and K. N. Venkatamaran, Gossiping in minimal time, *SIAM J. Comput.* **21** (1992) 111–139.
19. R. Labahn and I. Warnke, Quick gossiping by multi-telegraphs, in: R. Bodendiek and R. Henn (eds.), *Topics in Combinatorics and Graph Theory*, Physica-Verlag, Heidelberg, 1990, pp. 451–458.
20. F. Meyer auf der Heide, C. Scheideler, and V. Stemann, Exploiting storage redundancy to speed up randomized shared memory simulations, in: E. W. Mayr and C. Puech (eds.), Proc. 12th Annual Symposium on Theoretical Aspects of Computer Science (STACS 95), Lecture Notes in Computer Science 900, Springer, Berlin, 1995, pp. 267–278.
21. D. Richards and A. L. Liestman, Generalizations of broadcasting and gossiping, *Networks* **18** (1988) 125–138.
22. V. S. Sunderam and P. Winkler, Fast information sharing in a complete network, *Discrete Applied Math.* **42** (1991) 75–86.
23. G. Tel, *Introduction to Distributed Algorithms*, Cambridge University Press, Cambridge, 1994.
24. U. Vishkin and A. Wigderson, Trade-offs between depth and width in parallel computation, *SIAM J. Comput.* **14** (1985) 303–314.

On the Descriptive and Algorithmic Power of Parity Ordered Binary Decision Diagrams

Stephan Waack

Institut für Numerische
und Angewandte Mathematik
Georg–August–Universität Göttingen
Lotzestr. 16–18, 37083 Göttingen
Germany

Abstract. We present a data structure for Boolean functions, which we call Parity–OBDDs or POBDDs, which combines the nice algorithmic properties of the well–known ordered binary decision diagrams (OBDDs) with a considerably larger descriptive power.
Beginning from an algebraic characterization of the POBDD–complexity we prove in particular that the minimization of the number of nodes, the synthesis, and the equivalence test for POBDDs, which are the fundamental operations for circuit verification, have efficient deterministic solutions.
Several functions of pratical interest, i.e. the storage access function, have exponential ODBB–size but are of polynomial size if POBDDs are used.

1 Introduction

Formal circuit verification is a fundamantal task. The following approach for verification is often used (for a survey see [7] and [13]). A data structure for representing Boolean functions is chosen. It should allow compact representation of many important functions and efficient algorithms for certain operations. The fundamental operations are the *equivalence test* (decide whether two data objects represent one and the same Boolean function), the *synthesis* (compute, for a binary Boolean operation ω, a representation for $\omega(f,g)$, given representation for f and g) and the *reduction of the size* of a given data structure, preferably *down to a minimal representation.* The circuit to be verified and the specification are transformed step by step with the synthesis algorithm possibly combined with the reduction algorithm converting it into a smaller representation, which is in the best case minimal. Then the equivalence test algorithm is applied.

Ordered binary decision diagrams (OBDDs) provide the data structure applied in most cases. Their excellent algorithmic properties are the reason for this preference: All fundamental operations can be done in time linear in the maximum of the sizes of the input and the output (see [5], [12]). The restricted descriptive power of OBDDs is their main drawback. There are several methods to prove exponential lower bounds on the size of the OBDDs, which are nothing but oblivious read–once branching programs, for explicitly defined Boolean

Reischuk, Morvan (Eds.): STACS'97 Proceedings, LNCS 1200

functions. Let us consider some examples. The *hidden weighted bit function* $\mathrm{HWB}_n(x_1, \ldots, x_n) := x_\nu$, where $\nu := \sum_{i=1}^n x_i$, and $x_0 = 0$, was poved in [6] to have exponential size. The same was proved for the storage access function $\mathrm{SAF}_n(y_{k-1}, \ldots, y_0, x_{2^k-1}, \ldots, x_0) = x_{|y|}$, where $n = 2^k + k$, $|y| = \sum_{i=0}^{k-1} y_i 2^i$, in [4]. Ajtai et al. considered in [1] the function $\oplus\text{-}cl_{n,3}$, $n = \binom{n'}{2}$, which counts in an undirected graph the number of triangles mod2. They proved that $\oplus\text{-}cl_{n,3}$ has exponential size even if the more powerful model of read–once branching programs is used. For more lower bounds, see, for example, [14], [11], and [10].

MOD_2–representations of Boolean functions have been used for a long time. Probably best known is the ring–sum–expension. Gergov and Meinel ([8]) have presented a data structure which they called MOD-2-OBDDs. They have shown that the descriptive power of the MOD-2-OBDDs is larger than that of OBDDs by proving that the hidden weighted bit function and the function $\oplus\text{-}cl_{n,3}$ have MOD-2-OBDD–size $O(n^2)$, $O(n^3)$, respectively. Moreover, they have proved all fundamental operations for MOD-2-OBDDs *except the minimization* to have feasible running time. The exception concerning the minimization is a decisive drawback. This is because of the fact that the synthesis algorithm does not lead to a minimal MOD-2-OBDD in most cases even if the inputs are minimal.

We generalize Gergov's and Meinel's MOD-2-OBDDs to what we call Parity–OBDDs or POBDDs. We allow all branching nodes to have unbounded nondeterminism. We count, for each input, the number of accepting paths mod2 to calculate the function value. Then the local neighbourhood relation (see Figure 1) of branching nodes is nothing else than a linear equation in a vector space over the prime field of characteristic 2. Methods from linear algebra are applicable, in particular the Gaussian elimination. As to lower bounds, some techniques based on rank arguments even for the more general case of oblivious parity branching programs of restricted length are developed in [9].

On the basis of an algebraic charaterization (see Theorem 14) of the POBDD–complexity we efficiently solve in a deterministic manner the minimization problem for the number of nodes (see Theorem 16), the synthesis problem (see Theorem 19), the equivalence test problem (see Corollary 20), the replacement by functions problem (see Corollary 21), the replacement by constants problem (see Theorem 4), the quantification problem (see Corollary 22), the redundancy test problem (see Corollary 18), the satisfiability problem (see Corollary 17), and the evaluation problem (see Theorem 3). Moreover, we complement the upper bounds on the size of representations of [8] by showing that the storage access function SAF has POBDD–size $O(n^2)$, for any ordering of the variables (see Corallary 23).

2 The Model and First Results

We consider Boolean functions $f : \{0,1\}^n \to \{0,1\}$ to be mappings from $\mathbb{F}_2^n$ to $\mathbb{F}_2$, where $\mathbb{F}_2$ is the prime field of characteristic 2. Thus the set $\mathbb{B}_n \stackrel{def}{=} \mathrm{Map}(\{0,1\}^n, \{0,1\})$ of all Boolean functions depending on n Boolean variables becomes a $\mathbb{F}_2$-algebra by pointwise defined algebraic operations. Clearly, the addition is the exclusive-or operation, and the multiplication is the conjunction. In particular, we have that $\mathbb{B}_0 = \mathbb{F}_2$.

Definition 1. A Parity-OBDD $\mathcal{B}_n = \mathcal{B}$ on the Boolean variables $x = (x_1, \ldots, x_n)$ with respect to the permutation σ of the index set $\{1, \ldots, n\}$ consists of the following data. A set $\mathcal{N}^{\mathcal{B}} = \mathcal{N}$ of nodes which contains, if it is not empty, two not necessarily pairwise distinct special nodes q, s_1, called the source, the 1-sink, respectively. A set $\mathcal{A}^{\mathcal{B}} = \mathcal{A}$ of directed edges connecting nodes, which we call arcs. Two labelling functions both denoted by $\Lambda_{\mathcal{B}}$: $\Lambda_{\mathcal{B}} : \mathcal{N} \to \{x_1, \ldots, x_n\} \cup \{1\}$, and $\Lambda_{\mathcal{B}} : \mathcal{A} \to \{0, 1\}$.

The following conditions have to be satisfied. For each pair (u, v) of nodes, $u \neq v$, there are at most two arcs connecting them, at most one labeled with "0", and at most one labeled with "1". The directed multigraph $(\mathcal{N}, \mathcal{A})$ is acyclic, the indegree of the source and the outdegree of the 1-sink is zero, and the 1-sink is the only node labeled with "1". If we denote by $\mathcal{N}^{\mathcal{B}}_{n-k+1}$ the set $\Lambda_{\mathcal{B}}^{-1}(\{x_{\sigma(k)}\})$, the so-called $n-k+1$th level of $\mathcal{B}$, for $k = 1, \ldots, n$, and $\mathcal{N}^{\mathcal{B}}_0 = \{s_1\}$, then $(v_1, v_2) \in \mathcal{A}$ and $v_j \in \mathcal{N}_{k_j}$ $j = 1, 2$, implies $k_1 > k_2$. (Sometimes we call $\mathcal{B}_n$ a σ-POBDD.)

Define, moreover,

$$\mathrm{C}^{\sigma}_{\mathrm{POBDD}}(f) \stackrel{def}{=} \min \{\mathrm{SIZE}(\mathcal{B}_n) \mid \mathcal{B}_n \text{ is a } \sigma\text{-POBDD representing } f\},$$

for $f \in \mathbb{B}_n$.

Moreover, the following notions and notations are useful.

If $\Lambda_{\mathcal{B}}(v_1, v_2) = e$, then v_2 is called an *e-successor of* v_1 and v_1 an *e-predecessor of* v_2.

Let $x_{i_1}, \ldots, x_{i_k}$ be a sequence of Boolean variables, and let $e \in \{0,1\}^k$, $k \leq n$, be a Boolean vector, which we regard as an assignment to $x_{i_1}, \ldots, x_{i_k}$. Then we define the path $p : v \stackrel{*}{\to} w$ to be *a path under the input* e in the following way. If (u', u'') is an arc of the path p, and if $\Lambda_{\mathcal{B}}(u') = x_j$, then $j \in \{i_1, \ldots, i_k\}$, and the node u'' is an e_j-successor of the node u'. We write $p : v \stackrel{e}{\to} w$ in this case.

A node v such that $\Lambda_{\mathcal{B}}(v) \in \{x_1, \ldots, x_n\}$ is call a *branching node.*

The number of all branching nodes is called the *size* of the POBDD $\mathcal{B}$, abbreviated SIZE$(\mathcal{B})$. Why do we use this definition instead of the number of edges? The nodes of a POBDD can be regarded as elements of a vector space. There are algebraic methods for minimizing and transforming generating systems of spaces. The number of edges leaving a node v is related to what might be called the representation length of v in a basis. It seems to be much more complicated to minimize it. That's why this paper does not consider the number of edges, hence does not give exact results on the space needed to store a POBDD.

What we have defined up to now is only the syntax of σ-PBDDs. Let us turn to the semantics. Let $\mathcal{B}_n$ be a σ-PBDD. Then each node $v \in \mathcal{N}^{\mathcal{B}}_k$ represents a function

$$\mathrm{Res}^{\mathcal{B}}_v(e) \stackrel{def}{=} \#\left\{p \,|\, p : v \stackrel{e}{\to} s_1\right\} \bmod 2,$$

for all assignments e of the variables $x_{\sigma(n-k+1)}, \ldots, x_{\sigma(n)}$. We shall assume throughout this paper that each function $\phi \in \mathbb{B}_k$ depends on the Boolean variables $x_{\sigma(n-k+1)}, \ldots, x_{\sigma(n)}$. Then $\mathrm{Res}_v^{\mathcal{B}} \in \mathbb{B}_k$. In particular, we have $\mathrm{Res}_{s_1}^{\mathcal{B}} = 1 \in \mathbb{F}_2$.

Since a PBDD is acyclic, we can inductively calculate the functions $\mathrm{Res}^{\mathcal{B}}$ as follows.

Lemma 2. *Let $\mathcal{B}_n$ be a σ–PBDD, and let $v \in \mathcal{N}_{n-k+1}^{\mathcal{B}}$. If $v_1, \ldots, v_\lambda$ are the 0–successors and $w_1, \ldots, w_\mu$ the 1–successors of a node v, then $\mathrm{Res}_v^{\mathcal{B}} = (1 - x_{\sigma(k)}) \cdot \sum_{j=1}^{\lambda} \mathrm{Res}_{v_j}^{\mathcal{B}} + x_{\sigma(k)} \cdot \sum_{j=1}^{\mu} \mathrm{Res}_{w_j}^{\mathcal{B}}$, where we consider $\mathbb{F}_2$–operations.*

That means, POBDDs are based on the Shannon decomposition of Boolean functions. For a survey on various decomposition types see [2] and [3].

A POBDD is called to be *deterministic* if and only if each branching node has at most one e–successor, for each $e \in \{0,1\}$. Deterministic POBDDs are nothing but ordinary POBDDs defined in the literature. In particular, deterministic POBDDs can be identified with the ordered binary decision diagrams (OBDDs). Obviously, we have that $\mathrm{C}_{\mathrm{OBDD}}^{\sigma}(f) \geq \mathrm{C}_{\mathrm{POBDD}}^{\sigma}(f)$, for all $f \in \mathbb{B}_n$.

The results which follow are observations rather than theorems. Straightforward algorithms, which are linear in time and space, can be used.

Theorem 3 (Evaluation). *Let $\mathcal{B}_n$ be a σ–POBDD representing $f \in \mathbb{B}_n$. Then we can evaluate $f(e)$, for each input $e \in \{0,1\}^n$, in linear time.*

Theorem 4 (Replacement by constants). *Let $\mathcal{B}_n$ be a σ–POBDD representing $f \in \mathbb{B}_n$, and let $c \in \{0,1\}$ be a Boolean constant. Then a σ–POBDD of size less than or equal to $\mathrm{SIZE}(\mathcal{B})$ representing $f|_{x_i=c}$ can be computed in linear time.*

3 An Algebraic Charaterization of the σ–POBDD Complexity

First, we construct as the main tool appropriate $\mathbb{F}_2$–linear mappings $\mathbb{B}_k \to \mathbb{B}_l$.

- $\iota_{kl}^{\sigma e} : \mathbb{B}_k \to \mathbb{B}_l$, for $e \in \mathbb{F}_2^{l-k}$, $k \leq l$:
  If $k = l$, then $\iota_{kl}^{\sigma e} = \mathrm{id}_{\mathbb{B}_k}$. Otherwise

$$\iota_{kl}^{\sigma e}(\phi)\left(x_{\sigma(n-l+1)}, \quad \ldots \quad, x_{\sigma(n-k+1)} \quad \ldots \quad, x_{\sigma(n)}\right) = \\ = \begin{cases} \phi\left(x_{\sigma(n-k+1)}, \ldots, x_{\sigma(n)}\right) & \text{if } \left(x_{\sigma(n-l+1)}, \quad \ldots \quad, x_{\sigma(n-k)}\right) = e; \\ 0 & \text{otherwise.} \end{cases}$$

- $\pi_{lk}^{\sigma e} : \mathbb{B}_l \to \mathbb{B}_k$, for $e \in \mathbb{F}_2^{l-k}$, $k \leq l$:
  If $k = l$, then $\pi_{lk}^{\sigma e} = \mathrm{id}_{\mathbb{B}_k}$. Otherwise

$$\pi_{lk}^{\sigma e}(\phi)\left(x_{\sigma(n-k+1)}, \quad \ldots \quad, x_{\sigma(n)}\right) = \phi\,|_{x_{\sigma(n-l+1)}=e_1, \ldots, x_{\sigma(n-k)}=e_{l-k}} .$$

- $\Delta_{kl}^{\sigma} : \mathbb{B}_k \to \mathbb{B}_l$, for $k \leq l$: $\Delta_{kl}^{\sigma} \stackrel{def}{=} \sum_{e \in \{0,1\}^{l-k}} \iota_{kl}^{\sigma e}$.

In order to simplify notation, we omit the superscript σ from now on. The following lemmas are well–known or easy to prove.

Lemma 5. *For each $\phi \in \mathbb{B}_l$ there are uniquely determined functions $\phi_e \in \mathbb{B}_k$, for $e \in \{0,1\}^{l-k}$, and $k \leq l$, such that $\phi = \sum_{e\in\{0,1\}^{l-k}} \iota^e_{kl}(\phi_e)$, namely, $\phi_e = \pi^e_{lk}(\phi)$. In particular, it follows that the mappings Δ_{kl} are one–to–one.*

Lemma 6. *For $k \leq l \leq m$, $e' \in \{0,1\}^{l-k}$, $e'' \in \{0,1\}^{m-l}$, $e := e'e'' \in \{0,1\}^{m-k}$, we have $\pi^{e'}_{lk} \circ \pi^{e''}_{ml} = \pi^e_{mk}$, $\iota^{e''}_{lm} \circ \iota^{e'}_{kl} = \iota^e_{km}$, $\Delta_{lm} \circ \Delta_{kl} = \Delta_{km}$, $\pi^e_{ml} \circ \Delta_{km} = \Delta_{kl}$.*

Consequently, we may *identify along the mappings Δ_{kl}*, for $k \leq l$. We do so when we say that a function $\phi \in \mathbb{B}_k$ can be regarded as an element of $\mathbb{B}_l$, too, by agreeing that ϕ does not essentially depend on the additional variables $x_{\sigma(n-l+1)}, \ldots, x_{\sigma(n-k)}$. Thus the notation Δ_{kl} becomes superfluous. Moreover, let us abbreviate ι^e_{kl} by ι^e_l, and π^e_{lk} by π^e_l, for $e \in \mathbb{F}_2^{l-k}$, $k \leq l$.

Let us take the first step towards our goal.

Lemma 7. *Let $\mathcal{B}$ be a POBDD with respect to the permutation σ, let u be a branching node of the diagram $\mathcal{B}$ belonging to the level $\mathcal{N}^{\mathcal{B}}_l$. Then the neighbourhood relation in $\mathcal{B}$ shown in Figure 1, where we assume that $v_i \in \mathcal{N}^{\mathcal{B}}_{k'_i}$, $w_j \in \mathcal{N}^{\mathcal{B}}_{k''_j}$, $k'_i < l$, $k''_j < l$, for $i = 1, \ldots, \nu$, $j = 1, \ldots, \mu$, is equivalent to the equation $\mathrm{Res}^{\mathcal{B}}_u = \iota^0_l\left(\sum_{i=1}^{\nu} \mathrm{Res}^{\mathcal{B}}_{v_i}\right) + \iota^1_l\left(\sum_{j=1}^{\mu} \mathrm{Res}^{\mathcal{B}}_{w_j}\right)$, or, what is the same, to $\pi^1_l\left(\mathrm{Res}^{\mathcal{B}}_u\right) = \sum_{j=1}^{\mu} \mathrm{Res}^{\mathcal{B}}_{w_j}$ and $\pi^0_l\left(\mathrm{Res}^{\mathcal{B}}_u\right) = \sum_{i=1}^{\nu} \mathrm{Res}^{\mathcal{B}}_{v_i}$. (Observe that the sets $\{v_i \mid i = 1, \ldots, \nu\}$ and $\{w_j \mid j = 1, \ldots, \mu\}$ are not necessarily disjoint.)*

Proof. The claim follows directly from Lemma 2.

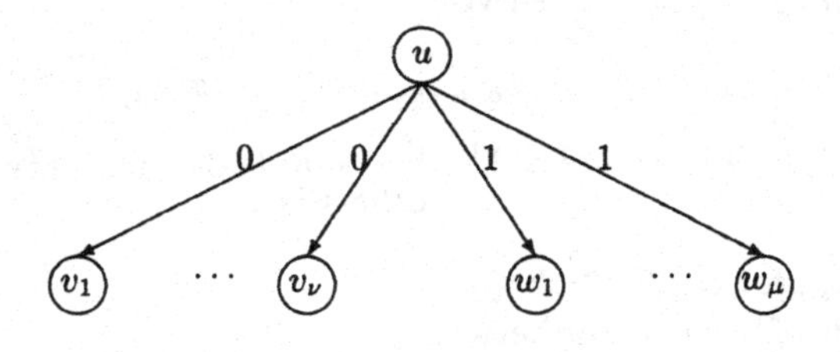

Fig. 1. The local neighbourhood relation in a POBDD

The second step towards the aim of this section is to define two subdiagrams of the diagram

$$\begin{array}{ccccccccc}
\mathbb{B}_0 & \overset{\pi_1^0}{\twoheadleftarrow} & \mathbb{B}_1 & \overset{\pi_2^0}{\twoheadleftarrow} & \dots & \overset{\pi_{n-1}^0}{\twoheadleftarrow} & \mathbb{B}_{n-1} & \overset{\pi_n^0}{\twoheadleftarrow} & \mathbb{B}_n \\
\uparrow \mathrm{id} & \circlearrowright & \uparrow \mathrm{id} & \circlearrowright & & \circlearrowright & \uparrow \mathrm{id} & \circlearrowright & \uparrow \mathrm{id} \\
\mathbb{B}_0 & \overset{\subseteq}{\rightarrow} & \mathbb{B}_1 & \overset{\subseteq}{\rightarrow} & \dots & \overset{\subseteq}{\rightarrow} & \mathbb{B}_{n-1} & \overset{\subseteq}{\rightarrow} & \mathbb{B}_n , \\
\downarrow \mathrm{id} & \circlearrowleft & \downarrow \mathrm{id} & \circlearrowleft & & \circlearrowleft & \downarrow \mathrm{id} & \circlearrowleft & \downarrow \mathrm{id} \\
\mathbb{B}_0 & \overset{\pi_1^1}{\twoheadleftarrow} & \mathbb{B}_1 & \overset{\pi_2^1}{\twoheadleftarrow} & \dots & \overset{\pi_{n-1}^1}{\twoheadleftarrow} & \mathbb{B}_{n-1} & \overset{\pi_n^1}{\twoheadleftarrow} & \mathbb{B}_n
\end{array}$$

which is commutative because of Lemma 6, the first one associated with the POBDD $\mathcal{B}$, the other one associated with the function f represented by $\mathcal{B}$. (The arrow "$\twoheadrightarrow$" indicates that the corresponding mapping is onto. A diagram

$$\begin{array}{ccc}
A & \overset{\alpha}{\rightarrow} & B \\
\downarrow \gamma & & \downarrow \beta \\
C & \overset{\delta}{\rightarrow} & D
\end{array}$$

is commutative if $\beta \circ \alpha = \delta \circ \gamma$. This notion generalizes in the obvious way to more complicated diagrams. The notion subdiagram means in our context, that we have subspaces, for each index, and the linear mappings are restrictions of the corresponding ones.) To do so, let us abbreviate for a moment the set of all subfunctions of the function $f \in \mathbb{B}_n$ depending on the k variables $x_{\sigma(n-k+1)}, \dots, x_{\sigma(n)}$ by $\tilde{\mathbb{B}}_k^f$. Then

$$\mathbb{B}_l^f \overset{def}{=} \sum_{\phi \in \tilde{\mathbb{B}}_k^f,\, k \le l} \mathbb{F}_2 \cdot \phi, \quad \mathbb{B}_l^{\mathcal{B}} \overset{def}{=} \sum_{v \in \mathcal{N}_k^{\mathcal{B}},\, k \le l} \mathbb{F}_2 \cdot \mathrm{Res}_v^{\mathcal{B}},$$

i.e. the subspace $\mathbb{B}_l^f$ of $\mathbb{B}_l$ is generated by the elements ϕ, whereas $\mathbb{B}_l^{\mathcal{B}}$ is generated by the functions $\mathrm{Res}_v^{\mathcal{B}}$.

The next two lemmas, whose proofs are trivial, justify Definition 10. Moreover, Lemma 8 shows that the diagrams defined there are fully determined by the spaces of their nth columns $\mathbb{B}_n^f$ and $\mathbb{B}_n^{\mathcal{B}}$, respectively.

Lemma 8. *For $k < l$, and $e \in \{0,1\}^{l-k}$, we have that $\pi_l^e\left(\mathbb{B}_l^f\right) = \mathbb{B}_k^f$, and $\pi_l^e\left(\mathbb{B}_l^{\mathcal{B}}\right) = \mathbb{B}_k^{\mathcal{B}}$.*

Lemma 9. *We have, for $k \le l$, that $\mathbb{B}_k^f \subseteq \mathbb{B}_l^f$, and $\mathbb{B}_k^{\mathcal{B}} \subseteq \mathbb{B}_l^{\mathcal{B}}$.*

Definition 10. The commutative diagram

$$\begin{array}{ccccccccc}
\mathbb{B}_0^f & \overset{\pi_1^0}{\twoheadleftarrow} & \mathbb{B}_1^f & \overset{\pi_2^0}{\twoheadleftarrow} & \dots & \overset{\pi_{n-1}^0}{\twoheadleftarrow} & \mathbb{B}_{n-1}^f & \overset{\pi_n^0}{\twoheadleftarrow} & \mathbb{B}_n^f \\
\uparrow \mathrm{id} & \circlearrowright & \uparrow \mathrm{id} & \circlearrowright & & \circlearrowright & \uparrow \mathrm{id} & \circlearrowright & \uparrow \mathrm{id} \\
\mathbb{B}_0^f & \overset{\subseteq}{\rightarrow} & \mathbb{B}_1^f & \overset{\subseteq}{\rightarrow} & \dots & \overset{\subseteq}{\rightarrow} & \mathbb{B}_{n-1}^f & \overset{\subseteq}{\rightarrow} & \mathbb{B}_n^f \\
\downarrow \mathrm{id} & \circlearrowleft & \downarrow \mathrm{id} & \circlearrowleft & & \circlearrowleft & \downarrow \mathrm{id} & \circlearrowleft & \downarrow \mathrm{id} \\
\mathbb{B}_0^f & \overset{\pi_1^1}{\twoheadleftarrow} & \mathbb{B}_1^f & \overset{\pi_2^1}{\twoheadleftarrow} & \dots & \overset{\pi_{n-1}^1}{\twoheadleftarrow} & \mathbb{B}_{n-1}^f & \overset{\pi_n^1}{\twoheadleftarrow} & \mathbb{B}_n^f
\end{array}$$

is called the σ–characteristic of the function f. The commutative diagram

$$\begin{array}{ccccccccc}
\mathbb{B}_0^{\mathcal{B}} & \overset{\pi_1^0}{\twoheadleftarrow} & \mathbb{B}_1^{\mathcal{B}} & \overset{\pi_2^0}{\twoheadleftarrow} & \dots & \overset{\pi_{n-1}^0}{\twoheadleftarrow} & \mathbb{B}_{n-1}^{\mathcal{B}} & \overset{\pi_n^0}{\twoheadleftarrow} & \mathbb{B}_n^{\mathcal{B}} \\
\uparrow \mathrm{id} & \circlearrowleft & \uparrow \mathrm{id} & \circlearrowleft & & \circlearrowleft & \uparrow \mathrm{id} & \circlearrowleft & \uparrow \mathrm{id} \\
\mathbb{B}_0^{\mathcal{B}} & \overset{\subsetneq}{\rightarrow} & \mathbb{B}_1^{\mathcal{B}} & \overset{\subsetneq}{\rightarrow} & \dots & \overset{\subsetneq}{\rightarrow} & \mathbb{B}_{n-1}^{\mathcal{B}} & \overset{\subsetneq}{\rightarrow} & \mathbb{B}_n^{\mathcal{B}} \\
\downarrow \mathrm{id} & \circlearrowright & \downarrow \mathrm{id} & \circlearrowright & & \circlearrowright & \downarrow \mathrm{id} & \circlearrowright & \downarrow \mathrm{id} \\
\mathbb{B}_0^{\mathcal{B}} & \overset{\pi_1^1}{\twoheadleftarrow} & \mathbb{B}_1^{\mathcal{B}} & \overset{\pi_2^1}{\twoheadleftarrow} & \dots & \overset{\pi_{n-1}^1}{\twoheadleftarrow} & \mathbb{B}_{n-1}^{\mathcal{B}} & \overset{\pi_n^1}{\twoheadleftarrow} & \mathbb{B}_n^{\mathcal{B}}
\end{array}$$

is defined to be the σ–characteristic of the POBDD $\mathcal{B}$.

The idea behind these two characteristics is the following. The σ–characteristic of $\mathcal{B}$ algebraically describes the bottom–up representation of the function f by the POBDD $\mathcal{B}$: $\mathbb{B}_k^{\mathcal{B}}$ is the space of all function represented by the diagram $\mathcal{B}$ from the bottom up to the k-th level. The projections $\pi_{k+1}^0, \pi_{k+1}^1 : \mathbb{B}_{k+1}^{\mathcal{B}} \twoheadrightarrow \mathbb{B}_k^{\mathcal{B}}$ constitute the algebraic counterpart of the edges of $\mathcal{B}$ starting from the level $k+1$ (see Lemma 7).

The σ–characteristic of a Boolean function is an invariant of this function. We show in Lemma 11 that the σ–characteristic of f is always a subdiagram of the σ–characteristic of a POBDD $\mathcal{B}$ representing f, and hence an "algebraic lower bound". In order to prove Theorem 14, we show by constructing an appropriate σ–POBDD that the σ–characteristic of a Boolean function f is, moreover, the "greatest lower bound".

We omit the proofs of the next three lemmas.

Lemma 11. $\mathbb{B}_l^{\mathcal{B}} \supseteq \mathbb{B}_l^f$, *for any σ–POBDD $\mathcal{B}$ representing a Boolean function $f \in \mathbb{B}_n$.*

Lemma 12. *If the Boolean function $f \in \mathbb{B}_n$ is represented by the σ–POBDD $\mathcal{B}$, then we have $\mathbb{B}_k^f \cap \mathbb{B}_{k-1} = \mathbb{B}_{k-1}^f$, and $\mathbb{B}_k^{\mathcal{B}} \cap \mathbb{B}_{k-1} = \mathbb{B}_{k-1}^{\mathcal{B}}$, for $k = 1, \dots, n$.*

Lemma 13. $\dim_{\mathbb{F}_2} \left(\mathbb{B}_k^{\mathcal{B}} / \mathbb{B}_{k-1}^{\mathcal{B}}\right) \geq \dim_{\mathbb{F}_2} \left(\mathbb{B}_k^f / \mathbb{B}_{k-1}^f\right)$, *for $k = 1, \dots, n$.*

Let us turn to the main theorem of this section. As to the proof we remark only that the foregoing three lemmas can be applied.

Theorem 14 (Algebraic Charaterization). *Let $f \in \mathbb{B}_n$ be a Boolean function. Then*

$$\mathrm{C}_{\mathrm{POBDD}}^{\sigma}(f) = \dim_{\mathbb{F}_2} \left(\mathbb{B}_n^f \,/\, \mathbb{B}_0^f\right).$$

Moreover, if $\mathcal{B}$ is any σ–POBDD representing f such that $\mathrm{C}_{\mathrm{POBDD}}^{\sigma}(f) = \mathrm{SIZE}(\mathcal{B})$, then

$$\#\mathcal{N}_k^{\mathcal{B}} = \dim_{\mathbb{F}_2} \left(\mathbb{B}_k^f / \mathbb{B}_{k-1}^f\right),$$

for $k = 1, \dots, n$.

We emphasise the fact that Theorem 14 characterizes the POBDD–complexity of a Boolean function by an algebraic invariant of this function. The situation is similar to that in the case of OBDDs, where we count the number of subfunctions obtained by fixing $x_{\sigma(1)}, \ldots, x_{\sigma(n-k+1)}$ that depend essentially on $x_{\sigma(n-k)}$. But while in OBDDs every "new" subfunction needs an extra node, in POBDDs it is sufficient to extend the basis so that every "new" subfunction can be linearly represented.

4 Minimizing the number of nodes in a σ–POBDD

Definition 15. Let $\mathcal{B}_n$ be a POBDD with respect to the permutation σ.

1. $\mathcal{B}_n$ is said to be σ–minimal if and only if $\mathrm{SIZE}(\mathcal{B}_n) = \mathrm{C}^{\sigma}_{\mathrm{POBDD}}(f)$, where f is the Boolean function represented by $\mathcal{B}_n$.
2. The k-th level of $\mathcal{B}_n$ is said to be algebraically reduced if and only if $\#\mathcal{N}_k^{\mathcal{B}} = \dim_{\mathbb{F}_2}\left(\mathbb{B}_k^{\mathcal{B}}/\mathbb{B}_{k-1}^{\mathcal{B}}\right)$. The whole diagram $\mathcal{B}_n$ is defined to be algebraically reduced if and only if all levels are algebraically reduced. It is called algebraically prereduced if and only if all of its levels but the top one are algebraically reduced.

Theorem 16 (Minimization of the number of nodes). *There is an algorithm which computes from a σ–POBDD $\mathcal{B}_n$ representing a Boolean function $f \in \mathbb{B}_n$ a σ–minimal POBDD representing f in time $O\left(n \cdot (\mathrm{SIZE}(\mathcal{B}_n))^3\right)$ and space $O\left((\mathrm{SIZE}(\mathcal{B}_n))^2\right)$.*

Proof. Let $\mathcal{B} = \mathcal{B}_n$ be the σ–POBDD under consideration. The algorithm computes in four global steps an algebraically reduced σ–POBDD representing f which fulfils, moreover, the property that, for each node $v \in \mathcal{N}_l$, Res_v belongs to $\mathbb{B}_l^f$. (Then it follows that this POBDD is σ–minimal.)

Global Step 1: The Bottom Up Reduction

It is our aim to compute an algebraically reduced σ–POBDD representing f level by level from the bottom up to the top. We shall eliminate all nodes which can be represented by the remaining ones. Define, for $k = 1, \ldots, n$, $d_k \overset{def}{=} \dim_{\mathbb{F}_2} \mathbb{B}_k^{\mathcal{B}}$. Assume that we have already reduced the diagram $\mathcal{B}$ up to level $\mathcal{N}_k^{\mathcal{B}}$. We discuss the reduction of level $\mathcal{N}_{k+1}^{\mathcal{B}}$.

First, we choose the local basis E_{k+1} consisting of the vectors

$$e_{\eta v} \overset{def}{=} \iota_{k+1}^{\eta}\left(\mathrm{Res}_v^{\mathcal{B}}\right), \text{ for } v \in \mathcal{N}_{k_v}^{\mathcal{B}},\ k_v \leq k,\ \eta \in \{0,1\},$$

where v ranges over all nodes of the part of $\mathcal{B}$ which we have already reduced. Clearly, $\#E_{k+1} = 2d_k$.

Second, we define the $(2d_k) \times (d_k + n_{k+1})$–matrix $M^{(k+1)}$, where $n_{k+1} \stackrel{def}{=} \#\mathcal{N}^{\mathcal{B}}_{k+1}$,

$$M^{(k+1)} \stackrel{def}{=} \begin{pmatrix} 1 & \dots & 0 & M^{(k+1)}_{1,d_k+1} & \dots & M^{(k+1)}_{1,d_k+n_{k+1}} \\ M^{(k+1)}_{2,d_k+n_{k+1}} & & & & & \\ \vdots & \ddots & \vdots & \vdots & & \vdots \\ 0 & \dots & 1 & M^{(k+1)}_{d_k,d_k+1} & \dots & M^{(k+1)}_{d_k,d_k+n_{k+1}} \\ 1 & \dots & 0 & M^{(k+1)}_{d_k+1,d_k+1} & \dots & M^{(k+1)}_{d_k+1,d_k+n_{k+1}} \\ \vdots & \ddots & \vdots & \vdots & & \vdots \\ 0 & \dots & 1 & M^{(k+1)}_{2d_k,d_k+1} & \dots & M^{(k+1)}_{2d_k,d_k+n_{k+1}} \end{pmatrix}.$$

The columns $M^{(k+1)}_{\cdot j}$,

- for $j = 1, \dots, d_k$, are the representations of the the functions $\mathrm{Res}^{\mathcal{B}}_v$ in the local basis E_{k+1}, where v ranges over the nodes of the part of $\mathcal{B}$ which we have already reduced.
- for $j = d_k + 1, \dots, d_k + n_{k+1}$, are the representations of the the functions $\mathrm{Res}^{\mathcal{B}}_v$, $v \in \mathcal{N}^{\mathcal{B}}_{k+1}$, in the local basis E_{k+1}, which can be obtained by Lemma 7.

This can be done in time $O\left((\mathrm{SIZE}(\mathcal{B}))^2\right)$.

Third, we select by the Gaussian elimination in time $O\left((\mathrm{SIZE}(\mathcal{B}))^3\right)$ the δ_{k+1} columns $M^{(k+1)}_{\cdot j}$, for $j \geq d_k + 1$, such that the rank of the matrix formed by these columns together with the first d_k ones is equal to the rank of the whole matrix $M^{(k+1)}$. Let us assume w.l.o.g. that the indices of the columns selected are $d_k + 1, \dots, d_k + \delta_{k+1}$. Obviously, we have that $d_{k+1} = d_k + \delta_{k+1}$.

Fourth, we use a simultaneous Gaussian elimination to solve all equations

$$\left(M^{(k+1)}_{\cdot 1} \quad \dots \quad M^{(k+1)}_{\cdot d_{k+1}}\right) \cdot \left(X_1, \dots, X_{d_{k+1}}\right)^T = M^{(k+1)}_{\cdot j},$$

for $j = d_{k+1} + 1, \dots, d_k + n_{k+1}$.

Thus we compute for each column $M^{(k+1)}_{\cdot j}$, for $j \geq d_{k+1} + 1$, a linear representation in the first d_{k+1} columns in time $O\left((\mathrm{SIZE}(\mathcal{B}))^3\right)$.

Last, we reconstruct the diagram, whose size is at most diminished. We delete all nodes from $\mathcal{N}^{\mathcal{B}}_{k+1}$ associated with columns $M^{(k+1)}_{\cdot j}$, $j \geq d_{k+1} + 1$. Their outgoing arcs are cancelled without compensation. The ingoing arcs are equivalently replaced by the help of the above representation of the corresponding columns and by Lemma 7 in time $O\left((\mathrm{SIZE}(\mathcal{B}))^3\right)$. Let us denote the resulting POBDD again by $\mathcal{B}$, and its depth by $\tilde{n}_{red}$.

Global Step 2: The Accessibility Reduction

Nodes which cannot be reached from the source are superfluous. That's why we compute by the help of depth first search the sub-POBDD of $\mathcal{B}$ induced by the subset of its nodes accessible from the source via a path in time $O\left((\text{SIZE}(\mathcal{B}))^2\right)$. Let us denote the resulting POBDD, which is reduced and represents f, again by $\mathcal{B}$.

Global Step 3: The Top Down Reduction

Why is this global step necessary? Up to now we have only selected a special basis of the space $\mathbb{B}_n^{\mathcal{B}}$ such that each space $\mathbb{B}_k^{\mathcal{B}}$, for $k = 0, \ldots, n$, is generated by a subset of this basis. Moreover, we have removed some superfluous basis vectors. The POBDD we have got so far is the representation of all necessary subfunctions of the function f in that basis. The problem is that it is possibly too large, or, what is the same, badly choosen. To overcome this problem, we carry out a level-by-level transformation constructing σ–POBDDs $\mathcal{B}_k$, for $k = \tilde{n}_{red}, \ldots, 1$. The resulting POBDD $\mathcal{B}_{\tilde{n}_{red}}$, denoted by $\mathcal{B}$, too, represents f in such a way that it is algebraically prereduced and all functions Res_v belong to $\mathbb{B}_{k_v}^f$, for all its nodes $v \in \mathcal{N}_{k_v}$.

The induction hypothesis for $\mathcal{B}_{k+1}$ is that all functions Res_v, $v \in \mathcal{N}_l$, $l \geq k+1$, belong to $\mathbb{B}_l^f$. Let us define $\mathcal{B}_{\tilde{n}_{red}}$ to be $\mathcal{B}$. Then the induction hypothesis is fulfilled at the very beginning.

Induction step down to $k < \tilde{n}_{red}$.

(i) For each $v \in \bigcup_{l \geq k+1} \mathcal{N}_l$, and each $\eta \in \{0,1\}$ we do the following. If v has no η–successor belonging to $\mathcal{N}_k$, then we skip to the next node. Otherwise we take $\text{Succ}(v, \eta, k)$ to be the set of all η–successors of v belonging to $\bigcup_{l \leq k} \mathcal{N}_l$. By the induction hypothesis and Lemma 12 we know that $\sum_{v' \in \text{Succ}(v,\eta,k)} \text{Res}_{v'} \in \mathbb{B}_k^f$. We create a new node $w(v)$, add it to the level $\mathcal{N}_k$, and create new arcs from $w(v)$ into the set $\bigcup_{l<k} \mathcal{N}_l$ such that $\text{Res}_{w(v)} = \sum_{v' \in \text{Succ}(v,\eta,k)} \text{Res}_{v'}$. We remove all arcs from v to the set $\text{Succ}(v, \eta, k)$ and connect v by a new arc labeled by η to $w(v)$.

(ii) Because the old elements of $\mathcal{N}_k$ are no longer accessible, we remove them.

(iii) We reduce the new level $\mathcal{N}_k$ in the same way as in Global Step 1.

The output of the Global Step 3, which can be done in time $O\left(n \cdot (\text{SIZE}(\mathcal{B}))^3\right)$, is $\mathcal{B}_1$.

Global Step 4: The Reduction of the Top Level

The output of Global Step 3 is nearly what we want. It remains to check whether the old top level is superfluous, and to remove it in the this case. We compute $k_4 := \max\{k < \tilde{n}_{red} \mid \#\mathcal{N}_k > 0\}$ in time $O\left((\text{SIZE}(\mathcal{B}))^2\right)$. If $\#\mathcal{N}_{k_4} > 1$, then we are done, because then $\mathcal{B}$ is algebraically reduced. Otherwise we carry out the reduction step for the top level of Global Step 1 again to algebraically reduce the top level of $\mathcal{B}$.

It is easy to see, that the entire algorithm works within space $O\left((\text{SIZE}(\mathcal{B}))^2\right)$.

Now it is no problem to solve the satisfiability and the redundancy problem, too.

Corollary 17 (Satisfiability). *Let $\mathcal{B}_n$ be a σ-POBDD representing $f \in \mathbb{B}_n$. Then it can be decided in time $O\left(n \cdot (\text{SIZE}(\mathcal{B}_n))^3\right)$ and space $O\left((\text{SIZE}(\mathcal{B}_n))^2\right)$ whether $f \neq 0$.*

Corollary 18 (Redundancy). *Let $\mathcal{B}_n$ be a σ-POBDD representing $f \in \mathbb{B}_n$. Then we can decide in time $O\left(n \cdot (\text{SIZE}(\mathcal{B}_n))^3\right)$ and space $O\left((\text{SIZE}(\mathcal{B}_n))^2\right)$ whether f depends essentially on the variable x_i.*

5 Synthesis, Equivalence and Applications

Theorem 19 (Synthesis). *1. If $\mathcal{B}$ is a POBDD representing a function f, then a POBDD of the same size representing $\neg f$ can be constructed in linear time.*

2. Let $\mathcal{B}^{(1)}, \ldots, \mathcal{B}^{(k)}$ be σ-POBDDs representing the Boolean functions $f^{(1)}, \ldots, f^{(k)} \in \mathbb{B}_n$. Then we can construct a σ-POBDD representing $f^{(1)} \oplus \ldots \oplus f^{(k)} \in \mathbb{B}_n$ of size less than or equal to $\sum_{i=1}^{k} \text{SIZE}\left(\mathcal{B}^{(i)}\right)$ in linear time.

3. Let $\mathcal{B}^{(1)}, \ldots, \mathcal{B}^{(k)}$ be σ-POBDDs representing the Boolean functions $f^{(1)}, \ldots, f^{(k)} \in \mathbb{B}_n$. Then we can construct a σ-POBDD representing $f^{(1)} \odot \ldots \odot f^{(k)} \in \mathbb{B}_n$ of size less than or equal to $\prod_{i=1}^{k} \text{SIZE}\left(\mathcal{B}^{(i)}\right)$ in time and space $\prod_{i=1}^{k} \left(\text{SIZE}\left(\mathcal{B}^{(i)}\right)\right)^2$, for $\odot \in \{\wedge, \vee\}$.

4. Let $\mathcal{B}'$ and $\mathcal{B}''$ be two σ-POBDDs representing the Boolean functions $f', f'' \in \mathbb{B}_n$. Then we can compute a σ-POBDD $\mathcal{B}$ of size less than or equal to $\text{SIZE}(\mathcal{B}') \cdot \text{SIZE}(\mathcal{B}'')$ representing the function $\omega(f', f'')$, where $\omega \in \mathbb{B}_2$ is a preassigned binary Boolean operation in time and space $O\left((\text{SIZE}(\mathcal{B}'_n)\,\text{SIZE}(\mathcal{B}''_n))^2\right)$.

Proof. The proof is similar to that for OBDDs and that's why omitted.

The next three corollaries can easily be verified.

Corollary 20 (Equivalence). *Let $\mathcal{B}'_n$ and $\mathcal{B}''_n$ be two σ-POBDDs representing the Boolean functions $f', f'' \in \mathbb{B}_n$. Then we can decide whether $f' = f''$ in time $O\left(n \cdot (\text{SIZE}(\mathcal{B}'_n) + \text{SIZE}(\mathcal{B}''_n))^3\right)$ and in space $O\left((\text{SIZE}(\mathcal{B}'_n) + \text{SIZE}(\mathcal{B}''_n))^2\right)$.*

Corollary 21 (Replacement by functions). *Let $\mathcal{B}'_n$ and $\mathcal{B}''_n$ be σ-POBDDs representing $f', f'' \in \mathbb{B}_n$. Then we can compute a σ-POBDD of size less than or equal to $(\text{SIZE}(\mathcal{B}'))^2\,\text{SIZE}(\mathcal{B}'')$ representing $f'|_{x_i = f''}$ in time and space $O\left((\text{SIZE}(\mathcal{B}'))^4\,(\text{SIZE}(\mathcal{B}''))^2\right)$.*

Corollary 22 (Quantification). *Let $\mathcal{B}$ be a σ-POBDD representing $f \in \mathbb{B}_n$. Then we can compute σ-POBDDs of size less than or equal to $(\text{SIZE}(\mathcal{B}))^2$ representing $(\exists x_i f), (\forall x_i f) \in \mathbb{B}_{n-k}$ in time and space $O\left((\text{SIZE}(\mathcal{B}))^2\right)$.*

Let us complement the upper bound results mentioned in the introduction now.

Corollary 23. *The storage access function SAF_n, for $n = 2^k + k$, has σ-POBDDs of size $O(n^2)$, for any σ.*

Proof. Clearly, $\text{SAF}_n = \bigoplus_{i=0}^{2^k-1} \delta(|y|, i) \wedge x_i$, where δ is Kronecker's function. Any POBDD for $\delta(|y|, i)$ is of size $O(n)$. The claim follows.

References

1. M. Ajtai, L. Babai, P. Hajnal, J. Komlos, P. Pudlak, V. Rödel, E. Semeredi, and G. Turan, *Two lower bounds for branching programs*, in: Proc. 18th ACM STOC 1986, pp. 30–38.
2. B. Becker, R. Drechsler, *How many decomposition types do we need?*, in: Proc. of the European Design and Test Conference, pp. 438–443, 1995.
3. B. Bollig, M. Löbbing, M. Sauerhoff, I. Wegener, *Complexity theoretical aspects of OFDDs*, in: Proc. of the Workshop on Applications of the Reed–Muller Expansion in Circuit Design, IFIP WG 10.5, pp. 198–205, 1995.
4. Y. Breitbart, H. B. Hunt, D. Rosenkrantz, *The size of binary decision diagrams representing Boolean functions*, preprint.
5. R. E. Bryant, *Graph-based algorithms for Boolean function manipulation*, IEEE Trans. on Computers 1986, **35**, pp. 677–691.
6. R. E. Bryant, *On the complexity of VLSI implementations of Boolean functions with applications to integer multiplication*, IEEE Trans. on Computers 1991, **40**, pp. 205–213.
7. R. E. Bryant, *Symbolic Boolean manipulation with ordered binary decision diagrams*, ACM Comp. on Surveys 1992, **24**, pp. 293–318.
8. J. Gergov, Ch. Meinel, *Mod-2-OBDDs — a data structure that generalizes EXOR-Sum-of-Products and Ordered Binary Decision Diagrams*, Formal Methods in System Design 1996, **8**, pp. 273–282.
9. M. Krause, *Separating ⊕L from L. NL, co-NL and AL (=P) for Oblivious Turing Machines of Linear Access Time* in: Proc. Mathematical Foundations of Computer Science 1990, Lecture Notes in Computer Science **452** pp. 385–391.
10. M. Krause, St. Waack, *On oblivious branching programs of linear length*, Information and Computation 1991, **94**, pp. 232–249.
11. K. Kriegel, St. Waack, *Lower bounds on the complexity of real-time branching programs*, RAIRO Theor. Inform. Appl. 1988, **22**, pp. 447–459.
12. D. Sieling, I. Wegener, *Reductions of OBDDs in linear time*, Information Processing Lettres 1993, **48**, pp. 139–144.
13. I. Wegener, *Efficient data structures for Boolean functions*, Discrete Mathematics 1994, **136**, pp. 347–372 .
14. S. Žák, *An exponential lower bound for read-once branching programs*, in: Proc. 11th MFCS 1984, Lecture Notes in Computer Sci. **176**, Springer–Verlag 1984, pp. 562–566.

A Reducibility Concept for Problems Defined in Terms of Ordered Binary Decision Diagrams *

Christoph Meinel[1 2] and Anna Slobodová[2]

[1] FB IV-Informatik, Universität Trier, D-54286 Trier, Germany
[2] ITWM-Trier, Bahnhofstr. 30-32, D-54292 Trier, Germany

Abstract. Reducibility concepts are fundamental in complexity theory. Usually, they are defined as follows: A problem Π is reducible to a problem Σ if Π can be computed using a program or device for Σ as a subroutine. However, this approach has its limitations if restricted computational models are considered. In the case of ordered binary decision diagrams (OBDDs), it allows merely to use the almost unmodified original program for the subroutine.
Here we propose a new reducibility concept for OBDDs: We say that Π is reducible to Σ if an OBDD for Π can be constructed by applying a sequence of elementary operations to an OBDD for Σ. In contrast to previous reducibility notions, the suggested one is able to reflect the real needs of a reducibility concept in the context of OBDD-based complexity classes: it allows to reduce those problems to each other which are computable with the same amount of OBDD-resources and it gives a tool to carrying over lower and upper bounds.

1 Introduction

Reducibility is one of the most basic notions in complexity theory. It provides a fundamental tool for comparing the computational complexity of different problems. The key idea is to use a program for a device that solves one problem Σ as a subroutine within the computation of another problem Π. If this is possible, Π is said to be reducible to Σ. Reductions provide the possiblity to conclude upper bound results on the computational complexity of problem Π and lower bounds for Σ, if one insists that the program for Π designed around the subroutine for Σ respects certain resource complexity bounds of interest.

In the past, a great variety of different reducibility notions has been investigated in order to get a better understanding of the different computational paradigms and/or resource bounds. Here we only mention polynomial-time Turing reducibility, log-space reducibility, polynomial projection reducibility, and NC^1-reducibility (see, e.g., [Lee90], [BDG88]). This great variety of different reducibility notions is a consequence of the fact that the computational power available to the reduction must not be stronger than the computational power of the complexity class under consideration. Otherwise, the possibility of hiding

* We are grateful to DAAD ACCIONES INTEGRADAS, grant Nr. 322-ai-e-dr

Reischuk, Morvan (Eds.): STACS'97 Proceedings, LNCS 1200

some essential computations within the reduction is a threat to the relevance of the obtained results. For example, polynomial-time reducibility does not give any insight into the computational complexity of logarithmic-space bounded computations. The computational power implementable in a reducibility notion for complexity classes defined in terms of very restricted computational models (e.g., eraser Turing machines [KMW88], real-time branching programs [KW87], or ordered binary decision diagrams [Bry92]) becomes extremely limited, since in general, almost all of the resources are consumed already by the programs which are used as subroutines in the reductions. Hence, the traditional approach results in reducibility concepts which enable to relate merely highly similar problems (e.g., [BW96]). Since complexity classes defined by such restricted models are interesting for the theory – they occur in connection with our limited abilities in proving lower bounds [e.g., KMW88, KW87, Mei89] – and of practical importance – ordered binary decision diagrams are the state of the art data structure for computer aided circuit design [Bry92, BCMD, Bry95] – it is highly desirable to develop more powerful reducibility concepts.

Here, we consider the case of complexity classes defined by ordered binary decision diagrams (OBDDs), (i.e. read-once binary decision diagrams with a fixed variable ordering). We attempt to overcome the difficulties mentioned above by introducing a new reducibility concept that is based on the following idea: A problem Π is reducible to a problem Σ if an OBDD for Π can be constructed from a given OBDD for Σ by applying a sequence of elementary operations (here 'elementary' means 'performable in constant time'). In contrast to previous reducibility notions the suggested one is able to reflect the real needs of a reducibility concept in the context of OBDD-based complexity classes: Firstly, it allows to reduce those problems to each other which are computable with the same amount of OBDD-resources, and, secondly, it allows to carry over lower and upper bounds.

Although well-motivated, a reducibility based on sequences of elementary operations is difficult to describe and to handle since it has to deal with permanently changing OBDDs. For this reason, we develop a formalism which allows a more 'static' description in terms of the so-called OBDD-transformer. We prove that the size of an OBDD which is obtained by the application of a sequence of elementary operations can be estimated in terms of the sizes of the original OBDD and of the corresponding OBDD-transformer. Hence, the formalism gives a solid basis for complexity theoretic investigations.

2 Notations and Preliminaries

Let X_n denote the set $\{x_1, x_2, ..., x_n\}$ of Boolean variables. A *variable ordering* on X_n is a total order on X_n and is described by a permutation of the index set $I_n = \{1, ..., n\}$, i.e. $x_i < x_j$ iff $\pi^{-1}(i) < \pi^{-1}(j)$. Throughout the paper, we will work with the extension of an ordering to constants *false* and *true* which are defined to be maximal (*false* and *true* become incomparable). Identity defines the so-called natural ordering. Two orderings (possibly defined on different variable sets) are said to be *consistent* if there is no pair (x_i, x_j) such that x_i precedes x_j in one ordering, and x_j precedes x_i in the other.

By functions, we mean Boolean functions $\{false, true\}^n \longrightarrow \{false, true\}$. The standard representation of *false* and *true* is 0 and 1, respectively. However, it will be more convenient for us to represent *false* by -1 and *true* by 1. $\cong$ is used for the isomorphism of labeled graphs.

Definition 1. An *ordered binary decision diagram (OBDD)* over X_n is a directed acyclic graph with the following properties:

(i) There is one distinguished node without incoming edges, called *root.*
(ii) Nodes without outgoing edges (called *sinks*) are labeled by -1 or 1.
(iii) Each non-sink node is labeled by a variable from X_n. The labeling fullfills the read-once property, i.e., on each root-to-sink path, any variable appears at most once.
(iv) Each non-sink node has two outgoing edges that are called *true*- and *false-edge* and are labeled by 1 and -1, respectively.
(v) Each node has a *negation mark* -1 or 1.
(vi) All variable orderings defined by the occurence of variables on root-to-sink paths are consistent.

The nodes that are labeled by the same variables form a *level.* Let π be a variable ordering. An ordered binary decision diagram is a πOBDD if all variable orderings defined by the occurence of variables on root-to-sink paths are consistent with π. The size of an OBDD P is defined as the number of its nodes and is denoted by *size*(P).

In order to make figures more compact, we omit the labels of the edges and use solid lines for the true-edges and dotted lines for the false-edges of an OBDD. The negation mark 1 is usually omitted.

Let v be a node of an OBDD. Each input assignment $\alpha = (a_1, \ldots, a_n)$ uniquely determines a v-to-sink path $p_v(\alpha)$ according to the following rule: At an inner node with label x_i, the outgoing edge with label a_i is chosen. Let $\text{sgn}_v(\alpha)$ denote the product of the negation marks on the nodes on $p_v(\alpha)$ and $sink_v(\alpha)$ be the label of the sink on $p_v(\alpha)$. The Boolean function which is represented by v is defined by $f_v(\alpha) = \text{sgn}_v(\alpha) \cdot sink_v(\alpha)$, for any input α. The function represented by an OBDD is the function represented by its root. We will not distinguish between an OBDD and its root as long as it introduces no ambiguity. In this sense, the successors of a node are the nodes reachable via edges starting in the node as well as the subOBDDs rooted in these nodes. An OBDD (respectively, a πOBDD) for a function f is denoted by OBDD(f) (respectively, πOBDD(f)).

The defined OBDD model slightly differs from the one usually used. The introduction of negated nodes is motivated by existing OBDD-implementations [e.g., BRB90, Lon93, Som96], where the use of negated edges allows to save up to half the size of the representation.

If P is an OBDD, then $\overline{P}$ denotes the OBDD obtained from P by multiplying the negation mark of the root by -1. Two OBDDs are *(functionally) equivalent* (denoted by $\equiv$) if they represent the same function. An OBDD is called *reduced* if all nodes have negation mark 1 and no two subgraphs represent equivalent OBDDs. We remark that OBDDs can be reduced in linear time [SW93]. For a fixed variable ordering π, the representation of a Boolean function in terms of reduced πOBDD is uniquely determined [Bry86].

The subject of this paper is the development of a reducibility concept usable for complexity investigation of OBDDs. Unfortunately, in the context of OBDDs, the notions 'reduce' and 'reduction' have a fixed meaning in the sense mentioned above. Speaking about reductions in the complexity theoretical sense, we avoid the terminological ambiguity by using the term 'OBDD-transformation' instead of 'OBDD-reduction'.

The non-uniformity of OBDDs and the sensitivity of the structure and the size of an OBDD to a variable ordering have to be taken into account in the definition of a 'problem' in this context. Due to the former attribute, we work with a *family of functions.* The latter makes differences among representations of a function with respect to different *orderings.* Indeed, the differences in sizes of the respective OBDDs may be exponential.

Definition 2. A *problem* (f, π) is a sequence of pairs $((f_n, \pi_n))_{n=1}^{\infty}$, where f_n is a n-ary Boolean function and π_n is a permutation which defines a variable ordering for f_n. For each n, (f_n, π_n) is called an *instance* of the problem.

Each problem (f, π) uniquely defines a sequence of reduced $\pi_n\text{OBDD}(f_n))_{n=1}^{\infty}$.

3 OBDD-Transformations

Our aim is to introduce a reducibility concept for complexity classes defined by OBDDs which reflects the expense that arises if one constructs an OBDD P' for one problem from a given OBDD P for another problem. This expense is measured in the number of 'elementary' operations that are necessary for constructing P' from P. Here, an operation is considered 'elementary' if it can be performed in constant time (under the unit cost measure).

Operation 1. Setting/deletion of the negation mark of a node.
Operation 2. Exchange of the true- and false-edge of a node.
Operation 3. Redirection of one outgoing edge towards the second one.
Operation 4. Replacing a node by a sink.
Operation 5. Node splitting – an edge of the considered node is redirected to a copy of the successor.
Operation 6. Introduction of a dummy node consistently with the given variable ordering.

Considering the usual representation of an OBDD, where each node stores its label, its mark and two pointers to its sons all operations are performable in constant time.

The operations defined above have clear semantics. The first operation negates the subfunction defined in a node, the second operation negates the evaluation of the checked variable, the third one corresponds to the restriction, and the fourth operation provides a replacement of a subfunction by a trivial function. Unlike the first four operations which may change the functionality of an OBDD, the last two allow to turn to unreduced OBDDs and to perform the subsequent operations merely on some distinguished subgraphs/subfunctions.

It is quite natural to have the possibility to rename variables. Among other things, it allows to move from one variable ordering to another. In order to preserve the 'read-once property', we have to insist that renaming does not identify any two different variables.

Definition 3. A problem (f,π) is an *(OBDD-)transformation* of a problem (g,σ) (written as $(f,\pi) \leq_{OBDD} (g,\sigma)$), if for each n, there is an m such that the reduced π_nOBDD(f_n) can be obtained from the reduced σ_mOBDD(g_m) by a sequence of elementary operations completed by reduction and a renaming of variables (i.e., application of a bijective mapping on the set of variables that occur in the reduced OBDD).

Example A. The following example illustrates an OBDD-transformation defined by means of the application of a sequence of elementary operations. We show that, for any variable ordering π,

$$\left(\bigvee_{i=1}^{n} x_i, \pi\right) \leq_{OBDD} \left(\bigwedge_{i=1}^{n} x_i, \pi\right) .$$

We start with an OBDD that computes $\bigwedge_{i=1}^{n} x_i$. First, applying Operation 2 on each node, we obtain an OBDD that computes $\bigwedge_{i=1}^{n} \overline{x_i}$. Then we apply Operation 1 to the root of this OBDD. We obtain an OBDD that computes $\overline{\bigwedge_{i=1}^{n} \overline{x_i}}$. Since, by DeMorgan's rules, $\overline{\bigwedge_{i=1}^{n} \overline{x_i}} = \bigvee_{i=1}^{n} x_i$ the constructed OBDD is the desired one. It is interesting to remark, that the functions considered in the example are not reducible by means of read-once projections [BW96].

4 Alternative Description of OBDD-Transformations

The desciption of more complex and interesting (OBDD) transformation in terms of sequences of elementary operations which have to be applied on permanently changing OBDDs may be quite cumbersome. In order to overcome these difficulties, we develop a formalism which allows a more 'static' and easier manipulatable desciption of transformations. The basic idea of this formalism is to encode a sequence of elementary operations by a certain OBDD-like graph structure with an extended labeling of the nodes, the so-called transformer. Then the application of a sequence of elementary operations to a given OBDD is realized by means of an algorithm *Derive* which computes the desired OBDD from the given one and a corresponding transformer.

Using this formalism it becomes much easier to get the transformations like *(MOD-2,π)* $\leq_{OBDD}$ *(MOD-2^i,π)*, *(MOD-2^i,π)* $\leq_{OBDD}$ *(MOD-2,π)*, or *(SQU,π)* $\leq_{OBDD}$ *(MUL,π')* which show the power of the introduced reduction concept in comparison with the formerly mentioned read-once projections.

Moreover, since we prove that the size of a transformed OBDD can be estimated in terms of the sizes of the original OBDD and of the transformer, our formalism open a way for complexity theoretical investigations.

Definition 4. For any variable set X_n we define $\tilde{X_n} = X_n \cup \{\bar{x} | x \in X_n\} \cup \{1_x | x \in X_n\} \cup \{-1_x | x \in X_n\} \cup \{-1, 1, \perp\}$. A mapping from $\tilde{X}_n$ to $X_n \cup \{-1, 1\}$ is defined by $x \mapsto |x|$, where

(i) $|x_i| = |\bar{x_i}| = |1_{x_i}| = |-1_{x_i}| = x_i$
(ii) $|-1| = -1$
(iii) $|1| = |\perp| = 1$

An *OBDD-transformer* (or simply *transformer*) T over X_n is a graph whose nodes are labeled by elements from $\tilde{X}_n$ such that the graph obtained from T by replacing each node label x by $|x|$ is an OBDD over X_n.
A πOBDD-transformer is an OBDD-transformer where the orderings of the variables associated with the labels of the nodes on the root-to-sink paths are consistent with π. Particularly, each πOBDD is a πOBDD-transformer. The *size* of a transformer T is defined as the number of its nodes and is denoted by *size*(T).

4.1 The Algorithm *Derive*

The algorithm *Derive* realizes OBDD-transformation defined in terms of OBDD-transformer. Let P be a πOBDD and T a πOBDD-transformer. The result of *Derive*(P, T) is denoted by $P \diamond T$. The algorithm starts in the roots of P and T, scans the graphs in parallel, and creates the result recursively. The information in the root of T describes the required changes in the root of P. Instead of modifying P, we create a new graph as the result. The mark -1 corresponds to Operation 1, the negative variable to Operation 2, labels 1_x and -1_x to Operation 3, and the sink node to Operation 4.

We explain this idea in detail. If x is a label (of the root) of T, then, for each constant $\delta \in \{-1, 1\}$, (the root of) $T_{|x|=\delta}$ denotes the δ-successor (of the root) of T. Similarly, $P_{x=\delta}$ is the δ-successor of P.

Trivial case:

(i) If T is a sink labeled by $\perp$, then $P \diamond T$ is a graph isomorphic to P but its mark equals to product the marks of P and T.
(ii) If T is a sink labeled by a constant, then $P \diamond T$ is isomorphic to T.

Nontrivial case:
Let x be the label of P and y the label of T. The program continues according to the ordering relation of x and $|y|$.

(i) If $x < |y|$, then a node is created with the same label and mark as P has and with true-successor $P_{x=1} \diamond T$ and false-successor $P_{x=-1} \diamond T$.

(ii) If $x > |y|$, then a node is created, labeled by $|y|$, with the same mark as T and with the true-successor $P \diamond T_{|y|=1}$ and false-successor $P \diamond T_{|y|=-1}$.

(iii) If $x = |y|$, then a node labeled by x is created. The mark of the node will be the product of the marks of P and T. The true- and false-successor of the node depend on y:

	true-successor	false-successor
$y = x$:	$P_{x=1} \diamond T_{x=1}$	$P_{x=-1} \diamond T_{x=-1}$
$y = \bar{x}$:	$P_{x=-1} \diamond T_{x=1}$	$P_{x=1} \diamond T_{x=-1}$
$y = 1_x$:	$P_{x=1} \diamond T_{x=1}$	$P_{x=1} \diamond T_{x=-1}$
$y = -1_x$:	$P_{x=-1} \diamond T_{x=1}$	$P_{x=-1} \diamond T_{x=-1}$

Particularly, if T is an OBDD, then for each OBDD P it holds $P \diamond T = T$.

The reduction running in parallel is easy to implement in the algorithm without changing its asymptotic time and space performance. W.l.o.g. we assume that the result of *Derive* is a reduced OBDD.

Proposition 5. *Let P be a πOBDD and T a πOBDD-transformer. The time as well as the space complexity of the algorithm Derive on an input (P, T) is bounded by $\mathcal{O}(size(P) \cdot size(T))$. The size of the output is bounded by $size(P) \cdot size(T)$.*

Proof. The main observation is that each pair of nodes (u, v), where $u \in P$ and $v \in T$, generates at most two recursive calls.

4.2 Composition of OBDD-Transformers

Let $T_1 \circ T_2$ denotes a transformer that realizes the composition of the OBDD-transformations represented by πOBDD-transformers T_1 and T_2, i.e., for any πOBDD P, $P \diamond (T_1 \circ T_2) \cong (P \diamond T_1) \diamond T_2$.

Proposition 6. *Let T_1 and T_2 be πOBDD-transformers. There is an algorithm that produces $T_1 \circ T_2$ of the size bounded by $size(T_1) \cdot size(T_2)$ within the time and space complexity $\mathcal{O}(size(T_1) \cdot size(T_2))$.*

4.3 Transformers vs. Sequences of Elementary Operations

We show that any sequence of elementary operations performed on a given OBDD P can be simulated by applying the algorithm *Derive* on P and a suitable transformer T. Moreover, the reverse is also true: any transformer describes a particular OBDD-transformation. Hence, transformer gives an equivalent 'static' description of the 'dynamical' transformation process.

Theorem 7. *A problem $(f, \pi) = (f_n, \pi_n)_{n=1}^{\infty}$ is an OBDD-transformation of a problem $(g, \sigma) = (g_n, \sigma_n)_{n=1}^{\infty}$, if and only if, for every n, there is an m and an OBDD-transformer T_n such that π_nOBDD(f_n) and σ_mOBDD$(g_m) \diamond T_n$ are equivalent up to a renaming of the variables.*

We say that the sequence of OBDD-transformers $(T_n)_{n=1}^{\infty}$ of the theorem *realizes* an OBDD-transformation $(f,\pi) \leq_{\mathrm{OBDD}} (g,\sigma)$. We remark that there is nothing said about the number of variables in T_n yet.

4.4 Examples of OBDD-Transformations

Example B: A transformer for the transformation in Example A has n nonterminal and one terminal node. i-th nonterminal node is labelled by $\overline{x_i}$ and its both outgoing edges enter $(i+1)$-st node. The sink is labeled by $\perp$. The only negation mark set to -1 is that of the root.

Example C: $(\text{MOD-}2,\pi) \leq_{OBDD} (\text{MOD-}4,\pi)$

$\text{MOD-}2_n(a_1,a_2,...,a_n) = \text{MOD-}4_{2n}(a_1,a_1,a_2,a_2,...,a_n,a_n)$, for any input. We can construct a transformer with $2n$ variables, where each variable with even index is forced to be set to the same value as the previous variable. For the natural ordering, there is one node labelled by x_i for each $i = 2j-1$. This node has the false-successor labelled by -1_{x_i} and the true-successor labelled by 1_{x_i}. The outgoing edges from these two successors enter the same node labelled by $x_{2(j+1)-1}$. The sink is labelled by $\perp$. This construction can be easily generalized to a transformation $\text{MOD-}2 \leq_{OBDD} \text{MOD-}2^i$, for any i.

Example D: $(\text{MOD-}4,\pi) \leq_{OBDD} (\text{MOD-}2,\pi)$

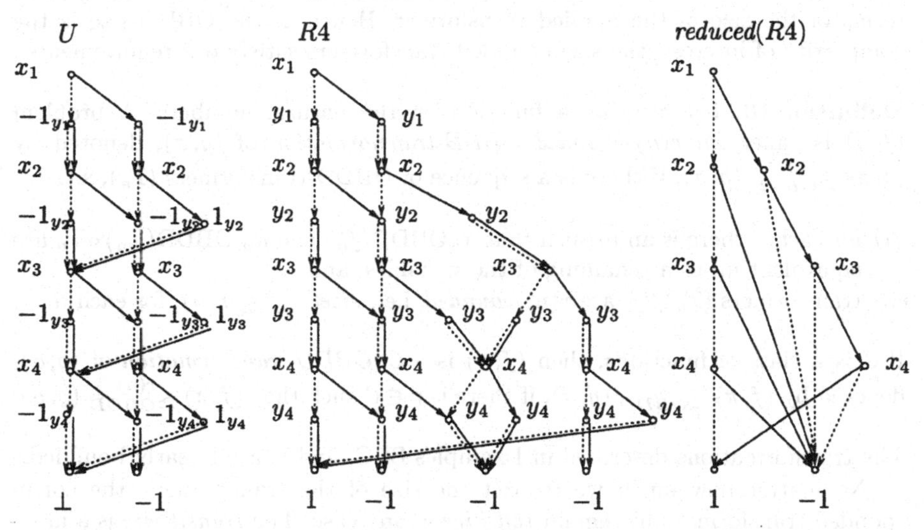

Fig. 1: Let R be an OBDD(MOD-2_4). $R4 = R \diamond U$ is an OBDD for MOD-4_4.

The transformation can be generalized to $\text{MOD-}2^i \leq_{OBDD} \text{MOD2}$, for any i.

5 Complexity-Bounded OBDD-Transformations

It is easy to see that the relation $\leq_{OBDD}$ defined via OBDD-transformations is *reflexive* and *transitive*. Even more, we can show that $\leq_{OBDD}$ is *symmetric*, too, and, hence, that all problems are equivalent with respect to *unbounded* OBDD-transformations. This is neither surprising nor disturbing, as long as we allow to apply any sequence of elementary operations, no matter how long it is.

Proposition 8. *Let (f, π) be any problem and σ any variable ordering. (f, π) and $(true, \sigma)$ are transformable to each other.*

Proof. Let P be any OBDD and $\mathbf{1}$ denote the reduced OBDD for the constant function 1. P can be seen as an OBDD-transformer that fulfills the relation $\mathbf{1} \diamond P = P$ and $P \diamond \mathbf{1} = \mathbf{1}$. □

Corollary 9. *Any two problems are OBDD-transformable to each other.*

5.1 Complexity Measure

In order to use OBDD-transformations for the investigation of complexity theoretic properties of OBDDs, we have to restrict their computational resources in an appropriate manner. Due to Propositions 5 and 6 together with Theorem 7, the size of the OBDD constructed by the transformation can be estimated in terms of the size of the applied transformer. Hence, if the OBDD-size is the complexity of interest, the size-bounded transformers satisfy our requirements.

Definition 10. Let $r(n)$ be a function on the natural numbers. A problem (f, π) is called an *$r(n)$-bounded OBDD-transformation* of (g, σ), denoted by $(f, \pi) \leq^{r(n)}_{OBDD} (g, \sigma)$, if there is a sequence of OBDD-transformers (T_n), s.t.

(i) for each n there is an m such that π_nOBDD(f_n) and σ_mOBDD$(g_m) \diamond T_n$ are equivalent up to a renaming of the variables, and
(ii) transformers $(T_n)_{n=1}^{\infty}$ are *$r(n)$-bounded*, i.e., *size*$(T_n) \leq r(n)$, for each n.

If C is a class of functions, then (f, π) is a *C–OBDD-transformation* of (g, σ), denoted by $(f, \pi) \leq^{C}_{OBDD} (g, \sigma)$, if there is $r \in C$ such that $(f, \pi) \leq^{r(n)}_{OBDD} (g, \sigma)$.

The transformations described in Examples B, C, and D are linearly bounded.

No matter how much we restrict the size of the transformers, the corresponded transformations remain *reflexive* in any case. The *transitivity* is a more sensitive property. Generally, we have

Proposition 11. *For every class of functions C, C–OBDD-transformations are* transitive *if and only if, for every $r_1, r_2 \in C$, there is $r \in C$, s.t. $r \geq r_1 r_2$.* □

According to Propositions 13, transitivity is fulfilled in the cases of e.g., constantly, polylogarithmically, or polynomially bounded transformations. A noteworthy phenomenon is that the *symmetry* property of unbounded transformations (Corollary 11) is no more true if one considers bounded transformations. For example, if one cuts out an exponentially large subOBDD by Operation 4, then an exponentially large transformer is needed to 'repair' this.

5.2 Polynomial OBDD-Transformations

Size bounded transformers provide a relevant reduction tool for the investigation of complexity classes defined in terms of OBDDs. For any class of functions $\mathcal{C}$, $\mathcal{C}_{OBDD}$ denotes all problems (f,π) for which there is some $r \in \mathcal{C}$ s.t. the size of the reduced $\pi_n\mathrm{OBDD}(f_n)$ is bounded by $r(n)$, for every n. Particularly, $\mathcal{P}_{OBDD}$ denotes the set of problems representable by polynomially bounded OBDDs, i.e.,

$$\mathcal{P}_{OBDD} = \{(f,\pi) \mid \exists \text{polynomial } p(n) \forall n \text{ with } size(\pi_n\mathrm{OBDD}(f_n)) \leq p(n)\}.$$

Let $\leq^p_{OBDD}$ denote the relation defined by means of polynomially bounded OBDD-transformations. Problems Π and Σ are said to be *polynomially equivalent* ($\Pi \equiv^p_{OBDD} \Sigma$) if $\Pi \leq^p_{OBDD} \Sigma$ and $\Sigma \leq^p_{OBDD} \Pi$.

Proposition 12. *Properties of polynomially bounded OBDD-Transformations.*

(i) Relation $\leq^p_{OBDD}$ is transitive.
(ii) Any two problems in $\mathcal{P}_{OBDD}$ are polynomially equivalent.
(iii) If $\Pi \leq^p_{OBDD} \Sigma$ then $\Sigma \in \mathcal{P}_{OBDD}$ implies $\Pi \in \mathcal{P}_{OBDD}$.

Polynomially bounded OBDD-transformations provide an adequate reduction tool for the investigation of the membership in the class $\mathcal{P}_{OBDD}$, which can be very useful in practical applications of OBDDs.

5.3 Polynomial OBDD-Transformations vs. Read-once Projections

According to the definition in [BW96], $f = (f_n)$ is a read-once projection of $g = (g_n)$ if there is a polynomially bounded sequence (p_n) such that for each n:

$$f_n(x_1, ..., x_n) = g_{p_n}(y_1, ..., y_{p_n}) \tag{1}$$

where

(i) $y_i \in \{-1,1\} \cup X_n \cup \{\overline{x_j} \mid x_j \in X_n\}$, for each $1 \leq i \leq p_n$, and
(ii) for any $i \neq j$: $y_i \in \{x_r, \overline{x_r}\}$ implies $y_j \notin \{x_r, \overline{x_r}\}$ (read-once property).

The problems considered in [BW96] are defined in terms of sequences of Boolean functions, unlike to this paper, where the variable ordering is taken in consideration as an important attribute of the problem. Roughly spoken, f is a read-once projection of g means that for each sequence $\pi = (\pi_n)$ of variable orderings for f there is a sequence $\sigma = (\sigma_n)$ of variable orderings for g such that $\pi_n\mathrm{OBDD}(f_n)$ and $\sigma_m\mathrm{OBDD}(g_m)$ are related as it is described in the definition. Hence, fine relations between the functions that are visible only with respect to some subset of orderings cannot be expressed in terms of read-once projections (see Example E). The extension of the notion of OBDD-transformations to problems defined as in [BW96] is possible, but not considered in this paper.

Proposition 13. *Let $f = (f_n)_{n=1}^{\infty}$ be a read-once projection of $g = (g_n)_{n=1}^{\infty}$ and $p(n)$ is the polynom from (1). For each sequence $\pi = (\pi_n)_{n=1}^{\infty}$ of orderings on variables in the support of f and g, it holds $(f, \pi) \leq_{\text{OBDD}}^{p(n)} (g, \pi)$.*

Proof. Omitted.

Corollary 14. *A read-once projection can be realized by means of a sequence of OBDD-transformers (T_n) with size$(T_n) = p_n + 1$.*

Additionaly to linearly bounded OBDD-transformation in Examples B, C and D, we can give an example of a constant OBDD-transformation that cannot be realized via any read-once projection.

Proposition 15. *There is an OBDD-transformation which can be realized by means of a single elementary operation, but cannot be obtained via any read-once projection.*

Proof. The constant functions *false* and *true* are reducible to each other by application of the one-node transformer labelled by -1 with negated mark 1. On the other hand, none of them is a read-once projection of the other. □

For the case of multivalued Boolean functions, OBDDs are generalized to multirooted OBDDs and transformers to multirooted ones. The most surprising result obtained for read-once projections in [BW96] was the non-reducibility of SQUaring to MULtiplication. We show that this result is not a witness of the higher complexity of the SQUaring. In contrary, it is an additional argument for the need of a more adequate notion of reducibility in the context of restricted complexity classes.

Example E: (SQU,π) $\leq_{OBDD}^{p}$ (MUL,π'),
where $\pi'_{2n} = (\pi(1), \pi(1)+n, \pi(2), \pi(2)+n, ..., \pi(n), \pi(n)+n)$.

Proof. For every n, let $\text{SQU}_n^i(x_1, x_2, ..., x_n)$ be the i-th output-bit of the square of the binary number $x_1x_2...x_n$ and $\text{MUL}_n^i(x_1, x_2, ..., x_{2n})$ be the i-th output-bit of the multiplication of the binary numbers $x_1x_2...x_n$ and $x_{n+1}x_{n+2}...x_{2n}$. We can construct a π'_{2n}OBDD-transformer that transforms a π'_{2n}OBDD for MUL_{2n}^i into a πOBDD for SQU_n^i, for each $i, 1 \leq i \leq n$. For each j, $1 \leq j \leq n$, the π'_{2n}OBDD-transformer contains one node labeled by x_{2j-1}. Each node labeled by x_{2j-1} has two successor nodes labeled by x_{2j}. The true-successor (respectively, the false-successor) has one outgoing edge labeled by 1 (resp., by -1). □

5.4 Derivation of an Exponential Lower Bound

Polynomial OBDD-transformations can be used for deriving exponential lower bounds on the OBDD-size.

Let $PERM_n$ be the test whether a given $n \times n$ matrix M is a permutation matrix, i.e., if there is exactly one 1 in each row and in each column. Let $MAG\text{-}2_n$ denote the test whether a given $n \times n$ matrix over $\mathbb{Z}_2$ is a magic square, i.e. whether the sum (XOR) of the elements in each row and the sum of each column agree. We show that

$$PERM \leq^{p}_{OBDD} MAG\text{-}2$$

The transformation is based on the observation that a permutation matrix is a magic matrix whose row sums over $\mathbb{Z}_2$ equal 1 and which overall sum of ones equals the number of rows. For every n, from the input we construct a transformer T^1_n that excludes from the the inputs the matrices with more or less than n ones. Transformer T^2_n assures that the sum in a row is equal to 1 in the accepted inputs. It looks like an OBDD for the parity of the variables in one (any chosen) row, where the 1-sink is replaced by the $\perp$-sink. The entire transformer is obtained as $T^1_n \circ T^2_n$.

In [KMW88], it was proven that the function *PERM* requires exponentially large OBDDs for any variable ordering. Due to this fact, the described OBDD-transformation proves an exponential lower bound for the magic square problem.

Corollary 16. *size*$(\pi OBDD(MAG\text{-}2)) = 2^{\Omega(n)}$*, for any variable ordering* π*.*

References

[BRB90] K. S. Brace, R. L. Rudell, R. E. Bryant: Efficient Implementation of a BDD Package, Proc. of 27th ACM/IEEE Design Automation Conference, 1990, 40-45.

[Bry92] R. E. Bryant: Symbolic Boolean Manipulation With Ordered Binary Decision Diagrams. *ACM Computing Surves, 24 (3), p. 293-218, 1992.*

[Bry95] R. E. Bryant: Binary Decision Diagrams and Beyond: Enabling Technologies for Formal Verification. *Proc. ICCAD'95.*

[BCMD] J.R. Burch, E.M. Clarke, K.L. McMillan, D.L. Dill: Symbolic Model Checking: 10^{20} states and beyond, Proc. of 5th IEEE Symposium on Logic in Computer Science, 46–51, 1990.

[BW96] B. Bollig, I. Wegener: Read-once Projections and Formal Circuit Verification with Binary Decision Diagrams. Proc. STACS'96 (1996), 491–502.

[BDG88] J. L. Balcázar, J. Díaz, J. Gabarró: Structural Complexity I., Springer 1988.

[Lee90] J. van Leeuwen: Handbook of Theoretical Computer Sci. MIT Press, 1990.

[KW87] K. Kriegel, S. Waack: Exponential Lower Bounds for Real-time Branching Programs. Proc. FCT'87,LNCS 278 (1987), 263-367.

[KMW88] M. Krause, Ch. Meinel, S. Waack: Separating the Eraser Turing Machine Classes $\mathcal{L}_\epsilon, \mathcal{NL}_\epsilon, co-\mathcal{L}_\epsilon$ and $\mathcal{P}_\epsilon$. Proc. MFCS'88, LNCS 324 (1988), 405-413.

[Lon93] D. Long: BDD-Package, CMU.

[Mei89] Ch. Meinel: Modified Branching Programs and Their Computational Power. LNCS 370, Springer Verlag, 1989.

[Som96] F. Somenzi: CUDD: CU Decision Diagram Package (Release 1.1.1). University of Colorado at Boulder. June 1996.

[SW93] D. Sieling, I. Wegener: Reduction of BDDs in linear time. Information Processing Letters 48 (1993) 139-144.

On the Classification of Computable Languages

John Case[1], Efim Kinber[2], Arun Sharma*[3], and Frank Stephan**[4]

[1] Computer and Information Sciences Department, 101A Smith Hall, University of Delaware, Newark, DE 19716-2586, U.S.A., case@cis.udel.edu.
[2] Computer Science, Sacred Heart University, 5151 Park Avenue, Fairfield, CT 06432-1000, U.S.A., kinber@shu.sacredheart.edu.
[3] School of Computer Science and Engineering, The University of New South Wales, Sydney, NSW, 2052, Australia, arun@cse.unsw.edu.au.
[4] Mathematisches Institut der Universität Heidelberg, Im Neuenheimer Feld 294, 69120 Heidelberg, Germany, fstephan@math.uni-heidelberg.de.

Abstract. A one-sided classifier for a given class of languages converges to 1 on every language from the class and outputs 0 infinitely often on languages outside the class. A two-sided classifier, on the other hand, converges to 1 on languages from the class and converges to 0 on languages outside the class. The present paper investigates one-sided and two-sided classification for classes of computable languages. Theorems are presented that help assess the classifiability of natural classes. The relationships of classification to inductive learning theory and to structural complexity theory in terms of Turing degrees are studied. Furthermore, the special case of classification from only positive data is also investigated.

1 Introduction

Consider the problem of determining whether a language A over $\mathbb{N}$, the set of natural numbers $\{0, 1, 2, \ldots\}$, satisfies a certain property. Let $\mathcal{A}$ denote the class of all languages over $\mathbb{N}$ that satisfy the given property. The question of classification can then be stated thus: if one is given data about A, can one determine if $A \in \mathcal{A}$.

We briefly discuss the various approaches to the study of classification in the literature. One of the earliest attempts was the design of finite automata to decide whether an infinite string (representing the characteristic function of a language) belongs to a given ω-language or not [5, 15, 17, 26]. But the restrictive computational ability of these finite automata led Büchi [5] and his successors to consider non-deterministic automata. The present paper takes the alternate approach of choosing Turing machines as classifiers. In fact this approach had already been initiated by Büchi and Landweber [6, 14].

Smith and Wiehagen [24] introduced a model of classification analogous to the Gold model of learning [4, 10, 19]. The (computable) classifier M sees longer

* Supported by a grant from the Australian Research Council.
** Supported by the Deutsche Forschungsgemeinschaft (DFG) grant Am 60/9-1.

Reischuk, Morvan (Eds.): STACS'97 Proceedings, LNCS 1200

and longer prefixes σ of the characteristic function of a language $A \in \mathcal{A}_1 \cup \mathcal{A}_2 \cup \ldots \cup \mathcal{A}_k$ and guesses on each input σ some number $h \in \{1, 2, \ldots, k\}$ to indicate that $A \in \mathcal{A}_h$. These guesses are supposed to converge, for each set $A \in \mathcal{A}_1, \mathcal{A}_2, \ldots, \mathcal{A}_k$, to a value h such that $A \in \mathcal{A}_h$. Smith, Wiehagen and Zeugmann [25] extended this study in various ways.

Ben-David [3] and Kelly [13] also interestingly studied classification. They call a class *classifiable* iff there exists a (not-necessarily-computable) functional that indicates in the limit for every A whether or not it belongs to a given class $\mathcal{A}$. They obtained topological conditions for classifiable classes. Gasarch, Pleszkoch, Stephan and Velauthapillai [9] extended this study and obtained relations between the Borel hierarchy on classes – which is induced by the space $\{0,1\}^\infty$ with product topology – and the query hierarchy obtained by allowing a certain number of quantifier-alternations during querying a teacher on the target set A.

Later Stephan [23] investigated the limits of (computable) classifiers. He considered classification of languages w.r.t. one single class $\mathcal{A}$ and used the following two natural models of classification: Two-sided classification which is the computable counterpart to Ben-David's classification in the limit and one-sided classification which is already implicit in the notion of reliable inference (on languages inside the class the learner converges to an index for the language and on languages outside the class the learner makes infinitely many mind changes). These two notions of classes are very natural and coincide with the Δ_2^0 and Σ_2^0 classes studied by recursion theorists [11, 18, 20, 21]. Our study derives from these models which we present next. But, first some notation.

We take a *classifier* to be an algorithmic device; M, N and H range over classifiers. Calligraphic letters range over classes, A, B range over sets and U ranges over oracles. We take σ, τ to range over prefixes of strings of characteristic functions of sets. $\sigma \preceq \tau$ means that $\tau(x)\downarrow = \sigma(x)$ for all $x \in dom(\sigma)$. $M(\sigma)$ denotes the guess issued by classifier M on a prefix $\sigma \preceq A$ of the input-set A.

Two-Sided Classification: For all languages A: $M(\sigma) = \mathcal{A}(A)$ for almost all $\sigma \preceq A$.

Here $\mathcal{A}(A)$ is 1 if $A \in \mathcal{A}$ and 0 otherwise, i.e., classes and sets are identified with their characteristic function. Two-sided classification may be considered to be too strong a requirement. In some applications it is sufficient if the classifier is able to signal the inclusion of a language in a given class, but only provides a weaker signal if the language is not in the class. Stephan [23] introduced the notion of one-sided classification to model this idea.

One-Sided Classification: For all languages A: if $A \in \mathcal{A}$, then $M(\sigma) = 1$ for almost all $\sigma \preceq A$; if $A \notin \mathcal{A}$, then $M(\sigma) = 0$ for infinitely many $\sigma \preceq A$.

We normally let M and N range over two-sided classifiers and let H range over one-sided classifiers. The notion of one-sided classification is reasonable since the classifier outputs 0 infinitely often thereby guaranteeing that the classifier never locks onto an incorrect conjecture.

In the present paper, we restrict our investigation to classification of computable languages. This restriction may be supported by the fact that practical examples are always computable and assuming an algorithmic view of the universe, it is unlikely that nature generates noncomputable languages. Thus, our classifiers can be relied upon if they are never expected to deliberate upon noncomputable languages. Hence, in the sequel, the statement "for all languages A" in the above two definitions is replaced by "for all computable languages A".[5]

The present paper may also be seen as closing the gap between Stephan's abstract work [23] and the more concrete approach of Smith, Wiehagen and Zeugmann [24, 25]. Before we begin a formal presentation of the results, we give an informal tour of the various sections in the paper.

In Section 2, we introduce the basic definitions and give preliminary results about two-sided and one-sided classification for classes of computable languages. We give concrete classes of languages that can be two-sided and one-sided classified. In particular we observe that one-sided classes are closed under finite monotone Boolean combinations and two-sided classes are closed under all finite Boolean combinations. We also show that every uniformly recursive family of languages is one-sided classifiable. Additionally, if the family is discrete, then it is also two-sided classifiable. As a consequence of this result, the class of pattern languages is two-sided classifiable. As a contrast, however, the class of regular languages is only one-sided classifiable.

In Section 3 we show that classes identifiable in the limit from informants can be *reliably* identified iff they are one-sided classifiable. We also investigate conditions under which reliable identification in the limit and two-sided classification are linked.

The characteristic function of a language conveys both positive and negative data about the language. In Section 4, we argue that it may not be realistic to assume the availability of both positive and negative data in practice. The experience from empirical studies of learning is that negative data is not always readily available and even when it is available, it is often tedious to obtain. Motivated by such concerns, we also investigate two-sided and one-sided classification from only positive data. Following the practice in inductive inference literature, we model positive data as texts. As expected, we show that classification from texts is very difficult. As a simple consequence of our result, the class of pattern languages is not even one-sided classifiable from texts.

Not deterred by the difficulty of classification from texts, we investigate a weaker version of classification for text presentation, called ***partial classification***, that yields some positive results. A class $\mathcal{A}$ is ***partially classifiable*** just in case there exists a machine that on texts for languages in $\mathcal{A}$ outputs exactly one guess infinitely often and on texts for nonmembers of $\mathcal{A}$ does not output any single guess infinitely often. The motivation here is that a partial classifier gives

[5] So, we ignore noncomputable sets everywhere. Accordingly, set-theoretic notions like the complement of classes are adapted to the computable universe: $\overline{\mathcal{A}} = \{\text{computable } A : A \notin \mathcal{A}\}$.

a weak signal if the language belongs to the class and refuses to give any signal if the language is not a member of the class being classified. We show that each uniformly recursive family of computable languages is partially classifiable. We also give a sufficient condition for partial classification from texts in terms of classification from both positive and negative data. We show that if a class is one-sided classifiable from both positive and negative data, then it is partially classifiable from texts. The converse, however, does not hold.

In Section 5, we investigate structurally the computational limits of classifying computable languages. In particular, we investigate the "computational distance" between one-sided and two-sided classification by determining the kind of noncomputable information that yields a two-sided classifier for a class that was otherwise only one-sided classifiable. This gives insight into what it takes for a class of interest to be two- *vs* one-sided classifiable. We show that access to a high oracle is sufficient to construct a two-sided classifier for a one-sided classifiable class. We also establish that in some cases the power of a high oracle is necessary as there are classes for which any two-sided classifier has high Turing degree. We adapt Post's notion of creative set to describe the one-sided classifiable classes that are effectively not two-sided classifiable. We call a one-sided classifiable class $\mathcal{A}$ *creative* just in case there is a uniformly computable sequence of languages $A_0, A_1, \ldots$ such that for each one-sided classifier H_e, the language A_e is a counterexample to the hypothesis "H_e classifies $\overline{\mathcal{A}}$". The analogy between the two notions of creativity turns out to be quite striking. We give examples of creative classes and show that a creative class is two-sided *only* relative to a high oracle. We discuss some interesting results about one-sided classifiable classes of intermediate complexity and compare our results with the more abstract study of classification by Stephan [23] in which a classifier has to behave correctly on noncomputable languages, too.

2 Basic Definitions and Results

We now proceed formally. Due to space restrictions, we only give a few representative proofs and omit results about classification with bounded mind changes and about inference of trial and error classifiers. The full paper is available as a technical report [8].

Definition 2.1 A classifier H is an algorithm that on every string σ outputs a number 0 or 1. It classifies a class $\mathcal{A}$ *one-sided* just in case

- if $A \in \mathcal{A}$, then $H(\sigma) = 1$ for almost all $\sigma \preceq A$;
- if $A \in \overline{\mathcal{A}}$, then $H(\sigma) = 0$ for infinitely many $\sigma \preceq A$.

The classifier H is furthermore *two-sided* iff the statement "for infinitely many" in the second clause can be strengthened to "for almost all". Note that in this definition the variable A ranges over only *computable* sets.

There is an effective list of classifiers H_e such that for each one-sided class there is some H_e classifying it one-sided and for each two-sided class there is some

H_e classifying it two-sided. Assuming an acceptable numbering φ_e of all partial computable functions, these classifiers are defined as follows:

$$H_e(\sigma) = \begin{cases} \varphi_e(\tau) & \text{for the longest } \tau \preceq \sigma \text{ such that} \\ & \varphi_e(\tau) \text{ outputs 0 or 1 within } |\sigma| \text{ steps;} \\ 0 & \text{if there is no such } \tau. \end{cases}$$

Now it is easy to verify that whenever φ_e is a one-sided classifier for $\mathcal{A}$, then so is H_e; and whenever φ_e is a two-sided classifier for $\mathcal{A}$, then so is H_e. This normalization has the advantage that now we can assume without loss of generality that all one-sided (two-sided) classes have a total and computable one-sided (two-sided) classifier. Therefore, in the sequel, we will consider H_e instead of the underlying φ_e.

One-sided classes are closed under finite monotone Boolean combinations and two-sided classes are closed under all finite Boolean combinations. These facts follow from the following theorem.

Theorem 2.2 *$\mathcal{A}$ is two-sided iff $\mathcal{A}$ and $\overline{\mathcal{A}}$ are one-sided. If $\mathcal{A}, \mathcal{B}$ are one-sided classes so are $\mathcal{A} \cup \mathcal{B}$ and $\mathcal{A} \cap \mathcal{B}$. If $\mathcal{A}$ is one-sided so is $\mathcal{B} = \{B : B$ is a finite variant of some $A \in \mathcal{A}\}$.*

Given a computable function $A(x, y)$, let $A_x = \{y : A(x, y) = 1\}$ and $\mathcal{A}$ be the class $\{A_0, A_1, \ldots\}$. Such a class $\mathcal{A}$ is called a *uniformly recursive family.* Angluin [2] initiated the study of learning uniformly recursive families from texts. After the introduction of monotonicity constraints many papers have considered the learnability of these families from texts and informants [12, 27, 28]. A class $\mathcal{A}$ is *closed* iff for each $A \notin \mathcal{A}$ there is a $\sigma \preceq A$ such that no $B \in \mathcal{A}$ extends σ.

Theorem 2.3 *Every uniformly computable family is one-sided. If it is also closed, then it is two-sided.*

Example 2.4 The immediately preceding results yield the following examples.

- $\mathcal{C} = \{A : A \text{ is cofinite}\}$ is one-sided, but not two-sided.
  The classifier is $H(\sigma w) = w$.
- $\mathcal{D} = \{1^\infty, 01^\infty, 001^\infty, 0001^\infty, \ldots\}$ is two-sided.
  The classifier M outputs 1 if $\sigma \in 0^*1^+$ and 0 otherwise.
- $\mathcal{E} = \{A : A \text{ has finite and even cardinality}\}$ is one-sided, but not two-sided.
  The classifier $H(\sigma)$ outputs 1 iff the number of 1s in σ is even and 0 iff this number is odd.
- $\mathcal{F}_\phi = \{A : \text{the formula } \phi(A) \text{ is true}\}$ is two-sided.
  Here $\phi(A)$ means that ϕ is a Boolean formula, such as $[5 \in A \vee [3 \notin A \wedge 4 \notin A]]$, with A being the only free variable representing the input-set A of the same name. Such formulas can be evaluated after having seen a sufficiently long part of the input and from then on the classifier outputs 1 if $\phi(A)$ holds and 0 if $\phi(A)$ does not hold.
- $\mathcal{G} = \{graph(p) : p \text{ is a polynomial}\}$ is one-sided, but not two-sided.
  $\mathcal{G}$ and $\mathcal{R}$ below are uniformly recursive families and, hence, have the one-sided classifier from Theorem 2.3.

- $\mathcal{P} = \{A : A \text{ is a pattern language}\}$ is two-sided.
  This is due to the fact that the class of the pattern languages is both closed and uniformly recursive.
- $\mathcal{R} = \{A : A \text{ is regular}\}$ is one-sided, but not two-sided.

There is also a prominent class which is not one-sided: the class $\{A : \varphi_{\min(A)}$ computes $A\}$ of the self describing sets. But this class has a one-sided complement.

3 Reliable Learning and One-Sided Classification

A learner is said to *reliably* Ex-identify a class of languages if it either either diverges or converges to a correct index, but it never converges to a false one (see Minicozzi for the definition and related results [16]). We denote by REx the class of language classes that can be reliably Ex-identified. The reader may observe that a reliably inferred class is also in some sense classified since convergence indicates membership in the class and divergence indicates membership in its complement. Hence, it might be expected that there are interesting links between reliable learning and classification.

Theorem 3.1 *Let $\mathcal{A}$ be in* Ex. *Then $\mathcal{A} \in$* REx *iff $\mathcal{A}$ is one-sided.*

Proof ($\Rightarrow$): Let $\mathcal{A}$ be reliably learnable. The classifier outputs 0 if the learner changes its mind and outputs 1 if there is no mind change. Whenever the learner converges to an index, then the classifier outputs only finitely many 0s and thus accepts the language. Whenever the learner does not converge to an index, i.e., the language does not belong to $\mathcal{A}$, then the classifier rejects the language by outputting infinitely many 0s. So the classifier accepts just the languages in $\mathcal{A}$ and is correct.

($\Leftarrow$): If $\mathcal{A}$ is learnable and one-sided classifiable, then a mind change can be introduced into the learning algorithm by padding at every place where the classifier outputs 0, i.e., if the learner outputs for σ and σw the same guess e, but the classifier outputs a 0 for σw, then the learner's output at σw is replaced by an equivalent but different index for the characteristic function computed by e. This does not effect convergence on $A \in \mathcal{A}$ since there these new mind changes are inserted only finitely often. But if $A \notin \mathcal{A}$, then the classifier outputs infinitely many 0s which induce infinitely many mind changes on the modified learner; so this modified learner diverges. Thus the modified learner is reliable, i.e., it converges on a computable A if and only if it learns A. ∎

The reader may have observed that in the above proof of Theorem 3.1, at no point are the guesses evaluated. Therefore this result can be translated to similar notions as long as the following two conditions are satisfied: padding is available and infinitely many mind changes (as in the notion of behaviorally correct learning) are not permitted. Case, Jain and Sharma [7] introduced the notion of learning limiting recursive programs or "trial-and-error-guesses" (LimEx). According to this criterion of learning, a learner has to converge on every language $A \in \mathcal{A}$ to a total program e in two variables which computes A in the limit:

$$(\forall x)\,(\exists y)\,(\forall z > y)\,[A(x) = \varphi_e(x,z)].$$

Reliable LimEx is the variant where the learner has to converge to a limiting program for every $A \in \mathcal{A}$ and has to make infinitely many mind changes for all computable $B \notin \mathcal{A}$. Since LimEx satisfies the two conditions above, Theorem 3.1 also holds for learning limiting recursive programs.

Theorem 3.2 *Let $\mathcal{A}$ be in* LimEx. *Then $\mathcal{A}$ is reliably* LimEx *learnable iff $\mathcal{A}$ is one-sided.*

4 Classification From Only Positive Data

Gold [10] introduced the notion of identification from text. A text for a language is an infinite sequence of numbers and the symbol "#" such that each element of the language appears at least once and no nonelement of the language ever appears in the sequence. Analogously to Gold's notion of inference, we can define classification from texts: a classifier for $\mathcal{A}$, upon being fed a text for some language A, converges to 1 iff $A \in \mathcal{A}$. As in the case of standard classification, one can define the obvious variants of one-sided and two-sided classification from texts.

Example 4.1 Every class $\mathcal{F}_\phi$ of all languages satisfying the Boolean formula ϕ is two-sided classifiable from text.

Theorem 4.2 *If $\mathcal{A}$ and $\mathcal{B}$ are both two-sided classifiable from text and a finite set belongs to $\mathcal{A}$ iff it belongs to $\mathcal{B}$, then $\mathcal{A} = \mathcal{B}$. Furthermore, if $\mathcal{A}$ is one-sided classifiable from text and contains only infinite languages, then $\mathcal{A}$ is void. In particular the class $\mathcal{P}$ of all pattern-languages is not classifiable from text.*

Furthermore some classes which are one-sided classifiable from text have a certain immunity-property:

Theorem 4.3 *The infinite class $\mathcal{A} = \{\{0, 1, \ldots, a\} : a \in \mathbb{N}\}$ is one-sided classifiable from text but every subclass $\mathcal{B} \subseteq \mathcal{A}$ which is two-sided classifiable from text is finite.*

The preceding theorems showed the limitations of classifying from text. So it is suitable to look for a weaker convergence criterion in order to make it possible to classify more realistic classes from text.

Definition 4.4 A machine H classifies a class $\mathcal{A}$ *partially* iff H on any text T for any set A outputs an infinite sequence of numbers such that $A \in \mathcal{A}$ iff exactly one number appears in the output infinitely often and $A \notin \mathcal{A}$ iff no number appears in the output infinitely often.

It is easy to see that every class which can be one-sided classified from texts can also be partially classified from texts. But there are classes which can be partially classified but cannot be one-sided classified from text.

Theorem 4.5 *If $\mathcal{A}$ is a uniformly recursive family $A_0, A_1, \ldots$, then $\mathcal{A}$ can be partially classified.*

Proof W.l.o.g. for every $A \in \mathcal{A}$ there is exactly one e with $A = A_e$. The algorithm H outputs each number e on text $T = w_0 w_1 \ldots$ for A at least n times iff A_e and T are "compatible at level n", i.e., iff $\{w_0, w_1, \ldots, w_n\} \subseteq A_e$ and each $x \in A_e$ with $x \leq n$ appears in T.

On one hand if the set A to be classified equals A_e, then H outputs e infinitely often. On the other hand if $A \neq A_e$, then there is an n such that either $w_n \in A - A_e$ or $n \in A_e - A$. In both cases, H outputs e less than n times. In the first case $A \in \mathcal{A}$ and there is an unique index e such that H outputs e infinitely often. In the second case $A \notin \mathcal{A}$ and H outputs no e infinitely often. ∎

Since the classes $\mathcal{C}$, $\mathcal{D}$, $\mathcal{E}$, $\mathcal{G}$, $\mathcal{P}$ and $\mathcal{R}$ (from Example 2.4) are uniformly recursive families, they can be partially classified. Furthermore, all classes $\mathcal{F}_\phi$ (from Example 2.4) can be partially classified since they are two-sided classifiable from text.

Theorem 4.6 *If $\mathcal{A}$ is one-sided classifiable from informant, then $\mathcal{A}$ is partially classifiable from text. The converse does not hold.*

So partial classification is very powerful. Nevertheless there are still classes which can even not be partially classified from text. Let $q_0, q_1, \ldots$ be an enumeration of the rational numbers and $<$ the ordering on them. Then a class which can not be partially classified is given by those languages A for which $\{q_x : x \in A\}$ is well-ordered w.r.t. "$<$".

While every class, which is one-sided classifiable from text, either contains or is disjoint to an infinite class which is two-sided classifiable from text, this does not longer hold for partial classification versus one-sided classification.

Theorem 4.7 *There is a class $\mathcal{A}$ partially classifiable from text such that any infinite class $\mathcal{B}$ which is one-sided classifiable from text is neither subclass of $\mathcal{A}$ nor of $\overline{\mathcal{A}}$.*

5 Structural Properties of Classification

Soare [22] contains an extensive study on the relation between recursively enumerable and computable sets. As Stephan [23] has already noted, the situation of one-sided versus two-sided classification is similar to that of recursively enumerable versus computable sets. This relationship not only holds in the setting of classifying all sets but also in setting of the present paper of classifying computable sets.

This section shows that if only computable sets are to be classified, then the analogy with recursively enumerable versus computable sets is even more striking. Turing degrees, an important tool for studying recursively enumerable sets, also turn out to be useful in analyzing the complexity of one-sided classification. The next result shows that every one-sided class is two-sided relative to a sufficiently complex oracle.

An oracle U is *Turing reducible* to V (written: $U \leq_T V$) iff U can be computed by a machine which has access to a database containing V by the membership-queries "Is $x \in V$?". For an oracle U the relativized halting problem U' to U is defined as $U' = \{e : \varphi_e^U(e) \downarrow\}$.[6] U is *high* iff $K' \leq_T U'$ [7]. An alternative characterization is that there is a function u computable relative to U which dominates every recursive function, i.e., which satisfies $(\forall^\infty x)\,[u(x) > f(x)]$ for all $f \in$ REC. Adleman and Blum [1] showed that high oracles play a significant role in inductive inference: REC can be Ex-identifiable relative to U iff U is high. Theorems 5.1 and 5.5 show that the high oracles play a similar special role in classification.

Theorem 5.1 *For each high oracle U, every one-sided class $\mathcal{A}$ has a two-sided classifier which is computable relative to U.*

Proof Let H be a one-sided classifier for a class $\mathcal{A}$ of computable sets. Furthermore let u be a function computable relative to U which dominates every computable function. Now the two-sided classifier is defined as follows where $n_H(\sigma)$ denotes the number of prefixes $\tau \preceq \sigma$ with $H(\tau) = 0$. The idea is now to repeat each 0 of H a large but finite number of times such that M still converges to 1 if H does but M converges to 0 if H only diverges.

If $u(n_H(\sigma)) > |\sigma|$, then let $M(\sigma) = 0$ else let $M(\sigma) = 1$.

If $A \in \mathcal{A}$, then there is only a finite number n of prefixes $\tau \preceq A$ with $H(\tau) = 0$. Almost all prefixes σ of A have length at least $u(n)$. So $|\sigma| \geq u(n) \geq u(n_H(\sigma))$ and $M(\sigma) = 1$ for these prefixes σ. If $A \notin \mathcal{A}$ and A is computable, then also the function $f_A(n) = \min\{m : n_H(A(0)A(1)\ldots A(m)) \geq n\}$ is computable and thus u dominates f_A. There is a n with $u(m) > f(m)$ for all $m \geq n$. In particular whenever a prefix $\sigma \preceq A$ has at least the length $u(n)$, then $u(n_H(\sigma)) > f_A(n_H(\sigma)) \geq |\sigma|$ and $M(\sigma) = 0$. So M converges on every computable set outside $\mathcal{A}$ to 0 and M is two-sided. ∎

Nevertheless there are hard problems, i.e., there are one-sided classes $\mathcal{A}$ which require that every two-sided classifier for $\mathcal{A}$ has high Turing degree.

[6] φ_e^U is the e-th partial recursive in U function.

[7] This differs slightly from Soare's definition [22, Definition IV.4.2]: Soare defined "$K' \equiv_T U'$" instead of "$K' \leq_T U'$" since he considers only oracles $U \leq_T K$.

Theorem 5.2 *If M two-sided classifies the class $\mathcal{C} = \{C : C$ is cofinite$\}$, then M is not computable and its Turing degree is high.*

A recursively enumerable set E is called *creative* [22, Definition II.4.3] iff there is an effective procedure which disproves for every e the hypothesis "$W_e = \overline{E}$" by a counterexample $f(e)$, i.e., either $f(e) \in \overline{E} - W_e$ or $f(e) \in W_e - \overline{E}$. The name "creative" derives from the fact that such an f creates a new element $f(e) \in \overline{E}$ outside W_e whenever $W_e \subseteq \overline{E}$. This concept is adapted to the context of classifying computable sets.

Definition 5.3 A one-sided classifiable class $\mathcal{A}$ is *creative* iff there is an uniformly computable array $A_0, A_1, \ldots$ such that for each one-sided classifier H_e the set A_e is a counterexample to the hypothesis "H_e classifies $\overline{\mathcal{A}}$".

The next theorem shows that there is a creative class, namely the class of all cofinite sets. So this class is effectively not two-sided.

Theorem 5.4 *The class $\mathcal{C}$ of all cofinite sets is creative.*

All creative sets are 1-equivalent to K and have in particular the same Turing degree as K, i.e., belong to the greatest recursively enumerable Turing degree. So it is natural to ask how complex are the creative classes. The next theorem shows that there is indeed an analogous result that only the high oracles allow them to be two-sided classified.

Theorem 5.5 *Every creative class, in particular $\mathcal{C}$, is two-sided only relative to high oracles.*

While the preceding results mainly dealt with creative classes, this one deals with several degrees of non-creativeness. First it is shown that there are one-sided classes of intermediate complexity: they are two-sided relative to some non-high oracle but not relative to the empty oracle. In particular they are also not creative by Theorem 5.5.

Theorem 5.6 *For each $U \geq_T K$ which is also enumerable relative to K there is a class $\mathcal{A}$ such that a Turing degree contains a classifier for $\mathcal{A}$ iff U is computable relative to its jump. In particular there are intermediate one-sided classes; these are neither two-sided nor creative.*

There are two kinds of immunity-properties for classes:

- For a class $\mathcal{A}$ there is no uniformly computable array $A_0, A_1, \ldots$ of pairwise different sets such that $\{A_0, A_1, \ldots\} \subseteq \mathcal{A}$.
- No infinite two-sided class $\mathcal{B}$ is contained in $\mathcal{A}$.

The following theorems investigate the extent to which one-sided classes and their complements satisfy these requirements. But, the first result shows that a one-sided class and its complement can never be simultaneously immune.

Theorem 5.7 *Let $\mathcal{A}$ be a one-sided class. Then there is an uniformly computable array $A_0, A_1, \ldots$ of pairwise distinct sets such that the class $\mathcal{B} = \{A_0, A_1, \ldots\}$ is two-sided and either $\mathcal{B} \subseteq \mathcal{A}$ or $\mathcal{B} \subseteq \overline{\mathcal{A}}$.*

Theorem 5.8 *There is an infinite two-sided class $\mathcal{A}$ which contains no subclass $\mathcal{B} = \{A_0, A_1, \ldots\}$ consisting of a uniformly computable array of pairwise distinct sets.*

It is well-known that every infinite recursively enumerable set has an infinite computable subset. Stephan [23] showed that this easy observation does not generalize to one-sided classification versus two-sided in his model which requires correct classification of non-computable sets. Furthermore Theorem 4.3 shows something similar for classification from texts. Since the classification of only computable sets from informants is more well-behaved than the two previously mentioned settings, the following problem might still have a positive solution.

Problem Does every infinite one-sided class have an infinite two-sided subclass?

References

1. Leonard Adleman and Manuel Blum: Inductive Inference and Unsolvability. *Journal of Symbolic Logic* 56 (1991) 891–900.
2. Dana Angluin: Inductive Inference of Formal Languages from Positive Data, *Information and Control* 45 (1980) 117–135.
3. Shai Ben-David: Can Finite Samples Detect Singularities of Real-Valued Functions? *Proceedings of the 24th Annual ACM Symposium on the Theory of Computer Science, Victoria, B.C.*, (1992) 390–399.
4. Lenore Blum and Manuel Blum: Toward a mathematical Theory of Inductive Inference. *Information and Control*, 28 (1975) 125–155.
5. J. Richard Büchi: On a decision method in restricted second order arithmetic. In *Proceedings of the International Congress on Logic, Methodology and Philosophy of Science*, Standford University Press, Standford, California, 1960.
6. J. Richard Büchi and Lawrence H. Landweber: Definability in the Monadic Second Order Theory of Successor. *Journal of Symbolic Logic* 34 (1969) 166–170.
7. John Case, Sanjay Jain and Arun Sharma: On Learning Limiting Programs. *International Journal of Foundations of Computer Science*, 3 (1992) 93–115.
8. John Case, Efim Kinber, Arun Sharma and Frank Stephan. On the Classification of Computable Languages. Technical Report No. 9603, School of Computer Science and Engineering, The University of New South Wales, Sydney NSW 2052, Australia.

9. William Gasarch, Mark Pleszkoch, Frank Stephan and Mahendran Velauthapillai: Classification Using Information. *To appear in: Annals of Mathematics and Artificial Intelligence.*
10. E. Mark Gold: Language Identification in the Limit. *Information and Control*, 10 (1967) 447–474.
11. Peter G. Hinman: *Recursion-Theoretic Hierarchies.* Springer-Verlag, Heidelberg, 1978.
12. Klaus Peter Jantke: Monotonic and Non-Monotonic Inductive Inference. *New Generation Computing* 8 (1991) 349–360.
13. Kevin Kelly: *The Logic of Reliable Inquiry.* Oxford University Press, Oxford, to appear.
14. Lawrence H. Landweber: Decision Problems for ω-Automata. *Mathematical Systems Theory*, 3 (1969) 376–384.
15. Robert McNaughton: Testing and Generating Infinite Sequences by a Finite Automaton. *Information and Control* 9 (1966) 434–448.
16. Eliana Minicozzi: Some Natural Properties of Strong Identification in Inductive Inference. *Theoretical Computer Science* 2 (1976) 345–360.
17. Maurice Nivat and Dominique Perrin (editors): *Automata on Infinite Words.* Lecture Notes to Computer Science 192, Springer-Verlag, Heidelberg, 1984.
18. Piergiorgio Odifreddi: *Classical recursion theory.* North-Holland, Amsterdam, 1989.
19. Daniel N. Osherson, Michael Stob and Scott Weinstein: *Systems that learn.* Bradford / MIT Press, London, 1986.
20. Hartley Rogers, Jr.: *Theory of Recursive Functions and Effective Computability.* McGraw-Hill Book Company, New York, 1967.
21. Gerald E. Sacks: *Higher Recursion Theory*, Perspectives in Mathematical Logic, Springer-Verlag, Heidelberg, 1990.
22. Robert I. Soare: *Recursively enumerable sets and degrees.* Springer-Verlag, Heidelberg, 1987.
23. Frank Stephan: On One-Sided Versus Two-Sided Classification. Forschungsbericht 25 / 1996 des Mathematischen Instituts der Universität Heidelberg, Heidelberg, 1996.
24. Rolf Wiehagen and Carl H. Smith: Generalization versus Classification. *Journal of Experimental and Theoretical Artificial Intelligence*, 7, 1995. Shorter version in *Proceedings 5th Annual Workshop on Computational Learning Theory*, (1992) 224–230. ACM Press, New York.
25. Carl H. Smith, Rolf Wiehagen and Thomas Zeugmann: Classifying Predicates and Languages. *To appear in:* International Journal of Foundations of Computer Science.
26. Boris A. Trakhtenbrot: Finite Automata and the Logic of One Place Predicates. *Siberian Mathematical Journal* 3 (1962) 103–131 [in Russian].
27. Thomas Zeugmann and Steffen Lange: A Guided Tour Across the Boundaries of Learning Recursive Languages. *Algorithmic Learning for Knowledge-Based Systems* (K. P. Jantke and S. Lange, Eds.), Lecture Notes in Computer Science 961 (1995) 193–262.
28. Thomas Zeugmann, Steffen Lange and Shyam Kapur: Characterizations of Monotonic and Dual Monotonic Language Learning, *Information and Computation* 120 (1995) 155–173.

A Conditional-Logical Approach to Minimum Cross-Entropy

Gabriele Kern-Isberner

FernUniversitaet Hagen,Fachbereich Informatik
P.O. Box 940, D-58084 Hagen, Germany
e-mail: gabriele.kern-isberner@fernuni-hagen.de

Abstract. The principle of minimum cross-entropy (ME-principle) is often used in the AI-areas of knowledge representation and uncertain reasoning as an elegant and powerful tool to build up complete probability distributions when only partial knowledge is available. The inputs it may be applied to are a prior distribution P and some new information $\mathcal{R}$, and it yields as a result the one distribution P^* that satisfies $\mathcal{R}$ and is closest to P in an information-theoretic sense. More generally, it provides a "best" solution to the problem "How to adjust P to $\mathcal{R}$?"
In this paper, we show in a rather direct and constructive manner that adjusting P to $\mathcal{R}$ by means of this principle follows a simple and intelligible conditional-logical pattern. The scheme that underlies ME-adjustment is made obvious, and in a generalized form, it provides a straightforward conditional-logical approach to the adaptation problem. We introduce the idea of a *functional concept* and show how the demands for *logical consistency* and *representation invariance* influence the functions involved. Finally, the ME-distribution arises as the only solution which follows the simple adaptation scheme given and satisfies these three assumptions. So a characterization of the ME-principle within a conditional-logical framework will have been achieved, and its logical mechanisms will be revealed clearly.

1 Introduction

Uncertain reasoning on the base of probability theory has a long tradition. From its early beginnings, probability theory has been regarded as an appropriate means to handle quantified degrees of (un)certainty in a logically consistent and sound way. In the fourties of this century, R.T.Cox [4] proved a fundamental connection between probability and conditionals: if any "measure of credibility" is to be attached to the conclusion B of a conditional statement "if A then B" presuming the antecedent A to be true, and if that measure is to follow two basic assumptions on commonsense inference, then it necessarily obeys the rules of probability.

Within the last decades, probabilistic inference has received increasing attention in the area of artificial intelligence. Its logical soundness provides a solid foundation for nonmonotonic reasoning methods (for a survey, cf. [6]; cf. [14], [2]), and probabilistic expert systems make use of the easy and consistent computability of (quantified) uncertainty (cf.[11], [15]). Usually, probabilistic knowledge is

Reischuk, Morvan (Eds.): STACS'97 Proceedings, LNCS 1200

represented by a probability distribution P or by a system of compatible distributions over a set of (discrete or continous) variables. If evidence E is present, the knowledge base is instantiated by calculating the conditional probability $P(\cdot|E)$, applying the principle of direct inference (cf. [2, ch.5.1]). Thus the notion of conditionals is central to probabilistic inference, and vice versa, as the results of Cox cited above show.

Given a (consistent) set $\mathcal{R}$ of conditionals, each equipped with a probability, however, there are generally arbitrary many distributions which all fulfill the probabilistic conditionals in $\mathcal{R}$. Which of them should be chosen to be the "most adequate" one, which of them yields "best" inferences?

From a ***statistical and information-theoretical*** point of view, the answer should be clear. Many authors (cf. e.g. [5], [7]) argued in eloquent words as well as by proving important statistical properties, that representation of probabilistic knowledge should be performed via the ***principle of maximum entropy***. The ***entropy*** $H(P) = -\sum_\omega p(\omega) \log p(\omega)$ (where the sum is taken over all elementary events ω) of a distribution P measures the uncertainty inherent to P. The idea of that principle is as follows: If the distribution P^* has maximum entropy under the constraints given, then P^* represents $\mathcal{R}$ most adequately, without any external information being added. This approach can be generalized to cases where prior knowledge P has to be taken into account. Here instead of (absolute) entropy the notion of ***cross-entropy*** (also called ***relative entropy***) $R(Q,P) = \sum_\omega q(\omega) \log \frac{q(\omega)}{p(\omega)}$ as a measure of dissimilarity between two distributions Q and P is used. The ***principle of minimum cross-entropy*** states that if P^* solves

$$\min R(Q,P) = \sum_\omega q(\omega) \log \frac{q(\omega)}{p(\omega)} \tag{1}$$

s.t. Q is a probability distribution which satisfies $\mathcal{R}$

then P^* constitutes the optimal adaptation of the prior P to the new conditional knowledge in $\mathcal{R}$. We refer to both principles as the ***ME-principle***, where the abbreviation ***ME*** stands both for ***M***inimum cross-***E***ntropy and for ***M***aximum ***E***ntropy.

Two articles [17] and [13] provide essential insights into the ***logical*** aspects of ME-inference. Both show that minimizing cross-entropy resp. maximizing entropy may serve as a logically consistent inference pattern. Shore and Johnson [17] [8] succeeded in characterizing (cross-)entropy as the only functional whose optimization satisfies four (resp. five) fundamental axioms of probabilistic inference. A similar result is proved for entropy in [13] by Paris and Vencovská without assuming that inference is performed by optimizing a functional but relying heavily on solving linear equational systems.

The aim of our paper is to develop a characterization of ME-inference within a completely logical framework. We are going to start from a rather ***simple and constructive conditional-logical scheme*** for adjusting a prior distribution P to new conditional information $\mathcal{R}$ to yield some posterior distribution P^* which satisfies $\mathcal{R}$. This approach makes things explicit and tangible, and there will be no need to make use neither of optimization theory nor of linear equational systems. The ME-solution is a special instance of that scheme, but generally there are a lot of other solutions of that type. Then we will assume that these solutions arise from an *implicit functional concept* which is to represent the underlying logical inference pattern, and we will refine our scheme slightly to make this functional concept explicit. Now only two further postulates - that of *representation invariance* and a fundamental *logical consistency* - will be necessary to uniquely determine the special type of the solution. And we will prove that the ME-solution is indeed the only solution of that type.

Thus the aim of this paper will be achieved. The essential and striking benefit of the present paper compared to the appreciated works of Shore and Johnson and of Paris and Vencovská is that it thoroughly lays bare the logical mechanisms at work in ME-inference. Using a constructive and intuitive approach, it develops step by step the road to the entropy-optimal solution, making all assumptions explicit and intelligible. Thus the conditional-logical point of view provides a most clear and concise characterization of ME-inferences.

2 Probabilistic conditionals

We consider probability distributions P over a finite set $\mathcal{V} = \{V_1, V_2, V_3, \ldots\}$ of propositional variables V_i which are assumed to be binary. The dotted *literal* $\dot{v}_i \in \{v_i, \bar{v}_i\}$ stands for one of the two possible outcomes of the corresponding variable: v_i is to symbolize "V_i is true", and negation is indicated by barring, i.e. $\bar{v}_i = \neg v_i$. P is uniquely determined by the values of its probability function p applied to the elementary events $\omega = \dot{v}_1\dot{v}_2\dot{v}_3\ldots, p(\omega) = p(\dot{v}_1\dot{v}_2\dot{v}_3\ldots) = P(V_1 = \dot{v}_1, V_2 = \dot{v}_2, V_3 = \dot{v}_3, \ldots)$.

A propositional language $\mathcal{L} = \mathcal{L}(\mathcal{V})$ is defined in the usual way, using the letters $\mathcal{V}$ and the classical connectives $\wedge$ (resp. juxtaposition) and $\neg$. Its formulas are denoted by capital roman letters A,B,.... To each propositional formula $\mathrm{A} \in \mathcal{L}$ a probability may be assigned via $p(\mathrm{A}) = \sum_{\omega:\mathrm{A}(\omega)=1} p(\omega)$, where the sum is taken over all elementary events ω, and $\mathrm{A}(\omega) = 1$ means that the complete conjunction corresponding to ω is a disjunct in the disjunctive normal form of A. Thus the correspondence between complete conjunctions and elementary events induces a probabilistic interpretation of $\mathcal{L}$ (on the base of the distribution P).

A ***probabilistic conditional*** (or a ***probabilistic rule***, both terms are used synonymously) is an expression $\mathrm{A} \rightsquigarrow \mathrm{B}[x]$ with ***antecedent*** $\mathrm{A} \in \mathcal{L}$, ***conclusion*** $\mathrm{B} \in \mathcal{L}$ and probability $x \in [0,1]$. It is to represent syntactically non-classical conditional assertions $\mathrm{A} \rightsquigarrow \mathrm{B}$ weighted with a degree of certainty x. A ***probabilistic fact*** has the form $\mathrm{B}[x]$, $\mathrm{B} \in \mathcal{L}$, $x \in [0,1]$ and is considered to be equivalent to the conditional $\top \rightsquigarrow \mathrm{B}[x]$, where $\top$ represents a tautology. Let $\mathcal{L}^*$ denote the set of all probabilistic conditionals over $\mathcal{L}$. A semantical interpretation of probabilistic conditionals is given by conditional probabilities: If P is a distribution, we write $P \models \mathrm{A} \rightsquigarrow \mathrm{B}[x]$ iff $p(\mathrm{B}|\mathrm{A}) = \frac{p(\mathrm{AB})}{p(\mathrm{A})} = x$. In general, we have $x = p(\mathrm{B}|\mathrm{A}) = \frac{1}{1+\frac{p(\mathrm{A}\bar{\mathrm{B}})}{p(\mathrm{AB})}}$ iff $\frac{p(\mathrm{AB})}{p(\mathrm{A}\bar{\mathrm{B}})} = \frac{x}{1-x}$, so the quotient $\frac{p(\mathrm{AB})}{p(\mathrm{A}\bar{\mathrm{B}})}$ determines the probability of the conditional A $\rightsquigarrow$ B . It represents the proportion of individuals or objects with property A which also have property B to those that B is not true of. Thus it is crucial for the acceptability of the conditional, not only within a probabilistic framework (cf. [12]).

The problem this paper is going to deal with can now be described in a formal manner:

Adjustment problem (*)
Given a prior distribution P and some set of probabilistic conditionals $\mathcal{R} = \{A_1 \rightsquigarrow B_1[x_1], \ldots, A_n \rightsquigarrow B_n[x_n]\} \subset \mathcal{L}^*$, how should P be modified to yield a posterior distribution P^* with $P^* \models \mathcal{R}$?

For (*) to be solvable at all, we now assume the prior P to be positive, i.e. $p(\omega) > 0$ for all complete conjunctions ω, and the set $\mathcal{R}$ to be consistent, that means there shall be some distribution Q that satisfies all rules in $\mathcal{R}$.

3 The principle of minimum cross-entropy

Cross-entropy is a well-known information-theoretic measure of dissimilarity between two distributions and has been studied extensively (for a brief, but informative introduction and further references cf. [16]; cf. [18]). It is also called ***directed divergence*** for it lacks symmetry, i.e. $R(Q,P)$ and $R(P,Q)$ differ in general, so it is not a metric. But cross-entropy is ***positive***, that means we have $R(Q,P) \geq 0$, and $R(Q,P) = 0$ iff $Q = P$ (cf. [5], [16]).

If the adjustment problem (*) is solvable at all, then it has a solution P_e with minimal relative entropy to the prior P (cf. [5]), i.e. P_e solves (1). The condition $Q \models \mathcal{R}$ imposed on a distribution Q can be transformed equivalently into a system of linear equality constraints for the probabilities $q(\omega)$. Using the Lagrangian techniques, we may represent P_e in the form

$$p_e(\omega) = \alpha_0 p(\omega) \prod_{i:A_iB_i(\omega)=1} \alpha_i^{1-x_i} \prod_{i:A_i\overline{B_i}(\omega)=1} \alpha_i^{-x_i} \tag{2}$$

with the α_i's being exponentials of the Lagrange multipliers, one for each conditional in $\mathcal{R}$, and $\alpha_0 = \exp(\lambda_0 - 1)$, where λ_0 is the Lagrange multiplier of the constraint $\sum_\omega q(\omega) = 1$.

By construction, P_e satisfies all conditionals in $\mathcal{R}$: $p_e(B_i|A_i) = x_i, 1 \leq i \leq n$. So $\alpha_1, \ldots, \alpha_n$ are solutions of the nonlinear equations

$$\alpha_i = \frac{x_i}{1-x_i} \frac{\sum\limits_{\omega:A_i\overline{B_i}(\omega)=1} p(\omega) \prod\limits_{\substack{j\neq i\\ A_jB_j(\omega)=1}} \alpha_j^{1-x_j} \prod\limits_{\substack{j\neq i\\ A_j\overline{B_j}(\omega)=1}} \alpha_j^{-x_j}}{\sum\limits_{\omega:A_iB_i(\omega)=1} p(\omega) \prod\limits_{\substack{j\neq i\\ A_jB_j(\omega)=1}} \alpha_j^{1-x_j} \prod\limits_{\substack{j\neq i\\ A_j\overline{B_j}(\omega)=1}} \alpha_j^{-x_j}}, \tag{3}$$

$1 \leq i \leq n$, and α_0 arises as a normalization factor. If $x_i = 1$ for some i, (2) and (3) may still be applied by setting $\infty^0 = 1$ and $\infty^{-1} = 0$.

In spite of the deterrent complexity of the formulas above, (2) shows rather clearly how the ME-adjustment to a rule is carried out: Apart from the normalization factor α_0, only the probabilities of those complete conjunctions ω are changed which satisfy the antecedent of this rule. Furthermore, the new probability depends also on whether ω satisfies the conclusion or not. In particular, the probabilities of all conditionals valid in P whose antecedents are not touched by any of the rules in $\mathcal{R}$ remain unchanged. This means that the ME-adaptation respects one of the most fundamental principles of conditional logics: Asserting a conditional should only affect the knowledge about states which the conditional may be applied to.

This intuitive and reasonable *principle of conditional preservation* will give rise to the adaptation scheme which will be presented in the next section.

But first, we are going to illustrate the use of the ME-approach and the benefits of the representation formulas (2) and (3) by two simple but informative examples. The first example shows knowledge processing in the case of conflicting information, whereas the second example deals with transitive inference. All numerical results were obtained by using the probabilistic expert system SPIRIT which realizes entropy-optimal knowledge processing (cf. [15]).

Example 1. A knowledge base is to be built up representing "Typically, students are adults", "Usually, adults are employed" and "Mostly, students are not employed" with degrees of uncertainty 0.99, 0.8 and 0.9, respectively. Let A, S, E denote the propositional variables A=Being an Adult, S=Being a Student, and E=Being Employed. The quantified conditional information may be written as $\mathcal{R} = \{s \rightsquigarrow a[x_1], a \rightsquigarrow e[x_2], s \rightsquigarrow e[x_3]\}$, $x_1 = 0.9, x_2 = 0.8, x_3 = 0.1$. No prior information is at hand, so we start from the uniform distribution. We are interested in the probability of the conditional $as \rightsquigarrow e$ the antecedent of which combines the evidences a and s conflicting with respect to E.

SPIRIT now tells us $p_e\,(e|as) = 0.1009$, so the more specific information s dominates a, as we should expect. If we set $x_1 = 1$, defining the set of students to be definitely a subset of the set of adults, this preference of the more specific knowledge conveyed by the third rule may be proved in general. If P_e' denotes the ME-solution to the same problem with $x_1 = 1$ instead of $x_1 = 0.99$, we obtain $p_e'\,(e|as) = x_3$ by using (2) and (3).

Example 2. Let denote Y, S, C denote the three propositional variables Y=Being Young, S= Being Single and C=Having Children. We know (or assume) that young people are usually singles (with probability 0.9) and that mostly, singles do not have children (with probability 0.85). Here we have $\mathcal{R} = \{y \rightsquigarrow s[x_1], s \rightsquigarrow \bar{c}[x_2]\}$ with $x_1 = 0.9, x_2 = 0.85$. Again we take the uniform distribution as prior information. A calculation with SPIRIT shows $y \rightsquigarrow \bar{c}[0.815]$, connecting both rules transitively. By use of the formulas (2) and (3), a more general transitive inference rule can be proved: For arbitrary x_1, x_2, we obtain for the ME-distribution P_e: $p_e\,(\bar{c}|y) = \frac{1}{2}\left(1 + 2x_1x_2 - x_1\right)$. Of course, the correctness of this formula is independent of the particular meanings of the propositional variables involved. So it states a general transitive inference rule for problems with an analogous knowledge structure.

4 A simple adaptation scheme preserving conditional structures

As we saw earlier in Section 2, ratios of probabilities determine the probabilities of conditionals. If the conditional knowledge inherent to P is to be preserved by the adjustment process (*) as far as possible, we have to preserve ratios of probabilities, the most fundamental of which being probabilities of elementary events.

Following Calabrese [3], a conditional $A \rightsquigarrow B$ can be represented as an *indicator function* $(B|A)$ on elementary events, setting $(B|A)(\omega) = \begin{cases} 1 & : \omega \in AB \\ 0 & : \omega \in A\overline{B} \\ u & : \omega \notin A \end{cases}$ where u stands for *undefined*. This definition captures excellently the non-classical character of conditionals within a probabilistic framework. According to it, a conditional is a function that polarizes AB and $A\overline{B}$, leaving $\overline{A}$ untouched. Now we are able to phrase the principle of conditional preservation in a more general form: The adjustment process (*) shall alter a probability ratio only if at least one conditional in $\mathcal{R}$ explicitly gives rise to do so. For two complete conjunctions ω_1, ω_2, this means that we shall have $\frac{p^*(\omega_1)}{p^*(\omega_2)} = \frac{p(\omega_1)}{p(\omega_2)}$ if $(B_i|A_i)(\omega_1) = (B_i|A_i)(\omega_2)$ for all i. This leads to the following simple approach: Let $\mathcal{R} = \{A_1 \rightsquigarrow B_1 [x_1], \ldots, A_n \rightsquigarrow B_n [x_n]\} \subset \mathcal{L}^*$ be a set of probabilistic conditionals. An adaptation $P * \mathcal{R}$ of the distribution P to the new conditional information $\mathcal{R}$ is constructionally defined by the scheme

$$p^*(\omega) = p * \mathcal{R}(\omega) = \alpha_0 p(\omega) \prod_{i:A_iB_i(\omega)=1} \alpha_i^+ \prod_{i:A_i\overline{B_i}(\omega)=1} \alpha_i^- \tag{4}$$

where α_i^+ and α_i^- are non-negative factors chosen in a suitable way so as to ensure that the rules of $\mathcal{R}$ are all valid in the posterior distribution P^* (as usual, the use of the small letter $p * \mathcal{R}$ indicates the probability function belonging to $P * \mathcal{R}$). Applying the constraints $p^*(B_i|A_i) = x_i$, $1 \leq i \leq n$, yields

$$\frac{\alpha_i^+}{\alpha_i^-} = \frac{x_i}{1 - x_i} \frac{\sum\limits_{\omega:A_i\overline{B_i}(\omega)=1} p(\omega) \prod\limits_{\substack{j \neq i \\ A_jB_j(\omega)=1}} \alpha_j^+ \prod\limits_{\substack{j \neq i \\ A_j\overline{B_j}(\omega)=1}} \alpha_j^-}{\sum\limits_{\omega:A_iB_i(\omega)=1} p(\omega) \prod\limits_{\substack{j \neq i \\ A_jB_j(\omega)=1}} \alpha_j^+ \prod\limits_{\substack{j \neq i \\ A_j\overline{B_j}(\omega)=1}} \alpha_j^-}, \tag{5}$$

with $\alpha_i^- = 0$ if $x_i = 1$, $1 \leq i \leq n$. Thus to each conditional $A_i \rightsquigarrow B_i [x_i]$ in $\mathcal{R}$, $1 \leq i \leq n$, a pair of "distortion factors" $\left(\alpha_i^+, \alpha_i^-\right)$ is associated which are solutions of the equations (5). α_0 is a normalization factor that makes P^* a probability distribution. $P * \mathcal{R}$ shall denote any distribution of the form (4) which solves the adaptation problem $P^* \models \mathcal{R}$.

In spite of its simplicity and arbitrariness, this adaptation scheme actually has some important properties in common with the ME-approach. In [10] it is shown that it satisfies ***system independence*** as well as ***subset independence*** which Shore and Johnson [17] proved to be characteristic of ME-inferences.

5 The functional concept

Generally a solution to our adaptation problem provided by the scheme (4) is far from being unique. If e.g. P is to be adjusted to a nontrivial probabilistic conditional $\mathrm{A} \rightsquigarrow \mathrm{B}\,[x]$ the different solutions (α^+, α^-) of the corresponding equation (5): $\frac{\alpha^+}{\alpha^-} = \frac{x}{1-x} \frac{p(\mathrm{A}\overline{\mathrm{B}})}{p(\mathrm{AB})}$ result in different distributions. Thus the scheme (4) describes how the posterior distribution has to be built up once a solution to the equations (5) is fixed, but up to now any solution is possible.

Of course, this situation is not very satisfactory because we intuitively feel that there are "good" solutions and "bad" solutions. For instance, if the prior distribution P already satisfies the information contained in $\mathcal{R}$ it would be reasonable to expect that the scheme does not alter any probability, yielding the prior as the posterior distribution.

Demanding uniqueness means to assume a ***functional concept*** that underlies the finding of a "best solution" so that a unique distribution of type (4) arises in dependence of the prior knowledge P and of the new conditional information $\mathcal{R}$.

Returning to the small example given at the beginning of this section, we see that all different solutions (α^+, α^-) have the same quotient $\frac{\alpha^+}{\alpha^-} = \frac{x}{1-x} \frac{p(\mathrm{A}\overline{\mathrm{B}})}{p(\mathrm{AB})}$ which is determined by the constraint (5). In general, however, this need not be the case, different solutions may have different quotients, as can easily be checked when $\mathcal{R}$ contains more than one rule. But in any case, equation (5) reflects for each pair (α_i^+, α_i^-) which is part of a solution the dependencies of its quotient on the other conditionals in $\mathcal{R}$, on the prior knowledge P and – in a distinguished way – on the probability x_i of the conditional it corresponds to. Thus the quotient $\frac{\alpha_i^+}{\alpha_i^-}$ plays a central role, representing the core of the impact that the corresponding conditional is to have on the posterior distribution. If we look for something that should guide us in finding "good" solutions, these quotients seem to be most obvious.

Therefore we now assume the factors α_i^+ and α_i^- to be functionally dependent on their common quotient $\alpha_i := \frac{\alpha_i^+}{\alpha_i^-}$: $\alpha_i^+ = F_i^+(\alpha_i), \alpha_i^- = F_i^-(\alpha_i)$. Of course the functions F_i^+ and F_i^- should not be arbitrary. They should be sufficiently regular and are to follow a pattern independent of the specific form of a rule $\mathrm{A}_i \rightsquigarrow \mathrm{B}_i\,[x_i]$, thus realizing a fundamental inference pattern. This last assumption is being expressed as

$$\alpha_i^+ = F_i^+(\alpha_i) = F^+(x_i, \alpha_i) \quad \text{resp.} \quad \alpha_i^- = F_i^-(\alpha_i) = F^-(x_i, \alpha_i) \tag{6}$$

so as to separate the numerical impact of the inferential strength x_i from the logical interdependencies of the rule represented by α_i. Under these assumptions, the equations (5) may now be transformed into

$$\alpha_i = \frac{x_i}{1-x_i} \frac{\sum\limits_{\omega:A_i\overline{B_i}(\omega)=1} p(\omega) \prod\limits_{\substack{j \neq i \\ A_jB_j(\omega)=1}} F^+(x_j,\alpha_j) \prod\limits_{\substack{j \neq i \\ A_j\overline{B_j}(\omega)=1}} F^-(x_j,\alpha_j)}{\sum\limits_{\omega:A_iB_i(\omega)=1} p(\omega) \prod\limits_{\substack{j \neq i \\ A_jB_j(\omega)=1}} F^+(x_j,\alpha_j) \prod\limits_{\substack{j \neq i \\ A_j\overline{B_j}(\omega)=1}} F^-(x_j,\alpha_j)}, \tag{7}$$

yielding a distribution P^* of the form

$$p^*(\omega) = \alpha_0 p(\omega) \prod_{i:A_iB_i(\omega)=1} F^+(x_i,\alpha_i) \prod_{i:A_i\overline{B_i}(\omega)=1} F^-(x_i,\alpha_i) \tag{8}$$

which solves the adaptation problem. The functions $F^+(x,\alpha)$ and $F^-(x,\alpha)$ are defined for $x \in [0,1]$ and for non-negative real α , and they are related by

$$F^+(x,\alpha) = \alpha F^-(x,\alpha). \tag{9}$$

To symbolize the presence of the functional concept, we will use $*_F$ instead of $*$.

6 Logical consistency

Surely, the adaptation scheme (8) will be considered sound only if the resulting posterior distribution can be used as a prior distribution for further adaptations. This is a very fundamental meaning of *logical consistency*.

In particular, if we first adjust P only to a subset $\mathcal{R}_1 \subseteq \mathcal{R}$ and then use this posterior distribution to perform another adaptation to the full conditional information $\mathcal{R}$, we should have the same distribution as if we adjusted P to $\mathcal{R}$ in only one step. More formally, the operator $*_F$ should satisfy the following equation for any positive distribution P and any sets $\mathcal{R}_1, \mathcal{R}_2 \subset \mathcal{L}^*$ (provided that all adaptation problems are solvable):

$$P *_F (\mathcal{R}_1 \cup \mathcal{R}_2) = (P *_F \mathcal{R}_1) *_F (\mathcal{R}_1 \cup \mathcal{R}_2) \tag{10}$$

(10) is the first postulate we demand the adjustment operator $*_F$ to satisfy. According to the reflections above, it will be called the *postulate for logical consistency*. Applying it will give us a first important result in determining the functions F^+ and F^-:

Theorem 1. *If the adjustment operator $*_F$ satisfies the postulate for logical consistency (10) then F^- necessarily fulfills the functional equation*

$$F^-(x,\alpha\beta) = F^-(x,\alpha)\, F^-(x,\beta) \tag{11}$$

for all $x \in [0,1]$, $\alpha, \beta \geq 0$.

Because of (9), F^+ satisfies (11) iff F^- does.

Proof. If (10) holds in principle for any adaptation carried out by $*_F$, it is surely valid for some special type of $P, \mathcal{R}_1$ and $\mathcal{R}_2$. So let P be any positive distribution over 3 variables A, B and C, let $\mathcal{R}_1 = \{a \leadsto c\,[x]\}$ and $\mathcal{R}_2 = \{b \leadsto c\,[y]\}$. Let $p_1, \ldots, p_8$ denote the prior probabilities of P in reverse order, i.e. $p_1 = p(abc), \ldots, p_8 = p(\bar{a}\bar{b}\bar{c})$.

The new distributions $P * (\mathcal{R}_1 \cup \mathcal{R}_2), P * \mathcal{R}_1$ and $(P * \mathcal{R}_1) * (\mathcal{R}_1 \cup \mathcal{R}_2)$ may be set up according to (8), with factors α, β (belonging to $P * (\mathcal{R}_1 \cup \mathcal{R}_2)$), $\tilde{\alpha}$ (belonging to $P * \mathcal{R}_1$) and α_1, β_1 (belonging to $(P * \mathcal{R}_1) * (\mathcal{R}_1 \cup \mathcal{R}_2)$). By equations (7), these factors may be calculated as follows:

$$\alpha = \frac{x}{1-x} \frac{p_2 F^-(y,\beta)+p_4}{p_1 F^+(y,\beta)+p_3}; \quad \beta = \frac{y}{1-y} \frac{p_2 F^-(x,\alpha)+p_6}{p_1 F^+(x,\alpha)+p_5}; \quad \tilde{\alpha} = \frac{x}{1-x} \frac{p_2+p_4}{p_1+p_3};$$

$$\alpha_1 = \frac{x}{1-x} (\tilde{\alpha})^{-1} \frac{p_2 F^-(y,\beta_1)+p_4}{p_1 F^+(y,\beta_1)+p_3}; \quad \beta_1 = \frac{y}{1-y} \frac{p_2 F^-(x,\tilde{\alpha}) F^-(x,\alpha_1)+p_6}{p_1 F^+(x,\tilde{\alpha}) F^+(x,\alpha_1)+p_5}.$$

For $P * (\mathcal{R}_1 \cup \mathcal{R}_2) = (P * \mathcal{R}_1) * (\mathcal{R}_1 \cup \mathcal{R}_2)$ to hold, as required, we must have $\beta = \beta_1$ and $\alpha = \tilde{\alpha}\alpha_1$. Comparing β to β_1 above, observing (9) and using $\alpha = \tilde{\alpha}\alpha_1$, we obtain $F^-(x,\alpha) = F^-(x,\tilde{\alpha}\alpha_1) = F^-(x,\tilde{\alpha}) F^-(x,\alpha_1)$. Because P, x and y can be chosen arbitrarily, this equation must hold for all non-negative real $\tilde{\alpha}, \alpha_1$ and all $x \in [0,1]$.

The example in the proof of Theorem 1 above was not chosen accidentally. The way in which two conditionals with common conclusion should interact is one of the main issues in conditional logic. Sometimes it is referred to as the *antecedent conjunction problem* (cf. [12]).

Assume x to be held fixed and let for a moment $F_x^-(\alpha) := F^-(x,\alpha)$ be regarded only as a function of α. For real α', β', (11) yields $F_x^- \circ \exp(\alpha' + \beta') = (F_x^- \circ \exp(\alpha'))\,(F_x^- \circ \exp(\beta'))$. Therefore according to [1, p. 140], $F_x^- \circ \exp$ must have the form $F_x^- \circ \exp(\alpha') = e^{c\alpha'}$ for some constant c, and again taking into consideration the dependency on x, we obtain

$$F^-(x,\alpha) = \alpha^{c(x)} \quad \text{and} \quad F^+(x,\alpha) = \alpha^{c(x)+1}, \tag{12}$$

for any positive real α, with a (sufficiently well-behaved) function $c(x)$.

7 Representation invariance

Up to now, we neglected completely how (conditional) knowledge is represented in $\mathcal{R}$. Of course, the way in which we build up the adjusted distribution P^* already allows for the *logical equivalence* of propositional formulas, by definitions (4) and (5). But what about *probabilistic equivalences*, i.e. equivalences that are due to elementary probability calculus? For instance, the sets of rules $\{\mathrm{A} \leadsto \mathrm{B}\,[x], \mathrm{A}\,[y]\}$ and $\{\mathrm{AB}\,[xy], \mathrm{A}\,[y]\}$ are equivalent in this respect because each rule in one set is derivable from rules in the other. We surely expect the result of our adjustment process to be independent of the syntactic representation of probabilistic knowledge in $\mathcal{R}$:

$$P *_F \mathcal{R} = P *_F \mathcal{R}' \quad \text{if} \quad \mathcal{R} \quad \text{and} \quad \mathcal{R}' \quad \text{are probabilistically equivalent.} \tag{13}$$

(13) is called the *postulate for representation invariance*. It is the second stipulation that we expect the adjustment operator $*_F$ to comply with.

Proposition 2. *If the adaptation operator $*_F$ with F^-, F^+ of the form (12) satisfies the postulate for representation invariance (13) then for all real $x, x_1, x_2 \in [0,1]$ the following equations hold:*

$$c(x) + c(1-x) = -1 \tag{14}$$

$$c(xx_1 + (1-x)x_2) = -c(x)c(x_1) - c(1-x)c(x_2) \tag{15}$$

where $c(x)$ is defined by (12).

Proof. As in the proof of Theorem 1, the treatment of special cases revealing fundamental relationships will imply the desired equations. This proof involves intricate calculations, so we only sketch it by giving the main ideas. A detailed proof may be found in [9].
The most obious probabilistic equivalence is that of each two rules $\mathrm{A} \rightsquigarrow \mathrm{B}\,[x]$ and $\mathrm{A} \rightsquigarrow \overline{\mathrm{B}}\,[1-x]$. Respecting this implies equation (14).
Because of $p\,(b|a) = p\,(b|ac)\,p\,(c|a) + p\,(b|a\bar{c})\,p\,(\bar{c}|a)$ for arbitrarily chosen variables A, B and C, the two sets of rules $\mathcal{R} = \{a \rightsquigarrow c\,[x],\ ac \rightsquigarrow b\,[x_1],\ a\bar{c} \rightsquigarrow b\,[x_2]\}$ and $\mathcal{R}' = \{a \rightsquigarrow b\,[y],\ ac \rightsquigarrow b\,[x_1],\ a\bar{c} \rightsquigarrow b\,[x_2]\}$ with $y = xx_1 + (1-x)x_2$ are probabilistically equivalent for all real $x, x_1, x_2 \in [0,1]$. Setting up the distributions $P *_F \mathcal{R}$ and $P *_F \mathcal{R}'$ according to (8) and postulating their equality shows the second equation (15) by comparing the corresponding factors.

The next theorem is now an easy consequence.

Theorem 3. *If the operator $*_F$ is to meet the fundamental demands for logical consistency (10) and for representation invariance (13), then F^+ and F^- necessarily must be*

$$F^+(x, \alpha) = \alpha^{1-x} \quad \textit{and} \quad F^-(x, \alpha) = \alpha^{-x}. \tag{16}$$

Proof. Logical consistency implies $F^-(x, \alpha) = \alpha^{c(x)}$ (cf. (12)) where $c(x)$ satisfies equations (14) and (15) because of representation invariance. Therefore $c(x)$ fulfills a Cauchy functional equation, and according to [1, p. 44f.], $c(x)$ must be of the form $c(x) = kx$ with $k = c(1) = -1$. This shows (16).

8 Uniqueness

So far we have proved that the demands for logical consistency and for representation invariance uniquely determine the functions which we assumed to underly the adjustment process $P * \mathcal{R}$ in the way we described in Section 5. Applying (16) to (8) and (7) we recognize that the posterior distribution necessarily is of the same type (2) as the ME-distribution if it is to yield sound and consistent inferences.

So we nearly have achieved the aim of this paper, the characterization of ME-inference within a conditional-logical framework. But one step is still missing: Is the distribution which arises as a solution to (2) and (3) uniquely determined?

Are there possibly several different solutions of type (2), only one of which being the ME-distribution? If we assume the functions F^- and F^+ to fulfill (16), is this*sufficient* to guarantee that the resulting operator $*_F$ satisfies logical consistency and representation invariance?

The question for uniqueness of the posterior distribution is at the centre of all these problems. If it can be answered positively, we will have reached our goal: The unique posterior distribution of type (2) must be the ME-distribution, $*_F$ then corresponds to ME-inference, and ME-inference is known to fulfill (10) and (13) as well as many other reasonable properties, cf. [13], [17], [18]. Indeed, this uniqueness will be affirmed by the next theorem. It is proved by making use of cross-entropy as an excellently fitting measure of distance for distributions of type (2).

Theorem 4. *There is at most one solution of the adaptation problem $P * \mathcal{R}$ of type (2).*

Proof. Let P_1^*, P_2^* be two distributions of type (2) with non-negative real factors $(\alpha_i)_{0 \leq i \leq n}$, $(\beta_i)_{0 \leq i \leq n}$ fulfilling equations (3). Calculating the cross-entropies between P_1^* and P_2^*, we see $R(P_1^*, P_2^*) = \log \alpha_0 - \log \beta_0$ and $R(P_2^*, P_1^*) = \log \beta_0 - \log \alpha_0$. Both equations together imply $R(P_1^*, P_2^*) = 0$ because cross-entropy is non-negative, and by using its positivity (cf. [16]), both distributions must be equal. This proves the theorem.

The following theorem summarizes our results in characterizing ME-adjustment within a conditional-logical framework:

Theorem 5. *Let $*_e$ denote the ME-adjustment operator, i.e. $*_e$ assigns to a prior distribution P and some set $\mathcal{R}$ of probabilistic conditionals the one distribution $P_e = P *_e \mathcal{R}$ which has minimal cross-entropy with respect to P among all distributions that satisfy $\mathcal{R}$ (provided the adjustment problem is solvable at all).*

Then $*_e$ is the only operator of type (4) that realizes a functional concept (6) and satisfies the fundamental postulates for logical consistency (10) and for representation invariance (13). $*_e$ is completely described by (2) and (3).

9 Conclusion

Starting from a conditional-logical point of view we found a completely logical characterization of the ME-solution to the adjustment problem (*) stated in Section 2. This characterization rests on three axioms: The existence of a functional concept, being described by (6), and the fundamental properties of logical consistency (10) and of representation invariance (13). All of these three principles are usually considered to be fundamental to any reasonable inference procedure. So the ME-solution arises in a very natural way, without imposing any external minimality demand. In fact, minimality is inherent to the idea of a functional concept and to logical consistency, and the simplicity of the scheme (4) may be regarded as minimizing the complextity of the conditional-logical approach.

Actually, there is a fourth axiom present, expressed by the approach (4). In this paper, we gave only an informative explanation for it realizing *conditional preservation.* In [9], a clear and exact formalization of this principle is given, and its equivalence to (4) is proved.

References

1. J. Aczél. *Vorlesungen ueber Funktionalgleichungen und ihre Anwendungen.* Birkhaeuser Verlag, Basel, 1961.
2. F. Bacchus. *Representing and Reasoning with Probabilistic Knowledge: a Logical Approach to Probabilities.* MIT Press, Cambridge, Mass., 1990.
3. P.G. Calabrese. Deduction and inference using conditional logic and probability. In I.R. Goodman, M.M. Gupta, H.T. Nguyen, and G.S. Rogers, editors, *Conditional Logic in Expert Systems*, pages 71–100. Elsevier, North Holland, 1991.
4. R.T. Cox. Probability, frequency and reasonable expectation. *American Journal of Physics*, 14(1):1–13, 1946.
5. I. Csiszár. I-divergence geometry of probability distributions and minimization problems. *Ann. Prob.*, 3:146–158, 1975.
6. M. Goldszmidt. Research issues in qualitative and abstract probability. *AI Magazine*, pages 63–66, Winter 1994.
7. E.T. Jaynes. *Papers on Probability, Statistics and Statistical Physics.* D. Reidel Publishing Company, Dordrecht, Holland, 1983.
8. R.W. Johnson and J.E. Shore. Comments on and correction to "Axiomatic derivation of the principle of maximum entropy and the principle of minimum cross-entropy". *IEEE Transactions on Information Theory*, IT-29(6):942–943, 1983.
9. G. Kern-Isberner. Characterizing the principle of minimum cross-entropy within a conditional logical framework. Informatik Fachbericht 206, FernUniversitaet Hagen, 1996.
10. G. Kern-Isberner. Conditional logics and entropy. Informatik Fachbericht 203, FernUniversitaet Hagen, 1996.
11. S.L. Lauritzen and D.J. Spiegelhalter. Local computations with probabilities in graphical structures and their applications to expert systems. *Journal of the Royal Statistical Society B*, 50(2):415–448, 1988.
12. D. Nute. *Topics in Conditional Logic.* D. Reidel Publishing Company, Dordrecht, Holland, 1980.
13. J.B. Paris and A. Vencovská. A note on the inevitability of maximum entropy. *International Journal of Approximate Reasoning*, 14:183–223, 1990.
14. J. Pearl. *Probabilistic Reasoning in Intelligent Systems.* Morgan Kaufmann, San Mateo, Ca., 1988.
15. W. Roedder and C.-H. Meyer. Coherent knowledge processing at maximum entropy by spirit. In *Proceedings 12th Conference on Uncertainty in Artificial Intelligence*, pages 470–476, 1996.
16. J.E. Shore. Relative entropy, probabilistic inference and AI. In L.N. Kanal and J.F. Lemmer, editors, *Uncertainty in Artificial Intelligence*, pages 211–215. North-Holland, Amsterdam, 1986.
17. J.E. Shore and R.W. Johnson. Axiomatic derivation of the principle of maximum entropy and the principle of minimum cross-entropy. *IEEE Transactions on Information Theory*, IT-26:26–37, 1980.
18. J.E. Shore and R.W. Johnson. Properties of cross-entropy minimization. *IEEE Transactions on Information Theory*, IT-27:472–482, 1981.

Undecidability Results on Two-Variable Logics*

Erich Grädel[1], Martin Otto[1], Eric Rosen[2]

[1] Mathematische Grundlagen der Informatik, RWTH Aachen,
{graedel,otto}@informatik.rwth-aachen.de
[2] Technion Haifa, erosen@csa.cs.technion.ac.il

Abstract. It is a classical result of Mortimer's that L^2, first-order logic with two variables, is decidable for satisfiability (whereas L^3 is undecidable). We show that going beyond L^2 by adding any one of the following leads to an undecidable logic:

- very weak forms of recursion, such as transitive closure or monadic fixed-point operations.
- cardinality comparison quantifiers.

In fact these extensions of L^2 prove to be undecidable both for satisfiability, and for satisfiability in finite models. Moreover the satisfiability problem for these systems is shown to be hard for Σ^1_1, the first level of the analytical hierarchy. They thereby exhibit a much higher degree of undecidability than first-order logic.

The case of monadic least fixed-point logic in two variables deserves particular attention, since this logic may be seen as the natural least common extension of two important decidable systems: first-order with two variables and propositional μ-calculus (propositional modal logic with a least fixed-point operator). It had been conjectured that this system might still be decidable.

1 Introduction

One of the most prominent problems of mathematical logic has been the *classical decision problem*: given a first-order sentence, determine whether it is satisfiable. After Church and Turing had proved that the classical decision problem is algorithmically unsolvable, the problem was transformed into a classification problem: which formula classes are decidable for satisfiability and which are not? Traditionally most attention has been given to fragments of first-order logic that are determined by the quantifier prefix and the vocabulary of relation and function symbols of the given formulae. With respect to such fragments the classical decision problem is completely solved. A comprehensive account of this classification is given in the forthcoming book [1]. An important variation is the decision problem for *finite satisfiability*, i.e. satisfiability in finite models. For most but not all of the decidable standard fragments of first-order logic, satisfiability and finite satisfiability coincide (*finite model property*).

* This work has been partially supported by the German-Israeli Foundation of Scientific Research and Development

Reischuk, Morvan (Eds.): STACS'97 Proceedings, LNCS 1200

In computer science a number of logic problems arose (e.g. for knowledge representation systems, automatic verification, concurrent systems) that are closely related to the classical decision problem. In most cases, however, the relevant formula classes are *not* those determined by quantifier prefix and vocabulary. More important are fragments determined by the number of variables. A number of logics used in computer science can be embedded into L^2, the class of first-order formulae with only two distinct variables. In particular this is the case for ***propositional modal logics*** and for a number of ***knowledge representation logics*** (or *concept logics*). Thus the classical result of Mortimer's [9] that L^2 is decidable and indeed has the finite model property immediately implies corresponding results for these logics. Of course, there are also many logics of interest in computer science that are *not* fragments of L^2. In fact they often contain constructors such as fixed-point operators, quantifiers over paths, counting constructs, etc. that are not even present in first-order logic. Nevertheless, many of them can still be seen as two-variable logics. For instance, the propositional μ-calculus L_μ can easily be embedded in the extension of L^2 with a fixed-point operator and several more powerful knowledge representation logics are indeed sublogics of C^2, the extension of L^2 with counting quantifiers, or sublogics of C^2 with transitivity.

It is therefore interesting to determine the borderline between decidability and undecidability for two-variables logics. We provide a number of results from the undecidable side of that borderline, namely that the following are undecidable for satisfiability as well as for finite satisfiability:
(a) two-variable transitive closure logic TC^2.
(b) two-variable monadic least fixed-point logic FP^2 (in a restricted form).
(c) the extension of L^2 by the Härtig (or equicardinality) quantifier.
Moreover these systems are hard for Σ_1^1, the first level of the analytical hierarchy, and thus have a much higher degree of undecidability than first-order logic.

In connection with (a), we even prove undecidability for the extension of L^2 by mere transitivity statements, which is a fragment of first-order logic. (b) is particularly interesting as a result on the least common extension of two important decidable systems, L^2 and the propositional μ-calculus. It had therefore been conjectured that this system itself might still be decidable. With respect to (c) it is interesting to note that another extension of L^2 that allows for certain cardinality statements, C^2, has recently been shown decidable in a companion paper to the present one [5].

In this extended abstract we explicitly treat (a) and (b) and only briefly address (c).

1.1 Recursive inseparability and conservative classes

Let X be a class of formulae. We write *sat*(X) for the the set of $\psi \in X$ that are satisfiable and *fin-sat*(X) for the set of $\psi \in X$ that have a finite model. Formulae in *sat*$(X) \setminus$ *fin-sat*(X) are called infinity axioms, *inf-axioms*(X) is the set of these.

A strong statement of the unsolvability of the classical decision problem is Trakhtenbrot's *Inseparability Theorem* which uses the concept of recursive inseparability. Two disjoint sets X, Y are called *recursively inseparable* if there is no recursive set R such that $X \subseteq R$ and $R \cap Y = \emptyset$. In particular neither X nor Y can then be decidable.

Theorem 1 (Trakhtenbrot). *The sets fin-sat*(FO), *inf-axioms*(FO) *and the complement of* sat(FO) *are pairwise recursively inseparable.*

A formula class X is a *conservative reduction class* if there exists a recursive function $g : \mathrm{FO} \rightarrow X$ that preserves (in the sense of if-and-only-if) satisfiability as well as finite satisfiability. For such X it follows from Trakhtenbrot's Theorem that *fin-sat*(X), *inf-axioms*(X), and the complement of *sat*(X) are pairwise recursively inseparable; in this case *fin-sat*(X) is r.e.-hard while *sat*(X) and *inf-axioms*(X) are co-r.e.-hard.

1.2 First-order logic with two variables

We denote by L^k the restriction of first-order logic to relational formulae (i.e. without function symbols) that contain only the variables $x_1, \ldots, x_k$. Note that interesting sentences in L^k are *not* in prenex normal form. Quite to the contrary, one extensively uses the possibility to re-use variables. It is well-known that the decision problem for L^k is unsolvable (even for formulae without equality) for all $k \geqslant 3$. However, first-order logic with two variables is decidable.

Theorem 2 (Mortimer). *fin-sat*(L^2) = *sat*(L^2) *and* sat(L^2) *is decidable.*

A more transparent proof of the finite model property with essentially optimal bounds on the model size has recently been provided in [3].

Throughout this paper we consider purely relational vocabularies (whose predicates have arities bounded by 2). It will be useful to introduce shorthand notation for the following basic predicate transformations that are L^2-definable. For binary E let dm(E), rg(E), E^{-1}, and E^* denote the domain and range, the converse and the symmetric reflexive closure of E, respectively:

$$\mathrm{dm}(E) = \{x : \exists y Exy\} \qquad E^{-1} = \{(x,y) : Eyx\}$$
$$\mathrm{rg}(E) = \{x : \exists y Eyx\} \qquad E^* = \{(x,y) : Exy \vee Eyx \vee x{=}y\}$$

1.3 Domino problems and local grids

Domino or tiling problems provide a simple and powerful method for proving undecidability results and lower complexity bounds for various systems of propositional logic, for subclasses of first-order logic and for decision problems in mathematical theories (see e.g. [1,2,6]).

Definition 3. A *domino system* $\mathcal{D}$ is a triple (D, H, V) where D is a finite set of dominoes and $H, V \subseteq D \times D$ are two binary relations.

If S is any of the spaces $\mathbb{Z} \times \mathbb{Z}$, $\mathbb{N} \times \mathbb{N}$ or $\mathbb{Z}/s\mathbb{Z} \times \mathbb{Z}/t\mathbb{Z}$, then a *tiling* of S with $\mathcal{D}$ is a mapping $\tau : S \to D$ such that for all $(x, y) \in S$: if $\tau(x, y) = d$, $\tau(x+1, y) = d'$, and $\tau(x, y+1) = d''$, then $(d, d') \in H$ and $(d, d'') \in V$.

A tiling of $\mathbb{Z} \times \mathbb{Z}$ is *periodic*, if it has non-trivial horizontal and vertical periods, i.e. if there are $s, t > 0$ such that for all $x, y \in \mathbb{Z}$: $\tau(x, y) = \tau(x+s, y) = \tau(x, y+t)$.

A tiling of $\mathbb{N} \times \mathbb{N}$ or $\mathbb{Z} \times \mathbb{Z}$ with $\mathcal{D}$ is *recurrent* with respect to some designated domino d if d occurs infinitely often in this tiling.

Theorem 4 (Berger, Gurevich-Koryakov). *The set of domino systems that admit, respectively, no tiling and a periodic tiling of* $\mathbb{N} \times \mathbb{N}$ *are recursively inseparable. In particular, these domino problems are both undecidable.*

Harel [6,7] showed that the existence of recurrent tilings is not only undecidable, but undecidable of very high degree. Recall that Σ^1_1 and Π^1_1 form the first level of the analytical hierarchy in recursion theory.

Theorem 5 (Harel). *The set of domino systems with designated domino* $(\mathcal{D}, d)$ *that admit a recurrent tiling of* $\mathbb{Z} \times \mathbb{Z}$ *(respectively of* $\mathbb{N} \times \mathbb{N}$*) is* Σ^1_1*-complete.*

Local grids. Two-dimensional grids form the basis of reductions from domino problems. Let $\mathcal{G}_{\mathbb{Z}}$ be the infinite standard grid on $\mathbb{Z} \times \mathbb{Z}$ with horizontal and vertical successor relations $H = \{((x, y), (x+1, y)) : x, y \in \mathbb{Z}\}$ and $V = \{((x, y), (x, y+1)) : x, y \in \mathbb{Z}\}$. We denote by $\mathcal{G}_{\mathbb{N}}$ the corresponding standard grid on $\mathbb{N} \times \mathbb{N}$, and by $\mathcal{G}_m$ the standard grid on $\mathbb{Z}/m\mathbb{Z} \times \mathbb{Z}/m\mathbb{Z}$ (with horizontal and vertical adjacency relations modulo m).

It will be convenient to deal with structures that only locally resemble grids.

Definition 6. A structure $\mathcal{G} = (G, H, V)$ is called a *local grid*, if H and V are the graphs of two injective functions h and v without fixed points, such that:
(i) $(H \cup H^{-1}) \cap (V \cup V^{-1}) = \emptyset$ (h and v are independent everywhere).
(ii) h and v commute, i.e. $h \circ v = v \circ h$.
A local grid *without boundary* is one in which h and v are bijections.

Thus $\mathcal{G}_{\mathbb{Z}}$ and the $\mathcal{G}_m$ are local grids without boundary, $\mathcal{G}_{\mathbb{N}}$ is a local grid with boundary. Thinking in terms of the neighbourhood of a point a, condition (ii) for local grids means that those points reachable by traversing at most one H- and one V-edge form a non-degenerate quadrangle which is closed at the far corner.

Towards the formulation, in Theorem 7 below, of concise criteria that embody our reduction to domino problems we review two standard notions from classical model theory.

A *projective characterization* of a class $\mathcal{C}$ of τ-structures is given by a sentence φ in some vocabulary $\tau' \supseteq \tau$, if $\mathcal{C}$ is the class of the τ-reducts of models of φ: $\mathcal{C} = \{\mathfrak{A}|\tau : \mathfrak{A} \text{ a } \tau'\text{-structure}, \mathfrak{A} \models \varphi\}$. A formula class L admits a projective characterization of $\mathcal{C}$ if there is a $\varphi \in L$ that projectively characterizes $\mathcal{C}$. This is a generalization of the standard notion of (finite) axiomatizability in allowing an extended vocabulary for the purposes of the axiomatization.

A formula class L has the *relativization property* if for any sentence φ of L and unary predicate U, there is another sentence φ^U of L, whose models are those structures in which U carries as an induced substructure a model of φ. In formulae: $(\mathfrak{A}, U^{\mathfrak{A}}) \models \varphi^U$ if and only if $\mathfrak{A}|U^{\mathfrak{A}} \models \varphi$. All logics considered in this paper satisfy this criterion.

Our technical criterion for undecidability of a formula class X centers on the possibility to characterize certain rich classes of local grids. A class $\mathcal{C}$ of local grids is here called *rich* if

(i) $\mathcal{C}$ contains at least one of the infinite standard grids, $\mathcal{G}_{\mathbb{Z}}$ or $\mathcal{G}_{\mathbb{N}}$.

(ii) For all $r \geqslant 1$ there exists a $k \geqslant 1$ such that $\mathcal{C}$ contains the grid $\mathcal{G}_{kr}$, the standard grid on $\mathbb{Z}/kr\mathbb{Z} \times \mathbb{Z}/kr\mathbb{Z}$.

Theorem 7. *Let L be a formula class that contains L^2 and is closed under conjunction. If L admits a projective characterization of a rich class of local grids then L is conservative. In particular,* sat(*L*) *is co-r.e. hard and* fin-sat(*L*) *is r.e.-hard. If L also has the relativization property and admits a projective characterization of one of the infinite standard grids, $\mathcal{G}_{\mathbb{N}}$ or $\mathcal{G}_{\mathbb{Z}}$, then* sat(*L*) *is even Σ^1_1-hard.*

Remark. The stronger requirement for the second claim, concerning a characterization of one of the infinite standard grids, serves to exclude those infinite local grids that have infinitely many connected components. First-order logic, for instance, does admit a projective characterization of the class of infinite local grids, but no first-order statement can force a particular domino to occur infinitely often in the same connected component of a corresponding tiling.

2 L^2 with transitivity

It is a well-known fact that L^2 cannot express transitivity of a binary relation. In fact transitivity is not even expressible (even over finite structures) in $L^2_{\infty\omega}$, two-variable infinitary logic. We shall consider the weak extension of L^2 that admits transitivity assertions about binary relations. In fact it suffices to admit these in positive form, and only for basic binary relations in the vocabulary. Formally, we may consider conjunctions $\varphi \wedge \theta$, where φ is an ordinary sentence of L^2 and θ is a statement trans$[R_1, \ldots, R_m]$ asserting the transitivity of some basic relations R_i in τ. Call the resulting class L^2 + trans. Note that L^2 + trans may be regarded as a fragment of first-order logic, as transitivity is FO-definable. We shall prove next that this weak system suffices for the projective characterization of a rich class of local grids.

2.1 Characterizing matchings, triangle matchings, and local grids

Matchings. A matching is a binary relation E which is the graph of an injective partial function with disjoint domain and range. For the projective characterization we use one extra binary predicate S whose intended interpretation will be the symmetric reflexive closure E^* of E. This intended connection between S and E, $S = E^*$, and the condition $\mathrm{dm}(E) \cap \mathrm{rg}(E) = \emptyset$ are directly expressible by an L^2-sentence φ_0. It is not difficult to verify that $\varphi_{\mathrm{match}}(E, S) := \varphi_0 \wedge \mathrm{trans}[S]$ provides a projective characterization of matchings E.

Triangle matchings. Call a triple of binary relations E_1, E_2, E_3 a triangle matching, if each E_i is a matching, $\mathrm{dm}(E_2) = \mathrm{rg}(E_1)$ and $E_3 = E_1 \cdot E_2$. In order to give a projective characterization of triangle matchings we use $\varphi_{\mathrm{match}}(E_i, S_i)$ together with one more transitivity statement concerning a new relation S_0 whose intended interpretation is $S_0 = E_1 \cup E_2 \cup E_3$. It is not hard to see that the following provides a projective characterization of triangle matchings in $L^2 + \mathrm{trans}$:

$$\begin{aligned}\varphi_\Delta(E_1, E_2, E_3, \overline{S}) = {} & \mathrm{dm}(E_2) = \mathrm{rg}(E_1) \wedge \mathrm{dm}(E_3) = \mathrm{dm}(E_1) \wedge \mathrm{rg}(E_3) = \mathrm{rg}(E_2) \\ & \wedge \textstyle\bigwedge_{i=1}^{3} \varphi_{\mathrm{match}}(E_i, S_i) \wedge S_0 = E_1 \cup E_2 \cup E_3 \wedge \mathrm{trans}[S_0].\end{aligned}$$

Local grids. We now axiomatize in $L^2 + \mathrm{trans}$ a rich class of local grids without boundary. Those conditions on H and V which are not right away expressible in L^2 alone, are that they are functional and bijective and that the induced functions commute. It will here be useful to restrict the periodicities in components that have the topological type of a cylinder or torus to even numbers.

Definition 8. Call a local grid without boundary *even*, if the two functions h and v associated with H and V satisfy the following: whenever some vertex is fixed by some composition $h^n \circ v^m$, then it must be the case that both n and m are even.

The resulting class of *even local grids* is clearly rich in the sense required for Theorem 7. It thus suffices to show the following.

Proposition 9. $L^2 + \mathrm{trans}$ *admits a projective characterization of the class of even local grids without boundary.*

Bijectivity and functionality of H and V are taken care of by applying the considerations about matchings to predicates obtained from H and V by splitting each of them into two disjoint parts, each of which becomes a matching. For commutativity of H and V, we introduce more auxiliary relations that serve to reduce the commutativity statement to a condition on triangle matchings. We shall work with the following binary predicates, apart from the original H and V: H_i, V_i, and H_{ij}, V_{ij}, D_{ij} for $i, j = 0, 1$. The intended interpretations are such that H becomes the disjoint union of H_0 and H_1 in such a way that horizontal edges indexed 0 alternate with those indexed 1; similarly for V and the V_i. To this end the standard grid $\mathcal{G}_{\mathbb{Z}}$ is first expanded with additional edge relations H_i and V_i according to the following (compare (a) in the figure below):

$$H_i = \{((u,v),(u+1,v)) : u \equiv i \bmod 2\},\ V_i = \{((u,v),(u,v+1)) : v \equiv i \bmod 2\}.$$

Note that the vertices are of four different types according to which indices their outgoing and incoming V- and H-edges carry. Obviously, there exist L^2-formulae

$\eta_{ij}(x)$ for $i,j = 0,1$ characterizing these four vertex classes. $\eta_{ij}(x)$ describes those vertices with outgoing edges H_i and V_j, and with incoming H_{i-1} and V_{j-1}. We further expand by the following to a structure $\mathcal{G}_{\mathbb{Z}}{}^*$ (cf. (b) in the figure):

$$H_{ij} = \{(x,y) \in H_i : \eta_{ij}(x)\},\ V_{ij} = \{(x,y) \in V_j : \eta_{ij}(x)\},\ D_{ij} = H_i \cdot V_j.$$

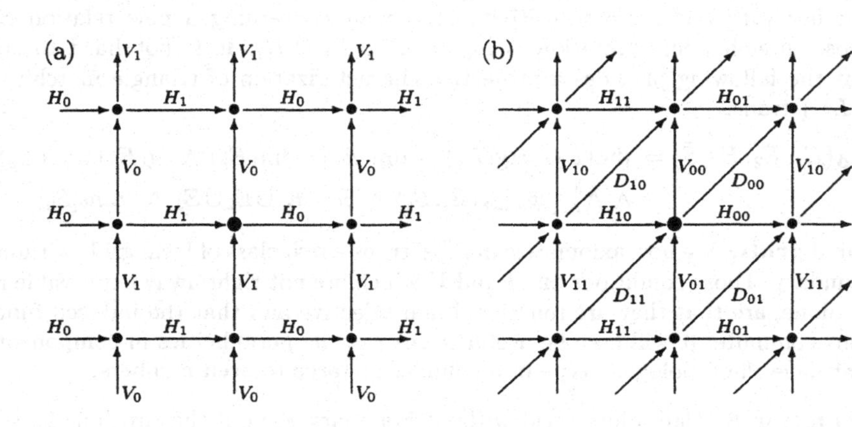

The following facts (1) – (4) about this expansion can be asserted in plain L^2. (5) uses the formulae φ_{match} and φ_Δ (and thereby actually a whole lot of extra auxiliary predicates, for instance for the symmetric reflexive closures of H_{ij},V_{ij}, D_{ij}, that we now suppress in the notation).

(1) H is the disjoint union of H_0 and H_1, H_i the disjoint union of H_{i0} and H_{i1}, V the disjoint union of V_0 and V_1, V_i the disjoint union of V_{0i} and V_{1i}.
(2) all vertices are of exactly one of the above four types: $\forall x \bigvee \eta_{ij}(x)$.
(3) the definition of the H_{ij} and V_{ij} in terms of the H_i and V_j and the η_{ij}.
(4) $\text{dm}(D_{ij}) = \text{dm}(H_{ij}) = \text{dm}(V_{ij})$, $\text{rg}(H_{ij}) = \text{dm}(H_{1-i\,j})$, $\text{rg}(V_{ij}) = \text{dm}(V_{i\,1-j})$, $\text{rg}(D_{ij}) = \text{dm}(H_{1-i\,1-j})$.
(5) the H_i, V_i and the D_{ij} are matchings, and the triples $(H_{ij}, V_{1-i\,j}, D_{ij})$ and $(V_{ij}, H_{i\,1-j}, D_{ij})$ form triangle matchings.

Let $\varphi \in L^2 + \text{trans}$ be a sentence asserting (1) – (5). We claim that φ provides a projective characterization of the class of even local grids without boundary. It is easy to see that $\mathcal{G}_{\mathbb{Z}}{}^*$ satisfies φ and in fact, every even local grid without boundary can similarly be expanded to a model of φ. It then remains to show that any model of φ must be an even local grid without boundary. The non-trivial parts of this claim are functionality and commutativity of V and H. But functionality follows from the fact that the relations H_i and V_i are matchings. For commutativity of H and V, consider without loss of generality a vertex $a \models \eta_{00}$. By (2) – (4) the neighbourhood of a must be as indicated in (a) of the following figure and we want to show that in fact $c_1 = c_2$.

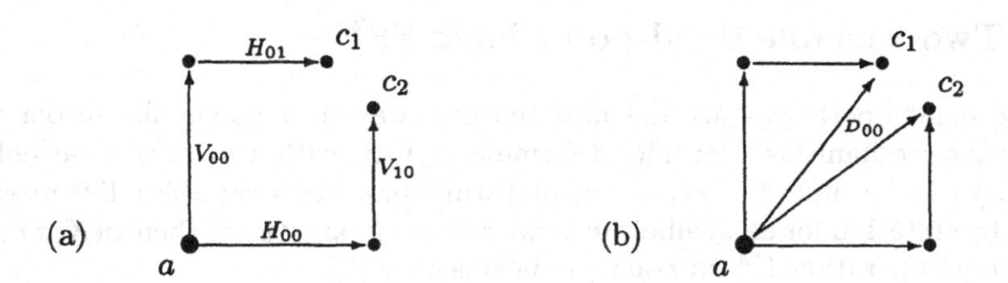

But by the triangle conditions in (5), both c_1 and c_2 are linked to a by D_{00}-edges. As D_{00} is a matching, these edges and therefore c_1 and c_2 must coincide.

We have thus shown that there is a projective characterization in $L^2 + \text{trans}$ of the even local grids without boundary. By Theorem 7 $L^2 + \text{trans}$ is a conservative reduction class. As $L^2 + \text{trans} \subseteq \text{FO}$ this implies the following.

Theorem 10. *sat*$(L^2 + \text{trans})$ *is co-r.e.-complete, fin-sat*$(L^2 + \text{trans})$ *is r.e.-complete.*

2.2 Two-variable transitive closure logic TC^2

First-order logic does not have the expressive means to model even very modest relational recursion. For instance it is well known that connectivity of graphs is not a first-order definable property. *Transitive closure logic* TC was introduced by Immerman to address this weakness of first-order logic. It augments first-order logic by allowing the formation of new formulae that define the transitive closure of definable binary relations. We can here in a natural way limit the application of the transitive closure operation to a framework which allows only two variables. Syntactically, we use a transitive closure operator TC, which may be applied to a formula φ to get a new formula $\text{TC}\varphi$ that is free in variables x and y. The semantics of the new formula is such that $\mathfrak{A} \models \text{TC}\varphi[a, b]$ if and only if the pair (a, b) is in the transitive closure of the binary relation that is defined by φ over $\mathfrak{A}$.

Definition 11. *Two-variable transitive closure logic* TC^2 is the simultaneous closure of L^2 under the first-order operations in L^2 and the TC-operator.

Obviously $L^2 + \text{trans} \subseteq \text{TC}^2$. TC^2 also allows to specify the overall topology of a local grid without boundary and thus admits a projective characterization of $\mathcal{G}_{\mathbb{Z}}$.

Theorem 12. TC^2 *is a conservative reduction class, and therefore undecidable for satisfiability as well as for satisfiability in finite models; sat*(TC^2) *is even* Σ^1_1*-hard.*

Slight modifications of the techniques presented here apply to the two-variable variant of *deterministic transitive closure logic.*

3 Two-variable fixed-point logic FP^2

Fixed-point operations are the most natural way to augment first-order with recursive mechanisms. Consider a formula $\varphi(X, \overline{x})$ with a predicate variable X of arity r and a tuple $\overline{x} = x_1, \ldots, x_r$ of distinct first-order variables. If $\mathfrak{A}$ provides an interpretation for all symbols in φ apart from X and the $\overline{x}$, then $\varphi(X, \overline{x})$ gives rise to an operation $F_\varphi^{\mathfrak{A}}$ on r-ary predicates over A:

$$F_\varphi^{\mathfrak{A}} : P \longmapsto \{\overline{a} \in A^r : \mathfrak{A} \models \varphi[P, \overline{a}]\}.$$

If φ is positive (and therefore monotone) in X then $F_\varphi^{\mathfrak{A}}$ possesses a ***least fixed-point*** $\mathrm{LFP}(F_\varphi^{\mathfrak{A}})$. By definition, $\mathrm{LFP}(F_\varphi^{\mathfrak{A}})$ is the $\subseteq$-minimal $P \subseteq A^r$ such that $F_\varphi^{\mathfrak{A}}(P) = P$. In procedural terms, $\mathrm{LFP}(F_\varphi^{\mathfrak{A}})$ may be defined as the limit of the sequence P_α of predicates obtained by iterating F_φ on the empty predicate:

$$\mathrm{LFP}(F_\varphi^{\mathfrak{A}}) = \bigcup_\alpha P_\alpha \qquad \text{where} \qquad \begin{array}{l} P_0 = \emptyset,\ P_{\alpha+1} = F_\varphi^{\mathfrak{A}}(P_\alpha), \\ P_\lambda = \bigcup_{\alpha<\lambda} P_\alpha, \ \text{for limits } \lambda. \end{array}$$

The formula $\psi = [\mathrm{LFP}_{X,\overline{x}}\varphi]\overline{x}$ is taken to describe the least fixed-point associated with F_φ. Its semantics is defined according to $\mathfrak{A} \models \psi[\overline{a}]$ iff $\overline{a} \in \mathrm{LFP}(F_\varphi^{\mathfrak{A}})$. Least fixed-point logic LFP is the extension of first-order logic obtained by closing with respect to the formation of least fixed-points for $\varphi(X, \overline{x})$ that are ***positive*** in X. Its importance in descriptive complexity is due to the well-known Theorem of Immerman and Vardi, that over linearly ordered finite structures LFP precisely captures PTIME.

There are several natural ways of restricting LFP to the context of two-variable logic. The straightforward notion would permit the application of LFP-operators $\mathrm{LFP}_{X,x}$ and $\mathrm{LFP}_{X,y}$ for monadic X, and $\mathrm{LFP}_{X,xy}$ (and $\mathrm{LFP}_{X,yx}$) for binary X. In fact TC^2 is a sublogic of this liberal variant of two-variable fixed-point logic: $\mathrm{TC}\varphi$ is equivalent with $\mathrm{LFP}_{X,y}(\varphi \vee \exists x(Xx \wedge \varphi))$. Note that x is a parameter in this fixed-point.

This motivates the introduction of the following much weaker (indeed weakest) form of adjoining a least fixed-point operation to L^2, which may actually also be regarded as the natural common extension of L^2 and propositional μ-calculus L_μ (which are both decidable).

Definition 13. Let FP^2 be the simultaneous closure of L^2 under L^2-constructors and monadic least fixed-points of the form $\mathrm{LFP}_{X,x}\varphi$ where y is not free in φ or $\mathrm{LFP}_{X,y}\varphi$ where x is not free in φ; X a unary second-order variable, in which φ is positive.

We remark that FP^2 and TC^2 are incomparable (even over finite structures).

3.1 Well-foundedness, order, successor, and triangle matchings

A typical example of the expressive power of FP^2 is the definability of the well-founded support of a binary predicate. If E is a binary predicate, then its *well-founded support* WF(E) is the set of those points, which have no infinite descending E-chain below them: $x \in \mathrm{WF}(E)$ if and only if there is no infinite sequence $(x_i)_{i \geqslant 0}$ such that $x_0 = x$ and $Ex_{i+1}x_i$ for all $i \geqslant 0$. It is easily checked that $\mathrm{LFP}_{X,x}(\forall y(\varphi(y,x) \to Xy))$ defines the well-founded support of the predicate defined by $\varphi(x,y)$. The real strength of FP^2 is brought out in the following lemma, which will also form the cornerstone in our proof of undecidability. It really seems intriguing that *well-orderings* are axiomatizable by a sentence of FP^2 (whereas orderings as such are not).

Lemma 14. *There are* FP^2*-sentences* $\psi_{\mathrm{wo}}(<)$ *and* $\psi_{\mathrm{wo,s}}(<,S)$ *whose models are exactly all well-orderings* $(A,<)$, *respectively all* $(A,<,S)$ *where* $(A,<)$ *is a well-ordering and* S *the associated successor relation.*

Proof. Let ψ_0 be the L^2-sentence $\forall x \forall y\big(x = y \lor x < y \lor y < x\big)$. Let ψ_{wo} be the conjunction of ψ_0 with the sentence $\psi_1 := \forall x\big[\mathrm{LFP}_{X,x}(\forall y(y < x \to Xy))\big]x$, which says that $<$ is well-founded. ψ_{wo} provides the desired characterization of well-orderings. ψ_1 forces $<$ to be irreflexive and antisymmetric in forbidding loops and 2-cycles. But $\psi_0 \land \psi_1$ forces transitivity as well: if $a_1 < a_2 < a_3$ in a model of ψ_{wo}, then $a_1 \neq a_3$ since $<$ is antisymmetric, and $a_3 \not< a_1$ since $<$ may not contain a 3-cycle. Therefore $a_1 < a_3$ follows by ψ_0.

Towards characterizing the successor relation S of a well-ordering $<$ consider the following formula, which is positive in X:

$$\xi(X,x) = \neg\exists y\Big(Sxy \land \exists x(x < y \land \exists y(y < x \land \neg Xy))\Big).$$

Suppose that $(A,<)$ is a well-ordering and consider the fixed-point generation for $\mathrm{LFP}_{X,x}\xi(X,x)$ over some expansion $(A,<,S)$. We assume that $(A,<,S)$ satisfies the L^2-sentence $\psi_2(<,S)$ which says that $S \subseteq <$, and that all elements apart from the first/last element (if the latter exists) have S-predecessors/-successors. An easy induction shows that, if S is the true successor relation for $<$, then stage X_α of the fixed-point generation consists of all elements of rank less than α in $(A,<)$. Therefore in this case, $\mathrm{LFP}_{X,x}\xi(X,x)$ evaluates to the full set A.

Any S that satisfies $\psi_2(<,S)$ can only fail to be the true successor because of the existence of S-edges that jump ahead in the sense that $(a_1,a_3) \in S$ and there is an a_2 such that $a_1 < a_2 < a_3$. It is easy to see that the least such a_1 will not enter the fixed-point, in fact $\mathrm{LFP}_{X,x}\xi(X,x)$ evaluates to the initial segment of $(A,<)$ below the least such a_1.

Putting these facts together, it is clear that the following characterizes well-orderings with successor: $\psi_{\mathrm{wo,s}}(<,S) = \psi_{\mathrm{wo}}(<) \land \psi_2(<,S) \land \forall x\big[\mathrm{LFP}_{X,x}\xi(X,x)\big]x$.

Proposition 15. *The class of all matchings* (V,E) *and the class of all triangle matchings* (V,E_1,E_2,E_3) *admit projective characterizations in* FP^2.

Proof. Let $\psi_0 \in L^2$ say that the domain and range of E are disjoint. For matchings it suffices to assert in addition that $E \subseteq S$ for an expansion $(V,E,<,S)$

where $<$ is a well-ordering with successor relation S. The desired projective characterization is given by the FP^2-sentence $\psi_{\mathrm{match}}(E,<,S) = \psi_{\mathrm{wo,s}}(<,S) \wedge \psi_0 \wedge \forall x \forall y (Exy \rightarrow Sxy)$.

For triangle matchings we firstly use a sentence $\psi_1 \in L^2$ to say that $\mathrm{dm}(E_2) = \mathrm{rg}(E_1)$, $\mathrm{dm}(E_3) = \mathrm{dm}(E_1)$, $\mathrm{rg}(E_3) = \mathrm{rg}(E_2)$, and that $\mathrm{dm}(E_i) \cap \mathrm{rg}(E_i) = \emptyset$ for $i = 1,2,3$.

Let ψ_Δ — with new binary $<$, S, S' and unary U — be the conjunction of ψ_1 with FP^2-sentences asserting the following:
(1) $U = \mathrm{dm}(E_1) \cup \mathrm{rg}(E_2)$.
(2) S is the successor relation related to the well-ordering $<$.
(3) S' is the successor relation associated with the well-ordering $<|U$.
(4) $E_1 \cup E_2 \subseteq S$ and $E_3 \subseteq S'$.
(2) and (3) use the sentence $\psi_{\mathrm{wo,s}}$ of Lemma 14 and its relativization to U, respectively. The rest is plain L^2. It is not difficult to show that ψ_Δ is as desired.

We can now translate the projective characterization in L^2 + trans of even local grids (and also of the infinite grid $\mathcal{G}_{\mathbb{Z}}$) into FP^2 and thus get the following result.

Theorem 16. *FP^2 is a conservative reduction class, fin-sat(FP^2) and sat(FP^2) are undecidable; sat(FP^2) is even Σ^1_1-hard.*

4 Two-variable logic with cardinality comparison

It is well-known that one of the major limitations of first-order logic (with respect to expressiveness) is its inability to count. For instance, there is no first-order sentence that defines parity, in the sense that its finite models are precisely the structures of even cardinality. In fact even much stronger languages like fixed-point logic or infinitary logic with bounded number of variables $L^{\omega}_{\infty\omega}$ cannot express parity. Since counting is a computationally simple task it is natural to investigate logics with counting constructs (such as counting quantifiers, counting terms or generalized quantifiers). We refer to [4,10] for a detailed discussion of logics with counting.

The Härtig quantifier. A classical presentation of logics with cardinality comparison involves a particular generalized quantifier, namely the equicardinality quantifier I, introduced by Härtig (see [8] for a survey). The extension $L[I]$ of a logic L by the Härtig quantifier is defined by adding to L the following rule for building formulae. Given two formulae φ_1, φ_2 and two (not necessarily distinct) variables z_1, z_2, the expression $(Iz_1, z_2\ \varphi_1, \varphi_2)$ is a formula, saying that the sets $\{z_1 : \varphi_1\}$ and $\{z_2 : \varphi_2\}$ have the same cardinality.

The power of the Härtig quantifier in the context of L^2 results from the fact that an application of H does not necessarily reduce the number of free variables. Indeed, if x, y are free in $\varphi(x,y)$ and $\psi(x,y)$, then both x and y are free also in $(Ix, y\ \varphi, \psi)(x,y)$. For instance, we can axiomatize in $L^2[I]$ the class of regular graphs $G = (V,E)$ by the formula $\psi_{\mathrm{reg}} := \forall x \forall y (\neg Exx \wedge (Exy \rightarrow Eyx) \wedge (Ix, y\ Eyx, Exy)(x,y))$. Functionality of a binary E is expressible through

$\forall x \forall y(Ix, y \; Eyx, x = y)$; injectivity and the matching property are similarly expressible.

Rather than give a proof of Theorem 17 below, we indicate the crucial idea by showing that the natural numbers with standard successor $(\mathbb{N}, S)$ admit a projective characterization in $L^2[I]$. Using an auxiliary binary E to co-ordinatize elements by their numbers of E-predecessors, we may assert the following in $L^2[I]$:

(1) E is irreflexive and no two distinct elements have E-predecessor sets of the same cardinality.
(2) S is the graph of a total injective function without fixed points.
(3) for all $(x, y) \in S$, the number of E-predecessors of y is one more than the number of E-predecessors of x.

For (3) it suffices (in the presence of (1) and (2)) to equate the cardinalities of $\{x : Exy\}$ with that of $\{y : Eyx \vee y = x\}$. It is checked that (1) – (3) provide a projective characterization as desired.

This idea may in fact be extended to a two-dimensional co-ordinatization of local grids or of the infinite standard grid $\mathcal{G}_{\mathbb{N}}$ to prove the following theorem. A detailed proof is given in the full paper.

Theorem 17. *$L^2[I]$ is a conservative reduction class. Thus fin-sat$(L^2[I])$ and sat$(L^2[I])$ are undecidable; sat$(L^2[I])$ is even Σ^1_1-hard.*

This does not carry over to C^2, the extension of L^2 by the monadic counting quantifiers $\exists^{\geqslant m}$. The two counting logics $L^2[I]$ and C^2 are incomparable with respect to expressive power, in fact we prove in [5] that C^2 is decidable.

References

1. E. Börger, E. Grädel, and Y. Gurevich, *The Classical Decision Problem*, Springer, 1996.
2. E. Grädel, *Dominoes and the complexity of subclasses of logical theories*, Annals of Pure and Applied Logic, 43 (1989), pp. 1–30.
3. E. Grädel, P. Kolaitis, and M. Vardi, *On the decision problem for two-variable first-order logic*. To appear in: Bulletin of the ASL (1997).
4. E. Grädel and M. Otto, *Inductive definability with counting on finite structures*, in Computer Science Logic, Selected Papers, E. Börger et al., eds., Lecture Notes in Computer Science No. 702, Springer, 1993, pp. 231–247.
5. E. Grädel, M. Otto, and E. Rosen, *Two-variable logic with counting is decidable*. Submitted for publication.
6. D. Harel, *Recurring dominoes: Making the highly undecidable highly understandable*, Annals of Discrete Mathematics, 24 (1985), pp. 51–72.
7. ———, *Effective transformations on infinite trees, with applications to high undecidability, dominoes and fairness*, Journal of the ACM, 33 (1986), pp. 224–248.
8. H. Herre, M. Krynicki, A. Pinus, and J. Väänänen, *The Härtig quantifier: a survey*, Journal of Symbolic Logic, 56 (1991), pp. 1153–1183.
9. M. Mortimer, *On languages with two variables*, Zeitschr. f. math. Logik u. Grundlagen d. Math., 21 (1975), pp. 135–140.
10. M. Otto, *Bounded Variable Logics and Counting — A Study in Finite Models*. Lecture Notes in Logic No. 9, Springer, 1997.

Methods and Applications of (max,+) Linear Algebra

Stéphane Gaubert and Max Plus*

INRIA, Domaine de Voluceau, BP105, 78153 Le Chesnay Cedex, France.
e-mail: Stephane.Gaubert@inria.fr

Abstract. Exotic semirings such as the "$(\max,+)$ semiring" $(\mathbb{R}\cup\{-\infty\},\max,+)$, or the "tropical semiring" $(\mathbb{N}\cup\{+\infty\},\min,+)$, have been invented and reinvented many times since the late fifties, in relation with various fields: performance evaluation of manufacturing systems and discrete event system theory; graph theory (path algebra) and Markov decision processes, Hamilton-Jacobi theory; asymptotic analysis (low temperature asymptotics in statistical physics, large deviations, WKB method); language theory (automata with multiplicities).
Despite this apparent profusion, there is a small set of common, non-naive, basic results and problems, in general not known outside the $(\max,+)$ community, which seem to be useful in most applications. The aim of this short survey paper is to present what we believe to be the minimal core of $(\max,+)$ results, and to illustrate these results by typical applications, at the frontier of language theory, control, and operations research (performance evaluation of discrete event systems, analysis of Markov decision processes with average cost).
Basic techniques include: solving all kinds of systems of linear equations, sometimes with exotic symmetrization and determinant techniques; using the $(\max,+)$ Perron-Frobenius theory to study the dynamics of $(\max,+)$ linear maps. We point out some open problems and current developments.

1 Introduction: the $(\max,+)$ and tropical semirings

The "max-algebra" or "$(\max,+)$ semiring" $\mathbb{R}_{\max}$, is the set $\mathbb{R}\cup\{-\infty\}$, equipped with max as addition, and $+$ as multiplication. It is traditional to use the notation $\oplus$ for max ($2\oplus 3=3$), and $\otimes$ for $+$ ($1\otimes 1=2$). We denote[1] by $\mathbb{0}$ the *zero* element for $\oplus$ (such that $\mathbb{0}\oplus a=a$, here $\mathbb{0}=-\infty$) and by $\mathbb{1}$ the *unit* element for $\otimes$ (such that $\mathbb{1}\otimes a=a\otimes\mathbb{1}=a$, here $\mathbb{1}=0$). This structure satisfies all the semiring axioms, i.e. $\oplus$ is associative, commutative, with zero element, $\otimes$ is associative, has a unit, distributes over $\oplus$, and zero is absorbing (all the ring axioms are satisfied, except that $\oplus$ need not be a group law). This semiring is *commutative* ($a\otimes b=b\otimes a$), *idempotent* ($a\oplus a=a$), and

* Max Plus is a collective name for a working group on $(\max,+)$ algebra, at INRIA Rocquencourt, comprising currently: Marianne Akian, Guy Cohen, S.G., Jean-Pierre Quadrat and Michel Viot.

[1] The notation for the zero and unit is one of the disputed questions of the community. The symbols ε for zero, and e for the unit, often used in the literature, are very distinctive and well suited to handwritten computations. But it is difficult to renounce to the traditional use of ε in Analysis. The notation $\mathbb{0},\mathbb{1}$ used by the Idempotent Analysis school has the advantage of making formulæ closer to their usual analogues.

Reischuk, Morvan (Eds.): STACS'97 Proceedings, LNCS 1200

non zero elements have an inverse for $\otimes$ (we call *semifields* the semirings that satisfy this property). The term *dioid* is sometimes used for an *idempotent* semiring.

Using the new symbols $\oplus$ and $\otimes$ instead of the familiar max and + notation is the price to pay to easily handle all the familiar algebraic constructions. For instance, we will write, in the (max, +) semiring:

$$ab = a \otimes b, \quad a^n = a \otimes \cdots \otimes a \ (n \text{ times}), \qquad 2^3 = 6 \ , \qquad \sqrt{3} = 1.5 \ ,$$
$$\begin{bmatrix} 2 & 0 \\ 4 & \mathbb{0} \end{bmatrix} \begin{bmatrix} 10 \\ 103 \end{bmatrix} = \begin{bmatrix} 2 \otimes 10 \oplus 0 \otimes 103 \\ 4 \otimes 10 \oplus \mathbb{0} \otimes 103 \end{bmatrix} = \begin{bmatrix} 103 \\ 14 \end{bmatrix},$$
$$(3 \oplus x)^2 = (3 \oplus x)(3 \oplus x) = 6 \oplus 3x \oplus x^2 = 6 \oplus x^2 \ \ (= \max(6, 2 \times x)) \ .$$

We will systematically use the standard algebraic notions (matrices, vectors, linear operators, semimodules — i.e. modules over a semiring—, formal polynomials and polynomial functions, formal series) in the context of the (max, +) semiring, often without explicit mention. Essentially all the standard notions of algebra have obvious semiring analogues, provided they do not appeal to the invertibility of addition.

There are several useful variants of the (max, +) semiring, displayed in Table 1. In the sequel, we will have to consider various semirings, and will universally use the

$\mathbb{R}_{\max}$	$(\mathbb{R} \cup \{-\infty\}, \max, +)$	(max, +) semiring max algebra	idempotent semifield
$\overline{\mathbb{R}}_{\max}$	$(\mathbb{R} \cup \{\pm\infty\}, \max, +)$	completed (max, +) semiring	$-\infty + (+\infty) = -\infty$, for $\mathbb{0} \otimes a = \mathbb{0}$
$\mathbb{R}_{\max,\times}$	$(\mathbb{R}^+, \max, \times)$	(max, ×) semiring	isomorphic to $\mathbb{R}_{\max}$ ($x \mapsto \log x$)
$\mathbb{R}_{\min}$	$(\mathbb{R} \cup \{+\infty\}, \min, +)$	(min, +) semiring	isomorphic to $\mathbb{R}_{\max}$ ($x \mapsto -x$)
$\mathbb{N}_{\min}$	$(\mathbb{N} \cup \{+\infty\}, \min, +)$	tropical semiring	(famous in Language Theory)
$\mathbb{R}_{\max,\min}$	$(\mathbb{R} \cup \{\pm\infty\}, \max, \min)$	bottleneck algebra	not dealt with here
$\mathbb{B}$	({false, true}, or, and)	Boolean semiring	isomorphic to $(\{\mathbb{0}, \mathbb{1}\}, \oplus, \otimes)$, for any of the above semirings
$\mathbb{R}_h$	$(\mathbb{R} \cup \{-\infty\}, \oplus_h, +)$ $a \oplus_h b = h \log(e^{a/h} + e^{b/h})$	Maslov semirings	isomorphic to $(\mathbb{R}^+, +, \times)$ $\lim_{h \to 0^+} \mathbb{R}_h = \mathbb{R}_0 = \mathbb{R}_{\max}$

Table 1. The family of (max, +) and tropical semirings ...

notation $\oplus, \otimes, \mathbb{0}, \mathbb{1}$ with a context dependent meaning (e.g. $\oplus = \max$ in $\mathbb{R}_{\max}$ but $\oplus = \min$ in $\mathbb{R}_{\min}$, $\mathbb{0} = -\infty$ in $\mathbb{R}_{\max}$ but $\mathbb{0} = +\infty$ in $\mathbb{R}_{\min}$).

The fact that $\oplus$ is idempotent instead of being invertible ($\mathbb{R}_h$ is an exception, for $h \neq 0$), is the main original feature of these "exotic" algebras, which makes them so different from the more familiar ring and field structures. In fact the idempotence and cancellativity axioms are exclusive: if for all a, b, c, $(a \oplus b = a \oplus c \Rightarrow b = c)$ and $a \oplus a = a$, we get $a = \mathbb{0}$, for all a (simplify $a \oplus a = a \oplus \mathbb{0}$).

This paper is not a survey in the usual sense. There exist several comprehensive books and excellent survey articles on the subject, each one having its own bias and motivations. Applications of $(\max, +)$ algebras are too vast (they range from asymptotic methods to decidability problems), techniques are too various (from graph theory to measure theory and large deviations) to be surveyed in a paper of this format. But there is a small common set of useful basic results, applications and problems, that we try to spotlight here. We aim neither at completeness, nor at originality. But we wish to give an honest idea of the services that one should expect from $(\max, +)$ techniques. The interested reader is referred to the books [15,44,10,2,31], to the survey papers listed in the bibliography, and to the recent collection of articles [24] for an up-to-date account of the maxplusian results and motivations. Bibliographical and historical comments are at the end of the paper.

2 Seven good reasons to use the $(\max, +)$ semiring

2.1 An Algebra for Optimal Control

A standard problem of calculus of variations, which appears in Mechanics (least action principle) and Optimal Control, is the following. Given a Lagrangian L and suitable boundary conditions (e.g. $q(0), q(T)$ fixed), compute

$$\inf_{q(\cdot)} \int_0^T L(q, \dot{q}) dt \ . \tag{1}$$

This problem is intrinsically $(\min, +)$ linear. To see this, consider the (slightly more general) discrete variant, with sup rather than inf,

$$\xi(n) = x, \ \xi(k) = f(\xi(k-1), u(k)), \ k = n+1, \dots, N, \tag{2a}$$

$$J_n^N(x, u) = \sum_{k=n+1}^{N} c(\xi(k-1), u(k)) + \Phi(\xi(N)) \ , \tag{2b}$$

$$V_n^N(x) = \sup_u J_n^N(x, u) \ , \tag{2c}$$

where the sup is taken over all sequences of *controls* $u(k), k = n+1, \cdots, N$, selected in a finite *set of controls* U, $\xi(k)$, for $k = n, \dots, N$, belongs to a finite set X of *states*, x is a distinguished *initial state*, $f : X \times U \to X$ is the *dynamics*, $c : X \times U \to \mathbb{R} \cup \{-\infty\}$ is the *instantaneous reward*, and $\Phi : X \to \mathbb{R} \cup \{-\infty\}$ is the *final reward* (the $-\infty$ value can be used to code forbidden final states or transitions). These data form a deterministic *Markov Decision Process* (MDP) with additive reward.

The function $V_n^N(\cdot)$, which represents the optimal reward from time n to time N, as a function of the starting point, is called the *value* function. It satisfies the backward dynamic programming equation

$$V_N^N = \Phi, \qquad V_k^N(x) = \max_{u \in U} \left\{ c(x, u) + V_{k+1}^N(f(x, u)) \right\} \ . \tag{3}$$

Introducing the *transition matrix* $A \in (\mathbb{R}_{\max})^{X \times X}$,

$$A_{x,y} = \sup_{u \in U, \ f(x,u) = y} c(x, u), \tag{4}$$

(the supremum over an empty set is $-\infty$), we obtain:

FACT 1 (DETERMINISTIC MDP = (max, +)-LINEAR DYNAMICS). *The value function V_k^N of a finite deterministic Markov decision process with additive reward is given by the* $(\max, +)$ *linear dynamics:*

$$V_N^N = \Phi, \qquad V_k^N = AV_{k+1}^N . \tag{5}$$

The interpretation in terms of paths is elementary. If we must end at node j, we take $\Phi = \mathbb{1}_j$ (the vector with all entries $\mathbb{0}$ except the j-th equal to $\mathbb{1}$), Then, the value function $V_0^N(i) = (A^N)_{ij}$ is the maximal (additive) weight of a path of length N, from i to j, in the graph canonically associated[2] with A.

Example 1 (Taxicab). Consider a taxicab which operates between 3 cities and one airport, as shown in Fig. 1. At each state, the taxi driver has to choose his next destination, with deterministic fares shown on the graph (for simplicity, we assume that the demand is deterministic, and that the driver can choose the destination). The taxi driver considers maximizing his reward over N journeys. The $(\max, +)$ matrix associated with this MDP is displayed in Fig. 1.

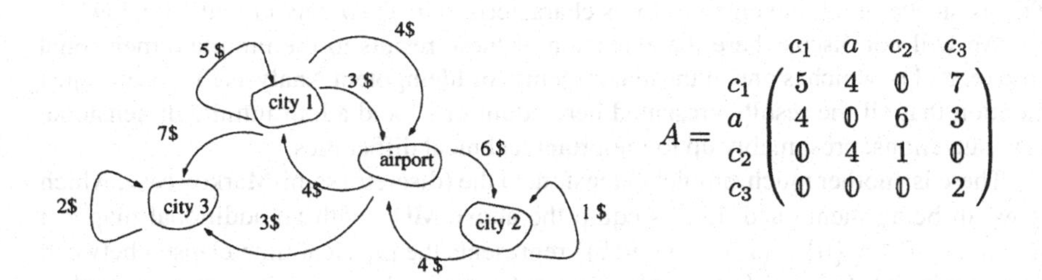

$$A = \begin{array}{c} \\ c_1 \\ a \\ c_2 \\ c_3 \end{array} \begin{array}{c} \begin{array}{cccc} c_1 & a & c_2 & c_3 \end{array} \\ \begin{pmatrix} 5 & 4 & \mathbb{0} & 7 \\ 4 & \mathbb{0} & 6 & 3 \\ \mathbb{0} & 4 & 1 & \mathbb{0} \\ \mathbb{0} & \mathbb{0} & \mathbb{0} & 2 \end{pmatrix} \end{array}$$

Fig. 1. Taxicab Deterministic MDP and its matrix

Let us consider the optimization of the average reward:

$$\chi(x) = \sup_u \limsup_{N \to \infty} \frac{1}{N} J_0^N(x, u) . \tag{6}$$

Here, the sup is taken over infinite sequences of controls $u(1), u(2), \dots$ and the trajectory (2a) is defined for $k = 0, 1, \dots$. We expect J_0^N to grow (or to decrease) linearly, as a function of the horizon N. Thus, $\chi(x)$ represents the optimal average reward (per time unit), starting from x. Assuming that the sup and lim sup commute in (6), we get:

$$\chi(x) = \limsup_{N \to \infty} \frac{1}{N} \times (A^N \Phi)_x \tag{7}$$

(this is an hybrid formula, $A^N \Phi$ is in the $(\max, +)$ semiring, $1/N \times (\cdot)$ is in the conventional algebra). To evaluate (7), let us assume that the matrix A admits an *eigenvector* v

[2] With a $X \times X$ matrix A we associate the weighted (directed) graph, with set of nodes X, and an arc (x, y) with weight $A_{x,y}$ whenever $A_{x,y} \neq \mathbb{0}$.

in the $(\max, +)$ semiring:

$$Av = \lambda v, \text{ i.e. } \qquad \max_j \{A_{ij} + v_j\} = \lambda_i + v_i \tag{8}$$

(the eigenvector v must be nonidentically $\mathbb{0}$, $\lambda \in \mathbb{R}_{\max}$ is the eigenvalue). Let us assume that v and Φ have only finite entries. Then, there exist two finite constants μ, ν such that $\nu + v \leq \Phi \leq \mu + v$. In $(\max, +)$ notation, $\nu v \leq \Phi \leq \mu v$. Then $\nu \lambda^N v = \nu A^N v \leq A^N \Phi \leq \mu A^N v = \mu \lambda^N v$, or with the conventional notation:

$$\nu + N\lambda + v \leq A^N \Phi \leq \mu + N\lambda + v. \tag{9}$$

We easily deduce from (9) the following.

FACT 2 ("EIGENELEMENTS = OPTIMAL REWARD AND POLICY"). *If the final reward Φ is finite, and if A has a finite eigenvector with eigenvalue λ, the optimal average reward $\chi(x)$ is a constant (independent of the starting point x), equal to the eigenvalue λ. An optimal control is obtained by playing in state i any u such that $c(i, u) = A_{ij}$ and $f(i, u) = j$, where j is in the* arg max *of (8) at state i.*

The existence of a finite eigenvector is characterized in Theorems 11 and 15 below.

We will not discuss here the extension of these results to the infinite dimensional case (e.g. (1)), which is one of the major themes of Idempotent Analysis [31]. Let us just mention that all the results presented here admit or should admit infinite dimensional generalizations, presumably up to important technical difficulties.

There is another much simpler extension, to the (discrete) semi-Markov case, which is worth being mentioned. Let us equip the above MDP with an additional map $\tau : X \times U \to \mathbb{R}^+ \setminus \{0\}$; $\tau(x(k-1), u(k))$ represents the physical time elapsed between decision k and decision $k+1$, when control $u(k)$ is chosen. This is very natural in most applications (for the taxicab example, the times of the different possible journeys in general differ). The optimal average reward per time unit now writes:

$$\chi(x) = \sup_u \limsup_{N \to \infty} \frac{\sum_{k=1}^N c(x(k-1), u(k)) + \Phi(x(N))}{\sum_{k=1}^N \tau(x(k-1), u(k))} . \tag{10}$$

Of course, the specialization $\tau \equiv 1$ gives the original problem (6). Let us define $\mathcal{T}_{ij} = \{\tau(i, u) \mid f(i, u) = j\}$, and for $t \in \mathcal{T}_{ij}$,

$$A_{t,i,j} = \sup_{u \in U, f(i,u) = j, \tau(i,u) = t} c(i, u) . \tag{11}$$

Arguing as in the Markov case, it is not too difficult to show the following.

FACT 3 (GENERALIZED SPECTRAL PROBLEM FOR SEMI-MARKOV PROCESSES). *If the generalized spectral problem*

$$\max_j \max_{t \in \mathcal{T}_{ij}} \{A_{t,i,j} - \lambda t + v_j\} = v_i \tag{12}$$

has a finite solution v, and if Φ is finite, then the optimal average reward is $\chi(x) = \lambda$, for all x. An optimal control is obtained by playing any u in the arg max *of (11), with j, t in the* arg max *of (12), when in state i.*

Algebraically, (12) is nothing but a generalized spectral problem. Indeed, with an obvious definition of the matrices A_t, we can write:

$$\bigoplus_{t\in\overline{\mathcal{T}}} \lambda^{-t} A_t v = v \ , \qquad \text{where } \overline{\mathcal{T}} = \bigcup_{i,j} \mathcal{T}_{ij} \ . \tag{13}$$

2.2 An Algebra for Asymptotics

In Statistical Physics, one looks at the asymptotics when the temperature h tends to zero of the spectrum of *transfer matrices*, which have the form

$$\mathcal{A}_h = (\exp(h^{-1}A_{ij}))_{1\le i,j\le n} \ .$$

The real parameters A_{ij} represent potential terms plus interaction energy terms (when two adjacent sites are in states i and j, respectively). The Perron eigenvalue[3] $\rho(\mathcal{A}_h)$ determines the free energy per site $\lambda_h = h\log\rho(\mathcal{A}_h)$. Clearly, λ_h is an eigenvalue of A in the semiring $\mathbb{R}_h$, defined in Table 1. Let $\rho_{\max}(A)$ denote the maximal $(\max,+)$ eigenvalue of A. Since $\lim_{h\to 0^+}\mathbb{R}_h = \mathbb{R}_0 = \mathbb{R}_{\max}$, the following result is natural.

FACT 4 (PERRON FROBENIUS ASYMPTOTICS). *The asymptotic growth rate of the Perron eigenvalue of $\mathcal{A}_h$ is equal to the maximal* $(\max,+)$ *eigenvalue of the matrix A:*

$$\lim_{h\to 0^+} h\log\rho(\mathcal{A}_h) = \rho_{\max}(A) \ . \tag{14}$$

This follows easily from the $(\max,+)$ spectral inequalities (24),(25) below. The normalized Perron eigenvector v_h of $\mathcal{A}_h$ also satisfies

$$\lim_{h\to 0^+} h\log(v_h)_i = u_i \ ,$$

where u is a special $(\max,+)$ eigenvector of A which has been characterized recently by Akian, Bapat, and Gaubert [1]. Precise asymptotic expansions of $\rho(\mathcal{A}_h)$ as sum of exponentials have been given, some of the terms having combinatorial interpretations.

More generally, $(\max,+)$ algebra arises almost everywhere in asymptotic phenomena. Often, the $(\max,+)$ algebra is involved in an elementary way (e.g. when computing exponents of Puiseux expansions using the Newton Polygon). Less elementary applications are WKB type asymptotics (see [31]), which are related to Large Deviations (see e.g. [17]).

2.3 An Algebra for Discrete Event Systems

The $(\max,+)$ algebra is popular in the Discrete Event Systems community, since $(\max,+)$ linear dynamics correspond to a well identified subclass of Discrete Event Systems, with only synchronization phenomena, called Timed Event Graphs. Indeed, consider a system with n repetitive tasks. We assume that the k-th execution of task i (firing of transition i) has to wait τ_{ij} time units for the $(k-\nu_{ij})$-th execution of task j. E.g. tasks represent the processing of parts in a manufacturing system, ν_{ij} represents an initially available stock, and τ_{ij} represents a production or transportation time.

[3] The Perron eigenvalue $\rho(B)$ of a matrix B with nonnegative entries is the maximal eigenvalue associated with a nonnegative eigenvector, which is equal to the spectral radius of B.

FACT 5 (TIMED EVENT GRAPHS ARE $(\max,+)$ LINEAR SYSTEMS). *The earliest date of occurrence of an event i in a Timed Event Graph, $x_i(k)$, satisfies*

$$x_i(k) = \max_j \left[\tau_{ij} + x_j(k - \nu_{ij})\right] \ . \tag{15}$$

Eqn 15 coincides with the value iteration of the deterministic semi-Markov Decision Process in § 2.1, that we only wrote in the Markov version (3). Therefore, the asymptotic behavior of (15) can be dealt with as in § 2.1, using $(\max,+)$ spectral theory. In particular, if the generalized spectral problem $v_i = \max_j[\tau_{ij} - \lambda\nu_{ij} + v_j]$ has a finite solution (λ, v), then $\lambda = \lim_{k\to\infty} k^{-1} \times x_i(k)$, for all i (λ is the *cycle time*, or inverse of the *asymptotic throughput*). The study of the dynamics (15), and of its stochastic [2], and non-linear extensions [11,23] (fluid Petri Nets, minmax functions), is the major theme of $(\max,+)$ discrete event systems theory.

Another linear model is that of *heaps of pieces*. Let $\mathcal{R}$ denote a set of *positions* or *resources* (say $\mathcal{R} = \{1,\dots,n\}$). A *piece* (or *task*) a is a rigid (possibly non connected) block, represented geometrically by a set of occupied positions (or requested resources) $R(a) \subset \mathcal{R}$, a lower contour (starting time) $\ell(a) : R(a) \to \mathbb{R}$, an upper contour (release time) $h(a) : R(a) \to \mathbb{R}$, such that $\forall a \in R(a)$, $h(a) \geq \ell(a)$. The piece corresponds to the region of the $\mathcal{R} \times \mathbb{R}$ plane: $P_a = \{(r,y) \in R(a) \times \mathbb{R} \mid \ell(a)_r \leq y \leq h(a)_r\}$, which means that task a requires the set of resources (machines, processors, operators) $R(a)$, and that resource $r \in R(a)$ is used from time $\ell(a)_r$ to time $h(a)_r$. A piece P_a can be translated vertically of any λ, which gives the new region defined by $\ell'(a) = \lambda + \ell(a)$, $h'(a) = \lambda + h(a)$. We can execute a task earlier or later, but we cannot change the differences $h(a)_r - \ell(a)_s$ which are invariants of the task. A *ground* or *initial condition* is a row vector $g \in (\mathbb{R}_{\max})^{\mathcal{R}}$. Resource r becomes initially available at time g_r. If we drop k pieces $a_1 \dots a_k$, in this order, on the ground g (letting the pieces fall down according to the gravity, forbidding horizontal translations, and rotations, as in the famous Tetris game, see Fig 2), we obtain what we call a *heap of pieces*. The upper contour $x(w)$ of the heap $w = a_1 \dots a_k$ is the row vector in $(\mathbb{R}_{\max})^{\mathcal{R}}$, whose r-th component is equal to the position of the top of the highest piece occupying resource r. The *height* of the heap is by definition $y(w) = \max_{r\in\mathcal{R}} x(w)_r$. Physically, $y(w)$ gives the *makespan* (= completion time) of the sequence of tasks w, and $x(w)_r$ is the release time of resource r.

With each piece a within a set of pieces $\mathcal{T}$, we associate the matrix $M(a) \in (\mathbb{R}_{\max})^{\mathcal{R}\times\mathcal{R}}$, $M(a)_{r,s} = h(a)_s - \ell(a)_r$ if $r, s \in R(a)$, and $M(a)_{r,r} = \mathbb{1}$ for diagonal entries not in $R(a)$ (other entries are $\mathbb{0}$). The following result was found independently by Gaubert and Mairesse (in [24]), and Brilman and Vincent [6].

FACT 6 (TETRIS GAME IS $(\max,+)$ LINEAR). *The upper contour $x(w)$ and the height $y(w)$ of the heap of pieces $w = a_1 \dots a_k$, piled up on the ground g, are given by the $(\max,+)$ products:*

$$x(w) = gM(a_1)\dots M(a_k), \qquad y(w) = x(w)\mathbb{1}_{\mathcal{R}},$$

($\mathbb{1}_X$ denotes the column vector indexed by X with entries $\mathbb{1}$).

In algebraic terms, the height generating series $\bigoplus_{w\in\mathcal{T}^*} y(w)w$ is *rational* over the (max,+) semiring ($\mathcal{T}^*$ is the free monoid on $\mathcal{T}$, basic properties of rational series can be found e.g. in [38]).

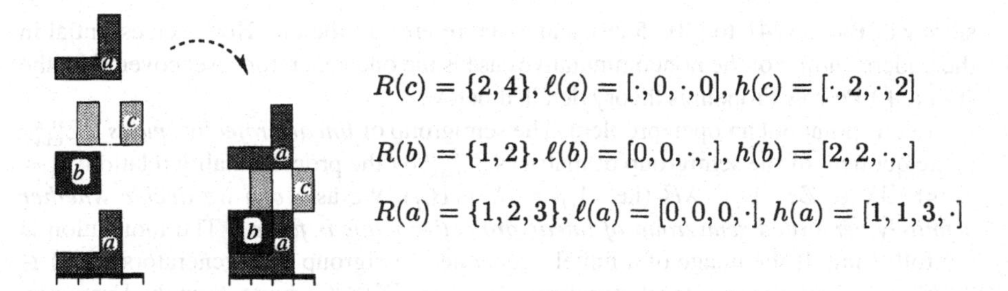

$R(c) = \{2,4\}, \ell(c) = [\cdot,0,\cdot,0], h(c) = [\cdot,2,\cdot,2]$
$R(b) = \{1,2\}, \ell(b) = [0,0,\cdot,\cdot], h(b) = [2,2,\cdot,\cdot]$
$R(a) = \{1,2,3\}, \ell(a) = [0,0,0,\cdot], h(a) = [1,1,3,\cdot]$
a
b
c

Fig. 2. Heap of Pieces

Let us mention an open problem. If an infinite sequence of pieces $a_1 a_2 \dots a_k \dots$ is taken at random, say in an independent identically distributed way with the uniform distribution on $\mathcal{T}$, it is known [14,2] that there exists an asymptotic growth rate $\lambda \in \mathbb{R}^+$:

$$\lambda = \lim_{k\to\infty} \frac{1}{k} y(a_1 \dots a_k) \quad \text{a.s.} \tag{16}$$

The effective computation of the constant λ (Lyapunov exponent) is one of the main open problems in (max,+) algebra. The Lyapunov exponent problem is interesting for general random matrices (not only for special matrices associated with pieces), but the heap case (even with unit height, $h(a) = 1 + \ell(a)$) is typical and difficult enough to begin with. Existing results on Lyapunov exponents can be found in [2]. See also the paper of Gaujal and Jean-Marie in [24], and [6].

2.4 An Algebra for Decision

The "tropical" semiring $\mathbb{N}_{\min} = (\mathbb{N}\cup\{+\infty\}, \min, +)$, has been invented by Simon [39] to solve the following classical problem posed by Brzozowski: *is it decidable whether a rational language L has the Finite Power Property (FPP):* $\exists m \in \mathbb{N},\ L^* = L^0 \cup L \cup \dots \cup L^m$. The problem was solved independently by Simon and Hashiguchi.

FACT 7 (SIMON). *The FPP problem for rational languages reduces to the finiteness problem for finitely generated semigroups of matrices with entries in* $\mathbb{N}_{\min}$, *which is decidable.*

Other (more difficult) decidable properties (with applications to the polynomial closure and star height problems) are the *finite section* problem, which asks, given a finitely generated semigroup of matrices S over the tropical semiring, whether the set of entries in position i, j, $\{s_{ij} \mid s \in S\}$ is finite; and the more general *limitation* problem, which asks whether the set of coefficients of a rational series in $\mathbb{N}_{\min}$, with noncommuting indeterminates, is finite. These decidability results due to Hashiguchi [25], Leung [29] and Simon [40] use structural properties of long optimal words in $\mathbb{N}_{\min}$-automata (involving multiplicative rational expressions), and combinatorial arguments. By comparison with basic Discrete Event System and Markov Decision applications, which essentially involve semigroups with a single generator ($S = \{A^k \mid k \geq 1\}$), these typically noncommutative problems represent a major jump in difficulty. We refer the reader to the survey of Pin in [24], to [40,25,29], and to the references therein. However, essential in the understanding of the noncommutative case is the one generator case, covered by the $(\max, +)$ Perron-Frobenius theory detailed below.

Let us point out an open problem. The semigroup of *linear projective maps* $\mathbb{PZ}_{\max}^{n\times n}$ is the quotient of the semigroup of matrices $\mathbb{Z}_{\max}^{n\times n}$ by the proportionality relation: $A \sim B \Leftrightarrow \exists\lambda \in \mathbb{Z},\ A = \lambda B$ (i.e. $A_{ij} = \lambda + B_{ij}$). We ask: *can we decide whether a finitely generated semigroup of linear projective maps is finite ?* The motivation is the following. If the image of a finitely generated semigroup with generators $M(a) \in \mathbb{Z}_{\max}^{n\times n}, a \in \Sigma$ by the canonical morphism $\mathbb{Z}_{\max}^{n\times n} \to \mathbb{PZ}_{\max}^{n\times n}$ is finite, then the Lyapunov exponent $\lambda = \text{a.s.} \lim_{k\to\infty} k^{-1} \times \|M(a_1)\dots M(a_k)\|$ (same probabilistic assumptions as for (16), $\|A\| = \sup_{ij} A_{ij}$, by definition) can be computed from a finite Markov Chain on the associated projective linear semigroup [19,20].

3 Solving Linear Equations in the $(\max, +)$ Semiring

3.1 A hopeless algebra?

The general system of n $(\max, +)$-linear equations with p unknowns $x_1, \dots, x_p$ writes:

$$Ax \oplus b = Cx \oplus d, \qquad A, C \in (\mathbb{R}_{\max})^{n\times p},\ b, d \in (\mathbb{R}_{\max})^n \ . \tag{17}$$

Unlike in conventional algebra, a square linear system ($n = p$) is not generically solvable (consider $3x \oplus 2 = x \oplus 0$, which has no solution, since for all $x \in \mathbb{R}_{\max}$, $\max(3 + x, 2) > \max(x, 0)$).

There are several ways to make this hard reality more bearable. One is to give general structural results. Another is to deal with natural subclasses of equations, whose solutions can be obtained by efficient methods. The *inverse* problem $Ax = b$ can be dealt with using *residuation*. The *spectral* problem $Ax = \lambda x$ (λ scalar) is solved using the $(\max, +)$ analogue of Perron-Frobenius theory. The *fixed point* problem $x = Ax \oplus b$ can be solved via rational methods familiar in language theory (introducing the "star" operation $A^* = A^0 \oplus A \oplus A^2 \oplus \cdots$). A last way, which has the seduction of forbidden things, is to say: "certainly, the solution of $3x \oplus 2 = x \oplus 0$ is $x = \ominus - 1$. For if this equation has no ordinary solution, the symmetrized equation (obtained by putting each occurrence of the unknown in the other side of the equality) $x' \oplus 2 = 3x' \oplus 0$ has the unique solution $x' = -1$. Thus, $x = \ominus - 1$ is the requested solution." Whether or not this argument is valid is the object of *symmetrization* theory.

All these approaches rely, in one way or another, on the *order* structure of idempotent semirings that we next introduce.

3.2 Natural Order Structure of Idempotent Semirings

An idempotent semiring $\mathcal{S}$ can be equipped with the following *natural* order relation

$$a \preceq b \iff a \oplus b = b. \tag{18}$$

We will write $a \prec b$ when $a \preceq b$ and $a \neq b$. The natural order endows $\mathcal{S}$ with a sup-semilattice structure, for which $a \oplus b = a \vee b = \sup\{a, b\}$ (this is the least upper bound of the set $\{a, b\}$), and $\mathbb{0} \preceq a$, $\forall a, b \in \mathcal{S}$ ($\mathbb{0}$ is the *bottom* element). The semiring laws preserve this order, i.e. $\forall a, b, c \in \mathcal{S}$, $a \preceq b \implies a \oplus c \preceq b \oplus c$, $ac \preceq bc$. For the $(\max, +)$ semiring $\mathbb{R}_{\max}$, the natural order $\preceq$ coincides with the usual one. For the $(\min, +)$ semiring $\mathbb{R}_{\min}$, the natural order is the opposite of the usual one.

Since addition coincides with the sup for the natural order, there is a simple way to define infinite sums, in an idempotent semiring, setting $\bigoplus_{i \in I} x_i = \sup\{x_i \mid i \in I\}$, for any possibly infinite (even non denumerable) family $\{x_i\}_{i \in I}$ of elements of $\mathcal{S}$, when the sup exists. We say that the idempotent semiring $\mathcal{S}$ is *complete* if any family has a supremum, and if the product distributes over infinite sums. When $\mathcal{S}$ is complete, $(\mathcal{S}, \preceq)$ becomes automatically a complete lattice, the greatest lower bound being equal to $\bigwedge_{i \in I} x_i = \sup\{y \in \mathcal{S} \mid y \leq x_i, \forall i \in I\}$. The $(\max, +)$ semiring $\mathbb{R}_{\max}$ is not complete (a complete idempotent semiring must have a maximal element), but it can be embedded in the complete semiring $\overline{\mathbb{R}}_{\max}$.

3.3 Solving $Ax = b$ using Residuation

In general, $Ax = b$ has no solution[4], but $Ax \preceq b$ always does (take $x = \mathbb{0}$). Thus, a natural way of attacking $Ax = b$ is to relax the equality and study the set of its subsolutions. This can be formalized in terms of *residuation* [5], a notion borrowed from ordered sets theory. We say that a monotone map f from an ordered set E to an ordered set F is *residuated* if for all $y \in F$, the set $\{x \in E \mid f(x) \leq y\}$ has a maximal element, denoted $f^\sharp(y)$. The monotone map $f^\sharp$, called *residual* or *residuated map* of f, is characterized alternatively by $f \circ f^\sharp \leq \mathrm{Id}$, $f^\sharp \circ f \geq \mathrm{Id}$. An idempotent semiring $\mathcal{S}$ is *residuated* if the right and left multiplication maps $\lambda_a : x \mapsto ax$, $\rho_a : x \mapsto xa$, $\mathcal{S} \to \mathcal{S}$, are residuated, for all $a \in \mathcal{S}$. A *complete* idempotent semiring is automatically residuated. We set

$$a \backslash b \stackrel{\text{def}}{=} \lambda_a^\sharp(b) = \max\{x \mid ax \preceq b\} \ , \quad b/a \stackrel{\text{def}}{=} \rho_a^\sharp(b) = \max\{x \mid xa \preceq b\} \ .$$

In the completed $(\max, +)$ semiring $\overline{\mathbb{R}}_{\max}$, $a \backslash b = b/a$ is equal to $b - a$ when $a \neq \mathbb{0} (= -\infty)$, and is equal to $+\infty$ if $a = \mathbb{0}$. The residuated character is transfered from scalars to matrices as follows.

Proposition 2 (Matrix residuation). *Let $\mathcal{S}$ be a complete idempotent semiring. Let $A \in \mathcal{S}^{n \times p}$. The map $\lambda_A : x \mapsto Ax, \mathcal{S}^p \to \mathcal{S}^n$, is residuated. For any $y \in \mathcal{S}^n$, $A \backslash y \stackrel{\text{def}}{=} \lambda_A^\sharp(y)$ is given by $(A \backslash y)_i = \bigwedge_{1 \leq j \leq n} A_{ji} \backslash y_j$.*

In the case of $\overline{\mathbb{R}}_{\max}$, this reads:

$$(A \backslash y)_i = \min_{1 \leq j \leq n} (-A_{ji} + y_j) \ , \tag{19}$$

[4] It is an elementary exercise to check that the map $x \mapsto Ax$, $(\mathbb{R}_{\max})^p \to (\mathbb{R}_{\max})^n$, is surjective (resp. injective) iff the matrix A contains a monomial submatrix of size n (resp. p), a very unlikely event — recall that a square matrix B is *monomial* if there is exactly one non zero element in each row, and in each column, or (equivalently) if it is a product of a permutation matrix and a diagonal matrix with non zero diagonal elements. This implies that a matrix has a left or a right inverse iff it has a monomial submatrix of maximal size, which is the analogue of a well known result for nonnegative matrices [4, Lemma 4.3].

with the convention dual to that of $\overline{\mathbb{R}}_{\max}$, $(+\infty) + x = +\infty$, for any $x \in \mathbb{R} \cup \{\pm\infty\}$. We recognize in (19) a matrix product in the semiring $\overline{\mathbb{R}}_{\min} = (\mathbb{R} \cup \{\pm\infty\}, \min, +)$, involving the transpose of the opposite of A.

Corollary 3 (Solving $Ax = y$). *Let $\mathcal{S}$ denote a complete idempotent semiring, and let $A \in \mathcal{S}^{n \times p}$, $y \in \mathcal{S}^n$. The equation $Ax = y$ has a solution iff $A(A \backslash y) = y$.*

Corollary 3 allows us to check the existence of a solution x of $Ax = y$ in time $O(np)$ (scalar operations are counted for one time unit). In the $(\max, +)$ case, a refinement (due to the total order) allows us to decide the existence of a solution by inspection of the minimizing sets in (19), see [15,44].

3.4 Basis Theorem for Finitely Generated Semimodules over $\mathbb{R}_{\max}$

A finitely generated semimodule $\mathcal{V} \subset (\mathbb{R}_{\max})^n$ is the set of linear combinations of a finite family $\{u_1, \dots, u_p\}$ of vectors of $(\mathbb{R}_{\max})^n$:

$$\mathcal{V} = \{ \bigoplus_{i=1}^{p} \lambda_i u_i \,|\, \lambda_1, \dots, \lambda_p \in \mathbb{R}_{\max} \} \ .$$

In matrix terms, $\mathcal{V}$ can be identified to the *column space* or *image* of the $n \times p$ matrix $A = [u_1, \dots, u_p]$, $\mathcal{V} = \operatorname{Im} A \stackrel{\text{def}}{=} \{Ax \mid x \in (\mathbb{R}_{\max})^p\}$. The *row space* of A is the column space of A^T (the transpose of A). The family $\{u_i\}$ is a *weak basis* of $\mathcal{V}$ if it is a generating family, minimal for inclusion. The following result, due to Moller [33] and Wagneur [42] (with variants) states that finitely generated subsemimodules of $(\mathbb{R}_{\max})^n$ have (essentially) a unique weak basis.

Theorem 4 (Basis Theorem). *A finitely generated semimodule $\mathcal{V} \subset (\mathbb{R}_{\max})^n$ has a weak basis. Any two weak bases have the same number of generators. For any two weak bases $\{u_1, \dots, u_p\}$, $\{v_1, \dots, v_p\}$, there exist invertible scalars $\lambda_1, \dots, \lambda_p$ and a permutation σ of $\{1, \dots, p\}$ such that $u_i = \lambda_i v_{\sigma(i)}$.*

The cardinality of a weak basis is called the *weak rank* of the semimodule, denoted $\mathrm{rk}_w \mathcal{V}$. The *weak column rank* (resp. weak row rank) of the matrix A is the weak rank of its column (resp. row) space. Unlike in usual algebra, the weak row rank in general differs from the weak column rank (this is already the case for Boolean matrices). Theorem 4 holds more generally in any idempotent semiring $\mathcal{S}$ satisfying the following axioms: $(a \succeq \alpha a$ and $a \neq \mathbb{0}) \implies \mathbb{1} \succeq \alpha$, $(a = \alpha a \oplus b$ and $\alpha \prec \mathbb{1}) \implies a = b$. The axioms needed to set up a general rank theory in idempotent semirings are not currently understood. Unlike in vector spaces, there exist finitely generated semimodules $\mathcal{V} \subset (\mathbb{R}_{\max})^n$ of arbitrarily large weak rank, if the dimension of the ambient space n is at least 3; and not all subsemimodules of $(\mathbb{R}_{\max})^n$ are finitely generated, even with $n = 2$.

Example 5 (Cuninghame-Green [15],Th. 16.4). The weak column rank of the $3 \times (i+1)$ matrix

$$A_i = \begin{bmatrix} 0 & 0 & \dots & 0 \\ 0 & 1 & \dots & i \\ 0 & -1 & \dots & -i \end{bmatrix}$$

is equal to $i+1$ for all $i \in \mathbb{N}$. This can be understood geometrically using a representation due to Mairesse. We visualize the set of vectors with finite entries of a semimodule $\mathcal{V} \subset (\mathbb{R}_{\max})^3$ by the subset of $\mathbb{R}^2$, obtained by projecting $\mathcal{V}$ orthogonally, on any plane orthogonal to $(1,1,1)$. Since $\mathcal{V}$ is invariant by multiplication by any scalar λ, i.e. by the usual addition of the vector $(\lambda, \lambda, \lambda)$, the semimodule $\mathcal{V}$ is well determined by its projection. We only loose the points with $\mathbb{0}$ entries which are sent to some infinite end of the $\mathbb{R}^2$ plane. The semimodules $\operatorname{Im} A_1, \operatorname{Im} A_2, \operatorname{Im} A_3$ are shown on Fig 3. The generators are represented by bold points, and the semimodules by gray regions. The broken *line* between any two generators u, v represents $\operatorname{Im}[u,v]$. This picture should make it clear that a weak basis of a subsemimodule of $(\mathbb{R}_{\max})^3$ may have as many generators as a convex set of $\mathbb{R}^2$ may have extremal points. The notion of weak rank is therefore a very coarse one.

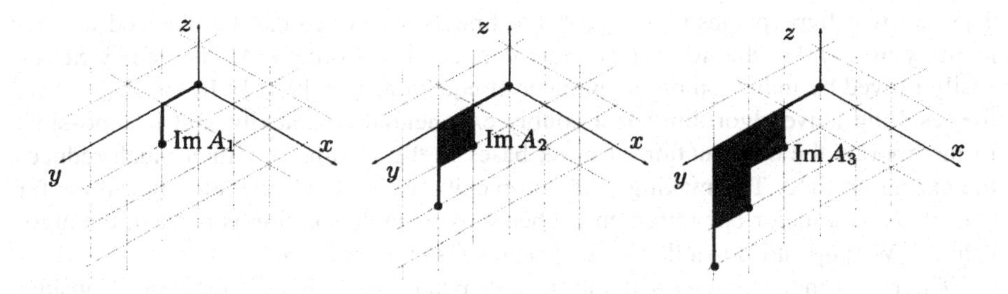

Fig. 3. An infinite ascending chain of semimodules of $(\mathbb{R}_{\max})^3$ (see Ex. 5).

Let $A \in (\mathbb{R}_{\max})^{n \times p}$. A weak basis of the semimodule $\operatorname{Im} A$ can be computed by a greedy algorithm. Let $A[i]$ denote the i-th column of A, and let $A(i)$ denote the $n \times (p-1)$ matrix obtained by deleting column i. We say that column i of A is *redundant* if $A[i] \in \operatorname{Im} A(i)$, which can be checked by Corollary 3. Replacing A by $A(i)$ when $A[i]$ is redundant, we do not change the semimodule $\operatorname{Im} A$. Continuing this process, we terminate in $O(np^2)$ time with a weak basis.

Application 6 (Controllability). The fact that ascending chains of semimodules need not stationnarize yields pathological features in terms of Control. Consider the controlled dynamical system:

$$x(0) = \mathbb{0}, \quad x(k) = Ax(k-1) \oplus Bu(k), \quad k = 1, 2, \ldots \tag{20}$$

where $A \in (\mathbb{R}_{\max})^{n \times n}$, $B \in (\mathbb{R}_{\max})^{n \times q}$, and $u(k) \in (\mathbb{R}_{\max})^q, k = 1, 2, \ldots$ is a sequence of control vectors. Given a state $\xi \in (\mathbb{R}_{\max})^n$, the *accessibility* problem (in time N) asks whether there is a control sequence u such that $x(N) = \xi$. Clearly, ξ is accessible in time N iff it belongs to the image of the *controllability* matrix $C_N = [B, AB, \ldots, A^{N-1}B]$. Corollary 3 allows us to decide the accessibility of ξ. However, unlike in conventional algebra (in which $\operatorname{Im} C_N = \operatorname{Im} C_n$, for any $N \geq n$, thanks to Cayley-Hamilton theorem), the semimodule of accessible states $\operatorname{Im} C_N$ may grow indefinitely as $N \to \infty$.

3.5 Solving $Ax = Bx$ by Elimination

The following theorem is due to Butkovič and Hegedüs [9]. It was rediscovered in [18, Chap. III].

Theorem 7 (Finiteness Theorem). *Let $A, B \in (\mathbb{R}_{\max})^{n \times p}$. The set $\mathcal{V}$ of solutions of the homogeneous system $Ax = Bx$ is a finitely generated semimodule.*

This is a consequence of the following universal elimination result.

Theorem 8 (Elimination of Equalities in Semirings). *Let $\mathcal{S}$ denote an arbitrary semiring. Let $A, B \in \mathcal{S}^{n \times p}$. If for any $q \geq 1$ and any row vectors $a, b \in \mathcal{S}^q$, the* hyperplane *$\{x \in \mathcal{S}^q \mid ax = bx\}$ is a finitely generated semimodule, then $\mathcal{V} = \{x \in \mathcal{S}^p \mid Ax = Bx\}$ is a finitely generated semimodule.*

The fact that hyperplanes of $(\mathbb{R}_{\max})^q$ are finitely generated can be checked by elementary means (but the number of generators can be of order q^2). Theorem 8 can be easily proved by induction on the number of equations (see [9,18]). In the $\mathbb{R}_{\max}$ case, the resulting naive algorithm has a doubly exponential complexity. But it is possible to incorporate the construction of weak bases in the algorithm, which much reduces the execution time. The making (and complexity analysis) of efficient algorithms for $Ax = Bx$ is a major open problem. When only a single solution is needed, the algorithm of Walkup and Borriello (in [24]) seems faster, in practice.

There is a more geometrical way to understand the finiteness theorem. Consider the following correspondence between semimodules of $((\mathbb{R}_{\max})^{1 \times n})^2$ (couples of row vectors) and $(\mathbb{R}_{\max})^{n \times 1}$ (column vectors), respectively:

$$\begin{array}{lcl} \mathcal{W} \subset ((\mathbb{R}_{\max})^{1 \times n})^2 & \longrightarrow & \mathcal{W}^\top = \{x \in (\mathbb{R}_{\max})^{n \times 1} \mid ax = bx, \ \forall (a,b) \in \mathcal{W}\} \ , \\ \mathcal{V}^\perp = \{(a,b) \in ((\mathbb{R}_{\max})^{1 \times n})^2 \mid ax = bx, \ \forall x \in \mathcal{V}\} & \longleftarrow & \mathcal{V} \subset (\mathbb{R}_{\max})^{n \times 1} \ . \end{array} \tag{21}$$

Theorem 7 states that if $\mathcal{W}$ is a finitely generated semimodule (i.e. if all the row vectors $[a, b]$ belong to the row space of a matrix $[A, B]$) then, its orthogonal $\mathcal{W}^\top$ is finitely generated. Conversely, if $\mathcal{V}$ is finitely generated, so does $\mathcal{V}^\perp$ (since the elements (a, b) of $\mathcal{V}^\perp$ are the solutions of a finite system of linear equations). The orthogonal semimodule $\mathcal{V}^\perp$ is exactly the set of *linear equations* $(a, b) : ax = bx$ satisfied by all the $x \in \mathcal{V}$. Is a finitely generated subsemimodule $\mathcal{V} \subset (\mathbb{R}_{\max})^{n \times 1}$ defined by its equations ? The answer is positive [18, Chap. IV,1.2.2]:

Theorem 9 (Duality Theorem). *For all finitely generated semimodules $\mathcal{V} \subset (\mathbb{R}_{\max})^{n \times 1}$, $(\mathcal{V}^\perp)^\top = \mathcal{V}$.*

In general, $(\mathcal{W}^\top)^\perp \supsetneq \mathcal{W}$. The duality theorem is based on the following analogue of the Hahn-Banach theorem, stated in [18]: *if $\mathcal{V} \subset (\mathbb{R}_{\max})^{n \times 1}$ is a finitely generated semimodule, and $y \notin \mathcal{V}$, there exist $(a, b) \in ((\mathbb{R}_{\max})^{1 \times n})^2$ such that $ay \neq by$ and $ax = bx, \ \forall x \in \mathcal{V}$.*

The *kernel* of a linear operator C should be defined as $\ker C = \{(x, y) \mid Cx = Cy\}$. When is the projector on the image of a linear operator B, parallel to $\ker C$, defined? The answer is given in [12].

3.6 Solving $x = Ax \oplus b$ using Rational Calculus

Let $\mathcal{S}$ denote a complete idempotent semiring, and let $A \in \mathcal{S}^{n\times n}, b \in \mathcal{S}^n$. The least solution of $x \succeq Ax \oplus b$ is A^*b, where the star operation is given by:

$$A^* \stackrel{\text{def}}{=} \bigoplus_{n\in\mathbb{N}} A^n \ . \tag{22}$$

Moreover, $x = A^*b$ satisfies the equation $x = Ax \oplus b$. All this is most well known (see e.g. [38]), and we will only insist on the features special to the $(\max, +)$ case. We can interpret A^*_{ij} as the *maximal weight* of a path from i to j of any length, in the graph[2] associated with A. We next characterize the convergence of A^* in $(\mathbb{R}_{\max})^{n\times n}$ (A^* is a priori defined in $(\overline{\mathbb{R}}_{\max})^{n\times n}$, but the $+\infty$ value which breaks the semifield character of $\mathbb{R}_{\max}$ is undesired in most applications). The following fact is standard (see e.g. [2, Theorem 3.20]).

Proposition 10. *Let $A \in (\mathbb{R}_{\max})^{n\times n}$. The entries of A^* belong to $\mathbb{R}_{\max}$ iff there are no circuits with positive weight in the graph[2] of A. Then, $A^* = A^0 \oplus A \oplus \cdots \oplus A^{n-1}$.*

The matrix A^* can be computed in time $O(n^3)$ using classical universal Gauss algorithms (see e.g. [21]). Special algorithms exist for the $(\max, +)$ semiring. For instance, the sequence $x(k) = Ax(k-1) \oplus b$, $x(0) = \mathbb{0}$ stationarizes before step n (with $x(n) = x(n+1) = A^*b$) iff A^*b is finite. This allows us to compute A^*b very simply. A complete account of existing algorithms can be found in [21].

3.7 The $(\max, +)$ Perron-Frobenius Theory

The most ancient, most typical, and probably most useful $(\max, +)$ results are relative to the spectral problem $Ax = \lambda x$. One might argue that 90% of current applications of $(\max, +)$ algebra are based on a complete understanding of the spectral problem. The theory is extremely similar to the well known Perron-Frobenius theory (see e.g. [4]). The $(\max, +)$ case turns out to be very appealing, and slightly more complex than the conventional one (which is not surprising, since the $(\max, +)$ spectral problem is a somehow degenerate limit of the conventional one, see §2.2). The main discrepancy is the existence of two graphs which rule the spectral elements of A, the weighted graph canonically[2] associated with a matrix A, and one of its subgraphs, called *critical* graph.

First, let us import the notion of *irreducibility* from the conventional Perron-Frobenius theory. We say that i *has access* to j if there is a path from i to j in the graph of A, and we write $i \stackrel{*}{\to} j$. The *classes* of A are the equivalence classes for the relation $i\mathcal{R}j \Leftrightarrow (i \stackrel{*}{\to} j$ and $j \stackrel{*}{\to} i)$. A matrix with a single class is *irreducible*. A class $\mathcal{C}$ is upstream $\mathcal{C}'$ (equivalently $\mathcal{C}'$ is downstream $\mathcal{C}$) if a node of $\mathcal{C}$ has access to a node of $\mathcal{C}'$. Classes with no other downstream classes are *final*, classes with no other upstream classes are *initial*.

The following famous $(\max, +)$ result has been proved again and again, with various degrees of generality and precision, see [37,41,15,44,22,2,31].

Theorem 11 ("$(\max,+)$ Perron-Frobenius Theorem"). *An irreducible matrix $A \in (\mathbb{R}_{\max})^{n\times n}$ has a unique eigenvalue, equal to the maximal circuit mean of A:*

$$\rho_{\max}(A) = \bigoplus_{k=1}^{n} \operatorname{tr}(A^k)^{\frac{1}{k}} \quad = \max_{1\le k\le n} \max_{i_1,\dots,i_k} \frac{A_{i_1 i_2} + \cdots + A_{i_k i_1}}{k} \ . \tag{23}$$

We have the following refinements in terms of inequalities [18, Chap IV], [3].

Lemma 12 ("Collatz-Wielandt Properties"). *For any $A \in (\mathbb{R}_{\max})^{n\times n}$,*

$$\rho_{\max}(A) = \max\{\lambda \in \mathbb{R}_{\max} \mid \exists u \in (\mathbb{R}_{\max})^n \setminus \{\mathbb{0}\},\ \ Au \succeq \lambda u\} \ . \tag{24}$$

Moreover, if A is irreducible,

$$\rho_{\max}(A) = \min\{\lambda \in \mathbb{R}_{\max} \mid \exists u \in (\mathbb{R}_{\max})^n \setminus \{\mathbb{0}\},\ \ Au \preceq \lambda u\} \ . \tag{25}$$

The characterization (25) implies in particular that, for an irreducible matrix A, $\rho_{\max}(A)$ is the optimal value of the linear program

$$\min \lambda \text{ s.t. } \forall i,j \ \ A_{ij} + u_j \le u_i + \lambda \ .$$

This was already noticed by Cuninghame-Green [15]. The standard way to compute the maximal circuit mean $\rho_{\max}(A)$ is to use Karp algorithm [27], which runs in time $O(n^3)$. The specialization of Howard algorithm (see e.g. [35]) to deterministic Markov Decision Processes with average reward, yields an algorithm whose average execution time is in practice far below that of Karp algorithm, but no polynomial bound is known for the execution time of Howard algorithm. Howard algorithm is also well adapted to the semi-Markov variants (12).

Unlike in conventional Perron-Frobenius theory, an irreducible matrix may have several (non proportional) eigenvectors. The characterization of the eigenspace uses the notion of *critical graph.* An arc (i,j) is *critical* if it belongs to a circuit $(i_1,\dots,i_k)$ whose mean weight attains the max in (23). Then, the nodes i,j are *critical.* Critical nodes and arcs form the *critical graph.* A *critical class* is a strongly connected component of the critical graph. Let $\mathcal{C}_1^c,\dots,\mathcal{C}_r^c$ denote the critical classes. Let $\tilde{A} = \rho_{\max}^{-1}(A)A$ (i.e. $\tilde{A}_{ij} = -\rho_{\max}(A) + A_{ij}$). Using Proposition 10, the existence of $\tilde{A}^* \ (\stackrel{\text{def}}{=} (\tilde{A})^*)$ is guaranteed. If i is in a critical class, we call the column $\tilde{A}^*_{\cdot,i}$ of $\tilde{A}^*$ *critical.* The following result can be found e.g. in [2,16].

Theorem 13 (Eigenspace). *Let $A \in (\mathbb{R}_{\max})^{n\times n}$ denote an irreducible matrix. The critical columns of $\tilde{A}^*$ span the eigenspace of A. If we select only one column, arbitrarily, per critical class, we obtain a weak basis of the eigenspace.*

Thus, the cardinality of a weak basis is equal to the number of critical classes. For any two i,j within the same critical class, the critical columns $\tilde{A}^*_{\cdot,i}$ and $\tilde{A}^*_{\cdot,j}$ are proportional.

We next show how the eigenvalue $\rho_{\max}(A)$ and the eigenvectors determine the asymptotic behavior of A^k as $k \to \infty$. The *cyclicity* of a critical class $\mathcal{C}_s^c$ is by definition the gcd of the lengths of its circuits. The *cyclicity* c of A is the lcm of the cyclicities of its critical classes. Let us pick arbitrarily an index i_s within each critical class $\mathcal{C}_s^c$, for $s = 1,\dots,r$, and let v_s, w_s denote the column and row of index i_s of $\tilde{A}^*$ (v_s, w_s are right and left eigenvectors of A, respectively). The following result follows from [2].

Theorem 14 (Cyclicity). *Let $A \in (\mathbb{R}_{\max})^{n\times n}$ be an irreducible matrix. There is an integer K_0 such that*

$$k \geq K_0 \implies A^{k+c} = \rho_{\max}(A)^c A^k \ , \tag{26}$$

where c is the cyclicity of A. Moreover, if $c = 1$,

$$k \geq K_0 \implies A^k = \rho_{\max}(A)^k P, \ \text{ where } P = \bigoplus_{s=1}^{r} v_s w_s \ . \tag{27}$$

The matrix P which satisfies $P^2 = P$, $AP = PA = \rho_{\max}(A)P$ is called the *spectral projector* of A. The cyclicity theorem, which writes $A_{ij}^{k+c} = \rho_{\max}(A) \times c + A_{ij}^k$ in conventional algebra, implies that $A^k x$ grows as $k \times \rho_{\max}(A)$, independently of $x \in (\mathbb{R}_{\max})^n$, and that a periodic regime is attained in finite time. The limit behavior is known a priori. Ultimately, the sequence $\rho_{\max}(A)^{-k}A^k$ visits periodically c accumulation points, which are $Q, AQ, \ldots, A^{c-1}Q$, where Q is the spectral projector of A^c. The length of the transient behavior K_0 can be arbitrarily large. In terms of Markov Decision, Theorem 14 says that optimal long trajectories stay almost all the time on the critical graph (Turnpike theorem). Theorem 14 is illustrated in Fig. 4, which shows the images of a cat (a region of the $\mathbb{R}^2$ plane) by the iterates of A (A, A^2, A^3, etc.), B and C, where

$$A = \begin{bmatrix} 0 & 0 \\ 0 & 2 \end{bmatrix}, \ B = \begin{bmatrix} 2 & 0 \\ 0 & 2 \end{bmatrix}, \ C = \begin{bmatrix} 0 & 2 \\ 2 & 0 \end{bmatrix} \ . \tag{28}$$

We have $\rho_{\max}(A) = 2$. Since A has a unique critical circuit, the spectral projector P is rank one (its column and row spaces are lines). We find that $\tilde{A}^2 = P$: every point of the plane is sent in at most two steps to the eigenline $y = 2 \otimes x = 2 + x$, then it is translated by $(2, 2)$ at each step. Similar interpretations exist for B and C.

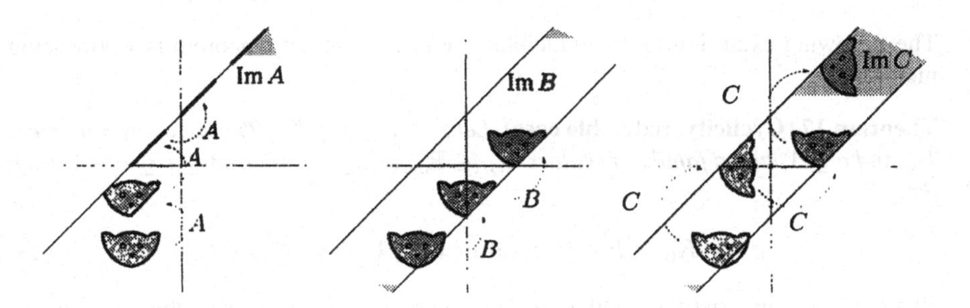

Fig. 4. A cat in a (max, +) dynamics (see (28))

Let us now consider a reducible matrix A. Given a class $\mathcal{C}$, we denote by $\rho_{\max}(\mathcal{C})$ the (max, +) eigenvalue of the restriction of the matrix A to $\mathcal{C}$. The *support* of a vector u is the set $\operatorname{supp} u = \{i \mid u_i \neq \mathbb{0}\}$. A set of nodes S is *closed* if $j \in S, i \xrightarrow{*} j$ implies $i \in S$. We say that a class $\mathcal{C} \subset S$ is *final in* S if there is no other downstream class in S.

Theorem 15 (Spectrum of reducible matrices). *A matrix $A \in (\mathbb{R}_{\max})^{n\times n}$ has an eigenvector with support $S \subset \{1,\dots,n\}$ and eigenvalue λ iff S is closed, λ is equal to $\rho_{\max}(\mathcal{C})$ for any class $\mathcal{C}$ that is final in S, and $\lambda \succeq \rho_{\max}(\mathcal{C}')$ for any other class $\mathcal{C}'$ in S.*

The proof can be found in [43,18]. See also [3]. In particular, eigenvalues of initial classes are automatically eigenvalues of A. The maximal circuit mean $\rho_{\max}(A)$ (given by (23)) is also automatically an eigenvalue of A (but the associated eigenvector need not be finite). A weak basis of the eigenspace is given in [18, Chap. IV,1.3.4].

Example 16 (Taxicab eigenproblem). The matrix of the taxicab MDP, shown in Fig 1, has 2 classes, namely $\mathcal{C}_1 = \{c_1, a, c_2\}, \mathcal{C}_2 = \{c_3\}$. Since $\rho_{\max}(\mathcal{C}_2) = 2 \prec \rho_{\max}(\mathcal{C}_1) = 5$, there are no finite eigenvectors (which have support $S = \mathcal{C}_1 \cup \mathcal{C}_2$). The only other closed set is $S = \mathcal{C}_1$, which is initial. Thus $\rho_{\max}(A) = \rho_{\max}(\mathcal{C}_1) = 5$ is the only eigenvalue of A. Let A' denote the restriction of A to $\mathcal{C}_1$. There are two critical circuits (c_1) and (a, c_2), and thus two critical classes $\mathcal{C}_1^c = \{c_1\}, \mathcal{C}_2^c = \{a, c_2\}$. A weak basis of the eigenspace of A' is given by the columns c_1 and (e.g.) c_2 of

$$(\tilde{A}')^* = \begin{array}{c} \\ c_1 \\ a \\ c_2 \end{array} \begin{array}{c} \begin{array}{ccc} c_1 & a & c_2 \end{array} \\ \begin{pmatrix} 0 & -1 & 0 \\ -1 & 0 & 1 \\ -2 & -1 & 0 \end{pmatrix} \end{array}$$

Completing these two columns by a $\mathbb{0}$ in row 4, we obtain a basis of the eigenspace of A. The non existence of a finite eigenvector is obvious in terms of control. If such an eigenvector existed, by Fact 2, the optimal reward of the taxicab would be independent of the starting point. But, if the taxi driver starts from City 3, he remains blocked there with an income of 2 \$ per journey, whereas if he starts from any other node, he should clearly either run indefinitely in City 1, either shuttle from the airport to City 2, with an average income of 5 \$ per journey (these two policies can be obtained by applying Fact 2 to the MDP restricted to $\mathcal{C}_1$, taking the two above eigenvectors).

The following extension to the reducible case of the cyclicity theorem is worth being mentioned.

Theorem 17 (Cyclicity, reducible case). *Let $A \in (\mathbb{R}_{\max})^{n\times n}$. There exist two integers K_0 and $c \geq 1$, and a family of scalars $\lambda_{ijl} \in \mathbb{R}_{\max}$, $1 \leq i,j \leq n$, $0 \leq l \leq c-1$, such that*

$$k \geq K_0, \quad k \equiv l \mod c \implies A_{ij}^{k+c} = \lambda_{ijl}^c A_{ij}^k \ , \tag{29}$$

Characterizations exist for c and λ_{ijl}. The scalars λ_{ijl} are taken from the set of eigenvalues of the classes of A. If i, j belong to the same class $\mathcal{C}$, $\lambda_{ijl} = \rho_{\max}(\mathcal{C})$ for all l. If i, j do not belong to the same class, the theorem implies that the sequence $\frac{1}{k} \times A_{ij}^k$ may have distinct accumulation points, according to the congruence of k modulo c (see [18, Chap. VI,1.1.10]).

The cyclicity theorems for matrices are essentially equivalent to the characterization of rational series in one indeterminate with coefficient in $\mathbb{R}_{\max}$, as a merge of ultimately geometric series, see the paper of Gaubert in [13] and [28]. Transfer series and rational algebra techniques are particularly powerful for Discrete Event Systems. Timed Event Graphs can be represented by a remarkable (quotient) semiring of series with Boolean coefficients, in two commuting variables, called $\mathcal{M}_{\min}^{\max}[[\gamma,\delta]]$ (see [2, Chap. 5]). The indeterminates γ and δ have natural interpretations as *shifts* in dating and counting. The complete behavior of the system can be represented by simple —often small— commutative rational expressions [2],[18, Chap. VII–IX] (see also [28] in a more general context).

3.8 Symmetrization of the $(\max,+)$ Semiring

Let us try to imitate the familiar construction of $\mathbb{Z}$ from $\mathbb{N}$, for an arbitrary semiring $\mathcal{S}$. We build the set of couples $\mathcal{S}^2$, equipped with (componentwise) sum $(x',x'')\oplus(y',y'')=(x'\oplus y',x''\oplus y'')$, and product $(x',x'')\otimes(y',y'')=(x'y'\oplus x''y'',x'y''\oplus x''y')$. We introduce the *balance* relation

$$(x',x'')\nabla(y',y'')\iff x'\oplus y''=x''\oplus y' \ .$$

We have $\mathbb{Z}=\mathbb{N}^2/\nabla$, but for an idempotent semiring $\mathcal{S}$, the procedure stops, since ∇ is not transitive (e.g. $(\mathbb{1},\mathbb{0})\nabla(\mathbb{1},\mathbb{1})\nabla(\mathbb{0},\mathbb{1})$, but $(\mathbb{1},\mathbb{0})\not\nabla(\mathbb{0},\mathbb{1})$). If we renounce to quotient $\mathcal{S}^2$, we may still manipulate couples, with the $\ominus$ operation $\ominus(x',x'')=(x'',x')$. Indeed, since $\ominus$ satisfies the sign rules $\ominus\ominus x=x$, $\ominus(x\oplus y)=(\ominus x)\oplus(\ominus y)$, $\ominus(xy)=(\ominus x)y=x(\ominus y)$, and since $x\nabla y\iff x\ominus y\nabla\mathbb{0}$ (we set $x\ominus y\stackrel{\text{def}}{=}x\oplus(\ominus y)$), it is not difficult to see that *all the familiar identities valid in rings admit analogues in* $\mathcal{S}^2$*, replacing equalities by balances*. For instance, if $\mathcal{S}$ is commutative, we have for all matrices (of compatible size) with entries in $\mathcal{S}^2$ (determinants are defined as usual, with $\ominus$ instead of $-$):

$$\det(AB)\ \nabla\ \det A\ \det B, \tag{30}$$

$$P_A(A)\ \nabla\ \mathbb{0}\qquad\text{where } P_A(\lambda)=\det(A\ominus\lambda\mathrm{Id})\ \text{(Cayley Hamilton)}. \tag{31}$$

Eqn 30 can be written directly in $\mathcal{S}$, introducing the positive and negative determinants $\det^+A=\bigoplus_{\sigma\text{ even}}\bigotimes_{1\le i\le n}A_{i\sigma(i)}$, $\det^-A=\bigoplus_{\sigma\text{ odd}}\bigotimes_{1\le i\le n}A_{i\sigma(i)}$ (the sums are taken over even and odd permutations of $\{1,\dots,n\}$, respectively). The balance (30) is equivalent to the ordinary equality $\det^+AB\oplus\det^+A\ \det^-B\oplus\det^-A\ \det^+B=\det^-AB\oplus\det^+A\ \det^+B\oplus\det^-A\ \det^-B$, but (30) is certainly more adapted to computations. Such identities can be proved combinatorially (showing a bijection between terms on both sides), or derived automatically from their ring analogues using a simple argument due to Reutenauer and Straubing [36, Proof of Lemma 2] (see also the *transfer principle* in [18, Ch. I]).

But in the $\mathbb{R}_{\max}$ case, one can do much better. Consider the following application of the Cayley-Hamilton theorem:

$$A=\begin{bmatrix}1&3\\4&1\end{bmatrix},\quad A^2\ominus\mathrm{tr}\,(A)A\oplus\det A\nabla\mathbb{0},\ \text{i.e}\qquad A^2\oplus2\mathrm{Id}=1A\oplus7\mathrm{Id}\ .$$

Obviously, we may eliminate the 2Id term which will never saturate the identity (since $2<7$), and obtain $A^2=1A\oplus7\mathrm{Id}$. Thus, to some extent $7\ominus2=7$. This can be formalized by introducing the congruence of semiring:

$$(x',x'')\ \mathcal{R}\ (y',y'') \Leftrightarrow (x' \neq x'', y' \neq y'' \text{ and } x'\oplus y'' = x''\oplus y') \text{ or } (x',x'') = (y',y'').$$

The operations $\oplus, \ominus, \otimes$ and the relation ∇ are defined canonically on the quotient semiring, $\mathbb{S}_{\max} = \mathbb{R}^2_{\max}/\mathcal{R}$, which is called the *symmetrized semiring* of $\mathbb{R}_{\max}$. This symmetrization was invented independently by G. Hegedüs [26] and M. Plus [34].

In $\mathbb{S}_{\max}$, there are three kinds of equivalence classes; classes with an element of the form $(a, \mathbb{0})$, identified to $a \in \mathbb{R}_{\max}$, and called *positive*, classes with an element of the form $(\mathbb{0}, a)$ denoted $\ominus a$, called *negative*, classes with a single element (a, a), denoted $a^\bullet$ and called *balanced*, since $a^\bullet \nabla \mathbb{0}$ (for $a = \mathbb{0}$, the three above classes coincide, we will consider $\mathbb{0}$ as both a positive, negative, and balanced element).

We have the decomposition of $\mathbb{S}_{\max}$ in sets of positive, negative, and balanced elements, respectively

$$\mathbb{S}_{\max} = \mathbb{S}^{\oplus}_{\max} \cup \mathbb{S}^{\ominus}_{\max} \cup \mathbb{S}^{\bullet}_{\max}\ .$$

This should be compared with $\mathbb{Z} = \mathbb{Z}^+ \cup \mathbb{Z}^- \cup \{0\}$. For instance, $3\ominus 2 = 3, 2\ominus 3 = \ominus 3$, but $3 \ominus 3 = 3^\bullet$. We say that an element is *signed* if it is positive or negative.

Obviously, if a system $Ax = b$ has a solution, the balance $Ax\nabla b$ has a solution. Conversely if $Ax\nabla b$ has a positive solution x, and if A, b are positive, it is not difficult to see that $Ax = b$. It remains to solve systems of linear balances. The main difficulty is that the balance relation is not transitive. As a result, $x\nabla a$ and $cx\nabla b$ do not imply $ca\nabla b$. However, when x is signed, the implication is true. This allows us to solve linear systems of balances by elimination, when the unknowns are signed.

Theorem 18 (Cramer Formulæ). *Let $A \in (\mathbb{S}_{\max})^{n\times n}$, and $b \in (\mathbb{S}_{\max})^n$. Every signed solution of $Ax\nabla b$ satisfies the Cramer condition $Dx_i\nabla D_i$, where D is the determinant of A and D_i is the i-th Cramer determinant[5]. Conversely, if D_i is signed for all i, and if D is signed and nonzero, then $x = (D^{-1}D_i)_{1\leq i\leq n}$ is the unique signed solution.*

The proof can be found in [34,2]. For the homogeneous system of n linear equations with n unknowns, $Ax\nabla\mathbb{0}$ has a signed non zero solution iff $\det A\nabla\mathbb{0}$ (see [34,18]), which extends a result of Gondran and Minoux (see [22]).

Example 19. Let us solve the taxicab eigenproblem $Ax = 5x$ by elimination in $\mathbb{S}_{\max}$ (A is the matrix shown in Fig 1). We have

$$5^\bullet x_1 \oplus 4x_2 \oplus 7x_4\ \nabla\ \mathbb{0} \tag{32a}$$
$$4x_1 \ominus 5x_2 \oplus 6x_3 \oplus 3x_4\ \nabla\ \mathbb{0} \tag{32b}$$
$$4x_2 \ominus 5x_3\ \nabla\ \mathbb{0} \tag{32c}$$
$$\ominus 5x_4\ \nabla\ \mathbb{0}\ . \tag{32d}$$

The only signed solution of (32d) is $x_4 = \mathbb{0}$. By homogeneity, let us look for the solutions such that $x_3 = 0$. Then, using (32c), we get $4x_2\nabla 5x_3 = 5$. Since we search a positive x_2, the balance can be replaced by an equality. Thus $x_2 = 1$. It remains to rewrite (32a),(32b): $5^\bullet x_1\nabla\ominus 5$, $4x_1\nabla 6^\bullet$, which is true for x_1 positive iff $0 \leq x_1 \leq 2$. The two extremal values give (up to a proportionality factor) the basis eigenvectors already computed in Ex. 19.

[5] Obtained by replacing the i-th column of A by b.

Determinants are not so easy to compute in $\mathbb{S}_{\max}$. Butkovič [8] showed that the computation of the determinant of a matrix with positive entries is polynomially equivalent (we have to solve an assignment problem) to the research of an even cycle in a (directed) graph, a problem which is not known to be polynomial. We do not know a non naive algorithm to compute the minor rank (=size of a maximal submatrix with unbalanced determinant) of a matrix in $(\mathbb{R}_{\max})^{n\times p}$. The situation is extremely strange: we have excellent polynomial iterative algorithms [34,18] to find a signed solution of the square system $Ax\nabla b$ when $\det A \neq \mathbb{0}$, but we do not have polynomial algorithms to decide whether $Ax\nabla\mathbb{0}$ has a signed non zero solution (such algorithms would allow us to compute $\det A$ in polynomial time). Moreover, the theory partly collapses if one considers rectangular systems instead of square ones. The conditions of compatibility of $Ax\nabla\mathbb{0}$ when A is rectangular cannot be expressed in terms of determinants [18, Chap. III, 4.2.6].

Historical and Bibliographical Notes

The $(\max, +)$ algebra is not classical yet, but many researchers have worked on it (we counted at least 80), and it is difficult to make a short history without forgetting important references. We will just mention here main sources of inspiration. The first use of the $(\max, +)$ semiring can be traced back at least to the late fifties, and the theory grew in the sixties, with works of Cuninghame-Green, Vorobyev, Romanovskiĭ, and more generally of the Operations Research community (on path algebra). The first enterprise of systematic study of this algebra seems to be the seminal "Minimax algebra" of Cuninghame-Green [15]. A chapter on dioids can be found in Gondran et Minoux [21]. The theory of linear independence using bideterminants, which is the ancester of symmetrization, was initiated by Gondran and Minoux (following Kuntzmann). See [22]. The last chapter of "Operatorial Methods" of Maslov [32] inaugurated the $(\max, +)$ operator and measure theory (motivated by semiclassical asymptotics). There is an "extremal algebra" tradition, mostly in East Europe, oriented towards algorithms and computational complexity. Results in this spirit can be found in the book of U. Zimmermann [44]. This tradition has been pursued, e.g. by Butkovič [7]. The *incline algebras* introduced by Cao, Kim and Roush [10] are idempotent semirings in which $a \oplus ab = a$. The tropical semiring was invented by Simon [39]. A number of language and semigroup oriented contributions are due to the tropical school (Simon, Hashiguchi, Mascle, Leung, Pin, Krob, Weber, ...). See the survey of Pin in [24], [40,25,29,28], and the references therein. Since the beginning of the eighties, Discrete Event Systems, which were previously considered by distinct communities (queuing networks, scheduling, ...), have been gathered into a common algebraic frame. "Synchronization and Linearity" by Baccelli, Cohen, Olsder, Quadrat [2] gives a comprehensive account of deterministic and stochastic (max,+) linear discrete event systems, together with recent algebraic results (such as symmetrization). Another recent text is the collection of articles edited by Maslov and Samborskiĭ [31] which is only the most visible part of the (considerable) work of the Idempotent Analysis school. A theory of probabilities in $(\max, +)$ algebra motivated by dynamic programming and large deviations, has been developed by Akian, Quadrat and Viot; and by Del Moral and Salut (see [24]). Recently, the $(\max, +)$ semiring has attracted attention from the linear algebra community (Bapat, Stanford, van den Driessche [3]). A survey with a very complete bibliography is the article of Maslov and Litvinov in [24]. Let us also mention the forthcoming book of Kolokoltsov and Maslov (an earlier version is in Russian [30]). The collection of articles edited by Gunawardena [24] will probably give the first fairly global overview of the different traditions on the subject.

References

1. M. Akian, R.B. Bapat, and S. Gaubert. Asymptotics of the Perron eigenvalue and eigenvector using max algebra. in preparation, 1996.
2. F. Baccelli, G. Cohen, G.J. Olsder, and J.P. Quadrat. *Synchronization and Linearity*. Wiley, 1992.
3. R.B. Bapat, D. Stanford, and P. van den Driessche. Pattern properties and spectral inequalities in max algebra. *SIAM Journal of Matrix Analysis and Applications*, 16(3):964–976, 1995.
4. A. Berman and R.J. Plemmons. *Nonnegative matrices in the mathematical sciences*. Academic Press, 1979.
5. T.S. Blyth and M.F. Janowitz. *Residuation Theory*. Pergamon press, 1972.
6. M. Brilman and J.M. Vincent. Synchronisation by resources sharing : a performance analysis. Technical report, MAI-IMAG, Grenoble, France, 1995.
7. P. Butkovič. Strong regularity of matrices — a survey of results. *Discrete Applied Mathematics*, 48:45–68, 1994.
8. P. Butkovič. Regularity of matrices in min-algebra and its time-complexity. *Discrete Applied Mathematics*, 57:121–132, 1995.
9. P. Butkovič and G. Hegedüs. An elimination method for finding all solutions of the system of linear equations over an extremal algebra. *Ekonomicko-matematicky Obzor*, 20, 1984.
10. Z.Q. Cao, K.H. Kim, and F.W. Roush. *Incline algebra and applications*. Ellis Horwood, 1984.
11. G. Cohen, S. Gaubert, and J.P. Quadrat. Asymptotic throughput of continuous timed petri nets. In *Proceedings of the 34th Conference on Decision and Control*, New Orleans, Dec 1995.
12. G. Cohen, S. Gaubert, and J.P Quadrat. Kernels, images and projections in dioids. In *Proceedings of WODES'96*, Edinburgh, August 1996. IEE.
13. G. Cohen and J.-P. Quadrat, editors. *11th International Conference on Analysis and Optimization of Systems, Discrete Event Systems*. Number 199 in Lect. Notes. in Control and Inf. Sci. Springer Verlag, 1994.
14. J.E. Cohen. Subadditivity, generalized products of random matrices and operations research. *SIAM Review*, 30:69–86, 1988.
15. R.A. Cuninghame-Green. *Minimax Algebra*. Number 166 in Lecture notes in Economics and Mathematical Systems. Springer, 1979.
16. R.A Cuninghame-Green. Minimax algebra and applications. *Advances in Imaging and Electron Physics*, 90, 1995.
17. A. Dembo and O. Zeitouni. *Large Deviation Techniques and Applications*. Jones and Barlett, 1993.
18. S. Gaubert. *Théorie des systèmes linéaires dans les dioïdes*. Thèse, École des Mines de Paris, July 1992.
19. S. Gaubert. Performance evaluation of (max,+) automata. *IEEE Trans. on Automatic Control*, 40(12), Dec 1995.

20. S. Gaubert. On the Burnside problem for semigroups of matrices in the (max,+) algebra. *Semigroup Forum*, 52:271–292, 1996.
21. M. Gondran and M. Minoux. *Graphes et algorithmes*. Eyrolles, Paris, 1979. Engl. transl. *Graphs and Algorithms*, Wiley, 1984.
22. M. Gondran and M. Minoux. Linear algebra in dioids: a survey of recent results. *Annals of Discrete Mathematics*, 19:147–164, 1984.
23. J. Gunawardena. Min-max functions. *Discrete Event Dynamic Systems*, 4:377–406, 1994.
24. J. Gunawardena, editor. *Idempotency*. Publications of the Newton Institute. Cambridge University Press, to appear in 1996 or 1997.
25. K. Hashiguchi. Improved limitedness theorems on finite automata with distance functions. *Theoret. Comput. Sci.*, 72:27–38, 1990.
26. G. Hegedüs. Az extremális sajátvektorok meghatározása eliminációs módszerrel — az általánosított Warshall algorithmus egy javítása. *Alkalmazott Matematikai Lapok*, 11:399–408, 1985.
27. R.M. Karp. A characterization of the minimum mean-cycle in a digraph. *Discrete Maths.*, 23:309–311, 1978.
28. D. Krob and A. Bonnier-Rigny. A complete system of identities for one letter rational expressions with multiplicities in the tropical semiring. *J. Pure Appl. Algebra*, 134:27–50, 1994.
29. H. Leung. Limitedness theorem on finite automata with distance function: an algebraic proof. *Theoret. Comput. Sci*, 81:137–145, 1991.
30. V. Maslov and V. Kolokoltsov. *Idempotent analysis and its applications to optimal control theory*. Nauka, Moskow, 1994. In Russian.
31. V. Maslov and S. Samborskiĭ, editors. *Idempotent analysis*, volume 13 of *Adv. in Sov. Math.* AMS, RI, 1992.
32. V.P. Maslov. *Méthodes Operatorielles*. Mir, Moscou, 1973. French Transl. 1987.
33. P. Moller. *Théorie algébrique des Systèmes à Événements Discrets*. Thèse, École des Mines de Paris, 1988.
34. M. Plus. Linear systems in $(\max,+)$-algebra. In *Proceedings of the 29th Conference on Decision and Control*, Honolulu, Dec. 1990.
35. M.L. Puterman. Markov decision processes. *Handbook in operations research and management science*, 2:331–434, 1990.
36. C. Reutenauer and H. Straubing. Inversion of matrices over a commutative semiring. *J. Algebra*, 88(2):350–360, June 1984.
37. I.V. Romanovskiĭ. Optimization and stationary control of discrete deterministic process in dynamic programming. *Kibernetika*, 2:66–78, 1967. Engl. transl. in Cybernetics 3 (1967).
38. A. Salomaa and M. Soittola. *Automata Theoretic Aspects of Formal Powers Series*. Springer, New York, 1978.
39. I. Simon. Limited subsets of the free monoid. In *Proc. of the 19th Annual Symposium on Foundations of Computer Science*, pages 143–150. IEEE, 1978.
40. I. Simon. On semigroups of matrices over the tropical semiring. *Theor. Infor. and Appl.*, 28(3-4):277–294, 1994.
41. N.N. Vorobyev. Extremal algebra of positive matrices. *Elektron. Informationsverarbeitung und Kybernetik*, 3, 1967. in Russian.
42. E. Wagneur. Moduloids and pseudomodules. 1. dimension theory. *Discrete Math.*, 98:57–73, 1991.
43. Chen Wende, Qi Xiangdong, and Deng Shuhui. The eigen-problem and period analysis of the discrete event systems. *Systems Science and Mathematical Sciences*, 3(3), August 1990.
44. U. Zimmermann. *Linear and Combinatorial Optimization in Ordered Algebraic Structures*. North Holland, 1981.

Regular Expressions and Context-Free Grammars for Picture Languages

Oliver Matz

Institut für Informatik und Praktische Mathematik
Christian-Albrechts-Univertität, Olshausenstraße 40, D-24098 Kiel
oma@informatik.uni-kiel.de

Abstract. We introduce a new concept of regular expression and context-free grammar for picture languages (sets of matrices over a finite alphabet) and compare and connect these two formalisms.
Keywords: formal languages, pictures languages, grammars, regular expressions.

1 Introduction

Many attempts have been made to generalize the definitions and results of the theory of formal word languages to other, more complex objects than words, e.g. traces, graphs, and trees. One possible generalization are *pictures* (matrices over a finite alphabet, e.g. two-dimensional words). Sets of pictures will be called *picture languages* or simply *languages.* Picture languages have been investigated by many authors; a comprehensive survey is [GR96]; another collection of references can be found for example in [Sir87].

It is a not finished subject to transfer results of the theory of word languages to picture languages. Here we present a new approach based on regular expressions and context-free grammars.

In [GRST96] the concept of *tiling systems* as a device for recognizing picture languages is investigated. Tiling systems are a possible generalization of the concept of finite automata for word languages. In [GRST96] it is shown that their expressive power is equal to that of formulas of existential monadic second order theory over pictures, so the close relation of automata and logic carries over from the theory of word languages to picture languages and gives a somewhat "robust" class of picture languages – the *recognizable picture languages.*

Regular expressions are a device for the definition of word languages that is more difficult to transfer to picture languages. A straightforward adaption of regular expressions to pictures — we will call them "simple" — is studied in [GR92,GRST96,GR96]. Simple expressions use two partial concatenations named row- and column concatenation (denoted by $\ominus$ and $\oplus$), which put their arguments vertically above each other (horizontally next to each other, resp.), provided they have the same width (height, resp.). Additionally, both of these concatenations may be iterated. But the expressive power of such expressions is much weaker than that of tiling systems, so the Kleene theorem cannot be carried over to the theory of picture languages.

Reischuk, Morvan (Eds.): STACS'97 Proceedings, LNCS 1200

In this paper we suggest a more powerful type of regular expressions for picture languages, the so-called *regular expressions with operators*. Inside these expressions we allow the Kleene star to range over more complex combinations of juxtapositions, unions, and even intersections of concatenations.

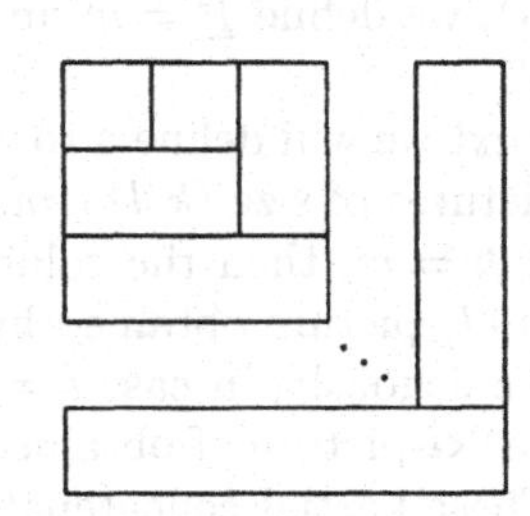

As an informal example, let us assume we have two expressions r and s that generate all columns, i.e. pictures of width 1, and all rows, i.e. pictures of height 1, resp. We consider the column-concatenation with r (let us denote it by $(\oplus r)$) as an individual object. It will enlarge a picture that it is applied to by one column. The row-concatenation with s (denoted $(\ominus s)$) will enlarge its argument by one row. Now, if we allow a Kleene star to iterate the juxtaposition of these two so-called operators, we get another operator, $((\oplus r)(\ominus s))^*$, which enlarges its argument alternatingly by one row and one column a finite number of times. If this operator is applied to the expression generating all 1×1-squares, we obtain a regular expression with operators that generates all squares, as illustrated by the figure.

We will distinguish different classes of expressions with operators, depending on how operators may be constructed. Given appropriate constraints, these expressions do not exceed the expressive power of tiling systems; but they are more powerful than simple regular expressions (the set of squares is not definable by a simple regular expression, see [GR92,GRST96,GR96]). For one particular class of expressions with operators it remains open whether it exhausts the class of recognizable languages.

We will also try to transfer the concept of context-free grammar from word languages to picture languages. Our approach for this differs very much from the one in [Sir87]. In our grammars, sentential forms are terms built by the two binary concatenation symbols $\ominus$, $\oplus$ as well as terminals and non-terminals, which are used as constant symbols. A rule has a non-terminal on the left hand side and a sentential form on the right hand side. The derivation proceeds as follows: One starts with the start symbol and replaces repeatedly a non-terminal A by the right hand side of an A-rule until a sentential form is reached that consists entirely of terminal symbols. If this can be evaluated to a picture, this picture is generated.

We characterize the expressive power of certain regular expressions with operators by context-free grammars with a certain constraint concerning the way recursion is allowed. This result corresponds in a way to the classical equivalence of regular word expressions to right-linear word grammars. This is why we think that our definition of "regular picture languages" by expressions with operators gives another natural and robust class of picture languages that is worth studying.

More detailed proofs can be found in [Mat95].

2 Basic Notions

Throughout the paper we consider a fixed finite alphabet Σ. A *picture* over Σ of *size* (m,n) (where $m,n \geq 1$) is a $m \times n$-matrix over Σ. For a picture P of size (m,n), we define $\overline{P} = m$ and $|P| = n$. We denote the set of all pictures over Σ by Σ^+_+.

Next we will define a row- and a column-concatenation for pictures. Let P, Q be pictures of size (k,l), (m,n) respectively.

If $k = m$, then the column concatenation $P \oplus Q$ of the two pictures is the $k \times (n+l)$-picture obtained by appending Q to the right of P.

Analogously, in case $l = n$ their row concatenation $P \ominus Q$ is defined as the $(k+m) \times l$-picture of obtained by appending Q to the bottom of P.

These partial concatenations can be extended to languages as usual, i.e. for $L, M \subseteq \Sigma^+_+$ we define $L \oplus M = \{P \oplus Q \mid P \in L, Q \in M\}$. These concatenations can be iterated: For a language $L \subseteq \Sigma^+_+$ we set $L^{\oplus 1} := L$ and $L^{\oplus(i+1)} := L^{\oplus i} \oplus L$. Now the *column closure* of L is defined as $L^{\oplus +} := \bigcup_{i \geq 1} L^{\oplus i}$. The *row closure* is defined analogously.

If no ambiguity arises, we denote the column concatenation $L \oplus M$ of two languages L, M by $(L\,M)$, and similarly use $\begin{pmatrix} L \\ M \end{pmatrix}$ instead of $L \ominus M$. The iterated column concatenation may be written as L^i and L^+ instead of $L^{\oplus i}$ and $L^{\oplus +}$, resp., whereas L_i and L_+ denote $L^{\ominus i}$ and $L^{\ominus +}$, resp. The latter notion will only be used in case no conflict with indices occurs.

3 Regular Expressions with Operators

The set $\cap$-$REG(\Sigma)$ of *simple regular expressions over* Σ with typical element r is defined by the following BNF-style rules:

$$r ::= a \mid (r_1 \cup r_2) \mid (r_1 \ominus r_2) \mid (r_1 \oplus r_2) \mid (r_1 \cap r_2) \mid r^{\oplus +} \mid r^{\ominus +}$$

Here a stands for an arbitrary letter from Σ. The language generated by such an expression is defined in a straightforward way: For all $a \in \Sigma$ let $\mathcal{L}(a) = \{a\}$ (the singleton of the 1×1-picture a), and for two expressions r and s we define $\mathcal{L}(r \ominus s) = \mathcal{L}(r) \ominus \mathcal{L}(s)$ an so on. The subset of $\cap$-$REG(\Sigma)$ of *monotonic* expressions (i.e. expressions without intersection symbol) will be denoted by $REG(\Sigma)$. The classes of languages definable by such expressions will be denoted by the corresponding calligraphic notations $\cap$-$\mathcal{REG}(\Sigma)$ and $\mathcal{REG}(\Sigma)$.

In these and similar cases we will omit the explicit mentioning of Σ if possible. We will omit brackets inside expressions following the usual conventions, i.e. $^{\oplus +}$ and $^{\ominus +}$ bind stronger than concatenation symbols, which bind in turn stronger than union and intersection symbols.

Example 1. Consider the language L that consists of the set of all pictures such that there is one row and one column (both not at the border) that hold b's and the remainder of the pictures is filled with a's. L is generated by the expression

$$\begin{pmatrix} a_+^+ \\ b^+ \\ a_+^+ \end{pmatrix} b_+ \begin{pmatrix} a_+^+ \\ b^+ \\ a_+^+ \end{pmatrix} \cap \begin{pmatrix} (a_+^+ b_+ a_+^+) \\ b^+ \\ (a_+^+ b_+ a_+^+) \end{pmatrix}.$$

The fact that the set of squares over a one-letter alphabet is not generated by a simple expression is an immediate consequence of the following characterization of the class $\cap$-$\mathcal{REG}(\Sigma)$.

Theorem 2. *(see [Mat95].) A language $L \subseteq \{a\}_+^+$ is in $\cap$-$\mathcal{REG}$ iff it is in $\mathcal{REG}$ iff the set $\{(m,n) \mid a_n^m \in L\}$ is a finite union of Cartesian products of ultimately periodic subsets of $\mathbb{N}_{\geq 1}$.*

The above theorem is an analogue to the known fact that a set N of integers is ultimately periodic iff the set $\{a^n \mid n \in N\}$ is a regular word language. (Note that this theorem remains true even when we allow also complementation symbols inside regular expressions.)

Theorem 2 shows that the expressive power of *REG* is very limited. But note that on the other hand any picture language that is recognizable by tiling systems in the sense of [GRST96,GR92] is the projection of a picture language that is generated by an expression from $\cap$-*REG* over a possibly larger alphabet. But we think that the use of intersection and projections disturb in a way the "assembling character" of regular expressions.

That is why we investigate another type of regular expressions whose main idea is to allow the Kleene star to range over more complex juxtapositions of concatenations. Before we give the syntax and semantics of these expressions, let us consider again the example of the introduction. The crucial point was to consider terms such as $(\oplus a_+)$, the concatenation with a column of a's to the right, as individual objects, which may be either applied to expressions or composed with each other by juxtaposition, iteration, and union.

With this intuitive idea of the class of regular expressions with operators, we will give a more formal definition. For the sake of maximal generality we will allow the intersection and union symbols as well, both in "expressions" and in "operators".

Definition 3. The set $\cap$-$REG^{UOP}(\Sigma)$ of regular expressions with typical element r and the set $\cap$-$UOP(\Sigma)$ of unrestricted regular operators with typical element ϱ are defined by the following BNF-style rules:

$$r ::= a \mid (r_1 \ominus r_2) \mid (r_1 \oplus r_2) \mid {r_1}^{\ominus+} \mid {r_1}^{\oplus+} \mid (r_1 \cup r_2) \mid (r_1 \cap r_2) \mid r\varrho$$
$$\varrho ::= (\ominus r) \mid (\oplus r) \mid (r\ominus) \mid (r\oplus) \mid (\varrho_1 \varrho_2) \mid \varrho^* \mid \varrho_1 \cup \varrho_2 \mid \varrho_1 \cap \varrho_2$$

Again, a stands for an arbitrary element from Σ.

As before, we drop the "$\cap$-" prefix in the respective notation to denote the classes of *monotonic* expressions and operators, i.e. ones that do not have intersection symbols in it.

Before we give the semantics of these type of expressions, let us consider another example. The expression $ab((a\oplus)(\oplus b))^*$ describes the set of all words (= pictures with one single row) that result from ab by appending repeatedly one a to the left and one b to the right, i.e. the language $\{a^i b^i \mid i \geq 1\}$. Since this word language is non-regular, we make the disappointing observation that we leave the class of recognizable languages if we allow the iteration in such an unrestricted way. Since our aim was to find suitable extensions of the concept of regular expressions to pictures, we shall put a certain constraint on the way operators may be juxtaposed in order to ensure that the resulting language is recognizable by tiling systems.

The crucial point for this constraint is that we make sure that an operator that "works to the right" (like $(\oplus b)$) is never juxtaposed, united, or intersected with another operator that "works to the left" (like $(a\oplus)$), and similarly top- and bottom-operators are seperated. On the other hand, we allow the combination of, say, bottom and right, so our example regular expression $a((\oplus a_+)(\ominus a^+))^*$ for the set of squares is still o.k. We will call operators that meet this constraint *restricted*. This is put formal by the following definition, in which the decoration symbols r, l, b and t (for "right", "left", "bottom" and "top", resp.) are used in order to indicate in which direction an operator potentially enlarges a picture that it is applied to.

Definition 4. The set $\cap$-$REG^{ROP}(\Sigma)$ (set of expression with restricted operators) with typical element r and the sets $ROP_{\mathfrak{d}}(\Sigma)$ for every $\mathfrak{d} \in \{br, bl, tl, tr\}$ (set of restricted $\cap$-regular operators for direction $\mathfrak{d}$) with typical element $\varrho^{\mathfrak{d}}$ are defined by the following BNF-style rules:

$$
\begin{aligned}
r &::= a \mid (r_1 \ominus r_2) \mid (r_1 \oplus r_2) \mid r^{\ominus +} \mid r^{\oplus +} \mid (r_1 \cup r_2) \mid (r_1 \cap r_2) \mid r\varrho \\
\varrho &::= \varrho^{br} \mid \varrho^{tr} \mid \varrho^{bl} \mid \varrho^{tl} \\
\varrho^{br} &::= (\ominus r) \mid (\oplus r) \mid (\varrho_1^{br} \varrho_2^{br}) \mid \varrho^{br\,*} \mid (\varrho_1^{br} \cup \varrho_2^{br}) \mid (\varrho_1^{br} \cap \varrho_2^{br}) \\
\varrho^{tr} &::= (r\ominus) \mid (\oplus r) \mid (\varrho_1^{tr} \varrho_2^{tr}) \mid \varrho^{tr\,*} \mid (\varrho_1^{tr} \cup \varrho_2^{tr}) \mid (\varrho_1^{tr} \cap \varrho_2^{tr}) \\
\varrho^{bl} &::= (\ominus r) \mid (r\oplus) \mid (\varrho_1^{bl} \varrho_2^{bl}) \mid \varrho^{bl\,*} \mid (\varrho_1^{bl} \cup \varrho_2^{bl}) \mid (\varrho_1^{bl} \cap \varrho_2^{bl}) \\
\varrho^{tl} &::= (r\ominus) \mid (r\oplus) \mid (\varrho_1^{tl} \varrho_2^{tl}) \mid \varrho^{tl\,*} \mid (\varrho_1^{tl} \cup \varrho_2^{tl}) \mid (\varrho_1^{tl} \cap \varrho_2^{tl})
\end{aligned}
$$

Again, a stands for a letter from Σ. The elements in $ROP_{\mathfrak{d}}(\Sigma)$ are called $\mathfrak{d}$*-regular operators* for every *direction* $\mathfrak{d} \in \{br, bl, tl, tr\}$. We also define $\mathfrak{d}$-regular operators for $\mathfrak{d} \in \{r, l, t, b\}$: For any $r \in \cap$-REG^{ROP}, the operator $(\oplus r)$ (or $(r\oplus)$, or $(\ominus r)$, or $(r\ominus)$, resp.) is an r-regular (or l-regular, or b-regular, or t-regular, resp.) elementary operator. For any $\mathfrak{d} \in \{r, l, t, b\}$, more complex $\mathfrak{d}$-regular operators are built by union, intersection, concatenation, or Kleene-star the way as above.

Now the definition of the semantics of expressions and operators is straightforward:

Definition 5. Two functions $[\![]\!] : \cap$-$REG^{UOP} \to 2^{\Sigma_+^+}$ and $[\![]\!] : \cap$-$UOP \to 2^{(\Sigma_+^+ \times \Sigma_+^+)}$ are defined simultaneously by induction over the structure of $\cap$-regular expressions and operators:

For $r, s \in REG^{UOP}$ and $\sigma, \varrho \in UOP$ let:

- $[\![a]\!] = \{a\}$ for all $a \in \Sigma$,
- $[\![(r \cup s)]\!] = [\![r]\!] \cup [\![s]\!]$, (similarly for $\cap$, $\oplus$ or $\ominus$ instead of $\cup$,)
- $[\![r^{\oplus+}]\!] = [\![r]\!]^{\oplus+}$, (similarly for $^{\ominus+}$ instead of $^{\oplus+}$,)
- $[\![r\varrho]\!] = \{R \in \Sigma_+^+ \mid \exists P \in [\![r]\!] : (P, R) \in [\![\varrho]\!]\}$,
- $[\![(\oplus r)]\!] = \{(P, R) \mid P \in \Sigma_+^+ \wedge R \in P \oplus [\![r]\!]\}$,
- $[\![(r\oplus)]\!] = \{(P, R) \mid P \in \Sigma_+^+ \wedge R \in [\![r]\!] \oplus P\}$, (similarly for $\ominus$ instead of $\oplus$,)
- $[\![(\sigma\varrho)]\!] = [\![\sigma]\!] \circ [\![\varrho]\!]$, where $\circ$ denotes the usual relational product,
- $[\![(\sigma \cup \varrho)]\!] = [\![\sigma]\!] \cup [\![\varrho]\!]$, (similarly for $\cap$ instead of $\cup$,)
- $[\![\sigma^*]\!] = \bigcup_{i \in \mathbb{N}} [\![\sigma]\!]^i$.

Binary relations on Σ_+^+ will be referred to as *operations*. For an expression r we denote the ***pictures language generated by*** r also with $\mathcal{L}(r)$ instead of $[\![r]\!]$. We we will denote the class of languages (operations, resp.) generated by elements from a class of expressions (operators, resp.) by the corresponding calligraphic notation. Thus $\cap$-$\mathcal{REG}^{\mathcal{ROP}}(\Sigma)$ denotes the class of languages definable by expressions with restricted operators and so on.

In the context of regular expressions with operators, the symbols $^{\oplus+}$ and $^{\ominus+}$ become superfluous because for every regular expression r one has $\mathcal{L}(r^{\oplus+}) = \mathcal{L}(r(\oplus r)^*)$, and similarly for $\ominus$ instead of $\oplus$.

A picture language or an operation will be called *(monotonic)* $\cap$*-regular* if it is the denotation of a (monotonic) $\cap$-regular expression or operation. We make the following simple observation:

Remark 6. All of the above mentioned classes of languages are closed under rotation and reflection. The classes defined by monotonic expressions are closed under projection.

For the set of regular expressions with restricted operators one can show that every picture language generated by such an expression is recognizable by a finite tiling system, as stated in the following theorem.

Theorem 7. *Every language in* $\cap$-$\mathcal{REG}^{\mathcal{ROP}}(\Sigma)$ *is recognizable by tiling systems as defined in [GR92].*

We conjecture that the converse of Theorem 7 is not true, but we cannot show this.

3.1 Two More Examples

We use an example of [GRST96] to show that $\cap$-$\mathcal{REG}^{\mathcal{ROP}}$ is not closed under complement. In fact, the complement of a $\cap$-regular language need not even be recognizable:

Example 8. Let Σ be a finite alphabet. Let $q = \Sigma((\ominus\Sigma^+)(\oplus\Sigma_+))^*$, and

$$r = \bigcup_{a \neq b} \begin{pmatrix} a \\ \Sigma_+ \\ b \end{pmatrix}, \; s = \Sigma_+^+ r \Sigma_+^+, \; t = s \cap \begin{pmatrix} q \\ \Sigma^+ \end{pmatrix}, \; u = \begin{pmatrix} \Sigma_+^+ \\ t \\ \Sigma_+^+ \end{pmatrix} \cap \begin{pmatrix} q \\ q \end{pmatrix}.$$

q generates the set of all squares; s generates all pictures whose top and bottom row are different; t all pictures which additionally have size $(n+1, n)$ for some n. Finally, u generates all pictures of size $(2n, n)$ for some $n \geq 2$, in which there are two different rows with a square in between, i.e. all pictures of the form $\begin{smallmatrix} P \\ Q \end{smallmatrix}$ where P and Q are different squares larger than 1×1. This language is not recognizable (see [GRST96]) and is thus (by Theorem 7) not in $\cap$-$\mathcal{REG}^{\mathcal{ROP}}$.

Example 9. Let q as above. The expression $\Sigma((\oplus q)(\ominus(q \oplus q)))^*$ generates the set of all squares whose side length is a power of two.

To see this, note that $(\oplus q)(\ominus(q \oplus q))$ is an operator that will enlarge a $m\times n$-picture P, where $m+n$ is even, to an $(m + \frac{1}{2}(m+n))\times(n+m)$-picture with P in its left upper corner. The expression Σ generates all 1×1-squares, so for all i, the expression $\Sigma((\oplus q)(\ominus(q \oplus q)))^i$ generates all $2^i\times2^i$-pictures.

4 Context-Free Grammars

We will introduce a concept for context-free grammars that we consider a straightforward adaption to pictures languages. For the classes $\mathcal{REG}^{\mathcal{ROP}}$, $\mathcal{REG}$, and $\mathcal{REG}^{\mathcal{UOP}}$, we will find certain subclasses of context-free grammars of the same expressive power (*rank-$\{br, bl, tl, tr\}$-linear*, *rank-$\{r, l, t, b\}$-linear grammars*, or *rank-linear grammars resp.* The first and second of these classes yield — when restricted to words rather than pictures — a class of grammars that characterizes the regular word languages.

$$\begin{aligned} S &\to \begin{pmatrix} A \\ C \end{pmatrix} \Big| \; a \\ A &\to (S\,B) \\ B &\to \begin{pmatrix} b \\ B \end{pmatrix} \Big| \; b \\ C &\to (c\,C) \mid a \end{aligned}$$

Before we introduce our notion of context-free grammar for picture languages, we give a toy example that might be self explaining. Here, S is the start symbol, and A, B, C are other non-terminal symbols, and a, b, c are the terminal symbols. B yields the set of all columns of b's, whereas C yields the set of rows in c^*a. The alternating recursion of S and A makes sure that this grammar produces the set of squares over $\{a, b, c\}$ that have a's on the diagonal, c's above, and b's below it. In order to obtain a grammar that produces all squares over the singleton a, one may replace all b's and c's by the letter a.

4.1 Sentential Forms, Grammars, Context-Free Languages

Trying to adapt the concept of (context-free) grammar from the theory of formal word languages to that of picture languages involves the following crucial problem: How can subpictures of a given picture be replaced by pictures of possibly different size? In order to avoid this problem, we do not use pictures as sentential forms, but terms built up by terminal and non-terminal symbols and the binary symbols $\ominus$ and $\oplus$.

Definition 10. Let V be any alphabet. A *sentential form over* V is an element from $REG(V)$ in which only the connectives $\ominus$ and $\oplus$ (but not $\cup$, ${}^{\oplus+}$ and ${}^{\ominus+}$) occur. $SF(V)$ denotes the set of all sentential forms over V.

Note that the language generated by a sentential form α can have at most one element, which we will denote by $[\![\alpha]\!]$.

For example, possible sentential forms over $\{a, b, c, d\}$ are $(a \ominus b) \oplus (c \ominus d)$ and $(a \ominus (b \oplus c) \ominus d)$. The first one generates the the picture $\begin{smallmatrix} a & c \\ b & d \end{smallmatrix}$, whereas the second does not generate a picture.

A context-free picture grammar will be defined very similarly to a word grammar. Derivation works the same way as for word grammars: One sentential form results from the preceding one by replacing a non-terminal with some corresponding right hand side, giving again a sentential form. The end of such a derivation is reached when there are only terminal symbols left. If this "terminal sentential form" can be evaluated to a picture, this picture is generated by the grammar.

Definition 11. A *context-free picture grammar* is a tuple $G = (N, \Sigma, \rightarrow, S)$, where N is a finite set of *non-terminal symbols*, disjoint from the set Σ of *terminal symbols*; $S \in N$ (the *start symbol*); and $\rightarrow \subseteq N \times SF(N \cup \Sigma)$ is a set of *rules*.

For a context-free grammar $G = (N, \Sigma, \rightarrow, S)$, we define the relation $\vdash_G$ on $SF(N \cup \Sigma)$ by $\beta \vdash_G \gamma$ iff there is some rule $(A, \alpha) \in \rightarrow$ such that γ results from β by replacing one occurrence of A by α. (We drop subscript G if possible.) We denote the reflexive and transitive closure of $\vdash$ by $\vdash^*$.

Two grammars G_1, G_2 with the same terminal symbol set Σ are called *strongly equivalent* iff for every terminal sentential form $\alpha \in SF(\Sigma)$ the equivalence $S_1 \vdash^*_{G_1} \alpha \iff S_2 \vdash^*_{G_2} \alpha$ holds.

We denote by $\mathcal{L}(G) = \{[\![\alpha]\!] \mid \alpha \in SF(\Sigma), S \vdash^* \alpha\}$ the picture language *generated by* G.

4.2 Limits of Context-Free Grammars

The following is proven similarly as the corresponding fact in the case of word languages:

Remark 12. Every context-free grammar is strongly equivalent to a context-free grammar in *Chomsky Normal Form (CNF)*, i.e. having only rules of the form $(A, B \ominus C)$, $(A, B \oplus C)$, and (A, a), where A, B, C are non-terminals and a is a terminal.

The Chomsky Normal Form can be used to prove the following.

Lemma 13. *The language of Example 1 is not context-free.*

Proof. Consider a context-free grammar G in CNF with $L \subseteq \mathcal{L}(G)$. We show $\mathcal{L}(G) \not\subseteq L$.

Provided that n is sufficiently large, among the $n-2$ different $n \times n$-pictures P_i over $\{a, b\}$ that have b's exactly in the i-th row and the i-th column ($i \in \{2, \ldots, n-1\}$) there are two different ones, say P_k and P_l, such that both of them can be derived from the same two-symbol sentential form α, say $\alpha = C \oplus D$. (The case $\alpha = C \ominus D$ is analogous.)

Now we can choose decompositions $P_k = P \oplus R$ and $P_l = P' \oplus R'$ with $P, R, P', R' \in \Sigma^+_+$ such that P, P' can be derived from C and R, R' can be derived from D. It is easy to see that $P \oplus R' \in \mathcal{L}(G) \setminus L$.

The above lemma shows that $\cap$-$\mathcal{REG} \not\subseteq \mathcal{CF}$ for two-letter alphabets. Together with Theorem 15 this shows that $\cap$-$\mathcal{REG}$ and $\mathcal{REG}^{\mathcal{ROP}}$ are incomparable.

4.3 Rank-Linear Grammars and Regular Expressions

Our aim is to find constraints for context-free grammars such that grammars with these constraints capture exactly the expressive power of monotonic expressions with unrestricted (or with restricted) operators.

These constraints are formalized in the definition of *rank-linear grammars*, which have some restriction on the way recursion is allowed. The main idea is to have some kind of ranking on the set of non-terminals and to require that the rank of the left hand side is larger than the rank of the non-terminals on the right hand side of a rule, except for at most one of them, which may have the same rank. If, additionally, this particular non-terminal is always "at the same place" – top-left, top-right, bottom-left or bottom-right – inside the right hand sides of all rules with "equally large" non-terminals on the left hand sides, then the grammar will be even *$\{br, bl, tl, tr\}$-rank-linear.* A corresponding definition applies for the directions from $\{r, l, t, b\}$.

Definition 14. Let $G = (N, \Sigma, \rightarrow, S)$ be a context-free grammar, $\leq$ be a pre-order (i.e. a reflexive, transitive relation) on N. The equivalence relation $\leq \cap \geq$ is denoted by $\equiv$. Let $< := \leq \setminus \equiv$.

A rule (A, α) is *linear wrt.* $\leq$ if there is at most one occurrence of some non-terminal B in α with $B \equiv A$ and all other non-terminals in α are $< A$.

If, additionally, there are $x \in \{l, r\}$ and $y \in \{t, b\}$ such that in subterms $\alpha_l \oplus \alpha_r$ of α the factor α_x does not contain this B and analogously for $\ominus$ and y, then the rule (A, α) is *yx-linear wrt.* $\leq$. If all non-terminals in α are $< A$, then the rule (A, α) is yx-linear for any x and y.

If a rule is tx- and bx-linear wrt. $\leq$, then it is *x-linear wrt.* $\leq$. If it is yr- and yl-linear wrt. $\leq$, then it is *y-linear wrt.* $\leq$.

The grammar G is *rank-linear wrt.* $\leq$ if all rules are linear wrt. $\leq$. The grammar G is called *$\{br, bl, tl, tr\}$-rank-linear wrt.* $\leq$ if for every $\equiv$-equivalence class $[A]$ there is a $\mathfrak{d} \in \{br, bl, tl, tr\}$ such that all the B-rules for $B \equiv A$ are $\mathfrak{d}$-linear.

$\{r, l, t, b\}$-rank-linear grammars are defined analogously.

A grammar is *rank-linear*, $\{br, bl, tl, tr\}$*-rank-linear*, or $\{r, l, t, b\}$-rank-linear iff there is a preorder $\leq$ such that it has the respective property wrt. $\leq$. A language is called *rank-linear* (or $\{br, bl, tl, tr\}$*-rank-linear* or $\{r, l, t, b\}$*-rank-linear*, resp.) if it is generated by some grammar with the respective property.

The notion on rank-linearity of context-free picture grammars compares to the linear word grammars as follows: Rank-linear grammars in which no $\ominus$-connective occurs may be viewed as word grammars; a word grammar is linear iff it is in this sense a rank-linear grammar *wrt. the universal relation on the set of nonterminals* as the preorder.

The example grammar from the beginning of this section is in CNF and $\{br, bl, tl, tr\}$-rank-linear wrt. a preorder with $S \equiv A$ and $B, C < A, S$. The S-rules and A-rules are br-linear wrt. this preorder, and the B-rules (C-rules) are t-linear (l-linear, resp.) wrt. this preorder.

Note the little incompatibility of notations that in our definition an r-linear rule wrt. the universal relation corresponds to what is known as a left-linear rule of a word grammar and vice versa.

Rank-linear word grammars can also generate non-linear word languages such as $\{a^i b^i \mid i \geq 1\}^+$. (Consider the grammar with the rules $S \to SA \mid A$, $A \to aAb \mid ab$ and a preorder such that $A < S$.)

The following theorem states that rank-linear grammars, $\{br, bl, tl, tr\}$-rank-linear grammars, and $\{r, l, t, b\}$-rank-linear grammars capture exactly the expressive power of REG^{UOP}, REG^{ROP}, and REG, resp.

Theorem 15. *For all languages $L \subseteq \Sigma^+_+$ we have*

- *L is rank-linear iff $L \in \mathcal{REG}^{\mathcal{UOP}}$.*
- *L is $\{br, bl, tl, tr\}$-rank-linear iff $L \in \mathcal{REG}^{\mathcal{ROP}}$.*
- *L is $\{r, l, t, b\}$-rank-linear iff $L \in \mathcal{REG}$.*

The proof uses the concept of *generalized operator grammars (GOGs)* as an intermediate stage. These are, roughly speaking, grammars in which the right hand side of rules need not be sentential forms but may be arbitrary monotonic expressions from REG^{UOP}. In a derivation step of a GOG, one non-terminal A is replaced by a sentential form α that is "below" a right hand side β of an A-rule in the sense that α results from β by replacing subexpressions in an outermost strategy, namely: Each union $\sigma \cup \varrho$ is replaced either with σ or with ϱ, each iteration σ^* is replaced with a finite number of compositions of σ, and so on.

The different notions of rank-linearity can be defined similarly for GOGs.

For a given monotonic regular expression r, there is the equivalent GOG with the only rule (S, r). So the construction of a grammar amounts to reducing the complexity of the right hand sides inductively by introducing new rules until a grammar is reached, whose right hand sides contains only sentential forms.

Conversely, a given grammar with one of the rank-linearity properties is also a GOG with this property. Now the construction of an expression means to reduce the number of rules and to replace the recursion of non-terminals with the iteration of operators. Formally, the latter is done by induction over the maximal length of a strictly decreasing chain wrt. to the given preorder.

The term "generalized operator grammar" has been chosen because of the similarity to the "generalized transition graphs" used to pass from automata to regular word expressions.

5 Summary of (Non-)Inclusion Results

The following table shows the mentioned non-inclusion results that hold over alphabets with at least two letters, and a witness for each.

$\cap\text{-}\mathcal{REG} \not\subseteq \mathcal{CF}$	language of lemma 13
$\mathcal{REG}^{\mathcal{ROP}} \not\subseteq \cap\text{-}\mathcal{REG}$	set of squares
$\mathcal{REG}^{\mathcal{UOP}} \not\subseteq \mathcal{REC}$	$\{a^i b^i \mid i \geq 1\}$

From the above facts, Theorem 7, and trivial inclusions like $\mathcal{REG} \subseteq \mathcal{REG}^{\mathcal{ROP}}$, one can infer the results presented in the figure for non-trivial alphabets. Here, the dotted lines indicate incomparability and the remaining lines show inclusions, which are marked by $\neq$ when known to be proper. It is open whether the inclusion $\mathcal{REG}^{\mathcal{ROP}} \subseteq \cap\text{-}\mathcal{REG}^{\mathcal{ROP}}$ is also proper in case of a one-letter alphabet, but it is known that the inclusion $\mathcal{REG}^{\mathcal{ROP}} \subseteq \mathcal{REC}$ is (see [Mat95]). Note that for a singleton alphabet both concatenations are commutative, from which one can deduce that any unrestricted operator can be transformed into an equivalent *br*-operator and therefore $\mathcal{REG}^{\mathcal{UOP}}(\{a\}) = \mathcal{REG}^{\mathcal{ROP}}(\{a\})$.

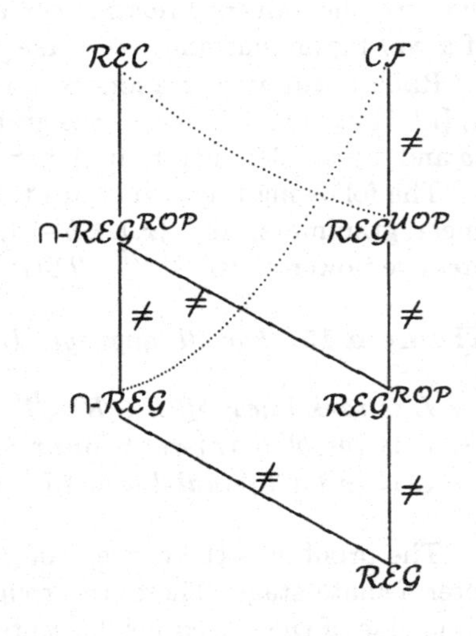

6 Conclusion

We have shown that besides the concept of tiling systems, which is an extension of recognizability to picture languages, there are interesting classes of regular expressions which allow to define a canonical analogon to regular word languages in the context of pictures. However, it seems to be impossible to define regular picture language in such a way that all characterization results and closure properties of the theory of words carry over to pictures.

Our characterization results may be viewed as a step towards a Kleene-like theorem for picture languages: In the same sense that a right-linear word grammar can be considered a non-deterministic finite automaton on words, a

$\{br, bl, tl, tr\}$-rank-linear grammar might be considered a kind of automaton on pictures (or rather on sentential forms). However, the translation into automata theoretic terms is not as immediate for $\{br, bl, tl, tr\}$-rank-linear grammars as for (right)-linear word grammars.

The following open problems are of particular interest:

- Do $\mathcal{CF}$ and $\mathcal{REG}^{\mathcal{ROP}}$ coincide in case of a one-letter alphabet? If so, this would be a nice correspondence to the situation of word languages.
- Do $\cap$-$\mathcal{REG}^{\mathcal{ROP}}$ and $\mathcal{REC}$ coincide? One candidate for a recognizable language not in $\cap$-$\mathcal{REG}^{\mathcal{ROP}}$ is the language of all pictures over $\{a\}$ for which $|P|$ is prime and $\overline{\underline{P}}$ is the length of an accepting run of a fixed LBA accepting the set of words of prime length.
- Is the emptiness problem for REG^{ROP} decidable? Note that it is undecidable for $\cap$-REG^{ROP}, even for a fixed alphabet of size 1; see [Mat95].

Besides these questions dealing with the picture language definition formalisms of this paper, one could imagine to transfer other concepts from word languages to pictures in such a way that some of the known results of the theory of word languages remain true. For example: Is there a reasonable analogon to the monoid recognizability of word languages?

Moreover, future research may deal with the more general case of n-dimensional arrays for arbitrary integers n instead of $n = 1$ or $n = 2$, e.g. words or pictures. It is easy and only a matter of notational inconvenience to redo all the definitions and proofs of this paper for arbitrary dimensions.

Acknowledgments. I thank Wolfgang Thomas for the help and advice he gave me during the preparation of both my diploma thesis and this paper.

References

[GR92] D. Giammarresi and A. Restivo. Recognizable picture languages. In *Proceedings First International Colloqium on Parallel Image Processing 1991*, number 6 in International Journal Pattern Recognition and Artificial Intelligence, pages 241–256, 1992.

[GR96] D. Giammarresi and A. Restivo. Two-dimensional languages. In G. Rozenberg and A. Salomaa, editors, *Handbook of Formal Language Theory*, volume III. Springer-Verlag, New York, 1996.

[GRST96] D. Giammarresi, A. Restivo, S. Seibert, and W. Thomas. Monadic second-order logic and recognizability by tiling systems. *Information and Computation*, 125:32–45, 1996.

[Mat95] Oliver Matz. *Klassifizierung von Bildsprachen mit rationalen Ausdrücken, Grammatiken und Logik-Formeln.* Diploma thesis, Christian-Albrechts-Universität Kiel, 1995. (German).

[Sir87] R. Siromoney. Advances in array languages. In Ehrig et al., editors, *Graph Grammars and Their Applications to Computer Science*, volume 291 of *Lecture Notes in Computer Science*, pages 549–563. Springer-Verlag, 1987.

Measuring Nondeterminism in Pushdown Automata

Jonathan Goldstine[1], Hing Leung[2], Detlef Wotschke*[3]

[1] Department of Computer Science and Engineering, The Pennsylvania State University, University Park PA 16802, USA
[2] Department of Computer Science, New Mexico State University, Las Cruces NM 88003, USA
[3] Fachbereich Informatik, Johann Wolfgang Goethe-Universität, Frankfurt, Germany

Abstract. The amount of nondeterminism that a pushdown automaton requires to recognize an input string can be measured by the minimum number of guesses that it must make to accept the string, where guesses are measured in bits of information. When this quantity is unbounded, the rate at which it grows as the length of the string increases serves as a measure of the pushdown automaton's "rate of consumption" of nondeterminism. We show that this measure is similar to other complexity measures in that it gives rise to an infinite hierarchy of complexity classes of context-free languages differing in the amount of this resource (nondeterminism) that they require. In addition, we show that there are context-free languages that can only be recognized by a pushdown automaton whose nondeterminism grows linearly, resolving an open problem in the literature. In particular, $\{ww^R : w \in \{a,b\}^*\}$ is such a language.

1 Introduction

Nondeterminism in pushdown automata can be treated as a consumable resource. In this paper, the amount of nondeterminism a pushdown automaton requires to recognize an input string is measured by the minimum number of guesses that it must make to accept the string, where guesses are measured in bits of information. This is the same measure used for finite automata in [2]. It is essentially what has been called the minmax-measure for pushdown automata [8, 9], except for the minor difference that measuring in bits of information gives a more precise measure by distinguishing between binary guesses and guesses that involve greater amounts of branching. When this quantity is unbounded, the rate at which it grows as the length of the string increases serves as a measure of the pushdown automaton's "rate of consumption" of nondeterminism. We show that this measure is similar to other complexity measures in that it gives rise to a hierarchy of complexity classes of context-free languages differing in the amount of this resource (nondeterminism) that they require.

* This author's research was supported in part by the Stiftung Volkswagenwerk under Grant No. II/62 325.

Reischuk, Morvan (Eds.): STACS'97 Proceedings, LNCS 1200

The difficult aspect of working with this measure is establishing lower bounds. In fact, there have previously been no proven examples of context-free languages requiring a linear amount of nondeterminism. So far as we know, there have not even been any tight lower bounds established for context-free languages that require unbounded nondeterminism. We resolve this situation by settling a question posed by Chandra Kintala in 1978. Kintala asked whether the language $\{ww^R\}$, where w ranges over an alphabet of at least two letters, requires at least $n/2$ binary guesses on inputs of length n for infinitely many inputs [6]. We show that the answer is no, but that it does infinitely often require a number of guesses linear in the length of the input. The best lower bound previously proved for this language was $\Omega(\log n)$ [9].

While proving lower bounds is difficult, finding examples of context-free languages that have interesting upper bounds can also be tricky. Salomaa and Yu give a simple example of a context-free language that requires unbounded nondeterminism but that can be accepted by a pushdown automaton that makes just $O(\sqrt{n})$ guesses [8]. We give a simple example of such a language that can be accepted by a nondeterministic counter automaton that makes only $O(\log n)$ guesses. In addition, we show that, for every unbounded monotone recursive function $f(n)$, there is a context-free language that requires unbounded nondeterminism but that can be accepted by a pushdown automaton that make only $O(f(n))$ guesses. These languages are more complicated, however, and they are no longer accepted by a counter automaton.

2 Preliminaries

We begin with the usual notion of a pushdown automaton, supplemented by some notation that can be useful for measuring nondeterminism.

Definition 1. A (nondeterministic) ***pushdown automaton*** or ***PDA*** is a 7-tuple

$$A = (Q, \Sigma, \Gamma, \delta, q_0, Z_0, F) , \tag{1}$$

where Q is a finite set of ***states***, Σ and Γ are the ***input*** and ***stack*** alphabets,

$$\delta \subset Q \times (\Sigma \cup \{\varepsilon\}) \times \Gamma \times \Gamma^* \times Q \tag{2}$$

is a finite set of ***moves***, and $q_0 \in Q$, $Z_0 \in \Gamma$ and $F \subseteq Q$ are the ***initial*** state, the ***initial*** stack symbol and the set of ***final*** states. A move $(p, x, Z, z, q) \in \delta$ is an ***ε-move*** if $x = \varepsilon$.

A ***configuration*** of A is $c = (q, w, z) \in Q \times \Sigma^* \times \Gamma^*$, where q is A's current state, w is the string of inputs not yet read, and z is the contents of the stack with the topmost symbol at the left. Each move $\mu = (p, x, Z, z, q)$ induces a binary relation on configurations, which is in fact a partial function,

$$(p, xw, Zv) \vdash_\mu (q, w, zv) \text{ for all } w \in \Sigma^* \text{ and } v \in \Gamma^* . \tag{3}$$

For $\delta' \subseteq \delta$, we write $c \vdash_{\delta'} c'$ if $c \vdash_\mu c'$ for some $\mu \in \delta'$, and we write $c \vdash_A c'$ if $c \vdash_\delta c'$. A (complete) ***move κ of A*** associated with a move $\mu \in \delta$ is a pair (c, c')

of configurations with $c \vdash_\mu c'$; we shall write this as $\kappa = c \vdash_\mu c'$, or simply as $\kappa = c \vdash_A c'$. A *partial computation* of A of *length* $t \geq 1$ *with input* w is a string of moves $\sigma = \mu_1\mu_2\cdots\mu_t \in \delta^*$, where $c_0 \vdash_{\mu_1} c_1 \vdash_{\mu_2} c_2 \vdash_{\mu_3} \cdots \vdash_{\mu_t} c_t$ for some configurations $c_0, c_1, \ldots, c_t$, with $c_0 = (p, w, z)$ and $c_t = (p', \varepsilon, z')$ for some $p, p' \in Q$, $z, z' \in \Gamma^*$. It is a *computation* of A *accepting* w if $c_0 = (q_0, w, Z_0)$ and $p' \in F$, in which case its *history* $H(\sigma)$ is the string of complete moves $\kappa_1 \cdots \kappa_t$, where $\kappa_i = c_{i-1} \vdash_{\mu_i} c_i$. (In addition, if $q_0 \in F$ then $\sigma = \varepsilon$ is considered a computation of length 0 accepting ε with history ε.) The set of computations of A is denoted by $C(A)$. The language *accepted* by A is

$$L(A) = \{w \in \Sigma^* : \text{there is a } \sigma \in C(A) \text{ accepting } w\} . \tag{4}$$

Two PDAs are *equivalent* if they accept the same language. A *context-free language* or *CFL* is a language accepted by a PDA.

Next, we define determinism, along with a normal form for nondeterministic PDAs that we shall discuss momentarily.

Definition 2. If $A = (Q, \Sigma, \Gamma, \delta, q_0, Z_0, F)$ is a PDA, a set of moves $\delta' \subseteq \delta$ is *deterministic* if $\vdash_{\delta'}$ is a partial function. The PDA A is a *deterministic* PDA or *DPDA* if δ is deterministic. A *deterministic context-free language* or *DCFL* is a language accepted by a DPDA.

The PDA A is *normal* if $\{\mu, \mu'\}$ is deterministic for all $\mu, \mu' \in \delta$ with μ an ε-move and μ' not an ε-move.

The PDA A is a (nondeterministic) *counter automaton* if $\Gamma = \{Z_0, c\}$ and its moves are of the form $(p, x, Z_0, c^i Z_0, q)$ or (p, x, c, c^i, q) for $i \geq 0$. A stack contents of $c^n Z_0$ represents a counter value of n, and we shall feel free to speak as though the counter automaton had a true nonnegative-integer-valued counter rather than a pushdown stack.

Finally, we specify notation for quantifying nondeterminism in terms of both branching and guessing, a distinction that we will discuss later in this section.

Definition 3. Let $A = (Q, \Sigma, \Gamma, \delta, q_0, Z_0, F)$ be a PDA. The *branching measure* β_A and the *guessing measure* γ_A of A are

$$\begin{aligned}
\beta_A(\mu) &= \#\{(p, x, Z, y, r) \in \delta : y, r\} && \text{for } \mu = (p, x, Z, z, q) \in \delta , \\
\beta_A(\kappa) &= \#\{c'' : c \vdash_A c''\} && \text{for } \kappa = c \vdash_A c' , \\
\beta_A(\pi) &= \textstyle\prod_{i=1}^{t} \beta_A(\kappa_i) && \text{for } \pi = \kappa_1 \cdots \kappa_t , \\
\beta_A(\sigma) &= \beta_A(\pi) && \text{for } \pi = H(\sigma) \text{ and } \sigma \in C(A) , \\
\beta_A(w) &= \min\{\beta_A(\sigma) : \sigma \text{ accepts } w\} && \text{for } w \in \Sigma^* , \\
\beta_A(n) &= \max\{\beta_A(w) : |w| \leq n\} && \text{for } n \geq 0 , \\
\beta_A &= \sup\{\beta_A(n) : n \geq 0\} ,
\end{aligned} \tag{5}$$

and, in each case, $\gamma_A(\cdot) = \log_2 \beta_A(\cdot)$, and $\gamma_A = \log_2 \beta_A$. In particular, $\gamma_A = \beta_A = \infty$ if $\{\beta_A(n) : n \geq 0\}$ is unbounded. For convenience, we define $\beta_A(w) = 1$ when $\{\beta_A(\sigma) : \sigma \text{ accepts } w\} = \emptyset$, i.e., when $w \notin L(A)$.

A word $x \in L(A)$ is γ_A-*minimal* if $|x| \leq |y|$ whenever $\gamma_A(x) \leq \gamma_A(y)$ for $y \in L(A)$.

If L is a CFL then $\gamma_L = \min\{\gamma_A : L(A) = L\}$. We shall not define $\gamma_L(n)$, but we shall write

$$\begin{array}{ll} \gamma_L(n) \leq O(f(n)) & \text{if } \gamma_A(n) = O(f(n)) \text{ for some PDA } A \text{ with } L(A) = L\ ; \\ \Omega(f(n)) \leq \gamma_L(n) & \text{if } \gamma_A(n) = \Omega(f(n)) \text{ for every PDA } A \text{ with } L(A) = L\ ; \\ \gamma_L(n) = \Theta(f(n)) & \text{if } \Omega(f(n)) \leq \gamma_L(n) \leq O(f(n))\ . \end{array} \tag{6}$$

When only one PDA A is under discussion, we may omit the subscript A from the notation.

There is a technical difficulty that arises from these definitions. The branching of a move in a computation may be less than the branching of the corresponding complete move in the history of the computation. This happens when an ε-move and a non-ε-move can be applied to the same configuration. The situation is particularly awkward in the case where the ε-move is the one that is chosen, since the PDA will not "know" the branching of the corresponding complete move until it reads the next input symbol, which may not occur until many ε-moves later. This creates difficulties when we want to construct a PDA to simulate another PDA while keeping track of the branching that is occurring. This problem does not arise for normal PDAs, since for them, ε-moves and non-ε-moves are never applicable to the same configuration. For this reason, it is helpful to have the following result. A proof, based on the predicting machine construction in [5], may be found in [4]. (Normal PDAs are called PDAs without ε-Σ-nondeterminism in [4].)

Lemma 4. *Each PDA A can be effectively converted to an equivalent normal PDA B with the property that $\beta_B(w) = \beta_A(w)$ for all input strings w.*

Note: *In view of this result, we shall henceforth assume that all of our PDAs are normal, so that $\beta(\mu_1 \cdots \mu_t) = \prod_{i=1}^{t} \beta(\mu_i)$ for each computation $\mu_1 \cdots \mu_t$.*

The guessing measure γ measures nondeterminism in units of bits of information: $\gamma(\kappa)$ is the number of bits of information needed to select the move κ from among the moves that can be chosen nondeterministically at a point in the history of a computation. This measure is additive over the individual moves in a history, and for PDAs in which every branch is binary, it simply counts the number of nondeterministic moves. It is the complexity measure that we shall study in this paper. (In [9], nondeterminism is measured by counting the number of nondeterministic moves in a history regardless of whether each branch is binary. This results in a less precise measure which is insensitive to multiplicative constants, but which is otherwise equivalent to our measure.)

A nondeterministic finite automaton can be considered a special case of a PDA, one which never changes its stack. The definition of $\gamma(n)$ for PDAs then agrees with that for nondeterministic finite automata studied in [2].

Note that $\beta(n)$ is equal to the maximum value of $\beta(w)$ over all γ-minimal strings w, $|w| \leq n$, if the maximum over the empty set is understood to be 1.

For if $|x| \leq n$ and x is not γ-minimal, then either $x \notin L(A)$ so that $\beta(x) = 1$, or $\gamma(y) \geq \gamma(x)$ for some y shorter than x. Either way, $\beta(x)$ does not affect the maximum value. Hence, in calculating $\gamma(n)$, we need only consider γ-minimal strings.

In this paper, we shall use guessing as the measure of nondeterminism in a PDA. This is a dynamic measure, based on the accepting computations of the PDA. Nondeterminism in a PDA may also be measured statically, as in the following definition of the nondeterministic depth of a PDA. (This definition is equivalent to the one in [9], which is a corrected version of the definition in [10].)

Definition 5. The *state graph* of the PDA $A = (Q, \Sigma, \Gamma, \delta, q_0, Z_0, F)$ is the digraph with set of vertices Q and set of arcs

$$\{(p, q) : (p, x, Z, z, q) \in \delta \text{ for some } x, Z, z\} . \tag{7}$$

A set $D \subseteq Q$ of states is *deterministic* if $\delta_D = \delta \cap (D \times (\Sigma \cup \{\varepsilon\}) \times \Gamma \times \Gamma^* \times D)$ is deterministic, and if $\{\mu, \mu'\}$ is deterministic for all ε-moves $\mu \in \delta_D$ and all moves $\mu' \in \delta$.

A ***deterministic partition*** of A is a partition of Q into deterministic subsets that induces an acyclic quotient digraph on the state graph of A. The ***depth*** of the partition is the length of the longest directed path in the quotient digraph. The (nondeterministic) *depth* of A is the minimum depth among all deterministic partitions of A. If A has no deterministic partitions, we say that A has infinite depth. The *depth* of a CFL L is the minimum of the depths of the PDAs that accept L.

Depth is a static measure of nondeterminism since it is not defined in terms of the computations of the PDA. Consequently, we cannot consider its rate of growth, as we can for guessing, but only whether it is equal to some finite value k or is infinite. The PDAs of depth zero are just the deterministic PDAs, and the PDAs with $\gamma = 0$ are equivalent to deterministic PDAs. The following example illustrates the difference between the two measures for values greater than zero. The obvious PDA for the language $\{ww^R : w \in \Sigma^*\}$ has depth one since it consists of two deterministic machines (one for the pushing phase and one for the popping phase) with a single nondeterministic transition point between them. However, this PDA has $\gamma(n) = \Theta(n)$, since it keeps passing through this transition point during its pushing phase. We shall examine this language more closely in a subsequent section.

Finally, we recall the definition of ambiguity for PDAs.

Definition 6. For a PDA $A = (Q, \Sigma, \Gamma, \delta, q_0, Z_0, F)$, the ***ambiguity*** of A on a string $w \in \Sigma^*$ is $\alpha_A(w) = \#\{\sigma \in C(A) : \sigma \text{ has input } w\}$. The ***ambiguity*** of A is defined to be $\alpha_A = \sup\{\alpha_A(w) : w \in \Sigma^*\}$, and A has ***infinite ambiguity, finite ambiguity***, or is ***unambiguous*** if $\alpha_A = \infty$, $2 \leq \alpha_A < \infty$, or $\alpha_A \leq 1$, respectively.

3 Reducing Nondeterminism

In this section, we show that the guessing performed by a PDA A having $\gamma_A = \infty$ can always be reduced by a linear amount by increasing the size of A, provided A has finite depth. That is why we have not attempted to define $\gamma_L(n)$ beyond the tolerances of O-notation. (It is, of course, impossible in any event to define $\gamma_L(n)$ at fixed values of n since any finite portion of L can be handled deterministically by some PDA.)

Definition 7. Call a PDA A *reducible* (*irreducible*) if it has (does not have) the following property. For every positive integer k, there exists a PDA B and a constant c such that $L(B) = L(A)$ and, for each string w accepted by A,

$$\gamma_B(w) \leq \frac{1}{k}\gamma_A(w) + c\,. \tag{8}$$

Theorem 8. *PDAs of finite depth are reducible.*

Proof. The proof is very simple: by guessing that A will not leave a deterministic component during its next k opportunities, B can replace k guesses of A with a single guess. We omit the details because of space limitations. □

Note that, because of the presence of the constant c, Theorem 8 is only of interest when $\gamma_A = \infty$.

4 CFLs Requiring Linear Nondeterminism

We now wish to show that there are CFLs that require linear nondeterminism. We begin with the following result.

Theorem 9. *Every PDA A has $\gamma_A(n) = O(n)$.*

Proof. The result that some PDA for $L(A)$ has this property follows from the fact that $L(A)$ can be generated by a grammar in Greibach normal form, and hence can be accepted by a PDA in which the length of every computation is linear in the length of its input. However, the stated result requires showing more: that, in every PDA, computations of more than linear length are superfluous. Although we cannot locate this precise result in the literature, it can be derived from the corresponding result for context-free grammars [3, 7]. We omit the details to conserve space. □

The following corollary is an immediate consequence.

Corollary 10. *For every CFL L, $\gamma_L(n) \leq O(n)$.*

Thus, a CFL L requires linear nondeterminism, $\Omega(n) \leq \gamma_L(n)$, if and only if $\gamma_L(n) = \Theta(n)$.

Salomaa and Yu show that $\Omega(\log n) \leq \gamma_L(n)$ for the language $L = \{ww^R : w \in \{a,b\}^*\}$ by an information-theoretic counting argument [9]. A somewhat similar technique can be used to show that the following language requires linear nondeterminism. This proves that there are such languages, which settles an open question in [8].

Theorem 11. *The language $L_0 = \left(\{a^ib^i, a^ib^{2i} : i \geq 0\}\$\right)^*$ requires linear nondeterminism.*

We omit the somewhat lengthy proof because of space limitations. In fact, the proof shows that PDAs accepting L_0 are irreducible. Hence, irreducible PDAs do exist, although, of course, by Theorem 8, they cannot have finite depth.

By a different method, we can show that the language $L = \{ww^R : w \in \{a,b\}^*\}$ also requires linear nondeterminism. This has been an open problem since 1978.[4] As we mentioned, the proof of Theorem 11 is based on a counting argument similar to that used in [9] to obtain the best previously known lower bound for $\gamma_L(n)$, $\Omega(\log n) \leq \gamma_L(n)$. Obtaining a linear lower bound is more difficult and requires something more than a counting argument. Perhaps it should not be surprising that it is more difficult to establish a linear lower bound for $\gamma_L(n)$ than for $\gamma_{L_0}(n)$ because, in some sense, the language L does require less nondeterminism than L_0. This is reflected in the fact that L has depth one, whereas L_0 has infinite depth. And in fact, L would require only logarithmic nondeterminism if a PDA had a binary counter available, so that it could guess the binary representation of the length of w. Thus, a proof of this result requires more than the information-theoretic argument that suffices for L_0, and we are forced into a detailed consideration of the behavior of the stack in an arbitrary PDA accepting L.

Theorem 12. *The language $L = \{ww^R : w \in \Sigma^*\}$ requires linear nondeterminism if Σ contains at least two symbols.*

Proof. The proof involves examining the structure of the computations of an arbitrary PDA accepting L. For this purpose, it is convenient to make use of the observation that the computations themselves constitute a context-free language, and therefore must conform to Ogden's Lemma.

Let $A = (Q, \Sigma, \Gamma, \delta, q_0, Z_0, F)$ be a PDA accepting L. We can construct a context-free grammar G generating the computations $C(A)$ of A as follows. The variables of G are $[qZ]$ and $[qZp]$, for $p, q \in Q$ and $Z \in \Gamma$, with $[q_0Z_0]$ as the starting variable. The rewriting rules enable $[qZ]$ to generate the partial

[4] "It would be interesting to know whether any pushdown automaton accepting $\{ww^R\}$ requires at least $n/2$ binary nondeterministic moves on inputs of length n, for infinitely many inputs." [6]. Since this language has depth one, Theorem 8 implies that this conjecture is false as stated, and that it should be phrased in terms of linear nondeterminism.

computations π such that $(q, w, Z) \vdash_\pi^* (p, \varepsilon, z)$ for some $w \in \Sigma^*$, $p \in F$ and $z \in \Gamma^*$, while $[qZp]$ generates the partial computations π such that $(q, w, Z) \vdash_\pi^* (p, \varepsilon, \varepsilon)$ for some $w \in \Sigma^*$. Specifically, the rules are the following, for all moves $\mu = (q, x, Z, Z_1 \cdots Z_n, q_1) \in \delta$, where $n \geq 0$ and $Z_i \in \Gamma$:

$$\begin{array}{rll} [qZ] & \to \varepsilon, & \text{if } q \in F \ ; \\ [qZ] & \to \mu, & \text{if } q_1 \in F \ ; \\ [qZ] & \to \mu[q_1 Z_1 q_2][q_2 Z_2 q_3] \cdots [q_{j-1} Z_{j-1} q_j][q_j Z_j], & 1 \leq j \leq n,\ q_i \in Q \ ; \\ [qZq_{n+1}] & \to \mu[q_1 Z_1 q_2][q_2 Z_2 q_3] \cdots [q_n Z_n q_{n+1}], & q_j \in Q \ , \end{array} \tag{9}$$

where the third line of (9) does not apply if $n = 0$. It is easily verified that the variables of G generate precisely the desired partial computations.

Let p be the pumping constant for the grammar G from Ogden's Lemma [1], and let $k = p \cdot \#Q \cdot \#\Gamma + 1$. It suffices to prove that, for each $n \geq 1$, if $w = (bba^p)^{kn+2}b$ and if $\pi \in C(A)$ is a computation accepting $ww^R = (bba^p)^{2kn+4}bb$, then $\gamma_A(\pi) \geq n$. To prove this, distinguish the p moves in π that consume the p a's in one of the a^p-terms of w. Then, by Ogden's Lemma, there is a derivation in G of the form

$$S \Rightarrow^* \pi_1 B \pi_5 \Rightarrow^* \pi_1 \pi_2 B \pi_4 \pi_5 \Rightarrow^* \pi_1 \pi_2 \pi_3 \pi_4 \pi_5 = \pi \ , \tag{10}$$

for some variable B of G, where S is the starting variable of G, and where π_1, π_2 and π_3 each contain a distinguished move of π, or π_3, π_4 and π_5 each contain a distinguished move of π. It follows that either π_2 or π_4 has as its input a nonempty substring a^i of the selected a^p-term of w. Since $\pi_1 \pi_2^j \pi_3 \pi_4^j \pi_5$ is generated by G, it must be a computation of A accepting a string in L, for all $j \geq 0$. It follows that the input to π_2 is a substring a^i of the selected a^p-term of w, and the input to π_4 is also a^i and is a substring of the corresponding a^p-term of w^R. Note that the variables of G derive partial computations that do not decrease the stack height below its starting value except perhaps on the very last move. Since $B \Rightarrow^* \pi_2 \pi_3 \pi_4$ and the input to $\pi_2 \pi_3 \pi_4$ extends from a point in the selected a^p-term of w to a point in the corresponding term of w^R, it follows that there is a point during the processing of the selected a^p-term of w when t a's of the a^p-term remain to be scanned, for some t, $1 \leq t \leq p$, and the stack height is as low as it will ever get during the processing of the rest of w. Call such a point a *low point* for the a^p-term.

Now consider a factorization $\pi = \sigma_1 \cdots \sigma_{kn+2}$, where σ_i begins at a low point for the i-th a^p-term of w, $2 \leq i \leq kn+2$. Let q_i be the state of A at that low point, let Z_i be the symbol on top of the stack, and let t_i be the number of a's in the i-th a^p-term that have not yet been scanned. To prove that $\gamma_A(\pi) \geq n$, it suffices to show that each partial computation $\sigma_{k(j-1)+2} \cdots \sigma_{kj+1}$, $1 \leq j \leq n$, contains a guessing move.

Suppose to the contrary that some $\sigma_{k(j-1)+2} \cdots \sigma_{kj+1}$ does not contain a guessing move. Since there can be at most $p \cdot \#Q \cdot \#\Gamma < k$ distinct (t_i, q_i, Z_i) triples, for some $k(j-1)+2 \leq r < s \leq kj+1$, $(t_r, q_r, Z_r) = (t_s, q_s, Z_s) = (t, q, Z)$, say. Let $\tau = \sigma_r \cdots \sigma_{s-1}$. Because the stack height during τ never dips below its starting value,

$$(q, a^t(bba^p)^{s-r-1}bba^{p-t}, Z) \vdash_\tau^* (q, \varepsilon, Zy) \ , \tag{11}$$

for some $y \in \Gamma^*$. Since no guessing move occurs during τ, the PDA A must keep repeating the partial computation τ; that is, the suffix $\sigma_r \cdots \sigma_{kn+2}$ of π is a prefix of τ^N for some N. Since π ends in a final state, τ must pass through a final state. Therefore, $\pi = \pi'\pi''$, where π' ends in a final state after scanning the input up to a point between the r-th a^p-term of w and the s-th a^p-term of w. As we have seen earlier, $\pi = \pi'\pi''$ can be factored as $\pi' = \pi_1'\pi_2'\pi_3'$ and $\pi'' = \pi_3''\pi_4''\pi_5''$, where the input to π_2' is a substring a^i of the first a^p-term of w for some i, $1 \leq i \leq p$; the input to π_4'' is a substring a^i of the last a^p-term of w^R; and $\pi_1'(\pi_2')^2\pi_3'\pi_3''(\pi_4'')^2\pi_5''$ is a computation of A. Since $\pi_1'(\pi_2')^2\pi_3'$ ends in a final state, it too is a computation of A, and hence its input x must be in L. However, the first a^*-term of x is a^{p+i}, while all the other a^*-terms of x are no larger than a^p. Hence, x is not in L, a contradiction. □

5 Nondeterministic Complexity Classes of CFLs

If nondeterminism truly behaves as a complexity measure in PDAs, we would expect there to be a hierarchy of complexity classes of CFLs with respect to this measure. In this section, we show that there is such a hierarchy.

We note first that there are PDAs that have slowly growing guessing measures.

Theorem 13. *There is a nondeterministic counter automaton of depth one with $\gamma(n) = \Theta(\log n)$.*

Proof. Consider the obvious nondeterministic counter automaton of depth one for $L = \{x10^i1y \in (0+1)^* : i \leq |x|\}$: it counts the length of its input until guessing (upon reading a 1) that it is time to count down against consecutive 0's. The γ-minimal words are

$$w_j = 10^1 10^3 10^7 10^{15} 1 \cdots 10^{2^{j-1}-1} 11, \quad j \geq 1\,. \tag{12}$$

Since $|w_j| = 2^j$ and $\gamma(w_j) = j$, $\gamma(n) = \lfloor \log_2 n \rfloor$ for $n \geq 1$. □

The PDA in the preceding proof has low depth but infinite ambiguity. We shall now construct nondeterministic counter automata of infinite depth but finite ambiguity whose guessing measures grow more slowly than $\log n$, in fact, more slowly than any prespecified recursive function. These PDAs are, however, artificial in the sense that they perform "unnecessary" guessing. The construction is based on the computations of deterministic multicounter machines: by a (deterministic) k-counter machine we mean a deterministic automaton having a finite-state control with a starting state and a set of final states, along with k counters, each of which can hold any nonnegative integer and each of which can be incremented by one, decremented by one (if it is greater than zero), and tested to see whether it is zero or nonzero. The multicounter machine begins its computation in its starting state with a zero in each counter. Recall that a machine with two counters can simulate a Turing machine [5]. Hence, for each recursive function $f(n)$, there is a two-counter machine that enters a final state infinitely often but that enters a final state for the n-th time only after it has performed at least $f(n)$ computational steps.

Theorem 14. *For each monotone, unbounded, recursive function $g(n)$, there is a nondeterministic counter automaton A with $\gamma_A(n)$ unbounded and $\gamma_A(n) \leq g(n)$.*

Proof. The proof resembles the construction by which the computations of a Turing machine can be represented as the intersection of two context-free languages. The minimization involved in the definition of $\gamma_A(w)$ plays the role of intersection. The role of the Turing machine is played by a two-counter machine that, running without input, enters a particular state infinitely often but very infrequently.

Since a PDA can always be modified to handle input strings up to a certain length deterministically by using an enlarged finite state control, it suffices to prove the condition $\gamma_A(n) \leq g(n)$ just for large values of n.

Let $f(n) = \min\{i : g(i) \geq n\}$. Since $f(n)$ is recursive, there is a two-counter machine M that enters a final state for the n-th time after performing at least $f(n)$ computational steps. If Q is M's set of states, then a configuration of M can be represented by a string in $C = Qc^*d^*$, where qc^id^j represents the configuration in which M is in state q with i in its first counter and j in its second counter. Let $\vdash$ denote the move relation of M on C and let q_0 be M's starting state. Let

$$P = \{x_0x_1 \cdots x_n : n \geq 0, x_0 = q_0, \text{ and } x_i \vdash x_{i+1} \text{ for } 0 \leq i < n\}, \tag{13}$$

and let $L = C^+ - P$.

The strings in P represent M's computation; hence, the final states of M occur within a string in P with a frequency determined by the function $f(n)$. A string in L represents a sequence of configurations which may form a valid computation for a while, but which eventually contains a "defect" in a configuration, a point at which the substring representing the configuration ceases to match the string that would represent the legitimate successor to the preceding configuration under a move of M (or, in the case of the first configuration, a point at which it ceases to match x_0). If a defect occurs in x_i for some odd (respectively, even) value of i, we shall say that the defect occurs in an odd (respectively, even) configuration.

The language L is accepted by a nondeterministic counter automaton A that uses its finite-state control to check that its input string is in C^+. It begins its computation by guessing whether to look for a defect involving the first or second counter of M, and whether to look for the defect in an odd or even configuration. It then searches for the defect. For example, if A guesses that there is a defect in the second counter at an even configuration, then A proceeds as follows. As A is verifying that its input has the form

$$q_0q_1c^{i_1}d^{j_1}q_2c^{i_2}d^{j_2} \cdots q_nc^{i_n}d^{j_n}, \tag{14}$$

it stores j_k in its counter if k is even; it determines from the value of q_k, and from whether or not $i_k = 0$ and $j_k = 0$, what q_{k+1} should be and whether j_{k+1} should equal $j_k - 1$, j_k, or $j_k + 1$; it then searches for a defect by checking whether q_{k+1} is what it should be, and by comparing j_k in its counter against j_{k+1} from its input.

Note that A, as described thus far, has $\gamma_A = 2$. We impose the following additional constraint on A. Each time that A scans a final state of M, except for the first two times, if A has not yet found a defect, we require that A guess whether to halt in a nonaccepting state or to proceed with the computation. The number of guesses made by a computation of A is now the number m of occurrences of final states that precede any defects found by that computation of A, for $m \geq 2$. This leads to the key observation in the proof: the minimal number of guesses on an input string $w \in L$ is made by whichever computation of A finds the first defect in w. Hence, $\gamma_A(w)$ is the number m of times that M enters a final state during the portion of its computation that is legitimately represented by a prefix of w, for $m \geq 2$. Since M enters a final state for the m-th time no earlier than the $f(m)$-th step of its computation, w contains at least $f(m)$ substrings in C, and so $|w| \geq f(m)$. Hence, $\gamma_A(w) = m \leq g(|w|)$ for $m \geq 2$, so that $\gamma_A(n) \leq g(n)$ for large n. □

The preceding result makes use of unnecessary guessing, but we can expand on this approach to obtain results that do not involve unnecessary guessing, and that therefore describe intrinsic properties of CFLs rather than artifacts of a particular PDA. The following result states that, no matter how slowly a recursive function tends to infinity, there is a context-free language that requires an amount of nondeterminism that tends to infinity at least that slowly.

Theorem 15. *For each monotone, unbounded, recursive function $f(n)$, there is a monotone, unbounded, recursive function $g(n)$ and a context-free language L such that*

$$\textit{some PDA } A \textit{ accepting } L \textit{ has } \gamma_A(n) \leq f(n) \textit{ for all } n\ , \tag{15}$$

and

$$\textit{every PDA } A \textit{ accepting } L \textit{ has } \gamma_A(n) \geq g(n) \textit{ for all large enough } n\ . \tag{16}$$

Proof. The proof involves inserting strings of the form $a^i b^j$, where $j = i$ or $j = 2i$, at each point where the previous counter automaton made unnecessary guesses, in order to force a PDA to guess whether to use its stack to check that $j = i$ or to check that $j = 2i$. This converts the unnecessary guesses to necessary ones, but at the cost of complicating the language so that it can no longer be accepted using just a counter. We omit the details for lack of space. □

From this result, we can establish an infinite hierarchy of complexity classes of CFLs, requiring ever diminishing amounts of nondeterminism.

Definition 16. For each real-valued function $f(x)$, let

$$\mathrm{CFL}(f(n)) = \{L(A) : A \text{ is a PDA with } \gamma_A(n) \le f(n) \text{ for all } n\} . \quad (17)$$

Theorem 17. *There exists an infinite hierarchy*

$$\mathit{CFL}(f_1(n)) \supset \mathit{CFL}(f_2(n)) \supset \cdots . \quad (18)$$

Proof. By the preceding theorem, for any monotone, unbounded, recursive function $f(n)$, there is another such function $g(n)$ and a CFL L such that $L \in \mathrm{CFL}(f(n))$ and such that every PDA A accepting L has $\gamma_A(n) \ge g(n)$ for large n. If $L \in \mathrm{CFL}(h(n))$ for some function h, then there is a PDA B accepting L with $\gamma_B(n) \le h(n)$. Hence,

$$g(n) \le \gamma_B(n) \le h(n) \quad \text{for large } n , \quad (19)$$

so that $h(n) = \Omega(g(n))$. However, we can always construct a monotone, unbounded, recursive function $h(n)$ that grows more slowly than $g(n)$, so that $h(n) \ne \Omega(g(n))$. Then $L \notin \mathrm{CFL}(h(n))$, and so $\mathrm{CFL}(h(n))$ is a proper subset of $\mathrm{CFL}(f(n))$. This process can be iterated indefinitely, proving the theorem. □

References

1. Aho, A., Ullman, J.: The Theory of Parsing, Translation, and Compiling. Prentice-Hall, Englewood Cliffs, NJ (1972)
2. Goldstine, J., Kintala, C., Wotschke, D.: On measuring nondeterminism in regular languages. Information and Computation **86** (1990) 179–194
3. Heilbrunner, S.: A parsing automata approach to LR theory. Theoretical Comp. Sci. **15** (1981) 117–157
4. Herzog, C.: Pushdown automata with bounded nondeterminism or bounded ambiguity. Proceedings LATIN '95, Lecture Notes in Computer Science **911** (1995) 358–370
5. Hopcroft, J., Ullman, J.: Introduction to Automata Theory, Languages, and Computation. Addison Wesley, Reading, MA (1979)
6. Kintala, C.: Refining nondeterminism in context-free languages. Math. Systems Theory **12** (1978) 1–8
7. Sippu, S., Soisalon-Soininen, E.: Parsing Theory. Springer-Verlag, Berlin (1988)
8. Salomaa, K., Yu, S.: Limited nondeterminism for pushdown automata. Bulletin of the EATCS **50** (1993) 186–193
9. Salomaa, K., Yu, S.: Measures of nondeterminism for pushdown automata. J. Comp. System Sci. **49** (1994) 362–374
10. Vermeir, D., Savitch, W.: On the amount of nondeterminism in pushdown automata. Fundamenta Informaticae **4** (1981) 401–418

On Polynomially $\mathcal{D}$-Verbose Sets

Arfst Nickelsen

Technische Universität Berlin
Fachbereich Informatik
10623 Berlin, Germany
nicke@cs.tu-berlin.de

Abstract. A general framework is presente for the study of complexity classes that are defined via polynomial time algorithms that compute partial information about the characteristic function of a given input. Given $n \in \mathbb{N}$ and a family $\mathcal{D}$ of sets $D \subseteq \{0,1\}^*$, a language A is polynomially $\mathcal{D}$-verbose (or: $A \in \mathrm{P}[\mathcal{D}]$) iff there is a polynomial time algorithm that on input $(x_1, \ldots, x_n)$ outputs a $D \in \mathcal{D}$ such that the characteristic string $\chi_A(x_1, \ldots, x_n)$ is in D. Also the variant where only pairwise distinct input words are allowed is studied. p-selective sets, p-verbose sets, easily p-countable sets, sets that allow a polynomial time frequency computation, and cheatable sets are special cases of this definition. It is shown that it suffices to study families that are in a certain normal form. An algorithm is presented that decides for given families $\mathcal{D}_1$, $\mathcal{D}_2$ whether $\mathrm{P}[\mathcal{D}_1] \subseteq \mathrm{P}[\mathcal{D}_2]$. The classes $\mathrm{P}[\mathcal{D}]$ are, except for trivial cases, not closed under union, intersection or join. The classes closed under complement are characterized, as well as those closed under $\leq_m^p$- and $\leq_{1\text{-}tt}^p$-reductions. For a given family of sets $\mathcal{D}$ the class of polynomially $\mathcal{D}$-verbose languages contains non-recursive languages iff it contains all p-selective languages. The families $\mathcal{D}$ for which a $\mathcal{D}$-verbose set can be non-recursive are fully characterized by a simple combinatorial property. It is also shown that for fixed n the classes form a distributive lattice. A diagram that shows this lattice for $n = 2$ is presented.

1 Introduction

Even if a language $A \subseteq \{0,1\}^*$ is not in P, some knowledge about the characteristic function of A may be computable in polynomial time. The idea is to consider n-tuples $(x_1, \ldots, x_n)$ of words as inputs and compute some partial information about $\chi_A(x_1, \ldots, x_n)$. This has lead (depending on the kind of "partial information" one is interested in) to the definition of various complexity notions.

This paper presents a general framework for the study of such classes. The goal is to show that many properties do not depend on specialities of the partial information considered. The focus here is on classes defined by polynomial time machines though some results also concern recursion theoretic questions. A definition that is general enough to capture all these types of classes appeared in [BGK95] (Definition 5.5.) where (strongly) $\mathcal{D}$-verbose sets were defined. In that paper this definition was used to formulate results about frequency computations. One can easily transfer this definition to the polynomial time case. For

Reischuk, Morvan (Eds.): STACS'97 Proceedings, LNCS 1200

given n and a family $\mathcal{D}$ of sets $D \subseteq \{0,1\}^*$ we call a language A polynomially $\mathcal{D}$-verbose iff there is a polynomially time bounded Turing machine M that for every input $(x_1, \ldots, x_n)$ outputs a set $D \in \mathcal{D}$ such that $\chi_A(x_1, \ldots, x_n) \in D$.

The best known and intensively studied examples of (polynomially) $\mathcal{D}$-verbose sets are the semi-recursive sets (introduced in [Joc68]) and their polynomial time analog, the p-selective sets (introduced in [Sel82]). In this case the information given by the algorithm for a set A on input $(x_1, \ldots, x_n)$ consists of a permutation of these words $(x_{i_1}, \ldots, x_{i_n})$ that is compatible with membership in A. The minimal non-zero information on $\chi(\bar{x})$ one can hope for is that at least one of the 2^n bitstrings that are possible for $\chi(\bar{x})$ is excluded. Sets for which a polynomial time algorithm with this property exists is called non-p-n-superterse (see [AG88]), appoximable, $(2^n - 1, n)_p$-verbose (see [BKS92]), or $P\text{-}mc(n)$(see [Ogi94]). In [ABG90] it was shown that non-p-superterse sets are in P/poly. In [BKS94] it was shown that if SAT is $(2^n - 1, n)_p$-verbose for some n then $P = NP$.

Two other prominent types of "partial information" are represented by cardinality computations, which lead to the definition of easily countable sets, and frequency computations. In the case of cardinality computations one tries to minimize the possible values for $|A \cap \{x_1, \ldots, x_n\}|$. In the case of frequency computations one tries on input $(x_1, \ldots, x_n)$ to find a bitstring $(b_1, \ldots, b_n)$ such that $b_i = \chi_A(x_i)$ for as many i as possible.

Should we demand the input words to be pairwise distinct? For selectivity and non-superterseness this makes no difference. But for frequency computations and cardinality computations these restrictions yields classes different from those in the unrestricted case. Therefore this paper treats both variants of definitions and answers the question in which cases this really leads to different classes.

2 Definitions and First Results

For basic notions and results in complexity theory see, e.g., [BDG88]. $\mathbb{N}$ denotes the set of positive integers. For a set X, its cardinality is denoted $|X|$. A language is a subset of $\{0,1\}^*$. For a language A the characteristic function $\chi_A : \{0,1\}^* \to \{0,1\}$ is defined by $\chi_A(x) = 1 \Leftrightarrow x \in A$. We extend this to tuples of words by setting $\chi_A(x_1, \ldots, x_n) = (\chi_A(x_1), \ldots, \chi_A(x_n))$. REC is the class of recursive languages and P the class of polynomial time decidable languages. For languages A, B the join is defined as $A \oplus B = \{x0 \mid x \in A\} \cup \{x1 \mid x \in B\}$.

Definition 1 (n-Families, Variants of $\mathcal{D}$-Verboseness). Let $n \in \mathbb{N}$ and let $\mathcal{D} = \{D_1, \ldots, D_r\}$ be a set of finite sets of bitstrings of length n, i.e., $D_i \subseteq \{0,1\}^n$ for all $i \in \{1, \ldots, r\}$. We call $\mathcal{D}$ an n-family if $\bigcup_{i=1}^r D_i = \{0,1\}^n$ and $D_i \not\subseteq D_j$ for all i, j with $i \neq j$. Now we introduce four types of complexity classes. For a given n-family $\mathcal{D} = \{D_1, \ldots, D_r\}$ a language A is

1. in REC $[\mathcal{D}]$ iff there is a deterministic Turing machine M that for every input $(x_1, \ldots, x_n)$ outputs an $i \in \{1, \ldots, r\}$ such that $\chi_A(x_1, \ldots, x_n) \in D_i$. Such languages are called $\mathcal{D}$-verbose.

2. in $\mathrm{REC}_{\mathrm{dist}}[\mathcal{D}]$ iff there is a deterministic Turing machine M that for every input $(x_1,\dots,x_n)$ of pairwise distinct words outputs an $i \in \{1,\dots,r\}$ such that $\chi_A(x_1,\dots,x_n) \in D_i$. Such languages are called weakly $\mathcal{D}$-verbose.
3. in $\mathrm{P}[\mathcal{D}]$ iff A is in $\mathrm{REC}[\mathcal{D}]$ via a polynomial time bounded machine M. Such languages are called polynomially $\mathcal{D}$-verbose.
4. in $\mathrm{P}_{\mathrm{dist}}[\mathcal{D}]$ iff A is in $\mathrm{REC}_{\mathrm{dist}}[\mathcal{D}]$ via a polynomial time bounded machine M. Such languages are called weakly polynomially $\mathcal{D}$-verbose.

We discuss shortly how classes that appear in the literature can be defined by n-families.

- Let $n \in \mathbb{N}$. Define $(k,n)\text{-SIZE} = \{D \subseteq \{0,1\}^n \mid |D| = k\}$. Then $P = \mathrm{P}[(1,n)\text{-SIZE}]$. All languages are in $\mathrm{P}[(2^n,n)\text{-SIZE}]$. A language A is non-p-n-superterse if $A \in \mathrm{P}[(2^n-1,n)\text{-SIZE}]$; A is cheatable if $A \in \mathrm{P}[(n,n)\text{-SIZE}]$ for some n.
- Call a set $B = \{b_0,\dots,b_n\} \subseteq \{0,1\}^n$ an ascending set if $b_1 = 0^n$, $b_n = 1^n$, and for all $i < n$, b_{i+1} is obtained from b_i by changing one 0 to 1. Define $(k,n)\text{-SORT} = \{D \subseteq \{0,1\}^n \mid, D \text{ is the union of } k \text{ ascending sets}\}$. Then a language A is easily sortable ([HN93]) if $A \in (\binom{n}{\lceil \frac{n}{2} \rceil} - 1, n)\text{-SORT}$. For $n = 2$ and $k = 1$ we get $(1,2)\text{-SORT} = \{D_1, D_2\}$ with $D_1 = \{00,01,11\}$ and $D_2 = \{00,10,11\}$. Therefore $\mathrm{REC}[(1,2)\text{-SORT}]$ consists of the semi-recursive sets and $\mathrm{P}[(1,2)\text{-SORT}]$ consists of the p-selective sets. We abbreviate $(1,n)$-SORT by n-SEL. Then the class of p-selective sets equals $\mathrm{P}[(1,n)\text{-SORT}]$ for all $n \geq 2$.
- For $b \in \{0,1\}^n$ let $\#_1(b)$ denote the number of 1's in b. Define $D_N \subseteq \{0,1\}^n$ by $b \in D_N \Leftrightarrow \#_1(b) \in N$. For $k \in \{1,\dots,n+1\}$ define $(k,n)\text{-CARD} = \{D_N \mid |N| = k\}$. Then $\mathrm{P}[(n,n)\text{-CARD}]$ are the easily countable sets in the sense of [HN93] and $\mathrm{P}_{\mathrm{dist}}[(n,n)\text{-CARD}]$ are the easily countable sets in the sense of [Rog92].
- For each $b \in \{0,1\}^n$ define $B_k(b) = \{b' \in \{0,1\}^n \mid b \text{ and } b' \text{ differ in at most } k \text{ bits}\}$. Define $(k,n)\text{-FREQ} = \{B_k(b) \mid b \in \{0,1\}^n\}$. Then a language A is (m,n)-recursive in the sense of [KS95] if $A \in \mathrm{REC}_{\mathrm{dist}}[(n-m,n)\text{-FREQ}]$.

In [BGK95] $\mathcal{D}$-verbose sets were called strongly $\mathcal{D}$-verbose. Observe that by definition $\mathrm{P}[\mathcal{D}] \subseteq \mathrm{P}_{\mathrm{dist}}[\mathcal{D}]$ and $\mathrm{REC}[\mathcal{D}] \subseteq \mathrm{REC}_{\mathrm{dist}}[\mathcal{D}]$. To compare the information that different n-families provide, we need the following definition.

Definition 2 (Refinement). Let $\mathcal{D}_1$ and $\mathcal{D}_2$ be n-families for some fixed n. $\mathcal{D}_1$ is a refinement of $\mathcal{D}_2$, $\mathcal{D}_1 \preceq \mathcal{D}_2$, if for each $D \in \mathcal{D}_1$ there is a $D' \in \mathcal{D}_2$ with $D \subseteq D'$. We write $\mathcal{D}_1 \prec \mathcal{D}_2$ iff $\mathcal{D}_1 \preceq \mathcal{D}_2$ and $\mathcal{D}_1 \neq \mathcal{D}_2$.

The relation $\preceq$ forms a partial ordering on n-families. Observe that if $\mathcal{D}_1 \preceq \mathcal{D}_2$ then $\mathrm{P}[\mathcal{D}_1] \subseteq \mathrm{P}[\mathcal{D}_2]$ and $\mathrm{P}_{\mathrm{dist}}[\mathcal{D}_1] \subseteq \mathrm{P}_{\mathrm{dist}}[\mathcal{D}_2]$. In the next section we show that if $\mathcal{D}_1$ and $\mathcal{D}_2$ are in certain normal forms also the converse holds. Section 4 answers the question in which cases inclusion can be strengthened to strict inclusion.

3 Normal Forms

If we know that a language A is in P$[\mathcal{D}]$ via a Turing machine M and we are given a tuple of input strings $(x_1, \ldots, x_n)$, we can try to get as much information about $\chi_A(x_1, \ldots, x_n)$ as possible by running M on modified inputs. There are essentially three ways to modify the input: We can change the order of the x_i. We can combine some of the x_i with some special strings y_j for which $\chi_A(y_j)$ is already known. Or we can leave out some of the x_i and put in some other x_j, possibly several times.

Definition 3 (Permutations, Projections, Replacements).

1. For $n \in \mathbb{N}$ let S_n be the group of permutations of $\{1, \ldots, n\}$. For $\sigma \in S_n$ and $(b_1, \ldots, b_n) \in \{0,1\}^k$ we define $\sigma(b_1, \ldots, b_n) = (b_{\sigma(1)}, \ldots, b_{\sigma(n)})$. For $D \subseteq \{0,1\}^n$ define $\sigma(D) = \{\sigma(b) \mid b \in D\}$. An n-family $\mathcal{D}$ is said to be closed under permutations iff $\sigma(D) \in \mathcal{D}$ for all $\sigma \in S_n$ and all $D \in \mathcal{D}$.
2. For $n \in \mathbb{N}$, $i \in \{1, \ldots, n\}$ and $c \in \{0,1\}$ define projections $\pi_i^c : \{0,1\}^n \to \{0,1\}^n$ by $\pi_i^c(b_1, \ldots, b_n) = (b_1, \ldots, b_{i-1}, c, b_{i+1}, \ldots, b_n)$. For $D \subseteq \{0,1\}^n$ define $\pi_i^c(D) = \{\pi_i^c(b) \mid b \in D\}$. An n-family $\mathcal{D}$ is said to be closed under projections iff for all projections π_i^c and all $D \in \mathcal{D}$ there is an $D' \in \mathcal{D}$ such that $\pi_i^c(D) \subseteq D'$.
3. For $n \in \mathbb{N}$, $i, j \in \{1, \ldots, n\}$ define a replacement operation $\rho_{i,j} : \{0,1\}^n \to \{0,1\}^n$ by $\rho_{i,j}(b_1, \ldots, b_n) = (b'_1, \ldots, b'_n)$ where $b'_k = b_k$ for $k \neq j$ and $b'_j = b_i$. For $D \subseteq \{0,1\}^n$ define $\rho_{i,j}(D) = \{\rho_{i,j}(b) \mid b \in D\}$. An n-family $\mathcal{D}$ is said to be closed under replacements iff for all replacements $\rho_{i,j}$ and all $D \in \mathcal{D}$ there is an $D' \in \mathcal{D}$ such that $\rho_{i,j}(D) \subseteq D'$.

Definition 4 (Normal Forms). An n-family $\mathcal{D}$ is in weak normal form if $\mathcal{D}$ is closed under permutations and projections. It is in normal form if it is additionally closed under replacements.

To simplify the proof of Theorem 6, and some of the following results we introduce a split-operation on n-families. This operation also plays a main role if one actually wants to compute the normal forms for a given n-family.

Definition 5 (Split-Operation). Let $\mathcal{D} = \{D_1, \ldots, D_r\}$ be an n-family. The operation split get as input the family $\mathcal{D}$ and a $D_i \in \mathcal{D}$ with $|D_i| \geq 2$ and outputs an n-family that is a proper refinement of $\mathcal{D}$. We define

$$\text{split}(D_i, \mathcal{D}) = \{D_j \mid j \neq i\} \cup \{D' \subset D_i \mid |D'| = |D_i| - 1 \text{ and } \forall j \neq i\ D' \not\subseteq D_j\} .$$

Basically, to obtain split$(D_i, \mathcal{D})$ from $\mathcal{D}$, we remove D_i from $\mathcal{D}$ and replace it by those subsets D' of D_i that contain one element less than D_i. Only if there is already a superset D_j of D' we leave out D'.

Theorem 6 (Normal Forms). *For all $n \in \mathbb{N}$ and all n-families $\mathcal{D}$ there is an n-family $\mathcal{D}'$ in weak normal form such that $\mathcal{D}' \preceq \mathcal{D}$ There also is an n-family $\mathcal{D}''$ such that $\mathcal{D}'' \preceq \mathcal{D}$ and $P[\mathcal{D}] = P[\mathcal{D}'']$.*

Proof. We prove the first claim. It suffices to show that for every $\mathcal{D}$ that is not closed under permutations and projections there is a strict refinement $\mathcal{D}' \prec \mathcal{D}$ with $\mathrm{P}_{\text{dist}}[\mathcal{D}] \subseteq \mathrm{P}_{\text{dist}}[\mathcal{D}']$. Thus we get a descending chain of n-families. Such a chain always reaches a family in weak normal form because the minimal n-family consisting only of one-element sets is in normal form.

a) Suppose $\mathcal{D}$ is not closed under permutations. Let $D \in \mathcal{D}$ a set of maximal size for which there is a permution σ such that $\sigma(D) \notin \mathcal{D}$. Define $\mathcal{D}' = \text{split}(D, \mathcal{D})$. $\mathcal{D}'$ is a strict refinement of $\mathcal{D}$. To show that $\mathrm{P}_{\text{dist}}[\mathcal{D}] \subseteq \mathrm{P}_{\text{dist}}[\mathcal{D}']$, let $A \in \mathrm{P}_{\text{dist}}[\mathcal{D}]$ via a machine M. A machine that witnesses $A \in \mathrm{P}_{\text{dist}}[\mathcal{D}']$ on input $(x_1, \dots, x_n)$ first computes $M(x_1, \dots, x_n)$. If $M(x_1, \dots, x_n) \neq D$ we are already done. Otherwise compute $M(x_{\sigma(1)}, \dots, x_{\sigma(n)})$. Then we know that $\chi_A(x_1, \dots, x_n) \in D \cap \sigma^{-1}(M(x_{\sigma(1)}, \dots, x_{\sigma(n)}))$. This is a strict subset of D. Output an $D' \in \mathcal{D}'$ with $D \cap \sigma^{-1}(M(x_{\sigma(1)}, \dots, x_{\sigma(n)})) \subseteq D'$.
b) Suppose $\mathcal{D}$ is closed under permutations but not under projections. W.l.o.g. suppose for a set $D \in \mathcal{D}$ that $\pi_1^1(D) \not\subseteq D'$ for all $D' \in \mathcal{D}$. Define again $\mathcal{D}' = \text{split}(D, \mathcal{D})$. Let $A \in \mathrm{P}[\mathcal{D}]$ via a machine M. Let a be a fixed word from A (w.l.o.g. $A \neq \emptyset$). A machine that witnesses $A \in \mathrm{P}[\mathcal{D}']$ on input $(x_1, \dots, x_n)$ first computes $M(x_1, \dots, x_n)$. If $M(x_1, \dots, x_n) \neq D$ we are already done. Otherwise compute $M(a, x_2, \dots, x_n)$. There is a string $b \in D$ such that $\pi_1^1(b) \notin M(a, x_2, \dots, x_n)$. Output $D' \in \mathcal{D}'$ with $D \setminus \{b\} \subseteq D'$.

The second claim can be proved in the same way. One has additionally to consider a case c) where $\mathcal{D}$ is closed under permutations and projections, but not under replacements. In all three cases one can show that $\mathrm{P}[\mathcal{D}] \subseteq \mathrm{P}[\mathcal{D}']$. □

The permutation group S_n is generated by two permutations σ_1 and σ_2 where $\sigma_1 = (1, 2)$ exchanges the first two elements and $\sigma_2 = (1, \dots, n)$ is a cyclic shift of all elements. Therefore, to check whether $\mathcal{D}$ is closed under permutations, it suffices to check whether $\mathcal{D}$ is closed under σ_1 and σ_2. If D is closed under permutations, to check closure under projections and replacements it suffices to check whether $\mathcal{D}$ is closed under the three operations π_1^0, π_1^1 (projections in the first component) and $\rho_{1,2}$ (replacing the second entry by the first one). The proof of Theorem 6 shows that a process of repeatedly splitting the sets $D \in \mathcal{D}$ that prevent $\mathcal{D}$ from being in normal form at last yields a normal form for $\mathcal{D}$. Thus the following algorithm can be used to compute the normal form for n-families. (To compute weak normal forms one has to omit the test for closure under $\rho_{1,2}$.)

Algorithm 7 (Computation of Normal Forms).

input $\mathcal{D}$
check for all $D \in \mathcal{D}$ *and all* $\tau \in \{\sigma_1, \sigma_2, \pi_1^0, \pi_1^1, \rho_{1,2}\}$
  whether there is a D' *with* $\tau(D) \subseteq D'$
  if no then D *and* τ *are found*
    such that $\tau(D) \not\subseteq D'$ *for all* $D' \in \mathcal{D}$*;*
    set $\mathcal{D} := \mathit{split}(D, \mathcal{D})$*;*
    start from the beginning
  else $\mathcal{D}$ *is in normal form;*
    return $\mathcal{D}$

4 The Inclusion Problem

It turns out that the inclusion problem for polynomial verboseness classes fully reduces to the refinement question for n-families:

Theorem 8 (Inclusion Reduces to Refinement).

For n-families $\mathcal{D}_1$, $\mathcal{D}_2$ in weak normal form, $P_{\text{dist}}[\mathcal{D}_1] \subseteq P_{\text{dist}}[\mathcal{D}_2]$ iff $\mathcal{D}_1 \preceq \mathcal{D}_2$.

For n-families $\mathcal{D}_1$, $\mathcal{D}_2$ in normal form, $P[\mathcal{D}_1] \subseteq P[\mathcal{D}_2]$ iff $\mathcal{D}_1 \preceq \mathcal{D}_2$.

Proof. We only give the proof of the first claim. The second can be proven similarly. Obviously $\mathcal{D}_1 \preceq \mathcal{D}_2$ implies $\mathrm{P}_{\text{dist}}[\mathcal{D}_1] \subseteq \mathrm{P}_{\text{dist}}[\mathcal{D}_2]$. It remains to show for $\mathcal{D}_1, \mathcal{D}_2$ in weak normal form that $\mathcal{D}_1 \not\preceq \mathcal{D}_2$ implies the existence of an $A \in \mathrm{P}[\mathcal{D}_1] \setminus \mathrm{P}[\mathcal{D}_2]$. If $\mathcal{D}_1 \not\preceq \mathcal{D}_2$ there is a $D_1 \in \mathcal{D}_1$ such that $D_1 \not\subseteq D_2$ for all $D_2 \in \mathcal{D}_2$. We construct step by step finite sets A_r such that $A = \bigcup_{r=1}^{\infty} A_r$. Let $\langle M_r \rangle_{r=1}^{\infty}$ be an enumeration of polynomial time Turing machines such that for all $B \in \mathrm{P}_{\text{dist}}[\mathcal{D}_2]$ there is an r such that $B \in \mathrm{P}_{\text{dist}}[\mathcal{D}_2]$ via M_r. Choose $l_1 \geq n$ and set $l_{r+1} = 2^{l_r}$. Let $x_1^r, \ldots, x_n^r$ be the n lexicographically first strings of length l_r. We will choose $A_r \subseteq \{x_1^r, \ldots, x_n^r\}$. Let s be the lexicographically first string in $D_1 \setminus M_r(x_1^r, \ldots, x_n^r)$. Define A_r such that $\chi_{A_r}(x_1^r, \ldots, x_n^r) = s$. This ensures that $A \notin \mathrm{P}_{\text{dist}}[\mathcal{D}_2]$. It remains to specify a machine M such that $A \in \mathrm{P}[\mathcal{D}_1]$ via M. Given an input $(x_1, \ldots, x_n)$ of pairwise distinct words M looks for the largest r such that $\{x_1, \ldots, x_n\} \cap \{x_1^r, \ldots, x_n^r\} \neq \emptyset$. If there is no such r then $\chi_A(x_1, \ldots, x_n) = (0, \ldots, 0)$ and M outputs a $D \in \mathcal{D}_1$ with $(0, \ldots, 0) \in D$. Otherwise M computes $\chi_A(x_i)$ for all $x_i \notin \{x_1^r, \ldots, x_n^r\}$ as follows. If there is no $r' < r$ with $x_i \in \{x_1^{r'}, \ldots, x_n^{r'}\}$ then $\chi_A(x_i) = 0$, otherwise M can simulate $M_{r'}$ on $(x_1^{r'}, \ldots, x_n^{r'})$ and thus compute $\chi_A(x_i)$. We know that $\chi_A(x_1^r, \ldots, x_n^r) \in D'$. Now we can compute from D_1 a set D with $\chi_A(x_1, \ldots, x_n) \in D$ by applying permutations and projections to D_1. □

Corollary 9. *Let $\mathcal{D}$ be an n-family in weak normal form. Then*

$$P[\mathcal{D}] = P_{\text{dist}}[\mathcal{D}] \text{ iff } \mathcal{D} \text{ is in normal form} .$$

Now the inclusion problem for fixed n is characterized. Next we show how to compare classes that are defined via different tuple lengths.

Theorem 10 (Changing the Tuple-Length).

1. *Let $\mathcal{D}$ be an n-family, $l \geq 1$. Let $\mathcal{D}' = \{D' \mid D \in \mathcal{D}\}$ where D' is the set of bitstrings obtained by extending the strings in D in all possible ways; i.e. $D' = \{(b_1, \ldots, b_{n+l}) \mid (b_1, \ldots, b_n) \in D\}$. Then $P[\mathcal{D}] = P[\mathcal{D}']$.*
2. *Let $n, l \geq 1$, $\mathcal{D}$ an $(n+l)$-family. Let $\mathcal{D}'' = \{D'' \mid D \in \mathcal{D}\}$ where D'' is the set of bitstrings of length n obtained by removing the last l bits from the strings in D. Then $P[\mathcal{D}] \subseteq P[\mathcal{D}'']$.*

We omit the proof. Obviously in 2. inclusion cannot be improved to equality. We combine the above results and give an algorithm that determines for given m-family $\mathcal{D}_1$ and n-family $\mathcal{D}_2$ the relation between $\mathrm{P}[\mathcal{D}_1]$ and $\mathrm{P}[\mathcal{D}_2]$.

Algorithm 11 (Comparing Two Polynomial Verboseness Classes).
input m-family $\mathcal{D}_1$ *and n-family* $\mathcal{D}_2$
if $m < n$ *then* $\mathcal{D}_1 := \{D' \mid D \in \mathcal{D}_1\}$ *where* $D' = \{(b_1, \ldots, b_n) \mid (b_1, \ldots, b_m) \in D\}$;
if $n < m$ *then* $\mathcal{D}_2 := \{D' \mid D \in \mathcal{D}_2\}$ *where* $D' = \{(b_1, \ldots, b_n) \mid (b_1, \ldots, b_m) \in D\}$;
$\mathcal{D}_1 :=$ *normal form*$(\mathcal{D}_1)$; $\mathcal{D}_2 :=$ *normal form*$(\mathcal{D}_1)$; *(with Algorithm 7)*
if $\mathcal{D}_1 \prec \mathcal{D}_2$ *output* " $P[\mathcal{D}_1] \subset P[\mathcal{D}_2]$ "; *if* $\mathcal{D}_2 \prec \mathcal{D}_1$ *output* " $P[\mathcal{D}_2] \subset P[\mathcal{D}_1]$ ";
if $\mathcal{D}_1 = \mathcal{D}_2$ *output* " $P[\mathcal{D}_1] = P[\mathcal{D}_2]$ ";
if $\mathcal{D}_1 \not\preceq \mathcal{D}_2$ *and* $\mathcal{D}_2 \not\preceq \mathcal{D}_1$ *output* " $P[\mathcal{D}_1]$ *and* $P[\mathcal{D}_2]$ *are uncomparable* "

In the partial ordering $\preceq$ on n-families $(2^n - 1, n)$-SIZE $= \{D \mid |D| = 2^n - 1\}$ is submaximal. We call an n-family $\mathcal{D}$ submaximal if for all $\mathcal{D}'$, except for the maximal family $\{D \mid |D| = 2^n\}$, it holds $\mathcal{D}' \preceq \mathcal{D}$. If we restrict ourselves to families in (weak) normal form, in both cases we also get an atomic n-family. We call $\mathcal{D}$ atomic if for all $\mathcal{D}'$, except for the minimal family $\{D \mid |D| = 1\}$, it holds $\mathcal{D} \preceq \mathcal{D}'$. We define these atomic families and state some easy properties.

Definition 12 (n-MIN, n-weakMIN). For $n \in \mathbb{N}$ define

1. n-weakMIN $= \{\{(b_1, \ldots, b_n), (b'_1, \ldots, b'_n)\} \mid \exists i\, b_i = 0 \wedge b'_i = 1 \wedge \forall j \neq i\, b_j = b'_j\}$
2. n-MIN $= \{\{(b_1, \ldots, b_n), (b'_1, \ldots, b'_n)\} \mid \exists I \subseteq \{1, \ldots, n\}\, \forall i \in I\, b_i = 0 \wedge b'_i = 1 \wedge \forall j \notin I\, b_j = b'_j\}$.

Lemma 13. *For all* $n \geq 2$, *n-weakMIN is atomic with respect to* $\preceq$ *for n-families in weak normal form; n-MIN is atomic with respect to* $\preceq$ *for n-families in normal form; and for all* $m \geq 2$ *is* $P_{\text{dist}}[n\text{-weakMIN}] = P_{\text{dist}}[m\text{-weakMIN}]$ *and* $P[n\text{-MIN}] = P[m\text{-MIN}]$.

5 Closure Properties

In this section closure of the classes $P_{\text{dist}}[\mathcal{D}]$ under boolean operations, many-one reductions and $1 - tt$-reductions is investigated. We first consider complement.

Definition 14 (Negation). For $n \geq 1$ let the negation operation $\nu : \{0,1\}^n \to \{0,1\}^n$ be defined by $\nu(b_1, \ldots, b_n) = (1 - b_1, \ldots, 1 - b_n)$. For $D \subseteq \{0,1\}^n$ define $\nu(D) = \{\nu(b) \mid b \in D\}$. An n-family $\mathcal{D}$ is called closed under negation if $D \in \mathcal{D}$ implies $\nu(D) \in \mathcal{D}$ for all D.

Theorem 15. *For all n and for all n-families* $\mathcal{D}$ *in weak normal form is* $P_{\text{dist}}[\mathcal{D}]$ *closed under complement iff* $\mathcal{D}$ *is closed under negation.*

Proof. "$\Rightarrow$" Let $\mathcal{D}$ be an n-family in weak normal form that is closed under complement, but not closed under negation. Choose $D \in \mathcal{D}$ such that $\nu(D) \not\subseteq D'$ for all $D' \in \mathcal{D}$. Define $\mathcal{D}' = \text{split}(\mathcal{D}, D)$. Clearly $\mathcal{D}'$ is a proper refinement of $\mathcal{D}$. By Theorem 8 there exists a language $A \in P_{\text{dist}}[\mathcal{D}] \setminus P_{\text{dist}}[\mathcal{D}']$. We will show that A is in $P_{\text{dist}}[\mathcal{D}']$ – a contradiction. Let B be the complement of A. Suppose A is in $P_{\text{dist}}[\mathcal{D}]$ via M_A and B is in $P_{\text{dist}}[\mathcal{D}]$ via M_B. The algorithm that witnesses $A \in P_{\text{dist}}[\mathcal{D}']$ works as follows. On input $(x_1, \ldots, x_n)$ if $M_A(x_1, \ldots, x_n) \neq D$

output $M(x_1,\ldots,x_n)$. If $M_A(x_1,\ldots,x_n) = D$ compute $M_B(x_1,\ldots,x_n)$. Then $\chi_A(x_1,\ldots,x_n) \in \nu(M_B(x_1,\ldots,x_n))$. Output $D \cap \nu(M_B(x_1,\ldots,x_n))$.

"$\Leftarrow$" Let $\mathcal{D}$ be an n-family in weak normal form that is closed under negation. Let $A \in \mathrm{P}_{\mathrm{dist}}[\mathcal{D}]$ and let B be the complement of A. Then $B \in \mathrm{P}_{\mathrm{dist}}[\mathcal{D}]$ because if $\chi_A(x_1,\ldots,x_n) \in D$ then $\chi_B(x_1,\ldots,x_n) \in \nu(D)$ and $\nu(D)$ is in $\mathcal{D}$. □

Theorem 16 (Closure under Join). *For all $n \geq 2$ there are sets $A_1,\ldots,A_n \in P_{\mathrm{dist}}[2\text{-}weakMIN]$ such that $A_1 \oplus \cdots \oplus A_n \notin P[(2^n-1,n)\text{-}SIZE]$.*

Sketch of Proof. To fix notation let $A_1 \oplus \cdots \oplus A_n = \{\langle w,i\rangle \mid 1 \leq i \leq n, w \in A_i\}$. We construct stepwise the A_i as tally sets. Consider only word lengths n_j where $n_1 = 1$ and $n_{j+1} = 2^{n_j}$. Let $\langle M_j\rangle_{j\in\mathbb{N}}$ be an effective enumeration of polynomial time Turing machines that possibly witness membership in $\mathrm{P}[(2^n-1,n)\text{-SIZE}]$. We can ensure that there is a polynomial p such that for inputs x up to length n_j a machine can simulate M_j on x in time less than $p(n_{j+1})$. Each A_i will be subsets of $\{0^n \mid n = n_j \text{ for some } j\}$. For each j there is there is a string $b = (b_1,\ldots,b_n)$ such that $b \notin M_j(\langle 0^{n_j},1\rangle,\ldots,\langle 0^{n_j},n\rangle)$. Put 0^{n_j} into A_i iff $b_i = 1$. Thus $A_1 \oplus \cdots \oplus A_n \notin \mathrm{P}[(2^n-1,n)\text{-SIZE}]$. On inputs of the form $0^{n_k},0^{n_j})$ with $k < j$ a machine can simulate M_k on $(\langle 0^{n_k},1\rangle,\ldots,\langle 0^{n_k},n\rangle)$ and determine $\chi_{A_i}(0^{n_k})$ for each i. Therefore each A_i is in $\mathrm{P}_{\mathrm{dist}}[\text{2-weakMIN}]$. □

Corollary 17. *All the classes $P_{\mathrm{dist}}[\mathcal{D}]$ except P and the class of all languages are not closed under join, union and intersection.*

Proof. They are not closed under join because of Theorem 16. The proof of Theorem 16 also shows that there are sets $A_1,\ldots,A_n \in \mathrm{P}_{\mathrm{dist}}[n\text{-weakMIN}]$ such that $A_1 \cup \ldots \cup A_n \notin \mathrm{P}[(2^n-1,n)\text{-SIZE}]$. The result for intersection follows because $\mathrm{P}_{\mathrm{dist}}[n\text{-weakMIN}]$ is closed under complement. □

Theorem 18 (Many-One Reductions). *For all n and for all n-families $\mathcal{D}$*

1. *$P[\mathcal{D}]$ is closed under polynomial time many-one reductions.*
2. *If $P_{\mathrm{dist}}[\mathcal{D}] \neq P[\mathcal{D}]$ then $P_{\mathrm{dist}}[\mathcal{D}]$ is not closed under polynomial time many-one reductions.*

Proof. To prove 1. let $A \leq_m^p B$ via a polynomial time computable function f and $B \in \mathrm{P}[\mathcal{D}]$ via a Turing machine M_B. On input $(x_1,\ldots,x_n)$ compute $j := M_b(f(x_1),\ldots,f(x_n))$. Because $\chi_A(x_1,\ldots,x_n) = \chi_B(f(x_1),\ldots,f(x_n))$ it is $\chi_A(x_1,\ldots,x_n) \in D_j \in \mathcal{D}$.

To prove 2. let $\mathcal{D}$ be in weak normal form, but not in normal form, and let $A \in \mathrm{P}_{\mathrm{dist}}[\mathcal{D}] \setminus \mathrm{P}[\mathcal{D}]$. Suppose $\mathrm{P}_{\mathrm{dist}}[\mathcal{D}]$ is closed under $\leq_m^p$. Then $A \oplus A \oplus \cdots A$ is in $\mathrm{P}_{\mathrm{dist}}[\mathcal{D}]$ because $A \oplus A \oplus \cdots A \leq_m^p A$. But if $A \oplus A \oplus \cdots A \in \mathrm{P}_{\mathrm{dist}}[\mathcal{D}]$ then $A \in \mathrm{P}[\mathcal{D}]$; a contradiction. □

Definition 19 (Bitflip). For $n \geq 1$ and $i \in \{1,\ldots,n\}$ define the bitflip operation $\nu_i : \{0,1\}^n \to \{0,1\}^n$ by $\nu_i(b_1,\ldots,b_n) = (b_1,\ldots,b_{i-1},1-b_i,b_{i+1},\ldots,b_n)$. For $D \subseteq \{0,1\}^n$ define $\nu_i(D) = \{\nu_i(b) \mid b \in D\}$. An n-family $\mathcal{D}$ is called closed under bitflip iff for all bitflip operations ν_i and all $D \in \mathcal{D}$ $\nu_i(D) \in \mathcal{D}$.

Obviously an n-family $\mathcal{D}$ is closed under negation if it is closed under bitflip. The converse does not hold as can be seen e.g. for $\mathcal{D} =$ 2-SEL.

Theorem 20 (1-tt-Reductions). *Let $\mathcal{D}$ be an n-family in normal form. $P[\mathcal{D}]$ is closed under polynomial time 1-tt-reductions iff $\mathcal{D}$ is closed under bitflip operations.*

Proof. "$\Leftarrow$" Let $\mathcal{D}$ be closed under bitflip operations. Let A, B be languages such that $A \leq^p_{1-tt} B$ via f and $B \in \mathrm{P}[\mathcal{D}]$ via a Turing machine M_B. For any input $(x_1, \ldots, x_n)$ for some x_i $x_i \in A \Leftrightarrow f(x_i) in B$ and for some x_i and for some x_i $x_i \in A \Leftrightarrow f(x_i) \notin B$; say $x_i \in A \Leftrightarrow f(x_i) in B$ for $i < j$ and $x_i \in A \Leftrightarrow f(x_i) \notin B$ for $i \geq j$ (otherwise permute the input words appropriately – $\mathcal{D}$ is closed under permutation). Compute $j := M_B(f(x_1), \ldots, f(x_n))$. Then $\chi_A(x_1, \ldots, x_n) \in \nu_{i_0} \circ \cdots \circ \nu_1(D_j)$ which is in $\mathcal{D}$ because $\mathcal{D}$ is closed under bitflips.

"$\Rightarrow$" Let $\mathcal{D}$ be not closed under bitflip operations. Then there is a $D \in \mathcal{D}$ such that $\nu_1(D) \notin \mathcal{D}$. Define $\mathcal{D}' = \mathrm{split}(D, \mathcal{D})$. Let B be a language that is in $\mathrm{P}[\mathcal{D}]$ via Turing machine M_B, but $B \notin \mathrm{P}[\mathcal{D}']$. Define $A = \{w0 \mid w \in B\} \cup \{w1 \mid w \notin B\}$. Then $A \leq^p_{1-tt} B$. It remains to show that $A \notin \mathrm{P}[\mathcal{D}]$. Suppose $A \in \mathcal{D}$. We argue that $B \in \mathrm{P}[\mathcal{D}']$. On input $(x_1, \ldots, x_n)$ compute $M_B(x_1, \ldots, x_n)$. If $M_B(x_1, \ldots, x_n) \neq D$ output $M_B(x_1, \ldots, x_n)$. If $M_B(x_1, \ldots, x_n) = D$ compute $M_B(x_1 1, x_2 0, \ldots, x_k 0)$. Then $\chi_B(x_1, \ldots, x_n) \in \nu_1(M_B(x_1 1, x_2 0, \ldots, x_k 0)) \cap D$ and $\nu_1(M_B(x_1 1, x_2 0, \ldots, x_k 0)) \cap D$ is a proper subset of D. Output the set $D' \in \mathcal{D}'$ with $\nu_1(M_B(x_1 1, x_2 0, \ldots, x_k 0)) \cap D \subseteq D'$. □

6 The Distributive Lattice

We define two operations $\sqcap$ and $\sqcup$ on n-families. It turns out that that the n-families for fixed n together with these operations form a distributive lattice.

Definition 21. For $\mathcal{D}_1$ and $\mathcal{D}_2$ let $\mathcal{D}_1 \sqcap \mathcal{D}_2 = \{D_1 \cap D_2 \mid D_1 \in \mathcal{D}_1 \text{ and } D_2 \in \mathcal{D}_2\}$ and $\mathcal{D}_1 \sqcup \mathcal{D}_2 = \{D \mid D \in \mathcal{D}_1 \cup \mathcal{D}_2, \text{ there is no } D' \in \mathcal{D}_1 \cup \mathcal{D}_2 \text{ with } D \subset D'\}$.

Lemma 22. *$P[\mathcal{D}_1] \cap P[\mathcal{D}_2] = P[\mathcal{D}_1 \sqcap \mathcal{D}_2]$ and $P[\mathcal{D}_1] \cup P[\mathcal{D}_2] \subseteq P[\mathcal{D}_1 \sqcup \mathcal{D}_2]$*

Lemma 23. *If $\mathcal{D}_1$, $\mathcal{D}_2$ are n-families in (weak) normal form then also $\mathcal{D}_1 \sqcap \mathcal{D}_2$ and $\mathcal{D}_1 \sqcup \mathcal{D}_2$ are n-families in (weak) normal form.*

Theorem 24 (Distributive Lattice). *For fixed n the partial ordering of n-families in normal form together with the operations $\sqcap$ and $\sqcup$ form a distributive lattice.*

Proof. For given $\mathcal{D}_1$, $\mathcal{D}_2$ it is easy to see that $\mathcal{D}_1 \sqcup \mathcal{D}_2$ is the least upper bound for $\mathcal{D}_1$, $\mathcal{D}_2$; and $\mathcal{D}_1 \sqcap \mathcal{D}_2$ is the greatest lower bound for $\mathcal{D}_1$, $\mathcal{D}_2$. It remains to show distributivity. We have to show that for n-families $\mathcal{D}_1$, $\mathcal{D}_2$, $\mathcal{D}_3$ it holds that a) $\mathcal{D}_1 \sqcap (\mathcal{D}_2 \sqcup \mathcal{D}_3) = (\mathcal{D}_1 \sqcap \mathcal{D}_2) \sqcup (\mathcal{D}_1 \sqcap \mathcal{D}_3)$ and b) $\mathcal{D}_1 \sqcup (\mathcal{D}_2 \sqcap \mathcal{D}_3) = (\mathcal{D}_1 \sqcup \mathcal{D}_2) \sqcap (\mathcal{D}_1 \sqcup \mathcal{D}_3)$. We here only give the proof of a).

We show that $\mathcal{D}_1 \sqcap (\mathcal{D}_2 \sqcup \mathcal{D}_3) \preceq (\mathcal{D}_1 \sqcap \mathcal{D}_2) \sqcup (\mathcal{D}_1 \sqcap \mathcal{D}_3)$. Let $D \in \mathcal{D}_1 \sqcap (\mathcal{D}_2 \sqcup \mathcal{D}_3)$. W.l.o.g. $D = D_1 \cap D_2$ for some $D_1 \in \mathcal{D}_1$, $D_2 \in \mathcal{D}_2$. Then $D \in \mathcal{D}_1 \sqcap \mathcal{D}_2$ and there is a $D' \in (\mathcal{D}_1 \sqcap \mathcal{D}_2) \sqcup (\mathcal{D}_1 \sqcap \mathcal{D}_3)$ with $D \subseteq D'$.

Now we show that $(\mathcal{D}_1 \sqcap \mathcal{D}_2) \sqcup (\mathcal{D}_1 \sqcap \mathcal{D}_3) \preceq \mathcal{D}_1 \sqcap (\mathcal{D}_2 \sqcup \mathcal{D}_3)$. Let $D \in (\mathcal{D}_1 \sqcap \mathcal{D}_2) \sqcup (\mathcal{D}_1 \sqcap \mathcal{D}_3)$. W.l.o.g. $D = D_1 \cap D_2$ for some $D_1 \in \mathcal{D}_1$, $D_2 \in \mathcal{D}_2$. There is a $D_2' \in \mathcal{D}_2 \sqcup \mathcal{D}_3$ with $D_2 \subseteq D_2'$. Then $D \subseteq D_1 \cap D_2' \in \mathcal{D}_1 \sqcap (\mathcal{D}_2 \sqcup \mathcal{D}_3)$. □

7 Recursive and Non-Recursive Classes

We now want to answer the question where exactly lies the borderline between recursive and non-recursive classes $\mathrm{P}_{\mathrm{dist}}[\mathcal{D}]$. (A class is called recursive if it contains recursive languages only.) Jockusch [Joc68] showed that there are non-recursive semi-recursive sets, Selman translated this to the poynomial time case showing that there are non-recursive p-selective sets. Kummer showed that $\mathrm{REC}_{\mathrm{dist}}[(n,n)\text{-CARD}]$ is recursive for all n; this result is called the Cardinality Theorem. For $n = 2$ these two results already give the complete answer to our question because in this case each of the $\mathrm{P}_{\mathrm{dist}}[\mathcal{D}]$ is either a subclass of $\mathrm{P}_{\mathrm{dist}}[(2,2)\text{-CARD}]$ $(= \mathrm{P}[(2,2)\text{-CARD}])$ or a superclass of $\mathrm{P}[2\text{-SEL}]$. But for $n \geq 3$ we need a stronger result than the Cardinality Theorem. We introduce the following notion:

Definition 25 (NONSEL). For each $n \geq 2$ we define the n-family

$$n\text{-NONSEL} = \{D \subseteq \{0,1\}^n \mid \text{there is no } D' \in n\text{-SEL with } D' \subseteq D\}.$$

It is easy to check that the family n-NONSEL is in normal form. n-NONSEL is maximal for those n-families in weak normal form for which n-SEL is not a refinement. This is the case because if a family $\mathcal{D}$ contains at least one D such that there is a $D' \in n$-SEL with $D' \subseteq D$ then for all $\sigma \in S_n$ $\sigma D' \subseteq \sigma D$. But n-NONSEL is closed under permutation and $n\text{-SEL} = \{\sigma D' \mid \Sigma \in S_n\}$. Therefore $n\text{-SEL} \preceq \mathcal{D}$. It is also not hard to see that $(n,n)\text{-CARD} \prec n\text{-NONSEL}$ for $n \geq 3$. Theorem 8 shows that $\mathrm{P}_{\mathrm{dist}}[(n,n)\text{-CARD}]$ is a proper subclass of $\mathrm{P}_{\mathrm{dist}}[n\text{-NONSEL}]$ for $n \geq 3$.

Theorem 26. *$REC_{\mathrm{dist}}[n\text{-}NONSEL] = REC$ for all $n \geq 2$.*

Sketch of Proof. We have to show that every $A \in \mathrm{REC}_{\mathrm{dist}}[n\text{-NONSEL}]$ is recursive. We only have to modify Kummer's proof of the Cardinality Theorem in its last step and therefore only give a sketch. For details see [Kum92]. Assume an ordering on $\{0,1\}^*$, e.g. lexicographic, and identify languages A with their infinite characteristic sequence χ_A. Let A be in $\mathrm{REC}_{\mathrm{dist}}[n\text{-NONSEL}]$ via the machine M. Define a tree T_M whose nodes are prefixes of characteristic sequences such that the infinite branches of T_M are the languages for which M witnesses membership in $\mathrm{REC}_{\mathrm{dist}}[n\text{-NONSEL}]$. Let B_n denote the full binary tree of height n. The rank of a tree T, denoted $rk(T)$, is the supremum

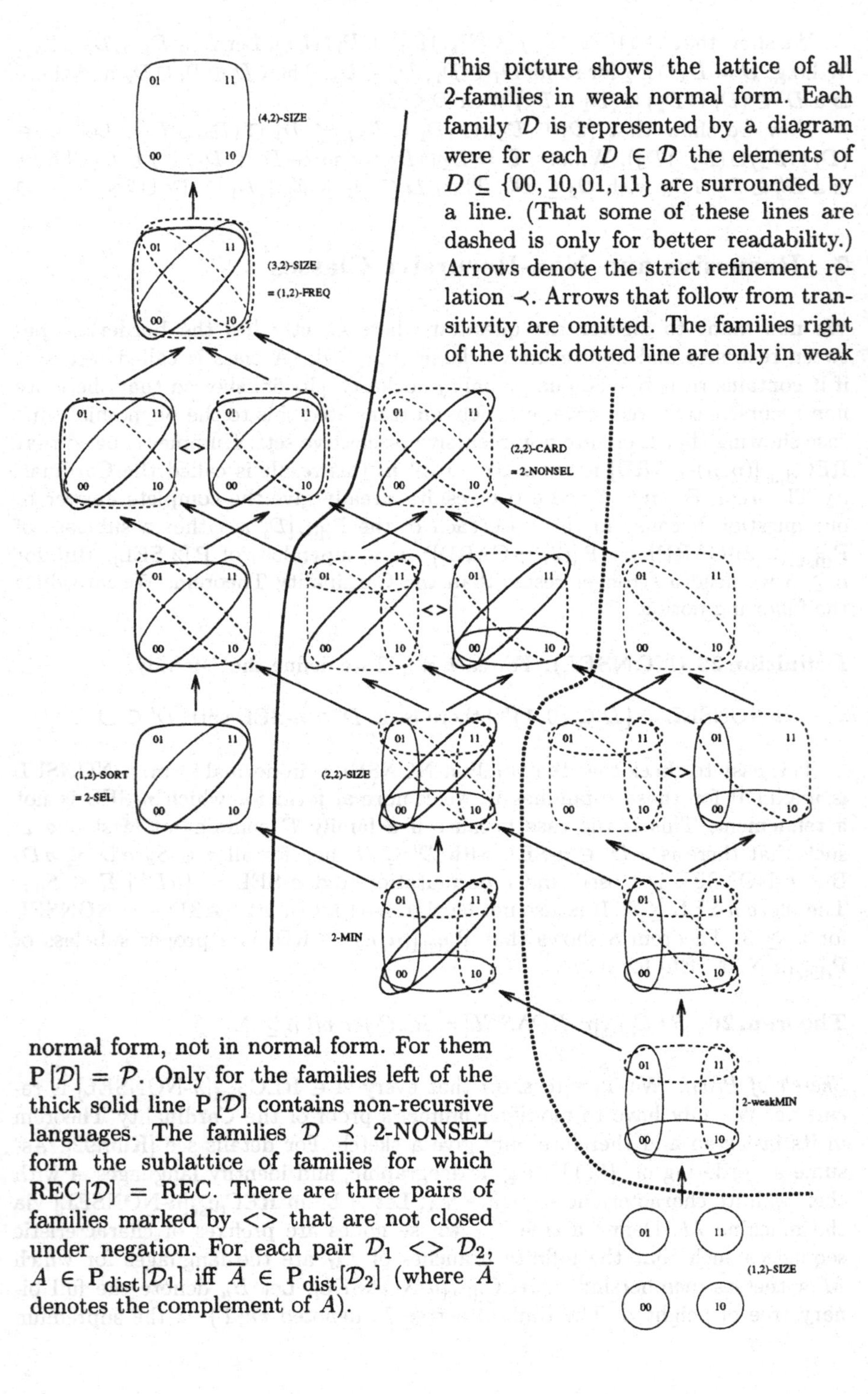

This picture shows the lattice of all 2-families in weak normal form. Each family $\mathcal{D}$ is represented by a diagram were for each $D \in \mathcal{D}$ the elements of $D \subseteq \{00, 10, 01, 11\}$ are surrounded by a line. (That some of these lines are dashed is only for better readability.) Arrows denote the strict refinement relation $\prec$. Arrows that follow from transitivity are omitted. The families right of the thick dotted line are only in weak normal form, not in normal form. For them $\mathrm{P}[\mathcal{D}] = \mathcal{P}$. Only for the families left of the thick solid line $\mathrm{P}[\mathcal{D}]$ contains non-recursive languages. The families $\mathcal{D} \preceq$ 2-NONSEL form the sublattice of families for which $\mathrm{REC}[\mathcal{D}] = \mathrm{REC}$. There are three pairs of families marked by <> that are not closed under negation. For each pair $\mathcal{D}_1$ <> $\mathcal{D}_2$, $A \in \mathrm{P}_{\mathrm{dist}}[\mathcal{D}_1]$ iff $\overline{A} \in \mathrm{P}_{\mathrm{dist}}[\mathcal{D}_2]$ (where $\overline{A}$ denotes the complement of A).

of all n such that B_n is embeddable in T. Lemma 1 of [Kum92] states that if T_M is of finite rank, then every branch of T_M, and especially χ_A, is recursive. It remains to show that T_M is of finite rank. From Lemma 3 of [Kum92] we can derive that if B_{4^n-2} is embeddable in T_M then there exist branches $A_0, \ldots, A_n$ and words $x_1, \ldots, x_n$ such that $\chi_{A_j}(x_i) = 0$ for $i < j$ and $\chi_{A_j}(x_i) = 1$ otherwise, for all $i \in \{1, \ldots, n\}$, $j \in \{0, \ldots, n\}$. But if we run M on input $(x_1, \ldots, x_n)$, at least one of the characteristic strings $0^j 1^{n-j}$ is excluded. If $0^j 1^{n-j}$ is excluded then A_j is not compatible with M, i.e. A_j is not in $\mathrm{REC}_{\mathrm{dist}}[n\text{-NONSEL}]$ via M, which contradicts the definition of T_M. □

Corollary 27. *Let $\mathcal{D}$ be an n-family. $P_{\mathrm{dist}}[\mathcal{D}]$ contains non-recursive sets iff n-SEL $\preceq \mathcal{D}$.*

Acknowledgements

I wish to thank Hans-Jörg Burtschick, Dieter Hofbauer, Wolfgang Lindner, Dirk Siefkes and Uli Hertrampf for helpful discussions and Birgit Schelm for implementing an algorithm for computing normal forms of n-families.

References

[ABG90] A. Amir, R. Beigel, and W. Gasarch. Some connections between bounded query classes and non-uniform complexity. In *Proc. 5th Structure in Complexity Theory*, 1990.

[AG88] A. Amir and W. Gasarch. Polynomial terse sets. *Information and Computation*, 77, 1988.

[BDG88] J. Balcázar, J. Díaz, and J. Gabarró. *Structural Complexity I.* 1988.

[BGK95] R. Beigel, W. Gasarch, and E. Kinber. Frequency computation and bounded queries. In *Proc. 10th Structure in Complexity Theory*, 1995.

[BKS92] R. Beigel, M. Kummer, and F. Stephan. Quantifying the amount of verboseness. In *Proc. Logical Found. of Comp. Sc.* LNCS 620, 1992.

[BKS94] R. Beigel, M. Kummer, and F. Stephan. Approximable sets. In *Proc. 9th Structure in Complexity Theory*, 1994.

[HN93] A. Hoene and A. Nickelsen. Counting, selecting, and sorting by query-bounded machines. In *Proc. STACS 93.* LNCS 665, 1993.

[Joc68] C. Jockusch, Jr. Semirecursive sets and positive reducibility. *Trans. Amer. Math. Soc.*, 131, 1968.

[KS95] M. Kummer and F. Stephan. The power of frequency computation. In *FCT 95.* LNCS 965, 1995.

[Kum92] M. Kummer. A proof of Beigel's cardinality conjecture. *J. of Symb. Logic*, 57(2), 1992.

[Ogi94] M. Ogihara. Polynomial-time membership comparable sets. In *Proc. 9th Structure in Complexity Theory*, 1994.

[Rog92] S. Rogina. Kardinalitätsberechnungen. Studienarbeit, Universtät Karlsruhe, 1992.

[Sel82] A. Selman. Analogues of semirecursive sets and effective reducibilities to the study of NP complexity. *Information and Control*, 1, 1982.

A Downward Translation in the Polynomial Hierarchy*

Edith Hemaspaandra[1]**, Lane A. Hemaspaandra[2]***, and Harald Hempel[3]†

[1] Department of Mathematics, Le Moyne College, Syracuse, NY 13214, USA.
[2] Department of Computer Science, University of Rochester, Rochester, NY 14627, USA.
[3] Inst. für Informatik, Friedrich-Schiller-Universität Jena, 07743 Jena, Germany.

Abstract. Downward translation (a.k.a. upward separation) refers to cases where the equality of two larger classes implies the equality of two smaller classes. We provide the first unqualified downward translation result completely within the polynomial hierarchy. In particular, we prove that, for $k > 2$,

$$\mathrm{P}^{\Sigma_k^p[1]} = \mathrm{P}^{\Sigma_k^p[2]} \iff \Sigma_k^p = \Pi_k^p,$$

where the "[1]" (respectively, "[2]") denotes that at most one query is (respectively, two queries are) allowed. We also extend this to obtain a more general downward translation result.

1 Introduction

The theory of NP-completeness does not resolve the issue of whether P and NP are equal. However, it does unify the issues of whether thousands of different problems—the NP-complete problems—have deterministic polynomial-time algorithms. The study of downward translation is similar in spirit. By proving downward translations, we seek to tie together central open issues regarding the computing power of complexity classes. For example, the displayed equation in the abstract, which is the main result of this paper, shows that (for $k > 2$) the issue of whether the kth level of the polynomial hierarchy is closed under complementation is identical to the issue of whether two queries give more power than one.

---

* A full version of this paper, including proofs of all claims, can be found as University of Rochester Department of Computer Science Technical Report TR-96-630, at `http://www.cs.rochester.edu/trs/theory-trs.html`.

** Email: `edith@bamboo.lemoyne.edu`. Supported in part by grant NSF-INT-9513368/DAAD-315-PRO-fo-ab. Work done in part while visiting Friedrich-Schiller-Universität Jena.

*** Email: `lane@cs.rochester.edu`. Supported in part by grants NSF-CCR-9322513 and NSF-INT-9513368/DAAD-315-PRO-fo-ab. Work done in part while visiting Friedrich-Schiller-Universität Jena.

† Email: `hempel@mipool.uni-jena.de`. Supported in part by grant NSF-INT-9513368/DAAD-315-PRO-fo-ab.

Reischuk, Morvan (Eds.): STACS'97 Proceedings, LNCS 1200

Informally, downward translation (a.k.a. upward separation) refers to cases in which the collapse of larger classes implies the collapse of smaller classes (for background, see, e.g., [All91,AW90]). For example, $\mathrm{NP}^{\mathrm{NP}} = \mathrm{coNP}^{\mathrm{NP}} \Rightarrow \mathrm{NP} = \mathrm{coNP}$ would be a (shocking, and inherently nonrelativizing [Ko89]) downward translation, the "downward" part referring to the well-known fact that $\mathrm{NP} \cup \mathrm{coNP} \subseteq \mathrm{NP}^{\mathrm{NP}} \cap \mathrm{coNP}^{\mathrm{NP}}$. No unqualified downward translation results are known. However, there are many fascinating near misses. Cases where the collapse of larger classes forces sparse sets (but perhaps not non-sparse sets) to fall out of smaller classes were found by Hartmanis, Immerman, and Sewelson ([HIS85], see also [Boo74]) and by others (e.g., Rao, Rothe, and Watanabe [RRW94], but in contrast see also [HJ95]). Existential cases have long been implicitly known (i.e., theorems such as "If PH = PSPACE then $(\exists k)\,[\mathrm{PH} = \Sigma_k^p]$"—note that here one can prove nothing about what value k might have). Hemaspaandra, Rothe, and Wechsung have given an example involving degenerate certificate schemes [HRW95], but that example does not satisfy the plausibly-strict-containment condition. Perhaps the most striking example to date is the result of Babai, Fortnow, Nisan, and Wigderson [BFNW93] that "If EH = E then P = BPP." Note, however, that this is not quite of the form of larger classes implying the collapse of smaller classes, as by E they mean $\mathrm{DTIME}[2^{\mathcal{O}(n)}]$, and it is not at all clear that $\mathrm{BPP} \subseteq \mathrm{E}$ (though under the hypothesis of their theorem this inclusion in fact clearly holds).

We provide an unqualified downward translation result that is not restricted to sparse or tally sets, whose conclusion does not contain a variable that is not specified in its hypothesis, and that deals with classes whose *ex ante* containments[4] are clear (and plausibly strict). Namely, as is standard, let $\mathrm{P}^{\mathcal{C}[j]}$ denote the class of languages computable by P machines making at most j queries to some set from $\mathcal{C}$. We prove that, for each $k > 2$, it holds that

$$\mathrm{P}^{\Sigma_k^p[1]} = \mathrm{P}^{\Sigma_k^p[2]} \Rightarrow \Sigma_k^p = \Pi_k^p = \mathrm{PH}.$$

(As just mentioned in footnote 4, the classes in the hypothesis clearly have the property that they contain both Σ_k^p and Π_k^p.) The best previously known results from the assumption $\mathrm{P}^{\Sigma_k^p[1]} = \mathrm{P}^{\Sigma_k^p[2]}$ collapse the polynomial hierarchy only to a level that contains Σ_{k+1}^p and Π_{k+1}^p [CK96,BCO93].

Our proof actually establishes a $\Sigma_k^p = \Pi_k^p$ collapse from a hypothesis that is even weaker than $\mathrm{P}^{\Sigma_k^p[1]} = \mathrm{P}^{\Sigma_k^p[2]}$. Namely, we prove that, for $i < j < k$ and $i < k - 2$, if one query each (in parallel) to the ith and kth levels of the polynomial hierarchy equals one query each (in parallel) to the jth and kth levels of the polynomial hierarchy, then $\Sigma_k^p = \Pi_k^p = \mathrm{PH}$.

In the final section of the paper, we generalize from 1-versus-2 queries to m-versus-$(m+1)$ queries. In particular, we show that our main result is in fact a reflection of an even more general downward translation. For each $k > 2$, if the truth-table hierarchy over Σ_k^p collapses to its mth level then the boolean hierarchy over Σ_k^p collapses one level further than one would expect.

[4] I.e., in the case of Theorem 1, $\Sigma_k^p \cup \Pi_k^p \subseteq \mathrm{P}^{\Sigma_k^p[1]} \cap \mathrm{P}^{\Sigma_k^p[2]}$ is well-known to be true (and most researchers suspect that the inclusion is strict).

2 Simple Case

Our proof works by extracting advice internally and algorithmically, while holding down the number of quantifiers needed, within the framework of a so-called "easy-hard" argument. Easy-hard arguments were introduced by Kadin [Kad88], and were further used by Chang and Kadin ([CK96], see also [Cha91]) and Beigel, Chang, and Ogihara [BCO93] (we follow the approach of Beigel, Chang, and Ogihara).

Theorem 1. *For each $k > 2$ it holds that*

$$\mathrm{P}^{\Sigma_k^p[1]} = \mathrm{P}^{\Sigma_k^p[2]} \Rightarrow \Sigma_k^p = \Pi_k^p = \mathrm{PH}.$$

Since the converse implication is trivial, we can equally well state that, for $k > 2$: $\mathrm{P}^{\Sigma_k^p[1]} = \mathrm{P}^{\Sigma_k^p[2]} \iff \Sigma_k^p = \Pi_k^p = \mathrm{PH}$. Theorem 1 follows immediately from Theorem 4 below, which states that, for $i < j < k$ and $i < k - 2$, if one query each to the ith and kth levels of the polynomial hierarchy equals one query each to the jth and kth levels of the polynomial hierarchy, then $\Sigma_k^p = \Pi_k^p = \mathrm{PH}$.

DPTM will refer to deterministic polynomial-time oracle Turing machines, whose polynomial time upper-bounds are clearly clocked, and are independent of their oracles. We will also use the following definitions.

Definition 2. 1. Let $M^{(A,B)}$ denote DPTM M making, simultaneously (i.e., in a truth-table fashion), at most one query to oracle A and at most one query to oracle B, and let

$$\mathrm{P}^{(\mathcal{C},\mathcal{D})} = \{L \subseteq \Sigma^* \mid (\exists C \in \mathcal{C})(\exists D \in \mathcal{D})(\exists \mathrm{DPTM}\ M)[L = L(M^{(C,D)})]\}.$$

2. (see [BCO93]) $A \tilde{\Delta} B = \{\langle x, y\rangle \mid x \in A \Leftrightarrow y \notin B\}$.

Lemma 3. *Let $0 \leq i < k$, let $L_{\mathrm{P}^{\Sigma_i^p[1]}}$ be any set $\leq_{\mathrm{m}}^{\mathrm{p}}$-complete for $\mathrm{P}^{\Sigma_i^p[1]}$, and let $L_{\Sigma_k^p}$ be any language $\leq_{\mathrm{m}}^{\mathrm{p}}$-complete for Σ_k^p. Then $L_{\mathrm{P}^{\Sigma_i^p[1]}} \tilde{\Delta} L_{\Sigma_k^p}$ is $\leq_{\mathrm{m}}^{\mathrm{p}}$-complete for $\mathrm{P}^{(\Sigma_i^p, \Sigma_k^p)}$.*

Theorem 4 contains the following two technical advances. First, it internally extracts information in a way that saves a quantifier. (In contrast, the earliest easy-hard arguments in the literature merely ensure that $\Sigma_k^p \subseteq \Pi_k^p/\mathrm{poly}$ and from that infer a weak polynomial hierarchy collapse. Even the interesting recent strengthenings of the argument [BCO93] still, under the hypothesis of Theorem 4, conclude only a collapse of the polynomial hierarchy to a level a bit above Σ_{k+1}^p.) The second advance is that previous easy-hard arguments seek to determine whether there exists a hard string for a length or not. Then they use the fact that if there is not a hard string, all strings (at the length) are easy. In contrast, we *never* search for a hard string; rather, we use the fact that the input itself (which we do not have to search for as, after all, it is our input) is either easy or hard. So we check whether the input is easy, and if so we can use it as an easy string, and if not, it must be a hard string so we can use it that way. This innovation is important in that it allows Theorem 1 to apply for all

$k > 2$ (as opposed to merely applying for all $k > 3$, which is what we would get without this innovation).

Theorem 4. *Let* $0 \leq i < j < k$ *and* $i < k-2$. *If* $\mathrm{P}^{(\Sigma_i^p,\Sigma_k^p)} = \mathrm{P}^{(\Sigma_j^p,\Sigma_k^p)}$ *then* $\Sigma_k^p = \Pi_k^p = \mathrm{PH}$.

Proof: Suppose $\mathrm{P}^{(\Sigma_i^p,\Sigma_k^p)} = \mathrm{P}^{(\Sigma_j^p,\Sigma_k^p)}$. Let $L_{\mathrm{P}^{\Sigma_i^p[1]}}$, $L_{\mathrm{P}^{\Sigma_{i+1}^p[1]}}$, and $L_{\Sigma_k^p}$ be $\leq_m^p$-complete for $\mathrm{P}^{\Sigma_i^p[1]}$, $\mathrm{P}^{\Sigma_{i+1}^p[1]}$, and Σ_k^p, respectively; such sets exist. From Lemma 3 it follows that $L_{\mathrm{P}^{\Sigma_i^p[1]}} \tilde{\Delta} L_{\Sigma_k^p}$ is $\leq_m^p$-complete for $\mathrm{P}^{(\Sigma_i^p,\Sigma_k^p)}$. Since (as $i < j$) $L_{\mathrm{P}^{\Sigma_{i+1}^p[1]}} \tilde{\Delta} L_{\Sigma_k^p} \in \mathrm{P}^{(\Sigma_j^p,\Sigma_k^p)}$, and by assumption $\mathrm{P}^{(\Sigma_j^p,\Sigma_k^p)} = \mathrm{P}^{(\Sigma_i^p,\Sigma_k^p)}$, there exists a polynomial-time many-one reduction h from $L_{\mathrm{P}^{\Sigma_{i+1}^p[1]}} \tilde{\Delta} L_{\Sigma_k^p}$ to $L_{\mathrm{P}^{\Sigma_i^p[1]}} \tilde{\Delta} L_{\Sigma_k^p}$. So, for all $x_1, x_2 \in \Sigma^*$: if $h(\langle x_1, x_2\rangle) = \langle y_1, y_2\rangle$, then $(x_1 \in L_{\mathrm{P}^{\Sigma_{i+1}^p[1]}} \Leftrightarrow x_2 \notin L_{\Sigma_k^p})$ if and only if $(y_1 \in L_{\mathrm{P}^{\Sigma_i^p[1]}} \Leftrightarrow y_2 \notin L_{\Sigma_k^p})$. Equivalently, for all $x_1, x_2 \in \Sigma^*$:

Fact 1:
if $h(\langle x_1, x_2\rangle) = \langle y_1, y_2\rangle$,
then

$$(x_1 \in L_{\mathrm{P}^{\Sigma_{i+1}^p[1]}} \Leftrightarrow x_2 \in L_{\Sigma_k^p}) \text{ if and only if } (y_1 \in L_{\mathrm{P}^{\Sigma_i^p[1]}} \Leftrightarrow y_2 \in L_{\Sigma_k^p}).$$

We can use h to recognize some of $\overline{L_{\Sigma_k^p}}$ by a Σ_k^p algorithm. In particular, following Kadin [Kad88], we say that a string x is *easy for length n* if there exists a string x_1 such that $|x_1| \leq n$ and $(x_1 \in L_{\mathrm{P}^{\Sigma_{i+1}^p[1]}} \Leftrightarrow y_1 \notin L_{\mathrm{P}^{\Sigma_i^p[1]}})$ where $h(\langle x_1, x\rangle) = \langle y_1, y_2\rangle$.

Let p be a fixed polynomial, which will be exactly specified later in the proof. We have the following Σ_k^p algorithm to test whether $x \in \overline{L_{\Sigma_k^p}}$ in the case that (our input) x is an easy string for $p(|x|)$. On input x, guess x_1 with $|x_1| \leq p(|x|)$, let $h(\langle x_1, x\rangle) = \langle y_1, y_2\rangle$, and accept if and only if $(x_1 \in L_{\mathrm{P}^{\Sigma_{i+1}^p[1]}} \Leftrightarrow y_1 \notin L_{\mathrm{P}^{\Sigma_i^p[1]}})$ and $y_2 \in L_{\Sigma_k^p}$. In light of Fact 1 above, it is clear that this is correct.

Again following Kadin [Kad88], we say that x is *hard for length n* if $|x| \leq n$ and x is not easy for length n, i.e., if $|x| \leq n$ and for all x_1 with $|x_1| \leq n$, $(x_1 \in L_{\mathrm{P}^{\Sigma_{i+1}^p[1]}} \Leftrightarrow y_1 \in L_{\mathrm{P}^{\Sigma_i^p[1]}})$, where $h(\langle x_1, x\rangle) = \langle y_1, y_2\rangle$.

If x is a hard string for length n, then x induces a many-one reduction from $\left(L_{\mathrm{P}^{\Sigma_{i+1}^p[1]}}\right)^{\leq n}$ to $L_{\mathrm{P}^{\Sigma_i^p[1]}}$, namely, $f(x_1) = y_1$, where $h(\langle x_1, x\rangle) = \langle y_1, y_2\rangle$. Note that f is computable in time polynomial in $\max(n, |x_1|)$.

We can use hard strings to obtain a Σ_k^p algorithm for $\overline{L_{\Sigma_k^p}}$. Let M be a Π_{k-i-1}^p machine such that M with oracle $L_{\mathrm{P}^{\Sigma_{i+1}^p[1]}}$ recognizes $\overline{L_{\Sigma_k^p}}$. Let the run-time of M be bounded by polynomial p, which without loss of generality satisfies $(\forall \widehat{m} \geq 0)[p(\widehat{m}+1) > p(\widehat{m}) > 0]$ (as promised above, we have now specified p). Then

$$\left(\overline{L_{\Sigma_k^p}}\right)^{=n} = L(M^{\left(L_{\mathrm{P}^{\Sigma_{i+1}^p[1]}}\right)^{\leq p(n)}})^{=n}.$$

If there exists a hard string for length $p(n)$, then this hard string induces a reduction from $\left(L_{\mathrm{P}^{\Sigma_{i+1}^p[1]}}\right)^{\leq p(n)}$ to $L_{\mathrm{P}^{\Sigma_i^p[1]}}$. Thus, with any hard string for length $p(n)$ in hand, call it w_n, $\widehat{M}$ with oracle $L_{\mathrm{P}^{\Sigma_i^p[1]}}$ recognizes $\overline{L_{\Sigma_k^p}}$ for strings of length n, where $\widehat{M}$ is the machine that simulates M but replaces each query to q by the first component of $h(\langle q, w_n\rangle)$. It follows that if there exists a hard string for length $p(n)$, then this string induces a Π_{k-1}^p algorithm for $\left(\overline{L_{\Sigma_k^p}}\right)^{=n}$, and therefore certainly a Σ_k^p algorithm for $\left(\overline{L_{\Sigma_k^p}}\right)^{=n}$.

However, now we have an $\mathrm{NP}^{\Sigma_{k-1}^p} = \Sigma_k^p$ algorithm for $\overline{L_{\Sigma_k^p}}$: On input x, the NP base machine of $\mathrm{NP}^{\Sigma_{k-1}^p}$ executes the following algorithm:

1. Using its Σ_{k-1}^p oracle, it deterministically determines whether the input x is an easy string for length $p(|x|)$. This can be done, as checking whether the input is an easy string for length $p(|x|)$ can be done by one query to Σ_{i+2}^p, and $i+2 \leq k-1$ by our $i < k-2$ hypothesis.

2. If the previous step determined that the input is not an easy string, then the input must be a hard string for length $p(|x|)$. So simulate the Σ_k^p algorithm induced by this hard string (i.e., the input x itself) on input x (via our NP machine itself simulating the base level of the Σ_k^p algorithm and using the NP machine's oracle to simulate the oracle queries made by the base level NP machine of the Σ_k^p algorithm being simulated).

3. If the first step determined that the input x is easy for length $p(|x|)$, then our NP machine simulates (using itself and its oracle) the Σ_k^p algorithm for easy strings on input x.

We need one brief technical comment. The Σ_{k-1}^p oracle in the above algorithm is being used for a number of different sets. However, as Σ_{k-1}^p is closed under disjoint union, this presents no problem as we can use the disjoint union of the sets, while modifying the queries so they address the appropriate part of the disjoint union.

Since $\overline{L_{\Sigma_k^p}}$ is complete for Π_k^p, it follows that $\Sigma_k^p = \Pi_k^p = \mathrm{PH}$. ∎

We conclude this section with three remarks. First, if one is fond of the truth-table version of bounded query hierarchies, one can certainly replace the hypothesis of Theorem 1 with $\mathrm{P}_{1\text{-tt}}^{\Sigma_k^p} = \mathrm{P}_{2\text{-tt}}^{\Sigma_k^p}$ (both as this is an equivalent hypothesis, and as it in any case clearly follows from Theorem 4). Indeed, one can equally well replace the hypothesis of Theorem 1 with the even weaker-looking hypothesis[5] $\mathrm{P}^{\Sigma_k^p[1]} = \mathrm{DIFF}_2(\Sigma_k^p)$ (as this hypothesis is also in fact equivalent to the hypothesis of Theorem 1—just note that if $\mathrm{P}^{\Sigma_k^p[1]} = \mathrm{DIFF}_2(\Sigma_k^p)$ then $\mathrm{DIFF}_2(\Sigma_k^p)$ is closed under complementation and thus equals the boolean hierarchy over Σ_k^p, see [CGH+88], and so in particular we then have $\mathrm{P}^{\Sigma_k^p[1]} = \mathrm{DIFF}_2(\Sigma_k^p) = \mathrm{P}^{\Sigma_k^p[2]}$).

[5] Where $\mathrm{DIFF}_2(\mathcal{C}) =_{\mathrm{def}} \{L \mid (\exists L_1 \in \mathcal{C})(\exists L_2 \in \mathcal{C})[L = L_1 - L_2]\}$, and $\mathrm{co}\mathcal{C} =_{\mathrm{def}} \{L \mid \overline{L} \in \mathcal{C}\}$ (see Definition 7 for background).

Of course, the two equivalences just mentioned—$\mathrm{P}^{\Sigma_k^p[1]} = \mathrm{P}^{\Sigma_k^p[2]} \Leftrightarrow \mathrm{P}_{1\text{-tt}}^{\Sigma_k^p} = \mathrm{P}_{2\text{-tt}}^{\Sigma_k^p} \Leftrightarrow \mathrm{P}^{\Sigma_k^p[1]} = \mathrm{DIFF}_2(\Sigma_k^p)$—are well-known. However, Theorem 4 is sufficiently strong that it creates an equivalence that is quite new, and somewhat surprising. We state it below as Corollary 6.

Theorem 5. *For each $k > 2$ it holds that:*

$$\mathrm{P}^{\Sigma_k^p[1]} = \mathrm{DIFF}_2(\Sigma_k^p) \cap \mathrm{coDIFF}_2(\Sigma_k^p) \Rightarrow \Sigma_k^p = \Pi_k^p = \mathrm{PH}.$$

Corollary 6. *For each $k > 2$ it holds that:*

$$\mathrm{P}^{\Sigma_k^p[1]} = \mathrm{DIFF}_2(\Sigma_k^p) \cap \mathrm{coDIFF}_2(\Sigma_k^p) \Leftrightarrow \mathrm{P}^{\Sigma_k^p[1]} = \mathrm{DIFF}_2(\Sigma_k^p).$$

Our second remark is that Theorem 1 implies that, for $k > 2$, if the bounded query hierarchy over Σ_k^p collapses to its $\mathrm{P}^{\Sigma_k^p[1]}$ level, then the bounded query hierarchy over Σ_k^p equals the polynomial hierarchy (this provides a partial answer to the issue of whether, when a bounded query hierarchy collapses, the polynomial hierarchy necessarily collapses to it, see [HRZ95, Problem 4]).

Third, in Lemma 3 and Theorem 4 we speak of classes of the form $\mathrm{P}^{(\Sigma_i^p,\Sigma_j^p)}$, $i \neq j$. It would be very natural to reason as follows: "$\mathrm{P}^{(\Sigma_i^p,\Sigma_j^p)}$, $i \neq j$, must equal $\mathrm{P}^{\Sigma_{\max(i,j)}^p[1]}$, as $\Sigma_{\max(i,j)}^p$ can easily solve any $\Sigma_{\min(i,j)}^p$ query "strongly" using the $\Sigma_{\max(i,j)-1}^p$ oracle of its base NP machine and thus the hypothesis of Theorem 4 is trivially satisfied and so you in fact are claiming to prove, unconditionally, that PH $= \Sigma_3^p$. This reasoning, though tempting, is wrong for the following somewhat subtle reason. Though it is true that, for example, $\mathrm{NP}^{\Sigma_q^p}$ can solve any Σ_q^p query and then can tackle any Σ_{q+1}^p query, it does not follow that $\mathrm{P}^{(\Sigma_{q+1}^p,\Sigma_q^p)} = \mathrm{P}^{\Sigma_{q+1}^p[1]}$. The problem is that the answer to the Σ_q^p query may *change the truth-table* the P transducer users to evaluate the answer of the Σ_{q+1}^p query.

We mention that H. Buhrman and L. Fortnow [BF96] have recently informed the authors that by extending our proof technique they can establish the $k = 2$ analog of Theorem 1. They also prove that there are relativized worlds in which the $k = 1$ analog of Theorem 1 fails.

3 General Case

We now generalize the results of Section 2 to the case of m-truth-table reductions. Though the results of this section are stronger than those of Section 2, the proofs are somewhat more involved, and thus we suggest the reader first read Section 2.

For clarity, we now describe the key differences between the proofs (not included here—see the full version mentioned in the footnote to the title) of the claims of this section and those of Section 2. (1) The completeness claims of Section 2 were simpler. Here, we now need Lemma 10, which extends [BCO93, Lemma 8] with the trick of splitting a truth-table along a simple query's dimension in such a way that the induced one-dimension-lower truth-tables cause no problems. (2) The proof of Theorem 11 is quite analogous to the proof of Theorem 4, except (i) it is a bit harder to understand as one continuously has to parse the deeply nested set differences caused by the fact that we are now working in the difference hierarchy, and (ii) the "input is an easy string" simulation is changed to account for a new problem, namely, that in the boolean hierarchy one models each language by a *collection* of machines (mimicking the nested difference structure of boolean hierarchy languages) and thus it is hard to ensure that these machines, when guessing an object, necessarily guess the same object (we solve this coordination problem by forcing them to each guess a lexicographically extreme object, and we argue that this can be accomplished within the computational power available).

The difference hierarchy was introduced by Cai et al. [CGH+88,CGH+89] and is defined below. Cai et al. studied the case $\mathcal{C} = \mathrm{NP}$, but a number of other cases have since been studied [BJY90,BCO93,HR].

Definition 7. Let $\mathcal{C}$ be any complexity class.

1. $\mathrm{DIFF}_1(\mathcal{C}) = \mathcal{C}$.
2. For any $k \geq 1$, $\mathrm{DIFF}_{k+1}(\mathcal{C}) = \{L \mid (\exists L_1 \in \mathcal{C})(\exists L_2 \in \mathrm{DIFF}_k(\mathcal{C}))[L = L_1 - L_2]\}$.
3. For any $k \geq 1$, $\mathrm{coDIFF}_k(\mathcal{C}) = \{L \mid \overline{L} \in \mathrm{DIFF}_k(\mathcal{C})\}$.

Note in particular that

$$\mathrm{DIFF}_m(\Sigma_k^p) \cup \mathrm{coDIFF}_m(\Sigma_k^p) \subseteq \mathrm{P}_{m\text{-tt}}^{\Sigma_k^p} \subseteq \mathrm{DIFF}_{m+1}(\Sigma_k^p) \cap \mathrm{coDIFF}_{m+1}(\Sigma_k^p).$$

Theorem 8. *For each $m > 0$ and each $k > 2$ it holds that:*

$$\mathrm{P}_{m\text{-tt}}^{\Sigma_k^p} = \mathrm{P}_{m+1\text{-tt}}^{\Sigma_k^p} \Rightarrow \mathrm{DIFF}_m(\Sigma_k^p) = \mathrm{coDIFF}_m(\Sigma_k^p).$$

Theorem 1 is the m equals one case of Theorem 8 (except the former is stated in terms of Turing access). Theorem 8 follows immediately from Theorem 11 below, which states that, for $i < j < k$ and $i < k - 2$, if one query to the ith and m queries to the kth levels of the polynomial hierarchy equals one query to the jth and m queries to the kth levels of the polynomial hierarchy, then $\mathrm{DIFF}_m(\Sigma_k^p) = \mathrm{coDIFF}_m(\Sigma_k^p)$. Note, of course, that by Beigel, Chang, and Ogihara [BCO93] the conclusion of Theorem 8 implies a collapse of the polynomial hierarchy. In particular, via [BCO93, Theorem 10], Theorem 8 implies that, for each $m \geq 0$ and each $k > 2$, it holds that: If $\mathrm{P}_{m\text{-tt}}^{\Sigma_k^p} = \mathrm{P}_{m+1\text{-tt}}^{\Sigma_k^p}$ then the polynomial hierarchy can be solved by a P machine that makes $m-1$ truth-table queries to Σ_{k+1}^p, and that in addition is allowed unbounded queries to Σ_k^p. This polynomial hierarchy collapse is about one level lower in the difference hierarchy over Σ_{k+1}^p than one could conclude from previous papers, in particular, from Beigel, Chang, and Ogihara. In fact, one can claim a bit more. The *proof* of [BCO93, Theorem 10] in fact proves the following: $\mathrm{DIFF}_m(\Sigma_k^p) = \mathrm{coDIFF}_m(\Sigma_k^p) \Rightarrow \mathrm{PH} = \mathrm{P}_{1,m-1\text{-tt}}^{(\Sigma_k^p,\Sigma_{k+1}^p)}$.

Thus, in light of Theorem 8, we have that for each $m \geq 0$ and each $k > 2$ it holds that $\mathrm{P}^{\Sigma_k^p}_{m\text{-tt}} = \mathrm{P}^{\Sigma_k^p}_{m+1\text{-tt}} \Rightarrow \mathrm{PH} = \mathrm{P}^{(\Sigma_k^p, \Sigma_{k+1}^p)}_{1,m-1\text{-tt}}$.

Definition 9. Let $M^{(A,B)}_{a,b\text{-tt}}$ denote DPTM M making, simultaneously (i.e., all $a + b$ queries are made at the same time, in the standard truth-table fashion), at most a queries to oracle A and at most b queries to oracle B, and let

$$\mathrm{P}^{(\mathcal{C},\mathcal{D})}_{a,b\text{-tt}} = \{L \subseteq \Sigma^* \mid (\exists C \in \mathcal{C})(\exists D \in \mathcal{D})(\exists \text{DPTM } M)[L = L(M^{(C,D)}_{a,b\text{-tt}})]\}.$$

Lemma 10. *Let $m > 0$, let $0 \leq i < k$, let $L_{\mathrm{P}^{\Sigma_i^p[1]}}$ be any set $\leq_m^p$-complete for $\mathrm{P}^{\Sigma_i^p[1]}$, and let $L_{\mathrm{DIFF}_m(\Sigma_k^p)}$ be any language $\leq_m^p$-complete for $\mathrm{DIFF}_m(\Sigma_k^p)$. Then $L_{\mathrm{P}^{\Sigma_i^p[1]}} \tilde{\Delta} L_{\mathrm{DIFF}_m(\Sigma_k^p)}$ is $\leq_m^p$-complete for $\mathrm{P}^{(\Sigma_i^p, \Sigma_k^p)}_{1,m\text{-tt}}$.*

Lemma 10 does not require proof, as it is a use of the standard mind-change technique, and is analogous to [BCO93, Lemma 8], with one key twist that we now discuss. Assume, without loss of generality, that we focus on $\mathrm{P}^{(\Sigma_i^p, \Sigma_k^p)}_{1,m\text{-tt}}$ machines that always make exactly $m + 1$ queries. Regarding any such machine accepting a set complete for the class $\mathrm{P}^{(\Sigma_i^p, \Sigma_k^p)}_{1,m\text{-tt}}$ of Lemma 10, we have on each input a truth-table with $m + 1$ variables. Note that if one knows the answer to the one Σ_i^p query, then this induces a truth-table on m variables; however, note also that the two m-variable truth-tables (one corresponding to a "yes" answer to the Σ_i^p query and the other to a "no" answer) may differ sharply. Regarding $L_{\mathrm{P}^{\Sigma_i^p[1]}} \tilde{\Delta} L_{\mathrm{DIFF}_m(\Sigma_k^p)}$, we use $L_{\mathrm{P}^{\Sigma_i^p[1]}}$ to determine whether the m-variable truth-table induced by the true answer to the one Σ_i^p query accepts or not when all the Σ_k^p queries get the answer no. This use is analogous to [BCO93, Lemma 8]. The new twist is the action of the $L_{\mathrm{DIFF}_m(\Sigma_k^p)}$ part of $L_{\mathrm{P}^{\Sigma_i^p[1]}} \tilde{\Delta} L_{\mathrm{DIFF}_m(\Sigma_k^p)}$. We use this, just as in [BCO93, Lemma 8], to find whether or not we are in an odd mind-change region *but now with respect to the m-variable truth-table induced by the true answer to the one Σ_i^p query.* Crucially, this still is a $\mathrm{DIFF}_m(\Sigma_k^p)$ issue as, since $i < k$, a Σ_k^p machine can first on its own (by its base NP machine making one deterministic query to its Σ_{k-1}^p oracle) determine the true answer to the one Σ_i^p query, and thus the machine can easily know which of the two m-variable truth-table cases it is in, and thus it plays its standard part in determining if the mind-change region of the m true answers to the Σ_k^p queries fall in an odd mind-change region *with respect to the correct m-variable truth-table.*

Theorem 11. *Let $m > 0$, $0 \leq i < j < k$ and $i < k - 2$. If $\mathrm{P}^{(\Sigma_i^p, \Sigma_k^p)}_{1,m\text{-tt}} = \mathrm{P}^{(\Sigma_j^p, \Sigma_k^p)}_{1,m\text{-tt}}$ then $\mathrm{DIFF}_m(\Sigma_k^p) = \mathrm{coDIFF}_m(\Sigma_k^p)$.*

Finally, remark that we have analogs of Theorem 5 and Corollary 6. The proof is analogous to that of Theorem 5; one just uses $\mathrm{P}^{(\mathrm{NP}, \Sigma_k^p)}_{1,m\text{-tt}}$ in the way $\mathrm{P}^{(\mathrm{NP}, \Sigma_k^p)}$ was used in that proof, and again invokes the relation between the difference and symmetric difference hierarchies (namely that $\mathrm{DIFF}_j(\Sigma_k^p)$ is exactly the class of sets L that for some $L_1, \cdots, L_j \in \Sigma_k^p$ satisfy $L = L_1 \Delta \cdots \Delta L_j$; this well-known equality is due to [KSW87, Section 3] in light of the standard equalities regarding boolean hierarchies (see [CGH+88, Section 2.1]); though both [KSW87,CGH+88] focus mostly on the $k = 1$ case, it is standard [Wec85, BBJ+89] that the equalities in fact hold for any class closed under union and intersection and containing $\emptyset$ and Σ^*).

Theorem 12. *Let* $m \geq 0$ *and* $k > 2$. *If* $\mathrm{P}^{\Sigma_k^p}_{m\text{-tt}} = \mathrm{DIFF}_{m+1}(\Sigma_k^p) \cap \mathrm{coDIFF}_{m+1}(\Sigma_k^p)$ *then* $\mathrm{DIFF}_m(\Sigma_k^p) = \mathrm{coDIFF}_m(\Sigma_k^p)$.

Corollary 13. *For each* $k > 2$ *and* $m \geq 0$, *it holds that:*

$$\mathrm{P}^{\Sigma_k^p}_{m\text{-tt}} = \mathrm{DIFF}_{m+1}(\Sigma_k^p) \cap \mathrm{coDIFF}_{m+1}(\Sigma_k^p) \Leftrightarrow \mathrm{P}^{\Sigma_k^p}_{m\text{-tt}} = \mathrm{DIFF}_{m+1}(\Sigma_k^p).$$

Acknowledgments

We thank Lance Fortnow for helpful email discussions and Jörg Rothe for helpful comments. The first two authors thank Gerd Wechsung's research group for its very kind hospitality during the visit when this research was performed.

References

[All91] E. Allender. Limitations of the upward separation technique. *Mathematical Systems Theory*, 24(1):53–67, 1991.

[AW90] E. Allender and C. Wilson. Downward translations of equality. *Theoretical Computer Science*, 75(3):335–346, 1990.

[BBJ+89] A. Bertoni, D. Bruschi, D. Joseph, M. Sitharam, and P. Young. Generalized boolean hierarchies and boolean hierarchies over RP. In *Proceedings of the 7th Conference on Fundamentals of Computation Theory*, pages 35–46. Springer-Verlag *Lecture Notes in Computer Science #380*, August 1989.

[BCO93] R. Beigel, R. Chang, and M. Ogiwara. A relationship between difference hierarchies and relativized polynomial hierarchies. *Mathematical Systems Theory*, 26:293–310, 1993.

[BF96] H. Buhrman and L. Fortnow. Two queries. Manuscript, September 1996.

[BFNW93] L. Babai, L. Fortnow, N. Nisan, and A. Wigderson. BPP has subexponential time simulations unless EXPTIME has publishable proofs. *Computational Complexity*, 3(4):307–318, 1993.

[BJY90] D. Bruschi, D. Joseph, and P. Young. Strong separations for the boolean hierarchy over RP. *International Journal of Foundations of Computer Science*, 1(3):201–218, 1990.

[Boo74] R. Book. Tally languages and complexity classes. *Information and Control*, 26:186–193, 1974.

[CGH+88] J. Cai, T. Gundermann, J. Hartmanis, L. Hemachandra, V. Sewelson, K. Wagner, and G. Wechsung. The boolean hierarchy I: Structural properties. *SIAM Journal on Computing*, 17(6):1232–1252, 1988.

[CGH+89] J. Cai, T. Gundermann, J. Hartmanis, L. Hemachandra, V. Sewelson, K. Wagner, and G. Wechsung. The boolean hierarchy II: Applications. *SIAM Journal on Computing*, 18(1):95–111, 1989.

[Cha91] R. Chang. *On the Structure of NP Computations under Boolean Operators.* PhD thesis, Cornell University, Ithaca, NY, 1991.

[CK96] R. Chang and J. Kadin. The boolean hierarchy and the polynomial hierarchy: A closer connection. *SIAM Journal on Computing*, 25(2):340–354, 1996.

[HIS85] J. Hartmanis, N. Immerman, and V. Sewelson. Sparse sets in NP−P: EXPTIME versus NEXPTIME. *Information and Control*, 65(2/3):159–181, 1985.

[HJ95] L. Hemaspaandra and S. Jha. Defying upward and downward separation. *Information and Computation*, 121(1):1–13, 1995.

[HR] L. Hemaspaandra and J. Rothe. Unambiguous computation: Boolean hierarchies and sparse Turing-complete sets. *SIAM Journal on Computing*. To appear.

[HRW95] L. Hemaspaandra, J. Rothe, and G. Wechsung. Easy sets and hard certificate schemes. Technical Report Math/95/5, Institut für Informatik, Friedrich-Schiller-Universität–Jena, Jena, Germany, May 1995.

[HRZ95] L. Hemaspaandra, A. Ramachandran, and M. Zimand. Worlds to die for. *SIGACT News*, 26(4):5–15, 1995.

[Kad88] J. Kadin. The polynomial time hierarchy collapses if the boolean hierarchy collapses. *SIAM Journal on Computing*, 17(6):1263–1282, 1988. Erratum appears in the same journal, 20(2):404.

[Ko89] K. Ko. Relativized polynomial time hierarchies having exactly k levels. *SIAM Journal on Computing*, 18(2):392–408, 1989.

[KSW87] J. Köbler, U. Schöning, and K. Wagner. The difference and truth-table hierarchies for NP. *RAIRO Theoretical Informatics and Applications*, 21:419–435, 1987.

[RRW94] R. Rao, J. Rothe, and O. Watanabe. Upward separation for FewP and related classes. *Information Processing Letters*, 52(4):175–180, 1994.

[Wec85] G. Wechsung. On the boolean closure of NP. In *Proceedings of the 5th Conference on Fundamentals of Computation Theory*, pages 485–493. Springer-Verlag *Lecture Notes in Computer Science #199* , 1985. (An unpublished precursor of this paper was coauthored by K. Wagner).

Strict Sequential P-completeness *

Klaus Reinhardt

Wilhelm-Schickhard Institut für Informatik, Universität Tübingen
Sand 13, D-72076 Tübingen, Germany
e-mail: reinhard@informatik.uni-tuebingen.de

Abstract. In this paper we present a new notion of what it means for a problem in P to be inherently sequential. Informally, a problem L is *strictly sequential P-complete* if when the best known *sequential* algorithm for L has polynomial speedup by parallelization, this implies that *all* problems in P have a polynomial speedup in the parallel setting. The motivation for defining this class of problems is to try and capture the problems in P that are truly inherently sequential. Our work extends the results of Condon who exhibited problems such that if a polynomial speedup of their best known *parallel* algorithms could be achieved, then all problems in P would have polynomial speedup. We demonstrate one such natural problem, namely the *Multiplex-select Circuit Problem* (MCP). MCP has one of the highest degrees of sequentiality of any problem yet defined. On the way to proving MCP is strictly sequential P-complete, we define an interesting model, the *register stack machine*, that appears to be of independent interest for exploring pure sequentiality.

1 Introduction

An important question in parallel complexity theory is whether problems in P have a speedup by parallelization. If we demand the speedup to be 'exponential', this means a speedup to polylogarithmic time, and if we allow a polynomial number of processors on a PRAM, we get the famous open problem 'P=NC?'. This problem can be instantiated to any P-complete problem as for example the circuit value problem or the context-free emptiness problem which means that P = NC iff one of those P-complete problems is in NC (see [Coo85,GHR95]).

We believe an even more realistic and interesting open question is whether problems in P have polynomial speedup by parallelization. This means for every sequential algorithm solving a problem in time $t(n)$ there is a parallel algorithm solving the problem in time $T(n) \in O((t(n))^{1-\varepsilon})$ for an $\varepsilon > 0$. For practical applications it is more important that the parallelization has a low inefficiency. The *inefficiency* is the quotient of the time processor product and the sequential time. Kruskal *et al.* defined the class EP (AP, SP, respectively) as the class of problems with polynomial speedup by parallelization and constant (polylogarithmic, polynomial, respectively) inefficiency [KRS90]. It is still open whether P⊆SP.

* This research has been supported by the DFG Project La 618/3-1 KOMET.

Reischuk, Morvan (Eds.): STACS'97 Proceedings, LNCS 1200

Unfortunately, the question P⊆EP (AP, SP, respectively) is not related to the notion of P-completeness in the sense that the membership of a P-complete problem in SP would force P⊆SP, since a lot of P-complete problems have polynomial speedup by parallelization (see [VS86]).

As a step towards addressing the issues above, Condon defined a problem as *strictly* $T(n)$ *complete* for P, if it has a parallel algorithm with running time $T(n)(\log(T(n)))^{O(1)}$ and if the existence of a parallel algorithm improving this by a polynomial factor, meaning to $O((T(n))^{1-\varepsilon})$, would imply that all problems in P have polynomial speedup [Con92]. She showed that the Square Circuit Value Problem SCVP is strictly $\sqrt{n}$ P-complete.

This was the first result showing that P⊆SP if a specific problem has a polynomial speedup relative to a known *parallel* running time, i.e., if SCVP has a polynomial speedup relative to $\sqrt{n}$. Since SCVP already has a polynomial speedup of $\sqrt{n}$ relative to its linear sequential time, we do not get the other direction. This means it may still be that P⊆SP but SCVP has no polynomial speedup relative to $\sqrt{n}$.

What we need is a P-complete candidate for not having polynomial speedup at all, which would otherwise force P to be contained in SP, that means a problem which is in P\SP iff P⊈SP. Thus in this paper we go one step further and look for a problem such that P⊂SP iff this specific problem has a polynomial speedup relative to a known *sequential* running time. This means a ***strictly sequential P-complete*** problem must be strictly $t(n)$ P-complete, where $t(n)$ is now the sequential time. Accordingly, we define a problem with a sequential running time $t(n)$ as ***strictly sequential efficient P-complete*** (***strictly sequential almost efficient P-complete***, respectively) if the existence of a parallel algorithm improving this by a polynomial factor, meaning to $O((t(n))^{1-\varepsilon})$, with a constant (polylogarithmic, respectively) inefficiency would imply that every problem in P, which has an efficient prediction of termination, is in EP (AP, respectively).

In this paper we use the multiplex select gate (see [FLR96]) in combination with a compressed representation of the wires connecting the gates. We show that the problem MCP to calculate the output of a circuit consisting of such gates which is given by a representing input, is strictly sequential P-complete. Furthermore we are able to prove that MCP is strictly sequential almost efficient P-complete, where we restrict our attention to those problems in P, which have an efficient prediction of termination.

As the sequential model we use the RAM which must read its complete input and needs at least linear time by definition. Additionally, we introduce the ***register stack machine*** as a new sequential model, which avoids the hidden parallelism in the storage of a RAM. Using this model for sequential algorithms, we can show that MCP is strictly sequential efficient P-complete.

As the parallel model we use the PRAM but the same results can be obtained if we use more restricted models such as grids of processors or linear arrays of processors, as long as this model is strong enough to calculate the reduction function $f(x)$ in time $|f(x)|^{1-\varepsilon}$ and an accordingly low inefficiency.

Thus far we have only been able to identify one natural strictly sequential P-complete problem. An interesting open question is to find more problems of this type.

2 Register stack machine

If we consider the simulation of a RAM by a circuit, in some sense the storage of a RAM contains a kind of parallelism on the circuit level: every cell must detect at every step, whether it is the one, where the processor it just writing on. To avoid this, we introduce a new model, where we have a write once storage and where we are able to simulate normal RAM's with only a logarithmic loss in time. Such a model is interesting on its own right.

We call this model register stack machine. The name stack is chosen in distinction to the name push-down because of the ability of the new model to read the entire stack without modifying it (see e.g. [HU79, chapter 14]).

A *register stack machine* (RSM) has a constant number c of registers and, in addition, a (non-erasing) stack of registers, i.e., a push-down like managed list of registers, which the machine is only allowed to read. In each step it simultaneously

- pushes the contents of the register number 2 on the stack,
- loads the contents of the stack indexed by the contents of register number 1 to register number 0 (if register number 1 is not negative) and
- all the registers with number >0 can be loaded by a constant or by the sum or the difference between other registers.

The next instruction can depend on a ≤ 0-test of register 3.

Formally an RSM M is a 5-tuple (c, Q, q_0, δ, Q_f) with the transition function $\delta : Q \times \{1,0\} \mapsto Q \times ((\mathbb{N} \cup \{+,-\} \times \{1,...c\}^2)^{\{1,\cdots c\}})$. A configuration is described by a sequence beginning with the contents of the stack, followed by the contents of the registers and ending with the current state. At the beginning the input is on the stack. The start configuration is $s_0, ..., s_n, r_0, ..., r_c, q_0$, where $s_0, ..., s_n$ contains the input, $r_4 = n$ and $r_0 = r_1 = r_2 = r_3 = r_5 = ... = r_c = 0$. The configuration transition is

$$s_0, ..., s_m, r_0, ..., r_c, q \vdash_{M} s_0, ..., s_{m+1}, r'_0, ..., r'_c, q'$$

where the test $t := 0$ if $r_3 \leq 0$ and $t := 1$ if $r_3 > 0$ determines the transition by $(q', a) := \delta(q, t)$. We set $s_{m+1} := r_2$, $r'_0 := s_{r_1}$, and $r'_i := a(i)$ if $a(i) \in \mathbb{N}$ and $r'_i := r_j t' r_k$ if $a(i) = (t', j, k) \in \{+,-\} \times \{1,...c\}^2$ for $i > 0$. The input $s_0, ..., s_n$ is accepted if a final state in Q_f is reached.

In the description of an algorithm for an RSM every line has the following shape:

q: if $r_3 \leq 0$ then r_1:=...,r_2:=... , ... r_c:=... goto q'
else r_1:=...,r_2:=... , ... r_c:=... goto q''

For convention we may add and subtract constants (which means add or subtract an additional register, which was loaded with this constant before), if a register r_i is not mentioned, it means $r_i := r_i$ (which means $r_i := r_i + 0$) and if a goto is omitted in line q_i it means goto q_{i+1}. Also 'if $r_3 \leq 0$ then ... else' may be omitted, if the same is done in both cases.

Proposition 1. *[Wie90] A RAM with running time $t(n)$ under logarithmic cost measure can be simulated in time $O(t(n))$ by a RAM using integers of length $O(\log(t(n)))$ and addresses of value in $O((t(n))^2)$.*

Lemma 2. *A RAM with running time $t(n)$ under logarithmic cost measure can be simulated by a RSM in time $O(t(n)\log(t(n)))$ using integers of length $O(\log(t(n)))$.*

Proof. The registers R_j which are accessed by the RAM directly (constantly many) are simulated by the RSM directly in registers $r_{j'}$. The indirectly addressed storage of the RAM is simulated as a binary tree. The address of the root node is stored in r_5, r_4 contains the address of the top of the stack (it is incremented in every step). Each inner node is represented as 3 consecutive register contents on the stack: one number containing the smallest address in the subtree[1] with the bigger addresses and two pointers to subtrees. The leaf nodes are represented as 2 consecutive register contents on the stack: one marking (by -1 or -2), which tells that this is a leaf node and whether this is one of the (constantly many) directly accessed cells (-2) and the contents of the register of the simulated RAM.

Because of Proposition 1, the addresses have a value in $O((t(n))^2)$. Thus the tree has logarithmic depth and we do not have to do balancing. Since the value of $t(n)$ may be unknown before the simulation, the binary tree can not be initialized completely at the beginning. Instead of this, each time the simulation addresses with a value which is out of the scope of the binary tree initialized so far, the rest of the binary tree with the size of the next power of the 2 is initialized[2] containing the old binary tree as a part.

In order to simulate an indirect reading of the RAM indexed by register R_j into register r_k, the RSM walks down the tree in the following way: first the pointer to the root node is loaded from r_5 to r_1, then, as long as r_0 does not contain a leaf marking, it sets $r_3 := r_{j'} - r_0$ which allows to set $r_1 := r_1 + 1$ or $r_1 := r_1 + 2$ in the two possible following states depending on whether $r_3 \leq 0$ and continue with $r_1 := r_0$. If the leaf node is found, it sets $r_1 := r_1 + 1$ (if R_{R_j} is not one of the directly accessed cells) and in the next step r_0 contains the contents of R_{R_j}.

The formal description of this looks like follows:

[1] In case of the root node this is always a power of the 2.

[2] The initialization can be done by copying the structure of the old tree with new leaf nodes having the value zero and with the values of the first registers of the inner nodes incremented by 2^k which was the size of the old tree. Then a new root node is introduced with 2^k in the first register and 2 pointers leading to the old and the new root node. This can be repeated until an appropriate power of the 2 is reached.

q_0: $r_1 := r_5$, $r_4 := r_4 + 1$
q_1: $r_3 := r_{j'} - r_0$, $r_4 := r_4 + 1$
q_2: if $r_3 \leq 0$ then $r_1 := r_1 + 1$, $r_4 := r_4 + 1$
else $r_1 := r_1 + 2$, $r_4 := r_4 + 1$
q_3: $r_1 := r_0$, $r_4 := r_4 + 1$
q_4: $r_3 := r_0 + 1$, $r_4 := r_4 + 1$
q_5: if $r_3 \leq 0$ then $r_3 := r_3 + 1$, $r_4 := r_4 + 1$
else $r_3 := r_{j'} - r_0$, $r_4 := r_4 + 1$, goto q_2
q_6: if $r_3 \leq 0$ then $r_k := r_0$, $r_4 := r_4 + 1$
else $r_k := r_{k'}$, $r_4 := r_4 + 1$

This takes $O(\log(t(n)))$ many steps.

In order to simulate an indirect writing of register r_k of the RAM indexed by register R_j, the RSM first checks whether R_{R_j} is not one of the directly accessed cells. Then the complete path to the corresponding node in the binary tree is copied on top of the stack in reverse order by replacing the pointers on the path by the new ones which are the current position in r_4 plus a constant. This constant is the number of steps until the RSM is copying the child node. Then it writes the new leaf node on top of the stack. Again this takes $O(\log(t(n)))$ many steps. ■

Remark 3. It is easy to see that an RSM with running time $t(n)$ can be simulated by a RAM in time $O(t(n))$.

In [LN93] K.-J. Lange and R. Niedermeier introduce the notion of write data-independence, which means that it is determined only by the length of the input, which memory cell is written at each time. This notion can be sharpened to *strict write data-independence*, which means that at any time for any memory cell the time, when this cell was written the last time, can be calculated in constantly many steps. It is easy to see that such a strict write data-independent RAM can be simulated by an RSM without a logarithmic loss of time. (Of course any RSM is strictly write data-independent.) In this maner for example Merge-sort can be performed on an RSM as fast as on a RAM.

3 The corresponding circuit

Analogously to the CVP we define MCP as the problem to calculate the output of a circuit, which has multiplex select gates like the ones which are used in [FLR96] to characterize OROW-PRAM's.

Such a multiplex select gate has two kinds of input signals: one bundle of $O(\log(n))$ steering signals and up to n bundles of $O(\log(n))$ data signals. The number which is encoded in binary by the steering signals is the number of the bundle of data signals which is switched through to the output.

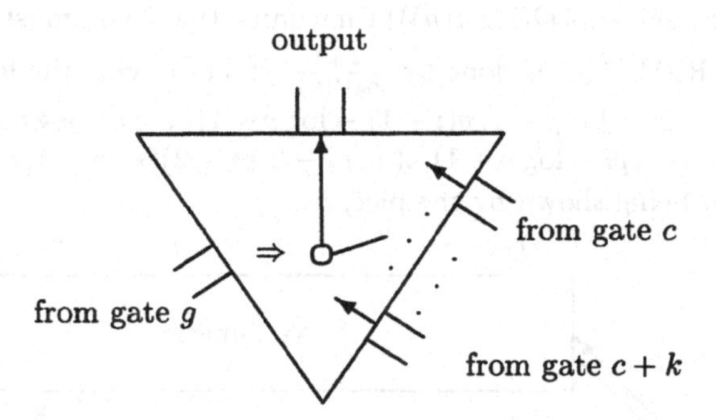

A multiplex select gate can be described by the (binary) encoding of $(\star, g, c, k)$ where g is the number of the gate whose output bundle is connected to the steering input bundle, k is the number of data input bundles and $c + j$ is the number of the gate whose output is connected to the j'th data input bundle for $j \leq k$. This means that the encoding can still have a logarithmic size although the number of input bundles can be linear.

Analogously we have gates for and ($\wedge$), or ($\vee$), addition (+) and subtraction (-) which are encoded by a symbol and two numbers of gates (for example $(+, d_1, d_2)$), gates for the >0 test and negation ($\neg$) which have one number of a gate in the encoding and gates having no input and a fixed output o, where the encoding is $(\$, o)$.

(All those gates could be simulated by polylogarithmically many multiplex select gates with independently connected steering signals and no bundles.)

MCP is the language of inputs encoding a circuit with multiplex select gates, where each gate has only inputs from gates with smaller numbers and where the last gate has the value 1.

Remark 4. To simulate one multiplex select gate with (unbounded fan-in) $\wedge$, $\neg$ and $\vee$ gates would need a number of gates, which is linear in the size of the circuit.

Lemma 5. *An RSM with running time $t(n)$ using only integers having the length $O(\log(t(n)))$ with a general log cost measure (that means that every step costs $O(\log(t(n)))$ time units) can be simulated by a uniform family of multiplex circuits of size $O(t(n) + n)$.*

Proof. Given an RSM and an input $s_0, ..., s_m$, we construct a circuit starting with m stages for the input having the form $(\$, 0)^{p-3}(\$, s_i)(\$, 0)(\$, 0)$ for $i < m$ and one stage having the form $(\$, 0)^{p-5}(\$, m)(\$, 0)(\$, s_m)(\$, 0)(\$, 0)$, where the period p of the circuit is the number of gates in one stage simulating one step of the RSM. Thus p is determined by the simulated RSM. We may assume that p is a power of 2. (Furthermore we may assume that the starting state is encoded by zeros.)

Since one step costs $O(\log(t(n)))$ time units, the circuit must simulate $\frac{t(n)}{\log(t(n))}$ steps of the RSM. This is done by $\frac{t(n)}{\log(t(n))}$ stages having the form

$...(+, p(i-1) - \log p - 1, p(i-1) - \log p - 1)(+, pi - \log p, pi - \log p)$
$(+, pi - \log p + 1, pi - \log p + 1)...(+, pi - 2, pi - 2)(\star, pi - 1, p - 2, p(i-1))$

for $i > m$ being shown by the picture

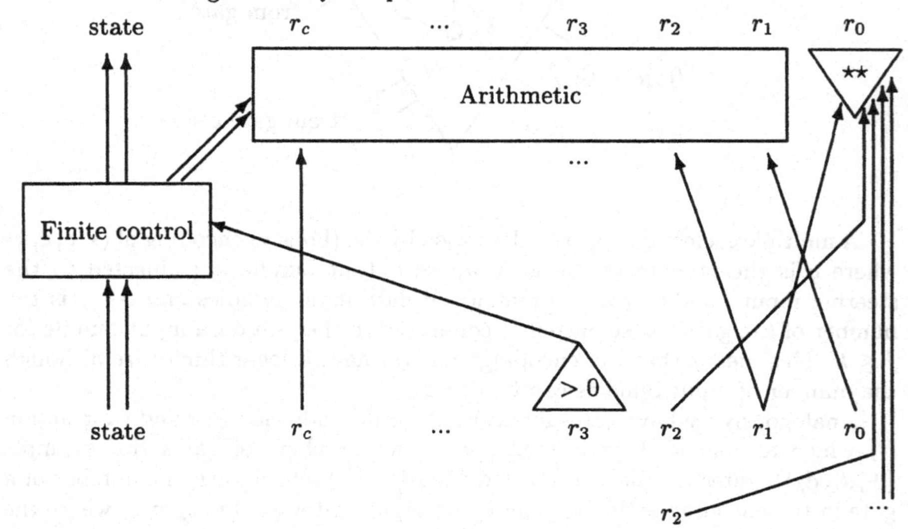

where all input wires to this stage come from the last stage except the bundles or wires from register 2, where gates simulating register 2 from all preceding stages are connected as data input to the multiplex select gate simulating register 0. The $\star\star$-gate in the picture is a shorthand for a $\star$-gate and $\log p$ addition-gates, which multiply the value for the steering input by p. Hence $(p-2) + jp$ is the number of the gate whose output is switched through to the output of gate number pi if $j < i$ is the output of gate number $p(i-1) - \log p - 1$, which simulates register r_1 in the previous stage. This simulates indexed reading from the stack which contains the input $s_0, ..., s_m$ as well as all contentses of register 2. For each register with a number $j > 0$, the arithmetic looks like

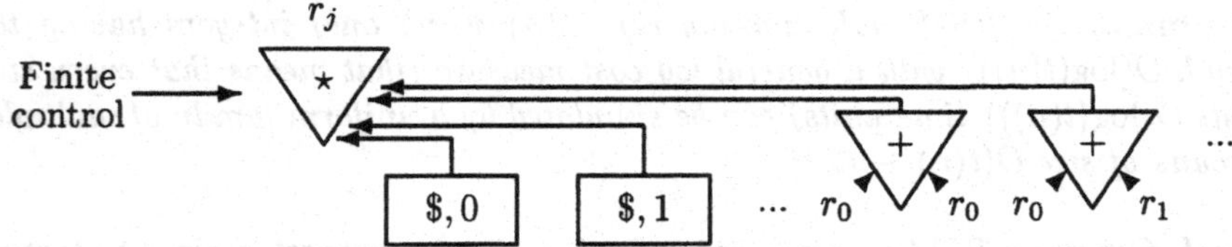

which preserves every possible assignment to the register depending on the state. The encodings of the stages look almost the same, except gate numbers appearing in them, which are linear dependent on the number of the stage. Thus the only difficulty to calculate the encoding of the circuit is to calculate $t(n)$.

The last stage contains only some logic gates such that the output of the last gate is 1 iff a final state in Q_f is reached. ■

Lemma 6. *MCP∈DTIME(O(n)) with log cost measure.*

Proof. The input $s_0, \dots, s_m$ on the stack of an RSM (or in the register of a RAM) contains the encoding of the circuit, where the specifying symbol and the numbers encoding one gate are contained in consecutive registers. An RSM can evaluate one gate in a constant h number of steps. (For example $h = 16$ should be sufficient.) Thus it can write the output of gate number j to the position $m + hj$ on the stack.

To evaluate a gate, the RSM first reads the specifying symbol to detect the kind of gate. To evaluate a multiplex select gate (encoded by $(\star, g, c, k)$), the machine first reads the number g, which is in the next position in the input, multiplies it by h in $\log h$ steps and loads $m+hg$ to register r_1. Then r_0 contains the number l of the input bundle, which is switched through to the output. The next position in the input contains c, which has to be added to get the number $l + c$ of the gate whose output is connected to this input bundle. Also the RSM has to check $l \leq k$, which is in the following position on the input. Now $m + h(l + c)$ is loaded to register r_1 (again in $\log h$ steps). Then r_0 contains the output value of the gate, which is loaded to r_2 to store it on the stack, if h steps took place since the evaluation of the last gate. The evaluation of the other kinds of gates works in analogous way.

In this way an RSM can evaluate the circuit in linear time. The same holds for a RAM. ∎

For proving the next theorem we need the following lemma which was proved by J. Hoover.

Lemma 7. *[Con94] Let l be an eventually non-decreasing function such that $l(n) \in \Omega(n)$ and $l(n) \in n^{O(1)}$. For all $\delta > 0$, there is a rational number σ and a natural number n_0 such that*

- $l(n) \leq n^\sigma$ *for* $n \geq n_0$ *and*
- $n^\sigma \in O(l(n) n^\delta)$.

Theorem 8. *MCP is strictly sequential P-complete.*

Proof. According to Lemma 6, MCP∈DTIME($O(n)$) with log cost measure. We have to show that if MCP has polynomial speedup, this means if it can be solved by a PRAM in time $O(n^{1-2\delta})$ for a $\delta > 0$, then every problem in P has polynomial speedup.

Let L be a language in P which is recognized by a RAM with running time $t(n) \in \Omega(n)$. According to Lemma 2, it can be recognized by a RSM in time $l(n) \in O(t(n)\log(t(n)))$ using only integers of length $O(\log(t(n)))$. For constructing a circuit, we need to know its size, but it is not known how to compute $l(n)$ in general. For that reason we apply Lemma 7 to get n^σ as a tight upper bound for the time for L to be recognized by a RSM. Thus according to Lemma 5, L can be reduced to MCP by a function f, which generates on input x the encoding of the circuit, which simulates the RSM on input x. For $|x| = n$ we have

$|f(x)| \in O(n^\sigma)$. Obviously the reduction $f(x)$ can be calculated by a PRAM in polylogarithmic time since n^σ can be calculated fast. The same PRAM can afterwards test in $O(n^{\sigma(1-2\delta)})$ steps whether $f(x) \in$MCP. Thus this PRAM can recognize L in time

$$O(n^{\sigma(1-2\delta)}) \subseteq O((l(n)n^\delta)^{1-2\delta}) \subseteq O((l(n)^{1+\delta})^{1-2\delta})$$
$$\subseteq O((l(n)^{1-\delta-2\delta^2}) \subseteq O((t(n))^{1-\delta}).$$

This means that $L \in$SP. ∎

Corollary 9. *If MCP∈SP then P⊆SP.*

4 Using prediction of termination

The inefficiency of the polynomial speedup of a problem in P in Theorem 8 is caused by the inefficiency for the speedup of MCP and by n^δ, which we have to sacrifice in order to get a a time limit, which can be calculated fast.

For practical reasons we are interested in those problems L in P having an optimal $t(n)$ time bounded sequential algorithm (this means no other algorithm recognizes L in time $o(t(n))$ with *efficient prediction of termination* which means $t(n)$ can be calculated in sublinear time.

Theorem 10. *MCP is strictly sequential almost efficient P-complete.*

This means if MCP∈AP, then every problem in P, which has an efficient prediction of termination, is in AP.

Proof. Let $\widetilde{O}(t(n)) := t(n)(\log(t(n)))^{O(1)}$. According to Lemma 6, we have MCP∈DTIME($O(n)$) with log cost measure. We have to show that if MCP has almost efficient polynomial speedup, this means if it can be solved by a PRAM in time $O(n^{1-\varepsilon})$ with $\widetilde{O}(n^\varepsilon)$ processors, then every problem in P, which has efficient prediction of termination, has almost efficient polynomial speedup.

Let L be a language in P which is recognized by a RAM with running time $t(n) \in \Omega(n)$, which can be calculated in sublinear time. According to Lemma 2 it can be recognized by a RSM in time $\widetilde{O}(t(n))$ using only integers of length $O(\log(t(n)))$. Thus according to Lemma 5 it can be reduced to MCP by a function f, which generates on input x the encoding of the circuit, which simulates the RSM on input x. For $|x| = n$ we have $|f(x)| \in \widetilde{O}(t(n))$. The reduction $f(x)$ can be calculated by a PRAM in $O(|f(x)|^{1-\varepsilon})$ steps and $O(|f(x)|^\varepsilon)$ processors. The same PRAM can afterwards test whether $f(x) \in$MCP in time $\widetilde{O}(|f(x)|^{1-\varepsilon})$ and $O(|f(x)|^\varepsilon)$ processors. This means that $L \in$AP. ∎

Nearly all computational models are equivalent for the definition of the class P, but this equivalence is not valid if we consider linear efficiency. Thus investigating efficiency, we have to make distinctions between the used models.

The inefficiency of the polynomial speedup of a problem in P in Theorem 8 is caused by the inefficiency for the speedup of MCP and by the $\log n$ factor for the simulation of a RAM by an RSM. Thus in analogous way we get the following result:

Theorem 11. *If we regard the RSM with a general log cost measure as the model to define sequential time complexity, then, MCP is strictly sequential efficient P-complete.*

This means if we regard the RSM with a general log cost measure as the model to define sequential time complexity and if MCP∈EP, then every problem in P, which has an efficient prediction of termination, is in EP.

Remark 12. We can use the uniformity of the stages in the construction of the circuit to obtain more compressed representations which can be padded afterwards. In this way we can create artificial variants of MCP which are also strictly sequential P-complete complete but which are in DTIME(n^{σ})\DTIME($n^{\sigma-\varepsilon}$) for $\sigma > 1$ and $\varepsilon > 0$.

Open Problems: Is CVP strictly sequential P-complete? Is CVP∈ SP? Are there other natural strictly sequential P-complete problems?

Acknowledgment: I thank Henning Fernau, Ray Greenlaw and Klaus-Jörn Lange for helpful remarks.

References

[Con92] A. Condon. A theory of strict P-completeness. In *Proceedings of the 9th Symposium on Theoretical Aspects of Computer Science*, number 577 in Lecture Notes in Computer Science, pages 33–44. Springer-Verlag, 1992.

[Con94] A. Condon. A theory of strict P-completeness. *Computational Complexity*, 4:220–241, 1994.

[Coo85] S. A. Cook. A taxonomy of problems with fast parallel algorithms. *Information and Control*, 64:2–22, 1985.

[FLR96] H. Fernau, K.-J. Lange, and K. Reinhardt. Advocating ownership. In V. Chandru, editor, *Proceedings of the 16th Conference on Foundations of Software Technology and Theoretical Computer Science*, volume 1180 of *Lecture Notes in Computer Science*, pages 286–297. Springer, December 1996.

[GHR95] R. Greenlaw, H. J. Hoover, and W. L. Ruzzo. *Limits to parallel computation: P-completeness theory.* New York u.a., Oxford Univ. Pr., 1995.

[HU79] J. E. Hopcroft and J. D. Ullman. *Introduction to Automata Theory, Languages and Computation.* Addison-Wesley, 1979.

[KRS90] C. P. Kruskal, L. Rudolph, and M. Snir. A complexity theory of efficient parallel algorithms. *Theoretical Computer Science*, 71:95–132, 1990.

[LN93] K.-J. Lange and R. Niedermeier. Data-independences of parallel random access machines. In R. K. Shyamasundar, editor, *Proceedings of the 13th Conference on Foundations of Software Technology and Theory of Computer Science*, number 761 in Lecture Notes in Computer Science, pages 104–113, Bombay, India, December 1993. Springer-Verlag.

[VS86] J. S. Vitter and R. A. Simons. New classes for parallel complexity: A study of unification and other complete problems for P. *IEEE Transactions on Computers*, C-35(5):403–418, 1986.

[Wie90] J. Wiedermann. Normalizing and accelerating RAM computations and the problem of reasonable space measures. In M.S. Paterson, editor, *Proceedings of the 17th ICALP (Warwick University, England, July 1990)*, LNCS 443, pages 125–138. EATCS, Springer-Verlag, 1990.

An Unambiguous Class Possessing a Complete Set

Klaus-Jörn Lange

Wilhelm-Schickard-Institut, Tübingen

Topics : Computational and structural complexity theory

Abstract

In this work a complete problem for an unambiguous logspace class is presented. This is surprising since unambiguity is a 'promise' or 'semantic' concept. These usually lead to classes apparently without complete problems.

1 Introduction

One of the most central questions of complexity theory is to compare determinism with nondeterminism. Our inability to exhibit the precise relationship between these two features motivates the investigation of intermediate features such as symmetry or unambiguity. In this paper we will concentrate on the notion of unambiguity.

Unfortunately, unambiguity of a device or of a language is in general an undecidable property. Unambiguous classes are not defined by a 'syntactical' machine property but rather by a 'semantical' restriction. A nasty consequence is the apparent lack of complete sets. In the case of time bounded computations there are relativizations of unambiguity which provably have no complete problem ([10]).

For space bounded computations the concept of unambiguity is not as uniform as in the time bounded case. There are several versions of it, which probably are different for space classes, while they coincide for time bounded computations. This is the case, since it is possible to keep track of the complete history of a computation without increasing the running time. It is remarkable, that this is precisely the same construction which shows the equivalence of nondeterminism with symmetry and of determinism with reversability in the time bounded case ([14,3]).

The main result of this work is to show that the unambiguous logspace class $RUSPACE(\log n)$ possesses a complete problem. The proof makes intensive use of the space-specific possibility to cycle through all configurations of a machine without increasing the resource bound. The proof doesn't seem to work for other versions of unambiguity or for other related semantic classes but nevertheless leaves the hope to exhibit complete problems for other semantic logspace classes defined by concepts like unambiguity and randomization.

As an interesting consequence of this result we get the first case of a single and explicit problem which can be solved by an unambiguous logspace algorithm

Reischuk, Morvan (Eds.): STACS'97 Proceedings, LNCS 1200

but which is not known to be solvable in logarithmic space deterministically. It was known before that the family *ULIN* of unambiguous linear context-free languages is contained in *USPACE*($\log n$) and it is still open whether *ULIN* $\subseteq$ *DSPACE*($\log n$) (see [4]). But there was no specific candidate known within *ULIN*, which was not known to be in *DSPACE*($\log n$).

2 Preliminaries

The reader is assumed to be familiar with the basic notions of complexity theory as they are contained in the standard text books on theoretical computer science.

2.1 Turing machines and configuration graphs

For a Turing machine T with input alphabet X we denote the graph of configurations over an input $w \in X^*$ by $G_T(w)$. If T is logarithmically space-bounded $G_T(w)$ has a size polynomial in the length of w. Without loss of generality we assume all machines considered here to be non-looping. Hence all configuration graphs are acyclic. In order to ease the presentation of our constructs we assume that in nondeterministic machines each non nonterminating configuration has exactly two successor configurations, which can be reached in one step. It should be remarked that this normal form can be reached without changing the number of accepting computations, i.e.: without changing properties like umambiguity.

Let $G = (V, E)$ be a directed acyclic graph. For two nodes x and y we denote by $N(x, y)$ the number of different paths leading from x to y in G. For $x = y$ we set $N(x, y) := 1$. For each pair of nodes (x, y) let $d(x, y)$ be the length of the shortest path between x and y. The length of a path is the number of its edges. If x and y are not connected, $d(x, y)$ is infinite. $d(x, x)$ is 0 for each $x \in V$. In the following we will work with complete binary graphs, that is, each node of G is either a leaf with no outgoing edges, or an inner node with two outgoing edges. This is determined by a mapping $\phi : V \longrightarrow \{i, l\}$, which takes the value l for leaves and i for inner nodes. The two successors of an inner node x will be denoted by $L(x)$ and $R(x)$. In the following we assume all graphs like $G_T(w)$ to be given in this form as (V, ϕ, L, R). Thus the edge set E would be $\{(x, L(x)) | \phi(x) = i\} \cup \{(x, R(x)) | \phi(x) = i\}$. Since G is acyclic, it contains no self-loop, i.e.: $L(x) \neq x \neq R(x)$ for $x \in V$. In addition, we assume without loss of generality that G contains no double edges (i.e.: $L(x) \neq R(x)$ for all x). If G has n nodes we assume V to be $\{v_1, v_2, \cdots, v_n\}$. We will be interested in the existence of paths between nodes v_1 and v_n. We will assume v_1 to be an inner node and v_n to be a leaf. Hence $n > 1$.

We call G *accepting* iff $N(v_1, v_n) > 0$, i.e.: if there is a path from v_1 to v_n. Otherwise we call G *rejecting.*

For $x \in V$ let $T(x) := \{y \in V | N(x, y) \geq 1\}$ be the set of all nodes reachable from x. For $d \geq 0$ we set $T_d(x) := \{y \in T(x) | d(x, y) \leq d\}$. Obviously, we have $T(x) = T_{n-1}(x)$ for every $x \in V$. Throughout the paper we will identify $T(x)$ and $T_d(x)$ with the subgraphs of G induced by $T(x)$ and by $T_d(x)$.

3 Unambiguity

A concept intermediate in power between determinism and nondeterminism is *Unambiguity.* A nondeterministic machine is said to be unambiguous, if for every input there exists at most one accepting computation. This leads for instance to the classes *UP* and *USPACE*$(\log n)$; we have $P \subseteq UP \subseteq NP$ and $DSPACE(\log n) \subseteq USPACE(\log n) \subseteq NSPACE(\log n)$ where none of these inclusions is known to be strict. The notion of unambiguity should be distinguished from that of *Uniqueness*, which uses the unique existence of an accepting path not as a restriction but as a tool. The resulting language classes $1NSPACE(\log n)$ and $1NP$ consists of languages defined by machines that accept their inputs if there is exactly one accepting path. Thus, the existence of two or more accepting computations is not forbidden, but simply leads to rejection. In the polynomial time case we have $Co\text{-}NP \subseteq 1NP$. In the logspace case inductive counting ([11,19]) shows $1NSPACE(\log n) = NSPACE(\log n)$.

3.1 Space bounded unambiguous classes

The concept of unambiguity of space bounded computations is not as uniform as that for time bounded classes. Instead we are confronted with a variety of probably different concepts of unambiguity. In the following we classify three notions of unambiguity for configuration graphs and complexity classes. For more varieties of unambiguities see [4].

A configuration graph $G = (V, \phi, L, R)$ is called *unambiguous* if there is at most one path from v_1 to v_n, i.e.: if $N(v_1, v_n) \leq 1$. G is called *reach-unambiguous* if for any x there is at most one path from v_1 to x, i.e.: if for each $x \in V$ $N(v_1, x) \leq 1$. G is called *strongly unambiguous* or a *mangrove* if for any pair (x, y) of nodes there is at most one path leading from x to y, i.e.: if $\forall_{x,y \in V} N(x, y) \leq 1$. Every mangrove is reach-unambiguous and every reach-unambiguous graph is unambiguous. Although a mangrove does not need to be a tree, for each x the subgraph $T(x)$ is indeed a tree and the same is true for the set of all nodes from which x can be reached. Some examples:

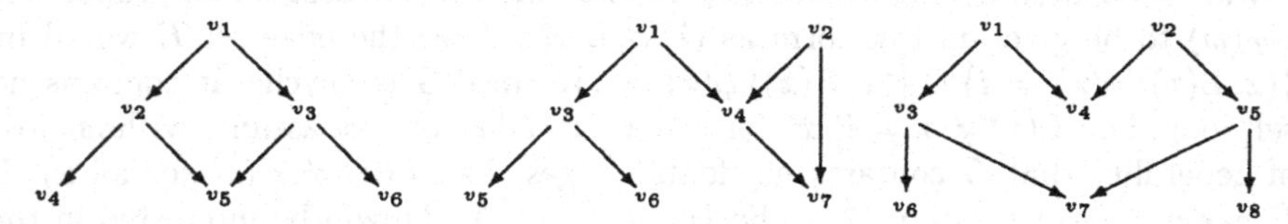

The left configuration graph is accepting and unambiguous since there is exactly one path from v_1 to v_6. It is not reach unambiguous since there exist two different paths between nodes v_1 and v_5. The second one is reach-unambiguous and accepting, but not a mangrove since there are two different paths between nodes v_2 and v_7. The third example is a mangrove, which is rejecting since there is no path from v_1 to v_8.

Let T be a nondeterministic Turing machine with input alphabet X. T is called *unambiguous*, if for all inputs $w \in X^*$ the configuration graph $G_T(w)$

is unambiguous. T is ***reach-unambiguous***, if for all $w \in X^*$ the configuration graph $G_T(w)$ is reach-unambiguous. Finally, T is ***strongly unambiguous***, if for all $w \in X^*$ the configuration graph $G_T(w)$ is a mangrove. Of course, both for the language and Turing machines each of this properties is undecidable. These concepts lead to the following classes:

Definition 1. a) *USPACE*$(\log n)$ is the class of all languages accepted by unambiguous logspace machines.
b) *RUSPACE*$(\log n)$ is the class of all languages accepted by reach-unambiguous logspace machines.
c) *StUSPACE*$(\log n)$ is the class of all languages accepted by strongly unambiguous logspace machines.

We mention in passing that in the case of time bounded classes these three concepts coincide.

By definition, these three classes fulfill: *StUSPACE*$(\log n) \subseteq$ *RUSPACE*$(\log n) \subseteq$ *USPACE*$(\log n) \subseteq$ *NSPACE*$(\log n)$. In addition, by [4] we know

Proposition 2. *i) RUSPACE*$(\log n) \subseteq$ *LOG(DCFL)*
ii)Both RUSPACE$(\log n)$ *and StUSPACE*$(\log n)$ *are closed under complement.*

Here *LOG(DCFL)* denotes the class of all languages reducible to deterministic context-free languages by deterministic many-one reductions. As a consequence of part i) both *RUSPACE*$(\log n)$ and *StUSPACE*$(\log n)$ are contained in $SC^2 :=$ *DTIMESPACE*$(pol, \log^2 n)$, the second level of the SC hierarchy ([6,7]).

The inclusion *StUSPACE*$(\log n) \subseteq$ *DSPACE*$(\log^2 n / \log\log n)$ was shown in [2]. But as remarked there, the proof uses only the fact, that the unfolding of the reachability graph, i.e.: the number of all paths starting in the root, is of polynomial size. While this is not true for *USPACE*$(\log n)$, this property is fulfilled by reach-unambiguous graphs. Hence we have:

Proposition 3. *RUSPACE*$(\log n) \subseteq$ *DSPACE*$(\log^2 n / \log\log n)$

It should be remarked, that nothing like Proposition 2 or 3 is known for *USPACE*$(\log n)$.

3.2 Sets of unambiguous reachability problems

The unambiguous classes defined in the previous subsection are semantic or promise classes. They are defined via machines which are subject to an undecidable restriction. As a consequence, they probably don't posses complete sets. There are relativizations in the time bounded case excluding the existence of complete sets ([10]).

For logarithmically space bounded classes the typical complete problems are reachability or connectivity problems in graphs. The question for existence of a connecting path in directed graphs is complete for *NSPACE*$(\log n)$, that in undirected graphs for *SymSPACE*$(\log n)$, and that in forests for *DSPACE*$(\log n)$.

In connection with the three unambiguous classes the corresponding connectivity problems would be:

$$L_u := \{G = (V, \phi, L, R) \,|N(v_1, v_n) = 1\},$$

$$L_{ru} := \{G = (V, \phi, L, R) \,|N(v_1, v_n) = 1 \,, \forall_{x \in V} \; N(v_1, x) \leq 1\},$$

$$L_{su} := \{G = (V, \phi, L, R) \,|N(v_1, v_n) = 1 \,, \forall_{x,y \in V} \; N(x, y) \leq 1\},$$

Obviously, these sets are hard for the corresponding complexity classes:

Proposition 4. *i) L_u is USPACE($\log n$)-hard, ii) L_{ru} is RUSPACE($\log n$)-hard, and iii) L_{su} is StUSPACE($\log n$)-hard.*

But to show the completeness of these languages we have to exhibit unambiguous logspace algorithms. But while the uniqueness of a computational path is used as a restriction in the definition of the complexity classes, this uniqueness is used as a tool (i.e.: as an acceptance criterion) in the definition of the connectivity problems. In fact, it turns out, that L_u seems to be harder than *USPACE*($\log n$), since it is complete for 1*NSPACE*($\log n$) = *NSPACE*($\log n$), as mentioned above:

Proposition 5. *L_u is NSPACE($\log n$)-complete.*

Proof: Obviously, $L_u \in$ *NSPACE*($\log n$) by the closure of *NSPACE*($\log n$) under complement. On the other hand, the usual *NSPACE*($\log n$)-complete GAP problem asking for the existence of a path from v_1 to v_n in an unrestricted graph is reduced to the complement of L_u by adding the edge (v_1, v_n). □

This seems to indicate that $L_u$is probably not a member of *USPACE*($\log n$). It is tempting in this situation to increase this appearence by trying to derive structural upper bounds as they were obtained for *StUSPACE*($\log n$) and *RUSPACE*($\log n$) ([4,2]). But by a result of Wigderson ([20]) this would immediately carry over to nonuniform *NSPACE*($\log n$).

4 Main Result

We will now show $L_{ru} \in$ *RUSPACE*($\log n$). The idea of proof will be as follows: assume $T(v_1)$ to be a tree and to perform a breadth first search for violations of this assumption while inductively using the fact that the parts searched so far are indeed trees. If no violation exists $T(v_1)$ is a tree and hence the input graph is reach-unambiguous. The problem is to traverse the tree since logspace doesn't suffice to keep track of the whole path leading from v_1 to the currently visited node. To traverse $T(v_1)$ as a tree we use nondeterminism. If in fact $T(v_1)$ is a tree, all paths are unique and guessing paths turns out to be reach-unambiguous. If the input graph is not a tree we will find the smallest counterexample in an reach-unambiguous way.

We will use the procedure *NEXT*(y, h, d), given in Table 1, which in case $T_d(v_1)$ is a tree computes the preorder successor of a node $y \in T_d(v_1)$ which

1 **IF** $\phi(y) = i$ **AND** $h < d$ **THEN**
2 $\quad (y, h) := (L(y), h + 1)$
ELSE
3 $\quad z := v_1$;
4 $\quad j := 1$;
5 $\quad$ **WHILE** $j < h$ **AND** $\phi(z) = i$ **DO**
6 $\quad\quad z := R(z)$;
7 $\quad\quad j := j + 1$
$\quad$ **OD**;
8 $\quad$ **IF** $R(z) = y$ **THEN**
9 $\quad\quad (y, h) := (v_1, 0)$
$\quad$ **ELSE**
10 $\quad\quad$ GUESSUNCLE(y,h,d)

Table 1. Procedure NEXT(y, h, d)

has depth h wrt v_1. Procedure NEXT first deterministically checks the easy cases that y is an inner node inside of $T_d(v_1)$ or that $y = R^h(v_1)$. If this is not successful NEXT uses the nondeterministic procedure GUESSUNCLE. It should be remarked, that in line 8 the condition $R(z) = y$ is not allways well-defined since not necessarily $\phi(z) = i$ holds. In this case we regard the condition as not fulfilled leading to a call of GUESSUNCLE. This procedure, given in Table 2 is invoked if on the path from v_1 to y an R-edge is used. Nondeterministically this path and in particular its last occurring R-edge are guessed. This situation is illustrated in Figure 1. Procedure GUESSUNCLE has two types of exits: either

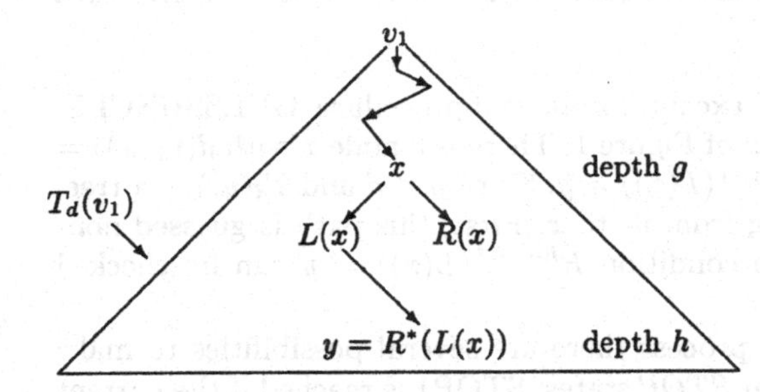

On the unique path from v_1 to y, x is the least predecessor of y which is followed by an L-successor. GUESSUNCLE guesses x and the path from v_1 to x. The rest is checked deterministically. As a result y is replaced by $R(x)$.

Fig. 1. The work of procedure GUESSUNCLE

it successfully computes the successor of y in inorder traversal of $T_d(v_1)$ or it ends its computation in a state STOPi for $i = 1, 2, 3$, or 4. To reach one of the STOP states means that the previous nondeterministic computation was not able to find the node y. If $T(v_1)$ is not a tree, the behaviour of GUESSUNCLE is ambiguous. But if $T(v_1)$ is a tree, we can show that GUESSUNCLE works reach-unambiguously. Further on, if it avoids a STOP state, the computed successor of (y, h) is uniquely determined.

```
 1 GUESS  g ∈ {0,1,···,h − 1};
 2 z := v1;
 3 FOR  j = 1,2,···,g DO
 4       GUESS  b ∈ {0,1};
 5       IF  b = 0 THEN
 6            z := L(z)
         ELSE
 7            z := R(z);
 8       IF  φ(z) ≠ i THEN  STOP1(d,y,h,g,z,j)
   OD;
 9 x := z;
10 z := L(z);
11 IF  g + 1 = hTHEN
12       IF  y ≠ z THEN STOP2(d,y,h,x);
13       (y,h) := (R(x),h)
   ELSE
14       FOR  j = g + 2,···,h DO
15            IF  φ(z) ≠ i THEN  STOP3(d,y,h,g,x,z,j);
16            z := R(z)
         OD;
17       IF  y ≠ z THEN STOP4(d,y,h,g,x,z);
18       (y,h) := (R(x),g + 1)
```

Table 2. Procedure GUESSUNCLE(y,h,d)

Proposition 6. *If for an input graph (V,ϕ,L,R) the subgraph $T_d(v_1)$ is a tree, then NEXT(y,h,d) works reach-unambiguously for each $0 \le h \le d$ and each $y \in T_d(v_1)$.*

Proof: NEXT is deterministic except for its subprocedure GUESSUNCLE, which is activated in the situation of Figure 1: There is a node x with $d(v_1,x)) = g$ for some $g < h$ such that $R^{h-g-1}(L(x)) = y$. Since $g < d$ and $T_d(v_1)$ is a tree, there is exactly one path leading from v_1 to x. First, this path is guessed nondeterministically. After that the condition $R^{h-g-1}(L(x)) = y$ can be checked deterministically.

During the nondeterministic process there are several possibilities to make wrong guesses which all end up in *STOP* states. STOP1 is reached if the current node is a leaf and thus has no outgoing edge. That is, while trying to guess a path of length g we got stuck after j steps. STOP2 indicates in the case $g = h-1$ that the guessed path didn't hit y. STOP3 is reached, if $R^{h-g-1}(L(x))$ does not exist. Finally, STOP4 means that we guessed a path of length h which doesn't lead to y.

In all these cases the actual values of the program variables uniquely determine the computational history which led to the STOP state, since $T_d(v_1)$ was assumed to be a tree. To make this more precise, with each STOP state the relevant variables are explicitly listed. Thus the computation is reach-unambiguous. □

It should be remarked that we couldn't replace NEXT and GUESSUNCLE by something using the Immerman-Szelepcséyi procedure ([11,19]) since this inherently admits an exponential number of possible computation before reaching an ACCEPT, REJECT, or STOP state. Thus it could only be unambiguous but never reach-unambiguous. The advantage of NEXT and GUESSUNCLE is that they stop early enough such that the complete computational history is uniquely determined by the program variables and the input. In this way we are able to avoid a superpolynomial number of (rejecting) computations.

With the help of the previous proposition we are now able to prove our main result.

Theorem 7. $L_{ru} \in RUSPACE(\log n)$.

Proof: The logspace algorithm for checking that $T(v_1)$ is a tree for a given input graph $G = (V, \phi, L, R)$ is given in Table 3. Obviously, this algorithm uses

```
1  reached:= false;
2  FOR  d = 1, 2, ..., n-2 DO
3       (y, h) := (v1, 0);
4       just-begun := true;
5       WHILE (y, h) ≠ (v1, 0) OR just-begun = true DO
6            just-begun := false;
7            IF v_n ∈ {L(y), R(y)} THEN reached:= true;
8            IF h = d AND φ(y) = i THEN
9                 (y', h') := (v1, 0);
10                just-begun' := true;
11                WHILE (y', h') ≠ (v1, 0) OR just-begun' = true DO
12                     just-begun' := false;
13                     IF y' ∈ {L(y), R(y)} THEN REJECT1(d, y, y', h');
14                     IF h' = d AND y ≠ y' AND φ(y') = i THEN
15                          IF {L(y'), R(y')} ∩ {L(y), R(y)} ≠ ∅ THEN
16                               REJECT2(d, y, y');
17                     NEXT(y', h', d)
                  OD;
18           NEXT(y, h, d)
       OD
   OD;
19 IF reached THEN ACCEPT ELSE REJECT3
```

Table 3. Program to check the tree property of $T(v_1)$

only $O(\log n)$ space. Further on, there exists an accepting computation if and only if $T(v_1)$ is a tree and $v_n \in T(v_1)$. That is, this a nondeterministic logspace algorithm recognizing L_{ru}. It is deterministic except for the calls of NEXT in lines 17 and 18. The cooperation of the main programm with its subroutines works in a way that the control is given back to the main program unless a STOP occurred. In that case the whole computation stops without acceptance.

If a REJECT is reached in the main program the computation is terminated as well, but with the knowledge that this input doesn't belong to L_{ru}.

We will now show that the algorithm sketched in the previous tables works reach-unambiguously. First, we may assume the graphs $T_0(v_1)$ and $T_1(v_1)$ to be trees, since by assumption our graphs contain neither double edges nor self-loops. (This could be tested deterministically in advance.)

The program searches in a breadth first way for ambiguities in the graphs $T_2(v_1), T_3(v_1), \cdots, T_{n-1}(v_1)$. It doesn't start to look over $T_{d+1}(v_1)$ unless $d \leq 1$ or $T_d(v_1)$ turned out to be a tree in the previous "stop-free" execution of the *FOR* loop in line 2. If this was the case, we traverse $T_d(v_1)$ with the help of NEXT in line 18. For every node $y \in T_d(v_1)$ on this tour with $d(v_1, y) = d$ and $\phi(y) = i$, that is for every leaf of $T_d(v_1)$ which is not a leaf in $T(v_1)$, $L(y)$ and $R(y)$ are compared with every $y' \in T_d(v_1)$. In case of any coincidence we end up in state REJECT1 which indicates the existance of two paths of different length from v_1 to y'. Otherwise, we compare $L(y)$ and $R(y)$ with $L(y')$ and $R(y')$ for every $y' \in T_d(v_1)$ with $d(v_1, y') = d, \phi(y') = i$, and $y \neq y'$. In case of any coincidence we end up in state REJECT2 which indicates the existance of two different paths of the same length leading from v_1 to either $L(y')$ or $R(y')$. If both REJECT1 and REJECT2 have been avoided, we know that $T_{d+1}(v_1)$ is a tree. After doing this for $d = 1, \cdots, n-2$ without reaching a STOP state we know that $T_{n-1}(v_1) = T(v_1)$ is a tree. Finally, we accept if a path from v_1 to v_n had been detected. Otherwise, we end up in state REJECT3.

Since procedure NEXT and hence GUESSUNCLE are only activated on subgraphs $T_d(v_1)$ which have been shown to be trees before, they work reach-unambiguously. Thus the whole program, being deterministic except for the call of NEXT, works reach-unambiguously, as well, since NEXT computes a single-valued function by either stopping or computing a successor value which is uniquely determined by the input. □

Corollary 8. *L_{ru} is $RUSPACE(\log n)$-complete.*

With [13] this implies that we can constructively enumerate the reach-unambiguous logspace languages:

Corollary 9. *$RUSPACE(\log n)$ is recursively presentable*

By changing the main program to check the tree property of $T(x)$ not only for $x = v_1$ but for every $x \in V$ gives us a reach-unambiguous program to recognize mangroves:

Corollary 10. $L_{su} \in RUSPACE(\log n)$

We remark, that this program is not strongly unambiguous.

We summarize the relations between unambiguous classes and hardest languages in the Figure 2.

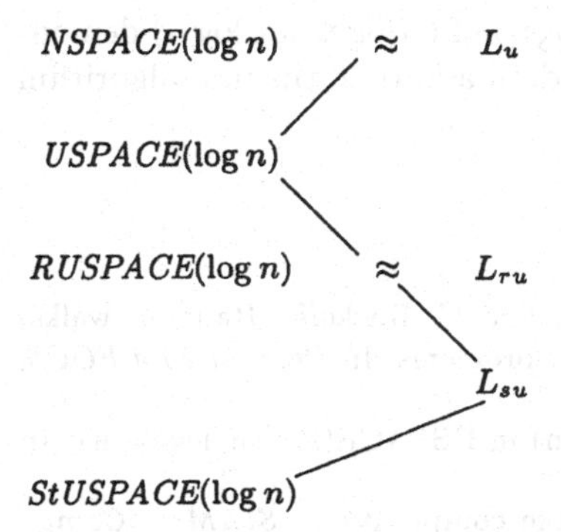

The three unambiguous classes seem to be weaker than the corresponding reachability problems. The $NSPACE(\log n)$-completeness of L_u indicates that we should not expect $USPACE(\log n)$-completeness. The more surprising is the $RUSPACE(\log n)$-completeness of L_{ru}

Fig. 2. Unambiguous logspace classes and reachability problems

5 Discussion and open questions

Considering the results of the previous section, it is natural to ask whether $StUSPACE(\log n)$ or even $RUSPACE(\log n)$ collapse down to $DSPACE(\log n)$. One fact speaking against equivalence, is that the unambiguous linear context-free languages are contained in $StUSPACE(\log n)$ but still are not known to be in $DSPACE(\log n)$.

Theorem 7 and Corollary 10 presented reach-unambiguous algorithms to recognize L_{ru} and L_{su}. None of the two works strongly unambiguous since the unreachable situation that GUESSUNCLE is activated on a nontree-like subgraph inherently results in an ambiguous behaviour. The open question here is, whether L_{su} could be $RUSPACE(\log n)$-complete, too.

The approach of Theorem 7 neither works for $USPACE(\log n)$ and L_u. But nevertheless we obtain with L_{ru} the first example of a single and explicit language which is known to be in $USPACE(\log n)$ and which is not known to be in $DSPACE(\log n)$. Compare this with the situation of the primality problem which is known to be in UP but not known to be in P ([9,15]). In view of the lack of complete problems for UP the primality problem and its relatives are the natural candidates for a problem in $UP \setminus P$. This role could now be played for $USPACE(\log n)$ vs. $DSPACE(\log n)$ by L_{ru}.

Finally, we would like to draw the attention of the reader to the structural similarity of reach-unambiguity and symmetry. Both $RUSPACE(\log n)$ and the symmetric logspace class $SymSPACE(\log n)$ possess complete problems and share nearly the same structural upper bounds, which seem to distinguish them from $NSPACE(\log n)$; they are contained in parity logspace, $DSPACE(o(\log^2 n))$, and SC^2 ([12,16,4,17,2]). Open questions here are: what is the relationship between $SymSPACE(\log n)$ and $RUSPACE(\log n)$. Can the inclusion of $SymSPACE(\log n)$ in randomized logspace ([1]) be extended to $RUSPACE(\log n)$? If so, the deterministic space bound of $O(\log^2 n / \log\log n)$ for $RUSPACE(\log n)$ could be improved to $O(\log^{1.5} n)$ ([18]). Can the inclusion $RUSPACE(\log n) \subseteq LOG(DCFL)$ be extended to $SymSPACE(\log n)$? If so, the currently best known

CREW-PRAM running time for $SymSPACE(\log n)$ of $O(\log n \cdot \log\log n)$ demonstrated in [5] would be improved to $O(\log n)$ and, in addition, the new algorithm would run on the simpler CROW-PRAM [8].

References

1. R. Aleliunas, R. Karp, R. Lipton, L. Lovasz, and C. Rackoff. Random walks, universal sequences and the complexity of maze problems. In *Proc. of 20rd FOCS*, pages 218–223, 1979.
2. E. Allender and K.-J. Lange. StUSPACE(log n) in DSPACE($\log^2$ n/loglog n). In *Proc. of the 17th ISAAC*, 1996. In print.
3. C. Bennet. Time/Space trade-offs for reversible computation. *SIAM J. Comp.*, 18:766–776, 1989.
4. G. Buntrock, B. Jenner, K.-J. Lange, and P. Rossmanith. Unambiguity and fewness for logarithmic space. In *Proc. of the 8th Conference on Fundamentals of Computation Theory*, number 529 in LNCS, pages 168–179, 1991.
5. Ka Wong Chong and Tak Wah Lam. Finding connected components in $O(\log n \log\log n)$ time on the EREW PRAM. In *Proceedings of the Fourth Annual ACM-SIAM Symposium on Discrete Algorithms, SODA*, pages 11–20, 1993.
6. S. Cook. Deterministic CFLs are accepted simultaneously in polynomial time and log squared space. In *Proc. of the 11th Annual ACM Symp. on Theory of Computing*, pages 338–345, 1979.
7. S. Cook. A taxonomy of problems with fast parallel algorithms. *Inform. and Control*, 64:2–22, 1985.
8. P. Dymond and W. Ruzzo. Parallel RAMs with owned global memory and deterministic context-free language recoginition. In *Proc. of 13th International Colloquium on Automata, Languages and Programming*, number 226 in LNCS, pages 95–104. Springer, 1986.
9. M. Fellows and N. Koblitz. Self-witnessing polynomial-time complexity and prime factorization. In *Proc. of the 7th IEEE Structure in Complexity Conference*, pages 107–110, 1992.
10. J. Hartmanis and L. Hemachandra. Complexity classes without machines: on complete languages for UP. *Theoret. Comput. Sci.*, 58:129–142, 1988.
11. N. Immerman. Nondeterministic space is closed under complementation. *SIAM J. Comp.*, 17:935–938, 1988.
12. M. Karchmer and A. Wigderson. On span programs. In *Proc. of the 8th IEEE Structure in Complexity Theory Conference*, pages 102–111, 1993.
13. W. Kowalczyk. Some connections between presentability of complexity classes and the power of formal systems of reasoning. In *Proc. of 10th Symposium on Mathematical Foundations of Computer Science*, number 201 in LNCS, pages 364–368. Springer, 1984.
14. P. Lewis and C.H. Papadimitriou. Symmetric space-bounded computation. *Theoret. Comput. Sci.*, 19:161–187, 1982.
15. G. Miller. Riemann's hypothesis and tests for primality. *J. Comp. System Sci.*, 13:300–317, 1976.
16. N. Nisan. $RL \subseteq SC$. In *Proc. of the 24th Annual ACM Symposium on Theory of Computing*, pages 619–623, 1992.
17. N. Nisan, E. Szemeredi, and A. Wigderson. Undirected connectivity in $O(\log^{1.5} n)$ space. In *Proc. of 33th Annual IEEE Symposium on Foundations of Computer Science*, pages 24–29, 1992.

18. M. Saks and S. Zhou. RSPACE(s) $\subseteq$ DSPACE($s^{3/2}$). In *Proc. of 36th FOCS*, pages 344–353, 1995.
19. R. Szelepcsényi. The method of forcing for nondeterministic automata. *Acta Informatica*, 26:279–284, 1988.
20. A. Wigderson. NL/poly $\subseteq$ $\bigoplus$L/poly. In *Proc. of the 9th IEEE Structure in Complexity Conference*, pages 59–62, 1994.

Deadlock-Free Interval Routing Schemes*

Michele Flammini[1,2]

[1] Dipartimento di Matematica Pura ed Applicata, University of L'Aquila, via Vetoio loc.Coppito, I-67100 L'Aquila, Italy. E-mail: flammini@univaq.it

[2] Project SLOOP I3S-CNRS/INRIA/Université de Nice–Sophia Antipolis, 930 route des Colles, F-06903 Sophia Antipolis Cedex, France

Abstract. k-Interval Labeling Schemes (k-ILS) are compact routing schemes on general networks which have been studied extensively and recently been implemented on the latest generation INMOS Transputer Router chips. In this paper we introduce an extension of the k-ILS to the $\langle k, s\rangle$-DFILS (Deadlock-Free ILS), where k is the number of intervals and s is the number of buffers used at each node or edge to prevent deadlock. Whereas k-ILS only compactly represents shortest paths between pairs of nodes, this new extension aims to represent those particular ones that give rise also to deadlock-free routing *controllers* which use a low number of buffers per node or per edge. In this paper we prove new NP-hardness results on the problem of devising low occupancy schemes, also for classical k-ILS. Moreover, while space complexity results are given for $\langle k, s\rangle$-DFILS in arbitrary networks, tight results are shown for specific topologies, such as trees, rings, grids, complete graphs and chordal rings. Finally, trade-offs are derived between the number of intervals k and the number of buffers s in Deadlock-Free Interval Routing Schemes for hypercubes, grids, tori and Cartesian products of graphs.

1 Introduction

Routing messages between pairs of processors is a fundamental task in a parallel or distributed computing system. In order to exchange messages between pairs of processors in such a way as to maintain a high throughput, it is important to route messages along paths of minimum cost (shortest paths). Moreover, the distributed nature of the system requires that path information be maintained somehow at each intermediate node.

The trivial solution is the one of storing, at each node v, a complete routing table which specifies, for each destination u, one incident link belonging to a shortest path between u and v. Such a table has size $\Theta(n \cdot \log \delta)$, where δ is the node degree and n the size of the network. Since in the general case this approach is too space-consuming for large networks, it is necessary to devise

* Work supported by the EU TMR Research Training Grant N. ERBFMBICT960861, by the EU ESPRIT Long Term Research Project ALCOM-IT under contract N. 20244 and by the Italian MURST 40% project "Algoritmi, Modelli di Calcolo e Strutture Informative".

Reischuk, Morvan (Eds.): STACS'97 Proceedings, LNCS 1200

routing schemes with smaller tables. This gives rise to a need of simple scalable and topology independent *compact routing* methods.

In the ILS (*Interval Labeling Scheme*, [21], [24], [25]) suitable labels are assigned to nodes and links. Node-labels belong to the set $\{1, \ldots, n\}$, while link-labels are pairs of node-labels representing disjoint intervals of $[1..n]$. To send a message m from a source v_i to a destination v_j, m is transmitted by v_i on the (unique) link $e = \{v_i, v_k\}$ such that the label of v_j belongs to the interval associated to e. With this approach, one always obtains an efficient memory occupation, while the problem is to choose node and link-labels in such a way that messages are routed along shortest paths.

In [21], [24], [25] it is shown how the ILS can be applied to optimally route messages on particular network topologies, such as trees, rings, etc., and in [25] the model has been extended to allow more than 1 interval to be associated to each link; in particular, a 2-ILS, i.e. a scheme associating at most 2 intervals per edge, is proposed for 2-dimensional tori. Other computational complexity and characterization results related to k-ILS and compact routing schemes can be found in [12,20,16,10,9,13] (see [26] for a survey).

Both in the packet and wormhole routing models (see [4,22,11,19] for a detailed description), in order to store messages during the routing process, a finite amount of buffers is assigned to each node or edge. Since messages are allowed to request buffers while holding others, deadlock configurations may occur in which no message can be delivered due to cyclic chains of buffer requests.

For every network, there exists a lower bound on the number of buffers which have to be maintained at each node or edge to allow deadlock-free routing mechanisms (or *controllers*) and several techniques have been developed to design minimal space routing controllers that avoid deadlocks by ordering buffers and allowing messages to use them in a monotonic fashion ([17,14,18,4,7,6,2,3,11] and others). This idea results in the generation of a directed acyclic resource dependencies graph (DAG), thus preventing deadlock configurations. DAG-based methods can be used with slight modifications both for packet and wormhole routing. All controllers considered in this paper are based on a standard DAG-based method in which the ordering of buffers is obtained by means of acyclic orientations of graphs (see for instance [23,5,1]).

k-ILS and wormhole routing have been implemented on the latest generation of INMOS Transputer C104 Router chips [15], thus the problem of reducing at the same time the number of intervals used by the scheme and the number of buffers used by the controller is a central and worth investigating issue.

In this paper we introduce an extension of the Interval Labeling Scheme (DFILS) which allows the integration of the interval model with the deadlock prevention techniques. Informally, given a graph G, a $\langle k, s \rangle$-DFILS consists of a k-ILS for G and a deadlock-free controller which allows the routing of messages along the paths represented by the k-ILS by maintaining at most s buffers per node (packet routing) or per edge (wormhole routing).

We show that with DFILS it is possible at the same time to efficiently represent shortest paths and to avoid deadlock configurations in some relevant interconnection networks such as trees, rings, grids, tori, complete graphs, chordal rings, hypercubes and Cartesian products of graphs.

Furthermore, new NP-completeness results are given concerning the problem of deciding the existence of 2-ILS for arbitrary networks. This extends the hardness results in [9] in two ways, as the number of intervals $k = 2$ is constant and the graph underlying the network is unweighted. Moreover, the NP-completeness holds also for linear and/or strict 2-ILS. Next, we show that deciding if a network admits a $\langle 2, 2\rangle$-DFILS is NP-complete, also in the linear and/or strict case.

Finally, we study the relationship between the number of intervals k and the number of buffers s: how it is possible to trade-off one parameter for the other in networks such as hypercubes, multi-dimensional grids and tori and Cartesian products of graphs.

The paper is organized as follows. The next section contains a description of the model and some definitions. In Sect. 3 we give some space complexity results for general networks and in Sect. 4 the NP-completeness ones. In Sect. 5 we show results on specific interconnection networks and present the above mentioned trade-offs between intervals and buffers for product graphs. Finally, we list some open problems in the last section.

2 The Model

The model we shall use is the ***point to point*** communication model, where each processor in the network has access only to its own local memory and communicates by sending messages via one of its neighbors. The network can then be represented as an undirected graph $G =< V, E >$ with processor set V of size n and link set E. Since we consider direct graphs obtained by orienting edges in E, in the following in order to avoid any ambiguity of notation we will denote undirected and directed edges between any couple of nodes u and v respectively as $\{u, v\}$ and (u, v).

An *Interval Labeling Scheme* (ILS) is a scheme of labeling each node in the graph with some unique integer in the set $\{1, ..., n\}$ and each link with a unique interval $[a, b]$, with $a, b \in \{1, ..., n\}$. Wrap-around of intervals is allowed, so if $a > b$ then $[a, b] = \{a, a+1, ..., n, 1, ..., b\}$. The set of all intervals associated with the edges of a node forms a partition of the set $\{1, ..., n\}$ (thus in reality each link needs to be labeled with only the left end-point of the interval). Messages to a destination node with label j are routed via the link labeled with the interval $[a, b]$ such that $j \in [a, b]$. An ILS is *valid* if, for all the nodes u and v of G, messages sent from u to v reach v correctly, eventually not routing shortest paths. A valid ILS is also called an *Interval Routing Scheme* (*IRS* for short).

In a k-ILS each link is labeled with up to k intervals, always under the assumption that at every node all intervals are disjoint. At any given node a message with destination node label j is routed on the link having the interval containing j. If no interval in the k-ILS wraps around, i.e. all intervals $[i, j]$ are such that $i \leq j$, the k-ILS is called *linear* or simply k-LILS. Valid k-ILS and k-LILS are called respectively k-IRS and k-LIRS. A k-IRS or k-LIRS is said *optimal* if it represents one shortest path between any couple of nodes. Finally, a k-IRS is strict if at each node u the intervals associated to its incident links must not contain the node-label of u. Thus for example, a node labeled 5 must not have an incident link with an interval containing 5, like for instance $[3,6]$.

For what concerns memory requirements, in a k-IRS both node-labels and intervals require $\lceil \log n \rceil$ bits of information, thus yielding at each node v of degree δ_v an overall space complexity of $(k+1) \cdot \delta_v \cdot \lceil \log n \rceil$ bits.

Concerning the buffering of messages, if we view each message simply as a couple source-destination and we denote as B the set of all buffers in the network, we can consider a controller $C : V \times V \times B \to 2^B$ as a function which specifies, given a message $m \equiv (u,v)$ and a buffer b containing m, the subset $C(u,v,b)$ of buffers which are allowed to store m in the next step along the path to its destination[1]. Thus, a controller C restricts the use of buffers and we say that C is deadlock-free if doesn't yield any deadlock configuration. This property can be guaranteed if the resulting buffer dependencies graph in which each node represents a buffer and there is a direct edge between b_i and b_j if there is at least one message $m \equiv (u,v)$ such that $b_j \in C(u,v,b_i)$, is acyclic [14,4,7,11].

In packet routing let us denote as s_u the number of buffers used by a controller C at node u and in wormhole routing as $s_{u,v}$ the number of buffers assigned at node u to the incident link $\{u,v\}$.

Definition 1. Given a network $G = (V,E)$ and a controller C for G, in packet routing (wormhole routing) we define the size s of C as $s = max_{u \in V}(s_u)$ (resp. $s = max_{\{u,v\} \in E}(s_{u,v} + s_{v,u})$).

In packet routing (wormhole routing) a controller of size s requires $s \cdot d_p$ bits per node ($s \cdot d_f$ bits per edge), where d_p (d_f) is the dimension of a packet (flit). In general, it is assumed $d_p = O(\log n)$ and $d_f = o(\log n)$.

Definition 2. Given a path $p = \langle v_1, \ldots, v_l \rangle$ connecting v_1 and v_l, in packet routing we say that C covers p if there exist l buffers $b_1, \ldots, b_l$ such that for each i, $1 \le i \le l$, b_i belongs to v_i and for each i, $1 \le i \le l-1$, $b_{i+1} \in C(v_1, v_l, b_i)$. Similarly, in wormhole routing, C covers p if there exist $l-1$ buffers $b_1, \ldots, b_{l-1}$ such that for each i, $1 \le i \le l-1$, b_i belongs at v_i to edge $\{v_i, v_{i+1}\}$ and for each i, $1 \le i \le l-2$, $b_{i+1} \in C(v_1, v_l, b_i)$.

We now extend the basic k-ILS.

Definition 3. A $\langle k,s \rangle$-DFILS (Deadlock-Free ILS) for a graph G is a k-ILS for G together with a deadlock-free routing controller of size s for G which covers the set of paths P represented by the k-ILS. The $\langle k,s \rangle$-DFILS is said valid (resp. optimal) if the k-ILS is valid (resp. optimal).

A valid $\langle k,s \rangle$-DFILS is also called a $\langle k,s \rangle$-*DFILS*, while linear $\langle k,s \rangle$-DFILS and $\langle k,s \rangle$-DFIRS are denoted respectively as $\langle k,s \rangle$-DFLILS and $\langle k,s \rangle$-DFLIRS. The space complexity of a $\langle k,s \rangle$-DFIRS is the sum of the memory space required by the k-IRS and by the controller.

[1] We assume $C(u,v,b) = \emptyset$ if b never stores a message $m \equiv (u,v)$

As already remarked, all routing controllers considered in this paper are based on the concept of acyclic orientation covering.

Definition 4. An acyclic orientation of a graph $G = (V, E)$ is an acyclic directed graph $DG = (V, \vec{E})$ obtained by orienting all edges in E.

Definition 5. Let $\mathcal{G} = \langle DG_1, \ldots, DG_s \rangle$ be a sequence of (not necessarily all different) acyclic orientations of a graph G and let $p =< v_1, \ldots, v_l >$ be a simple path in G. We say that $\mathcal{G}$ covers p if there exists a sequence of positive integers $j_1, \ldots, j_{l-1}$ such that $1 \leq j_1 \leq \ldots \leq j_{l-1} \leq s$ and for every i, $1 \leq i \leq l-1$, (v_i, v_{i+1}) belongs to DG_{j_i}.

Definition 6. Given a network G and a set of simple paths P, a sequence of orientations $\mathcal{G} = \langle DG_1, \ldots, DG_s \rangle$ is said to be an acyclic orientation covering for P of size $\mathcal{S}_{\mathcal{G}}(P) = s$ if it covers each $p \in P$. If P contains at least one path (resp. shortest path) between each couple of nodes, then $\mathcal{G}$ is said valid (resp. optimal) for G.

The importance of acyclic orientation coverings is stated by the following classical theorem (see for example [23]).

Theorem 7. *Given a network G and a set of simple paths P connecting couples of nodes, if an acyclic orientation covering of size s for P exists, then there exists a deadlock-free packet (wormhole) routing controller of size s for G which covers all paths $p \in P$.*

Motivated by Theorem 7, since in this paper we focus on deadlock-free controllers obtained by means of acyclic orientation coverings, in the following we will consider a specialized definition of $\langle k, s \rangle$-DFILS for a graph G as a k-ILS for G and an acyclic orientation covering of size s which covers the set of paths P represented by the k-ILS.

3 Deadlock-Free Interval Routing on General Networks

We now prove some results for arbitrary networks.

If are not interested in paths of minimal length, it is possible to obtain minimal space DFIRS.

Theorem 8. *For any graph G there exists a valid $\langle 1, 2 \rangle$-DFIRS.*

Proof. It suffices to route messages on any spanning tree T of G. Then, once chosen a root r of T, the acyclic orientation covering $\mathcal{G} = \langle DG_1, DG_2 \rangle$ of G is such that all edges are oriented from all the nodes toward the root r in DG_1 and from r toward the other nodes in DG_2, while all the other edges (not in T) are oriented in such a way to avoid cycles. A 1-IRS which routes messages on T can be easily constructed, for instance labeling nodes in preorder [21].

In order to show that $\mathcal{G}$ and the 1-IRS form a $\langle 1, 2 \rangle$-DFIRS for G, observe that given any two nodes u and v in G, a message from u to v follows edges in

T according to DG_1 until the first common ancestor of u and v is reached and then it descends to v along edges in T according to DG_2. □

Let us now consider paths of minimal length. In the optimal case, the set of shortest paths P represented by a k-IRS for G and the set of shortest paths P' covered by an acyclic orientation covering of size s for G can be different, so in general they do not imply the existence of a $\langle k, s\rangle$-DFIRS for G. In fact, weaker results can be proved.

Theorem 9. *For any graph G of n nodes having an optimal acyclic orientation covering $\mathcal{G}$ of size s there exist optimal $\langle\lfloor\frac{n}{2}\rfloor, s\rangle$-DFIRS and $\langle\lceil\frac{n}{2}\rceil, s\rangle$-DFLIRS.*

Theorem 10. *For any graph G having an optimal k-IRS there exists an optimal $\langle k, D+1\rangle$-DFIRS, where D is the diameter of G.*

Proof. Given any acyclic orientation DG for G, the acyclic orientation covering $\mathcal{G}$ of size $D+1$ obtained by alternating DG and its reversal covers all shortest paths in G and in particular those represented by the k-IRS, as any shortest path of length l is covered by the first $l+1$ orientations of $\mathcal{G}$. □

4 The NP-Completeness Results

We now consider the problem of the complexity of designing $\langle k, s\rangle$-DFIRS. In [9] it has been proved that given a weighted graph G (where each edge is assigned a non-negative weight) and an integer $k \geq 1$, deciding whether there exists an optimal k-IRS for G is NP-complete. As an extension of this result we obtain the following theorem.

Theorem 11. *Given a (unweighted) graph G, the problem of deciding whether there exists an optimal 2-IRS (resp. 2-LIRS, strict 2-IRS, strict 2-LIRS) for G is NP-complete.*

By exploiting the reductions used in the above theorem, the following results can be derived.

Theorem 12. *Given a (unweighted) graph G, the problem of deciding whether there exists an optimal $\langle 2,2\rangle$-DFIRS (resp. $\langle 2,2\rangle$-DFLIRS, strict $\langle 2,2\rangle$-DFIRS, strict $\langle 2,2\rangle$-DFLIRS) for G is NP-complete.*

5 Results for Specific Topologies

It still remains unresolved the question if, given a network G, optimal k-IRS and acyclic orientation covering of size s for G imply the existence of an optimal $\langle k, s\rangle$-DFIRS. In this section we show that in general this is indeed not the case by giving tight results for some known interconnection networks. Since we consider only optimal schemes, most of times for the sake of brevity we will drop the term "optimal" from the notation.

5.1 Some Interconnection Networks

We now show optimum results for some simple interconnection networks.

Theorem 13. *There exists a $\langle 1,2\rangle$-DFIRS for trees and complete graphs.*

The minimum size an acyclic orientation covering for a ring R_n with $n \geq 5$ nodes is 3 [23], that is there exist no deadlock-free routing controllers for R_n which use less than 3 buffers per node or per edge. If general controllers are concerned, that is not necessarily obtained by means of acyclic orientation coverings, this result still holds for rings R_n with $n > 12$ [3]. Thus, the bound established in the following theorem is optimum with respect to any deadlock-free routing controller.

Theorem 14. *For any ring R_n there exists a $\langle 1,3\rangle$-DFIRS.*

Let us consider now grid networks $G_{n\times m}$ $(n, m \geq 1)$, with node set $V = \{v_{i,j} \mid 1 \leq i \leq n, 1 \leq j \leq m\}$ and edge set $E = \{\{v_{i,j}, v_{i+1,j}\} \mid 1 \leq i \leq n-1, 1 \leq j \leq m\} \cup \{\{v_{i,j}, v_{i,j+1}\} \mid 1 \leq i \leq n, 1 \leq j \leq m-1\}$. We say that rows (columns) have *wrap-arounds* if further edges connecting the first and last node of each row (column) are added, i.e. if the new edge set $E' = E \cup \{\{v_{i,1}, v_{i,m}\} | 1 \leq i \leq n\}$ $(E' = E \cup \{\{v_{1,j}, v_{n,j}\} | 1 \leq j \leq m\})$. Doubly wrapped grids are also called *tori*.

1-IRS exist only for grids with no wrap-arounds or only rows (or columns, but not both) wrap-arounds, while there is a 2-IRS for tori [25,20]. Furthermore, the minimum size of an acyclic orientation covering for grids without any wrap-around is 2, while for grids with rows or columns (or both) wrap-arounds it is 3 (this is a direct implication of the result for rings in [3]).

The following results concerning $\langle k,s\rangle$-DFIRS can be proved.

Theorem 15. *There exist $\langle 1,2\rangle$-DFLIRS for grids, $\langle 1,3\rangle$-DFIRS for grids with either rows or columns wrap-arounds and $\langle 2,5\rangle$-DFLIRS for tori.*

Proof. We prove the claim only for the case of grids without any wrap-around.

Consider an $n \times m$ grid $G_{n\times m}$. The $\langle 1,2\rangle$-DFLIRS is constructed as follows. The acyclic orientation covering $\mathcal{G} = \langle DG_1, DG_2\rangle$ is such that $DG_1 = (V, \vec{E}_1)$ is the directed acyclic graph obtained by orienting edges in E in such a way that $\vec{E}_1 = \{(v_{i,j}, v_{i+1,j}) \mid 1 \leq i \leq n-1, 1 \leq j \leq m\} \cup \{(v_{i,j}, v_{i,j+1}) \mid i\, is\, odd, 1 \leq i \leq n, 1 \leq j \leq m-1\} \cup \{(v_{i,j}, v_{i,j-1}) \mid i\, is\, even, 1 \leq i \leq n, 2 \leq j \leq m\}$, while DG_2 is obtained by orienting edges in DG_1 in the opposite direction.

A trivial 1-IRS for $G_{n\times m}$ could be obtained by assigning label $l_{i,j} = (i-1)\cdot m+j$ to each node $v_{i,j}$ and assigning intervals in such a way that messages moves first vertically and then horizontally, but it would not form a $\langle 1,2\rangle$-DFLIRS with $\mathcal{G}$. In fact, given any two nodes $v_{i,j}$ and $v_{i',j'}$, the shortest path from $v_{i,j}$ and $v_{i',j'}$ which goes first vertically till node $v_{i',j}$ and then horizontally till $v_{i',j'}$ is not always covered by $\mathcal{G}$. In particular this holds if $i' < i$ and either j' is odd and $j' > j$ or j' is even and $j' < j$. In order to avoid such shortest paths, the 1-IRS can be modified as follows. Each node $v_{i,j}$ is assigned label $l_{i,j} = (i-1)\cdot m + j$ if i is odd and $l_{i,j} = i\cdot m - j + 1$ if i is even. At each node $v_{i,j}$ edge $\{v_{i,j}, v_{i+1,j}\}$

receives the interval $[i \cdot m + 1, n \cdot m]$, edge $\{v_{i,j}, v_{i-1,j}\}$ the interval $[1, l_{i-1,j}]$, edge $\{v_{i,j}, v_{i,j+1}\}$ the interval $[l_{i,j+1}, i \cdot m]$ if i is odd and $[l_{i-1,j+1}, l_{i,j+1}]$ if i is even and finally edge $\{v_{i,j}, v_{i,j-1}\}$ the interval $[1, j-1]$ if $i = 1$, $[l_{i-1,j-1}, l_{i,j-1}]$ if $i > 1$ and i is odd and $[l_{i,j-1}, i \cdot m]$ if i is even.

Consider now any two nodes $v_{i,j}$ and $v_{i',j'}$. Clearly if $i = i'$ or $j = j'$ the shortest path between $v_{i,j}$ and $v_{i',j'}$ is covered by DG_1 or DG_2. If $i' > i$ or if i' is odd and $j' < j$ or if i' is even and $j' > j$, then the path p routed by each message from $v_{i,j}$ to $v_{i',j'}$ is the unique shortest path through node $v_{i',j}$, otherwise it is the unique shortest path through nodes $v_{i'+1,j}$ and $v_{i'+1,j'}$. By construction, in every case p is covered by the two acyclic orientations DG_1 and DG_2, thus $\mathcal{G} = \langle DG_1, DG_2 \rangle$ and the 1-IRS form a $\langle 1, 2 \rangle$-DFLIRS for G. □

While it remains an open question the existence of a better DFIRS for tori, by the above observations the other results of Theorem 15 are optimal. Such an optimality holds also with respect to general controllers (i.e. not necessarily obtained by means of acyclic orientation coverings), since every deadlock-free routing controller has size at least 2 for grids (this is true for every network) and at least 3 for wrapped grids [3].

5.2 Chordal Ring Networks

We now consider chordal ring networks.

Definition 16. A chordal ring is an undirected graph $R_n(C) = (V, E)$, where: $C \subseteq \{2, ..., n-2\}$, $V = \{v_0, \ldots, v_{n-1}\}$ and $E = E_1 \bigcup E_2$ with
$E_1 = \{\{v_i, v_{(i+1) \bmod n}\} \mid i = 0, \ldots, n-1\}$,
$E_2 = \{\{v_i, v_{(i+j) \bmod n}\} \mid i = 0, ..., n-1,\ j \in C\}$. Edges in E_2 are called chords.

Chordal rings of type $R_n(\{l\})$ have a 1-IRS if $n \bmod l = 0$ [8], while the minimum k such that there is a k-IRS when $n \bmod l \neq 0$ is still unknown, thus we will always implicitly assume that the condition $n \bmod l = 0$ holds.

Throughout this subsection, addition and subtraction will be the appropriate modulo operations to allow for cyclic intervals and for consistency with the modulo notation we shall assume $\{0, \ldots, n-1\}$ as the set of node-labels, instead of $\{1, \ldots, n\}$. Moreover, for the sake of simplicity, when talking about 1-IRS for chordal rings we will denote as $I_{i,j}$ the (unique) interval associated at node v_i to the incident link $\{v_i, v_j\}$.

Consider an IRS for a given $R_n(\{l\})$ obtained by assigning integer labels in increasing order, starting from one node and traversing edges in E_1 along a chosen direction, i.e., by assigning label $(i + d) \bmod n$ to each node v_i or label $(d - i) \bmod n$, for some $0 \leq d \leq n-1$. A scheme derived by such an assignment is called *cyclic*.

We now completely characterize the class of labelings yielding a 1-IRS.

Definition 17. A k-IRS for any $R_n(\{l\})$ has rank r if it is obtained by a cyclic one by swapping labels of pair of nodes adjacent along the external face (i.e. connected by an edge in E_1) in such a way that each label participates to at most r swaps.

Theorem 18. *If $n/l \geq 5$ every 1-IRS for $R_n(\{l\})$ has rank at most* 1.

Notice that the bound established in the previous theorem is tight, since for $n/l < 5$ counterexamples of not cyclic 1-IRS for $R_n(\{l\})$ can be found.

As a corollary of Theorem 18, the paths represented by any 1-IRS for $R_n(\{l\})$ consist of a sequence of chords followed by a sequence of edges along the external face.

Corollary 19. *The sequence $e_1, \ldots, e_d$ of the edges traversed along any shortest path represented by a 1-IRS for a chordal ring $R_n(\{l\})$ with $n/l \geq 5$ is such that there exist i, $0 \leq i \leq d$, such that $e_j \in E_2$ for $j \leq i$ and $e_j \in E_1$ for $j > i$.*

The following lemma is useful to determine lower bounds for $\langle k, s\rangle$-DFIRS.

Lemma 20. *Given a ring R_m of $m \geq 5$ nodes and an acyclic orientation covering of size 3 for R_m, at least $m - 4$ nodes are not reachable by all the others in 2 orientations.*

We now prove a lower bound on the size of acyclic orientation coverings for chordal rings.

Lemma 21. *There exists no acyclic orientation covering of size 2 for $R_n(\{l\})$ if $l \geq 4$ or $n/l \geq 5$ and there exists an acyclic orientation covering of size 3 for any $R_n(\{l\})$.*

Notice that, by applying the result for rings in [3] to the subring of n/l nodes induced by $v_0, v_l, v_{2l}, \ldots, v_{n-l}$, if $n/l > 12$ the above result is tight for general controllers also.

The following lower bound concerning $\langle k, s\rangle$-DFIRS holds.

Theorem 22. *For n suitably large there exists no $\langle 1, 3\rangle$-DFIRS for $R_n(\{l\})$ if l is fixed and $l \geq 4$ or n/l is fixed and $n/l \geq 5$.*

Theorem 22 is relevant, as it shows the first counterexamples of networks having a 1-IRS and an acyclic orientation covering of size 3, but not a $\langle 1, 3\rangle$-DFIRS. From the previous theorem and Lemma 21 the following upper bounds are tight.

Theorem 23. *There exists a $\langle 1, 3\rangle$-DFIRS for $R_n(\{l\})$ with $l \leq 3$ or $n/l \leq 4$ and a $\langle 1, 4\rangle$-DFIRS for $R_n(\{l\})$, $l \geq 4$.*

5.3 Cartesian Products of Graphs

We now consider the *Cartesian product* of graphs. This class of graphs includes the topologies of some interconnection networks commonly used in parallel architectures, such as hypercubes, d-dimensional grids and tori.

Definition 24. Given two graphs $G_1 = (V_1, E_1)$ and $G_2 = (V_2, E_2)$, the *Cartesian product graph* $G_1 \times G_2$ is the graph $G = (V, E)$ where $V = V_1 \times V_2$ [2] and
$$E = \{\{v_{i_1,i_2}, v_{i_1,i'_2}\} \mid v_{i_1} \in V_1, v_{i_2} \in V_2, v_{i'_2} \in V_2, \{v_{i_2}, v_{i'_2}\} \in E_2\} \cup \{\{v_{i_1,i_2}, v_{i'_1,i_2}\} \mid v_{i_1} \in V_1, v_{i'_1} \in V_1, v_{i_2} \in V_2, \{v_{i_1}, v_{i'_1}\} \in E_1\}.$$

[2] For the sake of simplicity we will denote any node $(v_{i_1}, v_{i_2}) \in V$ by v_{i_1,i_2}

The definition states that the subgraph induced by all nodes v_{i_1,i_2} with the same i_1 (resp. i_2) value is isomorphic to G_2 (resp. G_1). Then an $n \times m$ grid (resp. torus) can be seen as the Cartesian product of two paths (resp. rings) respectively of n and m nodes. It is not difficult to show that the product operator is associative, so without any ambiguity it is possible to denote as $G_1 \times \ldots \times G_d$ the Cartesian product of $d > 2$ graphs $G_1, \ldots, G_d$ and as $v_{i_1,\ldots,i_d}$ a generic node of $G_1 \times \ldots \times G_d$.

We note here that the product of graphs with k-IRS has not necessarily a k-IRS [25,20], but this is true for Linear Interval Routing Schemes [16].

Theorem 25. *If each graph G_j in the set $\{G_1, \ldots, G_d\}$ has a $\langle k_j, s_j\rangle$-DFIRS ($\langle k_j, s_j\rangle$-DFLIRS), $1 \leq j \leq d$, then $G_1 \times \ldots \times G_d$ has a $\langle k+1, s\rangle$-DFIRS ($\langle k, s\rangle$-DFLIRS), with $k = \max(k_1, \ldots, k_d)$ and $s = s_1 + \ldots + s_d - d + 1$.*

If we define a d-grid $G_{n_1 \times \ldots \times n_d}$ as the Cartesian product of d paths $P_{n_1}, \ldots, P_{n_d}$ respectively of $n_1, \ldots, n_d$ nodes and analogously a d-torus $T_{n_1 \times \ldots \times n_d}$ as the Cartesian product of d rings $R_{n_1}, \ldots, R_{n_d}$, then by Theorem 10 it follows that $G_{n_1 \times \ldots \times n_d}$ admits a $\langle 1, n_1 + \ldots + n_d - d + 1\rangle$-DFLIRS and $T_{n_1 \times \ldots \times n_d}$ admits a $\langle 2, \lfloor \frac{n_1}{2} \rfloor + \ldots + \lfloor \frac{n_d}{2} \rfloor + 1\rangle$-DFIRS. Notice that a d-dimensional hypercube is a particular case of d-grid, namely $G_{2 \times \ldots \times 2}$, and admits a $\langle 1, d+1\rangle$-DFLIRS. By Theorem 25, these results can be refined as follows.

Corollary 26. *There exists a $\langle 1, \lceil \frac{d}{2} \rceil + 1\rangle$-DFLIRS for d-grids and a $\langle 2, 2d+1\rangle$-DFLIRS for d-tori.*

Even if it is not clear if the results established in the previous corollary are tight, the corresponding schemes exhibit a good space complexity. In fact, by the space considerations of Sect. 2, in packet routing at each node the ratio between the space required by buffers and the one needed to represent intervals is proportional to $\frac{d \cdot d_p}{d \log n} = O(1)$, i.e. constant, both for d-grids and for d-tori. Similarly, in wormhole routing, at each edge the ratio is proportional to $\frac{d \cdot d_f}{\log n} = o(d)$.

Given the d-th power G^d of a graph G (i.e. $G_1 \times \ldots \times G_d$ with each $G_i \equiv G$), by combining theorems 9 and 25, the following trade-off between number of intervals and number of buffers can be derived.

Theorem 27. *Given a network G of n nodes having an acyclic orientation covering of size s, for every integer i, $1 \leq i \leq d$, G^d has a $\langle \lceil \frac{n^i}{2} \rceil, (s-1) \cdot \lceil \frac{d}{i} \rceil + 1\rangle$-DFLIRS.*

Proof. For every $i \geq 1$ G^i has an acyclic orientation covering of size s [1]. Moreover, G^d is the product of $\lfloor \frac{d}{i} \rfloor$ graphs G^i and one graph $G^{d \bmod i}$. Such graphs have at most n^i nodes and the theorem follows from theorems 25 and 9. □

This shows how the increase of the number of buffers yields a corresponding decrease of the number of intervals needed and vice versa.

The following corollary is a direct consequence of the above theorem.

Corollary 28. *For every i and n ($i, n \geq 1$) there exist $\langle \lceil \frac{n^i}{2} \rceil, \lceil \frac{d}{i} \rceil + 1\rangle$-DFLIRS for d-grids $G_{n \times \ldots \times n}$ and $\langle \lceil \frac{n^i}{2} \rceil, 2 \cdot \lceil \frac{d}{i} \rceil + 1\rangle$-DFLIRS for d-tori $T_{n \times \ldots \times n}$.*

Notice that, since in d-grids and d-tori the number of nodes in each dimension j does not influence the existence of $\langle k, s\rangle$-DFIRS, the above corollary can be easily extended to any d-grid $G_{n_1 \times \ldots \times n_d}$ and d-torus $T_{n_1 \times \ldots \times n_d}$.

6 Conclusion and Open Problems

There are still many unresolved problems with $\langle k, s\rangle$-DFIRS.

First of all, there is the determination of tight space complexity results for wider classes of interconnection networks. Moreover, it would be worth to improve the trade-offs between the number of intervals k and the number of dimensions s for Cartesian products of graphs. A general result between k and s for arbitrary graphs would be quite desirable. Finally, it would be interesting to extend the NP-completeness results to 1-IRS and $\langle 1, 2\rangle$-DFIRS and to consider deadlock-free routing controllers based on other deadlock prevention techniques not necessarily related to acyclic orientation coverings.

References

1. J.C. Bermond, M. Di Ianni, M. Flammini, and S. Pérennès. Systolic acyclic orientations for deadlock prevention on usual networks. Manuscript, 1995.
2. J.C. Bermond and M. Syska. Routage wormhole et canaux virtuel. In M. Cosnard M. Nivat and Y. Robert, editors, *Algorithmique Parallèle*, pages 149–158. Masson, 1992.
3. Robert Cypher and Luis Gravano. Requirements for deadlock-free, adaptive packet routing. In *11th Annual ACM Symposium on Principles of Distributed Computing (PODC)*, pages 25–33, 1992.
4. W. J. Dally and C. L. Seitz. Deadlock-free message routing in multiprocessor interconnection networks. *IEEE Trans. Comp.*, C-36, N.5:547–553, May 1987.
5. M. Di Ianni, M. Flammini, R. Flammini, and S. Salomone. Systolic acyclic orientations for deadlock prevention. In *2nd Colloquium on Structural Information and Communication Complexity (SIROCCO)*. Carleton University Press, 1995.
6. J. Duato. Deadlock-free adaptive routing algorithms for multicomputers: evaluation of a new algorithm. In *3rd IEEE Symposium on Parallel and Distributed Processing*, 1991.
7. J. Duato. On the design of deadlock-free adaptive routing algorithms for multicomputers: theoretical aspects. In *2nd European Conference on Distributed Memory Computing*, volume 487 of *Lecture Notes in Computer Science*, pages 234–243. Springer-Verlag, 1991.
8. M. Flammini and G. Gambosi. Compact routing in chordal ring networks. Manuscript, 1996.
9. M. Flammini, G. Gambosi, and S. Salomone. Interval routing schemes. To appear on *Algorithmica*, 1996.
10. M. Flammini, J. van Leeuwen, and A. Marchetti Spaccamela. The complexity of interval routing on random graphs. In *20th Symposium on Mathematical Foundation of Computer Science (MFCS)*, volume 969 of *Lecture Notes in Computer Science*, pages 37–49. Springer-Verlag, 1995.

11. E. Fleury and P. Fraigniaud. Deadlocks in adaptive wormhole routing. Research Report, Laboratoire de l'Informatique du Parallélisme, LIP, École Normale Supérieure de Lyon, 69364 Lyon Cedex 07, France, March 1994.
12. G.N. Frederickson and R. Janardan. Designing networks with compact routing tables. *Algorithmica*, 3:171–190, 1988.
13. C. Gavoille and S. Pérennès. Lower bounds for interval routing on 3-regular networks. In *3rd Colloquium on Structural Information and Communication Complexity (SIROCCO)*. Carleton University Press, 1996.
14. K.D. Gunther. Prevention of deadlock in packet-switched data transport system. *IEEE Trans. on Commun.*, COM-29:512–514, May 1981.
15. *The T9000 Transputer Products Overview Manual*. INMOS, 1991.
16. E. Kranakis, D. Krizanc, and S.S. Ravi. On multi-label linear interval routing schemes. In *19th Workshop on Graph Theoretic Concepts in Computer Science (WG)*, volume 790 of *Lecture Notes in Computer Science*, pages 338–349. Springer-Verlag, 1993.
17. P.M. Merlin and P.J. Schweitzer. Deadlock avoidance in store-and-forward networks: Store and forward deadlock. *IEEE Trans. on Commun.*, COM-28:345–352, March 1980.
18. A.G. Ranade. How to emulate shared memory. In *Foundation of Computer Science*, pages 185–194, 1985.
19. Jean De Rumeur. *Communication dans les réseaux de processeurs*. Collection Etudes et Recherchers en Informatique. Masson, 1994.
20. P. Ruzicka. On efficiency of interval routing algorithms. In *Mathematical Foundations of Computer Science (MFCS)*, volume 324 of *Lecture Notes in Computer Science*, pages 492–500. Springer-Verlag, 1988.
21. N. Santoro and R. Khatib. Labeling and implicit routing in networks. *The Computer Journal*, 28:5–8, 1985.
22. A.S. Tannenbaum. *Computer Networks*. Englewood Cliffs, Prentice Hall, 1988.
23. Gerard Tel. *Introduction to Distributed Algorithms*. Cambridge University Press, Cambridge, U.K., 1994.
24. J. van Leeuwen and R.B. Tan. Routing with compact routing tables. In G. Rozemberg and A. Salomaa, editors, *The book of L*, pages 259–273. Springer–Verlag, 1986.
25. J. van Leeuwen and R.B. Tan. Interval routing. *The Computer Journal*, 30:298–307, 1987.
26. J. van Leeuwen and R.B. Tan. Compact routing methods: A survey. In *1st Colloquium on Structural Information and Communication Complexity (SICC)*, pages 71–93. Carleton University Press, 1994.

Power Consumption in Packet Radio Networks
(Extended Abstract)

Lefteris M. Kirousis[1], Evangelos Kranakis[2,4], Danny Krizanc[2,4], Andrzej Pelc[3,4]

[1] University of Patras, Department of Computer Engineering and Informatics, Rio 26500, Patras, Greece (kirousis@cti.gr).
[2] Carleton University, School of Computer Science, Ottawa, ON, K1S 5B6, Canada ({kranakis,krizanc}@scs.carleton.ca).
[3] Département d'Informatique, Université du Québec à Hull, Hull, Québec J8X 3X7, Canada (pelc@uqah.uquebec.ca).
[4] Research supported in part by NSERC (Natural Sciences and Engineering Research Council of Canada) grant.

Abstract. In this paper we study the problem of assigning transmission ranges to the nodes of a multi-hop packet radio network so as to minimize the total power consumed under the constraint that adequate power is provided to the nodes to ensure that the network is strongly connected (i.e., each node can communicate along some path in the network to every other node). Such assignment of transmission ranges is called complete. We also consider the problem of achieving strongly connected bounded diameter networks.
For the case of $n+1$ colinear points at unit distance apart (the unit chain) we give a tight asymptotic bound for the minimum cost of a range assignment of diameter h when h is a fixed constant and when $h \in \Omega(\log n)$. When the distances between the colinear points are arbitrary, we give an $O(n^4)$ time dynamic programming algorithm for finding a minimum cost complete range assignment.
For points in three dimensions we show that the problem of deciding whether a complete range assignment of a given cost exists, is NP-hard. For the same problem we give an $O(n^2)$ time approximation algorithm which provides a complete range assignment with cost within a factor of two of the minimum. The complexity of this problem in two dimensions remains open, while the approximation algorithm works in this case as well.

1 Introduction

A packet radio network is a network where the nodes consist of radio transmitter/receiver pairs distributed over a region. Communication takes place by a node broadcasting a signal over a fixed range (the size of which is proportional to the power expended by the node's transmitter). Any receiver within the range of the transmitter can receive the signal assuming no other nodes are transmitting signals that reach the receiver simultaneously. For a message to be sent to a node outside of the range of the message originator, multiple "hops" may be

Reischuk, Morvan (Eds.): STACS'97 Proceedings, LNCS 1200

required, whereby intermediate nodes pass on (re-broadcast) the message until the ultimate destination node is reached.

Such networks have applications in many situations, over many different scales, where traditional networks are too expensive or even impossible to build. Some examples include: (1) setting up a LAN in an historic building where adding wiring would destroy or obscure valuable features of the building; (2) battlefield or disaster situations where temporary WANs are required but the infrastructure for a traditional network doesn't exist; (3) networks which include nodes in outer space (e.g., satellites, space stations, the moon).

A key issue in setting up and running such a network is the amount of power required by each of the nodes for its transmission. It is well-established [5] that the power of the signal received at a node is inversely proportional to the distance the receiver is from the transmitter raised to an exponent known as the distance-power gradient, i.e., $P_r = \frac{P_o}{d^\alpha}$, where P_r is the power of the received signal, P_o is the power of the transmitted signal, d is the distance between the receiver and the transmitter, and α is the distance-power gradient. In an ideal situation $\alpha = 2$. However, due to various environmental factors such as building materials, street layouts, terrain characteristics, etc., the measured value of α may vary from less than two to more than six. (Here we will assume $\alpha = 2$ though all of our results are easily adjusted for any constant $\alpha \geq 1$.) This distance-power relationship implies there is a tradeoff between the power used by the nodes of the network (i.e., the size of the node ranges) and the diameter of the network (i.e., the number of hops in a path between communicating pairs of nodes if such a path exists).

In this paper we study the problem of assigning transmission ranges to the nodes of a multi-hop packet radio network so as to minimize the total power consumed under the constraint that adequate power is provided to the nodes to ensure that the network is strongly connected (i.e., each node can communicate along some path in the network to every other node). We also consider the problem of achieving strongly connected bounded diameter networks.

1.1 Terminology and Problem Statement

Let $V = \{x_1, \ldots, x_n\}$ be a set of n points in a Euclidean space. For two points $x_i, x_j \in V$, let $d(x_i, x_j)$ denote their Euclidean distance. We also refer to the points of V as *vertices.*

A *broadcasting range assignment* (or *range assignment*, for short) on the vertices in V is a function from V into the set of nonnegative real numbers.

If R is a range assignment on V, the *cost* of R is defined to be the sum $\sum_i (R(x_i))^2$. (Note that the exponent 2 was chosen for convenience and our results are easily adjusted for other choices of the distance-power gradient.)

The *communication graph* associated with a range assignment R (denoted by G_R) is a directed graph with V as its set of vertices and a directed edge from x_i to x_j iff $R(x_i) \geq d(x_i, x_j)$. In other words, a directed edge (x_i, x_j) indicates that x_j is within the range of x_i. A range assignment R has *diameter* h iff G_R has diameter h. R is called *complete* iff G_R is strongly connected.

The problems we consider in this work are those of finding a minimum cost complete range assignment and a minimum cost range assignment with a given diameter, for a given set of points.

1.2 Our Results

The results described in this paper deal with range assignment problems in one- and three-dimensional Euclidean space.

For the case of $n+1$ colinear points at unit distance apart (the unit chain) we give a tight asymptotic bound for the minimum cost of a range assignment of diameter h when h is a fixed constant and when $h \in \Omega(\log n)$ (section 2.1). When the distances between the colinear points are arbitrary, we give an $O(n^4)$ time dynamic programming algorithm for finding a minimum cost complete range assignment (section 2.2).

For points in three dimensions we show that the problem of deciding whether a complete range assignment of a given cost exists, is NP-hard (section 3.1). For the same problem we give an $O(n^2)$ time approximation algorithm which provides a complete range assignment with cost within a factor of two of the minimum (section 3.2). The complexity of this problem in two dimensions is open and we conjecture it remains NP-hard. Our approximation algorithm works in this case as well.

Many proofs are only outlined. Complete proofs will appear in the final version of the paper.

1.3 Related Results

Studies of multi-hop packet radio networks have mainly concentrated on the problem of scheduling communication so as to avoid simultaneous broadcast to the same receiver which results in a scrambled signal. Takagi and Kleinrock [9] consider the problem of assigning transmission radii so as to maximize the expected one-hop progress of a packet assuming randomly distributed packet radio terminals are broadcasting packets with fixed probability of transmission. A number of authors have shown that the problem of scheduling communications is NP-hard and have provided heuristics for it [1,2,6,7]. Sen and Huson [8] point out these previous authors assumed that the underlying graphs were arbitrary (and therefore the NP-completeness generally follows from known graph coloring problems). They show that the problem remains NP-hard when restricted to the domain of possible packet radio graphs and they give an $O(n \log n)$ time algorithm for the case of vertices located on a line. A survey of packet radio network technology appears in [4] and [5] contains useful background information on wireless networks in general.

2 Range Assignments in One Dimension

In this section we study range assignments when the points are arranged on a line.

2.1 The unit chain

Consider a set $N = \{0, 1, \ldots, n\}$ of $n + 1$ colinear points at unit distances. Let $Cost_h(n)$ be the minimum cost of a complete range assignment for N, of diameter at most h, for a positive integer h. From now on we consider only complete range assignments. We establish the exact order of magnitude of $Cost_h(n)$ for any fixed integer h and construct a range assignment of this cost. We omit the inductive proofs of the following two lemmas.

Lemma 1. *There exists a range assignment for N of diameter at most h, with cost $C_h(n) \in O(n^{E(h)})$, where $E(h) = \frac{2^{h+1}-1}{2^h-1}$, for any fixed positive integer h.*

Let $a < b$ be integers. Consider a set M of x colinear integer points in the segment (a, b). Call them *senders*. Let c be an integer point $b + y$, for $y \geq 1$. Let $C(h, x, y)$ be the minimum cost of a range assignment for (a, b), for which c can be reached from any sender in at most h hops.

Lemma 2. *$C(h, x, y) \in \Omega(xy^{e(h)})$, where $e(h) = \frac{2^h}{2^h-1}$, for any fixed positive integer h.*

Theorem 3. *$Cost_h(n) \in \Theta(n^{E(h)})$, where $E(h) = \frac{2^{h+1}-1}{2^h-1}$, for any fixed positive integer h.*

PROOF By lemma 1, $Cost_h(n) \in O(n^{E(h)})$. By lemma 2,

$$Cost_h(n) \geq C\left(h, \frac{n}{2}, \frac{n}{2}\right) \in \Omega(n^{e(h)+1}) = \Omega(n^{E(h)}).$$

■

The previous theorem deals with constant diameter range assignments. In the sequel we consider the case when the size of the diameter is $\Omega(\log n)$. We omit the proofs of the following two lemmas.

Lemma 4. *For any diameter h, $Cost_h(n) \geq n^2/h$.*

Lemma 5. *$Cost_h(n) \in O(n^2)$, where $h \in \Omega(\log n)$.*

Theorem 6. *For any $h \in \Omega(\log n)$, $Cost_h(n) \in \Theta(n^2/h)$.*

PROOF The lower bound of $\Omega(n^2/h)$ is immediate from Lemma 4. To prove the upper bound we need to construct the corresponding range assignments for the chain. Let $x \leq n$ be an integer such that $x \in \Omega(\log n)$. Construct a range assignment as follows. Divide the whole chain into x subchains by placing stations in locations $1+jn/x$, where $j \leq x$ each with range n/x. In each subchain use the range assignment of Lemma 5. The resulting diameter is $h = x+\log(n/x)$ and, by Lemma 5, the total cost is $Cost_h(n) \leq x\left(\frac{n}{x}\right)^2 + xCost_{\log(n/x)}(n/x) \leq 2 \cdot \frac{n^2}{x}$. However, it is clear that $h = x + \log(n/x) \in O(x)$. This shows that $Cost_h(n) \in O(n^2/h)$ and completes the proof of the theorem. ■

2.2 Arbitrary point-arrangements on a line

Theorem 7. *If the vertices in V lie on a line, then there is a $O(n^4)$ time algorithm that finds a minimum cost complete broadcasting range assignment.*

We will construct a minimum cost complete range assignment recursively. Suppose, without loss of generality, that the vertices $x_1, \ldots, x_n$ lie on the line from left to right in the order indicated by their subscripts. A natural first attempt towards a recursive definition of a minimum cost range assignment would be to assume that at a stage $k = 1, \ldots, n$ we know a minimum cost assignment R_k for $x_1, \ldots, x_k$ and try to extend R_k to include x_{k+1}. Unfortunately, this does not work because the range that will be assigned to x_{k+1} may render some of the ranges of R_k unnecessarily large. A second approach would be to assume that at stage k, we know for any given $x_l, l \geq k$ an assignment which is minimum cost among those that establish communication between any pair in $x_1, \ldots, x_k$ and, additionally, have the property that x_l is within the reach of at least one vertex from $x_1, \ldots x_k$. Then, in order to establish communication between any pair in $x_1, \ldots, x_{k+1}$, it would be sufficient to assign to x_{k+1} a range equal to $d(x_{k+1}, x_k)$. However, this also fails because for the recursive construction to be correct, it is necessary to examine the case that x_l is within the reach of x_{k+1}. The range of x_{k+1} that would guarantee this may, again, render some of the ranges of the $x_1, \ldots, x_k$ unnecessarily large. Fortunately, and despite the sometimes vicious circle of induction strengthening, an even stronger recursive assumption carries through: we assume that for any $l \geq k$ and any $i \leq k$, we have an assignment which is minimum cost among those such that (i) in the communication graph, there is a path between any pair from $x_1, \ldots, x_k$, (ii) x_l is within the reach of a vertex in $x_1, \ldots, x_k$, and (iii) in the communication graph, any backwards edge from x_k up to x_i is free of cost. Below we formalize and then prove the correctness of this approach.

We start with some definitions:

Let (V, E) be a directed (not necessarily strongly connected) graph with vertices V. Let x be an additional vertex which may or may not belong to V and which we call the *receiver* vertex. A range assignment R is called *total* for $((V, E); x)$ if (i) the graph on V obtained by adding to the set E the set of edges $\{(x_i, x_j) : R(x_i) \geq d(x_i, x_j)\}$ is strongly connected, and (ii) there is a vertex $x_i \in V$ such that $R(x_i) \geq d(x_i, x)$. The cost of such an assignment is $\sum_i (R(x_i))^2$, as usual. An optimal assignment with respect to $((V, E); x)$ is an assignment of minimum cost which is total for $((V, E); x)$. Intuitively, such an assignment has zero cost for the edges in E and establishes communication paths between any pair of vertices in V and also between a vertex in V and the receiver vertex x, in this direction only. We define $R \in \mathrm{Feas}((V, E); x)$ iff R is total for $((V, E); x)$ and $R \in \mathrm{Opt}((V, E); x)$ iff R is optimal with respect to $((V, E); x)$.

If the points $x_1, \ldots, x_n$ lie on a line from left to right in this order, and if x_i, x_j are any two of them with $j \geq i$ then $E_{i,j}$ is defined to be the set of edges $\{(x_s, x_r) : i \leq r < s \leq j\}$. Intuitively, $E_{i,j}$ is the set of edges from *right to left* which have their endpoints among $x_i, x_{i+1}, \ldots, x_j$.

We omit the proofs of the following technical lemmas:

Lemma 8. *Fix a k such that $1 \leq k \leq n$. Let j, m be such that $j \leq k+1 \leq m$ and let R be an assignment on $x_1, \ldots, x_{k+1}$. Finally, let $r = R(x_{k+1})$ and let R_k be the restriction of R on the set $\{x_1, \ldots, x_k\}$. Assuming that not both $r = 0$ and $j = k+1$ hold, then*
If $r < d(x_{k+1}, x_m)$ and $r < d(x_{k+1}, x_j)$, then

$$R \in \mathrm{Feas}((\{x_1, \ldots, x_{k+1}\}, E_{j,k+1}); x_m) \text{ iff } R_k \in \mathrm{Feas}((\{x_1, \ldots, x_k\}, E_{j,k}); x_m).$$

If $r \geq d(x_{k+1}, x_m)$ and $r < d(x_{k+1}, x_j)$, then

$$R \in \mathrm{Feas}((\{x_1, \ldots, x_{k+1}\}, E_{j,k+1}); x_m) \text{ iff } R_k \in \mathrm{Feas}((\{x_1, \ldots, x_k\}, E_{j,k}); x_{k+1}).$$

If $r < d(x_{k+1}, x_m)$ and $r \geq d(x_{k+1}, x_j)$, then if i is the least positive integer such that $r \geq d(x_{k+1}, x_i)$, we have that

$$R \in \mathrm{Feas}((\{x_1, \ldots, x_{k+1}\}, E_{j,k+1}); x_m) \text{ iff } R_k \in \mathrm{Feas}((\{x_1, \ldots, x_k\}, E_{i,k}); x_m).$$

If $r \geq d(x_{k+1}, x_m)$ and $r \geq d(x_{k+1}, x_j)$, then if i is the least positive integer such that $r \geq d(x_{k+1}, x_i)$, we have that

$$R \in \mathrm{Feas}((\{x_1, \ldots, x_{k+1}\}, E_{j,k+1}); x_m) \text{ iff } R_k \in \mathrm{Feas}((\{x_1, \ldots, x_k\}, E_{i,k}); x_{k+1}).$$

One more piece of notation: If $y \in V$ and r is a positive real, then

$$\mathrm{Opt}((V, E); x; (y, r))$$

is the class of assignments that have the least cost among the assignments R that (i) belong to $\mathrm{Feas}((V, E); x)$ and (ii) $R(y) = r$. The next lemma is an immediate corollary of the previous one.

Lemma 9. *Using the notation of the previous lemma, we have that if $r = 0$ and $j = k+1$ then*

$$\mathrm{Opt}((\{x_1, \ldots, x_{k+1}\}, E_{j,k+1}); x_m; (x_{k+1}, r)) = \emptyset,$$

otherwise we have:
If $r < d(x_{k+1}, x_m)$ and $r < d(x_{k+1}, x_j)$, then

$$R \in \mathrm{Opt}((\{x_1, \ldots, x_{k+1}\}, E_{j,k+1}); x_m; (x_{k+1}, r)) \text{ iff } R_k \in \mathrm{Opt}((\{x_1, \ldots, x_k\}, E_{j,k}); x_m).$$

If $r \geq d(x_{k+1}, x_m)$ and $r < d(x_{k+1}, x_j)$, then

$$R \in \mathrm{Opt}((\{x_1, \ldots, x_{k+1}\}, E_{j,k+1}); x_m; (x_{k+1}, r)) \text{ iff } R_k \in \mathrm{Opt}((\{x_1, \ldots, x_k\}, E_{j,k}); x_{k+1}).$$

If $r < d(x_{k+1}, x_m)$ and $r \geq d(x_{k+1}, x_j)$, then if i is the least positive integer such that $r \geq d(x_{k+1}, x_i)$, we have that

$$R \in \mathrm{Opt}((\{x_1, \ldots, x_{k+1}\}, E_{j,k+1}); x_m; (x_{k+1}, r)) \text{ iff } R_k \in \mathrm{Opt}((\{x_1, \ldots, x_k\}, E_{i,k}); x_m).$$

If $r \geq d(x_{k+1}, x_m)$ *and* $r \geq d(x_{k+1}, x_j)$, *then if* i *is the least positive integer such that* $r \geq d(x_{k+1}, x_i)$, *we have that*

$$R \in \mathrm{Opt}((\{x_1, \ldots, x_{k+1}\}, E_{j,k+1}); x_m; (x_{k+1}, r)) \text{ iff}$$
$$R_k \in \mathrm{Opt}((\{x_1, \ldots, x_k\}, E_{i,k}); x_{k+1}).$$

We are now in a position to give our recursive construction:

PROOF OF THEOREM 7 Assume that at stage k we know an assignment in

$$\mathrm{Opt}((\{x_1, \ldots, x_k\}, E_{i,k}); x_l),$$

for any i, l such that $i \leq k \leq l$. Under this assumption, for any j, m such that $j \leq k+1 \leq m$, we will recursively construct a range assignment

$$R \in \mathrm{Opt}((\{x_1, \ldots, x_{k+1}\}, E_{j,k+1}); x_m).$$

Of course, this will prove the theorem, since an optimal assignment for V is one in

$$\mathrm{Opt}((\{x_1, \ldots, x_n\}, E_{n,n}); x_n).$$

To construct the required R, we examine all possible values of $R(x_{k+1})$. There are $k+2$ of them as it only makes sense to have a range that extends from x_{k+1} to either one of the $x_1, \ldots, x_k$ or to x_m or one which is zero. For each such possible value r of $R(x_{k+1})$, find an assignment (if any) in

$$\mathrm{Opt}((\{x_1, \ldots, x_k\}, E_{j,k+1}); x_m; (x_{k+1}, r))$$

by making use of the recursion stack and Lemma 9. The one with least cost among all of them is obviously the required R. Notice that this algorithm takes time $O(n^4)$. ■

3 Range Assignments in Three Dimensions

3.1 NP-hardness

This section is devoted to the proof that the problem of finding a minimum cost complete range assignment for a given set of points in the 3-dimensional Euclidean space is NP-hard. We formulate the corresponding decision problem as follows.

Problem **RANGE**

Instance: A set of points A in the 3-dimensional space, an integer $q > 0$.

Question: Does there exist a complete range assignment for A of total cost at most q?

We will show a reduction from the vertex cover problem for connected planar cubic graphs, known to be NP-hard (cf. [3]). The decision version of it is formulated as follows.

Problem **COVER**
Instance: An undirected connected planar cubic graph G, an integer $k > 0$.
Question: Does G have a vertex cover of size at most k?

We first need some auxiliary notions, facts and constructions. A *subdivision* of a graph G is a graph H resulting from G by adding new vertices of order 2 on edges of G (every edge is replaced by a chain and distinct edges are replaced by vertex disjoint chains). The new vertices of order 2 are called *subdivision vertices.* An *even subdivision* is a subdivision in which an even number of vertices are added on every edge. We omit the proof of the following lemma.

Lemma 10. *Let G be a graph with edges $e_1, \ldots, e_k$. Let H be a subdivision of G such that $2x_i$ new vertices are added on edge e_i, for all $i \leq k$. Let $x = \sum_{i=1}^{k} x_i$. Then G has a vertex cover of size $\leq r$ iff H has a vertex cover of size $\leq r + x$.*

Until the end of this section, G denotes a connected planar cubic graph. For any such G consider its planar representation P. Edges incident to any vertex yield a division of the plane into 3 *sectors.* Take any vertex v_1 of G and any edge e_1 incident to it and assign to e_1 one of the two neighboring sectors. Consider the other end v_2 of e_1 and the other edge e_2 neighboring the chosen sector. Assign to e_2 the other neighboring sector and go to the other end v_3 of e_2. Proceed in this way, at each vertex trying to assign a free sector to the new edge neighboring the previously assigned sector. This can be done iff the graph G is bipartite. If G is not bipartite, call edges for which a conflict occurred - *special.* In Figure 1 we show a non-bipartite graph and a sector assignment depicted by arrows, with the special edges e_5 and e_6. For any planar representation of G this *sector assignment* can be constructed in polynomial time with at most one special edge incident to each vertex.

Consider a rectangular grid on the plane with grid points having integer coordinates. Call unit length grid edges - *grid lines.* Consider a planar representation of a graph G such that vertices are mapped to grid points and edges are mapped to polygonal lines composed of grid lines. Call this a *Valiant representation* of G. It follows from [10] that a Valiant representation of G always exists and can be constructed in polynomial time.

Take a Valiant representation R of G. Subdivide every grid line in R into 8 equal *segments* of length $d = \frac{1}{8}$. In the resulting figure every edge e_i of G is mapped to a rectangular polygonal line c_i consisting of an even number of segments. Take 2 middle segments of each chain c_i and replace them by a chain of 3 segments as shown in Figure 2.

Call the resulting figure a *normal picture of* G. Every edge of G is mapped to a polygonal line consisting of an *odd* number of segments, thus a normal picture of G yields a graph $P(G)$ which is an even subdivision of G.

Let P be a normal picture of G in the plane $\mathcal{P}$. Let X be the set of vertices of P and ϵ a small positive number. We define the *ϵ-envelope* of G as the set $E = X \cup Y$, where Y is a set of points in the 3-dimensional space constructed as follows.

First construct a sector assignment corresponding to P (P is homeomorphic to a planar representation of G). Next, for any segment S of P construct a

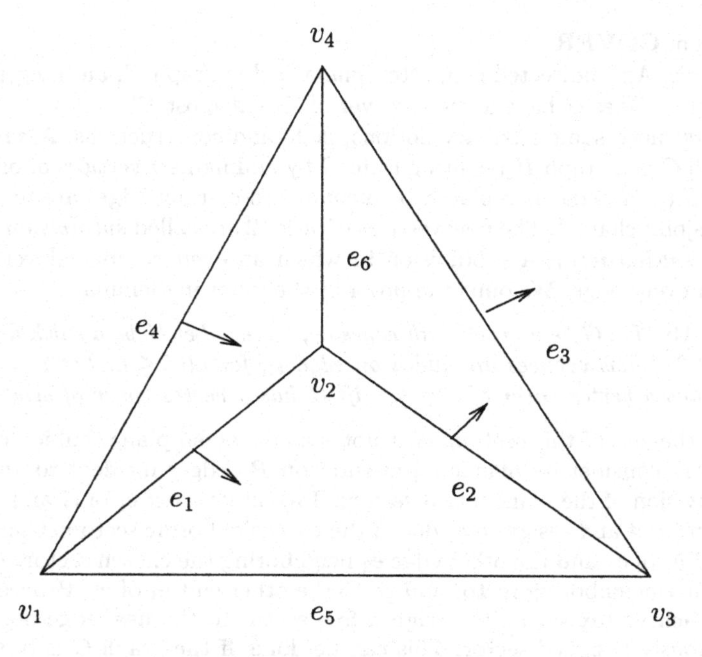

Fig. 1. Sector assignment.

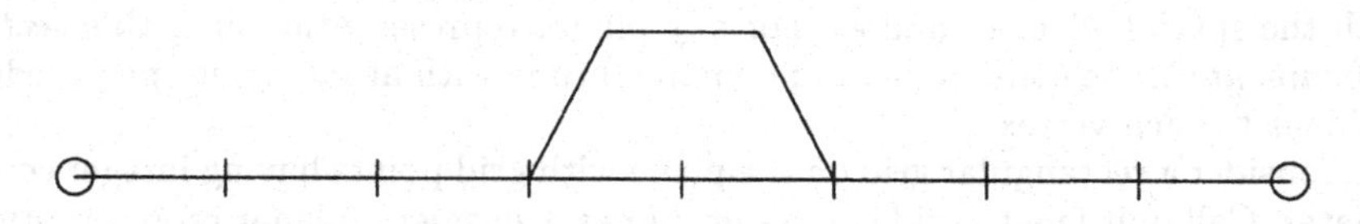

Fig. 2. Augmented chain in a normal picture.

companion point $c(S)$. This point is at distance $d+\epsilon$ from both ends of S. If S is a segment corresponding to a special edge in the sector assignment, $c(S)$ is in the plane perpendicular to $\mathcal{P}$, containing S. Otherwise, $c(S)$ is in one of the two sectors neighboring S, determined as follows. If S is one of the end segments of its chain then $c(S)$ is in the sector assigned to the corresponding edge. For other segments sectors alternate (cf. Figure 3). Recall that each chain has an odd number of segments. The set Y consists of companion points for all segments of P.

It follows from the construction that the distance between any two points in the ϵ-envelope of G is at least d. Moreover, every point $c(S) \in Y$ is at distance $d+\epsilon$ from the ends s_1, s_2 of the segment S. Consider any point t in E, different from $c(S)$, s_1 and s_2. Considering all possible cases it is easy to see that the distance between $c(S)$ and t is smallest when $t = c(S')$, segments S and S' are adjacent and perpendicular, and one of the points $c(S)$ or $c(S')$ is in the plane

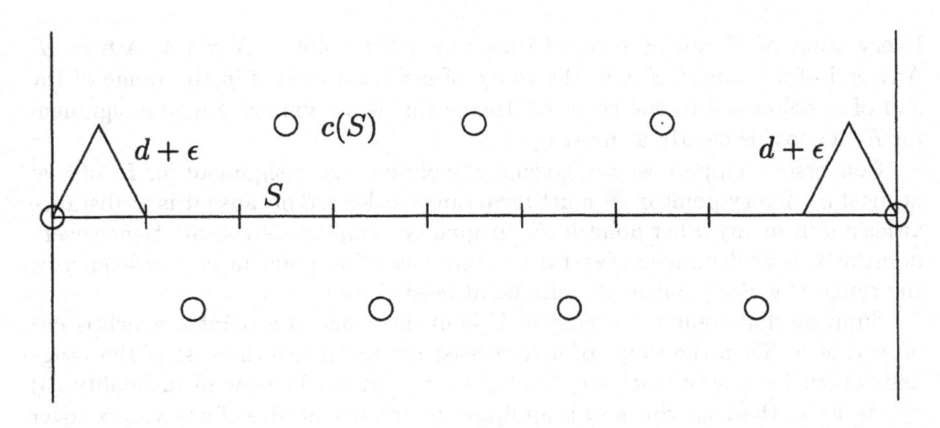

Fig. 3. Companion points in the ϵ-envelope. The figure depicts the sector assigned to e and the chain corresponding to e.

perpendicular to $\mathcal{P}$ (situated there because the corresponding edge was special). This smallest distance is then at least αd, where $\alpha = \sqrt{2 - \frac{\sqrt{3}}{2}} \approx 1.06$.

Theorem 11. *The problem* **RANGE** *is NP-hard.*

PROOF We show a polynomial reduction from **COVER**. Let G be a connected planar cubic graph and k a positive integer. Let N be the set of vertices of G. Let d and α be as before. Let P be a normal picture of G in the plane $\mathcal{P}$ and $H = P(G)$. Let X be the set of vertices of P, $|X| = m$, $X_1 = X \setminus N$, $|X_1| = 2x$. Let ϵ be a positive number satisfying the following two conditions:

$$\epsilon < (\alpha - 1)d, \tag{1}$$

$$\epsilon < \frac{(\alpha^2 - 1)d^2}{(m+1)(2d+1)}. \tag{2}$$

Inequality (1) implies $\epsilon < 1$. Hence inequality (2) implies $m(2d+\epsilon)\epsilon + 2d\epsilon + \epsilon^2 < (\alpha^2 - 1)d^2$ and consequently

$$md^2 + m(2d\epsilon + \epsilon^2) + y(d+\epsilon)^2 < md^2 + (y-1)(d+\epsilon)^2 + (\alpha d)^2. \tag{3}$$

By lemma 10 G has a vertex cover of size at most k iff H has a vertex cover of size at most $z = k + x$. Construct an ϵ-envelope $E = X \cup Y$ of G. The following claim concludes the proof of the theorem.

Claim. The graph H has a vertex cover of size at most z iff the envelope E has a complete range assignment of cost at most $q_z = md^2 + z(2d\epsilon + \epsilon^2) + y(d+\epsilon)^2$.

Proof of the claim. Suppose that H has a vertex cover C of size at most z. Assign range $d + \epsilon$ to all points in $C \cup Y$ and range d to all points in $X \setminus C$.

Every point of X can be reached from any other point of X via a path in H. Any end of a segment S is in the range of $c(S)$ and $c(S)$ is in the range of the end of S belonging to the cover C. Hence this is a complete range assignment for E. Its cost is clearly at most q_z.

Conversely, suppose we are given a complete range assignment for E, of cost at most q_z. Every point in X must have range at least d because it is at distance at least d from any other point in E. Inequality 1 implies $d+\epsilon < \alpha d$. Hence every point in Y is at distance at least $d+\epsilon$ from any other point in E. Consequently, the range of every point in Y must be at least $d+\epsilon$.

Suppose that some point $c(S)$ in Y is in the range of a point u which is not an end of S. Then the range of u is at least αd and hence the cost of the range assignment is at least $md^2 + (y-1)(d+\epsilon)^2 + (\alpha d)^2$. In view of inequality (3) this is larger than q_m. Since m is an upper bound on the size of any vertex cover of H, this yields a contradiction.

Hence every point $c(S)$ in Y is in the range of at least one of the ends of segment S. It follows that for every segment S of P one of its ends must have range at least $d+\epsilon$. Let $C \subset X$ be the set of vertices of P that have range at least $d+\epsilon$. Hence C is a vertex cover of H. The cost of the range assignment is at least

$$|C|(d+\epsilon)^2 + |X \setminus C|d^2 + y(d+\epsilon)^2 = md^2 + |C|(2d\epsilon + \epsilon^2) + y(d+\epsilon)^2 = q_{|C|}.$$

It follows that $q_{|C|} \leq q_z$ and consequently $|C| \leq z$, i.e., H has a vertex cover of size at most z. This concludes the proof of the claim and of the theorem. ■

3.2 An approximation algorithm

As was noted above, the minimum cost complete range assignment problem for a set of points in three dimensions is NP-hard. In this section we describe an $O(n^2)$ time approximation algorithm for this problem with a ratio bound of 2, i.e., the algorithm finds a solution within a factor of 2 of the optimal.

Given a set $V = \{x_1, \ldots, x_n\}$ of points in 3-space the algorithm proceeds as follows:

1. Construct an undirected weighted complete graph $G(V)$ with vertices V and where the weight of the edge between x_i and x_j is $d(x_i, x_j)^2$ for all i and j.
2. Find a minimum weight spanning tree T of $G(V)$.
3. For $i = 1, \ldots, n$ assign the range of x_i to be the maximum of $d(x_i, x_j)$ over j such that $\{x_i, x_j\}$ is an edge in T.

Clearly the algorithm runs in $O(n^2)$ time and the resulting range assignment is complete (since at the very least it contains all of the edges of the spanning tree in both directions). We omit the proof of the following theorem.

Theorem 12. *Let $OPT(V)$ be the minimum cost of a complete range assignment for V and let $APP(V)$ be the cost of the complete range assignment for V found by the above algorithm. Then $APP(V) < 2 \cdot OPT(V)$.*

4 Conclusions and Open Problems

In one dimension we gave asymptotically tight bounds on the minimum cost of a range assignment with diameter h on equi-distant points when h is constant and when $h \in \Omega(\log n)$. When h is between these ranges the bound is unknown. For the case of arbitrarily distributed points on a line we gave an $O(n^4)$ algorithm for finding the minimum cost complete range assignment. We believe our techniques may be extendable to find the minimum cost assignment of a given diameter.

In three dimensions we showed the problem of finding the minimum cost complete range assignment is NP-hard and gave an approximation algorithm optimal to within a factor of two. We conjecture the problem remains NP-hard in two dimensions. Note that the approximation algorithm works in two dimensions as well.

References

1. E. Arikan, "Some Complexity Results about Packet Radio Networks," *IEEE Transactions on Information Theory*, **IT-30** (1984), pp. 681–685.
2. A. Ephremides and T. Truong, "Scheduling broadcasts in multihop radio networks," *IEEE Transactions on Communications*, **30** (1990), pp. 456–461.
3. M.R. Garey and D.S. Johnson, "Computers and Intractability - A Guide to the Theory of NP-Completeness", Freeman and Co., New York, 1979.
4. R. Kahn, S. Gronemeyer, J. Burchfiel and R. Kunzelman, "Advances in packet-radio technology," *Proceedings of the IEEE*, **66** (1978), pp. 1468–1496.
5. K. Pahlavan and A. Levesque, "Wireless Information Networks," Wiley-Interscience, New York, 1995.
6. S. Ramanathan and E. Lloyd, "Scheduling broadcasts in multi-hop radio networks," *IEEE/ACM Transactions on Networking*, **1** (1993), pp. 166–172.
7. R. Ramaswami and K. Parhi, "Distributed scheduling of broadcasts in a radio network," *INFOCOM*, 1989, pp. 497–504.
8. A. Sen and M. Huson, "A New Model for Scheduling Packet Radio Networks," *INFOCOM*, 1996, pp. 1116–1124.
9. H. Takagi and L. Kleinrock, "Optimal Transmission Ranges for Randomly Distributed Packet Radio Terminals," *IEEE Transactions on Communications*, **COM-32** (1984), pp. 246–257.
10. L. Valiant, "Universality considerations in VLSI circuits," *IEEE Transactions on Computers*, **C-30** (1981), pp. 135–140.

The Complexity of Generating Test Instances

Christoph Karg*, Johannes Köbler, and Rainer Schuler

Abteilung Theoretische Informatik, Universität Ulm, 89069 Ulm, Germany

Abstract. Recently, Osamu Watanabe proposed a new framework for testing the correctness and average case behavior of algorithms that purport to solve a given NP search problem efficiently on average. The idea is to randomly generate certified instances in a way that resembles the underlying distribution μ. We discuss this approach and show that test instances can be generated for every NP search problem with non-adaptive queries to an NP oracle. Further, we introduce Las Vegas as well as Monte Carlo types of test instance generators. We show that these generators can be used to find out whether an algorithm is correct and efficient on average under μ. In fact, it is not hard to construct Monte Carlo generators for all RP search problems as well as Las Vegas generators for all ZPP search problems. On the other hand, we prove that (under the uniform distribution) Monte Carlo generators can only exist for problems in NP $\cap$ co-AM.

1 Introduction

The class NP plays a central role in computational complexity theory since it contains a large variety of problems that are of practical importance. Although, by definition, NP only contains decision problems (namely, to decide whether a given instance has a solution), in many applications it is necessary to solve the corresponding search problem (to actually find a solution in case one exists). On first glance, the search version of an NP problem seems to be less tractable than its decision version. However, in many cases, the decision and search versions are equivalent under polynomial-time Turing reductions (as, for example, in the case of NP-complete problems).

Since polynomial-time algorithms do not exist for NP-complete problems, unless P = NP, it seems reasonable to look for algorithms that are *efficient in the average case.* This idea has been formalized by Levin in [Lev86], where he introduced a framework for a theory of average-case complexity (see [Wan] for a survey). In practical applications, instances occur under some specific distribution. To simulate these distributions, we use probabilistic Turing machines that output instances in polynomial time (in the length of the generated instance) [BCGL92]. This model defines the class of polynomial-time samplable distributions.

Besides the design of efficient algorithms it is also important to find ways of testing the correctness (and performance) of a given algorithm. One way to

* Supported by the Deutsche Forschungsgemeinschaft, research grant Schö 302/4-2.

Reischuk, Morvan (Eds.): STACS'97 Proceedings, LNCS 1200

certify the purported quality of an algorithm is by giving a proof that it is correct and efficient. However, in many cases, these proofs may be long (if they can be found) and rather complicated. Thus, it is reasonable to test algorithms by feeding them with carefully chosen instances. A further reason to investigate the computational complexity of generating test instances is that it might give us more insight into the structure of the problem. In particular, we are interested in the relationship between the complexity of solving NP problems and the complexity of generating representative and certified test instances for them.

A framework for checking the correctness and average-case behavior of algorithms has been introduced by Watanabe [Wat94]. In this model, a probabilistic polynomial-time bounded Turing machine (from now on called test instance generator) outputs positive instances of the problem together with some solution. In contrast to other approaches [San89], the test instances have to be generated in accordance with a particular distribution, namely, with the *conditional* distribution on the set of all positive instances of fixed length. For some distributions, this requires to amplify the probability of the positive instances by an exponential factor. Hence, under such distributions test instance generators do not exist even for very simple search problems, unless RE = NE. We also show that on the other hand, non-adaptive queries to an NP-complete set are sufficient to generate test instances for any NP search problem. This result is tight since logarithmically many adaptive queries are shown to be useless.

In this paper we dispense with the condition that all generated test instances must be positive (i.e., possess a solution). In our model, a generator has to produce positive as well as negative instances. Moreover, we require that the probability that an instance is generated is polynomially related to the probability of that instance. This condition implies that any algorithm is efficient on the generated instances if and only if it is efficient under the given distribution.

We define two types of test instance generators providing test instances of different quality. Whereas a Monte Carlo generator is allowed to misclassify a generated instance with small probability, a Las Vegas generator has to classify all generated instances correctly. As in Watanabe's model we require that the generator outputs a witness for any positively classified instance. We show how these generators can be used to find out efficiently and with high confidence whether a given algorithm A for a distributional NP search problem is correct with respect to μ. Further, we construct Monte Carlo generators for all RP search problems as well as Las Vegas generators for all ZPP search problems.

Intuitively, a generator for an NP search problem only needs to solve the problem on instances that are (randomly) *generated by the generator itself*, whereas an algorithm has to solve the problem on instances that are (randomly) "generated from outside". Thus there might exist distributional NP search problems for which it is easier to find a generator than to find an algorithm that solves it. For example, it is not hard to construct a Las Vegas generator for the integer factorization problem. On the other hand, no efficient on average algorithm is known (under the standard distribution) that on input n outputs a factor of n in case n is composite, and outputs "prime" otherwise (in fact, this is the basis for the practical security of the RSA cryptosystem). Also, for certain random self-reducible problems [TW87,AFK89] it is possible to generate instances according to the distribution induced by the self-reduction. Examples are the *quadratic residue* and the *discrete logarithm* problems.

As pointed out above, there are distributions under which test instance generators don't even exist for very easy search problems. By an application of the universal hashing technique, we further show that the domain of any search problem that admits a Monte Carlo generator (under the standard probability distribution) necessarily belongs to co-AM. As a consequence, Monte Carlo generators don't exist for search problems with an NP-complete domain, unless the polynomial time hierarchy collapses.

Finally, we define a reduction between NP search problems which preserves the probabilities of the instances, similar to the reducibility introduced by Levin [Lev86,BG93]. Via this reducibility it is possible to transform the test instance generation problem between different NP search problems.

2 Preliminaries

In this paper we use the standard notations and definitions of computational complexity theory (see, e.g., [BDG95]). An introduction to the theory of computational average case complexity can be found in [Gur91,BCGL92,Wan].

All languages considered here are over the alphabet $\Sigma = \{0,1\}$. The length of a string $x \in \Sigma^*$ is denoted by $|x|$. For a set A of strings, let $A^{=n} = A \cap \Sigma^n$ and $A^{\leq n} = \bigcup_{k=0}^{n} A^{=k}$. We denote the cardinality of a set A by $\|A\|$. We use the pairing function $\langle \cdot,\cdot \rangle : \Sigma^* \times \Sigma^* \to \Sigma^*$, defined as $\langle x, y\rangle = d(x)01y$ where $d(x_1 \dots x_n) = x_1x_1 \dots x_nx_n$. Note that $\langle x,y\rangle$ is computable and invertible in polynomial time, and that for all $x, y \in \Sigma^*$, $|\langle x,y\rangle| = 2|x| + |y| + 2$.

Probability functions and distributions. A probability function μ is a total function from Σ^* to $[0,1]$ such that $\sum_{x\in\Sigma^*} \mu(x) = 1$. For example, we refer to $\mu_{st}(x) = 1/(|x|(|x|+1)2^{|x|})$ as the standard probability function. For any set $X \subseteq \Sigma^*$ let $\mu(X) = \sum_{x\in X} \mu(x)$. The distribution of μ, denoted by μ^*, is defined by $\mu^*(y) = \sum_{x:x\leq y} \mu(x)$. We denote by $\mu_n(x)$ the probability of x under the condition that $|x| = n$, i.e., $\mu_n(x) = \mu(x)/\mu(\Sigma^n)$ if $x \in \Sigma^n$ and $\mu(\Sigma^n) > 0$, and $\mu_n(x) = 0$ otherwise.

Let μ, ν be probability functions. μ is polynomially dominated by ν (in symbols: $\mu \leq_d^p \nu$) if for some polynomial p and all x, $\mu(x) \leq p(|x|) \cdot \nu(x)$. μ and ν are called equivalent (in symbols: $\mu \equiv_d^p \nu$) if $\mu \leq_d^p \nu$ and $\nu \leq_d^p \mu$.

A probability function μ is called p-computable if μ^* can be approximated via a polynomial-time algorithm in the sense of Ko and Friedman [KF82]. μ is called samplable [BCGL92] if there exists a probabilistic Turing machine M that on input λ outputs x with probability $\mu(x)$. If, in addition, M is polynomial-time bounded in the length of the output, then μ is called p-samplable.

Distributional NP search problems. An NP search problem is specified by a binary polynomial-time decidable relation R (called witness relation) and a polynomial q such that $R(x,w)$ implies $|w| = q(|x|)$. Any w with $R(x,w)$

is called a solution (witness) for x. W.l.o.g. we can assume that λ is never a solution. A witness relation R also specifies an NP decision problem, namely $D = \{x \mid \exists w : R(x, w)\}$. D is also called the domain of R.

A distributional NP search problem is a pair (R, μ) consisting of a witness relation R and a p-samplable probability function μ. An algorithm A solves (R, μ) if A computes a solution for all inputs x in the domain of R, and outputs λ on all other inputs.

3 Generating positive test instances

In this section we recall Watanabe's definition of a test instance generator and discuss his approach.

A test instance generator G for a distributional NP search problem (R, μ) is a probabilistic polynomial-time algorithm which on input 0^n, either generates some instance x of length n together with a solution w for x or outputs $\perp$ ("not successful"). In order to be useful, G should generate every positive instance x with sufficiently large probability.

Definition 1. [Wat94] Let (R, μ) be a distributional NP search problem and let D be the domain of R. A probabilistic polynomial-time Turing machine G is called a test instance generator for R under μ if there is a polynomial q such that for all n,

1. on every path, $G(0^n)$ outputs either $\perp$ or a pair $\langle x, w\rangle \in R$ with $x \in \Sigma^n$, and
2. for all $x \in D^{=n}$, $G(0^n)$ outputs with probability at least $\rho_n(x)/q(n)$ a pair of the form $\langle x, w\rangle$, $w \in \Sigma^*$, where

$$\rho_n(x) = \begin{cases} \mu(x)/\mu(D^{=n}), & \text{if } x \in D^{=n} \text{ and } \mu(D^{=n}) > 0, \\ 0, & \text{otherwise.} \end{cases}$$

is the probability of x under the condition that $|x| = n$ and x belongs to D.

As shown by Watanabe, generators of this type can be used to detect that an algorithm makes an error with probability ε (according to ρ_n). Furthermore, they can be used to find out that an algorithm is not polynomial on ρ-average [Wat94]. We note that even if an algorithm A is polynomial on ρ-average, A may take exponential time on average under the distribution induced by the generator.

The following result due to Watanabe shows that it is unlikely that there exists a test instance generator for every distributional NP search problem.

Theorem 2. [Wat94] *If every* NP *search problem has a test instance generator under the standard distribution, then* RE = NE.

In fact, the proof shows that randomly computing instance/witness pairs for all NP search problems with tally domain is impossible, unless RE = NE. We next show that generating test instances in this model may be impossible even for trivial search problems since for some distributions it is necessary to amplify the probability of the positive instances by an exponential factor.

Theorem 3. *Let $R = \{\langle x, 1\rangle \mid x \in \Sigma^* - \{1\}^*\}$. If for every p-samplable distribution μ there exists a test instance generator for (R, μ), then* RE = NE.

Proof. Since RE = NE if and only if NP $\cap$ TALLY $\subseteq$ RP [Boo74] it suffices to show that the assumption implies that every tally NP set T is contained in RP. Let R_T be a binary relation and p be a polynomial such that for all n, $0^n \in T$ if and only if there exists a string $w \in \Sigma^{p(n)}$ with $R_T(0^n, w)$.

Now consider the p-samplable distribution μ generated by the following probabilistic algorithm:

> First, guess a positive integer n with probability proportional to n^{-2}. Then uniformly guess a string $w \in \{0,1\}^{p(n)}$ and output 0^n in case $R_T(0^n, w)$ holds, else output 1^n.

Then any test instance generator G for (R, μ) has the property that for all n, $G(0^n)$ outputs $\langle 0^n, 1\rangle$ with probability $1/n^{O(1)}$ if $0^n \in T$, and with probability 0, otherwise. □

Before we give an upper bound on the complexity of generating test instances we observe that $O(\log n)$ many queries to an arbitrary oracle do not help in generating test instances in Watanabe's model.

Proposition 4. *If for a distributional* NP *search problem (R, μ), test instances can be generated with $O(\log n)$ queries to some oracle, then there exists a test instance generator for (R, μ).*

Proof. Let M be a probabilistic oracle Turing machine that generates test instances with $O(\log n)$ queries to some oracle. Let M' simulate M where oracle queries are answered randomly. If M outputs a pair $\langle x, w\rangle$ then M' outputs $\langle x, w\rangle$ only if $R(x, w)$ holds, and outputs $\perp$ otherwise. Then the probability that all oracle answers are guessed correctly is $1/n^{O(1)}$, implying that M' still fulfills the conditions of Definition 1. □

In the proof of the next theorem we make use of the universal hashing technique [CW79,Sip83]. Let $\mathcal{H}_{m,k}$ be the class of all linear hash functions from Σ^m to Σ^k. We will use the following well-known fact that can easily be derived from the universality of the class $\mathcal{H}_{m,k}$ (cf. [Pap94]).

Lemma 5. *Let S be a subset of Σ^m of cardinality $\|S\| = s$ such that $2^k \leq 3s \leq 2^{k+1}$ and let $x \in S$. Then, for a uniformly chosen hash function h from $\mathcal{H}_{m,k}$, it holds with probability at least $2/9s$ that x is the only element in S with $h(x) = 0^k$.*

Proof. Since $\mathcal{H}_{m,k}$ is a universal class of hash functions, it follows for a uniformly chosen h that $\text{Prob}[h(x) = 0^k] = 2^{-k}$ and that

$$\begin{aligned} &\text{Prob}[h(x) = 0^k \wedge \exists y \in S - \{x\} : h(y) = 0^k] \\ &\leq \sum_{y \in S - \{x\}} \text{Prob}[h(x) = 0^k \wedge h(y) = 0^k] \; < \; s2^{-2k}. \end{aligned}$$

Consequently, we get for the probability that x is the only preimage of 0^k in S (i.e., *h isolates x inside S*),

$$\begin{aligned} &\text{Prob}[h(x) = 0^k \wedge \forall y \in S - \{x\} : h(y) \neq 0^k] \\ &= \text{Prob}[h(x) = 0^k] - \text{Prob}[h(x) = 0^k \wedge \exists y \in S - \{x\} : h(y) = 0^k] \\ &> 2^{-k}(1 - s2^{-k}) \; \geq \; 2/9s. \end{aligned}$$

□

Theorem 6. *For every distributional* NP *search problem* (R, μ), *test instances can be generated with parallel queries to* SAT.

Proof. Let D be the domain of R and let q be a polynomial such that $|w| = q(|x|)$ for all $\langle x, w\rangle \in R$. Let M be a probabilistic Turing machine witnessing that μ is p-samplable and let p be a polynomial time bound for M, i.e., the time used by $M(\lambda)$ to output a string x is bounded by $p(|x|)$. Let $M_r(\lambda)$ denote the output of M (if any) when using sequence r as random source. The test instance generator G for (R_D, μ) is defined as follows.

> On input 0^n, G first guesses two integers $k \in \{1, \ldots, p(n) + 1\}$, $l \in \{1, \ldots, q(n) + 1\}$, and linear hash functions $h_1 \in \mathcal{H}_{p(n),k}$, $h_2 \in \mathcal{H}_{q(n),l}$. Then G asks in parallel the queries $\langle 0^n, i, h_1, h_2\rangle$, $1 \leq i \leq l(n) = 2p(n) + q(n) + 2$ to get a sequence $s = s_1 \ldots s_{l(n)}$ of oracle answer bits. Then G determines strings $r \in \Sigma^{p(n)}$, $w \in \Sigma^{q(n)}$, and $x = M_r(\lambda)$ such that $\langle r, w\rangle = s$. Finally, G outputs the pair $\langle x, w\rangle$, in case $|x| = n$ and $R(x, w)$, and $\perp$ otherwise.

For every x let $S_x = \{r \in \Sigma^{p(|x|)} \mid M_r(\lambda) = x\}$ denote the set of all random sequences r using which M outputs x, and let $S_n = \bigcup_{x \in D^{=n}} S_x$. For every $r \in S_n$ let α_r denote the probability that $G(0^n)$ guesses integers k, l and linear hash functions h_1, h_2 such that r is the only string in S_n with $h_1(r) = 0^k$ and there exists a unique $w \in \Sigma^{q(n)}$ with the property that $h_2(w) = 0^l$ and $R(M_r(\lambda), w)$. By Lemma 5 it follows that

$$\alpha_r \geq \frac{1}{(p(n)+1)(q(n)+1)} \cdot \frac{2}{9 \cdot \|S_n\|} \cdot \frac{2}{9}.$$

Hence, when using as oracle the set A defined as

$$\begin{aligned} \langle 0^n, i, h_1, h_2\rangle \in A \Leftrightarrow \exists\langle r, w\rangle : h_1(r) = 0^k,\ r \in S_n,\ h_2(w) = 0^l, \\ R(M_r(\lambda), w), \text{ and the } i\text{th bit of } \langle r, w\rangle \text{ is } 1, \end{aligned}$$

it follows that for every $x \in D^{=n}$, $G(0^n)$ outputs a pair of the form $\langle x, w\rangle$, $w \in \Sigma^*$, with probability at least

$$\sum_{r\in S_x} \alpha_r \geq \frac{4\,\|S_x\|}{81(p(n)+1)(q(n)+1)\,\|S_n\|} = \frac{4}{81(p(n)+1)(q(n)+1)} \cdot \rho_n(x).$$

□

We end this section by observing that the above theorem can easily be extended to the Monte Carlo type of generators that we will introduce in the next section.

4 Test instance generation without amplification

In this section we propose a slightly different model to test a given algorithm A that purports to solve a distributional NP search problem efficiently on average.

Definition 7. Let (R, μ) be a distributional NP search problem. A probabilistic Turing machine G is called a witness generator for R under μ if

1. on every halting computation, G outputs a pair $\langle x, w\rangle$ with the property that either $R(x, w)$ holds or $w = \lambda$ (note that G is not required to stop on all paths),
2. the time needed by G to output a pair of the form $\langle x, w\rangle$, $w \in \Sigma^*$, is polynomially bounded in $|x|$, and
3. μ_G is equivalent to μ, where $\mu_G(x)$ is the probability that G generates instance x, i.e., $\mu_G(x) = \sum_w \mu_G(x, w)$, where $\mu_G(x, w)$ is the probability that G outputs the pair $\langle x, w\rangle$.

Suppose that G is a witness generator for a distributional NP search problem (R, μ) and let A be an algorithm. Since the distributions μ and μ_G are equivalent, the running time of A is polynomial on μ-average if and only if it is polynomial on μ_G-average [Lev86].

We distinguish between two types of generators depending on the quality of the generated test instances. A Monte Carlo generator is allowed to err with small probability whereas a Las Vegas generator only generates test instances that are correctly classified.

Definition 8. Let G be a witness generator for a distributional NP search problem (R, μ) with domain D.

1. G is called a Monte Carlo generator for R under μ if for every polynomial q,

$$\mu_G(x, \lambda) \leq \frac{\mu_G(x)}{q(|x|)}$$

holds for all but finitely many instances x in D.
2. G is called a Las Vegas generator for R under μ if G never outputs a pair $\langle x, \lambda\rangle$ with $x \in D$.

We next show how a Monte Carlo generator can be used to find out efficiently and with high confidence whether a given algorithm A for a distributional NP search problem (R, μ) is correct with respect to μ. We can w.l.o.g. assume that A makes only errors on instances in the domain D of R. More formally, for any subset D' of D, we call the probability $\mu(\{x \in D' \mid A(x) = \lambda\})$ that $x \in D'$ and A does not find a witness for x the error rate of A on D'.

Theorem 9. *Let G be a Monte Carlo generator for a distributional* NP *search problem (R, μ) and let D be the domain of R. Then there exists a probabilistic polynomial time-bounded transducer T such that for every search algorithm A and all $n, m, l > 0$ the following holds:*

> *If the error rate of A on $D^{\leq l}$ is at least $1/m$, then with probability at least $1 - 2^{-n}$, $T(1^n, 1^m, 1^l)$ produces an instance $x \in D^{\leq l}$, together with a solution w such that $A(x) = \lambda$.*

Proof. Let G be a Monte Carlo generator for (R, μ) and let p be a polynomial such that for all instances x,

$$\mu(x)/p(|x|) \leq \mu_G(x) \leq p(|x|) \cdot \mu(x).$$

Then for all but finitely many $x \in D$, $\mu_G(x, \lambda) \leq \mu_G(x)/2$. Thus, w.l.o.g. we can assume that for all instances x,

$$\sum_{w \neq \lambda} \mu_G(x, w) \geq \mu_G(x)/2 \geq \mu(x)/2p(|x|).$$

Since the error rate of A on $D^{\leq l}$ is at least $1/m$, it follows that G outputs with probability at least $1/2mp(n)$ a pair $\langle x, w \rangle$ with $x \in D^{\leq l}$ and $A(x) = \lambda$. Therefore, during $2mnp(n)$ independent computations, G will output with probability at least $1 - (1 - 1/2mp(n))^{2mnp(n)} > 1 - 2^{-n}$ some pair $\langle x, w \rangle$ with $x \in D^{\leq l}$ and $A(x) = \lambda$. Note that since the running time of G is bounded by some polynomial q in the length of the output, T can suspend the simulation of G after $q(l)$ steps. □

It is interesting to note that under the following reasonable condition on μ it suffices to assume in the above theorem that A has error rate at least $1/m$ on D (instead of $D^{\leq n}$):

> For some polynomial q, $\sum_{x:|x|>q(n)} \mu(x) \leq 1/n$ holds for all n.

In fact, if under this condition the error rate of A on D is at least $1/m$, then it follows that $\mu(\{x \in D^{\leq q(2m)} \mid A(x) = \lambda\}) \geq 1/2m$. Hence, the above theorem implies that $T(1^n, 1^{2m}, 1^{q(2m)})$ produces with probability at least $1 - 2^{-n}$ an instance $x \in D^{\leq q(2m)}$ with $A(x) = \lambda$.

Next we consider distributional RP search problems (R, μ) (meaning that the binary relation R witnesses that its domain D belongs to RP). Since in this case instance/witness pairs can be easily generated by sampling a string x according to μ and randomly guessing a witness for x, we have the following positive result.

Proposition 10. *For every distributional* RP *search problem there exists a Monte Carlo generator.*

Intuitively, a generator for a distributional NP search problem only needs to solve the problem on instances that are (randomly) *generated by the generator itself*, whereas an algorithm has to solve the problem on instances that are (randomly) "generated from outside". Thus there might exist distributional NP search problems for which it is easier to construct a generator than to find an algorithm that solves it. For example, consider the factorization problem. The security of the RSA cryptosystem is based on the fact that no efficient on average algorithm is known (under the standard distribution) that on input n outputs a factor of n in case n is composite, and outputs "prime" otherwise. However, it is not hard to construct a Las Vegas generator for this search problem.

On the other hand, we show in the next theorem that the domain of every search problem that admits a Monte Carlo generator under the standard probability distribution necessarily belongs to co-AM. As a consequence, it cannot be NP-complete unless the polynomial time hierarchy collapses to the second level [BHZ87,Sch88]. For definitions of the class AM and of Arthur-Merlin games we refer the reader to [BM88] or some textbook, e.g., [BDG90,KST93].

Theorem 11. *If there exists a Monte Carlo generator for an* NP *search problem* R *under the standard distribution, then the domain of* R *is contained in* co-AM.

Proof. Let G be a Monte Carlo generator for (R, μ_{st}) and let p be a polynomial bounding the running time of G. From property (1) of Definition 8 we can conclude that for some polynomial r,

$$\frac{1}{r(|x|)2^{|x|}} \leq \mu_G(x) \leq \frac{r(|x|)}{2^{|x|}}.$$

From property (3) we know that for any polynomial q,

$$\mu_G(x, \lambda) \leq \frac{\mu_G(x)}{q(|x|)}$$

holds for all but finitely many x in the domain D of R. Now let S_x be the set of strings $r \in \Sigma^{p(n)}$ such that G outputs the pair $\langle x, \lambda\rangle$ when using r as random source. Let $q(n)$ be the polynomial $2^5 r(n)^2$. Then we get for all but finitely many x (letting $n = |x|$),

$$x \in D \Rightarrow \mu_G(x, \lambda) \leq \frac{r(n)}{2^n q(n)} \Rightarrow \|S_x\| \leq \frac{r(n)2^{p(n)}}{2^n q(n)} \Rightarrow \|S_x\| \leq \frac{2^{p(n)-n-5}}{r(n)},$$

$$x \notin D \Rightarrow \mu_G(x, \lambda) \geq \frac{1}{r(n)2^n} \Rightarrow \|S_x\| \geq \frac{2^{p(n)}}{r(n)2^n} \Rightarrow \|S_x\| \geq \frac{2^{p(n)-n}}{r(n)}$$

Now let $k(n) = p(n) - n - \lfloor \log r(n) \rfloor - 2$ and consider for a fixed instance x the random variable Z giving the number of strings $r \in S_x$ with $h(r) = 0^{k(n)}$, where

h is chosen uniformly at random from $\mathcal{H}_{p(n),k(n)}$. Then the expectation of Z is $E(Z) = 2^{-k(n)} \|S_x\|$ and the variance is $V(Z) = 2^{-k(n)}(1-2^{-k(n)}) \|S_x\|$ [VV86]. Thus, we have that

$$x \in D \Rightarrow \text{Prob}[Z \geq 1] \leq E(Z) \leq 1/4,$$

$$x \notin D \Rightarrow \text{Prob}[Z = 0] \leq \text{Prob}[|Z - E(Z)| \geq E(Z)] \leq V(Z)/E(Z)^2 \leq 1/4.$$

The Arthur-Merlin protocol for $\overline{D}$ proceeds as follows: On input x, $|x| = n$, Arthur randomly chooses a hash function $h \in \mathcal{H}_{p(n),k(n)}$ and asks Merlin to show him a string $r \in S_x$ with $h(r) = 0^{k(n)}$. Since Merlin succeeds with probability at least 3/4, if $x \notin D$, and with probability at most 1/4, otherwise, it follows that $\overline{D} \in$ AM. □

Corollary 12. *If there exists a Monte Carlo generator for some distributional* NP *search problem* (R, μ_{st}) *whose domain is* NP*-complete, then* PH $= \Sigma_2^P$.

We end this section by considering the Las Vegas type of generators that never output a wrongly classified instance. It is not hard to construct a Las Vegas generator for any ZPP search problem, i.e., for any RP search problem whose domain is in ZPP. On the other hand, Las Vegas generators can only exist for NP search problems whose domain belongs to NP ∩ co-NP.

Proposition 13.

1. *For every distributional* ZPP *search problem there exists a Las Vegas generator.*
2. *If there exists a Las Vegas generator for an* NP *search problem* R *under the standard distribution, then the domain of* R *is contained in* co-NP.

As a final remark, it is not hard to see that for every set L that has self-computable witnesses (in the sense of [BD76,Bal89]) there is a Las Vegas generator using L as an oracle. For example, since graph isomorphism (GI) and graph automorphism (GA) have self-computable witnesses and non-adaptively self-computable witnesses, respectively [Sch76,LT93], there exist Las Vegas generators for the corresponding search problems that ask adaptive (non-adaptive) queries to GI (GA, resp.).

5 Hard problems for test instance generation

In this section we consider the question of how the test instance generation problem for one NP search problem can be reduced to the test instance generation problem for another NP search problem. Inspired by [Wat94], we provide a way to transform a Monte Carlo (Las Vegas) generator for a distributional NP search problem (S, ν) to any problem (R, μ) that is reducible to (S, ν) by the following kind of reduction.

Definition 14. Let (R, μ) and (S, ν) be distributional NP search problems. Let D_R (D_S) be the domain of R (S, resp.). (R, μ) is **generator reducible** to (S, ν) if there exist functions $f, g \in \mathrm{FP}$ and a polynomial q such that:

1. For all y and r, if $f(y, r) \neq \perp$ then $f(y, r) \in D_R$ if and only if $y \in D_S$.
2. For all y, v and r, if $f(y, r) \neq \perp$ and $S(y, v)$ then $R(f(y, r), g(y, v, r))$.
3. μ is equivalent to ϕ, where ϕ is the distribution induced by ν and f, i.e., $\phi(x) = \sum_{y,r} \nu(y) 2^{-|r|}$, where the sum ranges over all strings y, r such that $|r| = q(|y|)$ and $f(y, r) = x$.
4. f is honest, i.e., for some polynomial p and all y, r, if $f(y, r) \neq \perp$ then $p(|f(y, r)|) \geq |y|$.

This reducibility is transitive, and, as we will show, the existence of generators is preserved under it.

Proposition 15. *Let (R, μ) be generator reducible to (S, ν). If there exists a Monte Carlo (Las Vegas) generator for (S, ν), then there exists a Monte Carlo (Las Vegas) generator for (R, μ).*

Proof. Assume that (R, μ) reduces to (S, ν) via the functions f and g, and let G be a generator for (S, ν). Then the following algorithm is a generator for (R, μ) (of the same type as G):

> Simulate G. If G produces a pair $\langle y, v \rangle$ then randomly guess a string $r \in \{0, 1\}^{q(|y|)}$. If $f(y, r) \neq \perp$, then output the pair $\langle f(y, r), g(y, v, r) \rangle$, otherwise loop forever. □

We finally observe that there are distributional NP search problems such that every other NP search problem (under any p-computable distribution) reduces to it. For example, this holds for the search version of all known distributional decision problems that are p-isomorphic to the randomized bounded halting problem [WB95].

References

[AFK89] M. Abadi, J. Feigenbaum, and J. Kilian. On hiding information from an oracle. *Journal of Computer and System Sciences*, 39:21–50, 1989.

[Bal89] J.L. Balcázar. Self-reducibility structures and solutions of NP problems. In *Revista Matematica*, volume 2, pages 175–184. Universidad Complutense de Madrid, 1989.

[BCGL92] S. Ben-David, B. Chor, O. Goldreich, and M. Luby. On the theory of average case complexity. *Journal of Computer and System Sciences*, 44(2):193–219, 1992.

[BD76] A. Borodin and A. Demers. Some comments on functional self-reducibility and the NP hierarchy. Technical Report 76–284, Dept. of Computer Science, Cornell University, 1976.

[BDG90] J.L. Balcázar, J. Díaz, and J. Gabarró. *Structural Complexity II*. Springer–Verlag, 1990.

[BDG95] J.L. Balcázar, J. Díaz, and J. Gabarró. *Structural Complexity I*. Springer-Verlag, second edition, 1995.

[BG93] A. Blass and Y. Gurevich. Randomized reductions of search problems. *SIAM Journal of Computing*, 22:949–975, 1993.

[BHZ87] R.B. Boppana, J. Hastad, and S. Zachos. Does co-NP have short interactive proofs? *Information Processing Letters*, 25:27–32, 1987.

[BM88] L. Babai and S. Moran. Arthur-Merlin games: A randomized proof system, and a short hierarchy of complexity classes. *Journal of Computer and System Sciences*, 36:254–276, 1988.

[Boo74] R.V. Book. Tally languages and complexity classes. *Information and Control*, 26:281–287, 1974.

[CW79] J.L. Carter and M.N. Wegman. Universal classes of hash functions. *Journal of Computer and System Sciences*, 18:143–154, 1979.

[Gur91] Y. Gurevich. Average case complexity. *Journal of Computer and System Sciences*, 42(3):346–398, 1991.

[KF82] K.-I. Ko and H. Friedman. Computational complexity of real functions. *Theoretical Computer Science*, 20:323–352, 1982.

[KST93] J. Köbler, U. Schöning, and J. Torán. *The Graph Isomorphism Problem: It's Structural Complexity*. Birkhäuser, 1993.

[Lev86] L.A. Levin. Average case complete problems. *SIAM Journal of Computing*, 15:285–286, 1986.

[LT93] A. Lozano and J. Torán. On the nonuniform complexity of the graph isomorphism problem. In K. Ambos-Spies, S. Homer, and U. Schöning, editors, *Complexity Theory—Current Research*, pages 245–271. Cambridge University Press, 1993.

[Pap94] C. Papadimitriou. *Computational Complexity*. Addison–Wesley, 1994.

[San89] L.A. Sanchis. Language instance generation and test case construction for NP-hard problems. Technical Report 296, University of Rochester, Dept. of Computer Science, 1989.

[Sch76] C.P. Schnorr. Optimal algorithms for self-transformable problems. In *Proc. 3th International Colloquium on Automata, Languages and Programming*, 1976.

[Sch88] U. Schöning. Graph isomorphism is in the low hierarchy. *Journal of Computer and System Sciences*, 37:312–323, 1988.

[Sip83] M. Sipser. A complexity theoretic approach to randomness. In *Proc. 15th ACM Symposium on the Theory of Computing*, pages 330–335, 1983.

[TW87] W. Tompa and H. Woll. Random self-reducibility and zero-knowledge interactive proofs of possession of information. In *Proc. 28st Symposium on Foundations of Computer Science*, pages 472–482, 1987.

[VV86] L.G. Valiant and V.V. Vazirani. NP is as easy as detecting unique solutions. *Theoretical Computer Science*, 47:85–93, 1986.

[Wan] J. Wang. Average-case computational complexity theory. In A. Selman and L. Hemaspaandra, editors, *Complexity Theory Retrospective II*. Springer-Verlag. (to appear).

[Wat94] O. Watanabe. Test instance generation for promised NP search problems. In *Proc. 9th Conference on Structure in Complexity Theory*, pages 205–216, 1994.

[WB95] J. Wang and J. Belanger. On the NP-isomorphism problem with respect to random instances. *Journal of Computer and System Sciences*, 50:151–164, 1995.

Efficient Constructions of Hitting Sets for Systems of Linear Functions

Alexander E. Andreev[1], Andrea E. F. Clementi[2], José D. P. Rolim[3]

[1] Dept. of Mathematics, University of Moscow,
andreev@matis.math.msu.su

[2] Dip. di Scienze dell'Informazione, University "La Sapienza" of Rome
clementi@dsi.uniroma1.it

[3] Centre Universitaire d'Informatique, University of Geneva, CH,
rolim@cui.unige.ch

Abstract. Given a positive number $\delta \in (0,1)$, a subset $\mathcal{H} \subseteq \{0,1\}^n$ is a *δ-Hitting Set* for a class $\mathcal{R}$ of boolean functions with n inputs if, for any function $f \in \mathcal{R}$ such that $\Pr(f=1) \geq \delta$, there exists an element $\mathbf{h} \in \mathcal{H}$ such that $f(\mathbf{h}) = 1$. Our paper presents a new deterministic method to efficiently construct *δ-Hitting Set* for the class of ***systems*** (i.e. logical conjunctions) ***of boolean linear functions***. Systems of boolean linear functions can be considered as the algebraic generalization of boolean ***combinatorial rectangular functions***, the only significative example for which an efficient deterministic construction of Hitting Sets were previously known. In the restricted case of boolean rectangular functions, our method (even though completely different) achieves equivalent results to those obtained in [11]. Our results also gives an upper bound on the minimum cardinality of ***solution covers*** for the class of systems of ***linear equations*** defined over a finite field. Furthermore, as preliminary result, we show a new upper bound on the circuit complexity of integer ***monotone*** functions generalizing previous results obtained in [12].

1 Introduction

General Motivations. This work is motivated by a recent result established by [4] in the theory of derandomization. Informally speaking, this result states that *quick Hitting Set Generators* can replace *quick Pseudorandom Generators* [14] in derandomizing BPP-algorithms. More precisely, in [4] it is proved that an efficient construction of a Hitting Set for the class of boolean functions having linear circuit-size complexity is sufficient to prove $P = BPP$. Consequently, a major, challenging goal in this area is now the development of algorithmic techniques to efficiently construct Hitting Sets for more and more general classes of boolean functions which can eventually culminate in the final efficient construction of Hitting Sets for boolean functions having linear circuit-size complexity.

Our paper presents a new method to efficiently construct Hitting Sets for an important natural class of boolean functions which is the algebraic generalization of the class of *combinatorial rectangle boolean functions* studied in [11].

Reischuk, Morvan (Eds.): STACS'97 Proceedings, LNCS 1200

We study the family of boolean functions $\mathcal{L} = \{\mathcal{L}(n,k), n > 0, 0 < k < n\}$, where $\mathcal{L}(n,k)$ consists of all such boolean functions, with n inputs, that can be expressed as *systems* (i.e. logical conjunctions) of boolean *linear* functions i.e.

$$f(x_1,\ldots,x_n) = \bigwedge_{j=1}^{k} \left(a_1^j x_1 \oplus a_2^j x_2 \oplus \ldots \oplus a_n^j x_n \oplus b^j\right) \; a_i^j, b^j \in \{0,1\}, k \leq n \,. \quad (1)$$

The complexity and the properties of linear functions have been the subject of several studies over the past few years [2,5,6,10]. Informally speaking, a major interest in linear functions is due to the fact that they have "small" (i.e. polynomial) circuit-size complexity [13] but they have a rather rich behavior recently used to approximate general boolean functions [5].

- *Previous Results on Hitting Sets.* In which follows, we consider the standard definition of circuit-size complexity of finite functions; moreover, given any boolean sequence $\mathbf{x} \in \{0,1\}^n$, we will use the term *complexity of* $\mathbf{x}$ to refer to the circuit-size complexity of the corresponding boolean function $x : \{0,1\}^{\lceil \log(n+1) \rceil} \to \{0,1\}$ where $x(i)$ is the i-th bit of $\mathbf{x}$. The circuit-size complexity of a finite function f (a finite sequence $\mathbf{x}$) will be denoted as $L(f)$ ($L(\mathbf{x})$).

Given a fixed positive number $\delta \in (0,1)$. A subset $\mathcal{H}_n \subseteq \{0,1\}^n$ is a *δ-Hitting Set* for a class $\mathcal{R}$ of boolean functions with n inputs if, for any function $f \in \mathcal{R}$ such that $\Pr(f=1) \geq \delta$, there exists an element $\mathbf{h} \in \mathcal{H}_n$ such that $f(\mathbf{h}) = 1$. A natural, well studied question concerning Hitting Sets is the *witness finding problem*: given a positive integer $n > 0$, and a positive number $\delta \in (0,1)$, find a subset $\mathcal{H}_n \subseteq \{0,1\}^n$ such that, for any *witness set* $\mathcal{W} \subseteq \{0,1\}^n$ with $|\mathcal{W}|/2^n \geq \delta$, we have $|\mathcal{W} \cap \mathcal{H}| > 0$. It is immediate to verify that the witness finding problem consists of finding a δ-Hitting Set for the class of all n-input boolean functions f.

Karp, Pippenger, and Sipser [9], and Sipser [15] introduced a randomized method to solve the witness finding problem that uses $O(n)$ random bits. Chor and Goldreich [7] derived a simpler algorithm that uses n random bits. This algorithm can be considered the best result in solving the witness finding problem for general witness sets. More recently, research has turned the attention also on non trivial restrictions of the problem where some combinatorial properties on the witness sets are imposed. In particular, we refer to Linial *et al*'s work [11] that will be described later.

Our results for systems of linear functions. In this paper, we study the witness finding problem for the family $\mathcal{L} = \{\mathcal{L}(n,k), n > 0, 0 < k < n\}$ of systems of linear functions (see Eq. 1). We also denote as $\mathcal{L}(n,k,q)$ ($q \leq n$) the class of boolean functions in $\mathcal{L}(n,k)$ having at most q non-zero columns in the matrix $A = [a_i^j]$ defined in Eq. (1) (q is commonly called the number of *essential* variables).

The definition given by Eq. (1) will be also written as $f = A_f\mathbf{x} \oplus \mathbf{b}$, where $A_f \in \{0,1\}^{n \times k}$, $\mathbf{x} \in \{0,1\}^n$, and $\mathbf{b} \in \{0,1\}^k$. The rank of a matrix $A \in \{0,1\}^{n \times k}$ (i.e. the maximum number of linearly independent rows) will be denoted as $\mathrm{r}(A)$.

Observe that for any non-zero function $f \in \mathcal{L}(n,k)$ (i.e. $f \neq \mathbf{0}$) we have $\Pr(f=1) \geq 2^{-k}$. Furthermore, we can consider the class $\mathcal{L}(n,k)$ as the class of non-zero functions which can be represented by Eq. (1) and such that the corresponding matrix A_f has maximum rank (i.e. $\mathrm{r}(A_f) = k$). Indeed, if $f = A'_f\mathbf{x} \oplus \mathbf{b}' \in \mathcal{L}(n,k')$ (with $f \neq \mathbf{0}$)) and A'_f has not maximum rank (i.e. $\mathrm{r}(A'_f) = k < k'$), then f can be represented as $f = A_f\mathbf{x} \oplus \mathbf{b}$, where A_f is a $k \times n$-matrix such that $\mathrm{r}(A_f) = k$, by choosing k linearly independent rows of A'_f.

With this assumption, given any function $f \in \mathcal{L}(n,k)$, we have $\Pr(f=1) = 2^{-k}$ and, consequently, the role of parameter δ in the definition of Hitting Sets is now replaced by the term 2^{-k}. In the case of class $\mathcal{L}$ of systems of linear functions, we can thus omit the parameter δ and simply use the term Hitting Set.

The main results of this paper can be stated in the following way.

Theorem 1. *Let $n > 0$ and ϵ be any positive constant such that $0 < \epsilon < 1/3$.*

- *If $k \geq n^{2/3+\epsilon}$, then it is possible to construct a Hitting Set $\mathcal{H}_n \subseteq \{0,1\}^n$ for $\mathcal{L}(n,k)$ such that $|\mathcal{H}_n| \leq 2^{O(k)}$.*
- *If $\log\log n \leq k \leq n^{2/3}$ and $k \geq q^{2/3+\epsilon}$, then it is possible to construct a Hitting Set $\mathcal{H}_n \subseteq \{0,1\}^n$ for $\mathcal{L}(n,k,q)$ such that $|\mathcal{H}_n| \leq 2^{O(k)}$.*
- *If $k \leq \log\log n$ and $k \geq q^{2/3+\epsilon}$, then it is possible to construct a Hitting Set $\mathcal{H}_n \subseteq \{0,1\}^n$ for $\mathcal{L}(n,k,q)$ such that $|\mathcal{H}_n| \leq 2^{O(k)}\log^2 n$.*
- *If $k \leq \min\{n^{2/3}, q^{2/3}\}$, then it is possible to construct a Hitting Set $\mathcal{H}_n \subseteq \{0,1\}^n$ for $\mathcal{L}(n,k,q)$ such that $|\mathcal{H}_n| \leq 2^{O(k)\log n}$*

In all cases, the time required by the construction is polynomially bounded in the size of the output.

Observe also that, for the first three cases, the logarithm of the size of the obtained Hitting Sets is optimal since a simple lower bound for the size of Hitting Sets for $\mathcal{L}(n,k)$ is 2^k.

A different interpretation of Theorem 1. Let $\mathcal{LS}(n,k)$ be the class of all feasible[1] linear systems of the form

$$A \times \mathbf{x} = \mathbf{b}\,, \quad \text{where } A \in \{0,1\}^{k\times n},\ \mathbf{x} \in \{0,1\}^n,\ \text{and } \mathbf{b} \in \{0,1\}^k. \tag{2}$$

Consider the following algebraic problem. Given $n > 0$ and $k \leq n$, find a *solution cover* for $\mathcal{LS}(n,k)$, i.e., a subset $\mathcal{H}_n \in \{0,1\}^n$ containing at least one solution for any system in $\mathcal{LS}(n,k)$. Clearly the goal here is to derive an efficient construction which minimizes the size of the obtained solution cover. It is easy to show that any Hitting Set for the function class $\mathcal{L}(n,k)$ represents a solution cover for $\mathcal{LS}(n,k)$ and viceversa. It follows that Theorem 1 provides also an efficient method to solve the algebraic problem defined above. Under this point of view, since our method is based only on algebraic properties of linear algebra

[1] we will consider only feasible systems and thus, in the following, we will omit the term feasible

in finite fields, we can derive equivalent constructive upper bounds for systems of linear equations on finite (non boolean) fields. In order to obtain the upper bounds for the case in which the $k \times n$ linear system is defined over the Galois field $GF(Q)$, it is sufficient to replace the basis 2 with Q in the formulas listed in Theorem 1. However, we will not mention this possible generalization further in this extended abstract.

- *Connections between combinatorial rectangles and systems of linear equations*
As mentioned above, Linial *et al* [11] studied the witness finding problem in the case of combinatorial rectangles. A *combinatorial rectangle* is any subset of the form $\mathcal{R} = R_1 \times R_2 \ldots \times R_n$ where $R_i \subseteq \{0,..,m-1\}$. The goal here is to generate a subset $\mathcal{H} \subseteq \{0,..,m-1\}^n$ that has non empty intersection with every combinatorial rectangle $\mathcal{R}$ whose size (also denoted as *volume*) is at least δm^n. Linial *et al*'s algorithm generates a Hitting Set $\mathcal{H}$ whose size is polynomial in $m \log n(1/\delta)$, and the running time is polynomial in $mn(1/\delta)$.

Observe that when $m = 2$ (i.e. the boolean case), then the characteristic function of a generic combinatorial rectangle $\mathcal{R} \subseteq \{0,1\}^n$ can be expressed as a system of simple boolean linear functions in which exactly one variable appears:

$$f(x_1, \ldots, x_n) = \bigwedge\nolimits_{j=1}^{k} (x_{i(j)} \oplus b^j) \tag{3}$$

Observe also that, given any boolean *rectangular* function $f^{\mathcal{R}}$ represented by Eq. 3, the size of the corresponding rectangle $\mathcal{R} \subseteq \{0,1\}^n$ easily verify the following equation $|\mathcal{R}| = 2^{-k} \cdot 2^n$. Thus, the role of the parameter δ in the definition of combinatorial rectangles is now replaced by the term 2^{-k}. More formally, the class of boolean rectangular functions having volume parameter $\delta = 2^{-k}$ is strictly contained in the class $\mathcal{L}(n,k,k)$ which always satisfies one of the first three cases of Theorem 1.

Boolean rectangle functions represent thus the intersection between Linial *et al*'s work and our work: while Linial *et al* provide an efficient construction of Hitting Set for general (i.e. non boolean) rectangular functions, our work gives an efficient solution of the same problem but for the class of boolean functions in which the "rectangular" condition is relaxed into the much more general condition expressed by Eq. 1. In the case of boolean rectangular functions, we thus give an another (completely different) method to construct Hitting Sets which has equivalent performances to those obtained by Linial *et al.*

Due to the lack of space, the proofs of lemmas and corollaries will be given in the full version of the paper (see also [1]). Furthermore the new result concerning monotone functions will be given in Appendix.

2 Hitting Sets for "large" linear systems

As observed in the Introduction, a subset $\mathcal{H} \subseteq \{0,1\}^n$ is a Hitting Set for $\mathcal{L}(n,k)$ if and only if $\mathcal{H}$ contains at least one solution for any feasible linear system. This equivalence will be strongly used in deriving our Hitting Sets.

Given a boolean (k,n)-matrix A, consider the following column partition. Let $n = n_1 + \ldots + n_s$ and let A_i be a boolean (k,n_i)-matrix, such that

$$A = (A_1, A_2, ..., A_s) . \tag{4}$$

Define $r_s = \mathrm{r}(A_s)$ where $\mathrm{r}(A)$ denotes the rank of A, and

$$r_i = \mathrm{r}((A_i, \ldots, A_s)) - \mathrm{r}((A_{i+1}, \ldots, A_s)) \ , \ i = 1, \ldots, s-1 .$$

Then consider the linear system

$$A\mathbf{x} = \mathbf{b} , \tag{5}$$

where $\mathbf{x} \in \{0,1\}^n$ and $\mathbf{b} \in \{0,1\}^k$. Using the above matrix representation, it is possible to show an interesting relation between the solutions of System (5) and the Hitting Sets for the classes $\mathcal{L}(n_i, r_i)$'s.

Lemma 2. *For any $i = 1, \ldots, s$, let $\mathcal{H}_i$ be a Hitting Set for $\mathcal{L}(n_i, r_i)$. Then, for any $\mathbf{b} \in \{0,1\}^k$, there exists a solution of System (5) which belongs to the set*

$$\mathcal{H}_1 \times \mathcal{H}_2 \times \ldots \times \mathcal{H}_s.$$

The following lemma states that the above matrix representation can be found satisfying a useful bound on the number s of submatrices.

Lemma 3. *Let $r, m \leq n$ and $r \leq m$. Then, given any (k,n)-matrix A, it is possible to construct Representation (4) of A that satisfies the following conditions.*

$$r_i \leq r \ , \ n_i \leq m \ , \qquad i = 1, 2, \ldots, s \ , \tag{6}$$

and

$$s \leq \frac{k}{r} + \frac{n}{m} . \tag{7}$$

Let $[\mathbf{a}]^j$ be the prefix of length j of sequence $\mathbf{a}$ and, for any set S of boolean sequences, define $[S]^j = \{[\mathbf{a}]^j \ : \ \mathbf{a} \in S\}$. Hitting Sets for systems of linear functions satisfy the following *monotone* property.

Lemma 4. *Let $r \leq m$ and $\mathcal{H}$ be a Hitting Set for $\mathcal{L}(m, r)$. Assume that Condition (6) is verified. Then there exists a solution of System (5) which belongs to the set $[\mathcal{H}]^{n_1} \times [\mathcal{H}]^{n_2} \times \ldots \times [\mathcal{H}]^{n_s}$.*

Given any class $\mathcal{R}$ of boolean functions, the function $\lambda(\mathcal{R})$ denotes the minimum size of any Hitting Set for $\mathcal{R}$.

Lemma 5. *For any $n > 0$ and $k \leq n$, we have $\lambda(\mathcal{L}(n,k)) \leq 2^k(n+1)k$.*

We can now prove the main result of this section.

Theorem 6. *Let ϵ be a positive constant such that $0 < \epsilon < 1/3$. If $k \geq n^{2/3+\epsilon}$ then any system of type (5) has at least one solution with complexity at most $O\left(\frac{k}{\log k}\right)$.*

Proof. From Lemma 3, we can construct the matrix representation in Eq. (4) that satisfies Conditions (6) and (7) (the choice of parameters r and m are given later). Then, Lemma 4 implies that there exists a solution of System (5) belonging to the set $[\mathcal{H}]^{n_1} \times [\mathcal{H}]^{n_2} \times \ldots \times [\mathcal{H}]^{n_s}$ where $\mathcal{H}$ is a Hitting Set for $\mathcal{L}(m,r)$. We now show a way to compute a sequence $\mathbf{a} = \mathbf{a}_1 \ldots \mathbf{a}_s$ where $\mathbf{a}_i \in [\mathcal{H}]^{n_i}$. We assume that $\mathcal{H} = \{\mathbf{h}_1, \ldots \mathbf{h}_{|\mathcal{H}|}\}$. For any $i = 1, \ldots, s$, $Q(i)$ denotes the index for which $\mathbf{a}_i = [\mathbf{h}_{Q(i)}]^{n_i}$. Define $NUM(u)$ as the function which gives, for any $u = 1, \ldots, n$, the index of the submatrix of A which contains column u. In other terms, $NUM(u)$ is uniquely determined by the following condition

$$\sum_{i=1}^{NUM(u)-1} n_i < u \leq \sum_{i=1}^{NUM(u)} n_i .$$

Define also $LEN(i) = \sum_{t=1}^{i-1} n_t$, and $SF(v) = \mathbf{h}_v$. Finally, let

$$SEL(i, \alpha_1, \ldots, \alpha_m) = \alpha_i \text{ if } : 1 \leq i \leq m \text{ and } 0 \text{ otherwise} .$$

We can derive the u-th bit of $\mathbf{a}$ using the following sequence of computations

$$i = NUM(u)\,;l = LEN(i)\,;j = u - l\,;p = Q(i)\,;\boldsymbol{\alpha} = SF(p)\,;\mathbf{a}(u) = SEL(j, \boldsymbol{\alpha}) .$$

The ranges of the parameters used in the above computations are the following

$$u, l \in \{1,2,..,n\},\ j \in \{1,2,..,m\},\ i \in \{1,2,..,s\},\ p \in \{1,2,..,|\mathcal{H}|\},\ \boldsymbol{\alpha} \in \{0,1\}^m .$$

It is then easy to prove the following bound for the complexity of $\mathbf{a}$:

$$L(\mathbf{a}) \leq L(NUM) + L(LEN) + O(\log n) + L(Q) + L(SF) + L(SEL) . \quad (8)$$

In which follows we give upper bounds for every element of the above sum. From Theorem 15 we have

$$L(NUM) \leq O\left(\frac{s}{\log s} \log \frac{n}{s} + (\log n)^{O(1)}\right) . \quad (9)$$

Since $LEN : \{1,2,..,s\} \to \{1,2,..,n\}$ then its output consists of $\log n$ bits; hence, by using Lupanov's result [12], we obtain

$$L(LEN) \leq (1 + o(1)) \frac{s}{\log s} \log n . \quad (10)$$

An equivalent argument holds for functions $Q : \{1,..,s\} \to \{1,..,|\mathcal{H}|\}$, and $SF : \{1,..,|S|\} \to \{0,1\}^m$:

$$L(Q) \leq (1 + o(1)) \frac{s}{\log s} \log |\mathcal{H}| \text{ and } L(SF) \leq (1 + o(1)) \frac{|\mathcal{H}|}{\log |\mathcal{H}|} m . \quad (11)$$

The function SEL can be easily constructed within the following circuit complexity

$$L(SEL) \leq O(m) . \tag{12}$$

By replacing Eq.s (9-12) in Eq. 8, we get

$$L(\mathbf{a}) \leq O\left(\frac{s}{\log s}(\log|\mathcal{H}| + \log n) + \frac{|\mathcal{H}|}{\log|\mathcal{H}|}m\right) . \tag{13}$$

Lemma 3 implies that $s \leq k/r + n/m$. If we choose $r = \epsilon \log n$ and $m = \lceil((nr)/k)\log n\rceil$, then we have $s \leq (1+o(1))(k/r)$. From Lemma 5, we have that $|\mathcal{H}| \leq 2^r(m+1)r$ and, consequently,

$$\frac{s}{\log s}(\log|\mathcal{H}| + \log n) \leq O\left(\frac{\frac{k}{r}}{\log\frac{k}{r}}(\log m + r + \log n)\right) \leq O\left(\frac{k}{\log k}\right) . \tag{14}$$

Furthermore,

$$\frac{|\mathcal{H}|}{\log|\mathcal{H}|}m \leq O\left(\frac{1}{\log n}2^r m^2\right) \leq O((\log n)^3 n^{2/3-\epsilon}) \leq O\left(\frac{k}{\log k}\right) . \tag{15}$$

Combining Eq.s (13), (14), and (15), we get $L(\mathbf{a}) \leq O(k/\log k)$. □

Let $\mathcal{F}(n,l)$ be the set of all sequences $\mathbf{a} \in \{0,1\}^n$ such that $L(\mathbf{a}) \leq l$.

Corollary 7. *Let ϵ be a positive constant such that $0 < \epsilon < 1/3$. If $k \geq n^{2/3+\epsilon}$ then*
1). *There exists a constant c (which can be efficiently derived from the proof of Theorem 6) such that $\mathcal{F}(n, c\frac{k}{\log k})$ is a Hitting Set for $\mathcal{L}(n,k)$.*
2). *Furthermore, the size of this Hitting Set is such that $|\mathcal{F}(n, c\frac{k}{\log k})| \leq 2^{O(k)}$.*
3). *It is possible to generate the Hitting Set $\mathcal{F}(n, c\frac{k}{\log k})$ for the class $\mathcal{L}(n,k)$ in polynomial time in $2^{O(k)}$.*

Note *(Connections with boolean rectangular functions).* It is easy to see that 2^k is a lower bound on the size of Hitting Sets for $\mathcal{L}(n,k)$. It follows that the logarithm of the size of our Hitting Set is optimal. In terms of rectangular boolean functions, since the volume parameter δ is equal to 2^{-k}, we thus generate a Hitting Set $\mathcal{H}$, for the class of boolean rectangular functions with n inputs and volume parameter δ, which has size polynomially bounded in $1/\delta$. Furthermore, the time to construct $\mathcal{H}$ is polynomially bounded in $(1/\delta)n$. For this case, the Hitting Set shown by Linial *et al* [11] has size polynomially bounded in $\log n(1/\delta)$, and the running time is polynomial in $(1/\delta)n$. Since $\delta = 2^{-k}$ and $k > n^{2/3}$, our method has equivalent performances to those of Linial *et al*'s method.

3 Hitting Sets for "small" linear systems

In this section, we describe a reduction technique whose goal is to extend the previous construction to the case of small (i.e. $k \leq n^{2/3}$) linear systems. This method works for the class $\mathcal{L}(n,k)$ with no restrictions on k but when a particular condition on the number of non-zero columns in the system matrix A is assumed. Let us now introduce the class of systems of linear functions determined by this new condition and its relation with boolean rectangular functions.

A function $f(x_1,..,x_n) \in \mathcal{L}(n,k)$ belongs to the subclass $\mathcal{L}(n,k,q)$ if it can be represented by Eq. (1) where matrix $A = [a_i^j]$ has at most q non-zero vertical columns. In which follows, we will consider the case in which k and q satisfy the following inequality: $k \geq q^{2/3+\epsilon}$. Since rectangular boolean functions are linear systems in which there is exactly one variable in every linear function (see Eq. 3) then it is easy to verify that the class of rectangular boolean functions with n variables and with volume $2^{-k}2^n$ is contained in the class $\mathcal{L}(n,k,k=q)$ which always satisfies the condition $k \geq q^{2/3+\epsilon}$.

It is possible to prove the existence of an integer function which has the following particular injectivity property. This function will be one of the key ingredients in the reduction shown in the next section.

Lemma 8. *Let $\mathcal{A} \subseteq \mathcal{B} \subseteq \{1,\ldots,n\}$. For any $s \geq 1$, there exists a function $f:\{1,\ldots,n\} \to \{1,\ldots,q\}$, with $q = |\mathcal{A}| + 2^s$, such that for any $a \in \mathcal{A}$ and $b \in \mathcal{B}$ ($a \neq b$) we have $f(a) \neq f(b)$, and $L(f) \leq O(|\mathcal{A}||\mathcal{B}|2^{-s}(s+\log n) + s\log n)$.*

3.1 Hitting Sets for case $k \geq \max\{\log^2 n, q^{2/3+\epsilon}\}$

The proof of the following theorem gives the main reduction which allows us to extend the results for large linear systems to the case of small linear systems.

Theorem 9. *Let ϵ be a constant such that $0 < \epsilon < 1/3$.*
If $k \geq \max\{\log^2 n, q^{2/3+\epsilon}\}$ then any system of type (5) has at least one solution with complexity at most $O\left(\frac{k}{\log k}\right)$.

Proof. Consider a system $A\mathbf{x} = \mathbf{b}$ where A is a boolean $k \times n$-matrix with $\mathrm{r}(A) = k, \mathbf{x} \in \{0,1\}^n$ and $\mathbf{b} \in \{0,1\}^k$. Assume also that A satisfies the conditions of the theorem. Let $\mathcal{A} \subseteq \{1,\ldots,n\}$ be the index subset determining a subset of A-columns of size and rank k. If $\mathcal{B}$ denotes the set of all indexes corresponding to non-zero columns of A, then we easily have that $\mathcal{A} \subseteq \mathcal{B}$. Let $s = \lfloor((2/3+\epsilon)/(2/3+\epsilon/2))\log q\rfloor$. From Lemma 8, there exists a function $f:\{1,..,n\} \to \{1,..,n'\}$ where $n' = 2^s + k$ such that for any $a \in \mathcal{A}$ and $b \in \mathcal{B}$, with $a \neq b$, we have $f(a) \neq f(b)$. Furthermore, $L(f) \leq O(|\mathcal{A}||\mathcal{B}|2^{-s}(s+\log n) + s\log n)$. Since $n' \leq q^{(2/3+\epsilon)/(2/3+\epsilon/2)} + k \leq 2 * k^{1/(2/3+\epsilon/2)}$, then it is not hard to prove that, for a.e. n, $k \geq (n')^{2/3+\epsilon/3}$. We now define a linear transformation for system $A\mathbf{x} = \mathbf{b}$ which leads us to the case of large systems described in Section 2. The linear transformation is defined by the following equations $x_i = y_{f(i)}$ for $i = 1,..,n$. The properties of function f in Lemma 8 implies that the new obtained system has still rank k. If $\mathbf{a} = (a_1, .., a_{n'})$ is a solution of the new system, then $\boldsymbol{\alpha} = (a_{f(1)}, .., a_{f(n)})$ is a solution for system $A\mathbf{x} = \mathbf{b}$. Furthermore, we have $L(\boldsymbol{\alpha}) \leq L(f) + L(\mathbf{a})$. We can now apply Theorem 6 thus proving that there exists a solution $\mathbf{a}$ of the new obtained system such that $L(\mathbf{a}) \leq O\left(\frac{k}{\log k}\right)$. From Lemma 8 and from the fact that $k = \Omega(\log^2 n)$, there exists a positive constant c for which

$$L(f) \leq O(k * q * 2^{-s}(s + \log n) + s \log n) = O\left(\frac{k}{\log k}\right),$$

Consequently, we have $L(\boldsymbol{\alpha}) = O\left(\frac{k}{\log k}\right)$.

□

The above theorem implies that the set $\mathcal{F}(n, c(k/\log k)$, for some constant $c > 0$, is a Hitting Set for the class $\mathcal{L}(n, k, q)$ when $k \geq \max\{\log^2 n, q^{2/3+\epsilon}\}$. We can thus obtain equivalent results to those in Corollary 7.

3.2 Hitting Sets for case $k < (\log n)^2$, $k \geq q^{2/3+\epsilon}$

Let us consider our linear system $A\mathbf{x} = \mathbf{b}$, where $\mathrm{r}(A) = k$. let $\mathcal{B}$ be the set of all indexes corresponding to non-zero columns of A. We consider some finite field $GF(Q)$ and the function $f_u : \{1, .., n\} \to GF(Q)$ with $u \in GF(Q)$ defined as follows. Let $m = \lceil \log(n + 1) \rceil$ and let $\mathbf{a} = (a_1, a_2, ..., a_m)$ be the standard binary representation of an integer $i \in \{1, .., n\}$. Then $f_u(\mathbf{a}) = \Sigma_{i=1}^{m} a_i * u^{i-1}$ where $+$ and $*$ are the operations defined in $GF(Q)$. Let $\mathbf{a}, \mathbf{b} \in \mathcal{B}$ such that $\mathbf{a} \neq \mathbf{b}$ (here $\mathcal{B}$ is considered as a set of boolean sequences of length $m = \lceil \log(n + 1) \rceil$). Then the equation

$$f_u(\mathbf{a}) = f_u(\mathbf{b}) . \tag{16}$$

is equivalent to the following $\sum_{i=1}^{m} (a_i - b_i) * u^{i-1} = 0$. The above equation can be true for at most $m - 1$ different u's. It follows that if $Q > m * q^2$ then there exists at least one element $u \in GF(Q)$ for which Eq. (16) is false for any pair $\mathbf{a}, \mathbf{b} \in \mathcal{B}$ such that $\mathbf{a} \neq \mathbf{b}$. Thus, we have the same property of Lemma 8 and, hence, we can apply the same method of the previous case (i.e. $k = \Omega(\log^2 n)$) in order to construct a solution for system $A\mathbf{x} = \mathbf{b}$. If $\mathcal{H}$ is a Hitting Set for the class $\mathcal{L}(Q, k, q)$, then the set of sequences

$$(a_{f_u(1)}, a_{f_u(2)}, ..., a_{f_u(n)}) , \quad u \in GF(Q) , \ \mathbf{a} \in \mathcal{H} , \tag{17}$$

will be a Hitting Set for $\mathcal{L}(n, k, q)$. Its size is at most $Q|\mathcal{H}|$.

Theorem 10. *Let ϵ be a constant such that $0 < \epsilon < 1/3$. If $k \geq \max\{(\log \log n)^3, q^{2/3+\epsilon}\}$, then a a Hitting Set for $\mathcal{L}(n, k, q)$ exists that has size bounded by $2^{O(k)}$. The time to construct the Hitting Set is $2^{O(k)} n$.*

Proof. We can choose Q such that $Q = O(q^2 \log n)$. In this case we have $k > (\log Q)^2$ and, from Theorem 9, we have that the obtained Hitting Set is such that $|\mathcal{H}| \leq 2^{O(k)}$. It follows that the size of the Hitting Set defined in Eq (17) is bounded by $O(q^2 \log n 2^{O(k)}) = O(2^{O(k)})$. □

Theorem 11. *Let ϵ be a constant, $0 < \epsilon < 1/3$. If $k \geq q^{2/3+\epsilon}$ then we can construct a Hitting Set for $\mathcal{L}(n,k,q)$ whose size is bounded by $2^{O(k)}(\log n)^2$. The time to construct the Hitting Set is $2^{O(k)}(\log n)^2 n$.*

Proof. If $k \geq \log\log n$ then we choose $Q = O(q^2 \log n$. Since $k > (\log\log Q)^3$ we can apply Theorem 10 and obtain a Hitting Set for the class $\mathcal{L}(Q,k,q)$. By considering the construction defined in Eq. 17, we derive a Hitting Set for the class $\mathcal{L}(n,k,q)$ whose size is bounded by $O(q^2 \log n 2^{O(k)}) = O(2^{O(k)} \log n)$. If $k \leq \log\log n$ then we can apply the construction of case $k = \lceil \log\log n \rceil$ and, consequently, the size of the obtained Hitting Set is bounded by $O(q^2 \log n)2^{O(k)}) = O(\log n)2^{O(k)} + O(\log^2 n)$. □

3.3 Hitting Sets for case $k \leq \min\{n^{2/3}, q^{2/3}\}$

Let $\{0,1\}^n_k$ be the set of all sequences in $\{0,1\}^n$ with at most k units. Unfortunately, when $k \leq q^{2/3}$ we are not able to construct Hitting Sets having the same (almost optimal) size of the previous cases.

Lemma 12. *The set $\{0,1\}^n_k$ is a Hitting Set for $\mathcal{L}(n,k)$.*

Lupanov [12] proved that the complexity of any sequence in $\{0,1\}^n_k$ is at most $(1+o(1))(k\log(n/k))/(\log k) + O(\log n)$.

We can thus repeat the same construction of the Set $\mathcal{F}(n,l)$ shown in Corollary 7, thus proving the following result.

Theorem 13. *If $k \leq n^{2/3}$ (and no restriction for q), then the set*

$$\mathcal{H} = \mathcal{F}(n, O(\frac{k\log\frac{n}{k}}{\log k} + \log n))$$

is a Hitting Set for the class $\mathcal{L}(n,k,q)$. Furthermore, we have $|\mathcal{H}| = O(2^{k\log n})$ and the time to construct $\mathcal{H}$ is polynomial in its size.

Appendix: The complexity of monotone functions

The construction of the Hitting Set for the class $\mathcal{L}(n,k)$ requires the use of integer *monotone* functions, i.e., $f : \{1,2,\ldots,n\} \to \{1,2,\ldots,s\}$, such that $f(i) \leq f(j)$ for any $i < j$. The complexity of these functions has been studied by Lupanov [12] in the restricted case $n = s$. However, in our construction we need an upper bound also for the case $s < n$.

Given any positive integer n and any positive integer $m \leq n$, the term $C(n,m)$ denotes the binomial coefficient. Let $1 \leq u_1 < \ldots < u_{ts} \leq n$ and consider the operator

$$F_U(i) = (u_{t(i-1)+1}, u_{t(i-1)+2}, \ldots, u_{t(i-1)+t}) . \quad (18)$$

Lemma 14. *If $s \leq n$ then*

$$L(F_U) \leq (1 + o(1))(\log C(n, ts)/(\log s) + O((t \log n)^4) .$$

The above Lemma can be used to obtain the generalization of Lupanov's result [12].

Theorem 15. *If $f : \{1, \ldots, n\} \rightarrow \{1, \ldots, m\}$ with $m < n$, then*

$$L(f) \leq (1 + o(1))(\log C(n, m))/(\log m) + (\log n)^{O(1)} .$$

Proof. Choose s and t as any pair of integers such that $m \leq st$; consider a monotone sequence $v_1, \ldots, v_{st}$ such that, if $v_{i-1} < j \leq v_i$ then $f(j) = i$. Now define $u_i = v_i + i - 1$ $(i = 1, \ldots, st)$, and consider the new monotone sequence $U = u_1, \ldots, u_{st}$ with the corresponding operator F_U defined in Eq. 18. Then, given any $j \in \{1, \ldots, n\}$, we can compute $f(j)$ by using sequence $F_U(j)$. It is not hard to prove that $L(f) \leq L(F_U) + (t \log n)^2$. The values s and t can be chosen in such a way that $\log st = (1 + o(1)) \log m$, $\log t = o(\log s)$, $t \geq (\log n)^2$. The theorem then follows by applying Lemma 14. □

References

1. Andreev A.E., Clementi A.E.F. and Rolim J.D.P. (1996), "Towards efficient constructions of hitting sets that derandomize BPP", Research Report in *ECCC*, TR-96-029.
2. Allender E, Beals R, and Ogihara M. (1996), "The complexity of matrix rank and feasible systems of linear equations", in Proc. of *28-th ACM STOC*, to appear. Also available by ftp/www in ECCC (Tech. Rep. 1996).
3. Andreev, A.E. (1989), On the complexity of the realization of partial Boolean functions by circuits of functional elements, Diskret. mat. 1, pp.36-45. (In Russian). English translation in Discrete Mathematics and Applications 1, pp.251-262.
4. Andreev A.E., Clementi A.E.F. and Rolim J.D.P. (1996), "Hitting Sets derandomize BPP", in Proc. of *23-th ICALP* LNCS, Springer-Verlag, to appear. Also available by ftp/www in ECCC (Tech. Rep. 1996).
5. Andreev A.E., Clementi A.E.F. and Rolim J.D.P. (1996), "Optimal bounds for the approximation of boolean functions and some applications", in Proc. of *13-th STACS*, LNCS, Springer-Verlag (1996). Also available by ftp/www in ECCC (Tech. Rep. 1995).
6. Andreev A.E., Clementi A.E.F. and Rolim J.D.P. (1996), "On the parallel computation of boolean functions on unrelated inputs", in Proc. of *4-th Israeli Symposium on Theory of Computing and Systems (ISTCS'96)*, to appear.
7. Chor B., and O. Goldreich (1989), "On the Power of Two-Point Based Sampling", *J. Complexity*, 5, 96-106.
8. Furedi, Z. (1988) "Matchings and Covers in Hypergraphs", *Graphs and Combinatorics*, 4, 115-206.
9. Karp R., Pippenger N., and Sipser M. (1982) "Time-Randomness, Tradeoff", presented at *AMS Conference on Probabilistic Computational Complexity*.

10. Karpinski, M., and Luby, M. (1993), "Approximating the number of solutions to a GF(2) Formula", *J. Algorithms*, 14, pp.280-287.
11. Linial N., Luby M., Saks M., and Zuckerman D. (1993), "Efficient construction of a small hitting set for combinatorial rectangles in high dimension", in *Proc. 25th ACM STOC*, 258-267.
12. Lupanov, O.B. (1965), "About a method circuits design - local coding principle", *Problemy Kibernet.* 14, pp.31-110. (in Russian). *Systems Theory Res.* v.14, 1966 (in English).
13. Nechiporuk, E.I. (1965), About the complexity of gating circuits for the partial boolean matrix, Dokl. Akad. Nauk SSSR 163, pp.40-42. (In Russian). English translation in Soviet Math. Docl.
14. Nisan N., and Wigderson A. (1994), "Hardness vs Randomness", *J. Comput. System Sci.* 49, 149-167 (also presented at the *29th IEEE FOCS*, 1988).
15. Sipser M. (1986), "Expanders, Randomness or Time vs Space", in *Proc. of 1st Conference on Structures in Complexity Theory*, LNCS 223, 325-329.

Protocols for Collusion-Secure Asymmetric Fingerprinting

(Extended Abstract)

Ingrid Biehl and Bernd Meyer

Universität des Saarlandes, Fachbereich Informatik
66041 Saarbrücken, Germany
email: ingi@cs.uni-sb.de, bmeyer@cs.uni-sb.de

Abstract. In [16] asymmetric fingerprinting of data is presented as a new method of copyright protection. The merchant of the data and each buyer interact in such a way that the buyer gets an individually fingerprinted version of the data, while the merchant does not know the version. If the merchant finds an illegally redistributed version of the data he can trace at least one dishonest buyer even if a collusion of c dishonest buyers created that version. Since the merchant himself never sees the buyers' versions he cannot cast suspicion on some innocent buyers by redistributing appropriate versions.
We present a general construction of an asymmetric fingerprinting scheme based on some arbitrary symmetric fingerprinting scheme. Moreover, we give a construction which is more suitable for broadcast data than the constructions presented in [16].

1 Introduction

The problem of copyright protection of digital data as pictures, texts, and software has gained increasing interest in the last years (see for example [8], [6], [2], [16], [15], [9], [11] and [14]). In recent literature, different approaches can be found which try to achieve different kinds of copyright protection. Fingerprinting of data is one of them. Each copy of the data is made unique in order to enable the merchant of the data to detect a dishonest buyer as soon as an illegal copy is found.

We investigate the following model for fingerprinting: errors are introduced into the data in order to detect them in an illegally distributed copy. We assume that there are a lot of possibilities for placing an error. We call such a modified part of the original data a *mark*. The form and the number of possible values for each individual mark may depend on the semantic of the data. For black-and-white pictures the change of a single bit in the map of the picture may serve as a good mark. For simplicity we assume that the data can be divided into t different blocks and that there are two versions for each block, an *unmodified version* and a *marked version*. The major task of a fingerprinting scheme is that the pattern according to which the blocks of the individual copy of a buyer are

Reischuk, Morvan (Eds.): STACS'97 Proceedings, LNCS 1200

marked or not, should be different for each buyer and enable the merchant to identify a dishonest buyer. Such a pattern is a binary string of length t and is called a *marking pattern.*

Obviously one is interested in schemes which are *secure against c-collusions.* That is, even if up to c buyers collude to modify their individual versions of the data, the fingerprinting scheme should enable the merchant to find at least one of the members of this collusion.

So far most of the publications on fingerprinting consider the *symmetric* case: the merchant modifies the data without interacting with the buyer. If the merchant finds an illegally distributed version of his data he can never decide whether it got distributed by some buyer or by one of his employees. For similar reasons the merchant has no means to prove to a third party (e.g. a court) that a buyer distributed the data illegally. In [16] this problem is solved by the invention of *asymmetric fingerprinting.* The idea is that merchant and buyer interactively compute a fingerprinted version of the data. The buyer sees the complete fingerprinted version of the data while the merchant only gets enough information to identify the buyer of a version of the data as soon as an illegal copy of it is found. In [16] a general model for this problem is presented. Moreover, they show how to construct an asymmetric fingerprinting scheme by means of secure multi-party computation given some *memory-less* symmetric fingerprinting scheme. Memory-less means that the fingerprinting algorithm has not to store individual information for each buyer in order to trace him if an illegal copy of the data redistributed by him is found.

The methods presented in [3] are the most efficient symmetric fingerprinting schemes published so far. But they are not memory-less, i.e. for each fingerprinted version of the data some information has to be stored which helps to assign an illegal copy of the data to a dishonest buyer. Thus the techniques developed in [16] cannot be used to create asymmetric fingerprinting schemes out of the symmetric fingerprinting schemes of [3].

In this paper we present a method for the construction of an asymmetric fingerprinting scheme based on an arbitrary symmetric fingerprinting scheme, not necessarily a memory-less scheme. We would like to stress here that a very similar construction was found independently by Birgit Pfitzmann and Michael Waidner (see [17]).

We divide the task of asymmetric fingerprinting data into a *codeword phase* and an *information-retrieval phase.* In the codeword phase merchant and buyer interact in such a way that the buyer gets a correct *personal marking pattern* without knowing the codebook and the merchant gets information which will enable him to identify a dishonest buyer if he finds an illegally distributed version of the data. In the information-retrieval phase the buyer gets some information which enables him to get the modified data which corresponds to the codeword computed in the codeword phase. This information can be the modified data itself or some key which the buyer can use to decrypt the data which is sold in encrypted form, for example, on a CD-ROM. By this technique, we get a method to asymmetrically fingerprint broadcast data.

In [8] the authors deal with the problem of distributing encrypted broadcast data: they encrypt the data blocks and store them for example on CD-ROMs. Then they give each buyer the appropriate decryption keys in a way which allows the tracing of dishonest buyers if one finds a set of decryption keys. While [8] studies the symmetric case [15] investigates the asymmetric case. In both cases: if the data is redistributed after decryption no dishonest person can be traced. This is a disadvantage compared to our construction.

In addition to general techniques for the construction of asymmetric fingerprinting schemes based on arbitrary symmetric fingerprinting schemes we study methods to minimize the number of interactions in the codeword and information-retrieval phase, since this is an important factor for the usability of these schemes. We introduce ***committed all-or-nothing-disclosure protocols*** (CANDOS) which allow us to compose the codeword phase and information-retrieval phase in an efficient way. In the codeword phase the buyer is committed to a marking pattern which automatically will be used in the information-retrieval phase.

2 Asymmetric Fingerprinting

2.1 General Model

We use the model proposed in [16]. We assume that marking is done in a way that it does not decrease the usefulness of the modified data for the buyer and that the chance to erase the marks is negligible if one does not know the original version of the data. Modifications of the data which make the errors undetectable by the merchant will, with overwhelming probability, disturb the data in a way that makes them useless. We call this the *Marking Assumption*.

In an *asymmetric fingerprinting* scheme (defined in [16]) merchant and buyer have to interact in a *fingerprinting protocol*. At the end of the interaction the buyer owns an individually fingerprinted version of the data and the merchant has some special information. If there is a coalition of at most c buyers which create a version of the data from their own versions and redistribute it, then there is an *identify algorithm* which identifies at least one dishonest buyer if the identify algorithm gets the afore mentioned special information. Moreover, there has to be a multi-party *dispute protocol* between the merchant, a third party called *arbiter*, and possibly up to c accused buyers, which guarantees: the probability that an innocent buyer is found guilty is bounded by δ and if a coalition of c buyers created an illegal copy of the data at least one member of the coalition is found guilty with probability $1 - \delta$.

2.2 Symmetric Fingerprinting

In the articles by Dan Boneh and James Shaw (see [3] and [4]) a formal model for the construction of the marking patterns is given and precisely investigated. They show that the marking patterns can be interpreted as codewords of so-called *c-secure* codes. Apart from some trivial cases, c-secure codes cannot be constructed which guarantee absolute security for innocent buyers. However, allowing that an innocent person comes under suspicion with probability ε, they construct *c-secure codes with ε-error* which demand polynomially in $\log(1/\varepsilon)$ and $\log(n)$ many different marking positions, where n is the number of possible buyers.

A *c-secure code with ε-error* is a code Γ of length ℓ with n codewords (without loss of generality a binary code) with the following properties: before the merchant starts to sell the data he randomly chooses a permutation $\pi \in S_\ell$ and keeps it secret. He assigns to each buyer u a codeword $w^{(u)}$ from Γ and uses the marking pattern $\pi(w^{(u)})$ to (symmetrical) fingerprint the data for u. Suppose a coalition C of at most c buyers constructs a version of the data which has a marking pattern x of length ℓ. According to the Marking Assumption x is a *reachable* word, that is if all marking patterns of the coalition members have the same bit in position $1 \leq i \leq \ell$, then the ith bit of x has the same value. There is a deterministic *tracing algorithm* A_Γ, which given x and the permutation π finds some $u \in C$ with probability $1-\varepsilon$, where the probability is taken over the random bits necessary to choose π and the random choices made by the coalition. The probability that some innocent person is traced is less than ε as well as the probability that the tracing algorithm finds nobody. We call the code Γ the *basis code* and the codebook resulting from Γ by applying a secret permutation a *c-secure permutation code with ε-error*. Notice that the basis code Γ may be publicly known. It gains its c-security by secretly and uniformly permuting the bit positions of the codewords.

A c-secure permutation code represents an example for a symmetric fingerprinting scheme which we define as follows:

Definition 1. A *symmetric (ℓ, n) fingerprinting scheme* (or in short *fingerprinting scheme*) is a function $\Phi(R, u, r)$ which maps a user number $1 \leq u \leq n$ and two strings of random bits R and r to a codeword in Σ^ℓ. The random bit strings R and r are kept hidden from the users.

Φ is *c-secure with ε-error* if there exists a tracing algorithm A with the following property: If R is chosen randomly then if a coalition of at most c users with codewords $\Phi(R, u_1, r_1), \ldots \Phi(R, u_c, r_c)$ create a reachable word $x \in \{0,1\}^\ell$ then A finds some $\Phi(R, u_i, r_i)$ $(1 \leq i \leq c)$ with probability of at least $1-\varepsilon$. (The probability is taken over the random choice of the bit strings R, r_i $(1 \leq i \leq c)$ and the random choices the coalition made.)

For each fixed R the set of all bit strings $\Phi(R, u, r)$ for all $1 \leq u \leq n$ and all random strings r is called the *codebook of $\Phi(R)$*. Its elements are the *codewords*.

Given a c-secure code Γ with ε-error and n codewords then one can construct secure codes for a set of buyers of size $N > n$ by concatenating $L \geq 1$ randomly chosen codewords from Γ. We generalize the construction of Theorem 17 in [3]:

Theorem 2. *Let $L \in \mathbb{N}$, $1 \geq \delta > 0$ and $\circ$ denote string concatenation. Let Γ be a d-secure symmetric fingerprinting scheme with ε-error, codeword length $\ell = \ell(n, d, \varepsilon)$, n codewords and tracing algorithm A. If $d \geq c$, $\varepsilon \leq \frac{\delta}{2L}$, $n \geq 2c$ and*

$L \geq 4(c-1)\log\left(\frac{4N}{\delta}\right)$ then one gets a c-secure symmetric fingerprinting scheme Γ' with δ-error, N codewords and codeword length $\ell = O(L\ell(n,d,\varepsilon))$ as follows: one has to choose random bit strings $R_1, \ldots, R_L$ identically for all users and random bit strings $r_1, \ldots, r_L$ individually for each user u and set

$$\Gamma'(R_1 \ldots R_L, u, r_1 \ldots r_L) = \Gamma(R_1, u, r_1) \circ \ldots \circ \Gamma(R_L, u, r_L).$$

We call Γ the *low-level code* of this fingerprinting scheme. The c-security is based on the secrecy of $R_1, \ldots, R_L$. Given a marking pattern found in an illegal version of the data the tracing algorithm A first traces each of the L pieces to a codeword in $\Gamma(R_i)$. The result is a sequence of codewords $\overline{w} = w_1, \ldots, w_\ell$. If a coalition of at most c dishonest users formed the whole pattern, there is a dishonest person which has a personal marking pattern which matches $\overline{w}$ in at least L/c low level codewords.

Given a c-secure code with ε-error and n codewords of length ℓ one gets a c-secure fingerprinting scheme Γ with ε-error as follows: one has to choose a permutation $\pi \in S_\ell$ randomly by some random bit string R. Then one gets $\Gamma(R, u, r)$ by permutating the codeword of the c-secure code which is randomly assigned to user u with respect to a random string r. Applying the construction of Theorem 2 leads to a fingerprinting scheme for a larger set of buyers: One has to secretly and randomly choose L permutations by means of random strings $R_1, \ldots, R_L$ and for each buyer, L random codewords from the c-secure code. Then the marking pattern for the buyer results from applying the permutations to the L codewords. The mentioned most efficient symmetric fingerprinting schemes (see [3]) are of this type.

3 Cryptographic Techniques

3.1 Bit Commitment Schemes

A *bit commitment scheme* consists of two protocols. The *commit protocol* allows Alice to commit to a bit b by sending an encryption $\mathrm{BC}(r, b)$, called *blob*, (in short $\mathrm{BC}(b)$) of that bit for some random binary string r to Bob (*completeness property*). Bob only has a negligible chance to determine the value of b from the blob (*security property*). In the *reveal protocol* Alice later can open the blob by sending r to convince Bob that b was the encrypted value. Again the probability is negligible that Alice constructed a blob she can open both as 0 and as 1 (*soundness property*) (see [10]).

In our more efficient constructions we need bit commitment schemes which have the so-called *blinding property*, i.e. there is some public information which enables Bob to *blind* blobs made by Alice: given a blob $\mathrm{BC}(b)$ for a bit b, Bob can modify randomly the blob to *blind*$(r, \mathrm{BC}(b))$ (with random string r) such that it is still a blob for b but randomly distributed in the set of all blobs for b. Moreover, there is some secret information in possession of Alice which allows her to correctly open each blob even if it was blinded by Bob.

We say a bit commitment scheme is *homomorphic* if there is some polynomial time computable operation $*$ such that for two blobs $\mathrm{BC}(b_0)$ and $\mathrm{BC}(b_1)$ of some bits b_0 and b_1 it follows $\mathrm{BC}(b_0) * \mathrm{BC}(b_1) = \mathrm{BC}(b_0 \oplus b_1)$ and Alice knows some secret information which allows her to open $\mathrm{BC}(b_0 \oplus b_1)$. (With $\oplus$ we denote the exclusive or.)

For example the *quadratic residue bit commitment scheme* has the blinding property and is homomorphic: Alice chooses two primes p and q with $p \equiv q \equiv 3 \bmod 4$ randomly, keeps them secret and publishes the Blum integer $n = pq$ and a quadratic non-residue y with Jacobi symbol $+1$. To commit to 0 she randomly chooses x and uses the quadratic residue x^2 as commitment, to commit to 1 she uses the quadratic non-residue x^2y. To open the commitment she publishes x.

Since Alice knows p and q she can find out whether an element of $\mathbb{Z}/n\mathbb{Z}$ is a quadratic residue or not. Bob can also encode 0 as x^2 and 1 as x^2y with some random x. Blinding of a blob by Bob can be achieved by multiplying the blob with x^2 for some random x. The homomorphism property can easily be seen by taking multiplication as the $*$ operation.

3.2 Zero-Knowledge Protocols

Zero-knowledge interactive proofs (see [13]) will be another primitive we use in our constructions. In our application merchant and buyer are Turing machines of bounded resources. Thus we consider the zero-knowledge proof model introduced by [18], called *zero-knowledge proof of the possession of information*, *zero-knowledge proofs of knowledge* or *zero-knowledge arguments*, where both the prover and verifier are probabilistic polynomial time Turing machines with auxiliary input.

A *move* is the act of sending a message either by the prover or by the verifier. A *round* is a pair of *moves*, i.e. a message from the prover followed by a message of the verifier or vice versa.

Theorem 3 ([10]). *If one-way functions exist then for every language in* NP *there is a 5-move zero-knowledge proof of the possession of information.*

Moreover, [10] presents protocols which take even less moves under some stronger assumptions.

The languages for which we have to construct zero-knowledge proofs of the possession of information in our protocols are *random self-reducible languages* $L \in \mathrm{NP}$ (see [10] and [18]) and *P-samplable*, i.e. there is a probabilistic polynomial time Turing machine which generates for a given integer n all pairs (x, w) of instances $x \in L$ of size n and witnesses w for x with the same probability. For these languages one can prove:

Theorem 4. *Let L be a random self-reducible language and P-samplable. If secure bit commitment schemes exist then there are 4-move zero-knowledge proofs of the possession of information for L.*

3.3 Committed ANDOS-Protocols

Assume Alice owns two secret t-bit strings s_0 and s_1. She allows Bob to learn one of these strings but Alice herself may not know which string Bob learns. A protocol which achieves this is called an *all-or-nothing disclosure of secrets protocol* (ANDOS) (see [5]). In a *committed ANDOS protocol* (CANDOS) Bob secretly has to commit at the beginning of the protocol to the secret he wants to learn, learns exactly this secret, and Alice does not know which secret he has learned.

CANDOS protocols can be easily constructed by means of a bit commitment scheme, an ANDOS protocol, a secret encryption scheme Enc, and zero-knowledge proofs: Bob has to commit to his choice b by a blob $\mathrm{BC}(b)$. Then Alice randomly chooses an encryption key k and encrypts both strings with this key $\mathrm{Enc}_k(s_0)$ and $\mathrm{Enc}_k(s_1)$ and plays the ANDOS protocol with Bob. At the end of this protocol Bob knows $\mathrm{Enc}_k(s_c)$. Then he has to show by means of a zero-knowledge protocol that he knows $\mathrm{Enc}_k(s_c)$ with $c = b$. If he can convince Alice, she sends k and Bob computes s_b.

We present a constant-round CANDOS protocol: Let BC_A be some homomorphic bit commitment scheme for Alice with blinding property and BC_B some homomorphic bit commitment scheme for Bob. Let $x_0 = x_{0,0} \dots x_{0,t}$ and $x_1 = x_{1,0} \dots x_{1,t}$ be the secrets. Bob commits to his choice by $\mathrm{BC}_B(b)$. Then Alice sends $\mathrm{BC}_A(x_0)$ and $\mathrm{BC}_A(x_1)$ to Bob. Moreover, she sends $\mathrm{BC}_A(r_0, 0)$ and $\mathrm{BC}_A(r_1, 1)$ for some random strings r_0 and r_1 and opens them. These strings combined with the blinding property of BC_A allow Bob to construct $\mathrm{BC}_A(y)$ for all strings $y \in \{0,1\}^*$. Bob chooses a bit c and two random strings w_0, w_1 of length t. Then he computes $y_0 = \mathrm{BC}(w_0)*\mathrm{BC}_A(x_0)$ and $y_1 = \mathrm{BC}_A(w_1)*\mathrm{BC}_A(x_1)$ and sends $(bc, d_0, d_1) = (\mathrm{BC}_B(c), y_c, y_{1-c})$ to Alice. The set of all strings of this form is a self-reducible and P-samplable language. Thus Bob can prove that this message has the correct form to Alice by means of a 4-move zero-konwledge proof of the possession of information (see Theorem 4). After that, Alice computes $\mathrm{BC}_B(b) * \mathrm{BC}_B(c)$ and Bob has to open this commitment. Alice gets the value b' and opens $d_{b'}$ for Bob. Since y_i is the product of the bit commitment of x_i and the bit commitment of some random string w_i it is the bit commitment of $w_i \oplus x_i$ which looks like a random string for Alice. Thus Alice gets no information about c. According to the security properties assumed for the zero-knowledge proof and bit commitment schemes we have $b' = b \oplus c$ with high probability. If $b = c$ then $b' = 0$, $d_{b'} = y_c$ and Bob gets $w_b \oplus x_b$. If $b \neq c$ then $b' = 1$, $d_{b'} = y_{1-c}$ and Bob gets $w_b \oplus x_b$. Bob learns only one secret, namely x_b.

Theorem 5. *If there exists a secure homomorphic bit commitment scheme and a bit commitment scheme with the blinding property then there is a 10-move CANDOS protocol.*

4 General Protocol Structure

In [16] a general protocol for asymmetric fingerprinting is constructed based on some memory-less symmetric fingerprinting scheme. We propose a method which can be based on every symmetric fingerprinting scheme. Our protocol consists of two phases. Let $0 \leq \delta \leq 1$. In the first phase, the codeword phase, merchant and buyer agree on some marking pattern in a way that even if up to c dishonest buyers collude, the merchant will still be able to identify a member of the coalition with probability $1 - \delta$. On the other hand the merchant may not know all bits of the pattern since we have to prevent him from generating a version of the data which could be derived from the version this buyer got. After the codeword phase the buyer is committed to a marking pattern. In the information-retrieval phase the buyer gets unmarked or marked blocks according to this marking pattern. There are several possibilities to make sure that the buyer indeed uses the marking pattern he is committed to. One can use zero-knowledge protocols to achieve this or use a CANDOS protocol where the buyer is committed to appropriate bits in the codeword phase. For black-and-white pictures one may use the idea presented in [16] in construction 2: the buyer has to use a homomorphic bit commitment scheme. The merchant commits to all bits of the picture and multiplies the bit commitments of the marking pattern with the bit commitments of the places of the picture which may be marked.

5 Codeword Phase

5.1 General Protocol

On the basis of an arbitrary symmetric fingerprinting scheme we construct a c-secure asymmetric fingerprinting scheme for at most N buyers, such that the probability to find a member of a cheating coalition is at least $1 - \delta$ and the probability that an innocent buyer is found guilty is at most δ.

The idea is that using a secret 2-party protocol (see for example [1] and [7]) buyer and merchant compute a random marking pattern consisting of $2t$ codewords of the symmetric fingerprinting scheme in such a way that the merchant knows only half of these codewords. They are sufficient for identifying a dishonest buyer if the merchant finds an illegal copy of the data and are called ***identification codewords***. In a trial the accused buyer has to show the information which determines the t codewords (called ***evidence codewords***) unknown to the merchant. If they also fit the marking pattern of the illegal copy the probability that the buyer is innocent is at most δ. In particular, the chance for a cheating merchant to guess the secret parts of the marking pattern and to cast suspicion on an innocent buyer is at most δ.

Our construction principle is as follows: we take a d-secure fingerprinting scheme Γ with ε-error and n codewords. If one randomly chooses $2t$ codewords of Γ and concatenates them one gets a c-secure fingerprinting scheme with δ-error for a larger set of buyers according to Theorem 2. The tracing algorithm of the construction in Theorem 2 is modified for our construction: at least $2t/c$ codewords have to match with the codewords in the personal marking pattern of a suspected buyer. Moreover, at least $t/2c$ codewords have to match with codewords in the personal marking pattern which are unknown to the merchant. The parameters d, n and ε have to be chosen appropriately. Furthermore, t has to be large enough that the probability for the merchant to guess $t/2c$ of the unknown codewords is low.

Let Sign_M (resp. Sign_B) be the combination of a secure hash algorithm and a signature algorithm of the merchant (resp. the buyer). Let BC_B be a bit commitment scheme of the buyer and Enc_M be a secure encryption algorithm of the merchant.

Initialization: Let $c \geq 2$, $t \geq 2(c-1)\log(\frac{8N}{\delta})$. The merchant secretly chooses a d-secure symmetric fingerprinting scheme Γ with ε-error and n codewords with $n \geq 40c$, $\varepsilon \leq \frac{\delta}{8t}$ and $d \geq c$. He randomly chooses a permutation π of the set of $\{1, \ldots, 2t\}$ and $2t$ random bit strings $R_1, \ldots, R_{2t}$. This choice is fixed for all buyers of the data and the merchant encrypts them and publicly commits to these encryptions.

Fingerprinting protocol:

1. The merchant randomly chooses t bit strings $r_{t+1}, \ldots, r_{2t}$ and forms t identification codewords $w_i = \Gamma(R_i, u, r_i)$ for $t+1 \leq i \leq 2t$.
2. The buyer forms some text *con* containing the details of the sale contract and secretly chooses random bit strings $r_1^u, \ldots r_t^u$. The merchant secretly chooses random bit strings $r_1^m, \ldots r_t^m$. Let $r_i = r_i^u \oplus r_i^m$ for $1 \leq i \leq t$, $r = r_1 \circ \ldots \circ r_t$, $r^u = r_1^u \circ \ldots \circ r_t^u$ and $r^m = r_1^m \circ \ldots \circ r_t^m$. Then the buyer and the merchant compute by means of a secure 2-party computation $\mathrm{Enc}_M(\mathrm{BC}_B(\Gamma(R_1, u, r_1)), \ldots, \mathrm{BC}_B(\Gamma(R_t, u, r_t)))$, *con*, $\mathrm{Sign}_B(\mathrm{BC}_B(r^u) \circ con)$ and $\mathrm{Sign}_M(\mathrm{Enc}_M(r^m) \circ con)$. Thus neither the merchant nor the buyer learn these evidence codewords, and only the merchant knows $f_i = \mathrm{BC}_B(\Gamma(R_i, u, r_i))$ for $1 \leq i \leq t$. For some concrete c-secure codes one may find more efficient ways to achieve that. We will explain such protocols in the next sections.
3. The merchant encrypts his identification codewords according to the buyers bit commitment scheme $f_{t+1}, \ldots, f_{2t} = \mathrm{BC}_B(w_1), \ldots, \mathrm{BC}_B(w_t)$ and permutes the bit commitments of the identification and the evidence codewords according to the fixed permutation π. The result is the bit commitment $\overline{f} = f_{\pi(1)}, \ldots, f_{\pi(2t)}$ of the marking pattern $\overline{m} = m_1 \ldots m_{2t}$ for this individual buyer. The merchant sends $\overline{f}$ and $\mathrm{Sign}_M(\overline{f} \circ con)$ to the buyer.
4. The buyer checks the signature, opens the blobs, gets his personal marking pattern and sends $S = \mathrm{Sign}_B(\overline{f} \circ con)$ as receipt to the merchant who checks this signature.

Identification algorithm: Having found an illegal copy of his data the merchant investigates its marking pattern $\tilde{p}$. Since the fingerprinting scheme is c-secure the part of the marking pattern which belongs to the identification codewords will enable him to trace members of a coalition.

Dispute protocol:

1. The merchant has to go to an arbiter and prove which persons are suspicious by showing the identification codewords and the result of the tracing algorithm to the arbiter.

2. Since the part corresponding to the identification codewords could be created by a dishonest merchant himself, the arbiter asks the merchant for S as defined in step 4. The suspicious person has to show his random string r^u and the marking pattern $\overline{m}$ by opening $BC_B(r^u)$ and $\overline{f}$. The arbiter checks the signatures and the correctness of the bit commitments. The merchant has to show to the arbiter how he has built the marking pattern out of correctly formed identification codewords and the committed evidence codewords according to the permutation π he publicly committed to. If the buyer is not able to pass these tests by the arbiter he is found guilty. If there are some inconsistencies in the information the merchant has to present, then obviously the merchant has tried to cheat.
3. Let $\overline{p}$ be the result of the tracing algorithm on input $\tilde{p}$. Then the arbiter compares $\overline{p}$ with $\overline{m}$. If he finds at least $2t/c$ matches and at least $t/2c$ of them correspond to evidence codewords this buyer is found guilty.

Theorem 6. *Let $c \geq 2$, $N \geq 1$, $\delta < 1$, $\varepsilon \leq \delta/8t$, $n \geq 40c$, $d \geq c$ and $t \geq 2(c-1)\log(\frac{8N}{\delta})$. Given a d-secure symmetric fingerprinting scheme with ε-error, n codewords and codeword length ℓ the above scheme in combination with a protocol for the information-retrieval phase is a c-secure asymmetric fingerprinting scheme. The probability that a dishonest coalition is not traced and the probability that some innocent person is found guilty are at most δ. The length of the marking patterns is $2tl$.*

Proof. Sketch:
At first notice that step 4 is necessary. Without step 4 a guilty buyer could claim that the merchant stopped the protocol after step 3 and that the merchant now created some marked version of the data with a marking pattern fitting to the identification codewords in order to cast suspicion on the buyer. If the buyer never got a marking pattern he cannot prove that the marking pattern of the data cannot be traced to his personal marking pattern. Thus the arbiter would have to find the buyer not guilty.

Now we estimate the probability that a dishonest coalition is not traced. According to Theorem 2 and the parameters chosen there is with probability $1 - \delta/2$ at least one dishonest buyer of a coalition of at most c persons such that the following is true: $\overline{p}$ matches $2t/c \leq s \leq 2t$ evidence or identification codewords in the personal marking pattern $\overline{m}$ of this buyer. Notice that the buyers cannot distinguish between identification and evidence codewords. We have to estimate the probability that $i < t/2c$ of these s matches are matches in the evidence codeword part. This probability is 0 for $s \geq t + t/2c$ and is $\frac{\binom{t}{s-i}\binom{t}{i}}{\binom{2t}{s}}$ for $2t/c \leq s < t + t/2c$. In the following let $s < t + t/2c$. Consider a random experiment in which a box is filled by t black and t red balls. The probability to get $s - i$ black balls and i red balls is less than $(1/2)^{s-2i} \leq (1/2)^{s-\frac{t}{c}+2} \leq (1/2)^{\frac{2t}{c}-\frac{t}{c}+2} = (1/2)^{\frac{t}{c}+2} \leq \delta/4$ for $t \geq 2(c-1)\log(8N/\delta) \geq 2(c-1)\log(8/\delta)$. To see this, consider a pair of scales with the black balls and the red balls on different sides. Consider for example the situation that the red side is up. This corresponds to the situation that there are more black balls in the box than red balls. The probability to get a ball of color C from the box is higher than 1/2 if and only if the scale of that color is down (not even) in our scales model. At the end of our random experiment in which we get $s-i$ black balls and i red balls, the black scale is up and one has to put $s-2i$ black balls on it to even the scales. Thus at least $s-2i$ times a ball was taken from the upper scale.

The probability that there is no match with an identification codeword is bounded by $\sum_{s=2t/c}^{t}\left(\frac{1}{2}\right)^{s} \leq \left(\frac{1}{2}\right)^{2t/c-1} < \frac{\delta}{4}$.

Thus the probability to trace and prove the guilt of at least one guilty buyer is at least $1-\delta/2-\delta/4-\delta/4=1-\delta$.

On the other hand we have to estimate the probability that an innocent person is found guilty. This may happen either by the choice of the marking pattern $\tilde{p}$ formed by a coalition of dishonest buyers or by an active attempt of a dishonest merchant to cast suspicion on a buyer. The probability of the first case is bounded by $\delta/2$ according to Theorem 2. In the second case the merchant has to guess at least $t/2c$ evidence codewords. These are random if the buyer has chosen its bit strings r^u randomly. Thus the probability for a codeword to be a fixed evidence codeword chosen by the buyer is $1/n$ for the merchant. Thus we can consider this as a random experiment with t samples and 0-1 random variables $X_1,\ldots,X_t$ which have the value 1 with probability $1/n$. By means of the Chernoff bound (see [12]) we estimate the probability p for the merchant to guess at least $t/2c$ evidence codewords and cast suspicion on the buyer. For minimal n the value of p is maximal. Since $n \geq 40c$ we get

$$p = Pr\left\{\sum_{i=1}^{t} X_i \geq \frac{t}{4c}\right\} \leq Pr\left\{\left|\sum_{i=1}^{t} \frac{X_i}{t} - \frac{1}{40c}\right| \geq \frac{9}{40c}\right\}$$

$$\leq 2e^{-\frac{(9/40c)^2}{2(1/40c)(1-(1/40c))}t} \leq 2e^{-\frac{81(c-1)}{40c-1}\log\left(\frac{8N}{\delta}\right)} \leq \frac{\delta}{2}.$$

Thus the probability for an innocent buyer to be found guilty is at most δ. □

Overhead of the asymmetric scheme: Theorem 6 shows that the length of the marking patterns used in an asymmetric fingerprinting scheme is about the same size it is in a symmetric fingerprinting scheme. Compare Theorem 6 and Theorem 2:

Theorem 7. *Let c, N, δ, d be as above, $n \geq 40c$ and $\varepsilon \leq \delta/\lceil 16(c-1)\log(8N/\delta)\rceil$. If there is a d-secure symmetric fingerprinting scheme Γ with ε-error and n codewords then there are a c-secure symmetric and a c-secure asymmetric fingerprinting scheme with δ-error and N codewords which use $L_{sym} = \lceil 4(c-1)\log(4N/\delta)\rceil$ resp. $L_{as} = \lceil 4(c-1)\log(8N/\delta)\rceil$ low-level codewords from Γ. Thus the overhead of the asymmetric scheme consists of $L_{as}-L_{sym} = 4(c-1)$ low-level codewords.*

The secure multi-party protocol for the computation of the evidence codewords may lead to a considerable amount of interaction. In the following section we will present a more efficient protocol based on permutation codes as symmetric fingerprinting schemes.

5.2 Asymmetric Fingerprinting Protocols based on c-secure Permutation Codes

We use the parameters of the previous section. Let Γ be a c-secure permutation code with ε-error. Then we prove by the following construction and Theorem 4:

Theorem 8. *If secure bit commitment schemes with blinding property and secure signature schemes exist, then there exist a 6-move codeword phase protocol.*

Initialization: The merchant randomly chooses $2t$ permutations $\pi_i \in S_\ell$, $1 \leq i \leq 2t$, a permutation $\pi \in S_t$, encrypts them and commits publicly to these encryptions.

Fingerprinting Protocol:

1. The buyer uses a bit commitment scheme with blinding property and sends the necessary public information signed to the merchant. In order to compute the evidence codewords the buyer creates t sets of the n different Γ codewords. The order of the codewords is permuted in each set and the codewords are encrypted by the bit commitment scheme $C_1 = \{\mathrm{BC}_B(c_{\pi_1(1)}), \ldots, \mathrm{BC}_B(c_{\pi_1(n)})\}$, ... , $C_t = \{\mathrm{BC}_B(c_{\pi_t(1)}), \ldots, \mathrm{BC}_B(c_{\pi_t(n)})\}$. The buyer sends them concatenated with *con* and signed to the merchant and proves by the following zero-knowledge technique that they have the right form: For each set some copies are made by permuting the set again and blinding the bit commitments. If the merchant asks 0 the permutation and blinding factors are shown otherwise the bit commitments are opened and the merchant can check whether the set is correct. This can be done for all copy sets in parallel and in a constant number of rounds (two rounds according to Theorem 4).
2. The merchant then chooses a committed codeword of each set C_i randomly, permutes its bit places according to the appropriate permutation and blinds the commitments. The result is the sequence of evidence codewords encrypted by the buyer's bit commitment scheme. Then the merchant randomly chooses t identification codewords and "encrypts" them by computing the bit commitments of them. He mixes the encrypted identification and evidence codewords according to the permutation π. The sequence of encrypted codewords concatenated with *con* is signed and sent to the buyer.
3. Since the blobs made by the buyer are blinded he cannot find out which blobs and codewords belong to evidence codewords and which belong to identification codewords. He opens all of them and uses this binary string as the marking pattern. Then he sends a receipt S similarly to the string S as in step 4 of the fingerprinting protocol in Sect. 5.1.

Notice that the evidence codewords are randomly chosen if either the merchant or the buyer correctly follows the protocol.

Moreover, notice that the merchant does not have to prove to the buyer that he has built the blobs correctly since the only problem which may arise is that the buyer has to prove to an arbiter that he is innocent. Suppose the merchant accuses the buyer. Then the merchant has to prove that he correctly followed the above protocol.

Generalization We would like to thank an anonymous referee for drawing our attention to the following generalization: It may be valuable to have different numbers of identification and evidence codewords. For example, if the probability of not tracing a dishonest coalition by the merchant may be much smaller than the probability of finding an innocent buyer guilty, this approach helps to get more efficient protocols which fit these requirements. Results analogous to Theorem 6 and Theorem 8 can be proven by using the same techniques as in the proofs of these theorems. For sake of simplicity we confined ourselves here to the case that the numbers of identification and evidence codewords are equal.

6 Information-Retrieval Phase

As stated earlier we suppose that the data can be partitioned into $2t$ blocks. For each unmarked block $B_{0,i}$ there is also a marked block $B_{1,i}$. Now the buyer is committed to a marking pattern $m_1 \ldots m_{2t}$ and he gets the unmarked ith block or the marked ith block according to the value of the ith bit of his marking pattern by means of a CANDOS protocol played in parallel for all $i = 1, \ldots, 2t$.

Instead of disclosing blocks of the data to the buyer, in the case of broadcast data the following method may be more efficient. The merchant chooses some publicly known secret key encryption scheme Enc and $4t$ secret keys $p_{0,1}, \ldots, p_{0,2t}$ and $p_{1,1}, \ldots, p_{1,2t}$ and encrypts the unmarked blocks with the $p_{0,i}$ keys and the marked block with the $p_{1,i}$ keys

$$\{\mathrm{Enc}_{p_{0,1}}(B_{0,1}), \mathrm{Enc}_{p_{1,1}}(B_{1,1}), \ldots, \mathrm{Enc}_{p_{0,2t}}(B_{0,2t}), \mathrm{Enc}_{p_{1,2t}}(B_{1,2t})\}$$

This sequence is stored, for example, on some CD-ROM. Then each buyer learns the keys corresponding to his codeword by means of a CANDOS protocol, that is $p_{m_1,1}, p_{m_2,2}, \ldots p_{m_{2t},2t}$. The technique to encode the data and marked data blocks and to distribute the appropriate key is well-known and can be found in [3].

By the above protocols, the construction of a CANDOS protocol in Sect. 3.3 and Theorem 4 we get protocols for the information-retrieval phase with a small number of interactions:

Theorem 9. *If secure homomorphic bit commitment schemes and bit commitment schemes with the blinding property exist then there are 10-move protocols for the information-retrieval phase.*

Acknowledgements

The authors would like to thank Dan Boneh and Birgit Pfitzmann for their help to understand the theory of c-secure codes and asymmetric fingerprinting systems. Susanne Wetzel, Christoph Thiel and Michael Jacobson supported our work by motivating discussions and proof-reading. We a very grateful to two anonymous referees for giving helpful hints.

References

1. D. Beaver, *Secure Multiparty Computation Protocols and Zero-Knowledge Proof Systems Tolerating a Faulty Minority*, Journal of Cryptology, 1991, pp. 75–122.
2. G.R. Blakeley, C. Meadows, G.B. Purdy, *Fingerprinting Long Forgiving Messages*, Proc. CRYPTO'85, Springer, 1986, pp. 180–189.
3. Dan Boneh, James Shaw, *Collusion-Secure Fingerprinting for Digital Data*, Proc. CRYPTO'95, Springer, 1995, pp. 452–465.
4. Dan Boneh, James Shaw, *Collusion-Secure Fingerprinting for Digital Data*, Princeton Computer Science Technical Report TR-468-94, 1994.
5. Gilles Brassard, Claude Crepeau, Jean-Marc Robert, *All-or-Nothing Disclosure of Secrets*, Proc. CRYPTO'86, Springer, 1987.
6. Germano Caronni, *Assuring Ownership Rights for Digital Images*, Proc. Verläßliche Informationssysteme'95, Wiesbaden, 1995, pp. 251–263.
7. D. Chaum, I. Damgard, J. van de Graaf, *Multiparty Computations Ensuring Privacy of each Party's Input and Correctness of the Result*, Proc. CRYPTO'87, Springer, 1988, pp. 88–119.
8. Benny Chor, Amos Fiat, Moni Naor, *Tracing Traitors*, Proc. CRYPTO'94, Springer, 1994, pp. 257–270.
9. C. Dwork, J. Lotspiech, M. Naor, *Digital Signets: Self-Enforcing Protection of Digital Information*, Proc. 28th STOC, 1996.
10. Uriel Feige, Adi Shamir, *Zero-Knowledge Proofs of Knowledge in Two Rounds*, Proc. of CRYPTO'89, Springer, 1990.
11. O. Goldreich,*Towards a Theory of Software Protection and Simulation by Oblivious RAMs*, Proc. 19th STOC, 1987, pp. 218–229.
12. O. Goldreich, *Foundations of Cryptography*, manuscript, available at http://www.wisdom.weizmann.ac.il/people/homepages/oded/frag.html.
13. S. Goldwasser, S. Micali, C. Rackoff, *The Knowledge Complexity of Interactive Proof Systems*, SIAM Journal on Computing, Vol. 18, No. 1, 1989, pp. 186–208.
14. Rafail Ostovsky, *An efficient Software Protection Scheme*, Proc. CRYPTO'89, Springer, 1990, pp. 610–611.
15. Birgit Pfitzmann, *Trials of Traced Traitors (Extended Abstract)*, Proc. Workshop on Information Hiding, Cambridge, UK, 1996, Proc. to appear in LNCS, Springer.
16. Birgit Pfitzmann, Mathias Schunter, *Asymmetric Fingerprinting*, Proceedings EUROCRYPT'96, Springer, 1996.
17. Birgit Pfitzmann, Michael Waidner, *Asymmetric Fingerprinting for Larger Collusions*, Proc. of the 4th ACM Conference on Computer and Communications Security, 1997, to appear; preliminary Version IBM Research Report RZ 2857 (#90805) 08/19/96, IBM Research Division, Zürich.
18. M. Tompa, H. Woll, *Random Self-Reducibility and Zero-Knowledge Interactive Proofs of Possession of Information*, Proc. 28th FOCS, 1987, pp. 472–482.

Minimal Transition Systems for History-Preserving Bisimulation*

Ugo Montanari and Marco Pistore

Computer Science Department, University of Pisa
Corso Italia 40, 56100 Pisa, Italy
{ugo,pistore}@di.unipi.it

Abstract. In this paper we propose a new approach to check history-preserving equivalence for Petri nets. Exploiting this approach, history-preserving bisimulation is proved decidable for the class of finite nets which are n-safe for some n (the approaches of [17] and of [8] work just for 1-safe nets). Moreover, since we map nets on ordinary transition systems, standard results and algorithms can be re-used, yielding for instance the possibility of deriving minimal realizations. The proposed approach can be applied also to other concurrent formalisms based on partial order semantics, like CCS with causality [4].

1 Introduction

Bisimulation is widely used to equip concurrent systems with an abstract semantics. A well-established theory and efficient algorithms have been developed for it. Automatic checking is successful in practice, since many interesting systems are finite state. One of the most used algorithms is the so-called partition refinement algorithm [10, 14]. It is particularly interesting since it allows for minimization, i.e., it can be used to find the minimal transition system in a class of bisimilar transition systems. Minimization is important both from a theoretical point of view — equivalent systems give rise to the same (up to isomorphism) minimal realization — and from a practical point of view — smaller state spaces can be obtained.

However, the standard definition of bisimulation — and most of the results and algorithms which have been developed for it — can be applied only to systems whose operational behavior is modeled by labeled transition systems. In this case computations are simply sequences of atomic actions and hence parallelism of actions is reduced to interleaving.

Many attempts have been made to overcome the limits of this interleaving approach and to allow the observer to discriminate systems via bisimulation also according to the degree of parallelism they exploit in their computations. Many

* Research supported in part by CNR Integrated Project *Metodi e Strumenti per la Progettazione e la Verifica di Sistemi Eterogenei Connessi mediante Reti di Comunicazione* and Esprit Working Group *CONFER2*.

Reischuk, Morvan (Eds.): STACS'97 Proceedings, LNCS 1200

of these approaches, however, do not allow for a full reusage of the existing theories and algorithms for standard bisimulation.

Consider for instance Petri nets. Processes have been defined in [7] to represent concurrent runs of nets. From a process, it is possible to derive a partial order of the events of the run, which represents the dependencies between them. A notion of bisimulation, called ***history-preserving bisimulation***, which takes into account the partial order behavior has been defined in [15] for event structures. The same notion has been introduced in [6] using mixed ordering observations. History-preserving bisimulation has been applied to Petri nets in [1]: for two nets to simulate, it is required not only that they can perform the same sequence of actions (with the same branching structure), as in ordinary bisimulation, but also that the partial orders corresponding to two matching computations are isomorphic, i.e., that the causal dependencies between the actions are the same in the two nets.

Notice however that the definition of bisimulation which is applied in this case is not the standard one on labeled transition systems. First of all, a bisimulation must deal with isomorphisms between partial orders and hence it cannot be simply a relation on states (which are processes in this case), but it should also refer to the isomorphism between the corresponding partial orders. Moreover, since processes and partial orders grow during a computation, it is possible to associate finite-state systems only to nets which cannot perform infinite sequences of actions. The above differences imply that theory and tools developed for standard bisimulation cannot be reused and, in particular, minimal models cannot be obtained.

Some alternative approaches have been proposed so that history-preserving bisimulation can be checked also for classes of nets with infinite behaviors, namely [17, 8]. However also in these approaches special definitions of bisimulation are used, so that reusage of standard tools and minimization are still not possible.

Notice that these problems are not limited to Petri nets, but are common to partial order semantics, i.e., to those semantics in which partial orders are used to remember the dependencies between the actions of a computation.

In this paper we try to address these problems. We first define ***causal automata*** as a general model for dealing with partial order semantics. In this model the dependencies between the actions are represented by means of names: each transition generates a new name which is then referenced in the labels of the transitions which depend causally from it. We then equip causal automata with a notion of bisimulation.

To show that causal automata are a good model for partial order semantics, we give a translation of Petri nets into causal automata, so that the obtained automata are bisimilar if and only if the nets are history-preserving bisimilar. Moreover the obtained automata are finite for a wide class of nets, more precisely for all the finite P/T nets which are n-safe for some n. It is important to remark that causal automata are a general model for dealing with partial order semantics also for other formalisms. For instance, CCS with causality [5, 4], can be mapped on causal automata.

Finally we show how, starting from causal automata, it is possible to build ordinary transition systems and to reuse ordinary bisimulation on them to decide bisimulation of causal automata. To obtain this, a notion of *active names* is exploited, where a name is active for a state if it appears in the label of a transition reachable from the state. Non-active names can be discarded, thus allowing for a static correspondence of names between bisimilar states. The idea of active names is not new: it has been used also in other contexts to obtain standard transition systems [9, 11, 13].

This translation into ordinary transition systems allows for the reusing of standard techniques and tools. In particular, it is possible to associate to each Petri net a transition system which is minimal w.r.t. those associated to history-preserving bisimilar nets. As far as we know, this is the first approach which leads to minimal realizations for Petri nets up to history-preserving bisimulation.

2 Background

In this section we present the basic definitions on Petri nets we use in the paper. Most of the definitions and of the notations are from [7].

Definition 1 (net). A *net* N is a tuple (S, T, F) where:

- S is a set of *places* and T is a set of *transitions*; we assume $S \cap T = \emptyset$;
- $F \subseteq (S \times T) \cup (T \times S)$ is the *flow relation.*

If $x \in S \cup T$ then ${}^\bullet x = \{y \mid (y, x) \in F\}$ and $x^\bullet = \{y \mid (x, y) \in F\}$ are called respectively the *pre-set* and the *post-set* of x. Let ${}^\circ N = \{x \in S \cup T \mid {}^\bullet x = \emptyset\}$ and $N^\circ = \{x \in S \cup T \mid x^\bullet = \emptyset\}$.
A net N is *finite* if S and T are finite sets.

Given a net $N = (S, T, F)$, we often write S_N, T_N, F_N for S, T, F. We will apply a similar convention also to the other structures we are going to define.

Definition 2 (P/T net). A *(labeled, marked) place/transition net* (or simply *P/T net*) N is a tuple (S, T, F, W, l, m_0), where (S, T, F) is a net and:

- $W : F \to \mathbf{N}^+$ assigns a positive *weight* to each arc of the net; we sometimes assume that W is defined on $(S \times T) \cup (T \times S)$ by requiring $W(x, y) = 0$ if $(x, y) \notin F$;
- $l : T \to \mathrm{Act}$ is the *labeling function*, where Act is a fixed set of action labels;
- $m_0 : S \to \mathbf{N}$ is the *initial marking.*

A *marking* is a mapping $m : S \to \mathbf{N}$. It represents a distribution of the *tokens* in the places of the net.
Transition $t \in T$ is *enabled* at marking m if $m(s) \geq W(s, t)$ for all $s \in {}^\bullet t$. In this case, the *firing* of t at m produces the marking m' with $m'(s) = m(s) + W(t, s) - W(s, t)$, and we write $m \xrightarrow{t} m'$.

Definition 3 (n-safe nets). A P/T net N is n-safe if for each reachable marking m (i.e., for each m such that $m_{0N} \to \cdots \to m$) we have:

$$m(s) \leq n \qquad \forall s \in S_N.$$

Definition 4 (occurrence net). An occurrence net is a net $K = (C, E, G)$ (in this case, states are also called *conditions* and transitions are also called *events*) such that:

- for all $c \in C$, $|{}^{\bullet}c| \leq 1$ and $|c^{\bullet}| \leq 1$ (conditions are not branching), and
- the transitive closure G^+ of G is irreflexive (the net is acyclic).

Definition 5 (process). A *process* π of a P/T net $N = (S, T, F, W, l, m_0)$ is a tuple (C, E, G, p), where $K = (C, E, G)$ is a finite occurrence net and $p : (C \cup E) \to (S \cup T)$ is such that:

- $p(C) \subseteq S$ and $p(E) \subseteq T$;
- $m_0(s) = |p^{-1}(s) \cap {}^{\circ}K|$ for all $s \in S$;
- $W(s, p(e)) = |\{c \in {}^{\bullet}e \mid p(c) = s\}|$ and $W(p(e), s) = |\{c \in e^{\bullet} \mid p(c) = s\}|$ for all $e \in E$ and all $s \in S$.

We write ${}^{\circ}\pi$ for ${}^{\circ}K$ and π° for K°.
The *initial process* of net N is the[1] process $\pi_0(N)$ with an empty set of events.
Let $\pi = (C, E, G, p)$ and $\pi' = (C', E', G', p')$ be two processes of N. If:

- $E' = E \cup \{\bar{e}\}$ for some $\bar{e} \notin E$; – $C' \supseteq C$; – $p'|_{C \cup E} = p$

then we write $\pi \xrightarrow{\bar{t}} \pi'$, where $\bar{t} = p'(\bar{e})$.

Now we define history-preserving bisimulation. We follow a classical characterization, as it appears in [1] under the name of *fully concurrent bisimulation*.

Definition 6 (event structure). The *(deterministic) event structure* for process $\pi = (C, E, G, p)$ of net N is the tuple $\mathbf{ev}(\pi) = (E, F^+|_E, l_N \circ p|_E)$. An *isomorphism* between two event structures is a bijective function between their events which respects ordering and labels.

Definition 7 (history-preserving bisimulation). A set $\mathcal{R}$ of triples is a *history-preserving bisimulation* for nets N_1 and N_2 if:

- whenever $(\pi_1, f, \pi_2) \in \mathcal{R}$ then π_1 is a process of N_1, π_2 is a process of N_2 and f is an isomorphism between $\mathbf{ev}(\pi_1)$ and $\mathbf{ev}(\pi_2)$;
- $(\pi_0(N_1), \emptyset, \pi_0(N_2)) \in \mathcal{R}$;
- whenever $(\pi_1, f, \pi_2) \in \mathcal{R}$ and $\pi_1 \xrightarrow{t_1} \pi_1'$ then $\pi_2 \xrightarrow{t_2} \pi_2'$ with $(\pi_1', f', \pi_2') \in \mathcal{R}$ and $f'|_{\mathbf{ev}(\pi_1)} = f$;
- whenever $(\pi_1, f, \pi_2) \in \mathcal{R}$ and $\pi_2 \xrightarrow{t_2} \pi_2'$ then $\pi_1 \xrightarrow{t_1} \pi_1'$ with $(\pi_1', f', \pi_2') \in \mathcal{R}$ and $f'|_{\mathbf{ev}(\pi_1)} = f$.

Two nets N_1 and N_2 are *history-preserving bisimilar*, written $N_1 \sim_{hp} N_2$, if there is a history-preserving bisimulation for them.

[1] Notice that the initial process of a net is unique only up to isomorphism of the set of initial conditions.

3 Causal Automata

In this section we define causal automata. They are a model for describing systems whose transitions may refer to previous transitions. Since these references can be used to represent dependencies and, hence, partial orders, it is clear that causal automata are an interesting operational model for partial order semantics. We also equip causal automata with an abstract semantics based on bisimulation.

In the next section we show how Petri nets can be mapped into causal automata so that two automata are equivalent if and only if the corresponding nets are history-preserving bisimilar.

Definition 8 (causal automaton). Let $\mathcal{N}$ be a fixed infinite denumerable set of event names.
A *causal automaton* is a tuple $A = \langle Q, w, \mapsto, q_0 \rangle$ where:

- Q is a set of *states*;
- $w : Q \to \mathcal{P}_{\text{fin}}(\mathcal{N})$ associates to each state a finite set of names;
- $\mapsto$ is a set of *transitions*; each transition has the form $q \overset{a}{\underset{M}{\mapsto}}_\sigma q'$, where:
  - $q, q' \in Q$ are the *source* and *target* states;
  - $a \in \text{Act}$ is the *label*;
  - $M \subseteq w(q)$ are the *dependencies* of the transition;
  - $\sigma : w(q') \hookrightarrow w(q) \cup \{\star\}$ is the injective (inverse) *renaming* for the transition; the special mark $\star \notin \mathcal{N}$ is used to recognize in the target state the name corresponding to the current transition;
- $q_0 \in Q$ is the *initial state*; we require that $w(q_0) = \emptyset$.

A causal automaton is hence an automaton particularly suited for dealing with dependencies between transitions. Each state q is labeled by the set $w(q)$ of names, which correspond to the past events that can still (but not necessarily will) be referenced in the future behaviors. These names have a meaning that is local, private to the state. Hence, the particular choice of event names cannot by itself make a distinction between two states of the causal automaton.

Each transition $\overset{a}{\underset{M}{\mapsto}}_\sigma$ depends on the past transitions identified by M. Due to the local meaning of names, each transition must also specify the correspondence between the names of the source and those of the target. This correspondence is obtained via the renaming σ, which permits also to deduce which names of the source are forgotten in the target; the name (if any) used in the target state to represent the current transition is mapped into the special mark $\star$.

On causal automata a bisimulation cannot simply be a relation on states: also a correspondence between the names of the states has to be specified and the same pairs of states can be in relation via more than one correspondence. This correspondence has to be partial in general, since two equivalent states can have a different number of private names.

Definition 9 (bisimulation on causal automata). A set $\mathcal{R}$ of triples is a *causal bisimulation* for causal automata A and B if:

- whenever $(p, \delta, q) \in \mathcal{R}$ then $p \in Q_A$, $q \in Q_B$ and δ is a partial injective function from $w_A(p)$ to $w_B(q)$;
- $(q_{0A}, \emptyset, q_{0B}) \in \mathcal{R}$;
- whenever $(p, \delta, q) \in \mathcal{R}$ and $p \overset{a}{\underset{M}{\mapsto}}_\sigma p'$ in A then there exist some $q \overset{a}{\underset{\delta(M)}{\mapsto}}_\rho q'$ in B and some δ' such that $(p', \delta', q') \in \mathcal{R}$ and $\delta'(m) = n$ implies $\sigma(m) = \star = \rho(n)$ or $\delta(\sigma(m)) = \rho(n)$;
- whenever $(p, \delta, q) \in \mathcal{R}$ and $q \overset{a}{\underset{M}{\mapsto}}_\sigma q'$ in B then there exist some $p \overset{a}{\underset{\delta^{-1}(M)}{\mapsto}}_\rho p'$ in A and some δ' such that $(p', \delta', q') \in \mathcal{R}$ and $\delta'(m) = n$ implies $\sigma(m) = \star = \rho(n)$ or $\delta(\sigma(m)) = \rho(n)$.

The causal automata A and B are ***bisimilar***, written $A \sim_{ca} B$, if there is some bisimulation for them.

Notice that if p and q correspond via δ in some bisimulation $\mathcal{R}$, then to each transition of p a transition of q must correspond, such that *i*) the two transitions perform the same action, *ii*) they depend on the same past events (via δ), and *iii*) the reached states correspond in $\mathcal{R}$ via some δ' which relates two names of the target states only if they both are the names corresponding to the current transitions or if they are related by δ in the source states.

4 From Petri Nets to Causal Automata

The classical definition of history-preserving bisimulation is based on processes. Not all the informations carried by processes, however, are used in the bisimulation. In fact both in [17] and in [8] processes are replaced by more compact structures — the *ordered markings* and the *pomset traces* respectively. These structures also allow part of the past history of a run to be discarded, so that a finite number of different states is sometimes sufficient to represent the behaviors of nets with infinite computations. However, these structures work just for 1-safe nets.

Now we introduce *configurations*, which are suitable to represent in a compact way the relevant part of the past history for generic P/T nets. We also show how processes can be mapped into configurations.

Definition 10 (configuration). Let $N = (S, T, F, W, l, m_0)$ be a P/T net. A *configuration* for N is a tuple $c = (E, \leq, \rho)$, where:

- E is a set of *events* and $\leq \subseteq E \times E$ is a partial ordering for E;
- $\rho : S \times (E \cup \{\texttt{init}\}) \to \mathbf{N}$.

We require that $\sum_{s \in S} \rho(s, e) > 0$ for each $e \in E$.
The ***initial configuration*** for N is the configuration $c_0(N) = (\emptyset, \rho_0, \emptyset)$, where $\rho_0(s, \texttt{init}) = m_0(s)$ for all $s \in S$.
Let $c = (E, \leq, \rho)$ and $c' = (E', \leq', \rho')$ be two configurations for N and $\bar{t} \in T$ be a transition of N. If[2]:

[2] For simplicity, in this definition we suppose that $\rho(s, e) = 0$ if $e \notin E$ and, similarly, $\rho'(s, e) = 0$ if $e \notin E'$.

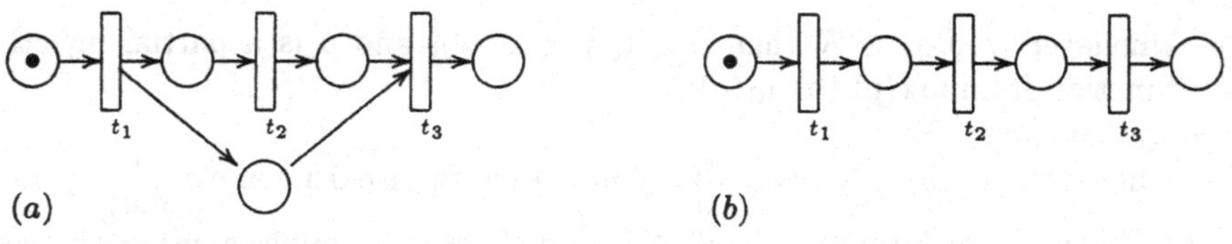

Fig. 1. Two bisimilar nets: transitions t_3 in the two nets have different sets of immediate causes.

- $E' \subseteq E \cup \{\bar{e}\}$ for some $\bar{e} \notin E$;
- $\leq' = (\leq \cup (\mathcal{IC}(c \xrightarrow{\bar{t}} c') \times \{\bar{e}\}))^* \cap (E' \times E')$;
- $\sum_{e \in E}(\rho(s,e) - \rho'(s,e)) = W(s,\bar{t})$ for all $s \in S$;
- $\rho'(s,\bar{e}) = W(\bar{t},s)$ for all $s \in S$;

then we write $c \xrightarrow{\bar{t}} c'$, where the set $\mathcal{IC}(c \xrightarrow{\bar{t}} c')$ of the *immediate causes* of the transition is:

$$\mathcal{IC}(c \xrightarrow{\bar{t}} c') = \{e \in E \mid \exists s \in S : \rho'(s,e) < \rho(s,e)\}.$$

The set $\mathcal{MC}(c \xrightarrow{\bar{t}} c')$ of the *maximal causes* of the transition contains the elements of $\mathcal{IC}(c \xrightarrow{\bar{t}} c')$ which are maximal w.r.t. $\leq$.

In a configuration, the set E of events represents the past events which are still referenced. Since we are interested in a partial order semantics, a partial order is defined on E, which represents the causal dependencies between the past events. Function ρ represents the current marking of the net; instead of simply defining how many tokens are in each place of the net, it also remembers which events generated these tokens (`init` is a special mark used for the tokens in the initial marking).

We require that in a configuration only the events are remembered which generated tokens still present in the net. This is important to obtain a finite number of different configurations also for certain classes of nets with infinite behaviors.

A transition between two configurations corresponds to the firing of a transition of the net. A new event $\bar{e}$ is generated: it directly depends on those events of the source configuration which correspond to the tokens consumed by the transition (these events are called the *immediate causes* of the transition) and the partial order in the target configuration respects these dependencies. The marking of the target configuration is obtained from the marking of the source by discarding tokens according to the pre-set of the transition and by adding new tokens according to the post-set (these tokens are associated to the new event $\bar{e}$). Events with no tokens in the marking are discarded and do not appear in the target configuration.

It is important to remark that corresponding events of history-preserving bisimilar nets can have different sets of immediate causes. In fact, if we consider the net in Figure 1(a), we see that both t_1 and t_2 are immediate causes of t_3, whereas in the net in Figure 1(*b*) t_1 is not a direct cause of t_3. It is possible to show, instead, that two matching events must have the same sets of maximal causes. In fact, notice that t_1 is not a maximal cause of t_3 in both nets of our example.

Definition 11 (from processes to configurations). Let $\pi = (C, E, G, p)$ be a process on net $N = (S, T, F, W, l, m_0)$. The configuration corresponding to π is $c_\pi = (E', \leq, \rho)$ which is defined as follows:

- $E' = \{e \in E \mid \pi^\circ \cap e^\bullet \neq \emptyset\}$ and $\leq = G^*|_{E'}$;
- $\rho(s, e) = |p^{-1}(s) \cap \pi^\circ \cap e^\bullet|$ for all $s \in S$ and $e \in E$; moreover $\rho(s, \mathtt{init}) = |p^{-1}(s) \cap \pi^\circ \cap {}^\circ\pi|$ for all $s \in S$.

Notice that the configuration corresponding to the initial process for N is precisely the initial configuration for N. The following proposition shows that the transitions on configurations exactly match the transitions on processes.

Proposition 12. *Let π be a process for net N and let c_π be the corresponding configuration. If $\pi \xrightarrow{t} \pi'$ then $c_\pi \xrightarrow{t} c_{\pi'}$ and, conversely, if $c_\pi \xrightarrow{t} c'$ then there exists some process π' such that $\pi \xrightarrow{t} \pi'$ and $c' = c_{\pi'}$.*

When a causal automaton is generated from a net, states of the automaton correspond to configurations of the net. However, to obtain a compact automaton, it is important to identify configurations which are isomorphic.

Definition 13 (isomorphic configurations). Let N be a P/T net. Two configurations $c = (E, \leq, \rho)$ and $c' = (E', \leq', \rho')$ for N are *isomorphic* if there exists some bijective function $i : E \to E'$ such that:

- $e \leq e'$ iff $i(e) \leq' i(e')$ for all $e, e' \in E$, and
- $\rho(e, s) = \rho'(i(e), s)$ for all $e \in E$, $s \in S_N$.

We assume to fix a representative for each class of isomorphic configurations and to have a function $\mathtt{norm}$ such that $\mathtt{norm}(c) = \langle c', i \rangle$ where c' is the representative of the class of configurations isomorphic to c and i is the bijection between $E_{c'}$ and E_c.

Now we are ready to show how, given a net, it is possible to build the causal automaton corresponding to it, by exploiting its behavior on configurations.

Definition 14 (from nets to causal automata). Let N be a P/T net. The causal automaton corresponding to N is $\mathtt{aut}(N) = \langle Q, w, \mapsto, c_0 \rangle$, where $c_0 \in Q$ is the initial configuration for N, and whenever $c \in Q$ then $w(c) = E_c$ and

- if $c \xrightarrow{t} c'$ and $\langle c'', \sigma \rangle = \mathtt{norm}(c')$ then $c'' \in Q$ and $c \overset{a}{\underset{M}{\mapsto}}_{[\star/\bar{e}]\circ\sigma} c''$, where:
  - $a = l(t)$,
  - $\{\bar{e}\} = E_{c'} \setminus E_c$ (if $E_{c'} \setminus E_c = \emptyset$ then we can assume $\bar{e} = \star$), and
  - $M = \mathcal{MC}(c \xrightarrow{t} c')$.

Notice that the renaming which corresponds to a transition on the causal automaton is obtained from the bijection defined by function norm: it is sufficient to re-direct the new name $\bar{e}$ to $\star$. Moreover, the maximal causes of the transition are used as dependencies in the automaton.

This construction generates finite causal automata for the finite nets which are n-safe for some n. We emphasize that it is decidable whether a finite P/T net is n-safe for some n. A possible procedure can be found in [16].

Proposition 15. *The causal automaton corresponding to the finite net N is finite iff net N is n-safe for some n.*

The general definition of bisimulation on causal automata exactly matches the classical definition of history-preserving bisimulation on nets.

Theorem 16. *Let N_1 and N_2 be two P/T nets. Then $N_1 \sim_{hp} N_2$ iff* $\mathbf{aut}(N_1) \sim_{ca} \mathbf{aut}(N_2)$.

Sketch of the proof. We have to show how to obtain a history-preserving bisimulation for N_1 and N_2 from a causal bisimulation for $\mathbf{aut}(N_1)$ and $\mathbf{aut}(N_2)$ and, conversely, how to obtain a causal bisimulation from a history-preserving bisimulation. To obtain this, we exploit Proposition 12 and the fact that matching transitions for history-preserving bisimilar nets have the same sets of maximal causes. □

5 From Causal Automata to Ordinary Automata

In the construction of the causal automata, we consider only names of past events which are referenced in the present state. In fact, the remaining names cannot for sure be relevant for the future computation. However it can happen that some of the names associated to a state are never referenced in future computations. These names can be safely discarded from the automaton, obtaining a more compact structure.

Definition 17 (active names). Given a causal automaton A, the sets of *active names* corresponding to the states of A, denoted by $\mathbf{an}(p)$ with $p \in Q_A$, are the smallest sets such that:

- if $p \overset{a}{\underset{M}{\mapsto}}_\sigma p'$ then $M \subseteq \mathbf{an}(p)$;
- if $p \overset{a}{\underset{M}{\mapsto}}_\sigma p'$, $m \in \mathbf{an}(p')$ and $\sigma(m) \neq \star$ then $\sigma(m) \in \mathbf{an}(p)$.

Definition 18 (irredundant automata). Let $A = \langle Q, w, \mapsto, q_0 \rangle$ be a causal automaton. The *irredundant reduction* of A is the causal automaton $\Downarrow\! A = \langle Q, \mathbf{an}, \mapsto', q_0 \rangle$ where $\mapsto'$ is obtained from $\mapsto$ by restricting the renamings to the active names of the target states.
We say that an automaton A is *irredundant* if $\Downarrow\! A = A$.

Proposition 19. *Let A be a causal automaton. Then $\Downarrow A \sim_{ca} A$.*

A causal automaton A can be visited beginning from the initial state. In this visit, the global meaning of the private names of the reached states is made explicit[3]. If the global meaning corresponding to the names of a reached state p is given by $\sigma : w(p) \hookrightarrow \mathcal{N}$ and transition $p \overset{a}{\underset{M}{\mapsto}}_\rho q$ is followed, the global meaning for q is given essentially by $\sigma \circ \rho$. However, a global meaning has to be associated also to the name created in the transition (the name of the target state mapped in $\star$ by the transition renaming). To this purpose we use a function new, which gets a transition $p \overset{a}{\underset{M}{\mapsto}}_\rho p'$ and a global meaning σ for the names of p and returns a new name. A possible definition of new is as follows:

$$\mathtt{new}(p \overset{a}{\underset{M}{\mapsto}}_\rho p', \sigma) = \min\{\mathcal{N} \setminus \sigma(\rho(w(p')))\}$$

This means that the first name is chosen, that is not already used in the target state. Other allocation strategies can be adopted by changing function new.

To formalize the idea of visiting a causal automaton A, we associate to A a standard labeled transition system (called the *unfolding* of A); each state of the unfolding is a pair ⟨state of the causal automaton, global meaning of its names⟩ and each transition has the form

$$\langle p, \sigma \rangle \overset{a}{\underset{m,M}{\rightarrow}} \langle p', \sigma' \rangle$$

where a is an action, M are the names the action depends from and m is the newly created name.

Definition 20 (unfolding). The *unfolding* corresponding to a causal automaton $A = \langle Q, w, \mapsto, q_0 \rangle$ is the labeled transition system $\mathtt{unf}(A) = \langle Q_u, \rightarrow, q_{0u} \rangle$ defined as follows:

- the initial state is $q_{0u} = \langle q_0, \emptyset \rangle \in Q_u$;
- if $\langle p, \sigma \rangle \in Q_u$ and $p \overset{a}{\underset{M}{\mapsto}}_\rho p'$ then $\langle p', \sigma' \rangle \in Q_u$ and $\langle p, \sigma \rangle \overset{a}{\underset{m,M'}{\rightarrow}} \langle p', \sigma' \rangle$, where $\sigma' = (\sigma \cup (\star, m)) \circ \rho$, $M' = \sigma(M)$ and $m = \mathtt{new}(p \overset{a}{\underset{M}{\mapsto}}_\rho p', \sigma)$.

It is easy to show that there are equivalent causal automata with non-equivalent unfoldings. This happens because two equivalent states of the causal automata can have a different number of names, and in the unfolding this can lead to different choices for the new names.

The following theorem expresses an important result of this paper: given two irredundant causal automata, then they are equivalent if and only if the corresponding unfoldings are equivalent. This allows us to apply a standard partitioning algorithm for checking the equivalence of two automata and to obtain minimal (standard) automata corresponding to them.

[3] A state can be visited more than once, with different meanings for its private names.

Theorem 21. *If A and B are irredundant causal automata then $A \sim_{ca} B$ iff* $\mathtt{unf}(A) \sim \mathtt{unf}(B)$.

Sketch of the proof. Assume p_A and p_B are states of A and of B, and $s_A = \langle p_A, \sigma_A \rangle$ and $s_B = \langle p_B, \sigma_B \rangle$ are states of $\mathtt{unf}(A)$ and of $\mathtt{unf}(B)$. Then s_A and s_B are equivalent states if and only if p_A and p_B are causal-equivalent via the name correspondence $\sigma_B^{-1} \circ \sigma_A$.

This is true only in the case A and B are irredundant, since only in this case we can guarantee that function `new` chooses the same new names for the transitions of s_A and of s_B. □

Corollary 22. *Given two P/T nets N_1 and N_2, $N_1 \sim_{hp} N_2$ iff* $\mathtt{unf}(\Downarrow\mathtt{aut}(N_1)) \sim \mathtt{unf}(\Downarrow\mathtt{aut}(N_2))$.

6 Partitioning Algorithm and Minimization

Corollary 22 suggests an algorithm for checking history-preserving equivalence of two P/T nets:

1. construct (separately) the causal automata corresponding to the nets;
2. discover (separately) the active names of the two automata and get the irredundant reductions: start marking the names that are active due to the first condition of Definition 17 and continue marking all the names reachable following the dependencies in the other conditions of Definition 17; at the end discard the unmarked names;
3. unfold (separately) the obtained irredundant automata;
4. use a standard algorithm for checking the equivalence of the obtained transition systems (for instance, partition refinement [10, 14]).

This algorithm works for all n-safe nets, since for these we are sure that finite causal automata (and hence finite unfoldings) can be built.

Corollary 23. *It is decidable whether two finite n-safe P/T nets are history-preserving bisimilar.*

The proposed approach can be used to obtain the minimal transition system corresponding to a net; to obtain this, the same procedure has to be applied by starting with just a net and, at the end, a minimization algorithm has to be applied.

We conclude this section with some comments on the complexity of the algorithm. Assume that the considered nets have (less than) p places and t transitions and that they are n-safe. Then there will be at most np tokens in all the reachable markings and hence at most np events in the reachable configurations. The partial orders on np elements are $2^{\mathcal{O}(np \cdot \log(np))}$ and hence also the reachable configurations are, up to isomorphism, $2^{\mathcal{O}(np \cdot \log(np))}$. This is also an upper bound for the number of states of the corresponding causal automata; the upper bound for the number of transitions is $2^{\mathcal{O}(np \cdot \log(np) + \log(t))}$.

It is possible that different states of the unfolding correspond to the same state of the causal automaton: to be more precise, to each state of the causal automaton can correspond at most $2^{\mathcal{O}(np \cdot \log(np))}$ states of the unfolding. Nevertheless, the number of states and transitions of the unfolding have the same bounds as those of the causal automaton[4].

Since the first two phases of the proposed algorithm require a time which is polynomial in the number of states and transitions of the causal automaton and the last two phases require a time which is polynomial in the number of states and transitions of the unfolding, we obtain an upper bound on the time to check the history-preserving bisimilarity of two P/T nets which is $2^{\mathcal{O}(np \cdot \log(np) + \log(t))}$. So, even in the case of 1-safe nets ($n = 1$), the time required by the algorithm is exponential in the dimension of the nets; this is not surprising, since in [8] it is proved that deciding history-preserving bisimilarity on 1-safe nets is DEXPTIME-hard in the size of the net[5]. In [8] no exact analysis of the degree of the exponent is reported — it is just proved that the algorithm is DEXPTIME. Nevertheless, also in that case, we guess, the exponent is of the form $x \cdot \log x$ in the size x of the net. Also the space complexity of the two algorithms should be similar. Thus the algorithm in [8] and ours should have the same complexity for 1-safe nets.

7 Concluding Remarks

In this paper we have extended to finite n-safe nets the decidability result for history-preserving bisimulation given in [17, 8] for finite 1-safe nets. Differently from the approaches of [17, 8], we reduce the checking of history-preserving bisimulation on P/T nets to the checking of ordinary bisimulation on ordinary transition systems. This allows us to reuse standard theory and tools and, in particular, minimal transition systems can be associated to the nets.

We have also introduced causal automata as an operational model suited for dealing with partial order semantics. Our claim is that the theory developed in this paper can be applied to other bisimulation semantics based on partial order. A confirmation in this sense comes from the possibility of mapping on causal automata the causal version of CCS [5, 4]; for lack of space we cannot present this encoding here; it can be found in the full version of this paper.

In [13] an analogous approach is applied to CCS with location equivalence [3] and a notion of location automata is introduced. Both location and causal automata are instances of a more general class of history-dependent automata, which are able to model also the ordinary and noninterleaving semantics of the π-calculus [11]. For history-dependent automata a categorical definition of bisimulation via open maps has been defined in [12]. We are convinced that this

[4] The upper bounds for the case graphs are instead $2^{\mathcal{O}(np)}$ for the states and $2^{\mathcal{O}(np + \log(t))}$ for the transitions. Thus the extra information we require in the case graph adds a logarithmic factor to the exponent.

[5] A DEXPTIME algorithm is a deterministic algorithm with a bound of the form $2^{\mathcal{O}(x^k)}$ where x is the size of the instance and k is fixed.

categorical definition coincides with the definition of bisimulation presented in this paper.

Some interesting directions of further research involve the study of causal automata: for instance it would be interesting to define minimization directly on the automata. Also the possibility of mapping back causal automata to Petri nets is of some interest.

References

1. E. Best, R. Devillers, A. Kiehn and L. Pomello. Fully concurrent bisimulation. *Acta Informatica* 28:231–264, 1991.
2. A. Bouali, S. Gnesi and S. Larosa. The integration project for the JACK environment. In *EATCS Bullettin*, 1994.
3. G. Boudol, I. Castellani, M. Hennessy and A. Kiehn. Observing localities. *Theoretical Computer Science* 114:31–61, 1993.
4. Ph. Darondeau and P. Degano. Causal trees. In *Proc. ICALP'89*, LNCS 372. Springer-Verlag, 1989.
5. P. Degano, R. De Nicola and U. Montanari. Observational equivalences for concurrency models. In *Proc. Formal Description of Programming Concepts – III*, 1986. North-Holland, 1987.
6. P. Degano, R. De Nicola and U. Montanari. Partial orderings descriptions and observations of nondeterministic concurrent processes. In *Proc. REX School/Workshop on Linear Time, Branching Time and Partial Orders in Logica and Models for Concurrency*, LNCS 354. Springer Verlag, 1989.
7. U. Goltz and W. Reisig. The non-sequential behaviour of Petri nets. *Information and Control* 57:125–147, 1983.
8. L. Jategaonkar and A. Meyer. Deciding true concurrency equivalences on finite safe nets. *Theoretical Computer Science* 154:107–143, 1996. Extended abstract in *Proc. ICALP'93*, LNCS 700. Springer Verlag, 1993.
9. B. Jonsson and J. Parrow. Deciding bisimulation equivalences for a class of non-finite-state programs. *Information and Computation* 107:272–302, 1993.
10. P. C. Kanellakis and S. A. Smolka. CCS expressions, finite state processes, and three problems of equivalence. *Information and Computation* 86:43–68, 1990.
11. U. Montanari and M. Pistore. Checking bisimilarity for finitary π-calculus. In *Proc. CONCUR'95*, LNCS 962. Springer Verlag, 1995.
12. U. Montanari and M. Pistore. History-dependent automata. Draft. Available as `ftp://ftp.di.unipi.it/pub/Papers/pistore/HDautomata.ps.gz`. Also: Technical Report, Università di Pisa, to appear.
13. U. Montanari, M. Pistore and D. Yankelevich. Efficient minimization up to location equivalence. In *Proc. ESOP'96*, LNCS 1058. Springer Verlag, 1996.
14. R. Paige and R. E. Tarjan. Three partition refinement algorithms. *SIAM Journal on Computing*, 16(6):973–989, 1987.
15. A. Rabinovich and B. A. Trakhtenbrot. Behaviour structures and nets. *Fundamenta Informaticae* 11:357–404, 1988.
16. R. Valk and M. Jantzen. The residue vector sets with applications to decidability problems in Petri nets. In *Proc. APN'84*, LNCS 188. Springer Verlag, 1984.
17. W. Vogler. Deciding history preserving bisimilarity. In *Proc. ICALP'91*, LNCS 510. Springer Verlag, 1991.

On Ergodic Linear Cellular Automata over $\mathbf{Z}_m$

Gianpiero Cattaneo[1], Enrico Formenti[1], Giovanni Manzini[2,3], and Luciano Margara[4]

[1] Dipartimento di Scienze dell'Informazione, Università di Milano, Milano, Italy.
[2] Dipartimento di Scienze e Tecnologie Avanzate, Università di Torino, Via Cavour 84, 15100 Alessandria, Italy.
[3] Istituto di Matematica Computazionale, Via S. Maria, 46, 56126 Pisa, Italy.
[4] Dipartimento Scienze dell'Informazione, Università di Bologna, Piazza Porta S. Donato 5, 40127 Bologna, Italy.

Abstract. We study the ergodic behavior of linear cellular automata over $\mathbf{Z}_m$. The main contribution of this paper is an easy-to-check necessary and sufficient condition for a linear cellular automaton over $\mathbf{Z}_m$ to be ergodic. We prove that, for general cellular automata, ergodicity is equivalent to topological chaos (transitivity and sensitivity to initial conditions). Finally we prove that linear CA over $\mathbf{Z}_p$ with p prime have dense periodic orbits.

1 Introduction

Cellular Automata (CA) are dynamical systems consisting of a regular lattice of variables which can take a finite number of discrete values. The state of the CA, specified by the values of the variables at a given time, evolves in synchronous discrete time steps according to a given *local rule.* CA can display a rich and complex temporal evolution whose exact determination is in general very hard, if not impossible. In particular, many properties of the temporal evolution of general CA are undecidable [4, 5, 12]. Despite their simplicity that allows a detailed algebraic analysis, linear CA over $\mathbf{Z}_m$ exhibit many of the complex features of general CA. Several important properties of linear CA have been studied during the last few years (see for example [2, 7, 8, 11]) and in some cases exact results have been carried out. As an example, in [11] the authors present criteria for surjectivity and injectivity of the global transition map of linear CA. The *qualitative* behavior of CA is a main subject in CA theory. Quoting from [11]: *"Criteria are desired for determining when the sequence of transations of a state-configuration of a cellular automata takes a certain type of dynamical behavior."* In this paper we study the dynamical behavior of linear CA over $\mathbf{Z}_m$ in the framework of ergodic theory. Ergodic theory has been recently applied to CA in a number of works. Some preliminary results can be found in [10, 13, 14, 16]. The main contribution of this paper is the solution to the two following open problems (Theorem 9).

(1) How to decide whether the global transition map of a given linear CA over $\mathbf{Z}_m$ is *ergodic.*

Reischuk, Morvan (Eds.): STACS'97 Proceedings, LNCS 1200

(2) How to find ergodic linear CA over $\mathbf{Z}_m$.

The solution of problem (1) generalizes a result presented in [13] (Corollary 2, page 406), while the solution of problem (2) answers a question raised in [14] (Question 2, page 605).
We also establish a connection between ergodic theory and topological chaos in the case of general CA.

Although a universally accepted definition of chaos does not exist, two properties are widely accepted as important features of chaotic behavior: topological transitivity and sensitivity to initial conditions. Sensitivity is recognized as a central notion in chaos theory because it captures the feature that in chaotic systems small errors in experimental readings lead to large scale divergence, i.e., the system is unpredictable. Transitivity guarantees that the system cannot be decomposed into two or more subsystems which do not interact under iterations of the map.

We prove (Theorem 6) that a CA is ergodic if and only if it is topologically transitive. Since in [3] one of the author proved that topologically transitive CA are sensitive to initial conditions, we conclude that, for CA, ergodicity is equivalent to topological chaos. In Theorem 7 we take advantage of the compactness of the space of the configurations on which CA are defined for proving that topologically transitive CA are surjective.

Another widely accepted feature of chaotic behavior is given by *denseness of periodic orbits* (see for example [6]). Here (Theorem 10) we prove that D-dimensional linear CA over $\mathbf{Z}_p$ with p prime have dense periodic orbits, which is a generalization of a result in [7] (Theorem 4).

The rest of this paper is organized as follows. In Section 2 we give basic definitions and notations. In Section 3 we list our results. Section 4 contains the proofs of the theorems stated in Section 3. Section 5 contains some indications for further works.

2 Basic definitions

Let $\mathbf{Z}$ be the set of integers. Let m and D be positive integers. Let $\mathcal{A} = \{0, 1, \ldots, m-1\}$ be a finite alphabet of cardinality m. Let f, $f\colon \mathcal{A}^s \to \mathcal{A}$, $s \geq 1$, be any map. We say that s is the size of the domain of f, or simply the size of f. A D-dimensional CA based on a *local rule* f of size s is a pair $(\mathcal{A}^{\mathbf{Z}^D}, F)$, where

$$\mathcal{A}^{\mathbf{Z}^D} = \{c \mid c : \mathbf{Z}^D \to \mathcal{A}\}$$

is the *space of configurations* and F, $F : \mathcal{A}^{\mathbf{Z}^D} \to \mathcal{A}^{\mathbf{Z}^D}$, is the *global transition map* defined as follows. For every $c \in \mathcal{A}^{\mathbf{Z}^D}$ and for every $\mathbf{v} \in \mathbf{Z}^D$

$$[F(c)](\mathbf{v}) = f\left(c(\mathbf{v} + \mathbf{nb}(1)), \ldots, c(\mathbf{v} + \mathbf{nb}(s))\right), \tag{1}$$

where $\mathbf{nb} : \{1, \ldots, s\} \to \mathbf{Z}^D$, is the *neighborhood structure map.*

Example 1. We describe a 2-dimensional binary CA whose evolution is governed by a local rule f which computes the sum modulo 2 of its 4 input values selected by the neighborhood map $\mathbf{nb}$. In this example $\mathbf{nb}$ selects the cells to the north, west, east, and south of the cell we are considering. Let $D = 2$, $\mathcal{A} = \{0,1\}$, and $s = 4$. Let $\mathbf{nb}$ be defined by $\mathbf{nb}(1) = \langle 0,1\rangle$, $\mathbf{nb}(2) = \langle -1,0\rangle$, $\mathbf{nb}(3) = \langle 1,0\rangle$, $\mathbf{nb}(4) = \langle 0,-1\rangle$. The local rule f is defined by $f(x_1,x_2,x_3,x_4) = (x_1 + x_2 + x_3 + x_4) \bmod 2$. The global transition map F is defined by

$$[F(c)](i,j) = (c(i,j+1) + c(i-1,j) + c(i+1,j) + c(i,j-1)) \bmod 2.$$

In the case of 1-dimensional CA, we use the following simplified notation. Let $f, f : \mathcal{A}^{2r+1} \to \mathcal{A}$, be any map. A 1-dimensional CA based on the local rule f is a pair $(\mathcal{A}^{\mathbf{Z}}, F)$, where $\mathcal{A}^{\mathbf{Z}}$ is the space of configurations and F, $F : \mathcal{A}^{\mathbf{Z}} \to \mathcal{A}^{\mathbf{Z}}$, is defined by

$$[F(c)](i) = f(c(i-r), \ldots, c(i+r)), \qquad c \in \mathcal{A}^{\mathbf{Z}},\ i \in \mathbf{Z}. \tag{2}$$

We say that r is the ***radius*** of f. Note that, even if f must depend on at least one between x_{-r} and x_r, in general f does not depend on all the $2r+1$ variables $x_{-r}, \ldots, x_r$.

Throughout the paper, $F(c)$ will denote the result of the application of the map F to the configuration c, $c(\mathbf{v})$ will denote the value of the entry with coordinates $\mathbf{v}$ of the configuration c, and $\mathbf{v}_i$ will denote the i-th component of the vector $\mathbf{v}$. We recursively define $F^n(c)$ by $F^n(c) = F(F^{n-1}(c))$, where $F^0(c) = c$.

We now give the definition of linear local rule over $\mathbf{Z}_m$.

Definition 1. Let $\mathcal{A} = \{0, 1, \ldots, m-1\}$. A map f, $f : \mathcal{A}^s \to \mathcal{A}$, is linear over $\mathbf{Z}_m$ if and only if it can be written as

$$f(x_1, \ldots, x_s) = \left(\sum_{i=1}^{s} \lambda_i x_i\right) \bmod m,$$

where $\lambda_i \in \mathbf{Z}$.

From now on, we say that a CA defined over a finite alphabet $\mathcal{A} = \{0, \ldots, m-1\}$ is linear over $\mathbf{Z}_m$ if the local rule on which it is based is linear over $\mathbf{Z}_m$. Note that for a linear D-dimensional CA, equation (1) becomes

$$[F(c)](\mathbf{v}) = \sum_{i=1}^{s} \lambda_i c(\mathbf{v} + \mathbf{nb}(i)) \bmod m. \tag{3}$$

For 1-dimensional linear CA of radius r, the notation can be further simplified by writing

$$f(x_{-r}, \ldots, x_r) = \sum_{i=-r}^{r} a_i x_i \bmod m,$$

so that equation (2) becomes

$$[F(c)](i) = \sum_{j=-r}^{r} a_j c(i+j) \mod m, \qquad c \in \mathcal{A}^{\mathbf{Z}},\ i \in \mathbf{Z}.$$

It is often useful to introduce a distance over the space of the configurations. Let $\Delta: \mathcal{A} \times \mathcal{A} \to \{0,1\}$ be such that

$$\Delta(i,j) = \begin{cases} 0, & \text{if } i = j, \\ 1, & \text{if } i \neq j. \end{cases}$$

Given $a, b \in \mathcal{A}^{\mathbf{Z}^D}$ the Tychonoff distance $d(a,b)$ is defined by

$$d(a,b) = \sum_{\mathbf{v} \in \mathbf{Z}^D} \frac{\Delta(a(\mathbf{v}), b(\mathbf{v}))}{2^{\max(\mathbf{v})}}, \tag{4}$$

where $\max(\mathbf{v})$ is the maximum of the absolute value of the components of $\mathbf{v}$. It is easy to verify that d is a metric on $\mathcal{A}^{\mathbf{Z}^D}$ and that the metric topology induced by d coincides with the product topology induced by the discrete topology of $\mathcal{A}$. With this topology, $\mathcal{A}^{\mathbf{Z}^D}$ is a compact and totally disconnected space and F is a (uniformly) continuous map.

We now recall the definition of three properties, namely topological transitivity, sensitivity to initial conditions, and denseness of periodic orbits which are widely accepted as important features of chaotic behavior for general discrete time dynamical systems. Here, we assume that the space of configurations X is equipped with a distance d and that the map F is continuous on X according to the topology induced by d. For CA the Tychonoff distance satisfies this property.

Definition 2. A dynamical system (X, F) is topologically transitive if and only if for all nonempty open subsets U and V of X there exists a natural number n such that $F^n(U) \cap V \neq \emptyset$.

Intuitively, a topologically transitive map has points which eventually move under iteration from one arbitrarily small neighborhood to any other. As a consequence, the dynamical system cannot be decomposed into two disjoint open sets which are invariant under the map.

Definition 3. A dynamical system (X, F) is sensitive to initial conditions if and only if there exists a $\delta > 0$ such that for any $x \in X$ and for any neighborhood $N(x)$ of x, there is a point $y \in N(x)$ and a natural number n, such that $d(F^n(x), F^n(y)) > \delta$. δ is called the sensitivity constant.

If a map possesses sensitive dependence on initial conditions, then, for all practical purposes, the dynamics of the map defies numerical approximation. Small errors in computation which are introduced by round-off may become magnified upon iteration. The results of numerical computation of an orbit, no matter how accurate, may be completely different from the real orbit.

In [3] it has been proved that, for CA, topological transitivity implies sensitivity. Thus, for CA, the notion of transitivity becomes central to chaos theory.

Definition 4. Let $P(F) = \{x \in X | \exists n \in \mathbf{N} : F^n(x) = x\}$ be the set of the periodic points of F. A dynamical system (X, F) has dense periodic orbits if and only if $P(F)$ is a dense subset of X, i.e., for any $x \in X$ and $\epsilon > 0$, there exists $y \in P(F)$ such that $d(x, y) < \epsilon$.

Denseness of periodic orbits is often referred to as the *element of regularity* a chaotic dynamical system must exhibit. The popular book by Devaney [6] isolates three components as being the essential features of chaos: transitivity, sensitivity to initial conditions and denseness of periodic orbits.
Finally, we recall the definition of ergodic map.

Definition 5. Let $(X, \mathcal{F}, \mu)$ be a probabilistic space. Let F, $F : X \to X$, be a measurable map which preserves μ, i.e., for every subset $E \in \mathcal{F}$ we have $\mu(E) = \mu(F^{-1}(E))$. Then F is ergodic with respect to μ if and only if for every $E \in \mathcal{F}$

$$(E = F^{-1}(E)) \Rightarrow (\mu(E) = 0 \text{ or } \mu(E) = 1)\,.$$

In the following we will adopt the Haar probability measure over $\mathcal{A}^{\mathbf{Z}^D}$ which is defined as the product measure induced by the uniform probability distribution over $\mathcal{A}$.

3 Statement of new results

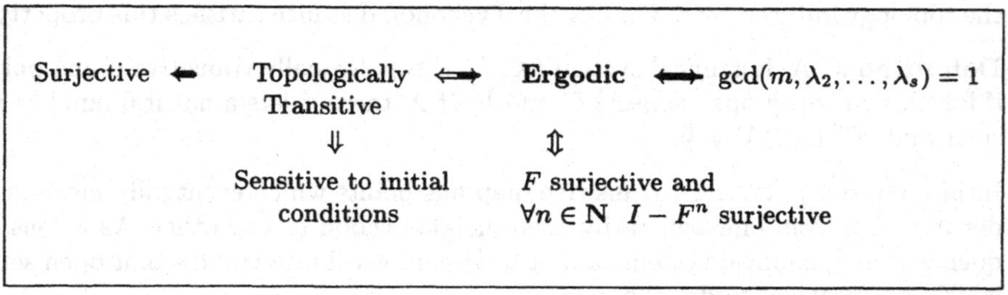

Surjective $\Leftarrow$ Topologically Transitive $\Longleftrightarrow$ **Ergodic** $\Longleftrightarrow$ $\gcd(m, \lambda_2, \ldots, \lambda_s) = 1$

Topologically Transitive $\Downarrow$ Sensitive to initial conditions

Ergodic $\Updownarrow$ F surjective and $\forall n \in \mathbf{N}$ $I - F^n$ surjective

Fig. 1. Implications among properties of global transition maps associated with D-dimensional linear CA over $\mathbf{Z}_m$. Solid arrows represent results proved in this paper

In this section we state the main results of this paper. The same results are summarized in the diagram of Fig. 1.

Our first result shows that transitive CA are ergodic with respect to the normalized Haar measure. Since the converse of this fact is already known (see Theorem 13), we establish that for CA the concepts of transitivity and ergodicity are indeed equivalent.

Theorem 6. *Topologically transitive cellular automata with respect to the metric topology induced by the Tychonoff distance are ergodic with respect to the normalized Haar measure.* □

Our next result establishes that transitive CA (hence also ergodic CA in view of Theorem 13) are surjective.

Theorem 7. *Topologically transitive CA with respect to the metric topology induced by the Tychonoff distance are surjective.* □

The following two theorems show that for linear CA there exists a simple characterization of ergodic maps based on the coefficients of the local rule. Note that Theorem 8 is indeed a special case of Theorem 9. We decided to state explicitly both theorems since the former has a much simpler proof and gives a clearer picture of the role of the coefficients.

Theorem 8. *Let F denote the global transition map of a linear 1-dimensional CA over $\mathbf{Z}_m$ with local rule $f(x_{-r},\ldots,x_r) = \sum_{i=-r}^{r} a_i x_i \bmod m$. The global transition map F is ergodic if and only if*

$$\gcd(m, a_{-r}, \ldots, a_{-1}, a_1, \ldots, a_r) = 1,$$

where gcd *denotes the greatest common divisor.* □

Theorem 9. *Let F denote the global transition map of a linear D-dimensional CA over $\mathbf{Z}_m$ defined by*

$$[F(c)](\mathbf{v}) = \sum_{i=1}^{s} \lambda_i c(\mathbf{v} + \mathbf{nb}(i)).$$

Assume $\mathbf{nb}(1) = \mathbf{0}$, that is, λ_1 is the coefficient associated to the null displacement. The global transition map F is ergodic if and only if

$$\gcd(m, \lambda_2, \lambda_3, \ldots, \lambda_s) = 1.$$

□

Theorem 9 generalizes a result given in [13] where the authors proved that 1-dimensional linear CA over $\mathbf{Z}_2$ based on a local rule different from the identity map are ergodic. In addition, Theorem 9 allows us to answer the following question raised in [14]. Are all surjective 1-dimensional CA defined over $\{0,1\}^{\mathbf{Z}}$ (with the exception of the identity and the inversion map) ergodic? Consider the subclass of linear CA over $\mathbf{Z}_m$. If m is prime, the answer is yes, otherwise one can easily construct a 1-dimensional non trivial linear CA over $\mathbf{Z}_m$ which is not ergodic. In the case of general CA, the answer is No. In fact, let F be the global transition map of the (global) injective binary 1-dimensional CA defined in [1]. Since $F = F^{-1}$ we have that F^2 is the identity map and then F is not ergodic.

Our final result shows that linear CA over an alphabet of prime cardinality have dense periodic orbits.

Theorem 10. *Let F denote the global transition map of a linear D-dimensional CA over $\mathbf{Z}_p$ with p prime. Then the set $P(F)$ of periodic orbits of F is dense in $\mathcal{A}^{\mathbf{Z}^D}$.* □

This result is a generalization of Theorem 4 of [7] where it has been proved that 1-dimensional linear CA over $\mathbf{Z}_2$ have dense periodic points.

4 Proof of the main theorems

We now prove the results stated in Section 3. In our proofs we make use of the following known facts about CA.

Theorem 11. *[11] Let $\mathcal{A} = \{0, 1, \ldots, m-1\}$. Let f, $f : \mathcal{A}^s \to \mathcal{A}$, be a linear map defined by*

$$f(x_1, \ldots, x_s) = \left(\sum_{i=1}^{s} \lambda_i x_i \right) \mod m.$$

The CA based on the local rule f is surjective if and only if $\gcd(m, \lambda_1, \ldots, \lambda_n) = 1$. □

Theorem 12. *[13] Let G be a compact abelian group with normalized Haar measure μ, and let θ be a continuous surjective endomorphism of G. Then, the following two statements are equivalent.*

(i) θ is ergodic.
(ii) For every $n \geq 1$, $I - \theta^n$ is surjective.

□

Theorem 13. *[16] Ergodic CA with respect to the normalized Haar measure are topologically transitive with respect to the metric topology induced by the Tychonoff distance.* □

Proof of Theorem 6. Let V_k, $k \geq 1$, be the set of all the D-dimensional vectors $\mathbf{v}_i$ such that $\|\mathbf{v}_i\|_\infty \leq k$. One can easily verify that the cardinality of V_k is $n_{kD} = (2k+1)^D$. Let $a_1, \ldots, a_{n_{kD}}$ be n_{kD} arbitrarily chosen elements of $\mathcal{A}$. We define a D-dimensional *cylinder* $Cyl(a_1, \ldots, a_{n_{kD}}) \subseteq \mathcal{A}^{\mathbf{Z}^D}$ as follows:

$$Cyl(a_1, \ldots, a_{n_{kD}}) = \{c \colon c(\mathbf{v}_i) = a_i,\ \mathbf{v}_i \in V_k,\ 1 \leq i \leq n_{kD}\}.$$

Assume that there exists a D-dimensional CA $(\mathcal{A}^{\mathbf{Z}^D}, F)$ which is topologically transitive but not ergodic. Then, there exists $E \subset \mathcal{A}^{\mathbf{Z}^D}$ such that

$$0 < \mu(E) < 1 \text{ and } F^{-1}(E) = E.$$

One can easily verify that since $\mu(E) > 0$, for $h \geq 0$ there exists a cylinder $Cyl(a_1, \ldots, a_{n_{hD}})$ entirely contained in E. In an analogous way, since $\mu(E) <$

1, there exists another cylinder $Cyl(b_1, \ldots, b_{n_{jD}})$, $j \geq 0$ entirely contained in $A^{\mathbf{Z}^D} \setminus E$. Since $Cyl(a_1, \ldots, a_{n_{hD}}) \subseteq E$ and $F^{-1}(E) = E$ we have that for every $n \in \mathbf{Z}$

$$F^n(Cyl(a_1, \ldots, a_{n_{hD}})) \subseteq F^n(E) \subseteq E.$$

Since $Cyl(b_1, \ldots, b_{n_{jD}}) \subseteq A^{\mathbf{Z}^D} \setminus E$ we have that

$$F^n(Cyl(a_1, \ldots, a_{n_{hD}})) \bigcap Cyl(b_1, \ldots, b_{n_{jD}}) = \emptyset,$$

i.e., F is not transitive which is a contradiction. □

Proof of Theorem 7 (sketch). Assume by contradiction that F is not surjective, that is, there exists $x \in X$ such that, for all $y \in X$, $F(y) \neq x$. Since F is transitive, we can find a sequence y_n such that $d(F(y_n), x) < 1/n$. Since (X, d) is a compact metric space, we can find a subsequence y_{n_i} which converges to $z \in X$. We have

$$d(F(z), x) \leq d(F(z), F(y_{n_i})) + d(F(y_{n_i}), x)\,.$$

Since the right-hand term goes to zero as $n_i \to \infty$ we have $F(z) = x$ as claimed. □

Let F denote the global transition map of a linear CA over $\mathbf{Z}_m$. In order to prove Theorems 8 and 9, we need to study the structure of F^n for $n \geq 1$. It is well known that F^n is a linear CA; in the following we derive a simple relationship between the coefficients of the local maps f and $f^{(n)}$ associated with F and F^n respectively. We first consider the simpler case of 1-dimensional CA. Let f denote a linear local map of radius r

$$f(x_{-r}, \ldots, x_{-1}, x_0, x_1, \ldots, x_r) = \sum_{i=-r}^{r} a_i x_i \bmod m.$$

By substitution we get

$$\begin{aligned} f^{(2)}(x_{-2r}, \ldots, x_{2r}) &= \sum_{i=-r}^{r} a_i \left(\sum_{j=-r}^{r} a_j x_{i+j} \right) \bmod m, \\ &= \sum_{k=-2r}^{2r} \left(\sum_{\substack{-r \leq i,j \leq r \\ i+j=k}} a_i a_j \right) x_k \bmod m. \end{aligned}$$

More in general, we can prove by induction that $f^{(n)}$ is a local map of radius nr such that

$$f^{(n)}(x_{-nr}, \ldots, x_{nr}) = \sum_{i=-nr}^{nr} b_i x_i \bmod m, \qquad b_i = \sum_{\substack{-r \leq i_1, \ldots, i_n \leq r \\ i_1 + \cdots + i_n = i}} a_{i_1} a_{i_2} \cdots a_{i_n}\,.$$

The fundamental observation is that the coefficient b_i is equal to the sum of all n-term products $a_{i_1}a_{i_2}\cdots a_{i_n}$ such that $i_1+\cdots+i_n=i$.

This formula can be easily generalized to dimensions greater than one. Assuming F is given by (3), we have

$$[F^n(c)](\mathbf{v})=\sum_{i=0}^{s(n)} b_i c(\mathbf{v}+\mathbf{nb}'(i)),$$

where

$$b_i=\sum_{\substack{1\le i_1,\ldots,i_n\le s\\ \mathbf{nb}(i_1)+\cdots+\mathbf{nb}(i_n)=\mathbf{nb}'(i)}} \lambda_{i_1}\lambda_{i_2}\cdots\lambda_{i_n}.$$

In other words, b_i is equal to the sum of all n-term products $\lambda_{i_1}\lambda_{i_2}\cdots\lambda_{i_n}$ such that the sum of the displacements $\mathbf{nb}(i_1)+\cdots+\mathbf{nb}(i_n)$ is equal to $\mathbf{nb}'(i)$. There is no simple way to determine the size $s(n)$ of the local rule $f^{(n)}$. However, it is clear that $s(n)<\infty$, that is, there is only a finite number of displacements $\mathbf{nb}'$ with a nonzero coefficient.

Proof of Theorem 8. Let $q=\gcd(m,a_{-r},\ldots,a_{-1},a_1,\ldots,a_r)$. We first prove that $q=1$ implies F is ergodic. By Theorem 11 we know that F is surjective. We prove that F is ergodic by showing that $I-F^n$ is surjective for $n\ge 1$ and using Theorem 12. For a fixed n let $b_{-nr},\ldots,b_{-1},b_0,b_1,\ldots,b_{nr}$ denote the coefficients of F^n. By Theorem 11 we know that $I-F^n$ is surjective if and only if

$$\gcd(m,-b_{-nr},\ldots,-b_{-1},1-b_0,-b_1,\ldots,-b_{nr})=1. \tag{5}$$

For $i=1,\ldots,r$, let

$$t_i=\gcd(m,a_{-r},\ldots,a_{-i},a_i,\ldots,a_r), \tag{6}$$

If $t_r=1$ the result is proven since

$$\gcd(m,a_{-r},a_r)=1 \quad\Longrightarrow\quad \gcd(m,a_{-r}^n,a_r^n)=\gcd(m,b_{-nr},b_{nr})=1$$

which implies (5). Assume now $t_r>1$, and let P denote the set of prime factors of t_r. To prove (5) we show that for all $p\in P$ there exists $j\neq 0$ such that $p\nmid b_j$. Given p, let i denote the greatest integer such that $p\nmid t_i$ but $p|t_{i+1}$. Such an integer must exist since by hypothesis $t_1=1$. We have that p divides all coefficients $a_{-r},\ldots,a_{-i-1},a_{i+1},\ldots a_r$, but p does not divide at least one between a_{-i} and a_i. Assume for simplicity that $p\nmid a_i$. We prove that, as a consequence, $p\nmid b_{ni}$. We know that b_{ni} is equal to the sum of the n-term products $a_{j_1}a_{j_2}\cdots a_{j_n}$ with $j_1+\cdots+j_n=ni$. Among the terms of the sum there is a_i^n which cannot be divided by p. We prove that $p\nmid b_{ni}$ by showing that every other term in the sum is a multiple of p. In fact, if $j_1+\cdots+j_n=ni$ and $\exists k: j_k\neq i$ at least one index j_h must be greater than i and the product $a_{j_1}a_{j_2}\cdots a_{j_n}$ can be divided by p. This proves that (5) holds which implies that F is ergodic.

Now we prove that $\gcd(m, a_{-r}, \ldots, a_{-1}, a_1, \ldots, a_r) = 1$ is a necessary condition for ergodicity. Assume by contradiction that F is ergodic and

$$q = \gcd(m, a_{-r}, \ldots, a_{-1}, a_1, \ldots, a_r) > 1.$$

Since ergodicity implies surjectivity (Theorems 13 and 7), by Theorem 11 we know that $\gcd(m, a_{-r}, \ldots, a_{-1}, a_0, a_1, \ldots, a_r) = 1$. Hence there exists a prime p such that $p|q$ and $p\not| a_0$. We establish our result by proving that $I - F^{p-1}$ is not surjective which is impossible by Theorem 12.

Let $n = p - 1$, and let $b_{-nr}, \ldots, b_{nr}$ denote the coefficients of $F^n = F^{p-1}$. We want to show that

$$p|\gcd(m, -b_{-nr}, \ldots, b_{-1}, 1 - b_0, b_1, \ldots, -b_{nr}). \tag{7}$$

Since $p|q$, p divides the coefficients $b_{-nr}, \ldots, b_{-1}, b_1, \ldots, b_r$. Consider now the coefficient b_0. We know that b_0 is equal to the sum of all n-term products $a_{j_1} a_{j_2} \cdots a_{j_n}$ with $j_1 + \cdots + j_n = 0$. The fundamental observation is that if some of the j_i's is $\neq 0$ the product $a_{j_1} \cdots a_{j_n}$ is a multiple of p. Hence

$$(1 - b_0) \bmod p \;=\; (1 - a_0^n) \bmod p \;=\; (1 - a_0^{p-1}) \bmod p \;= 0.$$

where the last equality follows from Fermat Theorem and the fact that by hypothesis $p\not| a_0$. Since $p|(1 - b_0)$, p satisfies (7) as claimed.

This completes the proof. □

Proof of Theorem 9 (sketch). The proof is a generalization of the previous one. Let $b_0, b_1, \ldots, b_{s(n)}$ denote the coefficients of F^n. The value b_i is associated to the displacement $\mathbf{nb}'(i)$, and we assume that $\mathbf{nb}'(0) = \mathbf{0}$. It suffices to show that

$$\gcd(m, \lambda_2, \ldots, \lambda_s) = 1 \implies \gcd(m, 1 - b_0, -b_1, \ldots, -b_{s(n)}) = 1, \tag{8}$$

since this implies that $I - F^n$ is surjective and we can apply Theorem 12. We reorder the coefficients $\lambda_2, \ldots, \lambda_s$ according to the euclidean norm of the displacements $\mathbf{nb}(i)$, (that is, $i < j \implies \|\mathbf{nb}(i)\| \leq \|\mathbf{nb}(j)\|$), and for $i = 2, \ldots, s$ we define

$$t_i = \gcd(m, \lambda_i, \lambda_{i+1}, \ldots, \lambda_s).$$

As in the proof of Theorem 8, we have that $t_s > 1$ implies (8). In addition, for each prime p such that $p|t_s$ we can find a coefficient b_j such that $p\not| b_j$. Hence (8) holds and the map F is ergodic.

The implication $\gcd(m, \lambda_2, \ldots, \lambda_s) = 1 \implies F$ ergodic is obtained by repeating verbatim the proof given for the 1-dimensional case. □

Proof of Theorem 10 (sketch). In order to prove that F has dense periodic points it is sufficient to prove that every cylindric subset of $A^{\mathbf{Z}^D}$ contains at least one periodic point. Let f and $\mathbf{nb}$ be the local rule and the neighborhood structure on which F is based.

We first prove this theorem in the 1-dimensional case. Let $a_1, \ldots, a_{2k+1}$ be $2k+1$ arbitrarily chosen elements of $\mathcal{A}$. Since F is topologically transitive, there exists a positive integer n such that

$$F^n(Cyl(a_1, \ldots, a_{2k+1})) \cap Cyl(a_1, \ldots, a_{2k+1}) \neq \emptyset.$$

Let $b \in F^n(Cyl(a_1, \ldots, a_{2k+1})) \cap Cyl(a_1, \ldots, a_{2k+1})$. Let $e_i \in \mathcal{A}^{\mathbf{Z}}$ be the configuration defined by

$$e_i(j) = \begin{cases} 1 \text{ if } j = i \text{ and} \\ 0 \text{ otherwise,} \end{cases}$$

Let

$$M = \max\{i : [F^n(b + e_i)](k) \neq [F^n(b)](k)\},$$
$$m = \min\{i : [F^n(b + e_i)](-k) \neq [F^n(b)](-k)\}.$$

When i is out of the interval $[-k, k]$, $b(i)$ may be different from $[F^n(b)](i)$. Let $b_1 \in \mathcal{A}^{\mathbf{Z}}$ be any configuration such that $b_1(i) = b(i)$, $i = m, m+1, \ldots, M$. It takes a little effort to verify that, since the cardinality of $\mathcal{A}$ is prime, it is possible to find suitable values for $b_1(m-1)$ and $b_1(M+1)$ such that $b_1(i) = [F^n(b_1)](i)$ inside the interval $[-k-1, k+1]$. By repeating the same process we may construct a Cauchy sequence of configurations b_h, $h \in \mathbf{N}$, such that $b_h(i) = [F^n(b_h)](i)$ for $-k-h \leq i \leq k+h$. Since $\mathcal{A}^{\mathbf{Z}}$ is complete b_h converges to a configuration c which is also a fixed point for F^n. Since $a_1, \ldots, a_{2k+1}$ can be arbitrarily chosen, we conclude that F has dense periodic orbits.
For simplicity of notation we prove this theorem in the 2-dimensional case by using a general decomposition technique which can be easily applied to the D-dimensional case.

A 2-dimensional additive CA can be seen as a 1-dimensional CA defined over the infinite alphabet $\mathcal{B} = \mathcal{A}^{\mathbf{Z}}$. The local rule g of the new CA is the sum of a finite number of surjective mappings (1-dimensional CA) g_i, $g_i : \mathcal{B} \to \mathcal{B}$, suitably defined according to f and **nb**.
The local rule f of Example 1 can be written as $f(x_1, x_2, x_3) = g_1(x_1) + g_2(x_2) + g_3(x_3)$, where g_i, $g_i : \{0,1\}^{\mathbf{Z}} \to \{0,1\}^{\mathbf{Z}}$, $1 \leq i \leq 3$, are defined as follows. For any $a \in \{0,1\}^{\mathbf{Z}}$, $g_1(a) = a$, $g_2(a) = b$, and $g_3(a) = a$, where $b(i) = a(i-1) + a(i+1)$, $i \in \mathbf{Z}$.

Since each g_i is surjective (being the local rule of an additive 1-dimensional CA defined on an alphabet of prime cardinality) we may prove the thesis by applying the proof technique used in the 1-dimensional case (with some little adjustments) to the new CA based on the local rule g.

A D-dimensional additive CA can be seen as a 1-dimensional CA definite over the alphabet $\mathcal{B} = \mathcal{A}^{\mathbf{Z}^{D-1}}$. The local rule g of the new D-dimensional CA is the sum of a finite number of surjective mappings from $\mathcal{B}$ to $\mathcal{B}$. The thesis follows from the surjectivity of $(D-1)$-dimensional additive CA defined on alphabets of prime cardinality.
We wish to emphasize that in the D-dimensional case, the surjectivity of the $(D-1)$-dimensional mappings which form the local rule is of fundamental importance and allows us to construct the Cauchy sequence of D-dimensional configurations whose limit is the fixed point for F^n. □

5 Further works

We are currently investigating the topological behavior of linear cellular automata over $\mathbf{Z}_m$ with the aim of providing easy-to-check characterizations of the following dynamical properties: expansivity, strongly transitivity, topological mixing, sensitivity to initial conditions, and denseness of periodic orbits (for composite m).

References

1. S. Amoroso and Y. N. Patt, Decision Procedures for Surjectivity and Injectivity of Parallel Maps for Tessellation Structures. *J. Comput. System. Sci. 6, 448-464*, 1972.
2. H. Aso and N. Honda, Dynamical Characteristics of Linear Cellular Automata. *J. Comput. System. Sci. 30, 291-317*, 1985.
3. B. Codenotti and L. Margara, Transitive Cellular Automata are Sensitive. *The Amer. Math. Monthly 103, 58-62*, 1996.
4. K. Culik, Y. Pachl, and S. Yu, On the limit sets of CA. *SIAM J. Comput., 18:831-842,*, 1989.
5. K.Culik and S. Yu, Undecidability of CA Classification Schemes. *Complex Systems 2(2), 177-190*, 1988.
6. R. L. Devaney, An Introduction to Chaotic Dynamical Systems. *Addison Wesley*, 1989.
7. P. Favati, G. Lotti and L. Margara, linear cellular Automata are chaotic According to Devaney's Definition of Chaos. *To appear on Theoret. Comp. Sci.*
8. P. Guan and Y. He, Exact Results for Deterministic Cellular Automata with Additive Rules. *Jour. Stat. Physics 43, 463-478*, 1986.
9. G. A. Hedlund, Endomorphism and Automorphism of the Shift Dynamical System. *Math. Sys. Th. 3(4), 320-375*, 1970.
10. M. Hurley, Ergodic Aspects of Cellular Automata. *Ergod. Th. & Dynam. Sys. 10, 671-685*, 1990.
11. M. Ito, N. Osato, and M. Nasu, Linear Cellular Automata over Z_m. *J. Comput. System. Sci. 27, 125-140*, 1983.
12. J. Kari, Rice's Theorem for the Limit Sets of Cellular Automata, *Theoret. Comp. Sci. 127, 229-254*, 1984.
13. M. Shirvani and T. D. Rogers, Ergodic Endomorphisms of Compact Abelian Groups. *Comm. Math. Physics 118, 401-410*, 1988.
14. M. Shirvani and T. D. Rogers, On Ergodic One-Dimensional Cellular Automata. *Comm. Math. Physics 136, 599-605*, 1991.
15. M. Vellekoop and R. Berglund On Intervals, Transitivity = Chaos. *The Amer. Math. Monthly 101, 353-355*, 1994.
16. S. J. Willson, On the Ergodic Theory of Cellular Automata. *Math. Sys. Th. 9(2), 132-141*, 1975.

Intrinsic Universality of a 1-Dimensional Reversible Cellular Automaton

Jérôme Olivier DURAND-LOSE *

LaBRI, Université Bordeaux I,
351, cours de la Libération,
33 405 TALENCE Cedex, FRANCE.

Abstract. This paper deals with simulation and reversibility in the context of Cellular Automata (CA). We recall the definitions of CA and of the Block (BCA) and Partitioned (PCA) subclasses. We note that PCA simulate CA. A simulation of reversible CA (R-CA) with reversible PCA is built contradicting the intuition of known undecidability results. We build a 1-R-CA which is intrinsic universal, *i.e.*, able to simulate any 1-R-CA.

1 Introduction

Cellular Automata (CA) model parallel computing as well as physical phenomena. They operate over regular infinite discrete lattices of finite dimension ($\mathbb{Z}^d$). Points are called cells and take a value from a finite set of state ($\mathcal{S}$). A CA iteration is the replacement of every state according to the states of the neighboring cells and a unique local function. This replacement is a local, uniform, parallel and synchronous update.

Reversibility is used for backtracking a phenomenon to its origin as well as for preserving information and energy. The possibilities of Reversible CA (R-CA) have been investigated from the 60s: the equivalence between bijectivity and injectivity by Moore and Myhill [11,14]; in the 70s: the equivalence of reversibility and bijectivity by Hedlung [5] and Richardson [15] and the decidability of reversibility in dimension 1 of Amoroso and Patt [1]; to its undecidability in higher dimension by Kari [6,7] in 1990.

The computing power of R-CA as well as their simulation powers was particularly investigated in [18]. Bennett [2] proved that reversible Turing machines could simulate any Turing Machine. In 1977, Toffoli [16] proved that R-CA of dimension d+1 are able to simulate any d-CA and thus are computationally universal. To built universal R-CA, Partitioned CA (PCA) and Block CA (BCA) were independently defined as special CA for which reversibility was decidable.

States of PCA, as defined by Morita [12,13], are partitioned according to the neighborhood and only the corresponding pieces of states are available to any cell. Sub-states are gathered to form one state to be updated. The local function operates over the finite set of states $\mathcal{S}$ rather that over two sets of different cardinality. Thus it can be bijective and this directly defines the reversibility.

* jdurand@labri.u-bordeaux.fr, http://www.labri.u-bordeaux.fr/~jdurand

Reischuk, Morvan (Eds.): STACS'97 Proceedings, LNCS 1200

Margolus [9,17] defines BCA to built the Billiard Ball Model (BBM), a computation universal 2-R-CA. For BCA, the underlying lattice, not the states, is partitioned into identical blocks regularly displayed. A transition is the replacement of all the blocks of a partition by their images by a unique local transition function. Again, since the local function operates over one finite set, it can be bijective. Originally, BCA were named: "Partitioning CA". Morita defined independently Partitioned CA. To avoid confusion in this article, we refer to Partitioning CA as Block CA because Kari [8] names Block Permutations bijective transitions.

A configuration is finite if all but a finite number of cells are in a define state. In 95, Morita [13] proved with PCA that any CA can be simulated with R-CA over finite configurations. This is enough for computing since it only treats finite information. But for physical modeling and as mathematical abstractions, there is no reason to restrict to such configurations. Moreover, finite configurations are a strict subset of recursive configurations (recursive mapping from $\mathbb{Z}^d$ to $\mathcal{S}$).

Trivially, BCA and PCA are CA. Although BCA and PCA are subclasses of CA, they are able to simulate any CA. It was proved in [3,4] that R-BCA can simulate R-CA. This was a 1990 conjecture by Toffoli and Margolus [18] was independently and partially proved in 1996 by Kari [8]. One of the results in the present paper is that R-PCA can also simulate R-CA.

In Sect. 2, we define the 3 CA models and prove the decidability of reversibility for PCA and BCA. In Sect. 3, we build simulations of CA with PCA, BCA with PCA and R-CA with R-PCA. In Sect. 4, we built U, a 1-R-PCA able to simulate any other 1-R-PCA. Sub-states are organized in 10 layers for identification, delimitation, table, signals, value, and translation of data. The dynamic is totally driven by signals which exchange values, test for equality, update when it should be done and move data. It is a signal like approach to computation with *a posteriori* tests to ensure reversibility.

The intrinsic universality of U as a 1-R-CA is proved from the existence of simulations between R-CA and R-PCA. This is one step ahead of the results in [3], where BBM is proven intrinsic universal for 2-R-CA, *i.e.*, able to simulate any 2-R-CA. The results and methods developed in [3] extend to higher dimensions, but do not hold for dimension 1. Our extension for dimension 1 is done differently, using PCA and an explicit R-PCA code which was not the case before. It extends to higher dimensions.

2 Definitions

Cellular automata define mappings over d-dimensional infinite arrays over a finite set of states $\mathcal{S}$. The set of configurations is $\mathcal{C}=\mathcal{S}^{\mathbb{Z}^d}$.

We denote $<$, $+$, mod, div and ., respectively the pointwise ordering, addition, modulo, Eulerian division and multiplication over $\mathbb{Z}^d$, *i.e.*: $\forall x, y \in \mathbb{Z}^d$; $x<y \Leftrightarrow \forall k, x_k<y_k$; $\forall k$; $(x+y)_k=x_k+y_k$; $(x \bmod y)_k=x_k \bmod y_k$; $(x \operatorname{div} y)_k=x_k \operatorname{div} y_k$; $(x.y)_k=x_k\, y_k$. For any configuration c and subset E of $\mathbb{Z}^d$, $c_{|E}$ is the restriction of c to E. For any $x \in \mathbb{Z}^d$, σ_x is the shift by x ($\forall c \in \mathcal{C}, \forall i \in \mathbb{Z}^d, \sigma_x(c)_i=c_{i+x}$). Periodicity is to be understood as in all directions.

A Cellular Automaton of dimension d (d-CA) is defined by $(\mathcal{S},\mathcal{N},f)$. The *neighborhood* $\mathcal{N}$ is a finite subset of $\mathbb{Z}^d$. The *local function* $f : \mathcal{S}^{\mathcal{N}} \to \mathcal{S}$ maps the states of a neighborhood into one state. The *global function* $\mathcal{G} : \mathcal{C} \to \mathcal{C}$ maps configurations into themselves as follows:

$$\forall c \in \mathcal{C},\ \forall x \in \mathbb{Z}^d,\ \mathcal{G}(c)_x \ = \ f\left((c_{x+\nu})_{\nu\in\mathcal{N}}\right) \ .$$

A new state of a cell depends only on the neighbor states as depicted by Fig. 1.

A Partitioned Cellular Automaton of dimension d (d-PCA) is defined by: $(\mathcal{S},\mathcal{N},\Phi)$. The set of states is a sets product indexed by the neighborhood: $\mathcal{S}=\prod_{\nu\in\mathcal{N}} \mathcal{S}^{(\nu)}$. The ν component of a state q is noted $q^{(\nu)}$. The local function f is defined with function $\Phi : \mathcal{S} \to \mathcal{S}$ as follows:

$$\forall c \in \mathcal{C},\ \forall x \in \mathbb{Z}^d,\ f(c)_x \ = \ \Phi\left(\prod_{\nu\in\mathcal{N}} c^{(\nu)}_{x+\nu}\right) \ .$$

Or equivalently, every state is the product of states to be sent around. The local function works only with what remains and what is received. Only partial information is accessible to a cell, even about its own state as depicted by Fig. 1.

A Block Cellular Automaton of dimension d (d-BCA) is defined by: $(\mathcal{S}, v, n, (o^{(j)})_{1\le j\le n}, t)$. The *size* v is an element of $\mathbb{Z}^d$ such that $0<v$. All $o^{(j)}$ are coordinates modulo v: $o^{(j)} \in \mathbb{Z}^d$ and $0\le o^{(j)}<v$. Block V is the subset $[\![0,v_1]\!]\times[\![0,v_2]\!]\times\ldots\times[\![0,v_d]\!]$ of $\mathbb{Z}^d$. The *local transition* t is a function over $\mathcal{S}^V$.

The *transition* T is the following mapping over $\mathcal{C}$: for any $c \in \mathcal{C}$ and $i \in \mathbb{Z}^d$, let $a=i \operatorname{div} v$, and $b=i \bmod v$, so that $i=a.v+b$, then $T(c)_i=t(c_{|a.v+V})_b$. In other words, the block containing i in the regular partition originated from 0 is updated according to t. The same happens for all the blocks of this partition. The transition of origin o, T_o is $\sigma_o \circ T \circ \sigma_{-o}$. It is the original one with the partition shifted by o. The global function is the composition of the transitions of origins $o^{(j)}$: $\mathcal{G}=T_{o^{(n)}} \circ T_{o^{(n-1)}} \circ \ldots T_{o^{(1)}}$. This is illustrated on the right part of Fig. 1 with 2 partitions and $v=(3)$. Since partitions come in a cycle, we assimilate $n+1$ with 1, and 0 with n from now on.

To see that BCA are indeed CA, consider the blocks of the first partition to be cells. At this scale, the global function commutes with any shift and is continuous for the product topology, according to a theorem by Richardson [15], it is a CA. A constructive proof can be found in [3].

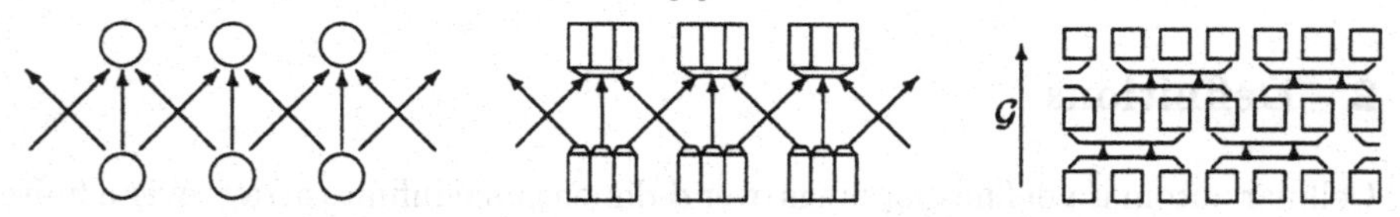

Fig. 1. Schematic CA, PCA and BCA updating.

2.1 Reversibility

A CA (PCA,BCA) is *reversible* if and only if its global function $\mathcal{G}$ is bijective and $\mathcal{G}^{-1}$ is the global function of some CA (PCA,BCA). We denote R-CA (R-BCA, R-PCA) the class of reversible CA (PCA,BCA). The main decidability result is:

Theorem 1. *The reversibility of* CA *is decidable in dimension* 1 (Amoroso and Patt 1972 [1]) *but it is undecidable for higher dimension* (Kari 1990 [6,7]).

Whereas for PCA and BCA, the following lemmas hold in any dimension and states that as far as reversibility is concerned, BCA and PCA fundamentally differ from CA. Recall that bijectivity for CA is equivalent to reversibility [5,15] and that Φ and t work over finite sets so that their bijectivities are decidable.

Lemma 2. (Morita) *A* PCA *is reversible iff its local function* Φ *is a permutation.*

Proof. If Φ is a permutation, then the inverse PCA is $\left(\prod_{\nu\in -\mathcal{N}} \mathcal{S}^{(-\nu)}, -\mathcal{N}, \Phi^{-1}\right)$ where $-\mathcal{N}=\{-\nu | \nu \in \mathcal{N}\}$, Φ is undone and pieces of states send back. Otherwise, since Φ works on a finite set, it is not one-to-one and it is easy to construct 2 configurations which have the same image. □

Lemma 3. (Margolus) *A* BCA *is reversible iff its local transition t is a permutation.*

Proof. If the local transition t is a permutation, by construction, any transition is reversible. The global transition as a composition of transitions, is reversible. Otherwise, t is not one-to-one, then neither is any transition, and neither is the global transition. □

3 Simulations

An automaton A *simulates* another B in linear time τ if there exist two functions α and β such that $\forall c \in \mathcal{C}_{\mathrm{B}}$, $\forall t \in \mathbb{N}$, $\mathcal{G}_{\mathrm{B}}^{t}(c)=\beta \circ \mathcal{G}_{\mathrm{A}}^{\tau t} \circ \alpha\ (c)$. If $\tau=1$, the simulation is *real time.* Since PCA and BCA (R-PCA and R-BCA) are CA (R-CA), they can obviously be simulated with CA (R-CA). In [3], it is proved that CA (R-CA) can be simulated with BCA (R-BCA) in real time. We built the missing simulations.

Proposition 4. *Any d-*CA *can be simulated with a d-*PCA *in real time.*

Proof. The idea is to duplicate the states in every part. Let $\mathsf{A}=(\mathcal{S},\mathcal{N},f)$ be any d-CA. It is simulated with the following d-PCA: $\mathsf{P}=\left(\mathcal{S}^{\mathcal{N}},\mathcal{N},\Phi\right)$, with:

$$\forall \mu \in \mathcal{N},\ \Phi\left(\prod_{\nu\in\mathcal{N}} s_{\nu}^{(\nu)}\right)^{(\mu)} = f\left(\left(s_{\nu}^{(\nu)}\right)_{\nu\in\mathcal{N}}\right) .$$

Any configuration c of $\mathcal{C}_{\mathrm{A}}$ is naturally mapped in d of $\mathcal{C}_{\mathrm{P}}$ with $\forall x \in \mathbb{Z}^d, \forall \nu \in \mathcal{N}, d_x^{(\nu)}=c_x$. Information is duplicated, resources are wasted. □

Since Φ only maps onto the diagonal of $\mathcal{S}^{\mathcal{N}}$, the simulating PCA is never reversible. Nevertheless, it is possible to simulate R-CA with R-PCA with the following result:

Lemma 5. *Any d-*R-BCA *can be simulated with a d-*R-PCA *in linear time n.*

Proof. The idea is to let one cell represent one block. This is a change of scale. Let $\mathsf{B}{=}(\mathcal{S}_{\mathrm{B}}, v, n, (o^{(j)})_{1\leq j\leq n}, t)$ be any d-BCA. Let $\mathsf{P} = \left(\prod_{\nu\in\mathcal{N}} \mathcal{S}_{\mathrm{P}}^{(\nu)}, \mathcal{N}, \Phi\right)$ where $\mathcal{N}{=}\{-1,0,1\}^d$, *i.e.*, coordinates which differ by at most one in any direction. The block of coordinates x (at block scale) of the jth partition is b_x^j. The block b_0^j holds the cell of coordinates 0. The sets of states are defined by:

$$\forall \nu \in \mathcal{N},\ \mathcal{S}_{\mathrm{P}}^{(\nu)} = \bigcup_{1\leq j\leq n} \left(\{j\} \times \mathcal{S}_{\mathrm{B}}^{\left(b_0^{j-1} \cap b_{-\nu}^{j}\right)} \right) .$$

It is the intersection of the block that holds the cell of coordinates 0 for a partitions and of the one of the next partition holding the cell 0 translated by $\nu.v$. Any intersection may be empty. Blocks are partitioned in function of the next partition so that every part is sent to the corresponding cells to form whole blocks of the next partition. Identically, each cell retrieves a full block, uses the local transition and sends the corresponding parts to the neighbors for the next transition. The local function is defined by:

$$\forall \mu \in \mathcal{N},\ \forall c \in \mathcal{C},\ \Phi\left(\prod_{\nu\in\mathcal{N}} x_{x+\nu}^{(\nu)}\right)^{(\mu)} = \Phi\left(k, b_0^k\right)^{(\mu)} = \left(k+1, t\left(b_0^k\right)_{|b_{-\mu}^{k+1}}\right) .$$

The first coordinates identify the current transition. They must match, otherwise Φ is undefined. Configurations are encoded by setting the first components to 1 and by putting states in the corresponding intersections between the last and the first partitions. On the first iteration of P, each cell gets one entire block of the first partition and make the first transition. Then every pieces are sent to the corresponding cells with 2 in the first component. Each iteration of P makes a successive transition of B. After n iterations of P, one iteration of B is made and the first component is 1 again.

This construction preserves reversibility: the partial definition of the PCA local function Φ is one-to-one if the local transition t of the BCA is reversible. □

It is proved in [3] that any R-CA can be simulated with a R-BCA in real time. From above Lemma and the transitivity of simulation comes:

Theorem 6. *Any d-R-CA can be simulated with a d-R-PCA in linear time.*

With the construction in [3], 2^{d+1}-1 partitions are needed and the number of B-states is quadratic. In dimension 2 and above, the size v is not bounded by any computable function (from the decidability results 1 and 3). Therefore, neither is the number of P-states.

4 Intrinsic Universality of 1-R-PCA

We built the following 1-R-PCA: $\mathsf{U}{=}(1, \mathcal{S}_{\mathrm{U}}, \{-1,0,1\}, \Phi_{\mathrm{U}})$. The states and the local function Φ_{U} are defined on Figs. 4 and 10. To avoid using $\{-1,0,1\}$, we denote l, c and r the left, center and right part. We prove that:

Theorem 7. U *is intrinsic universal, i.e., able to simulate any* 1-R-PCA.

For any 1-R-PCA P=($\mathcal{S}$,$\mathcal{N}$,Φ). With classical techniques of cells grouping, P can be simulated in real time with a 1-R-PCA with neighborhood $\{l, c, r\}$. From now on, we suppose that $\mathcal{N} = \{l, c, r\}$.

Macroscopic Level. Let B_x be the xth element of $\mathcal{S}$ modulo $|\mathcal{S}|$. We encode a configuration and the table of Φ in a configuration of U with the P-cell architecture of Figs. 2 and 6. The initial configuration given on Fig. 2 extends infinitely on both sides.

Index	...	B_{x-2}	B_{x-1}	B_x	B_{x+1}	B_{x+2}	...
Table	...	B_{x-2}	B_{x-1}	B_x	B_{x+1}	B_{x+2}	...
	...	$\Phi(B_{x-2})$	$\Phi(B_{x-1})$	$\Phi(B_x)$	$\Phi(B_{x+1})$	$\Phi(B_{x+2})$	...
Value	...	V_{x-2}	V_{x-1}	V_x	V_{x+1}	V_{x+2}	...
Mode	...	sma	sma	sma	sma	sma	...

Fig. 2. Initial configuration.

From PCA definition, all P-cells first exchange their l and r parts. The state is denoted V_x before the exchange and W_x after. The inner loop of the simulation is: shift the layers holding B_y and $\Phi(B_y)$ to the left, compare W_x to B_y and if they are equal, replace W_x by $\Phi(B_y)$. The table is fully scanned when the index B_x is equal to the B_y of the table, and then P-cells are updated and one P-iteration is done. The dynamics of the loop is given in Fig. 3.

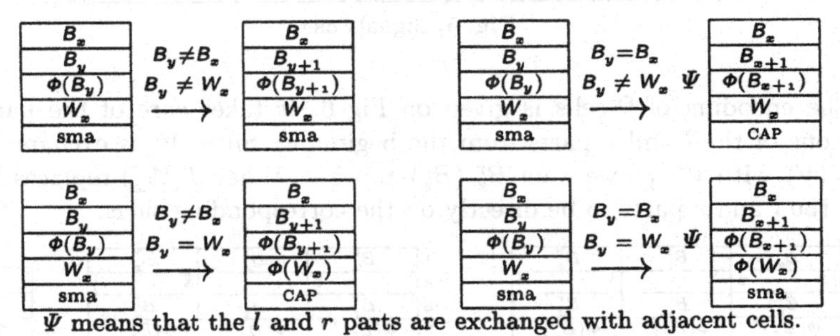

Fig. 3. Update of P-cells.

States, Layers and Initial Configuration at Microscopic Level. U-cells are organized in 10 layers as detailed in Fig. 4. Layer I holds an index to store where the reading of the table started. Architecture layer A holds delimiters for P-cells ([and]) and for l, c and r parts ($). Layers B and F hold one entry of the table B_y and its image $\Phi(B_y)$. Signals are found on layer S. The value of the P-cell (V_x or W_x) is stocked on layer V. Layers L_1 to L_4 work like conveyorbelts to transfer data. The values in layers I and A never change.

We use capital (B,$\Phi(B)$,W) to address the macroscopic level (P-cells) and small letters (b,c,f,v) for microscopic level (U-cells). All P-cells are binary encoded. For the exchange, the codes of l and r parts must have the same length (0's are added if necessary).

Layer	Name	States l c r	Use
1	I	0 1	P-cell identification B_x
2	A	_ [$]	Architecture: limits of cells and parts
3	B	0 1	Table entry B_y
4	F	0 1	Image of the table entry B_y, $\Phi(B_y)$
5	S	Σ Σ Σ	Control signals detailed on Fig. 5
6	V	0 1	Value of the P-cell (V_x or W_x)
7–10	L1–L4	0 1 0 1	Shift the table of Φ and exchange values (W_x^l & W_x^r)

Fig. 4. The 10 layers and corresponding sub-states.

Signals are 26 letters as described on Fig. 5. Small (sma) and capital (CAP) letters behave similarly. The sma or CAP mode distinguishes between before and after the replacement. During the simulation, signals are turn from CAP to sma when parts are exchanged and from sma to CAP when the value is replaced by its image. The number of states of U is $2^{13}.27^3 < 2^{28}$.

sma	CAP	Use
a-h	A-H	Loop which tests if [$W_x{=}B_y$ — $W_x{=}\Phi(B_y)$]
k	K	Write [$\Phi(B_y)$ — B_y] over W_x
m, n	M, N	Shift of the table: B_y and $\Phi(B_y)$
s, t	S, T	Exchange of Parts W_x^l and W_x^r

Fig. 5. Signals use.

The encoding of P-cells is given on Fig. 6. It takes care of the particular positions of the l and r parts from the beginning. Since V_x^l is exchanged with V_{x+1}^r (V_x^r with V_{x-1}^l), we want B_x^r (B_x^l) above it. When $\Phi(W_x)$ replaces W_x, we want the l and r parts to be directly on the corresponding sides.

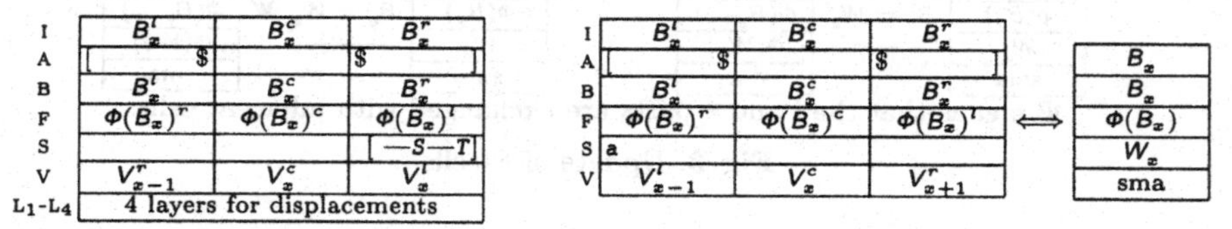

Fig. 6. Encoding of P-cell at coordinate x, before and after the exchange.

The Microscopic Algorithm is defined by space-time diagrams driven by signals. The corresponding rules are indicated on Fig. 10. The algorithm starts in CAP mode with [|S|T] signals in the rightmost cell. Signals in the different P-cells are always exactly synchronized.

First, the l and r parts of the P-cell are exchanged and the mode is switched to sma as depicted on Fig. 7. The initial value of the P-Cell is V_x. The bits of V_x^l and V_{x+1}^r are swapped on the layer L1 by signals S and T on they way from]. On crossing], the flows are transferred on L2 (for technical reasons explained below). Signals S and T turn back at $ and on their way back, they retrieve the bits from L2 and put them back on their destination slot. Synchronization is very important. Signals S and T finally get back together as h at], switch mode and go back to the left end of the P-cell. This is implemented with 11 rules of Fig. 10.

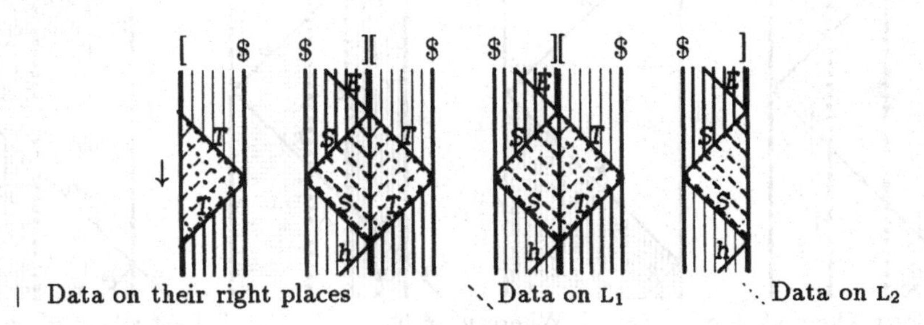

Fig. 7. Exchanging V_x^l and V_{x+1}^r.

Let us describe in details the inner loop. Value W_x and table entry B_y are in place, bit below bit, to be compared. Signal a crosses the whole P-cell to compare them. If they differ, a marker b is put on the first different bit, and a turns to b. On the way back, b marks d the last bit which differs and gets the marker b back as illustrated on the first column of Fig. 8. If W_x and B_y are equal, the signal reaches] as a, turns to k, writes $\Phi(B_y)$ over W_x on the way back and switch mode. This special behavior takes as much time as the regular one, keeping the synchronization. Equality is tested on the way back for reversibility: going backward in time, U must make the correct change at the adequate time, so it needs this test and it must change back $\Phi(W_x)$ into W_x.

To know that the table was completely scanned, signal must test whether B_x and B_y are equal. On the second crossing, signal d (or e) gets back the previous marker (if any) and turn to g and marks g the first different bit between B_x and B_y, as illustrated by Fig. 8. If they differ, g comes back and gets the marker. If there are equal, before returning e exchanges the l and r parts and switches mode and return as h.

On returning to [, h splits into m and n (right half of Fig. 8). These signals manage the shift of the table by one P-cell rightward using layers L1 to L4 as illustrated by Fig. 9. Signal m sets B_{y-1} and $\Phi(B_{y-1})$ on movement by swapping then on layers L1 and L3. On passing], bits go down a layer so as not to interfere with the moving ones of the next P-cell. On its way back, n sets B_{y-1} and $\Phi(B_{y-1})$ on their final places by swapping them from layers L2 and L4. Signals m and n gather and form a which starts the loop again. This corresponds to the last 12 rules of Fig. 10.

If the mode is CAP, it is exactly the same except that the equality tested is W_x with $\Phi(B_y)$ instead of B_y, and B_y is written over W_x instead of $\Phi(B_y)$. Since P is reversible, each value of $\Phi(B_y)$ appears once and only once in the table. After being copied $\Phi(B_y)$, it is never met again in the table as an image.

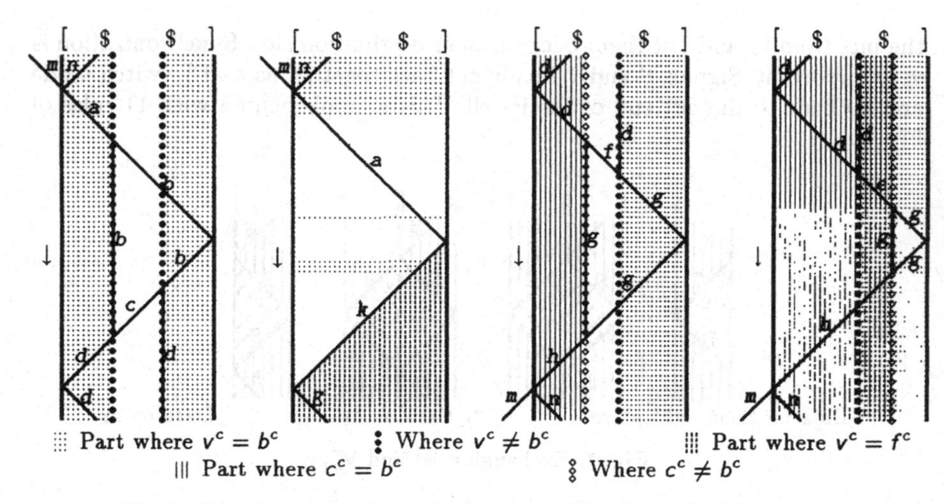

Fig. 8. Inner loop: test for replacement and for the end of P-iteration.

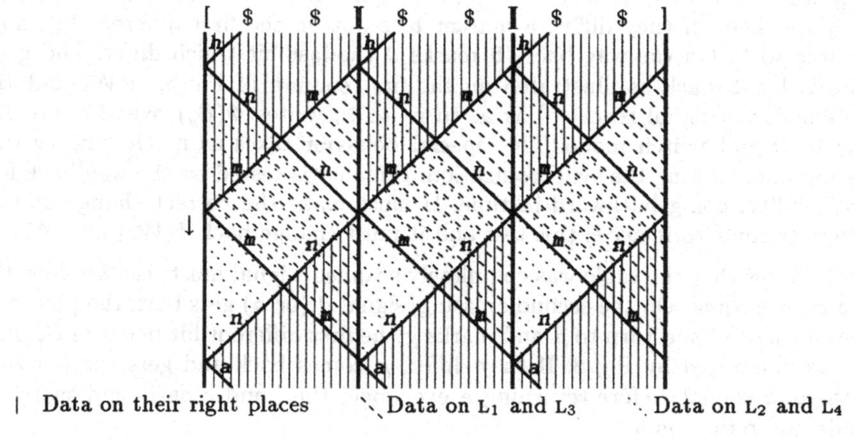

Fig. 9. Shifting the table.

Local Function of U. The necessary definitions of Φ_U are given in Fig. 10. Since they are one-to-one, Φ_U can be completed bijectively. The values of the layers that hold 0 and 1 are not indicated. These values are tested as requirement for rules and are not modified otherwise noted in the last column. These modifications are either swaps or of v^c, but in such case, the previous value is held somewhere else as indicated by a condition. For CAP signals, the differences are only for the lines with an '*': the test made is $v^c = f^c$, instead of $v^c = b^c$, and b^c, instead of f^c, is copied over v^c.

Rule (i) of Fig. 10 starts the replacement process: writing $\Phi(W_x)$ over W_x. Rules (ii) and (iii) start the shifting of the table. Rules (iv) and (v) start the

	Structure	Signals			Value Condition		Signals			Image Modification	
*	[		a		$v^c = b^c$				a		
					$v^c \neq b^c$			b	b		
*	' ',$	a			$v^c = b^c$				a		
					$v^c \neq b^c$			b	b		
*	]	a			$v^c = b^c$			k		$v^c \leftarrow f^c$	(i)
					$v^c \neq b^c$		d	d			
	' ',$	b							b		
	' ',$		b					b			
*	]	b			$v^c = b^c$		b				
					$v^c \neq b^c$		c	d			
*	' ',$			b	$v^c = b^c$		b				
					$v^c \neq b^c$		c	d			
*	' ',$		b	b	$v^c \neq b^c$		d	d			
*	[		b	b	$v^c \neq b^c$	$c^c = b^c$		e			
						$c^c \neq b^c$		g	g		
*	' ',$		b	c	$v^c \neq b^c$		d				
*	[		b	c	$v^c \neq b^c$	$c^c = b^c$			d		
						$c^c \neq b^c$		g	f		
	' ',$			c			c				
*	' ',$			d	$v^c = b^c$		d				
*	[			d	$v^c = b^c$	$c^c = b^c$			d		
						$c^c \neq b^c$		g	f		
*	' ',$	d			$v^c = b^c$	$c^c = b^c$			d		
						$c^c \neq b^c$		g	f		
*	' ',$	d	d		$v^c \neq b^c$	$c^c = b^c$			e		
						$c^c \neq b^c$		g	g		
*	]	d	d		$v^c \neq b^c$	$c^c = b^c$		s	t		(ii)
						$c^c \neq b^c$	h				
*	' ',$	e			$v^c = b^c$	$c^c = b^c$			e		
						$c^c \neq b^c$		g	g		
*	]	e			$v^c = b^c$	$c^c = b^c$		s	t		(iii)
						$c^c \neq b^c$	h				
	' ',$	f							f		
*	' ',$	f	d		$v^c \neq b^c$				g		
*	]	f	d		$v^c \neq b^c$		g				
*	' ',$	g			$v^c = b^c$				g		
*	]	g			$v^c = b^c$		g				
	[, ' ',$		g					g			
	' ',$			g			g				
	' ',$		g	g		$c^c \neq b^c$	h				
	]		g	g		$c^c \neq b^c$	m	n			(iv)
	' ',$			h		$c^c = b^c$	h				
	]			h		$c^c = b^c$	m	n			(v)
	]		s				s			swap(v^c, l_1^r)	
	' '			s			s			swap(v^c, l_1^r)	
	$			s				s		swap(v^c, l_1^r)	
	$		s						s	swap(v^c, l_2^r)	
	' '	s							s	swap(v^c, l_2^r)	
	]	s						s		swap(v^c, l_2^r)	
	[, ' '	t							t	swap(v^c, l_1^l)	
	$	t						t		swap(v^c, l_1^l)	
	$		t				t			swap(v^c, l_2^l)	
	[, ' '			t			t			swap(v^c, l_2^l)	
	]		s	t			H				(vi)
*	' ',$			k	$v^c = b^c$		o			$v^c \leftarrow f^c$	
*	]			k	$v^c = b^c$				E	$v^c \leftarrow f^c$	(vii)
	[, ' '			m			m			swap(b^c, l_1^r), swap(f^c, l_3^r)	
	' ',$			m			m			swap(b^c, l_1^l), swap(f^c, l_3^l)	
	' ',$	n		m			m		n	swap(b^c, l_1^l), swap(f^c, l_3^l)	
	]			m					m		
	' ',$]	m							m		
	]		n						n		
	' ',$	n							n		
	]	n						n			
	]		n				n			swap(b^c, l_2^l), swap(f^c, l_4^l)	
	' ',$			n			n			swap(b^c, l_2^l), swap(f^c, l_4^l)	
	[			n				n		swap(b^c, l_2^l), swap(f^c, l_4^l)	
	]	m	n						a		(viii)

Fig. 10. Table of Φ_U.

exchange of the l and r parts. Rules (*vi*) and (*vii*) switch the mode. The first corresponds to the end of the P-iteration, the second to the replacement of W_x by $\Phi(W_x)$. Rule (*viii*) is the end of the inner loop.

All rules are combined with the following: for the last 4 layers L_1 to L_4, the l and r parts are swapped so that l (r) parts move at speed 1 to the right (left). For all rules with] : layers L_1 and L_2 (L_3 and L_4) are swapped. This is technical for the flows of the table shift not to collide in the middle of Fig. 9 where 2 flows are traveling together.

Simulation Time. Let a be the width of a P-cell and b the width of the exchanged parts ($0 \leq 2b \leq a$ and $\lceil \log |\mathcal{S}| \rceil \leq a \leq 2\lceil \log |\mathcal{S}| \rceil + 2$). The inner loop needs $4(a-1)$ iterations for the tests and $2a$ for the shift of the table. It is done for every P-state, *i.e.*, $|\mathcal{S}|$ times. To make a P-iteration, values are exchanged between neighboring cells, this needs $2b+1$ iterations. All together, the simulation is in linear time and bounded by $12\,|\mathcal{S}| \log(|\mathcal{S}|) + o(|\mathcal{S}| \log(|\mathcal{S}|))$. This is fine for R-PCA. For R-BCA simulation, it have to be multiplied by $2^{d+1}-1$, here 3 and $\mathcal{S}$ is larger. For R-CA, $|\mathcal{S}|$ of the simulating R-BCA can be very big. With the construction of [3], it is $(|\mathcal{S}|^2 + |\mathcal{S}|)^{(4 \max(r_{\mathsf{A}}, r_{\mathsf{A}^{-1}}))^d}$ where r_{A} ($r_{\mathsf{A}^{-1}}$) is the radius of A (A^{-1}), *i.e.*, the maximum absolute value of any component of any ν in the neighborhood. Nevertheless, the simulation is still in linear time.

5 Conclusion

The inverse R-PCA is simulated by changing $[\ |S|T]$ by $[\ |s|t]$ in the initial configuration. When U runs with the code of an unreversible PCA, the Φ is not one-to-one so there are two states A and B such that $A \neq B$ and $\Phi(A) = \Phi(B)$. When A is encountered before B then A will be replaced by $\Phi(B)$ and in CAP it is then replaced by B and remains so until the end of the P-iteration. This is wrong.

The construction can be extended to greater dimension. The table and test are done in the first direction and sub-states exchanged in the other directions must be added.

Basic programming schemes can be embedded in R-PCA when conceived reversible. We have implemented with reversible local rules a global dynamic of move, test and replace which needs backward tests. The R-PCA is programmed: we make loops, tests and branches to subroutines.

The power of computations of d-CA, d-BCA and d-PCA over infinite configurations are the same. We have proved that the d-R-CA, d-R-BCA and d-R-PCA classes are also equivalent which is an important result since reversibility is decidable for BCA and PCA while it is not for CA.

In [3], it was proved the existence of intrinsic universal R-CA of dimension 2 and above. The construction was made using the R-CA simulation with R-BCA and then by constructing a simulation of any R-BCA with the Billiard Ball Model of Margolus [9,17]. We have proved using R-PCA and their 'source code' (*i.e.* local function) that this result still holds in dimension 1. Since any R-CA can be simulated with R-PCA (Th. 6) and U is able to simulate any R-PCA (Th. 7) and U is a R-CA:

Theorem 8. *There exist* 1-R-CA *able to simulate any* 1-R-CA *in linear time.*

There exist simulations of any Turing machines with R-CA [12] so that all partial recursive functions can be computed by R-PCA, so U is computation universal. The existence of a intrinsic universal R-PCA is proven here with the use of the source code of the R-PCA. So there should be some S-m-n theorem for R-PCA to prove that they form an acceptable programming system as proved for CA by Martin [10].

It is unknown whether the class of d-CA is strictly more powerful than the class of d-R-CA on infinite configurations. Nevertheless, if a 1-R-CA can simulate a nonreversible CA, then by transitivity, U is also able to do it, so that if U can not, none can.

References

1. S. Amoroso and Y. Patt. Decision procedure for surjectivity and injectivity of parallel maps for tessellation structure. *Journal of Computer and System Sciences*, 6:448–464, 1972.
2. C. H. Bennett. Logical reversibility of computation. IBM *Journal of Research and Development*, 6:525–532, 1973.
3. J. O. Durand-Lose. Reversible cellular automaton able to simulate any other reversible one using partitioning automata. In LATIN *'95*, number 911 in Lecture Notes in Computer Science, pages 230–244. Springer-Verlag, 1995.
4. J. O. Durand-Lose. *Automates Cellulaires, Automates à Partitions et Tas de Sable*. PhD thesis, LaBRI, 1996. In French.
5. G. A. Hedlung. Endomorphism and automorphism of the shift dynamical system. *Mathematical System Theory*, 3:320–375, 1969.
6. J. Kari. Reversibility of 2D cellular automata is undecidable. *Physica D*, 45:379–385, 1990.
7. J. Kari. Reversibility and surjectivity problems of cellular automata. *Journal of Computer and System Sciences*, 48(1):149–182, 1994.
8. J. Kari. Representation of reversible cellular automata with block permutations. *Mathematical System Theory*, 29:47–61, 1996.
9. N. Margolus. Physics-like models of computation. *Physica D*, 10:81–95, 1984.
10. B. Martin. A universal cellular automaton in quasi-linear time and its S-n-m form. *Theoretical Computer Science*, 123:199–237, 1994.
11. E. Moore. Machine models of self-reproduction. In *Proceeding of Symposium on Applied Mathematics*, volume 14, pages 17–33, 1962.
12. K. Morita. Computation-universality of one-dimensional one-way reversible cellular automata. *Information Processing Letters*, 42:325–329, 1992.
13. K. Morita. Reversible simulation of one-dimensional irreversible cellular automata. *Theoretical Computer Science*, 148:157–163, 1995.
14. J. Myhill. The converse of Moore's garden-of-eden theorem. In *Proceedings of the Symposium of Applied Mathematics*, number 14, pages 685–686, 1963.
15. D. Richardson. Tessellations with local transformations. *Journal of Computer and System Sciences*, 6:373–388, 1972.
16. T. Toffoli. Computation and construction universality of reversible cellular automata. *Journal of Computer and System Sciences*, 15:213–231, 1977.
17. T. Toffoli and N. Margolus. *Cellular Automata Machine - A New Environment for Modeling*. MIT press, Cambridge, MA, 1987.
18. T. Toffoli and N. Margolus. Invertible cellular automata: A review. *Physica D*, 45:229–253, 1990.

The Computational Complexity of Some Problems of Linear Algebra
(Extended Abstract)

Jonathan F. Buss*, Gudmund S. Frandsen**, and Jeffrey O. Shallit*

Abstract. In this paper we consider the computational complexity of some problems dealing with matrix rank. Let E, S be subsets of a commutative ring R. Let $x_1, x_2, \ldots, x_t$ be variables. Given a matrix $M = M(x_1, x_2, \ldots, x_t)$ with entries chosen from $E \cup \{x_1, x_2, \ldots, x_t\}$, we want to determine

$$\text{maxrank}_S(M) = \max_{(a_1,a_2,\ldots,a_t)\in S^t} \text{rank } M(a_1, a_2, \ldots, a_t)$$

and

$$\text{minrank}_S(M) = \min_{(a_1,a_2,\ldots,a_t)\in S^t} \text{rank } M(a_1, a_2, \ldots, a_t).$$

There are also variants of these problems that specify more about the structure of M, or instead of asking for the minimum or maximum rank, ask if there is some substitution of the variables that makes the matrix invertible or noninvertible.
Depending on E, S, and on which variant is studied, the complexity of these problems can range from polynomial-time solvable to random polynomial-time solvable to *NP*-complete to *PSPACE*-solvable to unsolvable.

1 Introduction

In this paper we consider the computational complexity of some problems of linear algebra — more specifically, problems dealing with matrix rank.

Our mathematical framework is as follows. If R is a commutative ring, then $\mathcal{M}_n(R)$ is the ring of $n \times n$ matrices with entries in R. The rows α_i of a matrix are said to be *linearly independent* over R if $\sum_i c_i\alpha_i = 0$ (with $c_i \in R$) implies $c_i = 0$ for all i, and similarly for the columns.

* Supported in part by grants from NSERC Canada. Address: Department of Computer Science, University of Waterloo, Waterloo, Ontario N2L 3G1, Canada. E-mail: jfbuss@plg.uwaterloo.ca, shallit@uwaterloo.ca.

** Supported in part by the ESPRIT Long Term Research Programme of the EU under project number 20244 (ALCOM-IT) and by BRICS, Basic Research in Computer Science, Centre of the Danish National Research Foundation. Address: BRICS, Department of Computer Science, University of Aarhus, Ny Munkegade, DK-8000 Aarhus C, Denmark. E-mail: gsfrandsen@daimi.aau.dk.

Reischuk, Morvan (Eds.): STACS'97 Proceedings, LNCS 1200

The *determinant* of $M = (a_{ij})_{1\le i,j\le n}$ is defined as follows:

$$\det M = \sum_{P=(i_1,i_2,\ldots,i_n)} (\operatorname{sgn} P) a_{1,i_1} a_{2,i_2} \cdots a_{n,i_n},$$

where $P = \binom{1\ 2\ \cdots\ n}{i_1\ i_2\ \cdots\ i_n}$ is a permutation of $\{1,2,\ldots,n\}$. We know that a matrix is invertible over R if and only if its determinant is invertible over R [8].

The *rank* of a matrix M is the maximum number of linearly independent rows. Rank can also be defined as the maximum number of linearly independent columns, and it is well-known [8] that these two definitions coincide. We denote the rank of M as rank M. An $n \times n$ matrix is invertible iff its rank is n.

A $k \times k$ *submatrix* of M is the array formed by the elements in k specified rows and columns; the determinant of such a submatrix is called a $k \times k$ *minor*. The rank of M can also be defined as the maximum size of an invertible minor.

The problems we consider are along the following lines: let E, S be two subsets of R. We are given an $n \times n$ matrix $M = M(x_1, x_2, \ldots, x_t)$ with entries chosen from $E \cup \{x_1, x_2, \ldots, x_t\}$, where the x_i are distinct variables. We want to compute

$$\operatorname{maxrank}_S(M) = \max_{(a_1,a_2,\ldots,a_t)\in S^t} \operatorname{rank} M(a_1,a_2,\ldots,a_t) \quad (1)$$

$$\operatorname{minrank}_S(M) = \min_{(a_1,a_2,\ldots,a_t)\in S^t} \operatorname{rank} M(a_1,a_2,\ldots,a_t). \quad (2)$$

Evidently there is no need to distinguish between column rank and row rank in this definition. Note also that we do not necessarily demand that we be able to exhibit the actual t-tuple that achieves the maximum or minimum rank.

There are several reasons for studying these problems. First, the problems seem — to us, at least — natural questions in linear algebra. Second, a version of the minrank problem is very closely related to determining the minimum rank rational series that approximates a given formal power series to a given order; see [6, 13]. Third, the maxrank problem is related to the problem of matrix rigidity which has recently received much attention [14, 5, 9], and may help explain why good bounds on matrix rigidity are hard to obtain.

Before describing our complexity results, we illustrate minrank and maxrank with some examples. First, consider the matrix

$$M = \begin{bmatrix} x_1 & x_2 & 2 \\ 4 & x_1 & 4 \\ 0 & 0 & x_3 \end{bmatrix}$$

with $\operatorname{minrank}_{\mathbb{Q}}(M) = 1$ attained at $(x_1, x_2, x_3) = (2, 1, 0)$ and $\operatorname{maxrank}_{\mathbb{Q}}(M) = 3$ attained at $(x_1, x_2, x_3) = (2, 2, 1)$.

Second, both $\operatorname{minrank}_S(M)$ and $\operatorname{maxrank}_S(M)$ may depend on S, as illustrated by the following examples. However, we show later that $\operatorname{maxrank}_S(M)$ is the same for all infinite S.

$$M = \begin{bmatrix} x & 1 \\ 1 & 2 \end{bmatrix}, \quad \det M = 2x - 1, \quad \begin{array}{l} \operatorname{minrank}_{\mathbb{Z}} M = 2 \\ \operatorname{minrank}_{\mathbb{Q}} M = 1 \text{ (using } x \mapsto \frac{1}{2}) \end{array}$$

$$T = \begin{bmatrix} x & 1 \\ 2 & x \end{bmatrix}, \ \det T = x^2 - 2, \quad \begin{array}{l} \text{minrank}_{\mathbb{Q}} T = 2 \\ \text{minrank}_{\mathbb{R}} T = 1 \ (\text{using } x \mapsto \sqrt{2}) \end{array}$$

$$U = \begin{bmatrix} x & 1 \\ -1 & x \end{bmatrix}, \ \det U = x^2 + 1, \quad \begin{array}{l} \text{minrank}_{\mathbb{R}} U = 2 \\ \text{minrank}_{\mathbb{C}} U = 1 \ (\text{using } x \mapsto i) \end{array}$$

$$V = \begin{bmatrix} x & x \\ 1 & x \end{bmatrix}, \ \det V = x^2 - x, \quad \begin{array}{l} \text{maxrank}_{GF(4)} V = 2 \ (\text{using } x \mapsto \\ \qquad\qquad \text{a generator of } GF(4)) \\ \text{maxrank}_{GF(2)} V = 1 \end{array}$$

2 Summary of Results

Most of our complexity results for the computation of minrank and maxrank are naturally phrased in terms of the decision problems given in Table 1. We have introduced two special problems, SING(ularity) and NONSING(ularity), which could possibly be easier than the more general minrank/maxrank problems.

Fixed: R, a commutative ring.
$E, S \subseteq R$.
Input: M is an $n \times n$ matrix with entries from $E \cup \{x_1, \ldots, x_t\}$.
k is a non-negative integer.

Problem	Input	Decide
MINRANK	M, k	$\min_{(a_1,\ldots,a_t)\in S^t} \text{rank } M(a_1,\ldots,a_t) \le k$?
MAXRANK	M, k	$\max_{(a_1,\ldots,a_t)\in S^t} \text{rank } M(a_1,\ldots,a_t) \ge k$?
SING	M	$\exists (a_1,\ldots,a_t) \in S^t$ such that $\det M(a_1,\ldots,a_t) = 0$?
NONSING	M	$\exists (a_1,\ldots,a_t) \in S^t$ such that $\det M(a_1,\ldots,a_t) \neq 0$?

Table 1. Decision problems.

Table 2 summarizes our results on the complexity of the four decision problems. The problems MAXRANK and NONSING are put together, since we have not been able to separate their complexities, although we do not know whether they have the same complexity in general. We have good evidence that the MINRANK and SING problems do not in general have the same complexity: over $\mathbb{C}$, MINRANK is *NP*-hard (Section 7) but SING has a randomized polynomial time solution (Section 3).

The exact value of E is not important for our bounds. All our lower bounds are valid for $E = \{0, 1\}$ and all our upper bounds are valid for E being $\mathbb{Q}$ or a finite-dimensional field extension of $\mathbb{Q}$ (respectively E being $GF(q)$ or a finite-dimensional field extension of $GF(q)$, when the characteristic is finite). For the upper bounds, we assume the input size to be the total number of bits needed to specify the matrix M, when using the standard binary representation

S	E	MAXRANK NONSING	SING	MINRANK
$GF(q)$	$\{0,1\} \subseteq E \subseteq GF(q)$	*NP*-complete		
$\mathbb{Z}$			r.e.; undecidable	
$\mathbb{Q}$			r.e.; *NP*-hard	
$\mathbb{Q}_p$	$\{0,1\} \subseteq E \subseteq \mathbb{Q}$	*RP*	*EXPEXPSPACE*; *NP*-hard	
$\mathbb{R}$			*PSPACE*; *NP*-hard	
$\mathbb{C}$				

Table 2. Complexity bounds for decision problems: the general case.

of numbers, representing a finite-dimensional algebraic extension by arithmetic modulo an irreducible polynomial, representing polynomials by coefficient vectors and listing the value of each entry in M. The upper bounds are also robust in another sense. We may allow entire multivariate polynomials (with coefficients from E) in a single entry of the matrix M and still preserve our upper bounds, provided such a multivariate polynomial is specified by an arithmetic formula using binary multiplication and binary addition but no power symbol, so that the representation length of a multivariate polynomial is at least as large as its degree.

S is significant for the complexity, as is apparent from Table 2. However, our upper and lower bounds for $S = C$ are valid for S being any algebraically closed field (in the case of S having finite characteristic, so must E of course).

The results of Table 2 fall in three groups according to the proof technique used. The randomized polynomial time upper bounds use a result due to Schwartz [12]. The undecidability result for $\mathbb{Z}$ uses a combination of Valiant's result that the determinant is universal [15] and Matiyasevich's proof that Hilbert's Tenth Problem is unsolvable [10]. All the remaining problems of the result table (those that are not marked either *RP* or *undecidable*) are equivalent (under polynomial-time transformations) to deciding the existential first-order theory over the field S. The equivalence immediately implies the *NP*-hardness of all these problems, and lets us use results by Egidi [3], Ierardi [7] and Canny [2] to obtain the doubly exponential space upper bound for a p-adic field $\mathbb{Q}_p$ and the *PSPACE* upper bounds for $\mathbb{C}$ and $\mathbb{R}$, respectively. Since it is presently an open problem whether the existential first-order theory over $\mathbb{Q}$ is decidable or not, we suspect it will be difficult to determine the decidability status of MINRANK and SING over $\mathbb{Q}$.

We also consider the special case when each variable in the matrix occurs exactly once. None of our lower bound proofs are valid under this restriction, and we have managed to improve some of the upper bounds. See Table 3 for a summary. The improved upper bounds all rely on the determinant polynomial being multi-affine when no variable occurs twice. In such case the *RP*-algorithm for singularity over $\mathbb{C}$ can be generalized to work for singularity over any field.

S	E	MAXRANK NONSING	SING	MINRANK
$GF(q)$	$GF(q)$			NP
$\mathbb{Z}$			r.e.	r.e.
$\mathbb{Q}$				
$\mathbb{Q}_p$	$\mathbb{Q}$	RP	RP	$EXPEXPSPACE$
$\mathbb{R}$				$PSPACE$
$\mathbb{C}$				

Table 3. Upper bounds when each variable occurs exactly once.

(Proof omitted from this extended abstract.)

Since minrank is at least *NP*-hard to compute over $\mathbb{Z}$ or a field, one might consider the existence of an efficient approximation algorithm. Suppose, however, that for some fixed S (S being $\mathbb{Z}$ or a field) and $E = \{0, 1\}$, there is a polynomial time algorithm that when given matrix $M = M(x_1, \ldots, x_t)$ always returns a vector $(a_1, \ldots, a_t) \in S^t$ satisfying rank $(M(a_1, \ldots, a_t)) \leq (1+\varepsilon) \cdot \text{minrank}_S(M)$. Then the assumption $P \neq NP$ implies $\varepsilon \geq \frac{7}{1755} \approx .0039886$. The proof uses a reduction from MAXEXACT3SAT; i.e., we use a known nonapproximability result for MAXEXACT3SAT [1] combined with a *MAXSNP*-hardness proof for our minrank approximation problem. (Proof omitted from this extended abstract.)

3 Randomized polynomial time algorithms

In this section we show that one can compute maxrank with a (Monte-Carlo) random polynomial-time algorithm over any infinite field and over the ring $\mathbb{Z}$, and one can similarly decide singularity over an algebraically closed field.

Our main tool is the following lemma, adapted from a paper of Schwartz [12]:

Lemma 1 *Let $p(x_1, x_2, \ldots, x_t)$ be a multivariate polynomial of total degree at most d which is not the zero polynomial, and let F be a field containing at least $2d$ distinct elements. Then if V is any set of $2d$ distinct elements of F, $p(a_1, a_2, \ldots, a_t) = p(\mathbf{a}) \neq 0$ for at least 50% of all $\mathbf{a} \in V^t$.*

Theorem 2 *Let $M = M(x_1, x_2, \ldots, x_t)$ be a $n \times n$ matrix with entries in $F \cup \{x_1, x_2, \ldots, x_t\}$. Let $V \subseteq F$ be a set of at least $2n$ distinct elements. Choose a t-tuple $(a_1, a_2, \ldots, a_t) \in V^t$ at random. Then with probability at least $1/2$, we have* $\text{maxrank}_F(M) = \text{rank } M(a_1, a_2, \ldots, a_t)$.

Proof. Suppose $\text{maxrank}_F(M) = k$. Then there exists some t-tuple $(a_1, a_2, \ldots, a_t) \in F^t$ such that rank $M(a_1, a_2, \ldots, a_t) = k$. Hence, in particular, there must be some $k \times k$ minor of $M(a_1, a_2, \ldots, a_t)$ with nonzero determinant. Consider the corresponding $k \times k$ submatrix M' of $M(x_1, x_2, \ldots, x_t)$. Then the determinant of M', considered as a multivariate polynomial p in the indeterminates $x_1, x_2, \ldots, x_t$, cannot be identically zero (since it is nonzero when $x_1 =$

$a_1, \ldots, x_t = a_t)$. It now follows from Lemma 1 that p is nonzero for at least half of all elements of V^t. Thus for at least half of all these t-tuples $(a_1, a_2, \ldots, a_t)$, the corresponding $k \times k$ minor of M must be nonzero, and hence $M(a_1, a_2, \ldots, a_t)$ has rank at least k. Since $\text{maxrank}_F(M) = k$, it follows that rank $M(a_1, a_2, \ldots, a_t) = k$ for at least half of the choices $(a_1, a_2, \ldots, a_t) \in V^t$. ■

The theorem justifies the following random polynomial-time algorithm to compute $\text{maxrank}_F(M)$ over an infinite field F: choose r t-tuples of the form $(a_1, a_2, \ldots, a_t)$ independently at random, and compute rank $M(a_1, a_2, \ldots, a_t)$ for each of them, obtaining ranks $b_1, b_2, \ldots, b_r$. Then with probability at least $1 - 2^{-r}$, we have $\text{maxrank}_F(M) = \max_{1 \le i \le r} b_i$.

It also follows from Theorem 2 that over an infinite field F, the quantity maxrank(M) cannot change when we consider an extension field F' with $F \subseteq F'$, or when we consider an infinite subset $S \subseteq F$. In particular, the decision problem MAXRANK is in the complexity class RP for $E = \mathbb{Q}$ and $\mathbb{Z} \subseteq S$.

For deciding singularity over an algebraically closed field, we need the following fact from algebra.

Lemma 3 *Let $p(x_1, x_2, \ldots, x_t)$ be a nonconstant multivariate polynomial over a field F. Then if F is algebraically closed, p takes on all values in F.*

Proof. By induction on t, the number of variables. Details omitted. ■

The lemma justifies the following random polynomial-time algorithm for deciding SING over an algebraically closed field F.

Let $V \subseteq F$ be a set of at least $2n$ distinct elements. Choose r t-tuples $\mathbf{a}_1$, $\mathbf{a}_2$, ..., $\mathbf{a}_r$ at random from V^t, and evaluate the determinant $\det M(\mathbf{a}_i)$ for $1 \le i \le r$. If at least two different values are obtained, return "yes". If all the values obtained are the same, and all are nonzero, return "no". If all the values are the same, and all are zero, return "yes".

4 Universality of the determinant

In this section, we prove a result that underlies all our lower bounds for the singularity and minrank problems: any multivariate polynomial is the determinant of a fairly small matrix. The result was first proven by Valiant [15], but since we need a slightly modified construction and the result is fundamental to our lower bound proofs, we feel it is best to make this paper self-contained and give the details of a construction.

To state the result, we need a few definitions. Let an *arithmetic formula* F be a well-formed formula using constants, variables, the unary operator $\{-\}$ and the binary operators $\{+, \cdot\}$. The *length* of a formula F (denoted by $|F|$) is defined as the total number of occurrences of constants, variables and operators. For example $|3(x + y - 4) + 5z| = |3 \cdot (x + y + (-(4))) + 5 \cdot z| = 12$. (Note that our definition of formula length is not the same as Valiant's.)

Proposition 4 *Let R be a commutative ring. Let F be an arithmetic formula using constants from $E \subseteq R$ and variables from $\{x_1, \ldots, x_t\}$.*

For some $n \leq |F|+2$, we may in time $n^{O(1)}$ construct an $n \times n$ matrix M with entries from $E \cup \{0,1\} \cup \{x_1, \ldots, x_t\}$ such that $p_F = \det M$ and $\operatorname{minrank}_R(M) \geq n-1$, where p_F denotes the polynomial described by formula F.

Proof. (Sketch) We use a modified version of Valiant's construction [15]. The main difference is that we insist that the rank of the constructed $n \times n$ matrix cannot be less than $n-1$ under any substitution for the variables. We also consider the negation operation explicitly, which allows us to avoid the use of negative constants in the formula, when wanted. Our construction is essentially a modification of Valiant's construction to take care of these extra requirements combined with a simplification that leads to matrices of somewhat larger size than Valiant's original construction. ■

5 The singularity problem over the integers

In this section we prove that the decision problem SING is unsolvable over $\mathbb{Z}$.

Theorem 5 *Given a matrix $M = M(x_1, \ldots, x_t)$ with entries from $\{0,1\} \cup \{x_1, \ldots, x_t\}$, it is undecidable whether there exist $a_1, \ldots, a_t \in \mathbb{Z}$ such that* $\det M(a_1, \ldots, a_t) = 0$

Proof. We reduce from Hilbert's Tenth Problem [10]. An instance of Hilbert's Tenth Problem is a Diophantine equation $p(x_1, \ldots, x_t) = 0$, where p is a multivariate polynomial with integer coefficients. We construct a formula for p using only $+, -, \cdot, 0, 1$ in addition to the indeterminates by replacing each integer constant $c \geq 2$ having binary representation $c = \sum_{i=0}^{l} b_i 2^i$ with the formula

$$b_0 + (1+1)[b_1 + (1+1)[b_2 + (1+1)[b_3 + \cdots + (1+1)[b_l] \cdots]]].$$

By the construction of Proposition 4, the resulting formula f_p for the polynomial $p(x_1, \ldots, x_t)$ is turned into a matrix $M = M(x_1, \ldots, x_t)$ such that $\det M(x_1, \ldots, x_t) = p(x_1, \ldots, x_t)$. The assertion of the theorem follows from the undecidability of Hilbert's Tenth Problem. ■

6 Existential first-order theories

In this section, we state some known bounds on the complexity of deciding existential first-order theories over specific fields. We will apply them later to our rank problems.

For a field F, the existential theory of F is $ETh(F) = \{ \phi : \phi$ is an existential sentence and $F \models \phi \}$. The decision problem for $ETh(F)$ is: On input ϕ, decide whether $F \models \phi$. (For definitions, see *e.g.* Enderton [4].)

The complexity of deciding $ETh(F)$ seems to depend on the field F. Table 4 summarizes some known upper bounds. $ETh(F)$ is *NP*-hard for any field F.

F	Upper bound on $ETh(F)$	reference
$GF(q)$	*NP*	
$\mathbb{Q}$	recursively enumerable	
$\mathbb{Q}_p$	*EXPEXPSPACE*	Egidi, 1993 [3]
$\mathbb{R}$	*PSPACE*	Canny, 1988 [2]; Renegar, 1992 [11]
$\mathbb{C}$	*PSPACE*	Ierardi, 1989 [7]

Table 4. Upper bounds on deciding $ETh(F)$

7 Lower bound for minrank over a field

In this section we prove that the decision problem MINRANK over a field is as hard as deciding the corresponding existential first-order theory. By a small extension of the proof, we also get that the decision problem SING over a field that is not algebraically closed and the decision problem NONSING over a finite field are as hard as deciding the corresponding existential first-order theories.

Lemma 6 *Let F be a field. Given an existential sentence $\exists x_1 \cdots \exists x_t .\, \phi(x_1, \ldots, x_t)$, of length m, we can in time $m^{O(1)}$ construct an equivalent existential sentence $\exists x_1 \cdots \exists x_{t'} .\, \psi(x_1, \ldots, x_{t'})$ such that ψ contains neither negation nor disjunction, i.e., ψ is a conjunction of atomic formulas,*

$$\psi(\mathbf{x}') \quad \equiv \quad p_1(\mathbf{x}') = 0 \ \wedge \cdots \wedge \ p_r(\mathbf{x}') = 0$$

for some arithmetic formulas p_i, $i = 1, \ldots, r$, and

$$F \models \exists \mathbf{x}.\ \phi(\mathbf{x}) \quad \textit{iff} \quad F \models \exists \mathbf{x}'.\ \psi(\mathbf{x}').$$

Proof. First we remove all negations from ϕ. This may be done using the rewriting rules in Table 5(a).

	Rewrite rules
Step 1	$\neg(F_1 \wedge F_2) \rightarrow (\neg F_1) \vee (\neg F_2)$
	$\neg(F_1 \vee F_2) \rightarrow (\neg F_1) \wedge (\neg F_2)$
Step 2	$\neg t(\mathbf{x}) = 0 \rightarrow 1 - z \cdot t(\mathbf{x}) = 0$

(a) Subconstruction for elimination of $\neg$

f_i	f_i'
$p_i(\mathbf{x}) = 0$	$p_i(\mathbf{x}) = z_i$
$f_j \vee f_k$	$z_j \cdot z_k = z_i$
$f_j \wedge f_k$	$z_j \cdot z_k = z_i \ \wedge \ z_j + z_k = z_i$

(b) Subconstruction for elimination of $\vee$

Table 5. Constructions for elimination of $\neg$ and $\vee$.

In step 1, we use de Morgan's laws to move all negations down so they are applied directly to the atomic formulas. In step 2, we replace each negated atomic formula with an unnegated formula. This requires the introduction of a new variable z for each such atomic formula (representing the inverse of the term $t(\mathbf{x})$). These new variables must be existentially quantified.

Without loss of generality, we may therefore assume that we are given the existential sentence $\exists x_1 \cdots \exists x_t \,.\, \phi(x_1, \ldots, x_t)$ where ϕ is an unquantified formula *without* negations using variables $x_1, \ldots, x_t$.

Let ϕ have s subformulas $f_1, \ldots, f_s$, each of which may be atomic or composite. For each such subformula f_i, we introduce a new (existentially quantified) variable z_i, and we construct a new formula f_i' that is either atomic or the conjunction of two atomic formulas. The $f_i's$ will be constructed such that

$$\begin{array}{l} \exists x_1 \cdots \exists x_t.\ \text{“}f_i \text{ is satisfied”} \\ \qquad\qquad \Updownarrow \\ \exists x_1 \cdots \exists x_t \exists z_1 \cdots \exists z_t.\ \text{“}z_i = 0 \text{ and } f_j' \text{ is satisfied} \\ \qquad \text{for all subformulas } f_j \text{ of } f_i \text{ (including } f_i\text{)”}. \end{array} \tag{3}$$

If the subformula f_1 corresponds to the entire formula ϕ, this implies that

$$\begin{array}{l} \exists \mathbf{x} \,.\, \phi(\mathbf{x}) \\ \qquad \Updownarrow \\ \exists \mathbf{x}, \mathbf{z}.\ \ z_1 = 0 \ \wedge\ f_1'(\mathbf{x}, \mathbf{z}) \ \wedge \cdots \wedge\ f_s'(\mathbf{x}, \mathbf{z}) \end{array}$$

For each original subformula f_i the new formula f_i' is constructed as described in Table 5(b). By induction in the structure of f_i, one may verify that this construction does satisfy (3), from which the theorem follows. ■

Lemma 7 *Let F be a field. Given an existential sentence ϕ of length m, we can in time $n^{O(1)}$ construct an integer k and an $n \times n$ matrix with entries from $\{0,1\} \cup \{x_1, x_2, \ldots, x_t\}$, where $n = O(m)$ such that*

$$\text{minrank}_F(M) \le k \quad \textit{iff} \quad F \models \phi$$

Proof. Let an existential sentence be given. First we remove all negations and disjunctions using the construction of Lemma 6.

Without loss of generality, we may therefore assume that we are given the existential sentence $\exists \mathbf{x}.\ p_1(\mathbf{x}) = 0 \wedge \cdots \wedge p_r(\mathbf{x}) = 0$ for some arithmetic formulas p_i, $i = 1, \ldots, r$.

By Proposition 4, we may for each $p_i(x_1, \ldots, x_t)$ find an $n_i \times n_i$ matrix M_i with entries from $\{0,1\} \cup \{x_1, x_2, \ldots, x_t\}$ such that $\det M_i = p_i(x_1, \ldots, x_t)$ and $\text{minrank}_F(M_i) \ge n_i - 1$.

Let $n = \sum_{i=1}^r n_i$, let $k = \sum_{i=1}^r (n_i - 1)$, and construct the $n \times n$ matrix M by placing $M_1, \ldots, M_r$ consecutively on the main diagonal and zeroes elsewhere. Clearly, $\text{minrank}_F(M) \ge k$ and rank $M = k$ only when all the polynomials p_i are simultaneously zero; therefore $\text{minrank}_F(M) \le k$ iff $F \models \phi$. ■

Lemma 6 can be extended to remove conjunction from fields that are not algebraically closed allowing us to prove a SING-version of Lemma 7 for these fields:

Lemma 8 *Let F be a fixed field that is not algebraically closed. Then there exists a finite set of constants $E \subseteq F$ such that given arithmetic formulas $p_1(\mathbf{x}), \ldots, p_r(\mathbf{x})$ of combined length m, we can in time $m^{O(1)}$ construct a single arithmetic formula $p(\mathbf{x})$ (using constants from E) such that*

$$F \models \exists \mathbf{x}.\ p_1(\mathbf{x}) = 0\ \wedge \cdots \wedge\ p_r(\mathbf{x}) = 0$$
$$\Updownarrow$$
$$F \models \exists \mathbf{x}.\ p(\mathbf{x}) = 0$$

The set of constants $E = \{0, 1\}$ suffices for any of the fields $\mathbb{Q}, \mathbb{R}, \mathbb{Q}_p, GF(q)$.

Proof. Since F is not algebraically closed, there exists a univariate polynomial $f(x) = \sum_{i=0}^{d} a_i x^i$ of degree d with $a_0, \ldots, a_d \in F$ such that f has no root in F. Define a new polynomial in two variables by $g(y, z) = z^d \cdot f(\frac{y}{z}) = \sum_{i=0}^{d} a_i y^i z^{d-i}$. Observe that $g(y, z)$ is nonzero except when $y = z = 0$. Using an arithmetic formula for g (of size $O(d^2)$) construct the formula $p(\mathbf{x})$ from $p_1(\mathbf{x}), \ldots, p_r(\mathbf{x})$ using the rewrite rule of table 6 repeatedly $\log r$ times. The size of $p(\mathbf{x})$ will be $O(d^{2 \log r} m) = O(r^{2 \log d} m) = m^{O(1)}$.

Rewrite rule
$p_1(\mathbf{x}) = 0 \wedge p_2(\mathbf{x}) = 0 \wedge p_3(\mathbf{x}) = 0 \wedge \cdots \wedge p_{2k-1}(\mathbf{x}) = 0 \wedge p_{2k}(\mathbf{x}) = 0$ $\rightarrow g(p_1(\mathbf{x}), p_2(\mathbf{x})) = 0 \wedge g(p_3(\mathbf{x}), p_4(\mathbf{x})) = 0 \wedge \cdots \wedge g(p_{2k-1}(\mathbf{x}), p_{2k}(\mathbf{x})) = 0$

Table 6. Construction for elimination of $\wedge$.

To see that $E = \{0, 1\}$ suffices for some special fields as claimed in the lemma, choose $f(x) = x^2 + 1$ for F being $\mathbb{Q}$ or $\mathbb{R}$, choose $f(x) = x^2 + x + 1$ for F being $\mathbb{Q}_2$ or $GF(2)$, choose $f(x) = x^2 + (p - a)$ for some quadratic nonresidue a modulo p when F is $\mathbb{Q}_p$ or $GF(p)$ and $p \neq 2$ is a prime (and use that $p - a = 1 + 1 + \cdots + 1$); finally, a suitable irreducible polynomial exists for any other specific finite field. ∎

Corollary 9 *Let F be fixed and one of the fields $\mathbb{Q}, \mathbb{R}, \mathbb{Q}_p, GF(q)$. Let an existential sentence $\exists \mathbf{x}.\ \phi(\mathbf{x})$ of length m be given. Then we can in time $m^{O(1)}$ construct an $n \times n$ matrix M with entries from $\{0, 1\} \cup \{x_1, \ldots, x_t\}$ such that*

$$F \models \exists \mathbf{x}.\ \phi(\mathbf{x}) \ \text{ iff } \ \exists (\mathbf{a}) \in F^t.\ \det M(\mathbf{a}) = 0.$$

If F is one of the finite fields $GF(q)$, we can also in time $m^{O(1)}$ construct an $n' \times n'$ matrix M' with entries from $\{0, 1\} \cup \{x_1, \ldots, x_t\}$ such that

$$F \models \exists \mathbf{x}.\ \phi(\mathbf{x}) \ \text{ iff } \ \exists (\mathbf{a}) \in F^t.\ \det M'(\mathbf{a}) \neq 0.$$

Proof. For the first part of the theorem combine Lemmas 6 and 8 and Proposition 4. For the second part make a similar argument, where the use of Lemma 8 is replaced by use of the fact that

$$GF(q) \models \exists \mathbf{x}.\ p_1(\mathbf{x}) = 0\ \wedge \cdots \wedge\ p_r(\mathbf{x}) = 0$$
$$\Updownarrow$$
$$GF(q) \models \exists \mathbf{x}.\ (1 - p_1(\mathbf{x})^{q-1}) \cdot \ldots \cdot (1 - p_r(\mathbf{x})^{q-1}) \neq 0$$

∎

8 Upper bounds for minrank over a field

In this section, we prove that the minrank problem over a field is no harder than deciding the corresponding existential first-order theory. Combined with our earlier results, this implies that the decision problem `MINRANK` is in fact equivalent (under polynomial-time transformations) to deciding the corresponding existential first-order theory. In addition, we inherit the upper bounds of Table 4.

In this extended abstract, we give the proof for matrices that use only the constants 0 and 1 and leave the proof of the general case to the full paper.

Lemma 10 *Let F be a field. Given an $n \times n$ matrix M with entries from $\{0,1\} \cup \{x_1, x_2, \ldots, x_t\}$, and some $k \leq n$, we may in time $n^{O(1)}$ construct an existential sentence ϕ such that* $\text{minrank}_F(M) \leq k$ *iff* $F \models \phi$.

Proof. Given $(n \times n)$ matrix M with variables $x_1, x_2, \ldots, x_t$ and constants from $\{0,1\}$, we express (in a first-order existential sentence) the assertion that some k columns of M span all columns of M. For this purpose we introduce n new variables $y_1, y_2, \ldots, y_n$ in addition to the variables already occurring in M. Define the modified matrix M', where $M'_{ij} = y_j \cdot M_{ij}$, i.e., each column of M' is a multiple (possibly zero) of the corresponding column in M. We also introduce n^2 new variables $z_{11}, \ldots, z_{nn}$ forming an $n \times n$ matrix Z. The assertion $\text{minrank}(M) \leq k$ is now equivalent to the following assertion: it is possible to choose the y_j's and z_{ij}'s in such a way that at most k of the y_j's are nonzero and the matrix equation $M' \cdot Z = M$ holds.

Our sentence will be an existential quantification of a conjunction of two formulas. The first one f_1 will assert that at most k of the y_j's are nonzero, and the second one f_2 will assert that the matrix equation $M' \cdot Z = M$ holds.

Construction of f_1: We use the elementary symmetric functions defined by

$$\sigma_j(y_1, \ldots, y_n) = \sum_{A \subseteq \{1,\ldots,n\} \wedge |A|=j} \prod_{i \in A} y_i, \qquad \text{for } j = 1, \ldots, n.$$

By considering the polynomial

$$\begin{aligned} p(z, y_1, y_2, \ldots, y_n) &= (z + y_1)(z + y_2) \cdots (z + y_n) \\ &= z^n + \sigma_1(y_1, \ldots, y_n) z^{n-1} + \cdots + \sigma_n(y_1, \ldots, y_n), \end{aligned}$$

it follows that there are at most k nonzero y_j's if and only if

$$\sigma_{k+1}(y_1, \ldots, y_n) = 0 \wedge \sigma_{k+2}(y_1, \ldots, y_n) = 0 \wedge \cdots \wedge \sigma_n(y_1, \ldots, y_n) = 0.$$

We need to find a short formula expressing that $\sigma_j(y_1, \ldots, y_n) = 0$. Applying the school method for multiplying out polynomials to the polynomial $p(z, y_1, y_2, \ldots, y_n)$ gives an arithmetic circuit of size $O(n^3)$ that computes $\sigma_j(y_1, \ldots, y_n)$ for all $j = 1, \ldots, n$ simultaneously. This circuit can be understood as a straight line program of length $s = O(n^3)$ using the operations $\{+, \cdot\}$. Let the atomic formula h_i be $w = u + v$ (respectively $w = u \cdot v$), if the i'th line of the straight line program is $w \leftarrow u + v$ (respectively $w \leftarrow u \cdot v$). Without loss of generality, we may assume that the variables used in the straight line program are $w_1, \ldots, w_s$ in addition to input variables $y_1, \ldots, y_n$ and output variables $s_1, \ldots, s_n$ computing $\sigma_1, \ldots, \sigma_n$. We may use the following conjunction of atomic formulas $f_1 \equiv h_1 \wedge \cdots \wedge h_s \wedge s_{k+1} = 0 \wedge \cdots \wedge s_n = 0$.

Construction of f_2*:* We need to express that $M' \cdot Z = M$. The ij'th entry in the matrix product is $\sum_{k=1}^{n} y_k M_{ik} z_{kj}$. Therefore construct the atomic formula $g_{ij} \equiv y_1 M_{i1} z_{1j} + y_2 M_{i2} z_{2j} + \cdots + y_n M_{in} z_{nj} = M_{ij}$ and let $f_2 \equiv g_{11} \wedge g_{12} \wedge \cdots \wedge g_{nn}$.

Combining, we get the existential sentence $\text{minrank}_F(M) \leq k \equiv \exists x_1 \cdots \exists x_t \, \exists y_1 \cdots \exists y_n \, \exists z_{11} \cdots \exists z_{nn} \, \exists w_1 \cdots \exists w_s \, \exists s_1 \cdots \exists s_n \, . \, f_1 \wedge f_2$. ■

Acknowledgement. The second author wishes to thank Igor Shparlinski for discussions leading to the proof of Lemma 8.

References

1. M. Bellare, O. Goldreich, and M. Sudan. Free bits, PCPs and non-approximability — towards tight results (3rd revision). Report Series 1995, Revision 02 of ECCC TR95-024, Electronic Colloqium on Computational Complexity, `http://www.eccc.uni-trier.de/eccc/`, December 1995. Earlier results appeared in *Proc. 36th Ann. Symp. Found. Comput. Sci.* (1995), 422–431.
2. J. Canny. Some algebraic and geometric computations in PSPACE. In *Proc. Twentieth ACM Symp. Theor. Comput.*, pp. 460–467, 1988.
3. L. Egidi. The complexity of the theory of p-adic numbers. In *Proc. 34th Ann. Symp. Found. Comput. Sci.*, pp. 412–421, 1993.
4. H. B. Enderton. *A Mathematical Introduction to Logic*. Academic Press, 1972.
5. J. Friedman. A note on matrix rigidity. *Combinatorica* **13** (1993), 235–239.
6. C. Hespel. Approximation de séries formelles par des séries rationnelles. *RAIRO Inform. Théor.* **18** (1984), 241–258.
7. D. Ierardi. Quantifier elimination in the theory of an algebraically-closed field. In *Proc. Twenty-first Ann. ACM Symp. Theor. Comput.*, pp. 138–147, 1989.
8. S. Lang. *Algebra*. Addison-Wesley, 1971.
9. S. V. Lokam. Spectral methods for matrix rigidity with applications to size-depth tradeoffs and communication complexity. In *Proc. 36th Ann. Symp. Found. Comput. Sci.*, pp. 6–16, 1995.
10. Y. V. Matiyasevich. *Hilbert's Tenth Problem*. The MIT Press, 1993.
11. J. Renegar. On the computational complexity and geometry of the first-order theory of the reals. part I: Introduction. preliminaries. the geometry of semi-algebraic sets. the decision problem for the existential theory of the reals. *J. Symbolic Comput.* **13** (1992), 255–299.
12. J. T. Schwartz. Fast probabilistic algorithms for verification of polynomial identities. *J. Assoc. Comput. Mach.* **27** (1980), 701–717.
13. J. O. Shallit. On approximation by rational series in noncommuting variables. Unpublished manuscript, in preparation, 1996.
14. L. Valiant. Graph-theoretic arguments in low-level complexity. In *6th Mathematical Foundations of Computer Science*, Vol. 197 of *Lecture Notes in Computer Science*, pp. 162–176. Springer-Verlag, 1977.
15. L. G. Valiant. Completeness classes in algebra. In *Proc. Eleventh Ann. ACM Symp. Theor. Comput.*, pp. 249–261, 1979.

Algebraic and Logical Characterizations of Deterministic Linear Time Classes

Thomas Schwentick
Universität Mainz

Institut für Informatik, Johannes–Gutenberg Universität Mainz, Germany

Abstract. In this paper an algebraic characterization of the class DLIN of functions that can be computed in linear time by a deterministic RAM using only numbers of linear size is given. This class was introduced by Grandjean, who showed that it is robust and contains most computational problems that are usually considered to be solvable in deterministic linear time.
The characterization is in terms of a recursion scheme for unary functions. A variation of this recursion scheme characterizes DLINEAR, the class which allows polynomially large numbers. A second variation defines a class that still contains DTIME(n), the class of functions that are computable in linear time on a Turing machine.
From these algebraic characterizations, logical characterizations of DLIN and DLINEAR as well as complete problems (under DTIME(n) reductions) are derived.

1 Introduction

Although *deterministic linear time* is a frequently used notion in the theory of algorithms it still does not have a universally accepted formalization in terms of a complexity class. This is merely because what can be computed in deterministic linear time strongly depends on the computational model and the representation of the input. In this paper we give an algebraic characterization, a logical characterization and a complete problem for some of the classes, that have been proposed to capture deterministic linear time, namely the classes DLIN and DLINEAR, that were introduced by Grandjean [13,12,14]. For other attempts to capture deterministic linear time see, e.g., [19,20,17,7].

We first describe the computational model that is used to define DLIN.

As mentioned before, the notion of deterministic linear time depends on the representation of the input. For graphs, as an example, it turns out that algorithms that are usually considered as *linear time* require the input graph to be represented by adjacency lists, which in turn can be easily represented by finite structures: the universe consists of the vertices and edges of G and the adjacency lists of G are encoded by two unary functions, in a natural way.

Grandjean [13,12,14] has pointed out that this is also possible for binary strings as inputs. He views a string x of length n as a sequence $x_0, \dots, x_m$ of words of size $\theta(\log n)$. Of course, every single word can be interpreted as a

Reischuk, Morvan (Eds.): STACS'97 Proceedings, LNCS 1200

number and hence we can represent x, via $f(i) := x_i$, as a unary function on $\{0,\dots,m\}$.

DLIN can be defined as the class of functions on finite structures (with unary functions) that are computable by a Random Access Machine (RAM), which has addition as only arithmetic operation, uses only numbers of size $O(n)$, and needs at most linearly many computation steps under the uniform cost measure. Grandjean [14] showed that DLIN contains most problems that are usually regarded as linear-time.

DLINEAR is defined similarly, but it allows numbers of polynomial size.

Grandjean also defined the nondeterministic counterparts NLIN and NLINEAR of these classes and showed that

- NLIN = NLINEAR [13];
- NLIN coincides with the class of problems that can be characterized by logical formulas of the form $\exists f_1,\dots,f_k\,\forall x\,\varphi$, where the f_i are unary function symbols and φ is a quantifier-free formula [10,11,13,15];
- there are natural complete sets for NLIN under DTIME(n)-reductions. These complete sets are derived from the mentioned logical characterizations. As DTIME(n) $\subset$ NTIME(n) $\subseteq$ NLIN, they are not in DTIME(n) [13].

Here DTIME(n) and NTIME(n) denote deterministic and non-deterministic linear time on multi-tape Turing machines, respectively.

Grandjean asked [13,14] for a natural complete problem and a logical characterization of DLIN. The main intermediate step in attacking these problems is an algebraic characterization of DLIN.

There is a large literature about characterizing computational complexity classes in terms of algebraic recursion schemes, e.g. [4,16,17,5,2,1,9,18,3].

In this paper, we invent a recursion scheme for unary functions that allows composition of functions and simultaneous recursion with the operations iteration, addition, case distinction, bounded composition and bounded search. By varying the power of the bounded search operation we obtain classes $\mathcal{F}_0$, $\mathcal{F}_1$, $\mathcal{F}_2$ for which we can show:

$$\mathrm{DTIME}(n) \subseteq \mathcal{F}_0 \subseteq \mathrm{DLIN} = \mathcal{F}_1 \subseteq \mathrm{DLINEAR} = \mathcal{F}_2.$$

From these results we derive logical characterizations of DLIN and DLINEAR and complete problems for DLIN and DLINEAR.

In Section 2 we give definitions of the considered classes and recursion schemes. In Section 3 we prove the algebraic characterizations of DLIN and DLINEAR and define complete problems for both classes. Section 4 gives the corresponding logical characterizations. In Section 5 we show DTIME(n) $\subseteq \mathcal{F}_0$ and a related result. Section 6 gives a short discussion.

I'd like to thank Clemens Lautemann, Etienne Grandjean, Judy Goldsmith, Ken Regan, Yuri Gurevich and two anonymous referees for a lot of valuable hints and suggestions.

2 Preliminaries

Finite Structures

We only consider signatures[1] σ of the form $(0, m, s, <, \dot{+}, f_0, \dots, f_k)$, where $0, m$ are constant symbols, $s, f_0, \dots, f_k$ are unary function symbols, $<$ is a binary relation symbol and $\dot{+}$ is a binary function symbol. In this paper, all σ-structures A have a universe, denoted $|A|$, of the kind $\{0, \dots, m\}$ and $<, s, 0, m, \dot{+}$ are interpreted as the usual linear order on $|A|$, the corresponding successor function, minimum element, maximum element and addition respectively. We set $s(m) = 0$ and $a \dot{+} b = 0$, if $a+b > m$. Unless otherwise stated, we always have $\sigma = (0, m, s, <, \dot{+}, f_0)$. Often, we do not notationally distinguish between the symbols of σ and their actual interpretations.

We are interested in *global σ-functions* f, i.e. functions that assign to every finite σ-structure A a unary function f^A on $|A|$ (c.f. [16]).

Let x be a binary string of length n. We set $l := \lfloor \frac{1}{2} \log n \rfloor$ and $m := \lfloor \frac{n}{l} \rfloor$. x can be written as $x_0 \circ x_1 \circ \cdots \circ x_m$, where, for every $i < m$, $|x_i| = l$ and $|x_m| < l$. We associate with x the σ-structure A_x with universe $\{0, \dots, m\}$ in which, for every i, $f_0(i)$ is the dyadic interpretation of x_i, i.e.

$$f_0(i) = \sum_{j=0}^{|x_i|-1} (x_i^j + 1) 2^j.$$

Here x_i^j denotes the j-th bit of x_i. We use this dyadic interpretation to assure that different strings are represented by different numbers.

Random Access Machines

In this paper, a simple $\{+\}$-RAM M has two accumulators A, B and registers R_i, for every $i \geq 0$. Its program is a sequence $I(1), \dots, I(r)$ of instructions of the following types.

- $A := 0$
- $A := 1$
- $A := A + B$
- $A := R_A$
- $B := A$
- $R_A := B$
- IF $A = B$ THEN $I(i_0)$ ELSE $I(i_1)$
- HALT

Here R_A denotes the register the number of which is given by the content of A. We require $I(r)$ to be HALT.

If $A = (\{0, \dots, m\}, 0, m, s, <, \dot{+}, f_0)$ is the input structure of M then in the initial configuration of M accumulator B contains m and R_i contains $f_0(i)$, for every $i \leq m$. All other registers and the accumulators have initial value 0. The program starts with instruction $I(1)$. The computation terminates when M

[1] For a general background in Finite Model Theory see [6]. Many of the definitions and notations in this section are essentially from [14].

reaches a HALT statement. Then, the output function g of M is given by the values $R_0, \ldots, R_m$ via $g(i) = R_i$, where we set $g(i) = 0$, whenever $R_i > m$. If we are interested in decision problems, we say that M accepts iff $R_0 \neq 0$.

In some of our proofs we will make use of Multimemory $\{+\}$-RAMs with several sequences $R^1, \ldots, R^l$ of registers and instructions $A := R^j_A$ and $R^j_A := B$, for every $j \leq l$. This does not change the power of the respective classes [12].

In [13,14] the following two classes of deterministic linear time computable problems are defined (Grandjean defines them as classes of functions of strings).

Definition 1.
- Let DLINEAR be the class of global σ-functions that are computed in time $O(m)$ by a $\{+\}$-RAM which uses only numbers (and addresses) of polynomial size in m.
- Let DLIN be the class of global σ-functions that are computed in time $O(m)$ by a $\{+\}$-RAM which uses only numbers (and addresses) of size $O(m)$.

The Algebraic Framework

In the following we define the recursion schemes. The main ingredient is a simultaneous recursion scheme with bounded composition, case distinction, iteration, addition and bounded search.

Definition 2. Let F be a set of global σ-functions and let $f_1, \ldots, f_k$ be unary function symbols. We write H for $\{f_1, \ldots, f_k\} \cup F$ and G_i for $\{f_1, \ldots, f_{i-1}\} \cup F$.

A *Linear Simultaneous Recursion Scheme (LSRS)* S for $f_1, \ldots, f_k$ is a tuple $e_1, \ldots, e_k$ of equations, where every e_i is of one of the following types.

- (*Bounded Composition*)
$$f_i(x) = \begin{cases} h(g(x)) & \text{if } g(x) < x \\ 0 & \text{otherwise} \end{cases}$$
- (*Case Distinction*)
$$f_i(x) = \begin{cases} g_1(x) & \text{if } P(x) \\ g_2(x) & \text{otherwise} \end{cases}$$
where $P(x)$ is a Boolean combination of expressions of the form $g(x) = g'(x)$,
- (*Iteration*)
$$f_i(0) = g(0)$$
$$f_i(x \dot{+} 1) = h(x)$$
- (*Addition*)
$$f_i(x) = g_1(x) \dot{+} g_2(x)$$
- (*Bounded Search*)
$$f_i(x) = \begin{cases} \max\{y < x \mid h(y) = g(x)\} & \text{if such } y \text{ exist} \\ 0 & \text{otherwise} \end{cases}$$

where always $g, g_j \in G_i$ and $h, h_j \in H$.

Lemma 3. *Let S be a LSRS which uses a set F of global σ-functions. Then, for every finite σ-structure A, there are unique functions $f_1^A, \ldots, f_k^A$ on $|A|$ which fulfil*[2] *the equations of S.*

The proof of the lemma is by a double induction on x and i. It relies on the fact that, for every i and x, the computation of $f_i(x)$ only needs values of the form $f_j(y)$ with either $y < x$ or $y = x$ and $j < i$ or of the form $g(y)$ with $g \in F$.

Hence every S defines unique global σ-functions $f_1, \ldots, f_k$. We say that $f_1, \ldots, f_k$ are LSRS-definable w.r.t. F.

Definition 4. Let $\mathcal{F}_1$ be the minimal set of global functions w.r.t. the following rules.

(1) For every constant symbol $c \in \sigma$ the constant function c^A is in $\mathcal{F}_1$.
(2) For every unary function symbol $f \in \sigma$, the global function which maps A to f^A is in $\mathcal{F}_1$.
(3) If $F \subseteq \mathcal{F}_1$ and f is LSRS-definable w.r.t. F then $f \in \mathcal{F}_1$.
(4) (*Composition*)
If $g, h \in \mathcal{F}_1$ and f is defined by $f(x) = h(g(x))$, then f is in $\mathcal{F}_1$.

It will turn out that the bounded search operator plays an important role in our characterizations. Let $\mathcal{F}_0$ be the class of global functions which is defined as $\mathcal{F}_1$ but without bounded search. Let $\mathcal{F}_2$ be the class of global functions which is defined as $\mathcal{F}_1$ but with the following extended rule for bounded search.
(*Extended Bounded Search*)

$$f_i(x) = \begin{cases} \max\{y < x \mid h_1(y) = g_1(x) \text{ and } h_2(y) = g_2(x)\} & \text{if such } y \text{ exist,} \\ 0 & \text{otherwise.} \end{cases}$$

As we will see later, allowing more than two comparisons does not increase the power of $\mathcal{F}_2$.

For every $j \in \{0, 1, 2\}$, we associate with every function $f \in \mathcal{F}_j$ a generation number $N(f)$ according to the following rules.

- If f is in $\mathcal{F}_j$ because of (1) or (2) then $N(f)$ is 0.
- If f is in $\mathcal{F}_j$ because of (3) or (4) then $N(f)$ is the minimal k such that f can be defined by $\mathcal{F}_j$ functions with generation numbers smaller than k.

The following lemma gives us another useful rule.

Lemma 5. *(Unbounded Search)*
If $g, h \in \mathcal{F}_1$ then the following global function f is in $\mathcal{F}_1$.

$$f(x) = \begin{cases} \max\{y \mid h(y) = g(x)\} & \text{if such } y \text{ exist,} \\ 0 & \text{otherwise,} \end{cases}.$$

The corresponding statement holds, with two comparisons, for $\mathcal{F}_2$.

The proof is relatively straightforward.

[2] In case of iteration $f_i(x+1) = h(x)$ needs only to hold for $x < m$.

3 Algebraic Characterizations

In this section we show that $\mathcal{F}_1$ = DLIN and $\mathcal{F}_2$ = DLINEAR. From this we derive complete sets for DLIN and DLINEAR.

Theorem 6. *(a) $\mathcal{F}_1$ = DLIN.*
(b) $\mathcal{F}_2$ = DLINEAR.

Proof. We have to show four inclusions.
$\mathcal{F}_1 \subseteq$ **DLIN:** Let $f \in \mathcal{F}_1$. The proof is by induction on the generation number, $N(f)$. If $N(f) = 0$ then g^A is either a constant function or an input function, which can, of course, be easily computed in $O(m)$ steps on a RAM.

Now let $N(f) > 0$. We are going to describe a Multimemory RAM M which computes f in $O(m)$ steps.

We have to distinguish two cases.

Case 1: f is defined by recursion via $f(x) = h(g(x))$

By induction there exist RAMs M^g and M^h which compute, respectively, g and h in linear time. M simulates M^g and M^h and stores the results in two arrays, R^g and R^h, respectively. Then it can easily compute f.

Case 2: f is defined by an LSRS

By definition there is a (finite) set $F \subseteq \mathcal{F}_1$ of global functions with generation number less than $N(f)$, and global functions $f_1, \ldots, f_k$, with $f_j = g$, for some j, such that $f_1, \ldots, f_k$ are LSRS-definable w.r.t. F by S. By induction, $F \subseteq$ DLIN, hence, for every $h \in F$, there is a RAM M^h which computes h in linear time. For every h, M simulates M^h and stores the result in an array R^h. Then, M computes the values $f_i(t)$, for every t and i, within two loops. In the outer loop, t is counted from 0 to m, in the inner loop i is counted from 1 to k. In the following, we describe how $f_i(t)$ is determined, given that $f_{i'}(t')$ has already been computed for all i', t' with $t' < t$ or $t' = t$ and $i' < i$. M uses one array of registers for every f_i.

Depending on the kind of equation that defines f_i, there are five cases, the most of which are straightforward. The definition of an LSRS assures that all function values that are needed are already computed. It only remains to show how $f_i(t)$ is computed, if f_i is defined by bounded search, i.e.

$$f_i(x) = \begin{cases} \max\{y < x \mid h(y) = g(x)\} & \text{if such } y \text{ exist,} \\ 0 & \text{otherwise,} \end{cases}$$

where $g \in G_i, h \in H$. In this case, M uses an extra array R'^i of registers. In the t-th iteration of the outer loop M stores in the $h(t)$-th register of R'^i the value t. Therefore, to determine $\max\{y \leq t \mid h(y) = g(t)\}$, M only has to read the $g(t)$-th register of R'^i. This can be done in constant time and the update of R'^i needs only constant time for every t.

DLIN $\subseteq \mathcal{F}_1$: Let M be a simple $\{+\}$-RAM, f the global function computed by M. Let M have running time bounded by cm with only using numbers $\leq dm$. In a first step let us assume $c = d = 1$. The proof is similar to the one given for

Theorem 1.3 in [15]. As in that proof, we define functions I, A, B, R_A, R'_A, where $I(t)$ is the number of the instruction that is performed in step t (e.g., $I(0) = 1$), $A(t), B(t)$ are the contents of accumulators A and B, respectively before step t is performed, $R_A(t)$, is the content of the register with number $A(t)$ **before** step t is performed, and $R'_A(t)$ is the content of the register with number $A(t)$ **after** step t has been performed.

In the following we give an LSRS for I, A, B, R_A, R'_A. As before, we make use of a somewhat more relaxed recursion scheme.

$$I(0) = 1.$$

$$I(t\dot{+}1) = \begin{cases} i_0 & \text{if } I(t) \text{ is } \mathbf{IF}\ \boldsymbol{A = B}\ \mathbf{THEN}\ \boldsymbol{I(i_0)}\ \mathbf{ELSE}\ \boldsymbol{I(i_1)} \text{ and } A(t) = B(t) \\ i_1 & \text{if } I(t) \text{ is } \mathbf{IF}\ \boldsymbol{A = B}\ \mathbf{THEN}\ \boldsymbol{I(i_0)}\ \mathbf{ELSE}\ \boldsymbol{I(i_1)} \text{ and } A(t) \neq B(t) \\ I(t) & \text{if } I(t) \text{ is HALT} \\ I(t)\dot{+}1 & \text{otherwise} \end{cases}$$

$$A(0) = 0.$$

$$A(t\dot{+}1) = \begin{cases} 0 & \text{if } I(t) \text{ is } \boldsymbol{A := 0} \\ 1 & \text{if } I(t) \text{ is } \boldsymbol{A := 1} \\ A(t)\dot{+}B(t) & \text{if } I(t) \text{ is } \boldsymbol{A := A + B} \\ R_A(t) & \text{if } I(t) \text{ is } \boldsymbol{A := R_A} \\ A(t) & \text{otherwise} \end{cases}$$

$$B(0) = 0.$$

$$B(t\dot{+}1) = \begin{cases} A(t) & \text{if } I(t) \text{ is } \boldsymbol{B := A} \\ B(t) & \text{otherwise} \end{cases}$$

$$R_A(t) = \begin{cases} R'_A(\max\{t' < t \mid A(t') = A(t)\}) & \text{if such } t' \text{ exist} \\ f_0(A(t)) & \text{otherwise} \end{cases}$$

$$R'_A(t) = \begin{cases} B(t) & \text{if } I(t) \text{ is } \boldsymbol{R_A := B} \\ R_A(t) & \text{otherwise} \end{cases}$$

We note that the function I only takes a fixed number of values. In the conditions of the above case distinctions, as an example, "$I(t)$ is $\boldsymbol{A := 0}$" is an abbreviation for "$I(t) = i_1$ or $I(t) = i_2$ or ... or $I(t) = i_q$", where the i_j are all numbers of instructions $\boldsymbol{A := 0}$. (Here, of course, i_j is an abbreviation for $s^{i_j}(0)$). The expression "$I(t)$ is $\mathbf{IF}\ \boldsymbol{A = B}\ \mathbf{THEN}\ \boldsymbol{I(i_0)}\ \mathbf{ELSE}\ \boldsymbol{I(i_1)}$" has to be interpreted similarly, depending on the possible values of i_0 and i_1.

The output function f of M is defined by (cf. Lemma 5)

$$g(t) = \begin{cases} R'_A(\max\{t' \mid A(t') = t\} & \text{if such } t' \text{ exist} \\ f_0(t) & \text{otherwise.} \end{cases}$$

In the following, we are going to sketch how this proof can be extended to the cases where $c > 1$ and/or $d > 1$.

In case $d > 1$, we replace A (and analogously B, R_A, R'_A) by functions $A^0, \ldots, A^{d-1}$ such that, for every t, it holds

- $A^i(t) = r$, if $A(t) = im + r$, with $0 < r \leq m$ (if such an i exists),
- $A^j(t) = 0$, for all other j.

We note that $A(t) > 0$ implies $A^i(t) > 0$ for at least one i.
In case $c > 1$ we proceed in c stages, each simulating m steps.

$\mathcal{F}_2 \subseteq$ **DLINEAR:** Compared to the case of DLIN only one modification is needed. If f is defined by bounded search with functions g_1, g_2, h_1, h_2 then, for every t, in the extra array the value t is stored in the register with number $g_1(t) + g_2(t)(m+1)$.

DLINEAR $\subseteq \mathcal{F}_2$: In [12] it was shown that, for every $\epsilon > 0$ any function in DLINEAR can be computed by a RAM M that uses only numbers of size $O(m^{1+\epsilon})$. Hence, any number z used by M can be encoded by a pair $(h(z), l(z))$ of numbers, where $z = h(z)(m+1) + l(z)$. Consequently, for every function g used in the proof of DLIN $\subseteq \mathcal{F}_1$ (with the exception of I) we use functions g^l, g^h that encode the high and low part of the values of g, respectively. Instead of bounded search we use extended bounded search. A further straightforward modification is needed for the simulation of $\boldsymbol{A} := \boldsymbol{A} + \boldsymbol{B}$.

Now we sketch the definition of complete problems for DLIN and DLINEAR. Let f be a function in $\mathcal{F}_1$. Let S be a complete recursion scheme for f. For a structure A we define the expanded recursion scheme $E(S, A)$ to be the sequence of equations $f_0(0) = f_0^A(0), \ldots, f_0(m) = f_0^A(m)$ followed by $S(0), \ldots, S(m)$, where every $S(j)$ consists of a copy of S in which every occurence of x is replaced by j.

Let Expand$_1$ be the set of all encodings of expanded recursion schemes $E(S, A)$ that fulfil $f(0) \neq 0$ for the function f with maximal index in S and let Expand$_2$ be defined analogously, with extended bounded search.

Then, Expand$_1$ turns out to be complete for the decision problems in DLIN under DTIME(n)-reductions, and Expand$_2$ is complete for the decision problems in DLINEAR under DTIME(n)-reductions.

4 Logical Characterizations

In this section we are going to sketch how one can easily get logical characterizations of DLIN and DLINEAR from the algebraic characterizations of Section 3. Grandjean [10,11,13] has shown that NLIN is the class of global σ-functions that can be defined by formulas of the form

$$\exists f_1, \ldots, f_k \forall x \varphi,$$

where the f_i are unary function symbols and φ is a quantifier-free formula. Our characterizations of DLIN and DLINEAR use formulas of almost the same type but with φ syntactically restricted in a way that allows the fast deterministic computation of the f_i. This approach should be compared with Grädel's characterization of **P** by Σ^1_1-HORN formulas [8].

Definition 7. Let $F = \{f_0, f_1, \ldots, f_k\}$ be a set of unary function symbols. Let every symbol f of F have a generation number $gn(f)$, such that $gn(f_i) \leq gn(f_j)$ whenever $i < j$. Let, for every $i \leq k$,

- $F_i = \{f_0, f_1, \ldots, f_{i-1}\}$,
- G_i the set of all function symbols f with $gn(f) \leq gn(f_i)$,
- H_i the set of all function symbols f with $gn(f) < gn(f_i)$.

We say a formula Φ of the form $\exists f_1, \ldots, f_k \forall x \bigwedge_{i=0}^{k} \varphi_i$ is 1-linear iff φ_0 is quantifier-free and, for every $i \geq 1$, φ_i is of one of the following types.

- *(Bounded Composition)*
  $[g(x) < x \wedge f_i(x) = h(g(x))] \vee [g(x) \geq x \wedge f_i(x) = 0]$, where $g \in F_i, h \in G_i$,
- *(Composition)*
  $f_i(x) = h(g(x))$, where $g \in F_i, h \in H_i$,
- *(Case Distinction)*
  $[P(x) \wedge f_i(x) = g(x)] \vee [\neg P(x) \wedge f_i(x) = g'(x)]$, where $P(x)$ is a boolean combination of formulas of type $h(x) = h'(x)$ and g, g' and all h, h' are in F_i,
- *(Iteration)*
  $[x = 0 \wedge f_i(x) = g(x)] \vee [s(x) \neq 0 \wedge f(s(x)) = h(x)]$, where $g \in F_i, h \in G_i$,
- *(Addition)*
  $f_i(x) = g(x) + h(x)$, where $g, h \in F_i$,
- *(Bounded Search)*
  $[f_i(x) = 0 \wedge \forall y(y < x \rightarrow g(y) \neq h(x))] \vee [f_i(x) < x \wedge g(f_i(x)) = h(x) \wedge \forall y(y \geq x \vee y \leq f_i(x) \vee g(y) \neq h(x))]$, where $g \in G_i, h \in F_i$,

It is easy to show that a set of finite σ-structures is in $\mathcal{F}_1$, if and only if it is definable by a 1-linear formula. Analogous characterizations can be given for $\mathcal{F}_0$ and $\mathcal{F}_2$.

Remark 8. (a) The additional universal quantifier in the subformulas for bounded search can be avoided by using more complicated formulas that define f_i and some additional functions.

(b) At the cost of even more complicated formulas it is also possible to get rid of the requirement that the input structure contains an addition relation and a linear order. The construction uses two projection functions p_h, p_l such that $x = p_h(x)\lfloor\sqrt(m)\rfloor + p_l(x)$, for every x.

5 $\mathcal{F}_0$ contains DTIME(n)

In this section we sketch the proof DTIME(n) $\subseteq \mathcal{F}_0$ and a related result. For a characterization for deterministic linear time on Turing machines see [3].

Theorem 9. *DTIME*(n) $\subseteq \mathcal{F}_0$

Proof. We can only give a sketch of the proof here. The proof is similar to that of DLIN $\subseteq \mathcal{F}_1$ but it makes essential use of the fact that the head of a Turing machine M cannot jump. At every time we only need to recover the contents of those segments of the tape that are immediately to the right and the left of the current segment.

Let $k := (\log n)/4$. In a first step it is shown that there are functions in $\mathcal{F}_0$ that describe the effect of M's computation on $\leq k$ cells of its tape within $\leq k$ steps. These functions depend only on the size of the input but not on its actual content. Afterwards the computation of M is simulated by several other functions. The two main ideas are as follows.

- We divide the tape into segments of length $k' := \frac{k}{3}$. By one "function time step" we simulate the computation of k' steps of M on three consecutive tape segments where, at the beginning of these k' steps, Ms head is on the middle segment (this assures that the head does not leave the three segments within k' steps).
- There are two functions which give the last time step at which the right, resp. left, neighbour of the current segment was visited.

We do not expect the opposite inclusion to hold.

In the following we sketch an apparently weaker result.

Let $\mathcal{F}_0^-$ be defined like $\mathcal{F}_0$ but without bounded composition. We note that $\mathcal{F}_0^-$ is still powerful enough to recognize palindromes.

Grandjean [14] invented the class DTIME$_{\text{SORT}}(n)$ which is the class of functions that are computed by a deterministic Turing machine M in time $O(n)$, where M is allowed to make use of a constant number of sorting operations. A sorting operation is performed if M enters a special state q_s. Whenever this happens the content of a special sorting tape T_S is interpreted as a sequence $S = x_1; x_2; \dots ; x_p$ of strings which is transformed within one step into a sequence $S' = y_1; y_2; \dots ; y_p$ such that both sequences contain exactly the same strings and S' is sorted lexicographically.

Grandjean has shown that DTIME$(n) \subseteq$ DTIME$_{\text{SORT}}(n) \subseteq$ DLIN.

For a σ-structure A with universe $\{0, \dots , m-1\}$, let the corresponding binary string x_A consist of the string $F(0) \circ F(1) \circ \cdots \circ F(m-1)$ where $F(i)$ is the binary representation of $f_0(i)$ of length $\lceil \log n \rceil$, hence, $|x| = \theta(m \log m)$.

Theorem 10. *If a global function f of finite σ-structures is in $\mathcal{F}_0^-$ then the corresponding function on binary strings is in DTIME$_{SORT}(n)$.*

Proof. Let $f \in \mathcal{F}_0^-$. The Turing machine M uses one tape T_g for every function g that occurs in the complete recursion scheme of f. M writes $(0, g(0))(1, g(1)) \dots ,$ $(m-1, g(m-1))$ onto T_g. If g is defined by case distinction, iteration or addition, then this is easy to do, because all function values that are needed are of the same argument and therefore the computation of $g(x)$ (or $g(x+1)$) can be done in logarithmic many steps.

If g is defined by $g(x) = h(h'(x))$ with functions that have already been computed then g is computed as follows. M writes $(h'(0), 0); (h'(1), 1); \dots ; (h'(m), m)$

on the extra tape T_s and enters the sorting state. Let $(y_0, z_0); \ldots ; (y_m, z_m)$ be the sorted sequence. M writes $(z_0, h(y_0)); \ldots ; (z_m, h(y_m))$ on T_s. This can be done in linear time as the y_i appear in increasing order. Finally, M enters the sorting state. The resulting sequence is $(0, h(h'(0))); \ldots ; (m, h(h'(m)))$. As the recursion scheme is fixed, M only needs a constant number of sorting operations.

6 Conclusion

Figure 1 shows the relations between computationally and algebraically defined classes that were established in this paper.

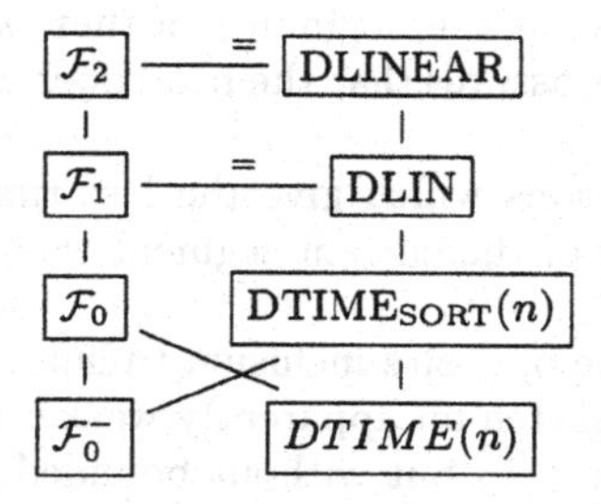

Fig. 1. Inclusion structure of the investigated classes.

One interesting aspect of these results is that the variation of one parameter, namely the power of the bounded search operator, reflects the apparent difference of DTIME(n), DLIN and DLINEAR. Furthermore we gave logical characterizations of DLIN and DLINEAR and derived complete problems for these classes. All these results support the robustness of DLIN and DLINEAR.

We conclude with some open problems:

- Can one find nicer logical characterizations ? Can one derive game characterizations to make lower bound proofs possible ?
- Is there a similar recursion scheme which characterizes deterministic linear-time Turing machines ? What is the computational counterpart of $\mathcal{F}_0$. Is there a corresponding Turing machine model ? (How can a TM handle bounded composition ?)
- Are there more natural complete problems for DLIN and DLINEAR?
- Are there similar algebraic characterizations of NLIN ?
- From the algebraic characterizations one can also derive normal forms for linear time algorithms. Does this give any new insights ?
- Can the algebraic characterizations be simplified, maybe avoiding simultaneous recursion ?

References

1. S. Bellantoni. Predicative recursion and the polytime hierarchy. In P. Clote and J.B. Remmel, editors, *Feasible Mathematics II*, pages 15–29. Birkhäuser, 1995.
2. S. Bellantoni and S. Cook. A new recursion-theoretic characterization of the polytime functions. *Comp. Complexity*, 2:97–110, 1992.
3. S.A. Bloch, J.F. Buss, and J. Goldsmith. Sharply bounded alternation within P. Technical Report TR96-011, Electronic Colloquium on Computational Complexity, 1996. http://ftp.eccc.uni-trier.de/pub/eccc/reports/1996/TR96-011/index.html.
4. A. Cobham. The intrinsic computational complexity of functions. In Y. Bar-Hillel, editor, *Proceedings of: Logic, Methodology, and the Philosophy of Science*, pages 24–30. North-Holland, 1964.
5. K. J. Compton and C. Laflamme. An algebra and a logic for NC^1. *Information and Control*, 87:241–263, 1990.
6. H.-D. Ebbinghaus and J. Flum. *Finite Model Theory*. Springer-Verlag, 1995.
7. E. Grädel. On the notion of linear time computability. *Int. J. Found. Comput. Sci.*, 1:295–307, 1990.
8. E. Grädel. The expressive power of second order horn logic. In *Proc. 8th Symposium on Theoretical Aspects of Computer Science STACS 91*, volume 480 of *LNCS*, pages 466–477. Springer-Verlag, 1991.
9. E. Grädel and Y. Gurevich. Tailoring recursion for complexity. *Journal of Symbolic Logic*, 60:952–969, 1995.
10. E. Grandjean. A natural NP-complete problem with a nontrivial lower bound. *SIAM Journal of Computing*, 17:786–809, 1988.
11. E. Grandjean. A nontrivial lower bound for an NP problem on automata. *SIAM Journal of Computing*, 19:438–451, 1990.
12. E. Grandjean. Invariance properties of RAMs and linear time. *Computational Complexity*, 4:62–106, 1994.
13. E. Grandjean. Linear time algorithms and NP-complete problems. *SIAM Journal of Computing*, 23:573–597, 1994.
14. E. Grandjean. Sorting, linear time and the satisfiability problem. *Annals of Mathematics and Artificial Intelligence*, 16:183–236, 1996.
15. E. Grandjean and F. Olive. Monadic logical definability of NP-complete problems. submitted, 1996.
16. Y. Gurevich. Algebras of feasible functions. In *Proc. 24th IEEE Symp. on Foundations of Computer Science*, pages 210–214, 1983.
17. Y. Gurevich and S. Shelah. Nearly linear time. In A. Meyer and M. Taitslin, editors, *Logic at Botik '89, Lecture Notes in Computer Science 363*, pages 108–118. Springer-Verlag, 1989.
18. D. Leivant. Ramified recurrence and computational complexity I: Word recurrence and poly-time. In P. Clote and J.B. Remmel, editors, *Feasible Mathematics II*, pages 320–343. Birkhäuser, 1995.
19. K. Regan. Machine models and linear time complexity. *SIGACT News*, 24:4, Fall 1993.
20. C. P. Schnorr. Satisfiability is quasilinear complete in NQL. *Journal of the ACM*, 25:136–145, 1978.

Finding the k Shortest Paths in Parallel

Eric Ruppert

Department of Computer Science
University of Toronto
Toronto, Ontario, Canada M5S 1A4

Abstract. A concurrent-read exclusive-write PRAM algorithm is developed to find the k shortest paths between pairs of vertices in an edge-weighted directed graph. Repetitions of vertices along the paths are allowed. The algorithm computes an implicit representation of the k shortest paths to a given destination vertex from every vertex of a graph with n vertices and m edges, using $O(m + nk \log^2 k)$ work and $O(\log^3 k \log^* k + \log n(\log\log k + \log^* n))$ time, assuming that a shortest path tree rooted at the destination is precomputed. The paths themselves can be extracted from the implicit representation in $O(\log k + \log n)$ time, and $O(n \log n + L)$ work, where L is the total length of the output.

Topics: parallel algorithms, data structures

1 Introduction

The problem of finding shortest paths in an edge-weighted graph is an important and well-studied problem in computer science. The more general problem of computing the k shortest paths between vertices of a graph also has a long history and many applications to a diverse range of problems. Many optimization problems may be formulated as the computation of a shortest path between two vertices in a graph. Often, the k best solutions to the optimization problem may then be found by computing the k shortest paths between the two vertices. A method for computing the k best solutions to an optimization problem may be useful if some constraints on the feasible solutions are difficult to specify formally. In this case, one can enumerate a number of the best solutions to the simpler problem obtained by omitting the difficult constraints, and then choose from among them a solution that satisfies the additional constraints. Knowledge of the k best solutions to an optimization problem can also be helpful when determining whether the optimal solution is sensitive to small changes in the input. If one of the best solutions is very different from the optimal solution but has a cost that is only slightly sub-optimal, it is likely that minor modifications to the problem instance would cause the sub-optimal solution to become optimal.

Sequential algorithms which compute the k best solutions to an optimization problem first compute an optimal solution using a standard algorithm. A number of candidates for the second best solution are then generated by modifying the optimal solution, and the algorithm outputs the best candidate as the second

Reischuk, Morvan (Eds.): STACS'97 Proceedings, LNCS 1200

best solution to the problem. In general, the kth best solution is chosen from a set of candidates, each one a modification of one of the best $k-1$ solutions. It seems that this approach cannot be used directly to obtain parallel algorithms with running times that are polylogarithmic in k, since the best $k-1$ solutions must be known before the algorithm can compute the kth best solution. A different technique is used here to produce a parallel algorithm for the k shortest paths problem with a running time that is polylogarithmic in k, and in the size of the problem instance.

A parallel algorithm will be developed in Sect. 3 to compute the k shortest paths to a given vertex t from every vertex of a weighted directed graph. It is assumed that the weights on the edges are positive, but the algorithm can easily be adapted to handle negative edge weights, as long as there are no negative cycles in the graph. The algorithm runs on a concurrent-read exclusive-write (CREW) PRAM. (See Karp and Ramachandran's survey [15] for definitions of PRAM models.) The algorithm finds the k shortest paths to t from every vertex in $O(\log^3 k \log^* k + \log n \log\log k + \log d \log^* d)$ time using $O(m+nk\log^2 k)$ work, where d is the maximum outdegree of any vertex in the graph, assuming that the shortest path to t from every other vertex is given. The algorithm computes an implicit representation of the paths, from which the paths themselves can be extracted in parallel using the techniques to be described in Sect. 3.2. New parallel algorithms for the weighted selection problem and the problem of selecting the kth smallest element in a matrix with sorted columns, which are used as subroutines, are outlined in Sect. 3.3. Some applications of the k shortest paths algorithm are described in Sect. 4.

Previous Work Dijkstra's sequential algorithm computes the shortest path to a given destination vertex from every other vertex in $O(m + n \log n)$ time [12]. In parallel, the shortest path between each pair of vertices can be found using a min/sum transitive closure computation in $O(\log^2 n)$ time and $O(n^3 \log n)$ work on an EREW PRAM [17]. More complicated implementations of the transitive closure computation run in $O(\log^2 n)$ time using $o(n^3)$ work on the EREW PRAM and in $O(\log n \log\log n)$ time on the CRCW PRAM [13]. There are no known polylogarithmic-time PRAM algorithms that find the shortest path from one particular vertex to another using less work than the all-pairs algorithm. This transitive closure bottleneck is not avoided by the algorithm presented here: the complexity bounds on the algorithm describe the amount of additional time and work to compute the k shortest paths, once the shortest paths are known.

The problem of finding the k shortest paths in sequential models of computation was discussed as early as 1959 by Hoffman and Pavley [14]. Fox presents an algorithm that can be implemented to run in $O(m + kn \log n)$ time [9]. Eppstein's recent sequential algorithm [7] is a significant improvement. It computes an implicit representation of the k shortest paths for a given source and destination in $O(m + n \log n + k)$ time. The k shortest paths to a given destination from every vertex in the graph can be found, using Eppstein's algorithm, in $O(m + n \log n + nk)$ time. The paths themselves can be extracted from the implicit representation in time proportional to the number of edges in the paths. A brief description of Eppstein's algorithm is given in Sect. 2.1. The k shortest paths problem has not previously been studied in parallel models of computation.

Sequential algorithms have been developed for other variations of the k shortest paths problem. Yen [23] gives an algorithm for the more difficult problem of finding the k shortest simple paths in $O(kn^3)$ time. Katoh, Ibaraki and Mine [16] describe an $O(kn^2)$ algorithm to find the k shortest simple paths in an undirected graph.

2 Preliminaries

Let $G = (V, E)$ be a directed graph with n vertices and m edges, where each edge (u, v) of E has a non-negative weight $w(u, v)$. The weight of a path in G is simply the sum of the weights of the edges that make up the path. The distance from vertex s to vertex t, $\text{dist}(s, t)$, is defined to be the weight of the path from s to t that minimizes this sum. A path that achieves this distance is a shortest path from s to t.

The problem of finding the k shortest paths from vertex s to vertex t is to find a set $\mathcal{P}$ of k s-t paths such that the weight of any path in $\mathcal{P}$ is no larger than the weight of any s-t path not in $\mathcal{P}$. There may be several paths in $\mathcal{P}$ with the same weight. If there are fewer than k distinct paths from s to t, the solution set $\mathcal{P}$ should consist of all s-t paths. Here, paths are not restricted to being simple; a vertex may appear more than once on a path.

Let T be a tree with root t that is a subgraph of G and is constructed so that the (unique) path in T to t from any vertex v is a shortest v-t path in G. The tree T is called a shortest path tree of G rooted at t. The edges of the graph which do not appear in T are called non-tree edges. Any path from a fixed vertex s to vertex t can be represented by the sequence of non-tree edges along the path. For example, in the graph shown in Fig. 1(a), edges of T are shown as solid lines, and non-tree edges are shown as dashed lines. In this graph, the path $s \dashrightarrow c \longrightarrow a \dashrightarrow f \longrightarrow b \longrightarrow t$ could be represented by the sequence of non-tree edges, $< (s, c), (a, f) >$; the rest of the path can be filled in by following the edges of T. If p is any path, let sidetracks(p) be the sequence of non-tree edges that occur along the path p.

For each edge $(u, v) \in E$, one can define a measure $\delta(u, v)$ of the extra distance added to the weight of a path from u to t if the edge (u, v) is used instead of taking the optimal path from u to t:

$$\delta(u, v) = w(u, v) + \text{dist}(v, t) - \text{dist}(u, t).$$

The following lemma describes some properties of this measure.

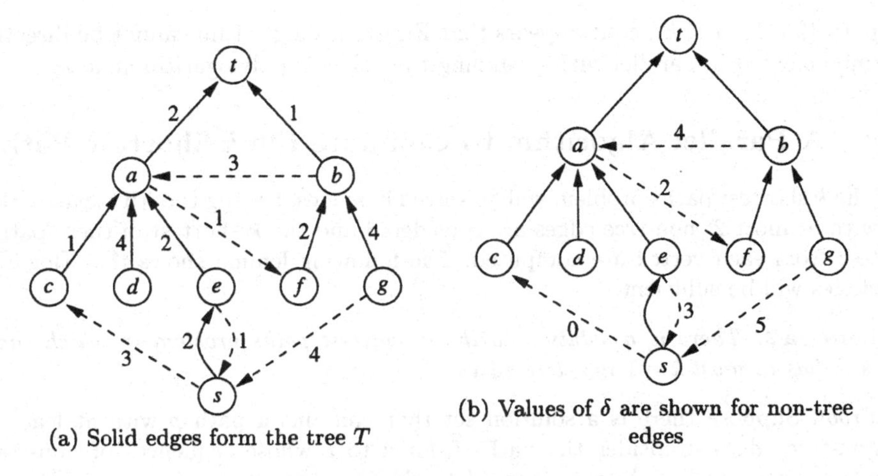

(a) Solid edges form the tree T

(b) Values of δ are shown for non-tree edges

Fig. 1. an example graph

Lemma 1 (Eppstein [7]).
(i) $\delta(u,v) \geq 0$ *for all* $(u,v) \in E$.
(ii) $\delta(u,v) = 0$ *for all* $(u,v) \in T$.
(iii) For any path p from s to t,

$$\text{weight}(p) = \text{dist}(s,t) + \sum_{(u,v)\in p} \delta(u,v) = \text{dist}(s,t) + \sum_{(u,v)\in \text{sidetracks}(p)} \delta(u,v).$$

To find the k shortest paths from s to t, it is therefore sufficient to find the paths p which yield the k smallest values of $\delta(p) = \sum_{(u,v)\in p} \delta(u,v)$. If δ is viewed as a weight function on the edges of G, a general instance of the k shortest paths problem has now been transformed into an instance where the distance to vertex t from any other vertex is 0. From now on, the weight function δ will be used instead of w.

2.1 Eppstein's Sequential Algorithm

Eppstein's sequential algorithm [7] computes an implicit representation of the k shortest paths. Each path's sidetracks sequence is represented as a modification of the sidetracks of a shorter path. Candidates for the kth shortest path are obtained from one of the shortest $k-1$ paths either by adding a non-tree edge to the sidetracks sequence or by replacing the last non-tree edge by another one. The algorithm constructs a new weighted directed graph G' in which each path starting at a fixed vertex s' (and ending at any other vertex) corresponds to an s-t path of G. This correspondence is bijective and weight-preserving, so the k shortest s-t paths of G can be found by computing the k shortest paths that begin at s' in G', using Frederickson's algorithm [10]. It is possible that the ith shortest path is obtained by adding a non-tree edge to the $(i-1)$th shortest path (for $2 \leq i \leq k$), so it appears that Eppstein's algorithm cannot be directly implemented in parallel with a running time that is polylogarithmic in k.

3 A Parallel Algorithm to Compute the k Shortest Paths

The k shortest paths problem will be solved in stages. During the ith stage, paths with at most 2^i non-tree edges are considered and the k shortest of these paths to t from each vertex are computed. The following lemma shows that $\lceil \log k \rceil$* stages will be sufficient.

Lemma 2. *There is a solution to the k shortest paths problem in which each path has at most $k-1$ non-tree edges.*

Proof. Suppose there is a solution set that contains a path p with at least k non-tree edges. Consider the paths from v to t whose sequences of non-tree edges are prefixes of the sequence sidetracks(p), where the prefixes are of length $0, 1, 2, \ldots, k-1$. There are k such paths, and each one has weight no greater than the weight of p, since δ is non-negative. So, the set of these paths is a correct solution to the problem, and each path has at most $k-1$ non-tree edges. □

The list of edges that make up each path will not be explicitly computed in each stage. Instead, each path is represented by a binary tree structure whose leaves represent the non-tree edges along the path. The ith stage constructs the implicit representations of paths by concatenating the sidetracks sequence of two paths that have been computed in previous stages. The result of such a concatenation is stored in the data structure by creating a new node, which will be the root of the tree representing the path, and setting its children pointers to point to the roots of the two smaller paths. Thus, the tree structures representing different paths the data structure may share common subtrees, and the height of the trees constructed during the ith stage have height at most i. Some additional information will be stored in each node of the tree structures to allow the the computations to be performed efficiently.

3.1 The Data Structure Used by the Algorithm

Let A_v^i be an array that will store the root nodes of the tree structures that represent the k shortest paths from v to t that have at most 2^i non-tree edges. If there are fewer than k such paths, some of the entries in the array will be nil. Elements of the array A_v^i will be formed by concatenating the sequences of non-tree edges of two paths with at most 2^{i-1} non-tree edges each.

Each array element stores the following information about the path p that it represents:

- pointers to the two previously computed paths whose sidetracks sequences were concatenated to form p, unless p contains only a single non-tree edge, in which case this edge is stored instead,

* All logarithms have base 2.

- the weight of the path (with respect to the weight function δ),
- the number of non-tree edges along the path,
- num, the number of edges on the path up to and including the last non-tree edge, and
- the head of the last non-tree edge on the path (this could be nil if all edges along the path are in T).

In the next section, a parallel algorithm is given for extracting the k shortest paths from this implicit representation. This is done by allocating processors to traverse the leaves of the tree structures that represent the paths to obtain the sidetracks sequences and filling in the rest of the edges along the paths by traversing branches of the tree T. The actual construction of the data structure is described in Sect. 3.3.

3.2 Extracting Information from the Data Structure

First, some preprocessing is done to the shortest path tree, T. Each vertex uses pointer jumping to locate a pointer to its ancestor 2^i levels above itself, for $i = 1, 2, \ldots, \lceil \log(n-1) \rceil$. Some of these pointers may be nil. This is done so that portions of the k shortest paths that are made up exclusively of tree edges can be traversed quickly. This computation uses $O(\log n)$ time and $O(n \log n)$ work.

Suppose that the jth shortest path from v to t contains l_j edges. The k shortest paths from v to t can then be explicitly stored one after another in an array P of size $L = \sum_{j=1}^{k} l_j$. The starting location of each path in the array P can be found by performing a prefix sum (see [15]) on $l_1, \ldots, l_k$.

Suppose $L/\log(kn)$ processors are available. Each processor is assigned the task of filling in a block of the output array P of length $\log(kn)$. The processor follows the pointers in the tree data structure that represents the path, starting from the root and going to the appropriate leaf to find the first edge it must write into P. At each node, the num field gives the number of edges in the subpath represented by the subtree rooted at that node, so that the processor can determine whether to go left or right at each node on its way to the leaf. When it reaches a leaf, the processor begins filling in entries of P sequentially. The portion of P that the processor must compute is made up of segments of branches of T, separated by non-tree edges. The processor can perform a linear traversal of the branches of T, copying the edges into P one by one. Whenever the processor reaches the end of a segment of tree edges, it traverses the tree data structure that represents the path, to the next leaf, which represents the next non-tree edge on the path. Once the non-tree edge has been entered into P, the processor can again start copying a segment of a branch of T into P.

If the first edge that the processor is required to enter into P is in the middle of a segment of tree edges in the path, the processor can jump to the correct point in T in $O(\log n)$ time using the ancestor pointers computed during the preprocessing of T. If a processor finishes entering one of the k shortest paths into P, it starts working on the next one. The total time to compute the output array

P, including the time to preprocess T, is $O(\log n + \log(kn)) = O(\log k + \log n)$ and the total work performed is $O(n \log n + L)$. The k shortest paths to t from every vertex v can be extracted in $O(\log k + \log n)$ time using $O(n \log n + L_{total})$ work, where L_{total} is the total length of the output for all starting vertices v, since the preprocessing of T need only be done once.

In fact, some properties of the paths can be computed without explicitly listing the edges in the path, as observed by Eppstein [7]. Suppose each edge in the graph is assigned a value from a semigroup, and the value of a path is defined as the product of the values of the edges along the path. If the associative semigroup operation can be evaluated in constant time by a single processor, the values of the k shortest paths can be computed in the same way as the num field of the data structure, without affecting the performance bounds.

3.3 Building the Data Structure

The construction of the data structure will be performed in stages. Stage i of the algorithm will compute A_v^i for each vertex v using the arrays computed in the previous stage.

First, the construction of A_v^0 is described. The first entry of A_v^0 will be the path from v to t in T. It has weight 0 and contains no non-tree edges. The rest of the paths in A_v^0 will each contain exactly one non-tree edge, so the tail of each of these edges must lie on the path from v to t in the tree T. Thus, A_v^0 can be computed by finding the $k-1$ shortest non-tree edges (with respect to δ) whose tails are on the path from v to t in T.

First, the $k-1$ shortest non-tree edges whose tails are at vertex v are selected, for each vertex v in the graph. This can be done using $O(\log d_v \log^* d_v)$ time and $O(d_v)$ work, where d_v is the outdegree of v, using Cole's selection algorithm [5]. In total, this requires $O(\log d \log^* d)$ time and $O(m)$ work, where d is the maximum outdegree of any vertex in the graph. In addition, $O(\log n)$ time and $O(n)$ work is used to allocate the appropriate number of processors to each vertex using a prefix sum computation. Cole's parallel merge sort [6] can be used to sort the $k-1$ smallest edges out of each vertex in $O(\log k)$ time using $n(k-1)$ processors.

Tree contraction is used to compute the array of the k shortest edges whose tails are on the path from each vertex v to the destination t. The tree contraction is similar to that described in [18] for computing the minimum weight ancestor of each node in a node-weighted tree. Instead of each primitive operation finding the minimum of two node weights, it computes the k smallest elements in a pair of sorted arrays, each containing k elements. This primitive operation is performed by merging the two sorted arrays, and then taking the first half of the resulting array. Ties between edge weights can be broken according to some arbitrary lexicographic order on the edges. Each merge step can be performed in $O(\log \log k)$ time and $O(k)$ work using Borodin and Hopcroft's merging algorithm [1], so the tree contraction takes $O(\log n \log \log k)$ time and $O(nk)$ work in total.

The num fields of the paths found during stage 0 can be filled in as follows. First, the depth of each vertex in T is computed using tree contraction. The algorithm is similar to the computation of minimum ancestors in [18], except that the minimization operations are replaced by additions, and each vertex in the tree is assigned a value of 1, except the root, which has value 0. This computation uses $O(\log n)$ time and $O(n)$ work. The value of num for any path from v to t found during stage 0 of the algorithm can then be computed easily: if the path contains the single non-tree edge (x, y), the value of num is depth(v) − depth$(x) + 1$.

Stage 0 of the algorithm uses $O(\log d \log^* d + \log k + \log n \log\log k)$ time in total, and performs $O(m + nk \log k)$ work.

Now, the computation of A_v^i, for $i > 0$ is described. The candidate paths for inclusion in A_v^i are those paths whose sidetracks sequence is obtained by concatenating sidetracks(p_1) and sidetracks(p_2), where p_1 is a path in the array A_v^{i-1}, p_2 is a path in A_w^{i-1}, and w is the head of the last non-tree edge of p_1. Any sidetracks sequence formed in this way represents a legal path, since there is a path in T from the head of the last non-tree edge of p_1 to the tail of the first non-tree edge in p_2. For example, in the graph of Fig. 1, combining the paths $p_1 = g \dashrightarrow s \dashrightarrow c \longrightarrow a \longrightarrow t$ and $p_2 = c \longrightarrow a \dashrightarrow f \longrightarrow b \longrightarrow t$ produces the path $g \dashrightarrow s \dashrightarrow c \longrightarrow a \dashrightarrow f \longrightarrow b \longrightarrow t$.

Paths p_1 containing l non-tree edges will be combined only with paths p_2 which have either l or $l-1$ non-tree edges. This ensures that each path considered is distinct, since any sequence of non-tree edges can be divided into such a pair p_1, p_2 in exactly one way. In fact, it is sufficient to consider only those pairs which, when concatenated, yield a path that has more than 2^{i-1} non-tree edges, since the k shortest paths with at most 2^{i-1} non-tree edges are already known.

From among these candidates, together with the paths in A_v^{i-1}, the shortest k are chosen to be the entries in A_v^i. To prove that this algorithm solves the problem correctly, it is helpful to define a total ordering $\prec_v$ on the paths from v to t so that an invariant describing the contents of A_v^i can be stated precisely. Paths are ordered by weight, and ties are broken in favour of the path with fewer non-tree edges. If two paths from v to t have the same weight and the same number of non-tree edges, then the lexicographic order of the sidetracks sequences is used to break the tie. This total ordering is used to select the k shortest paths for inclusion in A_v^i.

Lemma 3. *The paths in A_v^i are the k smallest paths (with respect to the order $\prec_v$) with at most 2^i non-tree edges. If there are fewer than k such paths, all of them appear in A_v^i.*

Proof. The computation of A_v^0 described above ensures that the invariant is true for $i = 0$.

Assume that the invariant is true for $i - 1$. Let p be any path from v to t with l_p non-tree edges, where $2^{i-1} < l_p \leq 2^i$. Let p_1 be the portion of p up to and including the $\lceil l_p/2 \rceil$ th non-tree edge. Let w be the last vertex on p_1, and let p_2 be the remaining portion of p from w to t. Let p_3 be the path from w to t consisting only of tree edges. The paths $p_1 p_3$ and p_2 each have at most 2^{i-1} non-tree edges.

Suppose p is one of the k smallest paths (with respect to $\prec_v$) from v to t with at most 2^i non-tree edges. To prove the invariant, it is sufficient to show that p_1p_3 appears in A_v^{i-1} and p_2 appears in A_w^{i-1}, since p will then be among the paths considered for inclusion in A_v^i.

Suppose p_1p_3 does not appear in A_v^{i-1}. By the inductive hypothesis, there are k paths from v to t that are smaller than p_1p_3 (with respect to $\prec_v$), each with at most 2^{i-1} non-tree edges. But $p_1p_3 \prec_v p_1p_2 = p$, since p_3 has no non-tree edges. This contradicts the fact that p is among the k smallest paths from v to t with at most 2^i non-tree edges. So, p_1p_3 must be in A_v^{i-1}.

Now suppose p_2 does not appear in A_w^{i-1}. Then there are k paths from w to t, each with at most 2^{i-1} non-tree edges, satisfying $r_1 \prec_w r_2 \prec_w \ldots \prec_w r_k \prec_w p_2$. But this implies that $p_1r_1 \prec_v p_1r_2 \prec_v \ldots \prec_v p_1r_k \prec_v p_1p_2 = p$ (by the definition of the orders $\prec_v$ and $\prec_w$). This is a contradiction, so p_2 must be in A_w^{i-1}. □

It follows from Lemmas 2 and 3 that $A_v^{\lceil \log k \rceil}$ will contain the k shortest paths from v to t. In the rest of this section, the implementation of the computation of A_v^i on a CREW PRAM is described, and the analysis of the performance of the algorithm is completed.

Using the notation introduced in the previous proof, the weight of the path formed by combining p_1 and p_2 is just the sum of the weights of p_1 and p_2, since the weight of each tree edge is 0. The num field is filled in similarly. Suppose that the paths of A_v^{i-1} are sorted according to the number of non-tree edges in the path, with ties broken in accordance with the ordering $\prec_v$. Then, the candidates for inclusion in A_v^i can be found by combining each path p_1 having l non-tree edges in A_v^{i-1} with the paths in two contiguous subarrays of A_w^{i-1}: the portions containing paths having l or $l-1$ non-tree edges. Each subarray will already be sorted by $\prec_w$, so the paths that result from combining the path p_1 with the elements of the subarray will be sorted according to $\prec_v$. Thus, the problem of picking the shortest k paths is now equivalent to selecting the k shortest paths from a set of $2k+1$ sorted arrays of paths, each with at most k elements (one of the arrays is A_v^{i-1}, and there are two arrays for each path p_1 in A_v^{i-1} whose elements are obtained by combining p_1 with other paths). If the arrays are considered to be the columns of a $k \times (2k+1)$ matrix, the computation of A_v^i amounts to the selection of the k smallest elements of the matrix. If ties between two paths with the same weight that appear in different columns are broken in favour of the element in the leftmost column, the elements selected will be the k smallest with respect to the ordering $\prec_v$.

A PRAM implementation of Frederickson and Johnson's sequential algorithm for selection in a matrix with sorted columns [11] may be used to find the required k paths.

Lemma 4. *There is an EREW PRAM algorithm that selects the kth smallest element of an $r \times r$ matrix with sorted columns using $O(\log r(\log k \log^* k + \log^* r) + \log k \log\log k)$ time and $O(r)$ work if $k \le r$.*

The implementation is straightforward except for the subproblem of weighted selection, which can be done by modifying Vishkin's parallel selection algorithm [22]. Further details of the parallel implementation of Frederickson and Johnson's algorithm may be found in [19].

Once the k elements of A_v^i are found using the algorithm of Lemma 4, they can be sorted in $O(\log k)$ time and $O(k \log k)$ work using Cole's parallel merge sort [6]. The ith stage of the algorithm that computes the arrays A_v^i (for all vertices v) therefore uses $O(\log^2 k \log^* k)$ time and $O(nk \log k)$ work.

Once the $\lceil \log k \rceil$ stages of the algorithm are complete, the k shortest paths are stored implicitly in the data structure. This completes the proof of the following theorem.

Theorem 5. *Let G be a directed graph with n vertices, m edges. Let d be the maximum outdegree of any vertex. Given the tree of shortest paths rooted at the destination vertex t, an implicit representation of the k shortest paths to t from every vertex can be computed on a CREW PRAM using $O(\log^3 k \log^* k + \log n \log\log k + \log d \log^* d)$ time and $O(m + nk \log^2 k)$ work.*

As described in Sect. 3.2, the paths can be explicitly listed using $O(\log k + \log n)$ time and $O(n \log n + L)$ work, where L is the length of the output.

The parallel algorithm developed here will also work on weighted directed multigraphs. The performance bounds proven above still apply unchanged. If some of the edge weights are negative, the algorithm will still work, provided there is no cycle in the graph whose total weight is negative. Lemma 1 still applies in this case. Therefore, the transformation used to reduce a general problem instance to one with non-negative edge weights (defined by δ), where the distance to vertex t from any other vertex is 0, also applies to graphs with negative edge weights that have no negative cycles.

4 Applications

Two applications of the k shortest paths problem will now be described. More detailed descriptions of these applications may be found in [19].

The Viterbi decoding problem is to estimate the state sequence of a discrete-time Markov process, given noisy observations of its state transitions. This problem has applications in communications (see [8]). The list Viterbi decoding problem is to compute the k state sequences that are most likely to have occurred, given a particular sequence of observations. Sequential algorithms exist that construct a weighted directed acyclic graph in which each path between two fixed vertices describes a possible state sequence [8]. The weight of the path corresponding to a state sequence is equal to the conditional probability that it occurred, given the observations. Sequential algorithms for this problem have appeared previously [20, 21], and a straightforward parallel implementation was described by Seshadri and Sundberg [20]. The parallel k shortest paths algorithm can be applied to this graph to solve the list Viterbi decoding problem on a CREW PRAM using $O(\log^2 s + \log(sT) \log\log k + \log^3 k \log k)$ time and $O(s^3T + sTk \log^2 k)$ work, where s is the size of the state space of the Markov process and T is the length of the observation sequence. This is the first parallel algorithm for this problem that runs in polylogarithmic time.

The quickest paths problem [4] is a generalization of the shortest path problem. It is used to model the problem of transmitting data through a computer network. Each edge of a directed graph is assigned a positive capacity and a non-negative latency. The capacity $c(p)$ of a path p is defined to be the minimum capacity of any edge on the path. The latency $l(p)$ is the sum of the latencies of the edges of p. The time to transmit σ bits of data along a path p is $l(p)+\sigma/c(p)$. The k quickest paths problem is to compute the k paths that require the least time to transmit a given amount of data between a given pair of vertices. If all edge capacities are equal, the problem reduces to the k shortest paths problem. Subpaths of quickest paths need not be quickest paths themselves, so the approaches used to solve shortest path problems are not directly applicable to quickest path problems. Sequential algorithms for the k quickest paths problem have been studied previously [2, 3, 19]. The problem can be solved by first finding the k shortest paths (with respect to latency) that have capacity at least c, for each edge capacity c, and then choosing from among them the k paths that have the shortest overall transmission time [3]. Repeated use of the parallel k shortest paths algorithm yields a CREW PRAM implementation of this approach that finds the k quickest paths to a given destination vertex t from every vertex v. For an n-node graph with m edges and r distinct edge capacities, the parallel algorithm runs in $O(\log^2 n \log r + \log^3 k + \log n \log\log k)$ time using $O(n^2 m \log n \log r + rnk \log^2 k)$ work.

Acknowledgements

I thank my advisor Faith Fich for many invaluable discussions throughout the development of this paper. Derek Corneil provided helpful comments on the presentation of this material. Financial support was provided by the Natural Sciences and Engineering Research Council of Canada.

References

1. A. Borodin and J. E. Hopcroft. Routing, merging and sorting in parallel models of computation. *Journal of Computer and System Sciences*, 30:130–145, 1985.
2. Gen-Huey Chen and Yung-Chen Hung. Algorithms for the constrained quickest path problem and the enumeration of quickest paths. *Computers and operations research*, 21(2):113–118, 1994.
3. Y. L. Chen. Finding the k quickest simple paths in a network. *Information Processing Letters*, 50(2):89–92, April 1994.
4. Y. L. Chen and Y. H. Chin. The quickest path problem. *Computers and Operations Research*, 17(2):153–161, 1990.
5. R. Cole. An optimally efficient selection algorithm. *Information Processing Letters*, 26:295–299, January 1988.
6. R. Cole. Parallel merge sort. *SIAM Journal on Computing*, 17(4):770–785, August 1988.

7. David Eppstein. Finding the k shortest paths. In *Proc. 35th IEEE Symposium on Foundations of Computer Science*, pages 154–165, 1994.
8. G. David Forney, Jr. The Viterbi algorithm. *Proceedings of the IEEE*, 61(3):268–278, March 1973.
9. B. L. Fox. Calculating kth shortest paths. *INFOR; Canadian Journal of Operational Research*, 11(1):66–70, 1973.
10. Greg N. Frederickson. An optimal algorithm for selection in a min-heap. *Information and Computation*, 104:197–214, 1993.
11. Greg N. Frederickson and Donald B. Johnson. The complexity of selection and ranking in $X + Y$ and matrices with sorted columns. *Journal of Computer and System Sciences*, 24:197–208, 1982.
12. Michael L. Fredman and Robert Endre Tarjan. Fibonacci heaps and their uses in improved network optimization algorithms. *Journal of the ACM*, 34(3):596–615, 1987.
13. Y. Han, V. Pan, and J. Reif. Efficient parallel algorithms for computing all pair shortest paths in directed graphs. In *4th Annual ACM Symposium on Parallel Algorithms and Architectures*, pages 353–362, 1992.
14. Walter Hoffman and Richard Pavley. A method of solution of the Nth best path problem. *Journal of the ACM*, 6:506–514, 1959.
15. Richard Karp and Vijaya Ramachandran. Parallel algorithms for shared-memory machines. In Jan van Leeuwen, editor, *Handbook of Theoretical Computer Science*, volume A, pages 871–941. Elsevier, 1990.
16. N. Katoh, T. Ibaraki, and H. Mine. An efficient algorithm for K shortest simple paths. *Networks*, 12:411–427, 1982.
17. Richard C. Paige and Clyde P. Kruskal. Parallel algorithms for shortest paths problems. In *Proceedings of the International Conference on Parallel Processing*, pages 14–20, 1985.
18. Margaret Reid-Miller, Gary L. Miller, and Francesmary Modugno. List ranking and parallel tree contraction. In John H. Reif, editor, *Synthesis of Parallel Algorithms*, chapter 3. Morgan Kaufmann, 1993.
19. Eric Ruppert. Parallel algorithms for the k shortest paths and related problems. Master's thesis, University of Toronto, 1996.
20. Nambirajan Seshadri and Carl-Erik W. Sundberg. List Viterbi decoding algorithms with applications. *IEEE Transactions on Communications*, 42(2/3/4 Part I):313–323, 1994.
21. Frank K. Soong and Eng-Fong Huang. A tree-trellis based fast search for finding the N best sentence hypotheses in continuous speech recognition. In *Proceedings of the International Conference on Acoustics, Speech and Signal Processing*, volume 1, pages 705–708, 1991.
22. Uzi Vishkin. An optimal parallel algorithm for selection. In *Advances in Computing Research*, volume 4, pages 79–86. JAI Press, 1987.
23. Jin Y. Yen. Finding the K shortest loopless paths. *Management Science*, 17(11):712–716, July 1971.

Sequential and Parallel Algorithms on Compactly Represented Chordal and Strongly Chordal Graphs

Elias Dahlhaus

Department of Computer Science,
University of Bonn,
Roemerstrasse 164,
53117 Bonn, Germany,
e-mail: dahlhaus@math.tu-berlin.de and dahlhaus@cs.uni-bonn.de

Abstract. For a given ordered graph $(G, <)$, we consider the smallest (strongly) chordal graph G' containing G with $<$ as a (strongly) perfect elimination ordering. We call $(G, <)$ a compact representation of G'. We show that the computation of a depth-first search tree and a breadth-first search tree can be done in polylogarithmic time with a linear processor number with respect to the size of the compact representation in parallel. We consider also the problems to find a maximum clique and to develop a data structure extension that allows an adjacency query in polylogarithmic time.

Keywords. Algorithms and Data Structures, Parallel Algorithms

1 Introduction

Chordal graphs are graphs with the property that every cycle of length greater than three has a *chord*, i.e. an edge that joins nonconsecutive vertices of the cycle. These graphs play an important role in sparse Gauss elimination [15] and in data base design [8]. Moreover, chordal graphs are exactly the intersection graphs of subtrees of trees [12, 3]. One can characterize chordal graphs also as those graphs having a *perfect elimination ordering*, i.e. for each vertex, all greater neighbors are pairwise joined by an edge. A subclass of chordal graphs are the strongly chordal graphs. They were introduced by M. Farber [9] for the purpose to get a subclass of chordal graphs that allows to solve the domination problem efficiently. Strongly chordal graphs are those graphs which have a strongly perfect elimination ordering, i.e.

1. $<$ is a perfect elimination ordering and
2. if $x < y < z$ and xy and xz are in E then $N[y] \cap \{u|u \geq v\} \subseteq N[z]$.

Here $N[y]$ is the set containing y and its neighbors (closed neighborhood).

Note that strongly chordal graphs are exactly those graphs that have a Γ-free adjacency matrix (we assume that each vertex is adjacent to itself).

Reischuk, Morvan (Eds.): STACS'97 Proceedings, LNCS 1200

Recently Spinrad [17] discussed the problem of the lowest nonredundant representation of Γ-free matrices, i.e. a 1-entry a_{ij} is not entered into the matrix if it is the consequence from other 1-entries. He proved that the size of an nonredundant representation of a Γ-free matrix is $O(n \log n)$.

Here we consider the following kind of problems. Given any graph $G = (V, E)$ and an ordering $<$ on the vertex set V, solve a problem (like maximum clique, coloring) on the smallest (strongly) chordal graph containing G as a subgraph with $<$ as a (strongly) perfect elimination ordering in linear or $O(n + m)polylog(n)$ time. Here n is the number of vertices and m is the number of edges of G. We mainly also consider parallel versions of these algorithms.

The background of this problem is the following. Where chordal graphs or strongly chordal graphs appear in practice (e.g. sparse Gauss elimination), they are not given in full graph representation but by some compact representation, e.g. by a graph G and an ordering on the vertices. We also would like to mention that Gilbert and Hafsteinsson [13] considered the problem, given a graph G and and ordering $<$, compute the smallest chordal graph G' containing G with $<$ as a perfect elimination ordering.

In section 3, we consider parallel solutions for elementary problems like breadth-first and depth-first search. In section 5, we show that we can determine the size of a maximum clique of a compactly represented chordal graph in $O(n + m) \log n$ time. It remains an open problem to get an efficient (parallel) solution of this problem for strongly chordal graphs.

2 Notation

A *graph* $G = (V, E)$ consists of a *vertex set* V and an *edge set* E. Multiple edges and loops are not allowed. The edge joining x and y is denoted by xy.

We say that x is a *neighbor* of y iff $xy \in E$. The set of neighbors of x is denoted by $N(x)$ and is called the *neighborhood*. The set of neighbors of x and x is denoted by $N[x]$ and is called the *closed neighborhood* of x.

Trees are always directed to the root. The notion of the *parent, child, ancestor*, and *descendent* are defined as usual.

A *subgraph* of (V, E) is a graph (V', E') such that $V' \subset V$, $E' \subset E$.

We denote by n the number of vertices and by m the number of edges of G. Note that in any chordal graph, the number of cliques is bounded by n and the number of pairs (x, c) such that x is in the clique c is bounded by m.

$(G, <) = (V, E, <)$ is called a *compact representation* of a (strongly) chordal graph $G' = (V, E')$ if E' is the smallest edge set containing E, such that $<$ is a (strongly) perfect elimination ordering. We implement the ordering $<$ on V as an enumeration of V. To solve a problem on a compactly represented (strongly) chordal graph, we have to solve the problem on G', e.g. if we would like to solve the problem to find a maximum clique on a compact representation $(G, <)$ of the strongly chordal graph G' then it means we want to find a maximum clique of G'.

The parallel computation models are the exclusive read exclusive write parallel random access machine (EREW-PRAM) and the concurrent read concurrent write parallel random access machine (CRCW-PRAM).

3 Breadth-first and Depth-first Search Trees and Related Topics

To get a depth-first search tree in a chordal graph in its full representation, we determine, for each vertex v, its smallest greater neighbor (see for example [18]). To get a breadth-first search tree, we determine for each vertex, its greatest larger neighbor (see for example [14]). The tree with the smallest greater neighbor as its parent function is also called the *minimum greater neighbor tree*, and the tree with the maximum greater neighbor as parent function is called the *maximum greater neighbor tree.*

Theorem 1. *For any compactly represented chordal graph, the minimum greater neighbor tree can be constructed in $O(\log n)$ time with $O(n+m)$ processors by a CRCW-PRAM.*

Proof: The key idea is the following observation.

Let G' be a chordal graph and G'_v be the subgraph of G induced by the vertices $\leq v$. Let C be the connected component of G'_v containing v. Then the greater neighborhood of v and the set of neighbors of C that are not in C coincide.

To get the minimum greater neighbor tree of a compactly represented chordal graph with the representation $(G, <)$, we first determine, for each vertex v, the connected component C_v of G_v that contains v. Note that the sets C_v form a tree like ordering with respect to the subset relation. Therefore we can represent all the sets C_v compactly by a tree of size $O(n)$. Moreover, the minimum greater neighbor of v is the vertex w with $C_v \subset C_w$, such that C_w is minimum. Therefore we get the minimum greater neighbor tree by computing the tree representation of the sets C_v.

We compute the sets C_v as follows. First we weight the edges vw by the maximum of v and w with respect to $<$. Then we compute a minimum spanning tree T in $O(\log n)$ time with $O(n+m)$ processors on a CRCW-PRAM [16] (The algorithm of [16] can be modified a little bit that a minimum spanning tree can be computed in the same time and processor bound. We allocate edges with smaller distance to smaller processors and let the processor of smallest number win).

Define as single link clusters of weight v the vertex sets of the trees that arise from deleting the edges of T of weight greater than v. Note that the set C_v is exactly the single link cluster of weight v that contains v and the single link clusters coincide exactly with the sets C_v.

The tree representation of the single link clusters can be computed in $O(\log n)$ time with $O(n)$ processors, provided the minimum spanning tree is given [4].

Q.E.D.

Theorem 2. *Let $(G,<)$ be an ordered graph. Then the minimum greater neighbor trees of the chordal graph G' and of the strongly chordal graph G'' represented by G and $<$ are identical.*

Sketch of Proof: We only have to show that the the connected components C_v of G_v do not change when we add a minimum number of edges that make $<$ a strongly perfect elimination ordering. Observe that when $x < y$, $x' < y'$, and $x < x'$ and if xy, xy', $x'y$ are in the edge set E' of the strongly chordal extension of G then $x'y' \in E'$. Note that x, x', and y are in the same connected component of G_v also if we leave out the edge $x'y'$.

Q.E.D.

Corollary 3. *The minimum greater neighbor tree of a compactly represented strongly chordal graph can be computed in $O(\log n)$ time with a linear processor number by a CRCW-PRAM.*

Theorem 4. *The maximum neighbor tree of a compactly represented chordal graph can be computed in $O(\log n)$ time with $O(n+m)$ processors on a CRCW-PRAM.*

Sketch of Proof: Note that the minimum greater neighbor tree T_m is a depth-first search tree, and therefore the maximum neighbor of v is an ancestor of v with respect to T_m. Moreover, the maximum neighbor of v in the chordal graph G' is the maximum neighbor of any descendent of v in T_m. Therefore, we first compute, for each vertex v, the maximum neighbor of v in the compact representation $(G,<)$ (in $O(1)$ time with $O(n+m)$ processors on a CRCW-PRAM). Afterwards, by tree contraction, we compute, for each vertex v the maximum neighbor of any descendent of v in $O(\log n)$ time with $O(n/\log n)$ processors [1].

Q.E.D.

Theorem 5. *The maximum neighbor tree of a compactly represented strongly chordal graph can be computed in $O(\log n)$ time with $O(n+m)$ processors on a CRCW-PRAM.*

Sketch of Proof: Here we use the following property of strongly chordal graphs.

If $x < y$ and x and y have at least one common neighbor then the maximum neighbor of x is a neighbor of y.

This follows immediately from the definition of a strongly perfect elimination ordering.

For each vertex v, we determine the next greater vertex $P(v)$ that has a common neighbor with v or is adjacent to v. Let T'_m be the tree with parent function P. T'_m is also called the *minimum greater common neighborhood tree.*

We can determine the tree T'_m in a similar way as T_m. We consider the bipartite graph B with two copies of V, say $V' = V \times \{1\} \cup V \times \{2\}$. $(x,1)(y,2) \in E'$ if and only if $x = y$ or $xy \in E$. We weight edges $(x,1)(y,2)$ with x. Let D_v

be the connected component of B restricted to $\{x|x \leq v\} \times \{1\} \cup V \times \{2\}$ that contains v. Then $P(v)$ is the smallest $w > v$, such that $(w, 1)$ is adjacent to some $(x, 2) \in D_v$, i.e. $D_v \subset D_w$ and there is no D_z between D_v and D_w. Again we use the single link method to get the tree T'_m.

To get the maximum greater neighbor of any vertex v, we determine, by tree contraction, the maximum greater neighbor of any descendent of v in T'_m.

Q.E.D.

4 Computation of the Full Strongly Chordal Graph from its Compact Representation

First we present a sequential algorithm to compute the full strongly chordal graph G', provided a compact representation $(G, <)$ of G' is known. We compute step by step a forest T_v with the parent function $P_v(x)$ = next greater y such that $N_{G'}[x] \cap N_{G'}[y] \cap \{z|z \leq v\}$ is not empty. Let $V = \{v_1, \ldots, v_n\}$ and $v_i < v_j$, for $i < j$. We compute the forests $T_i = T_{v_i}$ as follows.

Initialization: T_0 consists of the vertex set V and has no edges.

For $i = 1$ to n: 1. Let A_i be the set of ancestors of vertices $x \in N[v_i]$ in T_{i-1} including the vertices in $N[v_i]$.

2. T_i arises from T_{i-1} by making the vertices in A_i one path, such that with $x, y \in A_i$ and $x < y$, y is an ancestor of x, i.e. for any vertex $x \in A_i$, $P_i(x)$ = next greater vertex in A_i.

Theorem 6. *1. The closed neighborhood of v_i in G' is A_i.*

2. If $N_{G'}[x] \cap N_{G'}[y] \cap \{z|z \leq v_i\} \neq \emptyset$ then x is an ancestor of y in T_i or vice versa.

3. If y is the parent of x in T_i then $N_{G'}[x] \cap N_{G'}[y] \cap \{z|z \leq v_i\} \neq \emptyset$

The proof can be done by simultaneous induction on i.

To get the neighborhood of v_i in G', we only have to make the vertices in A_i adjacent with v_i.

The complexity of the algorithm is bounded by $\Sigma_{i=1}^{n} O(1 + |A_i|)$ and therefore by the number of vertices and edges of G'.

Theorem 7. *Let m' be the number of edges of the strongly chordal graph G' with a compact representation $(G, <)$. Then G' can be computed from $(G, <)$ in time $O(n + m')$.*

For a parallel algorithm, we use a divide and conquer strategy.

For simplicity, we assume that n is a power of 2, i.e. $n = 2^p$.

We define $T_1, \ldots, T_n$ as in the sequential algorithm.

The trees T_i can be computed in parallel as follows.

1. Initially we compute T_0 and T_n and the neighborhood of v_n.
2. For $q = p - 1, \ldots, 0$, for all i, $2^q | i$, $2^{q+1} \not| i$,

(a) compute T_i from T_{i-2^q} and the edges xv_j, $i - 2^q < j \leq i$;
(b) compute the neighborhood of v_i as the set of ancestors of the minimum neighbor of v_i in T_i.

The computation of T_i from T_{i-2^q} can be done as follows.

We proceed in a similar way as when we compute a depth-first search tree in a strongly chordal graph.

We define the graph $G_i = (V_i, E_i)$ as follows.

The vertex set is $\{(1, v_j)|j = 1, \ldots n\} \cup \{(2, v_j)|i - 2^q < j \leq i\}$. There is an edge $(1, v)(1, w)$ iff vw is an edge of T_{i-2^q}. The weight of this edge is the maximum index of v and w.
There is an edge $(1, v)(2, w)$ iff $vw \in E$. The weight of this edge is the index of v.

Let G_i' consist of the vertex set $U_i = \{(1, v)|v \in V\} \cup \{(2, v_j)|j \leq i\}$ and the weighted edges $(1, v)(2, w)$ with $vw \in E$ and the index of v as weight.

Lemma 8. *$(1, u)$ and $(1, v)$ can be joined by a path in G_i' with edges of weight $\leq k$ iff $(1, u)$ and $(1, v)$ can be joined by a path in G_i with edges of weight $\leq k$.*

Therefore T_i can be computed in parallel as follows.

1. Compute a minimum spanning forest S_i of G_i.
2. The parent of a vertex v_j in T_i is the next ancestor v_l of v_j in S_i with larger index l than j, i.e v_l is the neighbor of the connected component of edges of weight $\leq j$ v_j belongs to with smallest index $> j$.

We proceed here as in [4].

Obviously this algorithm runs in $O(\log^2 n)$ time with $O(n^2)$ processors on a CRCW-PRAM.

An $O(n+m')$ processor implementation can be realized as follows. Instead of keeping track of T_i, we only keep track of the subforest T_i' of T_i which contains all the ancestors of vertices of T_i that are in the closed neighborhood of a vertex v_j, $i < j \leq i + 2^q$, where q is the maximum, $2^q|i$.

Therefore:

Theorem 9. *Given a compact representation $(G, <)$ of the strongly chordal graph G', G' can be computed in $O(\log^2 n)$ time with $O(n + m')$ processors on an CRCW-PRAM.*

5 Problems in Compactly Represented Chordal and Strongly Chordal Graphs

We consider problems that have trivial solutions if the full chordal or strongly chordal graph is given. Examples are adjacency queries, degrees, and clique sizes (numbers of greater neighbors). But to get an efficient solution for these problems is more difficult if the chordal or strongly chordal graph is given by its compact representation.

5.1 Degrees and Clique Sizes in Compactly Represented Chordal Graphs

Theorem 10. *For all vertices of a compactly represented chordal graph simultaneously, the number of greater neighbors can be determined in $O(\log n)$ time with $O(n+m)$ processors on a CRCW-PRAM.*

Proof: The basic idea is that we can determine a compact tree representation of G' in $O(\log n)$ time with $O(n+m)$ processors. By a *compact tree representation*, we mean a tree representation $(T, \{T_v | v \in V\})$, such that each subtree T_v representing the vertex v is given by its leaves. Then we can determine a maximum clique in $O(\log n)$ time with $O(n+m)$ processors on an EREW-PRAM [7]. We can choose the minimum greater neighbor tree as tree T and first choose as T'_v the set $\{v\} \cup \{x | x < v \, and \, xv \in E\}$. We can erase in $O(\log n)$ time with $O(n+m)/\log n$ processors all vertices x in T'_v that are not equal v and are ancestors of some other $y \in T'_v$ (by Eulerian cycle techniques on T). This creates a representation of T_v, such that only the leaves of T_v are given.
Q.E.D.

Theorem 11. *For each chordal graph G' with its compact representation $(G, <)$, the degrees of all its vertices can be computed in $O(\log n)$ time with $O(n+m)$ processors.*

Proof: In the previous theorem we gave an algorithm that computes for each vertex v, the size of the clique c_v consisting of v and all its greater neighbors. It remains to compute the number of smaller neighbors of v. Note that the set of smaller neighbors of v is the full tree T_v, i.e. the set of ancestors of vertices in the compact representation T'_v of T_v that are not ancestors of v in the smallest greater neighbor tree T.

To get the size of T_v, we first compute the set T''_v of least common ancestors of vertices in T'_v. This can be done in $O(\log n)$ time with $\Sigma_v |T'_v| + n = O(n+m)$ processors, for all T_v simultaneously, ordering the vertices in T_v by postorder and determining the least common ancestors of neighbors.

By Eulerian cycle techniques, we also get, for each vertex x in T''_v, the next ancestor $P'(x)$ of x in T''_v. Let I_x be the path in T starting in x and ending one vertex before $P'(x)$. Note that the paths I_x are disjoint and cover $T_v \setminus \{v\}$. To get the size of $T_v \setminus \{v\}$ one only has to determine the length of the paths I_x, in $O(\log n)$ time with $|T''_v| = O(|T'_v|)$ processors, for each single v. Therefore the sizes of all T_v simultaneously can be determined in $O(\log n)$ time with $O(n+m)$ processors. Therefore also the degrees of all vertices can be determined simultaneously in $O(\log n)$ time with $O(n+m)$ processors.
Q.E.D.

Remark. The algorithms can be converted into linear time algorithms. The only step that needs $O(n+m)\log n$ workload was sorting. But given a sorted sequence of all the vertices, the neighbors of each vertex v can be simultaneously be sorted in $O(n+m)$ time, for all v.

5.2 Adjacency Queries for Compactly Represented Chordal graphs

Theorem 12. *There is an $O(n+m)$ data structure $\tilde{G}$ that can be computed in parallel in $O(\log n)$ time with $O(n+m)$ processors and sequentially in $O(n+m)$ time, such that the adjacency in G' of two vertices can be checked in $O(\log n)$ time if $\tilde{G}$ is known.*

Proof. Note that x and y are adjacent if and only if T_x and T_y intersect. That means that $x < y$ are adjacent if y is an ancestor of x and x is an ancestor of some vertex of the compact representation T'_y of T_y. To check that x is an ancestor of some vertex in T'_y, we assume that the vertices in T'_y are sorted with respect to preorder and select two vertices $u, v \in T'_y$, such that $u < x < v$ with respect to the preorder $<$ and no vertex $z \in T'_y$ is in between. It is sufficient to check whether x is an ancestor u or an ancestor of v. We only have to extend the compact tree representations T'_y by a preorder sorting and a binary search tree to check adjacency of two vertices in logarithmic time.
Q.E.D.

5.3 Adjacency Queries, Degrees, and Clique Sizes in Compactly Represented Strongly Chordal Graphs

Next we consider the problem to get the clique sizes, degrees and adjacency queries in compactly represented strongly chordal graphs. Due to a lack of space, we sketch only the ideas. Details will be described in the final version. Recall that T_v is the forest with the parent function $P_v(x) =$ next greater $y > x$ that has a common element with x in the closed neighborhoods $\leq v$.

We assume that an enumeration $v_1, \ldots, v_n$ of V is given and define $T_i := T_{v_i}$. We extend T_i to a data structure T'_i, such that the amortized complexity to get T'_{i+1} from T'_i is $O(n+m)polylog\, n$. We assume that $n = 2^p - 1$.

The idea is that we partition the branches of T_i into *binary search intervals* to speed up the updating procedure. For an x not divisible by two, $[x \cdot 2^i, (x+1) \cdot 2^i - 1]$ is called a *binary search interval* of length i. For $i \leq n$, let $I_1^i, \ldots, I_q^i$ be the increasing sequence of inclusion maximal binary search intervals $[a, b]$ with $i \leq a$ and $b \leq n$, called the interval sequence of i. The idea is that initially we store, for each vertex v_i, the interval sequence of i and consider intervals of the same range associated with different v_i as different. In the procedure, we identify intervals of the same range, say $I_l^i, I_{l'}^j$, if they represent the same path of T_k. We also keep track of intervals representing subpaths of a path represented by another interval.

Lemma 13. *If I_j^i is of length 2^l then I_{j+1}^i is of length $\geq 2^{l+1}$, and therefore the interval sequence of i if of length $O(\log n)$.*

Proof. Suppose $I_j^i = [a2^l, (a+1)2^l - 1]$. Then I_{j+1}^i starts with $(a+1)2^l$. Now, since a is odd, $a+1$ is even, and therefore $(a+1)2^l = b2^{l'}$ with $l' > l$ and b odd. Therefore I_{j+1}^i is of length $2^{l'} \geq 2^{l+1}$.

Q.E.D.

To have a compact representation of the subpath relation, we introduce also intervals $J_i^l = \bigcup_{j<i} I_j^l$. We call these intervals also *J-intervals*. For a binary search interval $[a, b]$, we keep track of the largest J-interval $[a', b]$ with $a' > a$ that represents a subpath of the path represented by $[a, b]$. We denote this J-interval by $largest_J[a, b]$. Note that initially $largest_J[a, b] = \emptyset$.

We also introduce pointers $larger(J_i^l) := I_i^l$ and $smaller(J_i^l) := J_{i-1}^l$, $i > 0$ and $smaller(J_0^l) := NIL$.

The main procedure is to get, for any v_i, all the neighbors in ancestor relation. Let $y_1, \ldots, y_k$ be the neighbors of v_i and $y_j > y_{j+1}$. Then it is sufficient to make step by step y_j an ancestor of y_{j+1}.

To make v_i an ancestor of v_j, $i > j$, we consider the maximum J-interval J_1 starting with i and the maximum J-interval J_2 starting with v_j. We unify all binary search intervals that appear in J_1 and in J_2 that have the same range. This is done by a procedure TRACK(J_1, J_2).

Procedure make v_i an ancestor of v_j
 J_1 := maximal J-interval starting with i;
 J_2 := maximal J-interval starting with j;
 TRACK(J_1, J_2)
 end **Procedure**
Procedure TRACK (J_1, J_2)
 If $range\, larger(J_1) = range\, larger(J_2)$ then
  begin
   UNIFY$(larger(J_1), larger(J_2)$; TRACK$(lower(J_1), lower(J_2))$
  end
 else (note that $range\, J_1 \subset range\, larger(J_2)$)
 if $largest_J(larger(J_2)) \subset J_1$ (the ranges) then
  begin
   TRACK$(largest_J(larger(J_2)), J_1)$;
   $largest_J(larger(J_2)) := J_1$
  end
 else
 TRACK$(J_1, larger(J_2))$
 end Procedure TRACK
Procedure UNIFY (I, I')
 If I and I' are not unified then
  begin
   if $largest_J(I) \subset largest_J(I')$ then
   begin
   TRACK$(largest_J(I), largest_J(I'))$;
   $largest_J(I) := largest_J(I')$
   end
   else
   begin
   TRACK$(largest_J(I'), largest_J(I))$;

$largest_J(I') := largest_J(I)$
end
identify I and I'
end **end Procedure UNIFY**

To get the forest T_k explicitly, we always also have to keep track of the largest element (of the path represented by) I_i^l and the smallest element of (the path represented by) I_i^l. We always make the smallest element of I_{i+1}^l the parent of the largest element of I_i^l.

Still we are not immediately able to support the number of ancestors query that is used to find out the degree of a vertex or the size of the greater neighbor clique of a vertex.

The basic idea would be to fix the size of I-intervals at each step. But this is difficult, because it can happen that I-intervals have a nonempty intersection but are not comparable with respect to inclusion. To solve this problem we introduce the notion of the *maximum descendent child* of a vertex v, i.e. a child of v that has a maximum number of descendents. All the other children are called *shunts*. We make it that all I-intervals are changed into I'-intervals that have, if they contain a shunt, they have its smallest shunt as largest element, and all greater vertices are cut away.

Lemma 14. *Each vertex has at most $O(\log n)$ shunts as ancestors.*

To get the number of ancestors, we have to keep track of the sizes of I'-intervals. Given a vertex v, we consider the sequence of maximal ancestor I'-intervals and add their sizes. The number of these intervals is bounded by $O(\log^2 n)$, because the number of shunts is bounded by $O(\log n)$ and the number of I'-intervals between two shunts is bounded by the number of maximal ancestor I-intervals of the smaller shunt, i.e. by $O(\log n)$.

To keep track of the number of descendents during the amalgamation process of T_i, i.e. unifying some paths, we keep track of the number of *side descendents* of each I'-interval I'. v is a side descendent of I' if it is a descendent of some vertex of I', v is not in I', and v is not a descendent of the maximum descendent child of the smallest element of I'.

Note that while we unify the descendents of v and w to one path the maximum descendent child relation can change. But it cannot happen that a non shunt vertex that is an ancestor of v or w becomes a shunt vertex. Therefore the maximum descendent child relation has only to be updated on places of the path of ancestors of v and w that are shunts of the tree T_i before unifying the ancestors of v and w to one path. They are bounded by $O(\log n)$. Again to count the number of descendents of a shunt x, we pick the maximum I'-interval that contains x and add the number of side descendents and the size. That can be done in constant time. To count the number of descendents of a non shunt vertex we also pick the maximum I'-interval I' containing x and select a sequence of maximal I'-intervals containing I' and being smaller than x. They are bounded by $\log n$, and therefore the time to find out the number of descendents of x

is bounded by $O(\log n)$. To change the I'-intervals if x becomes a non shunt, the I'-intervals containing x and the I' intervals containing the parent of x are involved, i.e. $O(\log n)$ intervals. Therefore to update the set of I'-intervals at one shunt needs $O(\log n)$ time. Note that also the size and the number of side descendents of these I'-intervals has to be updated (in $O(\log n)$ time). Therefore we can update the I'-intervals that are ancestors of v and w in $O(\log^2 n)$ time.

Details will be described in the final version of the paper.

We conclude this discussion with the following result.

Theorem 15. *The greater neighborhood clique sizes (and therefore a maximum clique) and the degrees of a compactly represented strongly chordal graph can be determined in* $O(n+m)\log^2 n$ *time.*

The last problem is to develop a data structure that supports the adjacency query.

To find out whether v_iv_j is an edge in the full graph, one only has to check whether in T_j, v_i is an ancestor the smallest neighbor of v_j in G. This can be done in $O(\log n)$ time, provided we have all T_i' available. What we really need are the intervals I_i^j and J_i^j and the moment when I_i^j intervals are identified and largest J-intervals are updated. This can be represented by a union-find tree structure [11].

Using the fact of Spinrad [17] that every strongly chordal graph has a compact representation of size $O(n \log n)$, we get the following.

Theorem 16. *There is a data structure of size* $O(n+m)\log n = O(n\log^2 n)$ *that allows to check the adjacency of two vertices in a compactly represented strongly chordal graph in* $O(\log n)$ *time. This data structure can be computed in* $O((n+m)\log^2 n) = O(n\log^3 n)$ *time.*

6 Conclusions

We did not parallelize many problems related to compactly represented strongly chordal graphs, e.g. to find out the degrees of a maximum clique. We believe that similar methods as in [2] might leed to a success. In principle we consider a more general problem. We have operations *unify the ancestors of x and y to one path, is x and ancestor of y*, and *determine the number of ancestors of x*, and we would like to develop a data structure that allows us to handle these operations with an amortized complexity that is as low as possible. We do not exclude that it can be done more efficiently. We can imagine that similar ideas as in [11] can be applied. One can also state the following problem. Given a (strongly) chordal graph in its full representation, find its smallest compact representation. We guess that this problem is NP-complete.

7 Acknowledgement

Sequential version of some algorithms have been implemented on the computer network of the Department of Mathematics, Technische Universität Berlin.

References

1. K. Abrahamson, N. Dadoun, D. Kirkpatrick, T. Przyticka, *A Simple Parallel Tree Contraction Algorithm*, Journal of Algorithms 10 (1989), pp. 287-302.
2. M. Atallah, M. Goodrich, S.R. Kosaraju, *Parallel Algorithms for Evaluating Sequences of Set Manipulation Operations*, Journal of the ACM 41 (1994), pp. 1049-1085.
3. P. Bunemann, *A Characterization of Rigid Circuit Graphs*, Discrete Mathematics 9 (1974), pp. 205-212.
4. E. Dahlhaus, *Fast parallel algorithm for the single link heuristics of hierarchical clustering*, Proceedings of the fourth IEEE Symposium on Parallel and Distributed Processing (1992), pp. 184-186.
5. E. Dahlhaus, P. Damaschke, *The Complexity of Domination Problems in Chordal and Strongly Chordal Graphs*, Discrete Applied Mathematics 52 (1994), pp. 261-273.
6. E. Dahlhaus, *A fast parallel algorithm to compute Steiner-trees in strongly chordal graphs*, Discrete Applied Mathematics 51 (1994), pp. 47-61.
7. E. Dahlhaus, *Efficient Parallel Algorithms on Chordal Graphs with a Sparse Tree Representation*, Proceedings of the 27-th Annual Hawaii International Conference on System Sciences, Vol. II (1994), pp. 150-158.
8. R. Fagin, *Degrees of Acyclicity and Relational Database Schemes*, Journal of the ACM 30 (1983), pp. 514-550.
9. M. Farber, *Characterizations of Strongly Chordal Graphs*, Discrete Mathematics 43 (1983), pp. 173-189.
10. M. Farber, *Domination, Independent Domination and Duality in Strongly Chordal Graphs*, Discrete Applied Mathematics 7(1984), pp. 115-130.
11. H. N. Gabow and R. E. Tarjan, *A linear-time algorithm for a special case of disjoint set union*, J. Comput. System Sci., 30 (1984), pp. 209–221.
12. F. Gavril, *The Intersection Graphs of Subtrees in Trees Are Exactly the Chordal Graphs*, Journal of Cobinatorial Theory Series B, vol. 16(1974), pp. 47-56.
13. J. Gilbert, H. Hafsteinsson, *Parallel Solution of Sparse Linear Systems*, SWAT 88 (1988), LNCS 318, pp. 145-153.
14. P. Klein, *Efficient Parallel Algorithms for Chordal Graphs*, 29. IEEE-FOCS (1988), pp. 150-161.
15. D. Rose, *Triangulated Graphs and the Elimination Process*, Journal of Mathematical Analysis and Applications 32 (1970), pp. 597-609.
16. Y. Shiloach, U. Vishkin, *An O(log n) Parallel Connectivity Algorithm*, Journal of Algorithms 3 (1982), pp. 57-67.
17. J. Spinrad, *Nonredundant 1's in Γ-free Matrices*, SIAM Journal on Discrete Mathematics 8 (1995), pp. 251-257.
18. R. Tarjan, M. Yannakakis, *Simple Linear Time Algorithms to Test Chordality of Graphs, Test Acyclicity of Hypergraphs, and Selectively Reduce Acyclic Hypergraphs*, SIAM Journal on Computing 13 (1984), pp. 566-579.
    Addendum: SIAM Journal on Computing 14 (1985), pp. 254-255.
19. K. White, M. Farber, W. Pulleyblank, *Steiner Trees, Connected Domination, and Strongly Chordal Graphs*, Networks 15 (1985), pp. 109-124.

Distance Approximating Spanning Trees

Erich Prisner *

Abstract. A spanning tree of a graph is *distance k-approximating* whenever the distance of every two vertices in the graph or in the tree differs by at most k. Variants or modifications of a simple approach yield distance k-approximating spanning trees for block graphs, interval graphs, distance-hereditary graphs, and cocomparability graphs, with $k = 1, 2, 2, 4$. On the other hand, there are chordal graphs without distance k-approximating spanning tree for arbitrary large k.

1 Introduction

Spanning trees are often used in applications where we want to save cost (edges) but maintain connectivity. In most of these applications, distance matters, so it is considered as a disadvantage that vertices of small distance in the original graph may have large distance in the spanning tree. We shall see in this paper that for certain tree-like graphs like interval graphs, distance-hereditary graphs, or cocomparability graphs, it is possible to find spanning trees keeping the size of this 'disadvantage' bounded.

The *distance* $d_G(x, y)$ of two vertices in a connected graph G is the length of a shortest path from x to y. Surely $d_G(x, y) \leq d_H(x, y)$ for all vertices x and y in a connected subgraph H of G. H is called *distance k-approximating* whenever $d_H(x, y) - d_G(x, y) \leq k$ for all vertices x, y of H. Such *spanning* subgraphs are called *additive spanners* in [6] — ordinary *k-spanners* are spanning subgraphs H where $d_H(x, y)/d_G(x, y) \leq k$ for all vertices x, y, see [2] for a survey of the various applications. Note also that distance 0-approximating subgraphs are just the isometric subgraphs.

In this paper we shall only deal with *spanning* sub*trees.* Surely trees are the only graphs with distance 0-approximating spanning subtrees. But for fixed $k \geq 1$, deciding whether a given graph has some distance k-approximating spanning tree is NP-complete [6].

In this paper we will show that certain well-known graph classes Γ allow some constant k_Γ such that every graph in Γ has some distance k_Γ-approximating spanning tree The simple approach is to consider certain breadth-first trees. For block graphs, and in a sense only for them, we get distance 1-approximating spanning trees in this way. For interval graphs and distance-hereditary graphs, variants of the approach yield distance 2-approximating spanning trees in linear

* Department of Mathematical Sciences, Clemson University, Clemson, SC 29631, U.S.A., on leave from the Mathematisches Seminar, Universität Hamburg, Bundesstr. 55, 20146 Hamburg, Germany; supported by the DFG under grant no. Pr 324/6-1.

Reischuk, Morvan (Eds.): STACS'97 Proceedings, LNCS 1200

time. For cocomparability graphs it doesn't suffice to consider breadth-first trees, instead the construction has to be modified. In this case we get distance 4-approximating spanning trees in time $O(n^2)$, where n is the number of vertices of the graph.

Some other classes Γ of graphs do not have such a constant k_Γ. This holds for instance if all cycles belong to Γ, since cycles of length $k+3$ do not have a distance k-approximating spanning tree. Even the class of chordal graphs does not have such a constant, as we shall see.

2 A Natural Approach

Let, for a vertex x in a graph G and any integer $i \geq 0$, $S_i(x)$ denote the set of all vertices y with $d(x, y) = i$. The *eccentricity* $ecc_G(x)$ of x is the largest integer i for which $S_i(x)$ is nonempty (i.e. the largest distance $d_G(x, y), y \in V(G)$). The most straightforward method of constructing some spanning tree in a graph G seems to be to start with some vertex x_0, attach all its neighbors, then attach all vertices in $S_2(x_0)$ towards the tree constructed so far (that is, towards vertices in $S_1(x_0)$), and so on.

Basic construction:
1. Choose some vertex x_0.
2. For $i = 0$ to $ecc_G(x) - 1$ do: Connect every $y \in S_{i+1}(x_0)$ with one of its neighbors $f(y) \in S_i(x_0)$, according to some **rule** f to be specified.

Note that every $y \in S_{i+1}(x_0)$ has exactly one neighbor in $S_i(x_0)$ on the tree constructed. So, what we do is construct a spanning tree T where all distances from x_0 towards the other vertices are identical in G and T. Whether the resulting tree is distance k-approximating depends only on whether $f^i(y) = f^j(z)$ is possible with small i, j, for every edge yz of G. So we do not have to compute all T-distances, but only T-distances between G-adjacent vertices.

Lemma 1. *A spanning tree T of G constructed by our basic construction is distance k-approximating whenever $d_T(y, z) \leq k+1$ for every edge yz of G.*

Proof: Necessity of this condition is obvious.

For sufficiency, assume that $d_T(y, z) \leq k+1$ for every edge yz of G. By induction over t we prove $d_T(v_0, v_t) \leq k+t$ for paths $v_0, v_1, \ldots, v_t$ of length t in G. The case $t = 1$ is just the assumption, now assume that $d_G(u, v) = t > 1$ and assume that $d_T(y, z) \leq d_G(y, z) + k$ whenever $d_G(y, z) < t$ (the induction hypothesis). Let p be a neighbor of v on the shortest $u-v$ path. Now $|d_G(x_0, p) - d_G(x_0, v)| \leq 1$, and $|d_G(x_0, p) - d_G(x_0, u)| \leq t-1$. By the construction of T there is some vertex q on the $p-v$ path $p = p_r, p_{r-1}, \ldots p_0 = q = v_0, v_1, \ldots, v_s = v$ of T such that $d_G(x_0, p_i) = d_G(x_0, q) + i$ and $d_G(x_0, v_j) = d_G(x_0, q) + j$. Moreover $r + s \leq k+1$ and $|r - s| \leq 1$.

The induction hypothesis implies $d_T(u, p) \leq k+t-1$. Let now $u = u_0, u_1, u_2, \ldots$ $\ldots u_\ell$ be the beginning of the $u-p$ path on T, with u_ℓ the first vertex on the path $p_r, p_{r-1}, \ldots, p_1, q, v_1, \ldots, v_s$. If $u_\ell = q$, then $d_T(u,v) = d_T(u,p) + s - r \leq d_T(u,p) + 1$ since $|r - s| \leq 1$. So we get $d_T(u,v) \leq k + t$ in this case. In case $u_\ell \neq q$ we obtain $d_G(u_{i+1}, x_0) + 1 = d_G(u_i, x_0)$ for every $0 \leq i \leq \ell - 1$, by the special shape of T. But $|d_G(u, x_0) - d_G(p, x_0)| \leq t - 1$, whence $d_T(u,q) \leq t - 1 + r$. Consequently $d_T(u,v) \leq t - 1 + r + s \leq k + t$ in this case also. □

For the algorithms we assume that the vertices of G are linearly ordered, and that the graph is given by means of its ordered neighborhood lists, i.e. for every vertex x there is some list NEIGH(x) containing its neighbors in increasing order.

The levels $S_i(x_0)$ can be computed in linear time $O(|V|+|E|)$ by breadth-first search. Moreover, it is also possible to compute the induced subgraphs $G[S_i(x_0)]$ in the levels in linear time — we simply check all neighborhood lists one after another, and delete in every list NEIGH(x) those neighbors that are in a different level than x. The resulting graph is $\bigcup_i G[S_i(x_0)]$.

How quickly T can be constructed surely depends on the rules, which themselves depend on the classes considered.

Although this approach seems natural, has the property of Lemma 1, and works for several graph classes, as we shall see, it has also one disadvantage. It doesn't appear to construct the optimum spanning tree—that which is k-approximating for the smallest possible k.

As an example, the graph of Figure 1 has some distance 1-approximating spanning tree, which, however, will not be found by our approach under *any* rule and *any* start vertex. The reason will be shown in the following subsection.

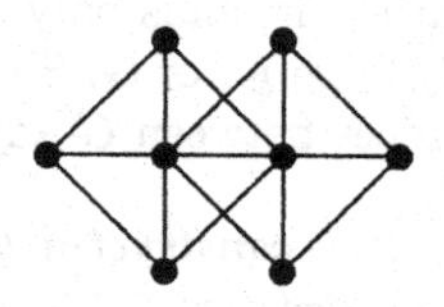

Fig. 1.

2.1 Block Graphs

In a *block graph*, every block must be complete. In other words, block graphs are just the chordal $(K_4 - e)$-free graphs, where $K_4 - e$ is obtained from the complete graph on 4 vertices by deleting one edge. It turns out that for block graphs, and only for them, the approach constructs a distance 1-approximating spanning tree for every start vertex.

Proposition 1 *If*

> *(A) the approach above yields, for every start vertex x_0 of the graph G (under some rule) a distance 1-approximating spanning tree,*

then G must be a block graph.

Proof: First we weaken the assumption. We assume that

> (B) there is *some* start vertex x_0 where the approach yields some distance 1-approximating spanning tree T.

(1) Then every $y \in S_i(x_0)$ has exactly one neighbor in $S_{i-1}(x_0)$. For, otherwise assume $f(y), z \in S_{i-1}(x_0)$ are adjacent to y. By construction, $f(y)$ and z cannot be adjacent in T, therefore $d_T(y,z) \geq 3$, a contradiction to T being distance 1-approximating.

(2) In the same way, if two vertices y_1 and y_2 of $S_i(x_0)$ are adjacent, then $f(y_1) = f(y_2)$.

(3) If G obeys (A) and contains nonadjacent vertices x and y and two adjacent vertices z_1, z_2, both adjacent to both x and y, then $d_G(x,y) = 2$ but $d_G(x,z_1) = d_G(x,z_2) = 1$, a contradiction to (1). Thus G must be $(K_4 - e)$-free.

(4) We prove by induction on the number of vertices that every connected $(K_4 - e)$-free graph obeying (B) must be a block graph. The start is trivial, since every connected graph with at most 3 vertices is a block graph. Now assume G obeys (B), is $(K_4 - e)$-free, and has more than one vertex. We choose some vertex x_0. Let Q be some component in the level $S_{ecc_G(x_0)}(x_0)$. By (1) and (2), there is some vertex $z \in S_{ecc_G(x_0)-1}(x_0)$ such that all vertices of Q are adjacent to z, and there are no further edges between Q and $V(G) \setminus Q$. By (3), Q must induce a complete graph. The approach yields $T - Q$ as spanning tree for $G - Q$, and obviously this is again distance 1-approximating. By induction hypothesis $G - Q$ is a block graph. Thus G is also a block graph. □

Theorem 2 *Every block graph has some distance 1-approximating spanning tree, that can be found by our basic construction and any rule in linear time.*

We postpone the proof until we have proven Theorem 3.

2.2 Distance-hereditary Graphs

A connected graph $G = (V, E)$ is *distance-hereditary* if every induced x-y path, $x, y \in V$, has length $d_G(x,y)$, see [1] or [4] for some characterizations. It is a superclass of the class of block graphs.

Let x_0 be any fixed vertex in a connected distance-hereditary graph G. Given the linear order of the vertices of G, we define the rule

> **Rule 1:** *Connect every $y \in S_{i+1}(x_0)$ with the smallest vertex $f(y) \in N(y) \cap S_i(x_0)$.*

Theorem 3 *Every connected distance-hereditary graph $G = (V, E)$ has some distance 2-approximating spanning tree which can be found by our basic construction and rule 1 in linear time $O(|V| + |E|)$.*

Proof: Let T be the tree constructed by the rule above, and let yz be some edge of G.

(1) If $d_G(y, x_0) = d_G(z, x_0) = i + 1$, then $N(y) \cap S_i(x_0) = N(z) \cap S_i(x_0)$ [4], thus $f(y) = f(z)$, and $d_T(y, z) \leq 2$.

(2) If w.l.g. $i+1 = d_G(y, x_0) = d_G(z, x_0)+1$, we assume $f(y) \neq z$ — otherwise we are already done. Then there are shortest y-x_0 paths in G, one going over $f(y)$, and one over z. In the terminology of [4], $f(y)$ and z are tied, and it has been shown in [4] that this implies $N(f(y)) \cap S_{i-1}(x_0) = N(y) \cap S_{i-1}(x_0)$. Therefore $f(f(y)) = f(z)$, and $d_T(y, z) = 3$, as desired.

Finally we apply Lemma 1.

Finding the tree T is very easy: We simply compute the levels, and then check for every vertex y the neighborhood list in increasing order until we find some vertex one level beyond the level of y — the resulting vertex is $f(y)$. □

Now Theorem 2 follows quite easily. Let x_0 be a vertex in some $(K_4 - e)$- and C_4-free distance-hereditary graph G, where C_4 is the cycle of length 4.

Then every vertex $y \in S_i(x_0)$ must have *exactly* one neighbor in $S_{i-1}(x_0)$.

For, assume otherwise $z_1, z_2 \in N_G(y) \cap S_{i-1}(x_0)$ with $z_1 \neq z_2$. Again z_1 and z_2 are tied, thus every neighbor w of z_1 in $S_{i-2}(x_0)$ is also adjacent to z_2. The four vertices y, z_1, z_2, and w induce a $K_4 - e$ or a C_4 in G, a contradiction.

Thus any rule yields identical results, and case (2) in the proof above cannot occur. Thus every spanning tree constructed by the approach is distance 1-approximating. But block graphs are $(K_4 - e)$- and C_4-free distance-hereditary graphs.

2.3 Interval Graphs

Interval graphs are intersection graphs of intervals of the real line. It is well-known that interval graphs are chordal, i.e. do not contain induced cycles of length 4 or more. Furthermore, interval graphs can be characterized as those chordal graphs without so-called asteroidal triple — three vertices where between any two of them there is a path avoiding the third and all neighbors of the third [5].

By a probably folklore result, chordal graphs have the nice feature that the neighbors of vertices or edges in the lower level look nice:

Lemma 2. *Let x_0 be a vertex in the chordal graph G.*

(a) For every $x \in S_{i+1}(x_0)$, $N(x) \cap S_i(x_0)$ induces a complete graph.
(b) For every edge $xv \in S_{i+1}(x_0)$, the two sets $N(x) \cap S_i(x_0)$ and $N(v) \cap S_i(x_0)$ must be comparable.

Proof: a) Assume $x \in S_{i+1}(x_0)$ had two nonadjacent neighbors y, z in $S_i(x_0)$. Then we choose some chordless y-z path that, except y and z, uses only vertices inside the levels $S_0(x_0)$ up to $S_{i-1}(x_0)$. Together with the edges zx and xy it forms an induced cycle of length 4 or more, a contradiction.

b) Again, assume to the contrary that there are vertices $y, z \in S_i(x_0)$ such that y is adjacent to x but not to v, and z is adjacent to v, but not to x. If y and z are adjacent, then we have some induced 4-cycle in G. Otherwise, again we find some induced y-z path where all internal vertices have distance less than i to x_0. Together with the edges zv, vx, xy this path yields an induced cycle of length at least 5 in G, a contradiction again. □

Lemma 3. *For every interval graph and every connected component Q of $G[S_{i+1}(x_0)]$, there is some vertex $f(Q) \in S_i(x_0)$ adjacent to all vertices of Q.*

Proof: We assume that there are two vertices y and z in some common connected component Q of $G[S_{i+1}(x_0)]$, whose sets of neighbors inside $S_i(x_0)$ are not comparable. By Lemma 2 (b), y and z are not adjacent. We find y' and z' in $S_i(x_0)$ such that y' is adjacent to y but not to z, and z' is adjacent to z but not to y.

Then the three vertices x_0, y, and z form an asteroidal triple: The y-z path inside $S_{i+1}(x_0)$ avoids x_0 and its neighbors, every shortest y-x_0 path going over y' avoids z and its neighbors, and every shortest z-x_0 path going over z' avoids y and its neighbors. Consequently the sets $N_G(y) \cap S_i(x_0), y \in V(Q)$, form a chain, thus some element is contained in all these sets. □

Let all the vertices $f(Q)$ for the components Q of the levels $G[S_{i+1}(x_0)]$ be chosen in advance, for instance we could choose the smallest element (in the given ordering of the vertices) in $\bigcap_{y \in Q} N_i(y) \cap S_i(x_0)$. The following rule fits into our general scheme:

> **Rule 2:** *Connect $y \in S_{i+1}(x_0)$ with $f(Q)$, where Q denotes the component of $G[S_{i+1}(x_0)]$ containing y.*

Theorem 4 *Every interval graph has some distance 2-approximating spanning tree, that can be found by our basic construction and rule 2 in linear time.*

Proof: By Lemma 1 it suffices to show that $d_T(y, z) \leq 3$ for every edge yz of G for the tree T constructed in this way. The case where both y and z have the same distance $d_G(y, x_0) = d_G(z, x_0) = i + 1$ towards x_0 is easy: Then both lie in the same component Q of $G[S_{i+1}(x_0)]$, whence they are connected over $f(Q)$ in T. So assume that $i + 1 = d_G(y, x_0) = d_G(z, x_0) + 1$, and let Q denote the component of $G[S_{i+1}(x_0)]$ containing y. Since we are done if $f(Q) = z$, assume $f(Q) \neq z$. Then both $f(Q)$ and z are neighbors of y in $S_i(x_0)$ — by Lemma 2 (a) they must be adjacent. By Lemma 3 they have distance two in T, whence $d_T(y, z) \leq 3$.

To find T, we first compute the levels, that is, mark every vertex x by some label $d_G(x, x_0)$. Then we compute $\bigcup_i G[S_i(x_0)]$ in linear time, as mentioned above. The components of the levels are just the components of this graph, and they can also be computed in linear time. We also create ordered lists LEVEL(i), $i \geq 0$, of vertices for the levels, simply be checking all vertices in the order and putting them in the appropriate lists.

All what remains to do is to find for every component Q of every level $S_i(x_0)$ some vertex $f(Q)$ in $S_{i+1}(x_0)$ adjacent to all vertices of Q. This is achieved by checking the vertices in the (ordered) list LEVEL($i-1$) one after another, and also browsing the neighborhood lists of the elements of Q from left to right in parallel, stopping the pointers every time the entry exceeds the actual entry of LEVEL($i-1$). Since the components partition G, every neighborhood list is traversed exactly once, giving a total time-complexity of $O(|V| + |E|)$. □

Both Theorems 4 and 3 are best possible. The graph in Figure 2 has no distance 1-approximating spanning tree, but it is both distance-hereditary and an interval graph.

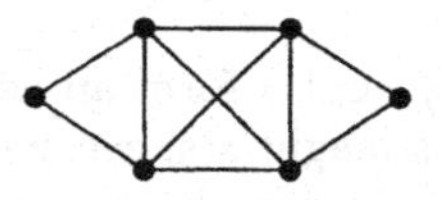

Fig. 2. A graph without distance 1-approximating spanning tree.

3 The Approach Modified for Cocomparability Graphs

A graph $G = (V, E)$ is a *cocomparability graph* if there is some poset $(V, <)$ such that distinct vertices are adjacent in G iff they are not comparable by $<$. Interval graphs are special cocomparability graphs. The approach presented in the preceding section does not work for cocomparability graphs. For the cocomparability graph in Figure 3, it would yield the bad tree consisting of the bold edges. Choosing another vertex as x_0 does not help too much.

Thus we have to allow edges of T inside the levels. But this means that the property of Lemma 1 is maybe no longer valid. Different from the approach in the preceding section, we now need a representation of the graph, and we have to choose a special vertex as start vertex.

Let $G = (V, E)$ be the cocomparabilty graph of the poset $(V, <)$, and let $p = ecc_G(x_0)$. Let x_0 be any minimal element of the poset. Each one of the sets $S_0(x_0) = \{x_0\}, S_1(x_0), \ldots, S_p(x_0)$ induces a poset, let for $i = 0, 1, \ldots, p$ by M_i denote the set of maximal elements of the poset $(S_i(x_0), <)$. Each M_i induces a complete graph in G.

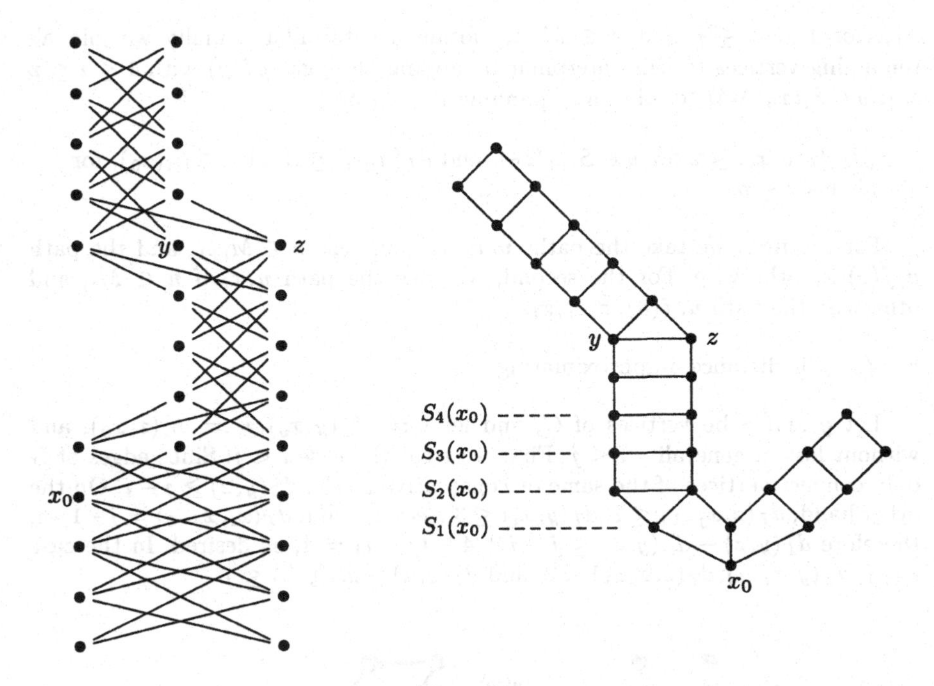

Fig. 3. A poset and its cocomparability graph

By the special choice of x_0, we get:

(1) If $y < z$, then $d_G(y, x_0) \leq d_G(z, x_0)$.

For, assume $y < z$ and $d_G(y, x_0) > d_G(z, x_0)$. Let $x_0, x_1, \ldots, x_t = z$ be some shortest x_0-z path. None of the vertices $x_0, x_1, \ldots, x_{t-1}$ can be adjacent to y, thus y is comparable with all these vertices, but also to $x_t = z$ by the assumption. Since x_0 is a minimal element, $x_0 < y$. Since $x_i < y < x_{i+1}$, $i \in \{0, 1, \ldots, t-1\}$, would contradict to x_i and x_{i+1} being adjacent, y must be greater than all these vertices, in particular $z = x_t < y$, a contradiction.

(2) For every $1 \leq i \leq p$, every $x \in S_i(x_0)$ has some neighbor $f(x) \in M_{i-1}$.

This is shown as follows: Assume this would not be the case for some $1 \leq i \leq p$ and some $x \in S_i(x_0)$. By (1) this would imply that all elements of M_{i-1} are smaller than x, consequently $y < x$ for all $y \in S_{i-1}(x_0)$. Then x would not be adjacent to any vertex of $S_{i-1}(x_0)$, a contradiction.

Here is the construction of the spanning tree T. We choose x_p in M_p. Defining $x_{p-1} := f(x_p), x_{p-2} := f^2(x_p)$, and so on, we obtain a path $x_0, x_1, \ldots, x_p$ with every $x_i \in M_i$. From this path we derive our tree T. First we add all edges

xx_i, for $1 \le i \le p$ and $x \in M_i$ to obtain a caterpillar. Finally we join all remaining vertices to this caterpillar by adding all edges $yf(y)$ with $1 \le i \le p$ and $y \in S_i(x_0) \setminus M_i$ to obtain a spanning tree T of G.

(3) $d_T(v, x_i) \le 2$ *for* $v \in S_{i+1}(x_0)$, *and* $d_T(u, x_i) \le 3$ *for* $u \in S_i(x_0)$, *for all* $1 \le i \le p$.

For the first, we take the paths v, x_{i+1}, x_i in case $v \in M_{i+1}$, and the path $v, f(v), x_i$ otherwise. For the second, we take the path u, x_i if $u \in M_i$, and otherwise the path $u, f(u), x_{i-1}, x_i$.

(4) T *is distance 4-approximating.*

Let y and z be vertices of G, and let $i := d_G(y, x_0), j := d_G(z, x_0)$, and without loss of generality $i \le j$. First we treat the case $i < j$: Since edges of G only connect vertices of the same or consecutive levels, $d_G(y, z) \ge j - i$. On the other hand, $d_T(z, x_{j-1}) \le 2$, $d_T(y, x_i) \le 3$ (see (3)), and $d_T(x_i, x_{j-1}) = j - 1 - i$, therefore $d_T(y, z) - d_G(y, z) \le j - i + 4 - (j - i) = 4$, as desired. In the case $i = j$, $d_T(y, x_{i-1}), d_T(z, x_{i-1}) \le 2$, and $d_T(y, z) - d_G(y, z) \le 4$.

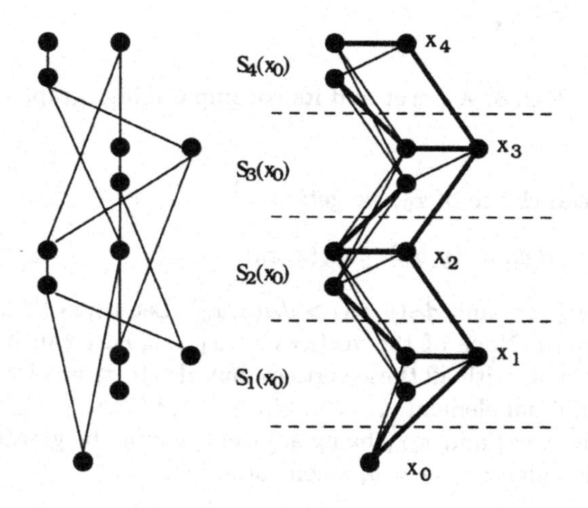

Fig. 4.

Finally, given the poset $(V, <)$ of the cocomparability graph, the construction of T is possible in time $O(|V|^2)$. We assume that the vertices are ordered, and that for every vertex y a ordered list SMALL(y) of vertices smaller than y is given. The total length of all these lists is $| < | \le |V|^2$. Any vertex x_0 with empty list SMALL(x_0) can serve as root. Now we construct the cocomparability graph $G = (V, E)$ (whose representation has size $|V|^2 - | < |$) and compute as usual the levels $S_i(x_0)$. Every vertex y gets its mark $d_G(x_0, y)$. To test whether a given vertex lies in some M_i, all we have to do is to compare the marks $d_G(x_0, z)$ of the entries z in SMALL(y) with $d_G(x_0, y)$—if $d_G(x_0, z) = d_G(x_0, y)$ then z can not lie in M_i. If we have excluded vertices in this way for all $y \in V$, the remaining vertices are in $\bigcup M_i$. Finding the vertices x_i, and the vertices $f(y)$ for $y \in S_i(x_0) \setminus M_i$ can be done by simply checking all neighborhood lists once.

Since a poset for a cocomparability graph $G = (V, E)$ can be computed in time $O(|V|^2)$ [7] we arrive at:

Theorem 5 *Every cocomparability graph has some distance 4-approximating spanning tree, which can be found in time $O(|V|^2)$.* □

Figure 4 may serve as an illustration for the construction of T. On the left, there is a poset, and on the right its cocomparability graph is given. The bold edges are the edges in T found by the rule. Note also that T is distance 4-approximating but not 3-approximating.

4 Chordal Graphs

Since block graphs, as well as interval graphs, are chordal, considering this class might seem promising. However, the following example, due to T.A. MCKEE, shows that for every fixed integer k there are chordal graphs without distance k-approximating spanning subtrees:

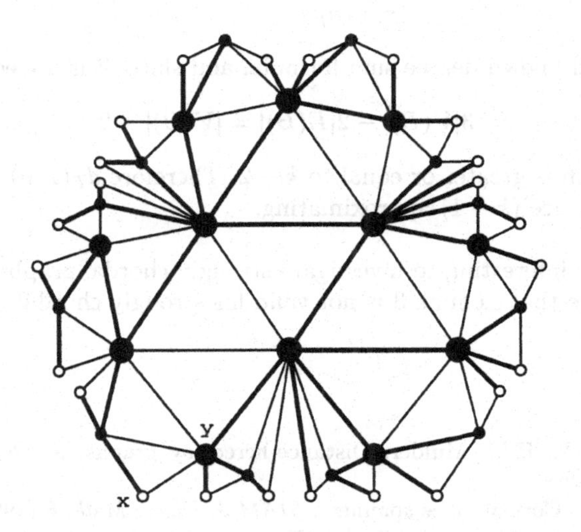

Fig. 5. A chordal snowflake graph.

Let the 1-snowflake graph be the triangle K_3, and let G_2 be the graph obtained from G_1 by adding three vertices, each one adjacent to distinct two vertices of

G_2. Let for every integer $s \geq 2$, the snowflake graph G_s be obtained from G_{s-1} and G_{s-2} by adding for every edge of $G_{s-1} \setminus E(G_{s-2})$ one new vertex, adjacent to the two vertices of the edge. These graphs are chordal, even 2-trees. G_5, together with some distance 5-approximating spanning tree is given in Figure 5. Look at vertices x and y to see that this particular tree is not distance 4-approximating

Proposition 6 *No distance $(k-1)$-approximating spanning tree is possible in G_k.*

Proof: Look at the canonical embedding of G_k in the plane. The first observation is, that the outer face F_0 of G_k has, as vertex in the dual G_k^*, eccentricity equal to k. In fact, all faces have the same eccentricity k in the dual graph for this example.

Let T be a spanning tree of G_k. The dual tree T^* contains all edges of G_k^* which cross edges of G_k that do *not* belong to T.

Let B be a largest connected component of the forest $T^* - F_0$, and let F_1 be the neighbor of F_0 in B. Note that B contains at least $ecc_{T^*}(F_0) \geq ecc_{G_k^*}(F_0) = k$ vertices.

The edge F_0F_1 in T^* crosses an edge xy on the outer cycle in G_k. Since F_0F_1 is an edge in T^*, x and y are not adjacent in T. Moreover $d_T(x,y)+1$ equals the number of edges in G_k^* that start in B and end outside B. Since all vertices except F_0 have degree 3 in G_k^*, this number equals

$$\sum_{F \in V(B)} (3 - d_B(F)),$$

but by the well known degree sum formula, and since B is a tree, this equals

$$3|V(B)| - 2|E(B)| = |V(B)| + 2$$

a number which is greater or equal to $k+2$. Therefore $d_T(x,y) \geq k+1$, and T cannot be distance $(k-1)$-approximating. □

It might be interesting to investigate strongly chordal graphs, see [3] for the definition. Note that Lemma 3 is not valid for strongly chordal graphs.

References

1. H.-J. Bandelt, H.M. Mulder, Distance-hereditay graphs, *J. Combin. Th. B* 41 (1986) 182-208.
2. L. Cai, D.G. Corneil, Tree spanners, *SIAM J. Discr. Math.* 8 (1995) 359-387.
3. M.C. Golumbic, Algorithmic Graph Theory and Perfect Graphs, Academic Press, London (1980).
4. P.L. Hammer, F. Maffray, Completely separable graphs, *Discrete Appl. Math.* 27 (1990) 85-99.
5. C. Lekkerkerker, J. Boland, Representation of a finite graph by a set of intervals on the real line, *Fund. Math.* 51 (1962) 45-64.

6. A.L. Liestman, T. Shermer, Additive graph spanners, *Networks* 23 (1993) 343-364.
7. J. Spinrad, On comparability and permutation graphs, *SIAM J. Computing* 14 (1985) 658-670.

A Better Upper Bound on the Bisection Width of de Bruijn Networks

(Extended Abstract)

Rainer Feldmann, Burkhard Monien, Peter Mysliwietz, and Stefan Tschöke

Department of Mathematics and Computer Science
University of Paderborn

Abstract. We approach the problem of bisectioning the de Bruijn network into two parts of equal size and minimal number of edges connecting the two parts (cross-edges). We introduce a general method that is based on required substrings. A partition is defined by taking as one part all the nodes containing a certain string and as the other part all the other nodes. This leads to good bisections for a large class of dimensions. The analysis of this method for a special kind of substrings enables us to compute for an infinite class of de Bruijn networks a bisection, that has asymptotically only $2 \cdot ln(2) \cdot 2^n/n$ cross-edges. This improves previously known bisections with $4 \cdot 2^n/n$ cross-edges.

1 Introduction

The graph-bisection-problem is one of the best studied problems in graph theory. It has various important applications, for example in VLSI-layout [Len94] and in parallel computing [Lei92]. Given a graph $G = (V, E)$ the bisection width $\beta(G)$ is the minimal number of edges that have to be removed from E in order to disconnect the graph into two parts of the same size.

The bisection width is one critical factor in determining whether a graph is able to serve as an efficient interconnection structure for parallel processing ([Lei92]). This is due to the fact that $|V|/\beta(G)$ is a lower bound for the complexity of some permutation routing problem. Therefore a bigger bisection width gives rise for a better interconnection network.

The de Bruijn network is one of the most prominent interconnection networks. It receives a lot of attention, because it assembles many desirable properties. It has constant number of edges per node, a low diameter, a fast shortest path routing and defines a natural connection structure for FFT applications. Especially this last property makes it a good candidate for the interconnection network of parallel computers. Many results concerning the de Bruijn network are given in [Lei92,SP89,BP89]. Nevertheless for the bisection width of the de Bruijn there is still a huge gap between known upper and lower bounds. If we define $dB(n)$ to be the de Bruijn network of dimension n (with 2^n nodes) we know ([Lei92]) that

$$2^n/n \leq \beta(dB(n)) \leq 4 \cdot 2^n/n.$$

Reischuk, Morvan (Eds.): STACS'97 Proceedings, LNCS 1200

The upper bound is derived by analyzing an embedding of the network into the complex plane [Lei92]. The lower bound can be obtained by simulating the complete graph with 2^n nodes on the de Bruijn network.

We use a different approach that is based on properties of words over a binary alphabet. It is not really surprising that such an approach can be successful. Note that the de Bruijn network got its name because every Hamiltonian path defines a de Bruijn sequence.

We partition the de Bruijn networks by looking at the nodes as binary strings. An initial partition is defined by taking as one part all the nodes containing a certain string and as the other part all the other nodes. For each such substring there are dimensions, where this results in an almost balanced partition with a low number of edges connecting the parts. To finally get a bisection we need to balance the sets by shifting some nodes that have two external edges into the other partition. Since the degree of each node is four, the shift of a node with two external edges does not increase the cost. We show that there are always enough of these nodes to get a bisection. By analyzing a special case that was suggested by Stéphane Perennes [Per95] (substrings are of the form 1^k), we are able to show for an infinite set of dimensions a new asymptotic upper bound of $2 \cdot ln(2) \cdot 2^n/n$ for $\beta(dB(n))$.

Although these results, almost closing the gap between upper and lower bound, are quite interesting, some even more interesting problems are yet to be solved. We conjecture that for the de Bruijn network our bisection induced by the string 1^k is optimal. This, and some other problems, will be discussed later in the paper.

A straightforward application of our method for the shuffle-exchange network gives an (in this case certainly non-optimal) upper bound of $3/2 \cdot ln(2) \cdot 2^n/n$ for $\beta(SX(n))$, which also significantly improves previous results.

We start with recalling the definitions of the considered networks and a description of our general method for the bisection. In section 4 we apply our method to a special case and analyze its asymptotics to finally derive the new bounds. Section 5 presents some extensions of the basic method and discusses further work.

Throughout the paper let $\mathbb{N}$ ($\mathbb{R}, \mathbb{C}$) denote the set of positive integers (reals, complex numbers, resp.). $ln(n)$ denotes the logarithm to the base e, $log(n)$ denotes the logarithm to base 2.

2 Description of the main idea

To fix the notation let us first define the de Bruijn and shuffle-exchange networks $dB(n)$ and $SX(n)$.

Let $\{0,1\}^n$ be the set of binary strings of length n. For $x \in \{0,1\}$ let $\bar{x}$ be the binary complement of x.

We define functions s, sx and ex on $\{0,1\}^n$ as follows:

$$\begin{aligned} s(v_0 \cdots v_{n-1}) &= v_1 \cdots v_{n-1} v_0 \\ sx(v_0 \cdots v_{n-1}) &= v_1 \cdots v_{n-1} \bar{v}_0 \text{ and} \\ ex(v_0 \cdots v_{n-1}) &= v_0 \cdots v_{n-2} \bar{v}_{n-1}. \end{aligned}$$

Let $dB(n)$ denote the graph where $\{0,1\}^n$ is the set of nodes and each node v has the neighbors $s(v), s^{-1}(v), sx(v), sx^{-1}(v)$. $SX(n)$ has the same set of nodes, and each node v has the neighbors $s(v), s^{-1}(v), ex(v)$.

For $w, u \in \{0,1\}^*$ w being substring of u is denoted by $w \prec u$. For $w \in \{0,1\}^*$ let

$$\begin{aligned} V_w(n) &= \{u \in \{0,1\}^n \mid w \prec u\} \text{ and} \\ \overline{V_w}(n) &= \{u \in \{0,1\}^n \mid w \not\prec u\} = \{0,1\}^n \setminus V_w(n). \end{aligned}$$

Let $g_w(n) := |\overline{V_w}(n)|$.

Identifying a word of the set $\{0,1\}^n$ with a node of the $dB(n)$, the sets $V_w(n)$ and $\overline{V_w}(n)$ define a partition of the $dB(n)$ into two parts, that are, in general, different in size. The number of cross-edges connecting the two parts is called the *cost* of the partition.

There is a good reason to assume that such partitions have low costs: If we take a node $v \in V_w(n)$ in $dB(n)$ and look at the neighbors of $s(v)$, $s^{-1}(v)$, $sx(v)$, and $sx^{-1}(v)$) of v we see that all these neighbors are also in $V_w(n)$ unless there is a single appearance of w in v at the beginning (or the end) of v that is destroyed by the application of a s or sx operation (s^{-1} or sx^{-1}, resp.).). Therefore only a small number of nodes located in $V_w(n)$ have neighbors in $\overline{V_w}(n)$ which results in a small number of cross-edges.

The main idea now is to take a word w and choose a dimension n_w , where the numbers of words that do respectively do not contain w are almost equal (i.e. $g_w(n_w) \approx 2^{n-1}$). To balance the partitions we use nodes that have exactly two internal and two external edges (later defined as B-nodes). A shift of these nodes does not increase the cost at all but may generate new B-nodes. By showing, that the number of B-nodes is big enough to close the gap to 2^{n-1} we have a bisection with the same cost as the previous partition.

For the exact description of our method we need some more definitions. For $w \in \{0,1\}^*$ we define

$$\begin{aligned} \mathcal{R}_w(n) &:= \{v \in V_w(n) \mid v \text{ has neighbors in } \overline{V_w}(n)\}, \\ \mathcal{B}_w(n) &:= \{v \in V_w(n) \mid v \text{ has exactly two neighbors in } \overline{V_w}(n)\}, \\ \mathcal{C}_w(n) &:= \mathcal{R}_w(n) \setminus \mathcal{B}_w(n) \end{aligned}$$

We denote by $R_w(n)$, $B_w(n)$ and $C_w(n)$ the cardinalities of $\mathcal{R}_w(n)$, $\mathcal{B}_w(n)$ and $\mathcal{C}_w(n)$ respectively. $\sigma_w(n)$ denotes the number of cross-edges connecting the sets $V_w(n)$ and $\overline{V_w}(n)$.

We call nodes corresponding to the above given set B-, C-, and R-nodes, resp. Note that $R_w(n) = B_w(n) + C_w(n)$ and $\sigma_w(n) = 2 \cdot B_w(n) + C_w(n)$ by construction.

Lemma 1 *Let* $n \geq 1$. *Then* $R_w(n) = 2 \cdot (2 \cdot g_w(n-1) - g_w(n))$.

Proof: Let $w = w_0 \cdots w_{k-1}$. Note that a node $v \in V_w(n)$ with neighbors in $\overline{V_w}(n)$ either starts or ends with a single appearance of w. Due to the symmetry of the operations in the de Bruijn network we have that

$$\frac{1}{2} \cdot R_w(n) = |\{u \in \{0,1\}^n \mid u = w_0 \cdots w_{k-1}v,\ |v| = n-k, w \not\prec w_1 \cdots w_{k-1}v\}|.$$

Note that a word xu where $x \in \{0,1\}$ and $u \in \overline{V_w}(n-1)$ either starts with a single appearance of w, or does not contain w. Therefore $g_w(n) = |\{0,1\}\overline{V_w}(n-1)| - \frac{1}{2} \cdot R_w(n)$. □

In order to compute $g_w(n)$ we need to study the structure of w to some detail. For a fixed $w = w_0 \cdots w_{k-1} \in \{0,1\}^k$ we call $l \in \{1, \ldots, k-1\}$ an *overlap-point*, iff $w_i = w_{l+i}$ for $0 \le i \le k-1-l$. If for example $w = 1011011$ then $l_1 = 3,\ l_2 = 6$ are the overlap-points, if $w = 1^k$ then $l_i = i$ for $1 \le i \le k-1$ are the overlap-points. Note that $l = 0$ is a trivial overlap-point that is excluded from our definition.

Lemma 2 *Let $l_1 \le l_2 \le \cdots \le l_r$ be the overlap-points of $w = w_0 \cdots w_{k-1}$. Then $g_w(n) = 2^n$ for $n < k$ and for $n \ge k$*

$$g_w(n) = 2 \cdot g_w(n-1) - g_w(n-k) + \sum_{i=1}^{r} (2 \cdot g_w(n-l_i-1) - g_w(n-l_i)).$$

Proof: By lemma 1 it is sufficient to show that

$$\frac{1}{2} \cdot R_w(n) = g_w(n-k) - \sum_{i=1}^{r} \frac{1}{2} \cdot R_w(n-l_i).$$

We do this by counting each word in $\Gamma := \{u \in \{0,1\}^n \mid u = w_0 \cdots w_{k-1}v,\ |v| = n-k, w \not\prec w_1 \cdots w_{k-1}v\}$ since this set has cardinality $\frac{1}{2} \cdot R_w(n)$. All words in Γ start with w followed by a word $u \in \overline{V_w}(n-k)$. However not each $u \in \overline{V_w}(n-k)$ corresponds to some $wu \in \Gamma$ because due to overlaps wu might very well contain words that have w as an interior substring starting at an overlap-point l_i. For each l_i we must therefore substract the nodes in the set $E_i := \{v \in \overline{V_w}(n-k) \mid w_{l_i} \cdots w_{k-1}v = wv', w \not\prec w_1 \cdots w_{k-1}v\}$. Note that by definition $w_{l_i} \cdots w_{k-1} = w_0 \cdots w_{k-l_i-1}$, therefore

$$\begin{aligned}
|E_i| &= |\{w_0 \cdots w_{k-l_i-1}v \mid w_0 \cdots w_{k-l_i-1}v = wv', |v| = n-k, w \not\prec w_1 \cdots w_{k-1}v\}| \\
&= |\{u \mid |u| = n-l_i,\ u = wv', w \not\prec w_1 \cdots w_{k-1}v'\}| \\
&= \frac{1}{2} \cdot R(n-l_i).
\end{aligned}$$

□

By looking at the extremal cases we are able to give bounds for $g_w(n)$. One of the extremal cases is the word 1^k since each position in the string is an overlap-point. The other extremal case is represented by the word $1^{k-1}0$ which has no overlap-point.

Lemma 3 *Let* $g_k(n) := g_{1^k}(n)$ *and* $\tilde{g}_k(n) := g_{1^{k-1}0}(n)$. *Then*

1. $g_k(n) = \sum_{i=1}^{k} g_k(n-i)$ *for* $n \geq k$ *and* $g_k(n) = 2 \cdot g_k(n-1) - g_k(n-k-1)$ *for* $n > k$.
2. $\tilde{g}_k(n) = 2 \cdot \tilde{g}_k(n-1) - \tilde{g}_k(n-k)$ *for* $n \geq k$
3. $\tilde{g}_k(n) \leq g_w(n) \leq g_k(n) \leq \tilde{g}_{k+1}(n)$ *for all* $w \in \{0,1\}^k$.

Proof: Omitted. □

The next lemma characterizes the cost of our bisection. It makes use of the line-digraph construction of the de Bruijn network: The $dB(n+1)$ can be constructed from $dB(n)$ by replacing an edge $\alpha v_1 \ldots v_{n-1} \to v_1 \ldots v_{n-1}\beta$ of the $dB(n)$ with the node $\alpha v_1 \ldots v_{n-1}\beta$ and connecting two such nodes $u_0 u_1 \ldots u_n$ and $w_0 w_1 \ldots w_n$ iff they correspond to a pair of consecutive edges in the $dB(n)$, i.e. $u_1 \ldots u_n = w_0 w_1 \ldots w_{n-1}$ or $u_0 u_1 \ldots u_{n-1} = w_1 \ldots w_n$.

Lemma 4 $\sigma_w(n) = R_w(n+1)$.

Proof: We count the number of pairs of the kind $(u, s(u))$ and $(u, sx(u))$ where $u \in V_w(n)$ and $s(u) \in \overline{V_w}(n)$ or $sx(u) \in \overline{V_w}(n)$. Due to the symmetries of the edges this count equals $\frac{1}{2} \cdot \sigma_w(n)$. Therefore $\frac{1}{2} \cdot \sigma_w(n) = |\{(u,v) \mid u = w_0 \cdots w_{k-1}u',\ |u'| = n-k,\ v = w_1 \cdots w_{k-1}u'x,\ x \in \{0,1\},\ w \not\prec v\}| = |\{w_0 \cdots w_{k-1}u'x \mid w \not\prec w_1 \cdots w_{k-1}u'x,\ x \in \{0,1\},\ |u'| = n-k\}| = \frac{1}{2} \cdot R_w(n+1)$.
□

For almost each R-node where the shuffle-edge is an external edge also the shuffle-exchange-edge is external. The same holds for the symmetric operations. Therefore it is easy to show the following

Lemma 5 *Let* $k \geq 2$, *let* $w \in \{0,1\}^k$ *be any string and let* $n \geq k+1$. *Then*

$$B_w(n) \geq \frac{1}{2} R_w(n).$$

Proof: Omitted. □

Now we are ready to finish the description of our method. For each w we choose a dimension n_w, such that the two sets $V_w(i)$ and $\overline{V_w}(i)$ have almost the same size. For $w \in \{0,1\}^*$ we define $n_w := min\{i \mid g_w(i) \leq 2^{i-1}\}$ to be the smallest dimension such that $|V_w(i)| \geq |\overline{V_w}(i)|$. $g_w(i) = 2^i$ for $i \leq k$ and the proportion of strings that do not contain a fixed substring approaches zero if we increase the size o the strings arbitrarily. Therefore n_w is well defined.

The following theorem shows that for dimension n_w the number of B-nodes is large enough to balance the partitions with B-nodes only. Since this balance does not change the number of cross-edges, the resulting bisection has a cost of $\sigma_w(n_w)$.

Theorem 1 *For $w \in \{0,1\}^*$ let $V_w(n)$, $\overline{V_w}(n)$, $g_w(n)$, $\sigma_w(n)$ and n_w be defined as above. Then the de Bruijn network of dimension n_w has a bisection with cost $\sigma_w(n_w)$.*

Proof: By definition we have that

$$2^{n-1} \geq g_w(n) = 2 \cdot g_w(n-1) - \frac{1}{2}R_w(n) > 2^{n-1} - \frac{1}{2}R_w(n).$$

Application of Lemma 5 for $n = n_w$ gives that $g_w(n_w) + B_w(n_w) \geq 2^{n_w - 1}$. Thus we know that there are enough type B nodes to balance the partitions. □

3 Analysis of the bisection cost for $w = 1^k$

In this section we will compute the partition cost for the special words $w = 1^k, k \in \mathbb{N}$. To simplify the notation we write $g_k(n)$ instead of $g_{1^k}(n)$. The other notations are simplified accordingly.

First we will show that for $w = 1^k, k \in \mathbb{N}$ the number of words in $\bar{V}_n^w$ is a generalized Fibonacci number $F_k(n+k)$. In lemmas 6 and 7, we present an approximation for $F_k(n+k)$. This allows us to determine the dimensions of the de Bruijn networks that can be bisected according to the method developed in section 2 using a word $w = 1^k, k \in \mathbb{N}$. Then, in theorem 2 we show that for the de Bruijn networks determined before the asymptotic bisection costs are $2 \cdot ln(2) \cdot \frac{2^n}{n}$, where n is the corresponding dimension of the network.

The proofs in this section apply standard methods from analysis and are mostly omitted.

By lemma 3, for $n \geq k$ we have that

$$g_k(n) = g_k(n-1) + g_k(n-2) + \ldots + g_k(n-k)$$

and for $n > k$

$$g_k(n) = 2 \cdot g_k(n-1) - g_k(n-k-1).$$

Together with lemma 4 we have that

$$\sigma_k(n) = R_k(n+1) = 2 \cdot (2 \cdot g_k(n) - g_k(n+1)) = 2g_k(n-k).$$

There is a nice correspondence between the function g_k and the generalized Fibonacci-sequence of order k which is defined as follows:

$F_k(n) = 0$, for $0 \leq n \leq k-2$
$F_k(k-1) = 1$
$F_k(n) = F_k(n-1) + \ldots + F_k(n-k)$ for $n \geq k$

The well known Fibonacci sequence 0, 1, 1, 2, 3, 5, 8, ... is the generalized Fibonacci-sequence of order 2. An easy induction gives us, that for $k, n \in \mathbb{N}$

$$g_k(n) = F_k(n+k).$$

In order to analyze the growth of $F_k(n)$ we need the roots of its characteristic polynomial which is given by

$$P_k(X) := X^k - X^{k-1} - \ldots - X - 1.$$

Let $z_{k,1}, z_{k,2}, \cdots, z_{k,k}$ (w.l.o.g. $|z_{k,1}| \leq |z_{k,2}| \leq \ldots \leq |z_{k,k}|$) be the roots of $P_k(X)$ in $\mathbb{C}$.

In [Mil60] it is shown that,

$z_{k,1}, z_{k,2}, \cdots, z_{k,k}$ are pairwise disjoint
$|z_{k,i}| < 1$ for $1 \leq i < k$
$z_{k,k} \in \mathbb{R}$, and $1 < z_{k,k} < 2$.

Therefore we know that there exist $\alpha_1, \alpha_2, \cdots, \alpha_k \in \mathbb{C}$ with

$$F_k(n) = \sum_{1 \leq i \leq k} \alpha_i z_{k,i}^n.$$

The determination of the $\alpha_i' s$ is described by the following

Lemma 6 *Let $k \in \mathbb{N}$, $k \geq 2$ and $P_k(X)$ and $z_{k,1}, z_{k,2}, \cdots, z_{k,k}$ be defined as above. Then*

$$\alpha_i = \frac{z_{k,i} \cdot (z_{k,i} - 1)}{(k+1) \cdot z_{k,i} - 2 \cdot k} \cdot \frac{1}{z_{k,i}^k}.$$

Proof: Omitted. □

Since there is only one root of $P_k(X)$ with absolute value greater than 1, it is clear that for increasing n $g_k(n) = \sum_{1 \leq i \leq k} \alpha_i z_{k,i}^{n+k}$ is heavily dominated by the term $\alpha_k \cdot z_{k,k}^{n+k}$. However, we are not interested in $g_k(n)$ for n going to infinity, but for a specific $n = n_k$. Therefore we need to be more precise about the quality of the approximation.

Lemma 7 *With the definitions given above, for $k \geq 2$ we have that*

$$|g_k(n) - \alpha_k \cdot z_{k,k}^{n+k}| < 1.$$

This lemma is essentially proven by showing that

$$|\alpha_i \cdot z_{k,i}^k| \leq \frac{2}{3k+1} \text{ for } 1 \leq i < k.$$

Since we know that the term $\alpha_k \cdot z_{k,k}^{n+k}$ is an almost exact approximation of $g_k(n)$, it is sufficient to determine this term. To simplify the notation we now write z_k instead of $z_{k,k}$.

Lemma 8 *Let $k > 2$. If $z_k \in \mathbb{R}$ is the unique root of $P_k(X)$ in the range $]1, 2[$ (by [Mil60] this root exists), then we have*

$$2^k - k/2 - \frac{k^2}{2^{k+1} - 2k} \leq z_k^k \leq 2^k - k/2 + \frac{k^2 - k}{2^{k+3}}$$

and

$$\lim_{k\to\infty} \alpha_k \cdot z_k^k = 1.$$

Proof: Omitted. □

Lemma 8 enables us to determine the asymptotics of n_k, giving the dimension of the de Bruijn network that has a bisection using substring 1^k.

Lemma 9

$$\lim_{k\to\infty} \frac{n_k}{2^k} = 2 \cdot ln(2).$$

Proof: Due to lemma 7 we do for convenience not distinguish between $g_k(n)$ and its approximation. For each k let $x_k \in \mathbb{R}, n_k - 1 < x_k \leq n_k$ with $g_k(x_k) = \alpha_k z_k^{x_k} = 2^{x_k - 1}$. Since $n_k = \lceil x_k \rceil$ it is sufficient to consider x_k. Let $\alpha'_k := \alpha_k \cdot z_k^k$, then by lemma 6 $\alpha_k z_k^{x_k + k} = \alpha'_k \cdot z_k^{x_k} = 2^{x_k - 1}$. It follows, that

$$x_k = \frac{1 + log(\alpha'_k)}{1 - log(z_k)}.$$

Let $h_k := (1 - log(z_k)) \cdot 2^k$. We first determine the asymptotics of h_k and finally the asymptotics of x_k.

We have that

$$h_k = (1 - log(z_k)) \cdot 2^k = (2^k - log(z_k^{2^k}) = 2^k - log((2 - \frac{1}{z_k^k})^{2^k})$$

$$= 2^k - log(2^{2^k} \cdot (1 - \frac{1}{2z_k^k})^{2^k}) = -log((1 - \frac{1}{2z_k^k})^{2^k})$$

Since $2^k - k \leq z_k^k \leq 2^k$, we know that $\lim_{k\to\infty}(1 - \frac{1}{2z_k^k})^{2^k} = e^{-\frac{1}{2}}$ and therefore

$$\lim_{k\to\infty} h_k = -log(e^{-\frac{1}{2}}) = \frac{log(e)}{2} = \frac{1}{2 \cdot ln(2)}.$$

By lemma 8 $\lim_{k\to\infty} \alpha'_k = 1$, therefore

$$\lim_{k\to\infty} \frac{n_k}{2^k} = \lim_{k\to\infty} \frac{x_k}{2^k} = \lim_{k\to\infty} \frac{1 + log(\alpha'_k)}{h_k} = \lim_{k\to\infty} \frac{1}{h_k} = 2 \cdot ln(2).$$

□

Lemma 9 enables us to derive the asymptotic cost of our bisection.

Theorem 2 *With $\sigma_k(n), n_k$ as defined in section 2, we have that*

$$\lim_{k\to\infty} \frac{\sigma_k(n_k) \cdot n_k}{2^{n_k}} = 2 \cdot ln(2),$$

therefore for the $dB(n_k)$ our method leads to an asymptotic bisection with costs

$$2 \cdot ln(2) \cdot \frac{2^{n_k}}{n_k}.$$

The proof of theorem 2 can be found in the technical report. The theorem shows that the bisection given in this paper has indeed asymptotically very low bisection costs, improving the factor to $2^n/n$ from 4 to 1.386. Table 1 gives the scaled cost for the first few k, and the dimensions n_k that are bisectioned.

k	3	4	5	6	7	∞
n_k	10	22	44	89	178	$2 \cdot ln(2)2^k$
$\sigma_k(n_k) \cdot n_k/2^{n_k}$	1.582	1.536	1.486	1.449	1.425	1.386

Table 1. Bisection costs

We have now analysed the asymptotic bisection costs obtained when using the string $w = 1^k$ to partition the de Bruijn networks. In theorem 1 we showed that for any $w \in \{0,1\}^*$ we are able to find a dimension which can be bisected using w. We will now show that for the special string $w = 1^k$ we are able to balance de Bruijn networks of k successive dimensions. The following theorem states this fact more precisely.

Theorem 3 *For all $k \geq 2$ and all $n \in \{n_k, \ldots, n_k+k-1\}$ the de Bruijn network $dB(n)$ can be bisectioned with costs $\sigma_k(n)$.*

Sketch of the proof: A bisection of $dB(n_k)$ can be found by balancing the initial partition with B-nodes. If $n > n_k$ there may not be enough B-nodes to get a balanced bisection. However, shifting all B-nodes creates new B-nodes. This process can be executed iteratively and we will call the B-nodes available after j steps of this process the j-th iterated B-nodes.

More formally, for $0 \leq j \leq k-2$ we define the sets

$$\begin{aligned}
\mathcal{B}_k^{(j)}(n) &= \{v1^k0u \mid v \in \{0,1\}^j, u \in \overline{V_k}(n-k-j), u \text{ does not end with } 1^{k-j-1}\} \\
&\cup \{u01^kv \mid v \in \{0,1\}^j, u \in \overline{V_k}(n-k-j), u \text{ does not start with } 1^{k-j-1}\} \\
\mathcal{C}_k^{(j)}(n) &= \{v1^k0u \mid v \in \{0,1\}^j,\ u \in \overline{V_k}(n-k-j), u \text{ ends with } 1^{k-j-1}\} \\
&\cup \{u01^kv \mid v \in \{0,1\},\ u \in \overline{V_k}(n-k-j), u \text{ starts with } 1^{k-j-1}\} \\
\mathcal{R}_k^{(j)}(n) &= \mathcal{B}_k^{(j)}(n) \cup \mathcal{C}_k^{(j)}(n).
\end{aligned}$$

For $n \geq 3k-3$ and $0 \leq j \leq k-2$ the union defining $\mathcal{B}_k^{(j)}(n)$ is disjoint. For fixed $k \in I\!N$ we will consider the de Bruijn networks of dimensions $n_k, \ldots, n_k + k - 1$ and for any $k \geq 3$ $n_k \geq 3k-3$. It is easy to show that, for general $n \in I\!N$, at least the nodes $v \in \underline{\mathcal{B}_k^{(j)}(n)}$ are B-nodes after all the sets $\mathcal{B}_k^{(i)}(n)$, $i < j$ have been shifted from $V_k(n)$ to $\overline{V_k}(n)$.

Let $B_k^{(0)}(n) = B_k(n)$ and $B_k^{(j)}(n)$ for $j > 0$ denote the number of B-nodes after shifting the $B_k^{(i)}(n)$, $i < j$, B-nodes to the smaller side. From the above fact we know that

$$B_k^{(j)}(n) \geq |\mathcal{B}_k^{(j)}(n)| \text{ for all } n \in I\!N.$$

If $n \geq 3k-3$ the cardinality of $\mathcal{B}_k^{(j)}(n)$ for $0 \leq j \leq k-2$ is given as

$$\begin{aligned}|\mathcal{B}_k^{(j)}(n)| &= 2 \cdot |\{\alpha_1 \ldots \alpha_j 1^k 0u \mid \alpha_i \in \{0,1\},\ u \in \overline{V_k}(n-k-j), \\ &\qquad\qquad u \text{ does not end with } 1^{k-j-1}\}| \\ &= 2^{j+1} \cdot \sum_{p=j}^{k-2} g_k(n-k-2-p).\end{aligned}$$

The sets $B_k^{(j)}(n)$, $j \in \{0, \ldots, k-2\}$ are disjoint since for any string v the innermost occurence of the string 1^k in v determines the index j uniquely. Therefore, the number of iterated B-nodes is given as

$$|\bigcup_{j=0}^{k-2} \mathcal{B}_k^{(j)}(n)| = \sum_{j=0}^{k-2} |\mathcal{B}_k^{(j)}(n)| = 2 \cdot \sum_{i=0}^{k-2} (2^{i+1}-1) g_k(n-k-2-i).$$

Now let $\Delta_k(n) := 2^{n-1} - g_k(n)$ be the number of nodes that have to be shifted from $V_k(n)$ to $\overline{V_k}(n)$ to get a balanced partition. Of course, $\Delta_k(n) \geq 0$ for all $n \geq n_k$. Using lemma 1 we obtain

$$\Delta_k(n_k) = 2^{n_k-1} - g_k(n_k) = \underbrace{2^{n_k-1} - 2 \cdot g_k(n_k-1)}_{=\Delta_k(n_k-1)<0} + \frac{1}{2} R_k(n_k) < \frac{1}{2} R_k(n_k).$$

Using lemma 1 again we obtain for all $n \geq n_k$

$$\Delta_k(n) = 2^{n-1} - g_k(n) = 2 \cdot \Delta_k(n-1) + \frac{1}{2} R_k(n)$$

Since $\Delta_k(n_k) < \frac{1}{2} R_k(n_k)$ we get for $i \leq k-1$

$$\Delta_k(n_k+i) < \frac{1}{2} \sum_{j=0}^{i} 2^{i-j} R_k(n_k+j) \leq \frac{1}{2} \sum_{j=0}^{k-1} 2^j R_k(n_k+i-j)$$

By lemma 1 $R_k(n) = 2(2 \cdot g_k(n-1) - g_k(n)) = 2 \cdot g_k(n-k-1)$ and therefore we get for all $n \in \{n_k, \ldots, n_k+k-1\}$

$$\Delta_k(n) < \sum_{j=0}^{k-1} 2^j g_k(n-k-1-j).$$

With this we are able to show that

$$\sum_{j=0}^{k-2} |\mathcal{B}_k^{(j)}(n)| - \Delta_k(n) \geq 0 \text{ for } k \geq 4 \text{ and } n \in \{n_k, \ldots n_k+k-1\}.$$

For $1 \leq k \leq 3$ the claim of the statement can be verified explicitely: if $k=1$ then $n_k = 2$. If $k=2$ the $dB(4)$ as well as the $dB(5)$ can be bisected by shifting B-nodes iteratively. If $k=3$ then the $dB(10), dB(11)$ and $dB(12)$ can be bisected. $\square$

4 Remarks and further work

4.1 Optimality

While our new asymptotic bound reduces the gap to the lower bound considerably, what we really would like to know is, whether the new bound is already optimal. We conjecture, that there is no bisection for a $dB(n_k)$ with cost less than $\sigma_k(n_k)$.

The only support for this claim, however, comes from experiments. For small dimensions n ($n < 10$) and for all k we have been able to compute an optimal cut of the de Bruijn network into a set of size $|V_n^{1^k}|$ and a set of size $2^n - |V_n^{1^k}|$ by solving the corresponding integer program. It turns out that $\sigma_k(n)$ is the optimal cost for each k where $g_k(n) >= 2^{n-1}$. We conjecture that this holds for all n and therefore as a special case for $n = n_k$ our bisection with $w = 1^k$ is optimal.

4.2 Bisectioning with other words

$\|w\|$	w	n	$\sigma_w(n)$	$\beta(n)$	$\delta_w(n)$	$r_w(n)$
3	100	7	40	30	10	1
3	101	8	56	54	2	1
		9	98	92	6	2
		10	162	162	0	1
3	111	11	298	298	0	1
		12	548	548	0	2
		12	652	548	104	1
4	1110	13	1200	1032	168	1
		14	2208	1928	280	2
		15	4062	3606	456	3
		14	2042	1928	114	1
4	1011	15	3812	3606	206	2
		16	7116	6646	470	3
		15	3690	3606	84	1
4	1010	16	6948	6646	302	1
		17	13086	12470	616	2
		22	294624	294624	0	1
4	1111	23	567906	567906	0	1
		24	1094674	1094674	0	2
		25	2110052	2110052	0	3

Although for fixed $|w|$ the word 1^k seems to be the best choice, we conjecture that each word leads to the same $2 \cdot ln(2)$ asymptotic bound. The table to the left gives an overview about the results with some words of length 3 and 4. For each word we give the resulting cut $\sigma_w(n)$, the distance $\delta_w(n)$ to the best cut known so far $\beta(n)$ and the number of rounds $r_w(n)$ of the iterative balancing process that have been applied. The best known cuts for the small dimensions $n \leq 17$ have been obtained by an algorithm using the "Helpful Set Heuristic" described in [DMP95] and [HM91], those for dimensions ≥ 18 with the method presented in this paper using more than one word w.

4.3 Bisectioning the shuffle-exchange network

Since the shuffle-exchange network has the same set of nodes and a structure quite similar to the de Bruijn network, it is natural to apply our method to the shuffle-exchange network. For $w = 1^k$ it is quite easy to count the number of cross-edges that result from a partition into $V_w(n)$ and $\overline{V_w}(n)$ and to show that in this case the cost is always less than or equal to 3/4 of the cost for the de Bruijn network. Since the shuffle-exchange network has degree 3, a shift of a node therefore either improves the cut by one (two external edges) or increases the cut by one (one external edge). It can be shown, that we can always shift a set of nodes that even improves the cost. Therefore we have bisection cost $\leq 3/2 \cdot ln(2) \cdot 2^n/n$.

5 Conclusions

We have given a new general method for the bisection of the de Bruijn network. This method uses the correspondence between the binary representation of the nodes and properties of words. An analysis of a special case of our method for an infinite set of dimensions gives an improved upper bound for the bisection width of the de Bruijn network. This improvement is significant, since it almost closes the gap to the best known lower bound. Moreover, we think that our new upper bound is already optimal, although we have no proof for this claim yet.

References

[BP89] J.C. Bermond and C. Peyrat. De Bruijn and Kautz networks: a Competitor for the Hypercube? In *Proceedings of the 1st European Workshop on Hypercubes and Distributed Computers*, pages 279–293. North-Holland, 1989.

[DMP95] R. Diekmann, B. Monien and R. Preis. Using Helpful Sets to Improve Graph Bisections. In *DIMACS Series in Discrete Mathematics and Theoretical Computer Science*, Am. Math. Society, 1995.

[FMMT96] R. Feldmann, B. Monien, P. Mysliwietz and S. Tschöke. A Better Upper Bound on the Bisection Width of de Bruijn Networks Technical report, University of Paderborn, 1996, to appear.

[HM91] J. Hromkovic and B. Monien. The Bisection Problem for Graphs of Degree 4 (configuring Transputer Systems). In *Proceedings of the 16th Math. Foundations of Computer Science (MFCS'91)*, Springer LNCS 520, pp 211-220, 1991.

[Lei92] F.T. Leighton. *Introduction to Parallel Algorithms and Architectures: Arrays, Trees, Hypercubes.* Morgan Kaufmann Publishers, 1992.

[Len94] T. Lengauer. *Combinatorial Algorithms for Integrated Circuit Layout.* Teubner, 1994.

[Mil60] E.P. Miles. Generalized Fibonacci Numbers and Associated Matrices. *Am. Math. Month.*, pages 745–752, 1960.

[Per95] S. Perennes. Personal communication. 1995.

[SP89] M.R. Samatham and D.K. Pradhan. The de Bruijn Multiprocessor Network: A Versatile Parallel Processing and Sorting Network for VLSI. *IEEE Transactions on Computers*, 38(4):567–581, 1989.

An Information-Theoretic Treatment of Random-Self-Reducibility*
(Extended Abstract)**

Joan Feigenbaum and Martin Strauss

AT&T Labs, Murray Hill, NJ 07974 USA
{jf,mstrauss}@research.att.com

Abstract. We initiate the study of random-self-reducibility from an information-theoretic point of view. Specifically, we formally define the notion of a random-self-reduction that, with respect to a given ensemble of distributions, leaks a limited number bits, *i.e.*, produces target instances $y_1, \ldots, y_k$ in such a manner that each y_i has limited mutual information with the input x. We argue that this notion is useful in studying the relationships between random-self-reducibility and other properties of interest, including self-correctability and NP-hardness. In the case of self-correctability, we show that the information-theoretic definition of random-self-reducibility leads to somewhat different conclusions from those drawn by Feigenbaum, Fortnow, Laplante, and Naik [13], who used the standard definition. In the case of NP-hardness, we use the information-theoretic definition to strengthen the result of Feigenbaum and Fortnow [12], who proved, using the standard definition, that the polynomial hierarchy collapses if an NP-hard set is random-self-reducible.

1 Introduction

Informally, a function f is *random-self-reducible* if the evaluation of f at any given instance x can be reduced in polynomial time to the evaluation of f at one or more *random* instances y_i.

Random-self-reducible functions have many applications, including:

Cryptography: The fact that certain number-theoretic functions are random-self-reducible is used extensively in the theory of cryptography — *e.g.*, to achieve *probabilistic encryption* [14] and *cryptographically strong pseudorandom number generation* [9]. Random-self-reductions also provide natural examples of *instance-hiding schemes* [1, 5, 6], in which a weak, private computing device uses the resources of a powerful, shared computing device without revealing its private data.

---

* Part of this work was done while the second author was at Iowa State University, supported by CCR-9157382.

** A full version of this paper has been submitted for journal publication and is available as AT&T Technical Report 96.13.2.

Reischuk, Morvan (Eds.): STACS'97 Proceedings, LNCS 1200

Interactive proof systems and program checkers, self-testers, and self-correctors: Random-self-reductions are crucial ingredients in many of the original examples of interactive proof systems and program checkers, self-testers, and self-correctors [7, 8, 15]. Intuitively, this is because the verifier, checker, tester, or corrector interrogates the prover or program by comparing its output on the specific input of interest to its outputs on other related random instances. These ideas play a crucial role in the characterization of the language-recognition power of interactive proof systems [4, 18, 19]. A very active theme of current research in this area is the question of whether NP-complete sets have checkers, self-testers, or self-correctors [8, 12, 13]. We explore this theme further in this paper, using an information-theoretic perspective for the first time.

Average-case complexity: A random-self-reduction maps an arbitrary, worst-case instance x in the domain of f to a set of random instances $y_1, \ldots, y_k$ in such a way that $f(x)$ can be computed in polynomial-time, given x, $f(y_1), \ldots, f(y_k)$, and the coin-toss sequence used in the mapping. Thus the average-case complexity of f, where the average is taken with respect to the induced distribution on instances y_i, is the same, up to polynomial factors, as the worst-case randomized complexity of f. One important example of this connection between average-case complexity and worst-case complexity is the result of Lipton [17] that the PERM (permanent of integer matrices) function is random-self-reducible. The PERM function is also #P-complete [21]; thus, if PERM could be computed efficiently *on average* (with respect to the target distribution of the reduction), then *every* function in #P could, with a randomized algorithm, be computed efficiently in the *worst case*. Furthermore, the random-self-reduction for PERM is very simple, whereas average-case hardness proofs are often complicated.

Lower bounds: The random-self-reducibility of the parity function is used in [2] to obtain a simple proof that a random oracle separates the polynomial hierarchy (PH) from PSPACE. (An earlier proof of this result in [10] does not use random-self-reducibility.)

In this work we investigate a new notion, *random-self-reducibility with respect to an ensemble* $D = \{D_n\}_{n\geq 1}$, *leaking* $l(n)$ *bits*. Roughly, a function has this property if $f(x)$ can be recovered with high probability from x, $y_1, \ldots, y_k$ and $f(y_1), \ldots, f(y_k)$, where each y_i, although chosen at random from a distribution that may depend on x, has no more than $l(|x|)$ bits of mutual information with x. In particular, we may require that, for all c, fewer than $1/n^c$ bits of information are leaked. We assume that the random-self-reduction can sample from D — this is justified because, in practice, the reduction can sample from D by requesting more input. This definition is a small modification of the previous definition, but it has implications for the application areas listed above.

Cryptography: One reason that the standard definition of a random-self-reduction is useful in cryptographic constructions is that it naturally provides a notion of privacy of the input x: The recipient of a single target instance y_i receives no information about x except its length. Our new definition provides a robust way to *quantify* the extent to which the privacy of x may be compromised. We assume that x is drawn from an ensemble $D = \{D_n\}_{n\geq 1}$; this ensemble will be determined by the particular cryptographic scenario in which the reduction is used and is assumed to be known by all of the parties in that scenario. A target instance y_i will also be drawn from a distribution, and this target distribution will depend both on x and the algorithm used in the reduction. The mutual information of the two random variables x and y_i thus measures what a party who receives y_i knows about x that he did not already know from D before receiving y_i. A reduction may be considered sufficiently privacy-preserving if it divulges fully an x that is assigned high probability by D but conceals almost fully an x that is assigned low probability by D.

Self-correction: Blum, Luby, and Rubinfeld [8] defined program self-correction in order to address the following question. Let P be any program that purports to compute f, and suppose that one can determine that, with respect to an ensemble $D = \{D_n\}_{n\geq 1}$, the measure of inputs on which P errs, while not necessarily zero, is limited. Is it possible to write an auxiliary program C that corrects the errors of any such P with high probability? More precisely, on any input $x \in Dom(f)$, C should produce the correct answer $f(x)$ with high probability, and C may call the (potentially faulty) program P several times in the course of its computation. If such a C exists, then the function f is said to be self-correctable with respect to the ensemble D. Blum, Luby, and Rubinfeld observed that every f that has a standard random-self-reduction is also self-correctable, and it is a well-known open question whether the two properties are equivalent, *i.e.*, whether every self-correctable function has a standard random-self-reduction. Recently, Feigenbaum, Fortnow, Laplante, and Naik [13] provided a partial answer to this question by exhibiting, under a plausible complexity theoretic hypothesis, a function that is self-correctable but does not have a standard random-self-reduction.

In Section 3 below, we give evidence that our new definition of random-self-reducibility is better suited than the standard definition to a direct comparison with self-correctability. Specifically, we claim that, because inputs to a self-corrector are drawn from an ensemble D, inputs to a random-self-reduction should also be drawn from such an ensemble if the notions are to be compared meaningfully. To justify this claim, we show that the function f that is proven in [13] to be self-correctable with respect to a particular ensemble D, but not random-self-reducible according to the standard definition, *is* in fact random-self-reducible (leaking 0 bits) according to our definition, in which the random-self-reduction is given the ability to sample from D.

Average-case complexity: If we exhibit a standard random-self-reduction for f, we show that the worst-case complexity of f is no worse than the average-case complexity with respect to an ensemble *of our own construction*, namely the target ensemble of the reduction. In practice, however, we might want to know the average-case complexity of f with respect to an instance ensemble $D = \{D_n\}_{n\geq 1}$ arising in a particular scenario or application. In such a context, we should only consider random-self-reductions that take inputs x from D and produce queries y_i from D (or from a distribution D_x that is close enough to D to allow the average-case complexity with respect to D_x to tell us something about the average-case complexity with respect to D). Restricting random-self-reductions in such a way may make them considerably harder to construct than they are to construct under the standard definition. Our new definition of a leaky random-self-reduction with respect to D allows us to take advantage of the peculiarities of D and so may make it easier to exhibit reductions that give bounds on the average-case complexity with respect to ensembles that arise naturally.

Some results and all proofs have been omitted from this extended abstract because of space limitations. They can be found in our journal submission, which is available as AT&T Technical Report 96.13.2.

2 Preliminaries

Throughout this paper, f is a function on $\{0,1\}^*$, and x is an arbitrary input for which we would like to determine $f(x)$. Reductions will make a polynomial number of queries, denoted $k(n)$, on inputs of length n. We use r to denote a sequence of fair coin tosses; if $|x| = n$, then $|r| = w(n)$, where w is a polynomially bounded function of n. $D = \{D_n\}_{n\geq 1}$ denotes an ensemble of distributions; there is a constant c such that, for all n, any y to which D_n assigns positive probability satisfies $n^{-c} \leq |y| \leq n^c$.

2.1 Previous Definitions

We start by recalling the standard definitions. For a more in-depth review of these definitions and of their role in complexity theory, see [11].

Definition 1.

- A **basic reduction** of f to g is a pair of polynomial-time computable functions (ϕ, σ) such that, for all n, all $x \in \{0,1\}^n$, and at least 3/4 of the r's in $\{0,1\}^{w(n)}$, f satisfies

$$f(x) = \phi(x, r, g(\sigma(1, x, r)), \ldots, g(\sigma(k(n), x, r))).$$

(Note that this is equivalent to the statement that there is a nonadaptive, probabilistic polynomial-time reduction from f to g; we use the term "basic reduction" for brevity.)
- A basic reduction (ϕ, σ) of f to f is a **self-corrector with respect to an ensemble D of distributions** if it is also a basic reduction of f to P for each program P that, for all n, computes f correctly on a set Y_n such that $\Pr_{D_n}(Y_n) \geq 3/4$.
- A basic reduction (σ, ϕ) of f to f is called a **random-self-reduction** of f if, for each i, for all x_1, x_2 such that $|x_1| = |x_2|$, the random variables $\sigma(i, x_1, r)$ and $\sigma(i, x_2, r)$ are identically distributed. (Note that, because r is chosen uniformly at random from $\{0,1\}^{w(n)}$, $\sigma(i, x, r)$ is a random variable, for fixed i and x.)

- Let L be a function on $Dom(f)$. The basic reduction (σ, ϕ) is a **random-self-reduction leaking at most** L if $\sigma(i, x_1, r)$ and $\sigma(i, x_2, r)$ are identically distributed for all x_1, x_2 such that $L(x_1) = L(x_2)$. Thus, "random-self-reduction" is really shorthand for "random-self-reduction leaking at most $|x|$."

Note that we are restricting attention to ***nonadaptive*** reductions, *i.e.*, those in which the query y_i does not depend on the answers to the queries y_1 through y_{i-1}; see, *e.g.*, [11] for a discussion of the more general, adaptive versions of these reductions. Because of this restriction, we may assume that the random variables $\sigma(i_1, x, r)$ and $\sigma(i_2, x, r)$ are identically distributed, for all i_1, i_2, as pointed out by Szegedy and reported in [12], and thus we use $\sigma(x, r)$ to denote this random variable and σ_x for its distribution. The shorthand "rsr" is used for both "random-self-reduction" and "random-self-reducibility."

As usual, we may reduce the error probability of any of these reductions from 1/4 to 2^{-n} by running a polynomial number of independent copies of the reduction and returning the plurality answer. One may also parameterize self-correction in such a way that the program P is assumed to err on D_n with probability at most $\epsilon(n)$ (as Blum, Luby, and Rubinfeld did in their original paper [8]).

2.2 Mutual Information

The **mutual information** $I(a; b)$ between two events a, b is defined to be

$$\log \frac{\Pr(ab)}{\Pr(a)\Pr(b)} = \log \frac{\Pr(a \mid b)}{\Pr(a)}.$$

Conventions differ about the mutual information between two events if (at least) one of these events has probability zero. In the context that we use information theory, there is a natural definition, but we defer discussion of that definition to the end of this section.

The **self-information** of the event a is the mutual information between a and a or, equivalently, the maximum, over all random variables B and reals b, of the mutual information between a and the event $B = b$.

2.3 Information-Theoretic Definition of RSR

We introduce three variations on the standard definitions given in Section 2.1. The first concerns the requirement that the distribution on queries be exactly the same for all x's of the same length, the second concerns the sampleability of inputs, and the third concerns a connection between distributions on queries and the natural distributions on inputs (that we introduce).

First, we introduce rsr's that leak at most $l(n)$ bits, as opposed to those that leak at most some function L.

Definition 2. For $n \geq 1$, the **(information-theoretic) leakage** of a basic reduction (ϕ, σ) for f with respect to the ensemble D is

$$\max_{\{x \in \{0,1\}^n, r\}} I(x; \sigma(x,r)).$$

A basic reduction (ϕ, σ) is an **rsr with respect to D, leaking l bits** if, for all $n \geq 1$, its leakage is at most $l(n)$.

Let $\sigma \circ D$ be the ensemble of distributions on y produced by the experiment of sampling x from D then sampling y from σ_x. Thus, if we are given x from D and we query y, the mutual information between x and y is

$$\log \frac{\Pr_{\sigma_x}(y)}{\Pr_{\sigma \circ D}(y)} = \log \frac{\Pr(y \mid x)}{\Pr(y)},$$

and the leakage of a basic reduction is bounded by l if and only if, for all x and y, $\Pr_{\sigma_x}(y) \leq 2^l \Pr_{\sigma \circ D}(y)$.

Both self-correctors and leaky rsr's have associated ensembles of instances. We are motivated by the case in which these ensembles arise in nature, and thus we assume that additional data points are abundant:

Convention 3 *We assume the functions ϕ and σ of a self-corrector can sample from D_n (by requesting more input). We assume samples from D_n are independently and identically distributed.*

Thus, in our model, sampling from $\sigma \circ D$ is feasible.

The following definition is useful in the context of average-case complexity.

Definition 4. An rsr with respect to D is **reciprocating** if there is a polynomial $p(n)$ such that, for all x, y, $\Pr_{\sigma_x}(y) \leq p(n) \Pr_D(y)$.

That is, the rsr is reciprocating if the distribution on queries conditioned on x is approximated from above by the distribution on inputs. If f has a reciprocating rsr with respect to D leaking (the function) $|x|$, then the distributions σ_x are actually the one common distribution σ, and the worst-case running time of f is at most polynomially worse than the average-case running time with respect to the natural ensemble D.

An important special case of a reciprocating rsr is that in which D_n and σ_x are uniform distributions. Traditional rsr theory allows two arbitrary distributions, D_n and a common distribution σ for all x, but requires no connection between D_n and σ; we allow different distributions for each x but require σ_x to be loosely bounded from above by both D and $\sigma \circ D$.

It should now be clear why our definition of leakage captures the notion of the (worst-case) amount of information about the private instance x that is given away by a basic reduction. Some further comments on the definition are in order:

- The phrase "... leaking l *bits*" (as opposed to "... leaking l") is used to distinguish information-theoretic leakage from the standard notion of functional leakage (which is referred to as "... leaking L").

- It is $I(x;\sigma(x,r))$ that is bounded and not the absolute value $|I(x;\sigma(x,r))|$: There is no limit on the amount of negative (*i.e.*, misleading) information about x that may be "divulged" by the queries.
- As in the literature on standard rsr's, we assume that we make our queries of powerful players that are not allowed to collude. Thus the mutual information between an input x and a single query y_i is limited in the worst case, but that between x and a pair of queries y_i, y_j is not.

We now argue that leaking $O(\log n)$ bits is reasonable. Standard rsr theory focuses on functional leakage of $n = |x|$, and we wish to develop an analogous notion of a "reasonable amount" of information-theoretic leakage. First we argue that divulging $n = |x|$ is equivalent to leaking at least $\log n + \log\log n$ bits, information-theoretically. Suppose that x is chosen from a distribution on $\{0,1\}^*$ rather than a distribution on $\{0,1\}^{|x|}$; so $|x|$ is initially unknown. Because $\sum_n \Pr(|x| = n) = 1 < \infty$, for infinitely many x's, $\Pr(|x| = n) < 1/(n \log n)$; therefore, for arbitrarily large n, the self-information, $-\log \Pr(|x| = n)$ bits, contained in $n = |x|$ is at least $\log n + \log\log n$ bits. (If only finitely many instances x are possible, then no complexity-theoretic treatment is meaningful.) Furthermore, the next lemma shows that leaking $O(\log n)$ bits is a robust notion for rsr's, in the same way that erring with probability at most $1/2 - \epsilon$ is a robust notion for probabilistic algorithms.

Lemma 5. *If f is rsr with respect to D, leaking $O(\log n)$ bits, then, for any polynomial $p(n)$, f is rsr with respect to D, leaking $1/p(n)$ bits.*

This argument that $O(\log n)$ bits is the "right" amount of leaking justifies the following definition.

Definition 6. A function has a **leaky rsr with respect to** D if it is rsr with respect to D leaking at most $O(\log n)$ bits.

Another useful way to view the information-theoretic leakage of a basic reduction is in relationship to the entropy of D. For example, suppose that D_n is the uniform distribution on $\{0,1\}^n$ and that a basic reduction leaks $n/2$ bits. Each x drawn from D_n contains n bits of self-information, and therefore the basic reduction preserves $n/2$ bits of privacy. This should be considered more privacy-providing, for example, than any basic reduction (ϕ, σ) with respect to an ensemble D in which D_n is uniform on $2^{n/3}$ of the strings in $\{0,1\}^n$, even if (ϕ, σ) is a standard rsr, because the latter provides only $n/3$ bits of privacy even *before* queries are made.

2.4 Zero Probability

One final issue remains. Suppose that D_n is a distribution such that $\Pr_D(x_0) = 0$. How should this be handled? Ideally, our analysis should be continuous, *i.e.*, the case in which $\Pr_D(x_0)$ is equal to zero should be treated similarly to the case in which $\Pr_D(x_0)$ is small.

First, should an rsr with respect to D be required to recover $f(x_0)$? Because we regard an rsr as a transformation from worst case to average case, we answer "yes."

A related question is whether there should be a limit to the leakage when $\Pr_D(x_0) = 0$. The leakage is defined in terms of the mutual information $I(x_0; y)$, which in turn is defined in terms of a conditional probability $\Pr(y|x_0)$. In many contexts, there is no sensible way to define the conditional probability when $\Pr(x_0) = 0$. We, however, use mutual information in just one context, in which there is a natural and meaningful definition. The mutual information between x and y is taken to be

$$\log \frac{\Pr_{\sigma_x}(y)}{\Pr_{\sigma \circ D}(y)}.$$

Even if $\Pr_D(x_0) = 0$, it still makes sense to talk about σ_{x_0} and about recovering $f(x_0)$ from $f(y)$ with y chosen from σ_{x_0}. Also, if the denominator is zero but the numerator is non-zero, which can only happen if $\Pr_D(x_0) = 0$, then we define the mutual information to be $+\infty$ — if we query such a y, then we have (completely) leaked the fact that the input is one to which D assigns probability zero. Similarly, if the denominator is non-zero but the numerator is zero then the mutual information is defined to be $-\infty$ (not a constraint). Finally, if both numerator and denominator are zero (whether $\Pr_D(x_0) = 0$ or not), then the mutual information is defined to be zero (*i.e.*, not a constraint), because we could easily define a new rsr that, with small probability, queries from the uniform ensemble and disregards the answers; in this new rsr, the numerator and denominator would be equal, non-zero quantities. These conventions make our analysis continuous.

3 RSR versus Self-correction

Feigenbaum, Fortnow, Laplante, and Naik [13] consider the question of whether all self-correctible functions are rsr. They show that the question cannot be settled with current techniques, because there are currently unrefutable hypotheses that support both answers. Nonetheless, one can interpret the results of [13] as evidence that there is a function f and an ensemble D such that f does not have a standard rsr but is self-correctable with respect to D — [13] gives a plausible hypothesis that implies the existence of such an f and D and an implausible hypothesis that implies the nonexistence of such an f and D.

Here we revisit the question, using our notion of rsr's that can sample from D, which we have argued in Section 1 is more suitable for a fair comparison of rsr with self-correction. While we cannot show definitively that all self-correctible functions have such rsr's, our results lend themselves to the opposite interpretation of those in [13].

3.1 Inequivalence

Feigenbaum, Fortnow, Laplante, and Naik [13] observe that certain functions f are self-correctible with respect to a singleton ensemble D, *i.e.*, one that, for each

n, puts all the weight on one string in $\{0,1\}^n$: The crux is that any program that computes f correctly with probability 3/4 with respect to D already computes f correctly on the one positive-probability string. While the peculiarities of D can make a function self-correctible, they do not necessarily endow that function with a standard rsr, as shown in [13]. We show here that, for such peculiar ensembles D, all self-correctible functions have rsr's that sample D. In fact, the rsr that we exhibit leaks zero bits beyond $n = |x|$; the proof relies on Convention 3 only.

Theorem 7. *Suppose $p(n)$ is a polynomial and $T \subseteq 1^* \cap \mathrm{UP}$. For each n such that $1^n \in T$, let u_n be the unique witness of this, and without loss of generality assume that $|u_n| = n$. Let W be the set of all such u_n's. Let δ_{u_n} denote the distribution on $\{0,1\}^n$ that puts all the weight on u_n, and define D_n by*

$$D_n = \begin{cases} \delta_{u_n} & 1^n \in T \\ \textit{uniform} & \textit{otherwise} \end{cases}$$

If L is a subset of W, then L has a leaky rsr with respect to D.

We conclude that the self-correctible function exhibited in [13] fails to be rsr simply because the standard definition of rsr does not allow the reduction to sample from the ensemble D. The fact that a standard rsr leaks at most n is not really used.

3.2 Equivalence

In [13], it is shown that, if PF = #P, then any function that is self-correctible with respect to a P-sampleable ensemble also has a standard rsr. In this section, we proceed along similar lines, drawing a weaker conclusion from a weaker hypothesis: If P = NP, then any function that is self-correctible with respect to a P-sampleable D has a leaky rsr with respect to D. The proof in [13] uses the assumption PF = #P to conclude that a P-sampleable ensemble is simply a function in P. Instead, we use the weaker hypothesis P = NP and the technique of universal hashing [20] merely to approximate the ensemble. This is sufficient to produce a leaky rsr.

Theorem 8. *Let $D = \{D_n\}_{n \geq 1}$ be a P-sampleable ensemble. If P = NP and f has a self-corrector with respect to D, then f has a leaky, reciprocating rsr with respect to D.*

Proof. (sketch) Our proof builds on [13]. Suppose the self-corrector (ϕ, σ) makes $k(n) = n^{O(1)}$ queries and fails with probability at most 2^{-n} to correct any program that correctly computes f on $\{0,1\}^n$ with probability at least $\epsilon(n) = 1/n^{O(1)}$, with respect to D_n.

Fix c. A query z is called *superfluous* with respect to x if

$$\Pr_{\sigma_x}(z) \geq \frac{k}{\epsilon^2} \Pr_{D_n}(z),$$

and *very superfluous* if

$$\Pr_{\sigma_x}(z) \geq \frac{k}{\epsilon^2}\left(1+\frac{1}{n^c}\right)\Pr_{D_n}(z).$$

The term superfluous originated in [13], where it is used to describe z's with the property that no self-corrector can rely heavily on $f(z)$ to recover $f(x)$. The term has additional meaning here: If we promise that our rsr queries from D (a goal that originated in [13] and is related to reciprocity), then a query z is superfluous precisely when it has more than $\log(k/\epsilon^2)$ bits of mutual information with x.

In [13], it is shown how to construct a standard rsr with target ensemble D, by preparing samples z from σ_x and another distribution $\Pr''(x,z)$, then querying the non-superfluous samples from σ_x mixed with the $\Pr''$ samples. The non-superfluous queries z do not reveal much about x so burying them among a modest number queries from $\Pr''$ that are not needed to reconstruct $f(x)$ results in a target ensemble that reveals sufficiently little about x.

The construction of $\Pr''$ depends on identifying superfluous queries. We can do this approximately, using universal hashing. The construction is given in full detail in our journal submission and is omitted here because of space limitations. For each x, the reduction we construct queries an instance z with probability $\Pr_{\tilde{D}_x}(z) = \Pr_{D_n}(z)(1 \pm O(1/n^c))$.

What is leaked? For a fixed x, the mutual information between x and a query z from $\tilde{D}_x$ is

$$\log \frac{\Pr_{\tilde{D}_x}(z)}{\Pr_{\tilde{D}\circ D_n}(z)}.$$

$\tilde{D}_x$ is close to D uniformly in x, and so

$$\Pr_{\tilde{D}\circ D_n}(z) = (1 \pm O(1/n^c)) \Pr_{D_n}(z).$$

Also $\Pr_{\tilde{D}_x}(z) = (1 \pm O(1/n^c)) \Pr_{D_n}(z)$, so the leakage is $\log(1 \pm O(1/n^c)) = O(1/n^c)$.

We conclude that f has a leaky reciprocating rsr.

4 Random-self-reducibility of SAT

Feigenbaum and Fortnow [12] show that SAT is not random-self-reducible unless the PH collapses at the third level. In this section, we show an analogous result for our notion of leaky rsr.

The language class AM^{poly} is defined in [12]; it consists of languages with Arthur-Merlin protocols [3] in which Arthur gets advice and the subsequent AM protocol need only be valid on the correct advice. In [12], it is shown that, if S is in NP and has a standard rsr, then the complement of S is in AM^{poly}. Since it is known [3, 16] that $AM^{poly} = NP/poly$, membership of $\overline{SAT}$ in AM^{poly} implies that co-NP $\subseteq$ NP/poly, whence the PH collapses [22].

In Section 2, we argued that $O(\log n)$ bits of leakage is the "right" amount to allow. That is, one should require that $\Pr_{\sigma_x}(z) \leq n^{O(1)} \Pr_{\sigma \circ D_n}(z)$. For the results of this section, we need tighter control on the leakage. Relatively tight one-sided bounds will allow us to deduce the two-sided bounds necessary for the results of this section.

Theorem 9. *Let S be a set in* NP. *If χ_S has an rsr making $k(n) \geq 4$ queries and leaking $1/k^5$ bits with respect to an ensemble D, then $\overline{S}$ is in* $\mathrm{AM}^{\mathrm{poly}}$.

Corollary 10. *If SAT has an rsr making $k(n) \geq 4$ queries and leaking $1/k^5$ bits, then the PH collapses.*

5 Conclusions and Open Problems

We have modified the standard notion of random-self-reducibility. We have argued that our modification is natural in many of the contexts in which random-self-reducibility arises, including cryptography, average-case complexity, and program self-correction.

The results in Section 3.1 raise the following question. Can one exhibit, perhaps under a plausible complexity theoretic hypothesis, a self-correctable function that has no leaky rsr, not even an rsr that samples from the ensemble D?

In Section 4, we show that SAT has no rsr making k queries and leaking $1/k^5$ bits, unless the PH collapses. Can this result be improved to show that SAT has no leaky rsr (leakage independent of k), unless the PH collapses? A more useful and possibly easier task would be to show that SAT has no leaky, reciprocrating rsr.

Can our information-theoretic formulation be used to analyze adaptive rsr's and to resolve some of the questions about adaptiveness that have gone unresolved for standard rsr's?

References

1. M. Abadi, J. Feigenbaum, and J. Kilian. On Hiding Information from an Oracle. *Journal of Computing and System Sciences*, 39:21–50, 1989.
2. L. Babai. Random oracles separate PSPACE from the polynomial-time hierarchy. *Information Processing Letters*, 26:51–53, 1987.
3. L. Babai and S. Moran. Arthur-Merlin games: A randomized proof system and a hierarchy of complexity classes. *Journal of Computing and System Sciences*, 36:254–276, 1988.
4. L. Babai, L. Fortnow, and C. Lund. Nondeterministic exponential time has two-prover interactive protocols. *Computational Complexity*, 1:3-40, 1991.
5. D. Beaver and J. Feigenbaum. Hiding instance in multioracle queries. In *Proc. 7th Symposium on Theoretical Aspects of Computer Science*, Lecture Notes in Computer Science, vol. 415, pages 37–48. Springer, Berlin, 1990.

6. D. Beaver, J. Feigenbaum, J. Kilian, and P. Rogaway. Locally random reductions: Improvements and applications. *Journal of Cryptology*, to appear. Preliminary version in *Proc. Crypto '90*, pages 62–76, under the title "Security with low communication overhead."
7. M. Blum and S. Kannan. Designing programs that check their work. *Journal of the ACM*, 42:269–291, 1995.
8. M. Blum, M. Luby, and R. Rubinfeld. Self-testing/correcting, with applications to numerical problems. *Journal of Computing and System Sciences*, 59:549–595, 1993.
9. M. Blum and S. Micali. How to generate cryptographically strong sequences of pseudo-random bits. *SIAM Journal on Computing*, 13:850–864, 1984.
10. J. Cai. With probability one, a random oracle separates PSPACE from the polynomial hierarchy. *Journal of Computing and System Sciences*, 38:68–85, 1989.
11. J. Feigenbaum. Locally random reductions in interactive complexity theory. *DIMACS Series in Discrete Mathematics and Theoretical Computer Science*, vol. 13, pages 73–98. American Mathematical Society, Providence, 1993.
12. J. Feigenbaum and L. Fortnow. Random-self-reducibility of complete sets. *SIAM Journal on Computing*, 22:994–1005, 1993.
13. J. Feigenbaum, L. Fortnow, S. Laplante, and A. Naik. On coherence, random-self-reducibility, and self-correction. In *Proc. 11th Conference on Computational Complexity,* pages 59–67. IEEE Computer Society Press, Los Alamitos, 1996.
14. S. Goldwasser and S. Micali. Probabilistic encryption. *Journal of Computing and System Sciences*, 28:270–299, 1984.
15. S. Goldwasser, S. Micali, and C. Rackoff. The knowledge complexity of interactive proof systems. *SIAM Journal on Computing*, 18:186–208, 1989.
16. S. Goldwasser and M. Sipser. Private coins versus public coins in interactive proof systems. *Advances in Computing Research*, vol. 5, pages 73–90. JAI Press, Greenwich, 1989.
17. R. Lipton. New directions in testing. *DIMACS Series in Discrete Mathematics and Theoretical Computer Science*, vol. 2, pages 191–202. American Mathematical Society, Providence, 1991.
18. C. Lund, L. Fortnow, H. Karloff, and N. Nisan. Algebraic methods for interactive proof systems. *Journal of the ACM*, 39:859–868, 1992.
19. A. Shamir. IP = PSPACE. *Journal of the ACM*, 39:869–877, 1992.
20. M. Sipser. A complexity theoretic approach to randomness. In *Proc. 15th Symposium on Theory of Computation*, pages 330–335. ACM Press, New York, 1983.
21. L. Valiant. The complexity of computing the permanent. *Theoretical Computer Science*, 8:189–201, 1979.
22. C. Yap. Some consequences of nonuniform conditions on uniform classes. *Theoretical Computer Science*, 26:287–300, 1983.

Equivalence of Measures of Complexity Classes *

Josef M. Breutzmann[1] and Jack H. Lutz[2]

[1] Department of Mathematics
and Computer Science
Wartburg College
Waverly, Iowa 50677
U.S.A.

[2] Department of Computer Science
Iowa State University
Ames, Iowa 50011
U.S.A.

Abstract. The resource-bounded measures of complexity classes are shown to be robust with respect to certain changes in the underlying probability measure. Specifically, for any real number $\delta > 0$, any uniformly polynomial-time computable sequence $\vec{\beta} = (\beta_0, \beta_1, \beta_2, \ldots)$ of real numbers (biases) $\beta_i \in [\delta, 1-\delta]$, and any complexity class $\mathcal{C}$ (such as P, NP, BPP, P/Poly, PH, PSPACE, etc.) that is closed under positive, polynomial-time, truth-table reductions with queries of at most linear length, it is shown that the following two conditions are equivalent.

(1) $\mathcal{C}$ has p-measure 0 (respectively, measure 0 in E, measure 0 in E_2) relative to the coin-toss probability measure given by the sequence $\vec{\beta}$.

(2) $\mathcal{C}$ has p-measure 0 (respectively, measure 0 in E, measure 0 in E_2) relative to the uniform probability measure.

The proof introduces three techniques that may be useful in other contexts, namely, (i) the transformation of an efficient martingale for one probability measure into an efficient martingale for a "nearby" probability measure; (ii) the construction of a *positive bias reduction*, a truth-table reduction that encodes a positive, efficient, approximate simulation of one bias sequence by another; and (iii) the use of such a reduction to *dilate* an efficient martingale for the simulated probability measure into an efficient martingale for the simulating probability measure.

1 Introduction

In the 1990's, the measure-theoretic study of complexity classes has yielded a growing body of new, quantitative insights into various much-studied aspects of computational complexity. Benefits of this study to date include improved bounds on the densities of hard languages [15]; newly discovered relationships among circuit-size complexity, pseudorandom generators, and natural proofs [21];

* This research was supported in part by National Science Foundation Grant CCR-9157382, with matching funds from Rockwell, Microware Systems Corporation, and Amoco Foundation. E-mail: lutz@@cs.iastate.edu

Reischuk, Morvan (Eds.): STACS'97 Proceedings, LNCS 1200

strong new hypotheses that may have sufficient explanatory power (in terms of provable, plausible consequences) to help unify our present plethora of unsolved fundamental problems [18, 15, 7, 16, 11]; and a new generalization of the completeness phenomenon that dramatically enlarges the set of computational problems that are provably strongly intractable [14, 6, 2, 7, 8, 1]. See [13] for a survey of these and related developments.

Intuitively, suppose that a language $A \subseteq \{0,1\}^*$ is chosen according to a random experiment in which an independent toss of a fair coin is used to decide whether each string is in A. Then *classical* Lebesgue measure theory (described in [5, 20], for example) identifies certain *measure 0* sets X of languages, for which the probability that $A \in X$ in this experiment is 0. *Effective* measure theory, which says what it means for a set of decidable languages to have measure 0 as a subset of the set of all such languages, has been investigated by Freidzon [4], Mehlhorn [19], and others. The *resource-bounded* measure theory introduced by Lutz [12] is a powerful generalization of Lebesgue measure. Special cases of resource-bounded measure include classical Lebesgue measure; a strengthened version of effective measure; and most importantly, measures in $\mathrm{E} = \mathrm{DTIME}(2^{\mathrm{linear}})$, $\mathrm{E}_2 = \mathrm{DTIME}(2^{\mathrm{polynomial}})$, and other complexity classes. The *small* subsets of such a complexity class are then the measure 0 sets; the *large* subsets are the measure 1 sets (complements of measure 0 sets). We say that *almost every* language in a complexity class $\mathcal{C}$ has a given property if the set of languages in $\mathcal{C}$ that exhibit the property has measure 1 in $\mathcal{C}$.

All work to date on the measure-theoretic structure of complexity classes has employed the resource-bounded measure that is described briefly and intuitively above. This resource-bounded measure is based on the *uniform* probability measure, corresponding to the fact that the coin tosses are fair and independent in the above-described random experiment. The uniform probability measure has been a natural and fruitful starting point for the investigation of resource-bounded measure (just as it was for the investigation of classical measure), but there are good reasons to also investigate resource bounded measures that are based on other probability measures. For example, the study of such alternative resource-bounded measures may be expected to have the following benefits.

(i) The study will enable us to determine which results of resource-bounded measure are particular to the uniform probability measure and which are not. This, in turn, will provide some criteria for identifying contexts in which the uniform probability measure is, or is not, the natural choice.
(ii) The study is likely to help us understand how the complexity of the underlying probability measure interacts with other complexity parameters, especially in such areas as algorithmic information theory, average case complexity, cryptography, and computational learning, where the variety of probability measures already plays a major role.
(iii) The study will provide new tools for proving results concerning resource-bounded measure based on the uniform probability measure.

The present paper initiates the study of resource-bounded measures that are based on nonuniform probability measures.

Let $\mathbf{C}$ be the set of all languages $A \subseteq \{0,1\}^*$. (The set $\mathbf{C}$ is often called *Cantor space.*) Given a probability measure ν on $\mathbf{C}$ (a term defined precisely below), the full version of this paper describes the basic ideas of resource-bounded ν-measure, generalizing definitions and results from [12, 14, 13] to ν in a *natural and straightforward* way. This full version of the paper specifies formally what it means for a set $X \subseteq \mathbf{C}$ to have p-ν-measure 0 (written $\nu_{\mathrm{p}}(X) = 0$), p-ν-measure 1, ν-measure 0 in E (written $\nu(X|\mathrm{E}) = 0$), ν-measure 1 in E, ν-measure 0 in E_2, or ν-measure 1 in E_2, but the intuitive remarks in this introduction should suffice for evaluating this extended abstract.

Most of the results in the present paper concern a restricted (but broad) class of probability measures on $\mathbf{C}$, namely, coin-toss probability measures that are given by P-computable, strongly positive sequences of biases. These probability measures are described intuitively in the following paragraphs (and precisely in the full paper).

Given a sequence $\boldsymbol{\beta} = (\beta_0, \beta_1, \beta_2, \ldots)$ of real numbers (biases) $\beta_i \in [0,1]$, the *coin-toss probability measure* (also call the *product probability measure*) *given by* $\boldsymbol{\beta}$ is the probability measure $\mu^{\boldsymbol{\beta}}$ on $\mathbf{C}$ that corresponds to the random experiment in which a language $A \in \mathbf{C}$ is chosen probabilistically as follows. For each string s_i in the standard enumeration $s_0, s_1, s_2, \ldots$ of $\{0,1\}^*$, we toss a special coin, whose probability is β_i of coming up heads, in which case $s_i \in A$, and $1 - \beta_i$ of coming up tails, in which case $s_i \notin A$. The coin tosses are independent of one another.

In the special case where $\boldsymbol{\beta} = (\beta, \beta, \beta, \ldots)$, i.e., the biases in the sequence $\boldsymbol{\beta}$ are all β, we write μ^{β} for $\mu^{\boldsymbol{\beta}}$. In particular, $\mu^{\frac{1}{2}}$ is the uniform probability measure, which, in the literature of resource-bounded measure, is denoted simply by μ.

A sequence $\boldsymbol{\beta} = (\beta_0, \beta_1, \beta_2, \ldots)$ of biases is *strongly positive* if there is a real number $\delta > 0$ such that each $\beta_i \in [\delta, 1-\delta]$. The sequence $\boldsymbol{\beta}$ is P-*computable* (and we call it a P-*sequences of biases*) if there is a polynomial-time algorithm that, on input $(s_i, 0^r)$, computes a rational approximation of β_i to within 2^{-r}.

In section 2, we prove the Summable Equivalence Theorem, which implies that, if $\boldsymbol{\alpha}$ and $\boldsymbol{\beta}$ are strongly positive P-sequences of biases that are "close" to one another, in the sense that $\sum_{i=0}^{\infty} |\alpha_i - \beta_i| < \infty$, then for every set $X \subseteq \mathbf{C}$,

$$\mu_{\mathrm{p}}^{\boldsymbol{\alpha}}(X) = 0 \Longleftrightarrow \mu_{\mathrm{p}}^{\boldsymbol{\beta}}(X) = 0.$$

That is, the p-measure based on $\boldsymbol{\alpha}$ and the p-measure based on $\boldsymbol{\beta}$ are in absolute agreement as to which sets of languages are small.

In general, if $\boldsymbol{\alpha}$ and $\boldsymbol{\beta}$ are not in some sense close to one another, then the p-measures based on $\boldsymbol{\alpha}$ and $\boldsymbol{\beta}$ need not agree in the above manner. For example, if $\alpha, \beta \in [0,1]$, $\alpha \neq \beta$, and

$$X_\alpha = \left\{ A \in \mathbf{C} \middle| \lim_{n\to\infty} 2^{-n} |A \cap \{0,1\}^n| = \alpha \right\},$$

then a routine extension of the Weak Stochasticity Theorem of [15] shows that $\mu_{\mathrm{p}}^{\alpha}(X_\alpha) = 1$, while $\mu_{\mathrm{p}}^{\beta}(X_\alpha) = 0$.

Notwithstanding this example, many applications of resource-bounded measure do not involve *arbitrary* sets $X \subseteq \mathbf{C}$, but rather are concerned with the measures of *complexity classes* and other closely related classes of languages. Many such classes of interest, including P, NP, co-NP, R, BPP, AM, P/Poly, PH, PSPACE, etc., are closed under positive, polynomial-time truth-table reductions ($\leq^{\mathrm{P}}_{\mathrm{pos-tt}}$-reductions), and their intersections with E are closed under $\leq^{\mathrm{P}}_{\mathrm{pos-tt}}$-reductions with linear bounds on the lengths of the queries ($\leq^{\mathrm{P,lin}}_{\mathrm{pos-tt}}$-reductions).

The main theorem of this paper is the Bias Equivalence Theorem. This result, proven in section 5, says that, for every class $\mathcal{C}$ of languages that is closed under $\leq^{\mathrm{P,lin}}_{\mathrm{pos-tt}}$-reductions, the p-measure of $\mathcal{C}$ is somewhat robust with respect to changes in the underlying probability measure. Specifically, if $\boldsymbol{\alpha}$ and $\boldsymbol{\beta}$ are strongly positive P-sequences of biases and $\mathcal{C}$ is a class of languages that is closed under $\leq^{\mathrm{P,lin}}_{\mathrm{pos-tt}}$-reductions, then the Bias Equivalence Theorem says that

$$\mu^{\boldsymbol{\alpha}}_{\mathrm{p}}(\mathcal{C}) = 0 \Longleftrightarrow \mu^{\boldsymbol{\beta}}_{\mathrm{p}}(\mathcal{C}) = 0.$$

To put the matter differently, for every strongly positive P-sequence $\boldsymbol{\beta}$ of biases and every class $\mathcal{C}$ that is closed under $\leq^{\mathrm{P,lin}}_{\mathrm{pos-tt}}$-reductions,

$$\mu^{\boldsymbol{\beta}}_{\mathrm{p}}(\mathcal{C}) = 0 \Longleftrightarrow \mu_{\mathrm{p}}(\mathcal{C}) = 0.$$

This result implies that most applications of resource-bounded measure to date can be immediately generalized from the uniform probability measure (in which they were developed) to arbitrary coin-toss probability measures given by strongly positive P-sequences of biases.

The Bias Equivalence Theorem also offers the following new technique for proving resource-bounded measure results. If $\mathcal{C}$ is a class that is closed under $\leq^{\mathrm{P,lin}}_{\mathrm{pos-tt}}$-reductions, then in order to prove that $\mu_{\mathrm{p}}(\mathcal{C}) = 0$, it suffices to prove that $\mu^{\boldsymbol{\beta}}_{\mathrm{p}}(\mathcal{C}) = 0$ for some conveniently chosen strongly positive P-sequence $\boldsymbol{\beta}$ of biases. (The Bias Equivalence Theorem has already been put to this use in the forthcoming paper [17].)

The plausibility and consequences of the hypothesis $\mu_{\mathrm{p}}(\mathrm{NP}) \neq 0$ are subjects of recent and ongoing research [18, 15, 7, 16, 11, 3, 17]. The Bias Equivalence Theorem immediately implies that the following three statements are equivalent.

(H1) $\mu_{\mathrm{p}}(\mathrm{NP}) \neq 0$.
(H2) For every strongly positive P-sequence $\boldsymbol{\beta}$ of biases, $\mu^{\boldsymbol{\beta}}_{\mathrm{p}}(\mathrm{NP}) \neq 0$.
(H3) There exists a strongly positive P-sequence $\boldsymbol{\beta}$ of biases such that $\mu^{\boldsymbol{\beta}}_{\mathrm{p}}(\mathrm{NP}) \neq 0$.

The statements (H2) and (H3) are thus new, equivalent formulations of the hypothesis (H1).

The proof of the Bias Equivalence Theorem uses three main tools. The first is the Summable Equivalence Theorem, which we have already discussed. The second is the Martingale Dilation Theorem, which is proven in section 3. This result concerns martingales (defined in the full version of this paper), which are the betting algorithms on which resource-bounded measure is based. Roughly speaking, the Martingale Dilation Theorem gives a method of transforming ("dilating") a martingale for one coin-toss probability measure into a martingale for another, perhaps very different, coin-toss probability measure, provided that the former measure is obtained from the latter via an "orderly" truth-table reduction.

The third tool used in the proof of our main theorem is the Positive Bias Reduction Theorem, which is presented in section 4. If $\boldsymbol{\alpha}$ and $\boldsymbol{\beta}$ are two strongly positive sequences of biases that are exactly P-computable (with no approximation), then the *positive bias reduction* of $\boldsymbol{\alpha}$ to $\boldsymbol{\beta}$ is a truth-table reduction (in fact, an orderly $\leq_{\mathrm{pos-tt}}^{\mathrm{P,lin}}$-reduction) that uses the sequence $\boldsymbol{\beta}$ to "approximately simulate" the sequence $\boldsymbol{\alpha}$. It is especially crucial for our main result that this reduction is efficient and positive. (The circuits constructed by the truth-table reduction contain AND gates and OR gates, but no NOT gates.)

The Summable Equivalence Theorem, the Martingale Dilation Theorem, and the Positive Bias Reduction Theorem are only developed and used here as tools to prove our main result. Nevertheless, these three results are of independent interest, and are likely to be useful in future investigations.

Most of the proofs have been omitted in the conference version of this paper.

2 Summable Equivalence

If two probability measures on $\mathbf{C}$ are sufficiently "close" to one another, then the Summable Equivalence Theorem says that the two probability measures are in absolute agreement as to which sets of languages have p-measure 0 and which do not. In this section, we define this notion of "close" and prove this result.

Definition 1. Let ν be a positive probability measure on $\mathcal{C}$, let $A \subseteq \{0,1\}^*$, and let $i \in \mathbb{N}$. Then the ith *conditional ν-probability along A* is

$$\nu_A(i+1|i) = \nu(\chi_A[0..i] \mid \chi_A[0..i-1]).$$

Definition 2. Two positive probability measures ν and ν' on $\mathbf{C}$ are *summably equivalent*, and we write $\nu \approx \nu'$, if for every $A \subseteq \{0,1\}^*$,

$$\sum_{i=0}^{\infty} |\nu_A(i+1|i) - \nu'_A(i+1|i)| < \infty.$$

Theorem 3 Summable Equivalence Theorem. *If ν and ν' are strongly positive, p-computable probability measures on $\mathbf{C}$ such that $\nu \approx \nu'$, then for every set $X \subseteq \mathbf{C}$,*

$$\nu_{\mathrm{p}}(X) = 0 \Longleftrightarrow \nu'_{\mathrm{p}}(X) = 0.$$

Definition 4. A P-*exact sequence of biases* is a sequence $\boldsymbol{\beta} = (\beta_0, \beta_1, \beta_2, \ldots)$ of (rational) biases $\beta_i \in \mathbb{Q} \cap [0,1]$ such that the function $i \longmapsto \beta_i$ is computable in time polynomial in $|s_i|$.

Definition 5. If $\boldsymbol{\alpha}$ and $\boldsymbol{\beta}$ are sequences of biases, then $\boldsymbol{\alpha}$ and $\boldsymbol{\beta}$ are *summably equivalent*, and we write $\boldsymbol{\alpha} \approx \boldsymbol{\beta}$, if $\sum_{i=0}^{\infty} |\alpha_i - \beta_i| < \infty$.

It is clear that $\alpha \approx \beta$ if and only if $\mu^\alpha \approx \mu^\beta$.

Lemma 6. *For every P-sequence of biases β, there is a P-exact sequence of biases β' such that $\beta \approx \beta'$.*

3 Martingale Dilation

In this section we show that certain truth-table reductions can be used to *dilate* martingales for one probability measure into martingales for another, perhaps dissimilar, probability measure on $\mathbf{C}$. Our terminology and notation for truth-table reductions are standard (e.g., see Rogers [22] or the full version of this paper). Given a truth-table reduction (f,g), we use the notation $F_{(f,g)}$ interchangeably for the induced functions

$$F_{(f,g)} : \mathbf{C} \longrightarrow \mathbf{C} \quad \text{and} \quad F_{(f,g)} : \{0,1\}^* \longrightarrow \{0,1\}^* \cup \mathbf{C}.$$

If ν is a probability measure on $\mathbf{C}$ and (f,g) is a $\leq_{\mathrm{tt}}$-reduction, then we use the notation $\nu^{(f,g)}$ for the *probability measure induced by ν and (f,g)*, i.e.,

$$\nu^{(f,g)}(z) = \nu(F^{-1}_{(f,g)}(\mathbf{C}_z)).$$

In this paper, we only use the following special type of $\leq_{\mathrm{tt}}$-reduction.

Definition 7. A $\leq_{\mathrm{tt}}$-reduction (f,g) is *orderly* if, for all $x,y,u,v \in \{0,1\}^*$, if $x < y$, $u \in Q_{(f,g)}(x)$, and $v \in Q_{(f,g)}(y)$, then $u < v$. That is, if x precedes y (in the standard ordering of $\{0,1\}^*$), then every query of (f,g) on input x precedes every query of (f,g) on input y.

The following is an obvious property of orderly $\leq_{\mathrm{tt}}$-reductions.

Lemma 8. *If ν is a coin-toss probability measure on $\mathbf{C}$ and (f,g) is an orderly $\leq_{\mathrm{tt}}$-reduction, then $\nu^{(f,g)}$ is also a coin-toss probability measure on $\mathbf{C}$.*

Note that, if (f,g) is an orderly $\leq_{\mathrm{tt}}$-reduction, then $F_{(f,g)}(w) \in \{0,1\}^*$ for all $w \in \{0,1\}^*$. Note also that the length of $F_{(f,g)}(w)$ depends only upon the length of w (i.e., $|w| = |w'|$ implies that $|F_{(f,g)}(w)| = |F_{(f,g)}(w')|$). Finally, note that for each $m \in \mathbb{N}$ there exists $l \in \mathbb{N}$ such that $|F_{(f,g)}(0^l)| = m$.

Definition 9. Let (f,g) be an orderly $\leq_{\mathrm{tt}}$-reduction.

1. An *(f,g)-step* is a positive integer l such that $F_{(f,g)}(0^{l-1}) \neq F_{(f,g)}(0^l)$.
2. For $k \in \mathbb{N}$, we let *step*(k) be the least (f,g)-step l such that $l \geq k$.

The following construction is crucial to the proof of our main theorem.

Definition 10. Let ν be a positive probability measure on $\mathbf{C}$, let (f,g) be an orderly $\leq_{\mathrm{tt}}$-reduction, and let d be a $\nu^{(f,g)}$-martingale. Then the (f,g)-*dilation* of d is the function

$$(f,g)\hat{\ }d : \{0,1\}^* \longrightarrow [0,\infty)$$

$$(f,g)\hat{\ }d(w) = \sum_{u\in\{0,1\}^{l-k}} d(F_{(f,g)}(wu))\nu(wu|w),$$

where $k = |w|$ and $l = step(k)$.

In other words, $(f,g)\hat{\ }d(w)$ is the conditional ν-expected value of $d(F_{(f,g)}(w'))$, given that $w \sqsubseteq w'$ and $|w'| = step(|w|)$. We do not include the probability measure ν in the notation $(f,g)\hat{\ }d$ because ν (being positive) is implicit in d.

Intuitively, the function $(f,g)\hat{\ }d$ is a strategy for betting on a language A, assuming that d itself is a strategy for betting on the language $F_{(f,g)}(A)$. The following theorem makes this intuition precise.

Theorem 11 Martingale Dilation Theorem. *Assume that ν is a positive coin-toss probability measure on $\mathbf{C}$, (f,g) is an orderly $\leq_{\mathrm{tt}}$-reduction, and d is a $\nu^{(f,g)}$-martingale. Then $(f,g)\hat{\ }d$ is a ν-martingale. Moreover, for every language $A \subseteq \{0,1\}^*$, if d succeeds on $F_{(f,g)}(A)$, then $(f,g)\hat{\ }d$ succeeds on A.*

A very special case of the above result (for strictly increasing $\leq^{\mathrm{P}}_{\mathrm{m}}$-reductions under the uniform probability measure) was developed by Ambos-Spies, Terwijn, and Zheng [2], and made explicit by Juedes and Lutz [8]. Our use of martingale dilation in the present paper is very different from the simple padding arguments of [2, 8].

4 Positive Bias Reduction

In this section, we define and analyze a positive truth-table reduction that encodes an efficient, approximate simulation of one sequence of biases by another.

Intuitively, if $\boldsymbol{\alpha}$ and $\boldsymbol{\beta}$ are strongly positive sequences of biases, then the positive bias reduction of $\boldsymbol{\alpha}$ to $\boldsymbol{\beta}$ is a $\leq_{\mathrm{tt}}$-reduction (f,g) that "tries to simulate" the sequence $\boldsymbol{\alpha}$ with the sequence $\boldsymbol{\beta}$ by causing $\mu^{\boldsymbol{\alpha}}$ to be the probability distribution induced by $\mu^{\boldsymbol{\beta}}$ and (f,g). In general, this objective will only be approximately achieved, in the sense that the probability distribution induced by $\mu^{\boldsymbol{\beta}}$ and (f,g) will actually be a probability distribution $\mu^{\boldsymbol{\alpha}'}$, where $\boldsymbol{\alpha}'$ is a sequence of biases such that $\boldsymbol{\alpha}' \approx \boldsymbol{\alpha}$. This situation is depicted schematically in Figure 1, where the broken arrow indicates that (f,g) "tries" to reduce $\boldsymbol{\alpha}$ to $\boldsymbol{\beta}$, while the solid arrow indicates that (f,g) actually reduces $\boldsymbol{\alpha}'$ to $\boldsymbol{\beta}$.

The reduction (f,g) is constructed and analyzed precisely in the full version of the paper.

The following result presents the properties of the positive bias reduction that are used in the proof of our main theorem.

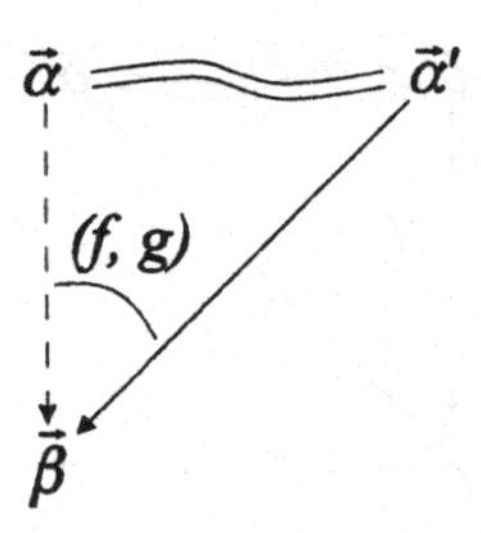

Fig. 1. Schematic depiction of positive bias reduction

Theorem 12 Positive Bias Reduction Theorem. *Let α and β be strongly positive,* P*-exact sequences of biases, and let (f, g) be the positive bias reduction of α to β. Then (f, g) is an orderly $\leq^{\mathrm{P,lin}}_{\mathrm{pos-tt}}$-reduction, and the probability measure induced by μ^{β} and (f, g) is a coin-toss probability measure $\mu^{\alpha'}$, where $\alpha \approx \alpha'$.*

5 Equivalence for Complexity Classes

Many important complexity classes, including P, NP, co-NP, R, BPP, AM, P/Poly, PH, PSPACE, etc., are known to be closed under $\leq^{\mathrm{P}}_{\mathrm{pos-tt}}$-reductions, hence certainly under $\leq^{\mathrm{P,lin}}_{\mathrm{pos-tt}}$-reductions. The following theorem, which is the main result of this paper, says that the p-measure of such a class is somewhat insensitive to certain changes in the underlying probability measure. The proof is now easy, given the machinery of the preceding sections.

Theorem 13 Bias Equivalence Theorem. *Assume that α and β are strongly positive P-sequences of biases, and let $\mathcal{C}$ be a class of languages that is closed under $\leq^{\mathrm{P,lin}}_{\mathrm{pos-tt}}$-reductions. Then*

$$\mu^{\alpha}_{\mathrm{p}}(\mathcal{C}) = 0 \Longleftrightarrow \mu^{\beta}_{\mathrm{p}}(\mathcal{C}) = 0.$$

Proof. Assume the hypothesis, and assume that $\mu^{\alpha}_{\mathrm{p}}(\mathcal{C}) = 0$. By symmetry, it suffices to show that $\mu^{\beta}_{\mathrm{p}}(\mathcal{C}) = 0$.

The proof follows the scheme depicted in Figure 2. By Lemma 6, there exist P-exact sequences α' and β' such that $\alpha \approx \alpha'$ and $\beta \approx \beta'$. Let (f, g) be the positive bias reduction of α' to β'. Then, by the Positive Bias Reduction Theorem (Theorem 12), (f, g) is an orderly $\leq^{\mathrm{P,lin}}_{\mathrm{pos-tt}}$-reduction, and the probability measure induced by μ^{β} and (f, g) is $\mu^{\alpha''}$, where $\alpha' \approx \alpha''$.

Since $\alpha \approx \alpha' \approx \alpha''$ and $\mu^{\alpha}_{\mathrm{p}}(\mathcal{C}) = 0$, the Summable Equivalence Theorem (Theorem 3) tells us that there is a p-α''-martingale d such that $\mathcal{C} \subseteq S^{\infty}[d]$. By the Martingale Dilation Theorem (Theorem 11), the function $(f, g)\hat{}d$ is then a β'-martingale. In fact, it easily checked that $(f, g)\hat{}d$ is a p-β'-martingale.

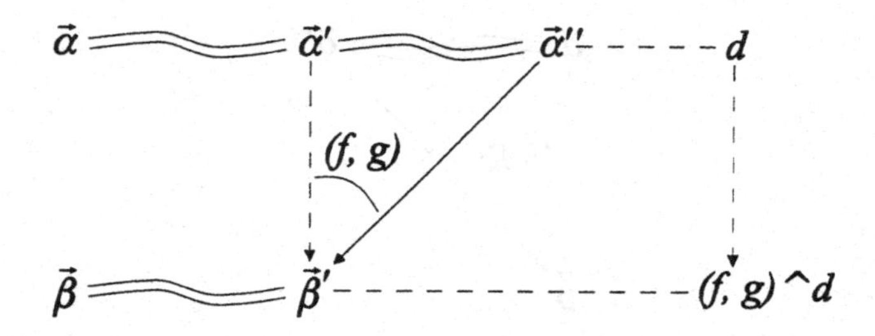

Fig. 2. Scheme of proof of Bias Equivalence Theorem

Now let $A \in \mathcal{C}$. Then, since $\mathcal{C}$ is closed under $\leq^{\mathrm{P,lin}}_{\mathrm{pos-tt}}$-reductions, $F_{(f,g)}(A) \in \mathcal{C} \subseteq S^\infty[d]$. It follows by the Martingale Dilation Theorem that $A \in S^\infty[(f,g)\hat{}d]$. Thus $\mathcal{C} \subseteq S^\infty[(f,g)\hat{}d]$. Since $(f,g)\hat{}d$ is a p-$\boldsymbol{\beta}'$-martingale, this shows that $\mu_{\mathrm{p}}^{\boldsymbol{\beta}'}(\mathcal{C}) = 0$. Finally, since $\boldsymbol{\beta} \approx \boldsymbol{\beta}'$, it follows by the Summable Equivalence Theorem that $\mu_{\mathrm{p}}^{\boldsymbol{\beta}}(X) = 0$.

It is clear that the Bias Equivalence Theorem remains true if the resource bound on the measure is relaxed. That is, the analogs of Theorem 13 for p_2-measure, pspace-measure, rec-measure, constructive measure, and classical measure all immediately follow. We conclude by noting that the analogs of Theorem 13 for measure in E and measure in E_2 also immediately follow.

Corollary 14. *Under the hypothesis of Theorem 13,*

$$\mu^{\boldsymbol{\alpha}}(\mathcal{C}|\mathrm{E}) = 0 \Longleftrightarrow \mu^{\boldsymbol{\beta}}(\mathcal{C}|\mathrm{E}) = 0$$

and

$$\mu^{\boldsymbol{\alpha}}(\mathcal{C}|\mathrm{E}_2) = 0 \Longleftrightarrow \mu^{\boldsymbol{\beta}}(\mathcal{C}|\mathrm{E}_2) = 0.$$

Proof. If $\mathcal{C}$ is closed under $\leq^{\mathrm{P,lin}}_{\mathrm{pos-tt}}$-reductions, then so are the classes $\mathcal{C} \cap \mathrm{E}$ and $\mathcal{C} \cap \mathrm{E}_2$.

6 Conclusion

Our main result, the Bias Equivalence Theorem, says that every strongly positive, P-computable, coin-toss probability measure ν is *equivalent* to the uniform probability measure μ, in the sense that

$$\nu_{\mathrm{p}}(\mathcal{C}) = 0 \Longleftrightarrow \mu_{\mathrm{p}}(\mathcal{C}) = 0$$

for all classes $\mathcal{C} \in \Gamma$, where Γ is a family that contains P, NP, co-NP, R, BPP, P/Poly, PH and many other classes of interest. It would be illuminating to learn more about which probability measures are, and which probability measures are not, equivalent to μ in this sense.

It would also be of interest to know whether the Summable Equivalence Theorem can be strengthened. Specifically, say that two sequences of biases $\boldsymbol{\alpha}$ and $\boldsymbol{\beta}$ are *square-summably equivalent*, and write $\boldsymbol{\alpha} \approx^2 \boldsymbol{\beta}$, if $\sum_{i=0}^{\infty}(\alpha_i - \beta_i)^2 < \infty$. A classical theorem of Kakutani [9] says that, if $\boldsymbol{\alpha}$ and $\boldsymbol{\beta}$ are strongly positive sequences of biases such that $\boldsymbol{\alpha} \approx^2 \boldsymbol{\beta}$, then for every set $C \subseteq \mathbf{C}$, X has (classical) $\boldsymbol{\alpha}$-measure 0 if and only if X has $\boldsymbol{\beta}$-measure 0. A constructive improvement of this theorem by Vovk [28] says that, if $\boldsymbol{\alpha}$ and $\boldsymbol{\beta}$ are strongly positive, computable sequences of biases such that $\boldsymbol{\alpha} \approx^2 \boldsymbol{\beta}$, then for every set $X \subseteq \mathbf{C}$, X has constructive $\boldsymbol{\alpha}$-measure 0 if and only if X has constructive $\boldsymbol{\beta}$-measure 0. (The Kakutani and Vovk theorems are more general than this, but for the sake of brevity, we restrict the present discussion to coin-toss probability measures.) The Summable Equivalence Theorem is stronger than these results in one sense, but weaker in another. It is stronger in that it holds for p-measure, but it is weaker in that it requires the stronger hypothesis that $\boldsymbol{\alpha} \approx \boldsymbol{\beta}$. We thus ask whether there is a "square-summable equivalence theorem" for p-measure. That is, if $\boldsymbol{\alpha}$ and $\boldsymbol{\beta}$ are strongly positive, p-computable sequences of biases such that $\boldsymbol{\alpha} \approx^2 \boldsymbol{\beta}$, is it necessarily the case that, for every set $X \subseteq \mathbf{C}$, X has p-$\boldsymbol{\alpha}$-measure 0 if and only if X has p-$\boldsymbol{\beta}$-measure 0? (Note: Kautz [10] has very recently answered this question affirmatively.)

Acknowledgments. We thank Giora Slutzki, Martin Strauss, and other participants in the ISU Information and Complexity Seminar for useful remarks and suggestions. We especially thank Giora Slutzki for suggesting a simplified presentation of a key lemma.

References

1. K. Ambos-Spies, E. Mayordomo, and X. Zheng. A comparison of weak completeness notions. In *Proceedings of the Eleventh IEEE Conference on Computational Complexity*. IEEE Computer Society Press, 1996. To appear.
2. K. Ambos-Spies, S. A. Terwijn, and X. Zheng. Resource bounded randomness and weakly complete problems. *Theoretical Computer Science*, 1996. To appear. See also *Proceedings of the Fifth Annual International Symposium on Algorithms and Computation*, 1994, pp. 369–377. Springer–Verlag.
3. J. Cai and A. L. Selman. Fine separation of average time complexity classes. In *Proceedings of the Thirteenth Symposium on Theoretical Aspects of Computer Science*, pages 331–343. Springer-Verlag, 1996.
4. R. I. Freidzon. Families of recursive predicates of measure zero. translated in *Journal of Soviet Mathematics*, 6(1976):449–455, 1972.
5. P. R. Halmos. *Measure Theory*. Springer-Verlag, 1950.
6. D. W. Juedes. Weakly complete problems are not rare. *Computational Complexity*, 5:267–283, 1995.
7. D. W. Juedes and J. H. Lutz. The complexity and distribution of hard problems. *SIAM Journal on Computing*, 24(2):279–295, 1995.
8. D. W. Juedes and J. H. Lutz. Weak completeness in E and E_2. *Theoretical Computer Science*, 143:149–158, 1995.
9. S. Kakutani. On the equivalence of infinite product measures. *Annals of Mathematics*, 49:214–224, 1948.

10. S. M. Kautz. Personal communication, 1996.
11. J. H. Lutz. Observations on measure and lowness for Δ_2^{P}. *Mathematical Systems Theory*. To appear. See also *Proceedings of the Thirteenth Symposium on Theoretical Aspects of Computer Science*, pages 87–97. Springer-Verlag, 1996.
12. J. H. Lutz. Almost everywhere high nonuniform complexity. *Journal of Computer and System Sciences*, 44:220–258, 1992.
13. J. H. Lutz. The quantitative structure of exponential time. In *Proceedings of the Eighth Annual Structure in Complexity Theory Conference*, pages 158–175, 1993. Updated version to appear in L.A. Hemaspaandra and A.L. Selman (eds.), *Complexity Theory Retrospective II*, Springer-Verlag, 1996.
14. J. H. Lutz. Weakly hard problems. *SIAM Journal on Computing*, 24:1170–1189, 1995.
15. J. H. Lutz and E. Mayordomo. Measure, stochasticity, and the density of hard languages. *SIAM Journal on Computing*, 23:762–779, 1994.
16. J. H. Lutz and E. Mayordomo. Cook versus Karp-Levin: Separating completeness notions if NP is not small. *Theoretical Computer Science*, 164:141–163, 1996.
17. J. H. Lutz and E. Mayordomo. Genericity, measure, and inseparable pairs, 1996. In preparation.
18. E. Mayordomo. Almost every set in exponential time is P-bi-immune. *Theoretical Computer Science*, 136(2):487–506, 1994.
19. K. Mehlhorn. The "almost all" theory of subrecursive degrees is decidable. In *Proceedings of the Second Colloquium on Automata, Languages, and Programming*, pages 317–325. Springer Lecture Notes in Computer Science, vol. 14, 1974.
20. J. C. Oxtoby. *Measure and Category*. Springer-Verlag, second edition, 1980.
21. K. W. Regan, D. Sivakumar, and J. Cai. Pseudorandom generators, measure theory, and natural proofs. In *36th IEEE Symposium on Foundations of Computer Science*, pages 26–35. IEEE Computer Society Press, 1995.
22. H. Rogers, Jr. *Theory of Recursive Functions and Effective Computability*. McGraw - Hill, New York, 1967.
23. C. P. Schnorr. Klassifikation der Zufallsgesetze nach Komplexität und Ordnung. *Z. Wahrscheinlichkeitstheorie verw. Geb.*, 16:1–21, 1970.
24. C. P. Schnorr. A unified approach to the definition of random sequences. *Mathematical Systems Theory*, 5:246–258, 1971.
25. C. P. Schnorr. Zufälligkeit und Wahrscheinlichkeit. *Lecture Notes in Mathematics*, 218, 1971.
26. C. P. Schnorr. Process complexity and effective random tests. *Journal of Computer and System Sciences*, 7:376–388, 1973.
27. M. van Lambalgen. *Random Sequences*. PhD thesis, Department of Mathematics, University of Amsterdam, 1987.
28. V. G. Vovk. On a randomness criterion. *Soviet Mathematics Doklady*, 35:656–660, 1987.
29. A. K. Zvonkin and L. A. Levin. The complexity of finite objects and the development of the concepts of information and randomness by means of the theory of algorithms. *Russian Mathematical Surveys*, 25:83–124, 1970.

Better Algorithms for Minimum Weight Vertex-Connectivity Problems*

Vincenzo Auletta and Mimmo Parente

Dipartimento di Informatica ed Applicazioni, "R.M. Capocelli", Università di Salerno, 84081 Baronissi (Italy). e-mail: {auletta,parente}@dia.unisa.it

Abstract. Given a k vertex-connected graph with weighted edges, we study the problem of finding a minimum weight spanning subgraph which is k vertex-connected, for small values of k. The problem is known to be NP-hard for any k, even when edges have no weight.
In this paper we provide a 2 approximation algorithm for the cases $k = 2, 3$ and a 3 approximation algorithm for the case $k = 4$. The best approximation factors present in literature are 2, $3 + \frac{2}{3}$ and $4 + \frac{1}{6}$, respectively.

1 Introduction

Let $G = (V, E)$ be an undirected graph and for each edge e, let $w(e)$ be the non negative weight of e. The *Minimum Weight k Vertex-Connectivity* problem, in short k-MWVC, consists in finding a minimum weight subgraph of G that remains connected in presence of up to $k - 1$ vertex failures.

The computation of minimum weight graphs having a given connectivity degree is a fundamental issue in the design of communication networks. In fact, we can think of the weights of the edges as the costs of the realization of connections between any two nodes of the network. In this case the connectivity degree can be seen as a measure of the tolerance to failures occurring in the nodes of the network. For more motivations see e.g. [8].

The k-MWVC problem is NP-hard [7], even in presence of unweighted edges or even when the weights can be just 1 or 2. Various approximation algorithms for its solution have been proposed in [9, 12, 5]. An approximation algorithm with factor α, is a polynomial time algorithm that returns a solution whose cost is no greater than α times the cost of an optimal solution.

The best approximation algorithm for the general case is due to Ravi and Williamson [12] that obtains a factor $2H(k)$, where $H(k) = \sum_{i=1}^{k} 1/i$. In [9] an approximation algorithm that obtains a factor not greater than $2 + 2(k-1)/n$ when the edge weights satisfy the triangle inequality is given. Better algorithms have been proposed for particular values of k. In [12] and [5] two algorithms are described that obtain a factor 3 for the case $k = 2$. In [9], Khuller and

* Work partially supported by the Italian Ministry of University and Scientific Research in the framework of the "Algoritmi, Modelli di Calcolo e Strutture Informative" project.

Reischuk, Morvan (Eds.): STACS'97 Proceedings, LNCS 1200

Raghavachari improved on this result obtaining a factor of $2 + 1/n$, where n is the number of vertices in the graph.

In this paper we study the k-MWVC problem for small values of k. We do not make any assumption on the weights of the edges of the graph. We give a 2 approximation algorithm for the cases $k = 2, 3$ and a 3 approximation algorithm for the case $k = 4$, thus improving on the results found in the literature.

Recently, independently from us, similar results have been obtained: in [11] a 2 approximation algorithm for the case $k = 2$ and a 3 approximation algorithm for the case $k = 3$ have been presented; in [2] a 2 approximation algorithm for $k = 3$ and a 3 approximation algorithms for $k = 4, 5$ are proposed.

The rest of the paper is organized as follows: in the next section we recall some basic definitions on graphs; in section 3 we describe the algorithm *Conn* that solves the cases $k = 2, 3$; in section 4 we describe the algorithm *4-Conn*, that uses *Conn* as a subroutine and solves the case $k = 4$. Finally section 5 contains some concluding remarks.

2 Preliminaries and definitions

Throughout all the paper we will refer to graphs having nonnegative weighted edges. Let G be a graph having n vertices and m edges. The weight of G, denoted by $w(G)$, is the sum of the weights of all its edges. For each subgraph H of G we denote by $G \backslash H$ the graph obtained from G by deleting all the vertices of H and the edges incident to vertices of H. If H contains all the vertices of G, then we say that H is a *spanning* subgraph.

Two paths connecting vertices u and v are *independent* if they have only the vertices u and v in common. As in [9], we can generalize this notion and consider paths between u and a set of vertices A: we say that there are k independent paths between u and A if the paths have only the vertex u in common and each path starts from u and ends to a distinct vertex of A. A graph G is said to be *connected* if for each pair of vertices u and v there exists a path in G between u and v. If G is not connected, then it is said to be *disconnected*. A connected component C of G is a connected subgraph of G. We say that C is a maximal connected component if adding any vertex to C we obtain a disconnected graph. A graph is *k-vertex connected* (or simply k-connected) if deleting up to $k - 1$ vertices, the remaining graph is connected. Equivalently, a graph is k-connected if for each pair of vertices u and v, there exist k independent paths that connect u to v. A set of vertices whose removal disconnects the graph is said a *cut-set*. A cut set of cardinality k is said a k-cut-set.

The *Minimum Weight k Vertex-Connectivity Problem* for a graph G consists in finding a k-connected spanning subgraph of G of minimum weight.
An equivalent formulation of the problem can be given in terms of minimal *augmentation sets*, see [3]. Let $G = (V, E)$ be a k-connected graph, G_0 be a spanning subgraph of G (possibly not k-connected) and let $F \subseteq E$. If the graph obtained by adding the edges of F to G_0 is k-connected, then F is said a *k-augmentation set* of G_0. An *augmentation set* of G_0 is a set of edges that added to G_0 increases

its connectivity by one. Given G and G_0, the *minimum k-augmentation problem* consists in finding a k-augmentation set for G_0 of minimum weight among the k-augmentation sets contained in G.

The *directed version* of G is the directed graph obtained by replacing every edge e by two anti-parallel edges, each with the same weight of e. Symmetrically, given a directed graph D with the same weight on the anti-parallel edges, the *undirected version* of D is the undirected graph obtained by taking a single copy for each pair of anti-parallel edges and by ignoring the orientation of all the other edges.

Let D be a directed graph and let F be a set of edges of D outgoing from a vertex v: we define the *reduced graph* of D with respect to F, denoted by $D_v(F)$, as the graph obtained from D by deleting all the edges outgoing from v except those in F.

The Frank and Tardos Algorithm. Following the strategy used in [9], our algorithms use as a subroutine a polynomial time algorithm *FT*, proposed in [4] by András Frank and Éva Tardos, to compute an approximated solution of the k-MWVC problem. The *FT* algorithm takes in input a weighted directed graph D, a distinguished vertex *root* and an integer k and returns a minimum weight subgraph of D (if it exists) that, for each vertex u, has k independent paths from *root* to u.

3 Two and Three Connectivity

Let G be a k-connected graph and let H^* be a minimum weight k-connected spanning subgraph of G. In this section we describe the algorithm *Conn* that computes a spanning subgraph H of G that is at least $((\lceil \frac{k}{2} \rceil + 1))$-connected and it is such that $w(H) \leq 2w(H^*)$. Clearly *Conn* gives a 2 approximated solution to the k-MWVC problem, when $k = 2, 3$. Moreover, *Conn* will be used as a subroutine by the algorithm *4-Conn* that solves the case $k = 4$ and that will be presented in the next section.

The algorithm *Conn* is described in Fig. 1. It takes in input a k-connected graph G and returns a distinguished vertex *root* and a subgraph H that has k independent paths between *root* and each other vertex.

We notice that the running time of *Conn* depends on the time required to execute the $O(n^{k+1})$ calls of the *FT* algorithm. Using the implementation of the *FT* algorithm given in [6], that requires time $O(kn^2m)$, it can be proved that, for any constant k, *Conn* has a polynomial running time. We will prove that the graph returned by *Conn* is at least $((\lceil \frac{k}{2} \rceil + 1))$-connected and has weight no greater than $2w(H^*)$.

We say that a directed k-connected graph D is *minimally* connected if the graph obtained from D by removing any edge is not k-connected. Observe that each minimum weight k-connected spanning subgraph of D is still a minimally connected graph.

Input: A weighted k-connected graph $G = (V, E)$.
Output: A k degree vertex *root* and a $((\lceil \frac{k}{2} \rceil + 1))$-connected spanning subgraph H such that $w(H) \leq 2w(H^*)$.

begin
1) Let D be the directed version of G.
2) **for** every vertex v of V **do**
3) **for** every set of k edges $F = \{e_1, e_2, \ldots, e_k\}$, outgoing from v in D **do**
4) Run the *FT* algorithm on the graph $D_v(F)$, the vertex v and the integer k;
end
end
5) Let D_{FT} be the minimum weight subgraph returned by the *FT* algorithm and let *root* be its distinguished vertex.
6) **return** the graph H, that is the undirected version of D_{FT} and the vertex *root*.
end

Fig. 1. The algorithm *Conn*

Lemma 1. *For each k-connected directed graph D there exist a vertex v and a set F of k edges outgoing from v such that in $D_v(F)$ there are k independent paths from v to every other vertex.*

Proof. Let D^{min} be a minimally connected subgraph of D. From a classical result of the Graph Theory we have that there is at least a vertex of D^{min} that has out-degree k, (see e.g. [1]). Let v be such a vertex and let F be the set of edges outgoing from v. Obviously, in D^{min} there are k independent paths between v and all the other vertices. Thus, there is a subgraph of $D_v(F)$ that has k independent paths from v to each other vertex.

By the previous Lemma we have that *Conn* always returns a subgraph of G. To prove that this subgraph is at least $((\lceil \frac{k}{2} \rceil + 1))$-connected we use the following result from [9].

Lemma 2 [9]. *For each degree k vertex v of H, if there are k independent paths from v to each other vertex in H, then the graph $H \backslash v$ is at least $(\lceil \frac{k}{2} \rceil)$-connected.*

Lemma 3. *If $H \backslash$root is λ-connected then H is $(\lambda + 1)$-connected.*

Proof. The proof is by contradiction. Suppose that H is not $(\lambda + 1)$-connected and let $X = \{x_1, x_2, \ldots, x_\lambda\}$ be a cut set of H. We notice that *root* cannot be a vertex of the cut set. In fact if *root* $\in X$ then the other vertices of X should form a cut set of $H \backslash$*root*. But this is a contradiction since, by Lemma 2, $H \backslash$*root* is λ-connected. Suppose, now, that *root* does not belong to X. Since *root* is connected to v in H with k independent paths and X can cut at most λ of these paths, we have that each vertex of $H \backslash X$ is connected to *root*. Thus, each pair of vertices of $H \backslash X$ is connected through a path going through *root* and X is not a cut set for H.

Corollary 4. *H is at least* $((\lceil \frac{k}{2} \rceil + 1))$*-connected.*

We can state now the main result of the section.

Theorem 5. *The algorithm* Conn, *on input a weighted k-connected graph G, returns a* $((\lceil \frac{k}{2} \rceil + 1))$*-connected spanning subgraph H such that* $w(H) \leq 2w(H^*)$, *where* H^* *is the minimum weight k-connected spanning subgraph of G.*

Proof. By Corollary 4, H is at least $((\lceil \frac{k}{2} \rceil + 1))$-connected. Let D be the directed version of G and let D^{FT} the directed graph computed in step 5 of *Conn*. Since H is the undirected version of D^{FT}, we have that $w(H) \leq w(D^{FT})$.

Observe now, that D^{FT} is a spanning subgraph of D of minimum weight among the subgraphs of D containing a degree k vertex v such that there are k independent paths from v to each other vertex. Thus

$$w(D^{FT}) \leq w(D^{min})$$

where D^{min} is a minimum weight minimally k-connected spanning subgraph of D. On the other hand, we have that $w(D^{min}) \leq 2w(H^*)$, since H^* is minimally connected. Therefore we can conclude that

$$w(H) \leq w(D^{FT}) \leq w(D^{min}) \leq 2w(H^*).$$

Corollary 6. *The algorithm* Conn *solves the k-MWVC problem, for* $k = 2, 3$, *with an approximation factor* 2.

4 Four Connectivity

In this section we describe the algorithm *4-Conn* that computes a minimum weight 4-connected spanning subgraph of a graph G with an approximation factor 3. The algorithm calls *Conn* to find a distinguished vertex *root* and a spanning subgraph H that contain 4 independent paths between *root* and each other vertex of G. By Corollary 4, H is at least 3-connected. Thus, if H is 4-connected the algorithm halts; if instead, H is 3-connected, then *4-Conn* computes an augmentation set for H. The augmentation algorithm is based on some properties of the graph H.

Define a k–rooted graph as a graph that contains k distinct vertices $A = \{v_1, \cdots, v_k\}$, called root vertices, such that for each vertex $u \notin A$ there are k independents paths from u to A. Moreover, there are $k-1$ independent paths from the root vertex v_i to the other vertices of A, for $i = 1, \ldots, k$.

Let J be the graph obtained from H by removing the distinguished vertex *root*. It can be easily seen that J is a 4–rooted graph, having the 4 neighbours of *root* as root vertices. Moreover, by Lemma 3, each augmentation set that takes the connectivity of J from 2 to 3 is also an augmentation set that takes the connectivity of H from 3 to 4. Therefore, in the sequel we describe how our algorithm computes an augmentation set of the 4–rooted graph J.

4.1 Properties of 4–rooted graphs

In the sequel we call *cut set* a 2–cut set and we call *block* a set of vertices that is either a maximal 3-connected component or it has no more than 4 vertices. We also refer to a *component* of J meaning either a block or a cut set. We say that two cut sets X and Y are adjacent if and only if each vertex of Y either belongs to X or it is adjacent to a vertex of X. Moreover, we say that a block B is adjacent to the cut set X if and only if $X \subset B$.

In this subsection we first prove (in Lemma 7) that each cut set of J partitions the 4–rooted graph into two connected components each one containing two root vertices. As a consequence we obtain that the root vertices of J do not belong to any cut set. We also prove that all the cut sets partition the set A in the same two subsets, denoted by A_L and A_R, where A_L is the subset containing v_1. Thus, we can define the left side of a cut set X, denoted by L_X, as the connected component of $J \backslash X$ that contains A_L, and the right side of X, denoted by R_X, as the connected component containing A_R. Similarly, Lemma 8 states that each block B not containing root vertices partitions J into two connected components: a left side L_B containing A_L and a right side R_B containing A_R. It can be easily seen that if B contains A_L then L_B is empty and R_B is equal to $J \backslash B$; if B contains A_R, instead, we have that L_B is equal to $J \backslash B$ and R_B is empty. In Fig. 2 an example of a decomposition of a 4–rooted graph into its components is given.

We start by proving that each component of J that does not contain root vertices partitions the 4-rooted graph into exactly two connected components, each one containing exactly two root vertices. The proofs of the last two lemma of the subsection are given in the full version of the paper.

Lemma 7. *For each cut-set X of J the graph $J \backslash X$ consists of two connected components, each one containing two root vertices of J. Moreover, all the cut sets partition the set A of the root vertices in A_L and A_R.*

Proof. Let C be a connected component of $J \backslash X$ and let y be a vertex of C. Obviously X can cut at most two of the paths between y and A. Therefore, if $y \notin A$, then y is connected in $J \backslash X$ to at least two root vertices; if $y \in A$, instead, y it is connected in $J \backslash X$ to at least another root vertex. Thus, C contains at least two root vertices. On the other hand, C cannot contain more than two root vertices, otherwise it would be the unique connected component of $J \backslash X$, contradicting the hypothesis that X is a cut set.

Suppose now that $X = \{x_1, x_2\}$ and $Y = \{y_1, y_2\}$ are two cut sets that partition A in different ways and assume, without loss of generality, that $A_L(X) = \{v_1, v_2\}$ and $A_L(Y) = \{v_1, v_3\}$. Without loss of generality assume that $x_1 \neq y_1$ and $y_1 \in R_X$. We observe that in J there are two independent paths between y_1 and v_1 (resp. v_2) that go through vertices of X and do not contain any vertex of R_Y (resp. L_Y). Thus $x_1 \in L_Y \cap R_Y$. However this is a contradiction since L_Y and R_Y are disjoint.

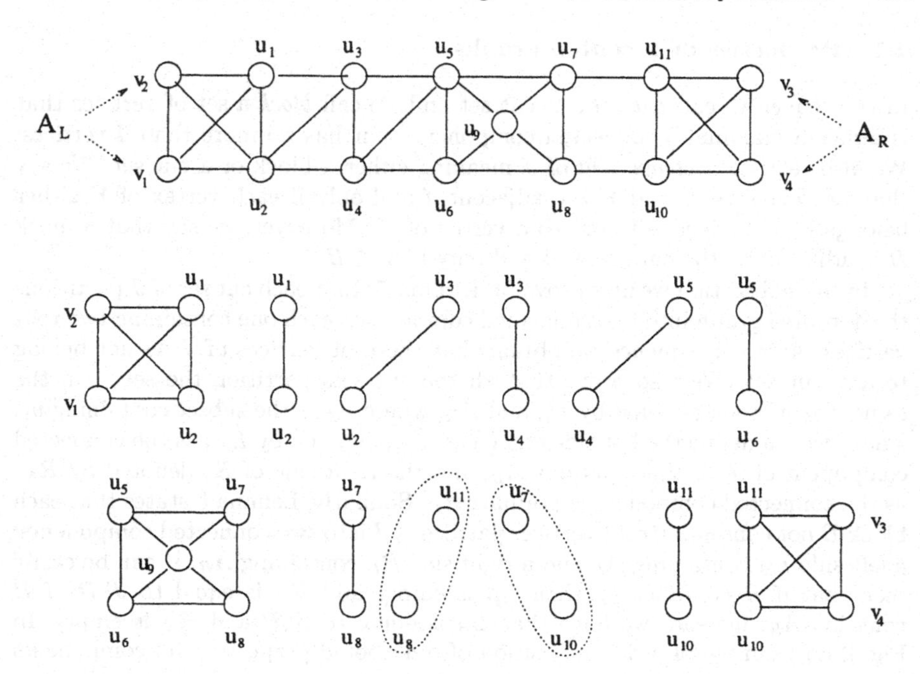

Fig. 2. An example of a decomposition of a 4–rooted graph J in its components.

Lemma 8. *For each block B that does not contain root vertices, it holds that the graph $J \backslash B$ consists of two connected components L_B and R_B, containing A_L and A_R, respectively.*

We notice that if B contains root vertices then it contains A_R or A_L. In this case the graph $J \backslash B$ either is empty, if $A \subseteq B$, or it is connected, if it contains only one of the subsets of root vertices.

Consider now the problem of listing all the components of J. We observe that a vertex v can belong to several components. However, if v belongs to a block at least one of the cut sets containing v is adjacent to this block. Moreover, we will prove that all the cut sets containing v are adjacent to each other. Thus, we can obtain the list of all the components of J by simply starting from the component containing A_L and, iteratively, computing for each component C the components adjacent to C and contained in $R_C \cup C$.

Lemma 9. *If $X = \{x, z\}$ and $Y = \{y, z\}$ are two cut sets of J, then x and y are adjacent.*

4.2 The block-cut graph

The block-cut graph $\Gamma(J)$ is a weighted directed graph associated to the 4-rooted graph J. It has a cut vertex for each vertex of J that belongs to a cut set and a block vertex for each block. We say that a vertex x of J is represented in $\Gamma(J)$ by the vertex $\gamma(x)$, where $\gamma(x)$ is the cut vertex associated to x, if x belongs to a cut set, or to the block vertex associated to the block containing x. If x belongs both to a block and to a cut set its representative is a cut vertex. We say that a vertex x in J corresponds to a vertex v in $\Gamma(J)$ if $v = \gamma(x)$. Let s and t be the block vertices representing the vertices of A_L and A_R, respectively.

For each edge (x, y) in J there is an edge in $\Gamma(J)$ between $\gamma(x)$ and $\gamma(y)$ if and only if $\gamma(x) \neq \gamma(y)$ and $\{x, y\}$ is not a cut set. If x is in the left (resp. right) side of a component containing y then the edge is directed from $\gamma(x)$ to $\gamma(y)$, (resp. from $\gamma(y)$ to $\gamma(x)$). We remark that if x belongs to a component A and y is to the right of A, then there does not exist any component B containing x such that y is to the left of B, thus the direction of the edge $(\gamma(x), \gamma(y))$ is well defined. The block cut graph has also edges not corresponding to edges in J. For each cut set X if there exist two cut vertices y and z adjacent to both the vertices of X such that $\{y, z\}$ is not a cut set and $y \in L_X$ and $z \in R_X$, then there exists an edge in $\Gamma(J)$ from $\gamma(y)$ to $\gamma(z)$. All the edges of the block-cut graph have weight 0. In Fig. 3 an example of a block-cut graph is given. It can be easily seen that the construction of the block-cut graph $\Gamma(J)$ takes polynomial time.

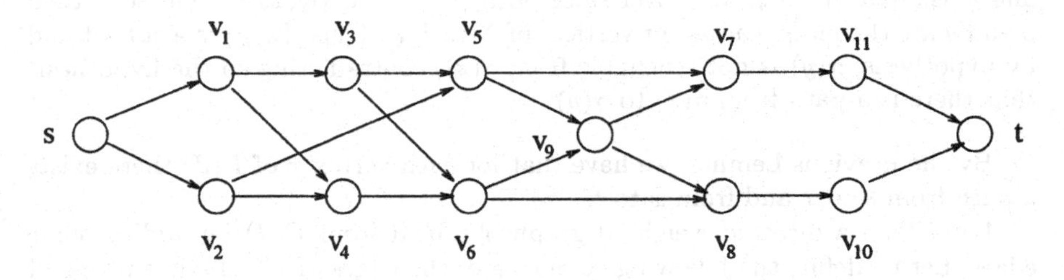

Fig. 3. The block-cut graph associated to the graph J.

Lemma 10. *For each pair of vertices $x, y \in J$, with $\gamma(x) \neq \gamma(y)$, there exists a path in $\Gamma(J)$ from $\gamma(x)$ to $\gamma(y)$ if and only if $\{x, y\}$ is not a cut set and $y \in R_X$, where X is a component containing x.*

Proof. It can be easily seen that if $y \in X \cup L_X$ then there is no path in $\Gamma(J)$ from $\gamma(x)$ to $\gamma(y)$, since there is no edge connecting representatives of vertices of $X \cup R_X$ to representatives of vertices of $X \cup L_X$. Therefore, we have only to prove that if y belongs to R_X, then there is a path from $\gamma(x)$ to $\gamma(y)$ if and only if $\{x, y\}$ is not a cut set.

The lemma trivially holds for the vertices of the block containing A_R. In fact its right side is empty and no vertex of $\Gamma(J)$ is reachable from t.

Let Z be a component such that for each pair of vertices in $Z \cup R_Z$ the lemma holds and let X be adjacent to Z and contained into $Z \cup L_Z$. We will show that for each pair of vertices in $X \cup R_X$, the lemma holds. We will consider only the case in which X and Z are cut sets and they have a vertex in common. All the other cases are similar. Let $X = \{x, w\}$ and $Z = \{z, w\}$ (observe that $R_X = R_Z \cup \{z\}$). It is sufficient to show that for each $y \in X \cup R_X$ there exists a path from $\gamma(x)$ to $\gamma(y)$ if and only if $\{x, y\}$ is not a cut set.

Suppose that $\{x, y\}$ is not a cut set. By Lemma 9, x and z are adjacent and there is an edge from $\gamma(x)$ to $\gamma(z)$. Thus, if $\{y, z\}$ is not a cut set or if $y = z$ then there is a path from $\gamma(x)$ to $\gamma(y)$ that goes through $\gamma(z)$. If $\{y, z\}$ is a cut set instead, then $\gamma(y)$ is not reachable from $\gamma(z)$. However, by Lemma 9 we have that x is adjacent to z and y is adjacent to w. Moreover, since $\{w, z\}, \{y, z\}$ and $\{w, x\}$ are cut sets, we have that x is adjacent to w and z is adjacent to y. Thus, since $x \in L_Z$ and $y \in R_Z$, we have that there is an edge in $\Gamma(J)$ that connects $\gamma(x)$ to $\gamma(y)$.

Suppose now that $\gamma(y)$ is reachable from $\gamma(x)$. We observe that, since $x \in L_Z$, then x is not adjacent to any vertex of R_Z. Thus, the path from $\gamma(x)$ to $\gamma(y)$ either consists of the unique edge $(\gamma(x), \gamma(y))$ or it goes through $\gamma(z)$. In the first case it can be easily seen that $\{x, y\}$ is not a cut set, since no edge of $\Gamma(J)$ connects representatives of the same cut set. It remains to prove that there is no vertex $\gamma(y)$ reachable from $\gamma(z)$ that represents a vertex y that forms a cut set together with x. But, if such a vertex exists, then by Lemma 9 z is adjacent to x and w is adjacent to y. Moreover since both $\{x, y\}$ and $\{w, z\}$ are cut sets, then z and y are the unique adjacent vertices of X in R_X. Thus, $\{z, y\}$ is a cut set and by hypothesis, $\gamma(y)$ is not reachable from $\gamma(z)$, contradicting so the hypothesis that there is a path from $\gamma(z)$ to $\gamma(y)$.

By the previous Lemma, we have that for each vertex x of $\Gamma(J)$ there exists a path from s to x and from x to t.

Let Γ^* be a directed weighted graph obtained from $\Gamma(J)$ by adding some edges. Let us define the following colouring of the edges of Γ^*: if e is an edge of $\Gamma(J)$ then it is coloured white, otherwise it is black. We say that a *b-w path* is a path that contains two consecutive black edges $\{u, v\}$ and $\{v, z\}$ only if v is a not cut vertex. We will prove that the problem of computing an augmentation set for J can be reduced to the problem of constructing from $\Gamma(J)$ a graph that has a b-w path from t to s.

We define the following relation between edges added to J and the black edges of Γ^*. For each pair of vertices x and y in J, let X and Y be two distinct components containing x and y, respectively, and such that $y \in L_X$. If (x, y) is an edge added to J we say that its representative in Γ^* is the black edge $(\gamma(x), \gamma(y))$. For each black edge (u, v) of Γ^* the edge corresponding to (u, v) in the augmentation set of J is the edge (x', y') that has minimum weight among the edges that are represented by (u, v). The weight of the edge (u, v) is equal to the weight of its corresponding edge. Notice that if there are no two distinct components X and Y that contain x and y, respectively, then the edge (x, y) has no representative in Γ^*. However, it can be easily seen that this edge does not belong to any minimum weight augmentation set of J. In fact this edge do not cross any cut set of J and thus it can be removed from an augmentation set, still having an augmentation set.

Lemma 11. *Let AUG be a set of edges added to J and let AUG^* be the set of representatives of AUG. Denote by Γ^* the graph obtained by adding the edges of AUG^* to $\Gamma(J)$. Then, AUG is an augmentation set for J if and only if there is a b-w path in Γ^* from t to s.*

Proof. It can be easily seen that if there is in Γ^* a b-w path p from t to s, then the graph J' obtained by augmenting J with the edges of AUG is 3-connected. In fact, for each cut set X, the b-w path p connects the representative of a vertex of R_X to the representative of a vertex of L_X. Then, by Lemma 10 and by the definition of b-w path, there is at least a black edge e of p that connects the representative of a vertex of R_X to a representative of a vertex of L_X. The edge corresponding to e crosses the cut set X and, thus, X is not a cut set for J'.

Suppose now that AUG is an augmentation set of J, we prove that there is a b-w path in Γ^* from t to s. Let B_r be the connected component containing A_R and let X_0 be the unique cut set adjacent to B_r. In AUG there is at least one edge that crosses X_0 and connects a vertex of B_r to a vertex $x_1 \in L_{X_0}$. The corresponding black edge in AUG^* connects t, that is the representative of the vertices of B_r, to $\gamma(x_1)$. Let X_1 be the component containing x_1 such that no vertex of R_{X_1} forms a cut set together with x_1. By Lemma 10, we have that there is a b-w path in Γ^* from t to any representative of a vertex in R_{X_1}. Furthermore, if X_1 is a cut set then in AUG there is an edge that connects a vertex of R_{X_1} to a vertex $x_2 \in L_{X_1}$. The corresponding black edge connects a vertex reachable from $\gamma(x_1)$ (using only white edges) to $\gamma(x_2)$. Thus, there is a b-w path from t to all the representatives of R_{X_2}, where X_2 is the component containing x_2 such that no vertex of R_{X_2} forms a cut set together with x_2. If X_1 is a block, instead, then by Lemma 8 there is a unique cut set adjacent to X_1 that separates X_1 from A_L and we can apply to this cut set the previous argument. Iterating the previous argument we can prove that there is a b-w path from t to any other vertex of Γ^*.

By the previous Lemma we obtain that the minimum weight augmentation set for J can be found computing the minimum weight set of edges that, added to $\Gamma(J)$, produce a b-w path from t to s. The algorithm *4-Conn* is given in Fig. 4. It can be easily seen that *4-Conn* takes polynomial time to compute a 4-connected spanning subgraph of G. Next Theorem proves that the weight of this subgraph is not greater than three times the weight of the minimum weight 4-connected spanning subgraph of G.

Theorem 12. *The algorithm* 4-Conn *solves the 4-MWVC problem with an approximation factor* 3.

Proof. Let H^* be the minimum weight 4-connected spanning subgraph of G. From Theorem 5, we have that the weight of the subgraph H obtained in the step 1 of the algorithm *4-Conn* is at most $2w(H^*)$. Consider now the weight

Input: A weighted 4-connected graph G.
Output: A 4-connected spanning subgraph H' of G such that $w(H') \leq 3w(H^*)$.

begin
1) **Run** algorithm *Conn* on input G and let H and *root* be the graph and the distinguished vertex returned, respectively.
2) **if** H is 4-connected, **then** stop.
3) Construct the block-cut graph associated to $J = H \backslash root$.
4) For each edge e of G not contained in H add to $\Gamma(J)$ the edge corresponding to e. Let Γ^* be the graph obtained.
5) Compute the minimum weight b-w path p in Γ^* that connects t to s. Let AUG^* be the set of black edges contained in p and let AUG be the set of edges corresponding to the edges of AUG^*.
6) **return** the graph H' obtained by augmenting H with the edges of AUG.
end

Fig. 4. Algorithm 4-*Conn*.

of the edges of the augmentation set AUG computed in step 5. Since H^* is 4-connected it exists a subset B of edges of H^* that is an augmentation set for J. On the other hand, the set AUG is a minimum weight augmentation set for J. Thus, we can conclude that

$$w(H \cup AUG) = w(H) + w(AUG) \leq 3w(H^*).$$

The arguments used for the case $k = 4$ can be applied to solve the 5-MWVC problem to obtain an algorithm with an approximation factor of 3 as well. The technical details involved are quite tedious.

5 Concluding Remarks

In this paper we have given approximation algorithms for the k-MWVC problem that obtain better approximation factors for small values of k. It is an open problem if there is a constant approximation algorithm for any k. Unfortunately the techniques known so far do not seem to suggest any way to generalize the current results.

However, our algorithms (as well as the other algorithms in literature) are quite inefficient. We think the problem of designing efficient approximation algorithms for the problem deserves some further investigation.

Acknowledgements. We thank R. Ravi and S. Khuller for telling us about [11] and M. Penn for pointing us towards the work of Dinitz and Nutov. We also greatly acknowledge one of the referees for his helpful suggestions. Finally we are glad to thank Ugo Vaccaro for his continue flow of suggestions.

References

1. B. Bollobás *Extremal Graph Theory*, Academic Press, London, 1978.
2. Y. Dinitz, Z. Nutov, *Finding minimum weight k-vertex connected spanning subgraphs: approximation algorithms with factor 2 for k* = 3 *and factor 3 for k* = 4, 5, TR-CS0886, Technion, Israel, 1996. (Also to appear in the proceedings of CIAC '97.)
3. K.P. Eswaran, R.E. Tarjan, *Augmentations Problems*, SIAM Jour. on Computing, (5), 4, 653–665, (1976).
4. A. Frank, É. Tardos, *An application of Submodular Flows*, Linear Algebra and its Applications 114/115, 329–348, (1989).
5. G. N. Frederickson, J. JáJá, *On the relationship between the biconnectivity augmentation and travelling salesman problem* Theoretical Computer Science, 19(2), 189–201, (1982).
6. H. N. Gabow, *A representation for Crossing Set Families with Applications to Submodular Flow Problems*, in Proc. of Symposium On Discrete Algorithms, *SODA '93*, 202–211, (1993).
7. M. R. Garey, D.S. Johnson, *Computers and Intractability*, Freeman, New York, 1979.
8. M. Grötschel, C. Monma, M. Stoer, *Design of survivable networks*, Handbook in Operations Research and Management Science, Volume on Networks, 1993.
9. S. Khuller, B. Raghavachari, *Improved Approximation Algorithms for Uniform Connectivity Problems*, in Proc. of Symposium on the Theory of Computing, *STOC '95*, 1–10, (1995).
10. S. Khuller, R. Thurimella, *Approximation Algorithms for Graph Augmentation*, J. of Algorithms 14, 214–225, (1993).
11. M. Penn, H. Shasha-Krupnik, *Improved Approximation Algorithms for Weighted 2 & 3 Vertex Connectivity Augmentation Problems*, Manuscript. (to appear on J. of Algorithms).
12. R. Ravi, D.P. Williamson, *An Approximation Algorithm for Minimum-Cost Vertex-Connectivity Problems*, in Proc. of Symposium On Discrete Algorithms, *SODA '95*, 332–341, (1995).

RNC-Approximation Algorithms for the Steiner Problem

Hans Jürgen Prömel*[1] and Angelika Steger[2]

[1] Institut für Informatik, Humboldt Universität zu Berlin, 10099 Berlin, Germany; *proemel@informatik.hu-berlin.de*

[2] Institut für Informatik, TU München, 80290 München, Germany; *steger@informatik.tu-muenchen.de*

Abstract. In this paper we present an $\mathcal{RNC}$-algorithm for finding a minimum spanning tree in a weighted 3-uniform hypergraph, assuming the edge weights are given in unary, and a fully polynomial time randomized approximation scheme if the edge weights are given in binary. From this result we then derive $\mathcal{RNC}$-approximation algorithms for the Steiner problem in networks with approximation ratio $(1+\epsilon)\,5/3$ for all $\epsilon > 0$.

1 Introduction

In recent years, the Steiner tree problem in graphs attracted considerable attention, as well from the theoretical point of view as from its applicability, e.g., in VLSI-layout. It is rather easy to see and has been known for a long time that a minimum Steiner tree spanning a given set of terminals in a graph or network can be approximated in polynomial time up to a factor of 2, cf. e.g. Choukhmane [6] or Kou, Markowsky, Berman [14]. After a long period without any progress Zelikovsky [23], Berman and Ramaiyer [2], Zelikovsky [24], and Karpinski and Zelikovsky [13] improved the approximation factor step by step from 2 to 1.644.

In this paper we present $\mathcal{RNC}$-approximation algorithm for the Steiner problem with approximation ratio $(1+\epsilon)\,5/3$ for all $\epsilon > 0$. The running time of these algorithms is polynomial in ϵ^{-1} and n. Our algorithms also give rise to conceptually much easier and faster (though randomized) sequential approximation algorithms than the Karpinski-Zelikovsky approach which almost match their approximation factor.

The core of our algorithm is a new $\mathcal{RNC}$-algorithm for finding a minimum spanning tree in 3-uniform hypergraphs. The problem of finding a minimum spanning tree in a given graph is well studied and known to be sequentially solvable almost in linear time [10]. Moreover, this problem can also be solved efficiently in parallel [21]. On the other hand, the problem of finding a minimum spanning tree in a k-uniform hypergraph is known to be $\mathcal{NP}$-hard whenever $k \geq 4$. The status of the case $k = 3$, however, is not yet completely decided. For

* Supported in part by DFG grant Pr 296/4

Reischuk, Morvan (Eds.): STACS'97 Proceedings, LNCS 1200

the unweighted case Lovász [16] provided a very complicated $\mathcal{O}(n^{17})$ algorithm, which was later improved to a $\mathcal{O}(n^4)$ algorithm by Gabow and Stallmann [11]. Here, n denotes the number of vertices of the hypergraph. Also, Lovász [17] presented a conceptionally simple randomized algorithm for the unweighted case by reducing it to a sequence of computations of determinants of appropriate matrices. For the weighted case Camerini, Galbiati and Maffioli [5] generalized the approach of Lovász [17] to obtain a randomized pseudo-polynomial time algorithm.

In this paper we present an $\mathcal{RNC}$-algorithm for finding a minimum spanning tree in a weighted 3-uniform hypergraph with edge-weights in *unary*. This algorithm simplifies to an $\mathcal{RNC}$-algorithm for deciding the existence of a spanning tree in an unweighted 3-uniform hypergraph and implies a fully polynomial approximation scheme if the edge-weights are given in binary.

To achieve this result, we combine ideas from Mulmuley, Vazirani, Vazirani [18] which they used to obtain an $\mathcal{RNC}$-algorithm for finding a perfect matching in a given graph with ideas from Lovász [17] resp. Camerini, Galbiati and Maffioli [5], where they presented a random pseudo-polynomial time algorithm for the general problem of finding a base of specified value in a weighted represented matroid subject to parity conditions.

2 Minimum Spanning Trees in 3-Uniform Hypergraphs

A *hypergraph* $H = (V, F)$ is a generalization of graphs where F is an arbitrary family of subsets of V (and not just a family of 2-element subsets). An *r-uniform* hypergraph is a hypergraph all of whose edges have cardinality exactly r.

Many notions and results of graph theory generalize to hypergraphs. Here we just need cycles and trees. A *cycle* (of length $l \geq 2$) in H is a sequence $x_1, e_1, \dots, x_{l-1}, e_{l-1}, x_l, e_l$ of vertices and edges such that the x_i are distinct vertices, the e_i are distinct edges, $x_1 \in e_1 \cap e_l$ and $x_i \in e_{i-1} \cap e_i$ for all $i = 2, \dots, l$. A hypergraph H is a tree iff H is connected and contains no cycles. A *spanning tree* of a hypergraph H is a subhypergraph T of H that is a tree and satisfies $V(T) = V(H)$. (In contrary to graphs, not every connected hypergraph contains a spanning tree!)

MINIMUM SPANNING TREE PROBLEM (MST)
Input: A weighted hypergraph $H = (V, F; w)$, where $w : F \to \mathbb{N}_0$,
Output: Find a spanning tree of H of minimum weight.

Restricted to 2-uniform hypergraphs (that is, to graphs) the minimum spanning tree problem is easily solved efficiently sequentially as well as in parallel. On the other hand, a trivial reduction from EXACT COVER BY 3-SETS shows that for unweighted 4-uniform hypergraphs even deciding whether there *exists* a spanning tree is $\mathcal{NP}$-complete. The status of the minimum spanning tree problem for 3-uniform hypergraphs is not yet completely resolved. Lovász [16] showed that the problem is in $\mathcal{P}$ for unweighted 3-uniform hypergraphs. His algorithm, however, is quite complicated and not very efficient. Later, Gabow and Stallmann [11] reduced the complexity from $\mathcal{O}(n^{17})$ to $\mathcal{O}(n^4)$. In [17] Lovász presented a conceptionally simple randomized algorithm for the unweighted case by reducing it to a sequence of computations of determinants of appropriate matrices. Camerini, Galbiati and Maffioli [5] generalized this approach of Lovász to obtain a randomized pseudo-polynomial time algorithm for the weighted case[3].

In this section we develop a randomized parallel algorithm for the minimum spanning tree problem in 3-uniform hypergraphs. It is also based on the algebraic approach of Lovász [17]. The parallelization is based on ideas from Mulmuley, Vazirani, Vazirani [18]. We start with some definitions.

Let $A = (a_{ij})$ be a skew-symmetric matrix (*i.e.*, $A^T = -A$) of size $2n \times 2n$ and let $\mathcal{P}$ be the set of all partitions of $\{1, \ldots, 2n\}$ into two element sets. For an element $p = \{\{i_1, i_2\}, \ldots, \{i_{2n-1}, i_{2n}\}\}$ of $\mathcal{P}$ we denote by $\sigma(p)$ the sign of the permutation

$$\begin{pmatrix} 1 & 2 & \cdots & 2n-1 & 2n \\ i_1 & i_2 & \cdots & i_{2n-1} & i_{2n} \end{pmatrix}$$

and by $\rho(p)$ the product

$$\rho(p) := \prod_{j=1}^{n} a_{i_{2j-1}\, i_{2j}}.$$

One easily verifies that $\sigma(p) \cdot \rho(p)$ is independent of the order of the classes and the order within the classes of p. Therefore

$$\mathrm{pf}(A) = \sum_{p \in \mathcal{P}} \sigma(p) \cdot \rho(p)$$

is well defined. It is called the *pfaffian* of A. A well-known result from linear algebra (cf. e.g. [15]) says:

Lemma 1. *If A is a skew-symmetric matrix A of size $2n \times 2n$ and B an arbitrary $2n \times 2n$ matrix, then*

$$\det(A) = [\mathrm{pf}(A)]^2 \qquad \textit{and} \qquad \mathrm{pf}(BAB^T) = \det(B) \cdot \mathrm{pf}(A). \qquad \square$$

Lemma 2. [5, 17] *Let $m \geq n$ and $a_1, b_1, \ldots, a_m, b_m$ be vectors in $\mathbb{R}^{2n}$, and let $x_1, \ldots, x_m$ be m indeterminants. Then the $2n \times 2n$ matrix*

$$A = \sum_{i=1}^{m} x_i (a_i b_i^T - b_i a_i^T)$$

is skew-symmetric and satisfies

$$\mathrm{pf}(A) = \sum_{1 \leq i_1 < \cdots < i_n \leq m} x_{i_1} \cdot \cdots \cdot x_{i_n} \cdot \det(a_{i_1} | b_{i_1} \cdots | a_{i_n} | b_{i_n}).$$

(With the notation | we just mean the concatenation of the column vectors a_{i_j} and b_{i_j}.)

[3] In fact, neither Lovász [17] nor Camerini, Galbiati and Maffioli [5] study the minimum spanning tree problem directly. They both consider the matroid parity problem which contains the minimum spanning tree problem in 3-uniform hypergraphs as a special case.

Let $H = (V, F)$ be a 3-uniform hypergraph on $2n+1$ vertices. For every edge $f = \{i, j, k\}$ in F we pick one vertex arbitrarily, say i, and let $e_f = \{i, j\}$ and $\tilde{e}_f = \{i, k\}$. Let $G = (V, E)$ be a (multi-)graph on the same vertex set as H and with edge set $E = \{e_f, \tilde{e}_f \mid f \in F\}$.

Fact 3. *A set $\{f_1, \ldots, f_n\} \subseteq F$ forms a spanning tree in H if and only if $\{e_{f_1}, \tilde{e}_{f_1}, \ldots, e_{f_n}, \tilde{e}_{f_n}\}$ forms a spanning tree in G.* □

That is, the problem of finding a minimum spanning tree in a 3-uniform hypergraph is equivalent to the problem of finding a minimum spanning tree in a (multi-)graph, where the edges are "paired", *i.e.*, either both are in the tree or none.

For every pair of edges e_f, $\tilde{e}_f$ we define two $2n$-dimensional vectors a_f and b_f as follows:

$$(a_f)_i = \begin{cases} 1 & \text{if } i \in e_f, \\ 0 & \text{otherwise,} \end{cases} \qquad (b_f)_i = \begin{cases} 1 & \text{if } i \in \tilde{e}_f, \\ 0 & \text{otherwise.} \end{cases}$$

(Note that the a_f's and b_f's are essentially the incidence vectors of e_f and $\tilde{e}_f$ – except that the $(2n+1)$st component has been cut off.) The following fact resembles a well-known property of the incidence matrix of a graph.

Fact 4. *Let $f_1, \ldots, f_n$ be n edges in F. Then*

$$|\det(a_{f_1}|b_{f_1}|\ldots|a_{f_n}|b_{f_n})| = \begin{cases} 1 & \textit{if } f_1, \ldots, f_n \textit{ form a spanning tree in } H, \\ 0 & \textit{otherwise.} \end{cases} \qquad \square$$

Finally, let A denote the $2n \times 2n$ matrix

$$A = \sum_{f \in F} 2^{w(f)} (a_f b_f^T - b_f a_f^T). \tag{1}$$

With the above facts and notation at hand it is now relatively straightforward to design an algorithm for constructing a minimum spanning tree whenever this tree is unique.

Lemma 5. *Assume H is a 3-uniform hypergraph on $2n+1$ vertices and $w : F \to \mathbb{N}_0$ is a weight-function such that H has a unique spanning tree T_0 of minimum weight, say, w_0. Then $\det(A) \neq 0$ and 2^{2w_0} is the highest power of 2 that divides $\det(A)$. Moreover, if we let $A_f = A - 2^{w(f)}(a_f b_f^T - b_f a_f^T)$ for all $f \in F$, then*

$$f \in T_0 \qquad \textit{if and only if} \qquad \frac{\det(A_f)}{2^{2w_0}} \quad \textit{is even.}$$

Proof. Combining Lemma 1, 2, and Fact 4 we deduce that

$$\det(A) = [\mathrm{pf}(A)]^2 = \left[\sum_T 2^{w(T)} \cdot \delta_T\right]^2,$$

where the sum is over all spanning trees T of H and $\delta_T \in \{-1,+1\}$ for all such trees. Hence,

$$\det(A) = \sum_{i\geq 0} c_i 2^{2w_0+i} \qquad \text{with } c_0 = 1 \text{ and appropriate } c_1, c_2, \dots \in \mathbb{Z}.$$

The first part of the theorem follows. For the second part just observe that A_f is the matrix corresponding to the hypergraph $H - f$. The reasoning above therefore implies that

$$\det(A_f) = \begin{cases} 2^{2w_0} + c_f \cdot 2^{2w_0+1} & \text{if } f \notin T_0, \\ c_f \cdot 2^{2w_0+2} & \text{if } f \in T_0, \end{cases}$$

for appropriate constants $c_f \in \mathbb{Z}$. □

To achieve the uniqueness of a minimum spanning tree in general hypergraphs we use randomization.

Lemma 6. [18] *Let $H = (V, F)$ be a hypergraph on n vertices. For every vertex $v \in V$ choose uniformly and independently at random an integer $r(v)$ from $[1, 2n]$ and define the weight $w(f)$ of an edge $f \in F$ as $w(f) = \sum_{v \in f} r(v)$. Then*

$$\text{Prob}[\textit{There exists a unique edge of minimum weight}] \geq \frac{1}{2}. \qquad \square$$

Corollary 7. *Let $H = (V, F; w)$ be a weighted 3-uniform hypergraph on $2n+1$ vertices containing at least one spanning tree. For every edge $f \in F$ choose uniformly and independently at random an integer $r(f)$ from $[1, 2\binom{2n+1}{3}]$ and define a weight $w' : E \to \mathbb{N}$ as follows:*

$$w'(f) := 3n^4 \cdot w(f) + r(f).$$

Then

$$\text{Prob}[\textit{There exists a unique minimum spanning tree with respect to } w'] \geq \frac{1}{2}.$$

Proof. Construct a hypergraph H' as follows. The vertex set of H' consists of all edges of H, and the edges of H' correspond to all spanning trees of H. Apply Lemma 6 with respect to the hypergraph H'. Then the weight of an edge in H' is at most $n \cdot 2\binom{2n+1}{3} \leq \frac{8}{3}n^4$ and with probability at least $\frac{1}{2}$ there exists a minimum weight edge.

On the other hand, after scaling the weights $w(f)$ of the edges in H by $3n^4$ the weight of a spanning tree in H is a multiple of $3n^4$. That is, by adding the values $r(f)$ we maintain the order relation on the spanning trees of H of different weight. We just disturb the order of the spanning trees in H which have the same weight "a little" – just enough to reach uniqueness with respect to w'. □

Algorithm 8 (MINIMUM SPANNING TREES IN 3-UNIFORM HYPERGRAPHS)
Input: A weighted 3-uniform hypergraph $H = (V, F; w)$ on $2n + 1$ vertices.
Output: A spanning tree T of H or FAILURE.

1. Compute the weight function w' as defined in Corollary 7;
2. Compute the matrix A as defined in (1) and let w_0 be the largest integer such that 2^{2w_0} divides $\det(A)$;
   Let $T := \emptyset$;
3. For every edge $f \in F$ do in parallel:
   Compute $\frac{\det(A_f)}{2^{2w_0}}$;
   if $\frac{\det(A_f)}{2^{2w_0}}$ is even **then** $T := T \cup \{f\}$;
4. **if** T is a spanning tree of H **then** **return** T
   else **return** FAILURE.

Observe that step 3 involves most of the computational effort. Here, one can use e.g. Pan's [19] randomized matrix-inversion algorithm which requires $\mathcal{O}(\log^2 n)$ time and $\mathcal{O}(n^{3.5} \cdot l)$ processors for computing the determinant of an $n \times n$ matrix whose entries are l bit integers. As the entries of the matrix A are of size exponential in $\mathcal{O}(n^4 \max_{f \in F} w(f))$, step 3 needs thus in total $\mathcal{O}(\log^2 n)$ time and $\mathcal{O}(m \cdot n^{7.5} \cdot \max_{f \in F} w(f))$ processors. Combining this observation with Lemma 5 and Corollary 7 we obtain the following result.

Theorem 9. *For all 3-uniform hypergraphs which contain at least one spanning tree Algorithm 8 returns a minimum spanning tree with probability at least $\frac{1}{2}$. The running time of the algorithm is $\mathcal{O}(\log^2 n)$ and it uses at most $\mathcal{O}(m \cdot n^{7.5} \cdot max_{f \in F} w(f))$ processors. In particular, if the weight function w is polynomially bounded in n or if the weights are given in unary, this yields an $\mathcal{RNC}$-algorithm for finding a minimum spanning tree in a 3-uniform hypergraph.* □

Corollary 10. *For every $\epsilon > 0$ there exists a randomized parallel algorithm with running time $\mathcal{O}(\log^2 n)$ and $\mathcal{O}(\frac{1}{\epsilon} \cdot m^2 \cdot n^{8.5})$ processors that returns, for all weighted 3-uniform hypergraphs H with at least one spanning tree, a spanning tree T such that $w(T) \leq (1+\epsilon) mst(H)$ with probability at least $\frac{1}{2}$. (Here $mst(H)$ denotes the length of a minimum spanning tree in the hypergraph H.)*

Proof. We apply the usual scaling technique. Let $H = (V, F; w)$ be a weighted 3-uniform hypergraph on n vertices and let $w_{\max} := \max_{f \in F} w(f)$. For a given $\epsilon > 0$ we set $t := \epsilon \cdot w_{\max}/n$ and define a new hypergraph $H' = (V, F; w')$ with the same vertex and edge set and weight function w' given by

$$w'(f) := \left\lceil \frac{w(f)}{t} \right\rceil \qquad \text{for all } f \in F.$$

Observe that, by construction,

$$mst(H') \leq \frac{1}{t} mst(H) + \frac{n-1}{2}$$

and that $w'_{\max} = \max_{f \in F} w'(f)) = \mathcal{O}(\frac{n}{\epsilon})$. Hence, by Theorem 9, Algorithm 8 returns a minimum spanning tree T in H' with probability at least $\frac{1}{2}$, in $\mathcal{O}(\log^2 n)$

time, using $\mathcal{O}(\frac{1}{\epsilon} \cdot m \cdot n^{8.5})$ processors. If T is indeed a minimum spanning tree in H' we can bound its weight in H as follows:

$$w(T) \leq t \cdot w'(T) = t \cdot mst(H') \leq mst(H) + \frac{1}{2}tn \leq mst(H) + \frac{1}{2}\epsilon w_{\max}.$$

That is, if we could guarantee that $mst(H) \geq w_{\max}$ we would be home. Unfortunately this is not true in general. But another trick helps here.

Let $w_1 < \cdots < w_s$ be the occurring weights in H sorted in increasing order. For every $1 \leq i \leq s$ we define a hypergraph $H_i = (V, F_i; w)$ by deleting all edges from H which have weight larger than w_i. That is, we let $F_i := \{f \in F \mid w(f) \leq w_i\}$. Furthermore, let T_{opt} be an (arbitrary) minimum spanning tree in H and let i_0 be defined such that w_{i_0} is the weight of an edge of maximum weight in T_{opt}. Then, clearly, $w_{i_0} \leq mst(H_{i_0}) = mst(H)$. That is, if use the scaling technique outlined above to compute (in parallel) a minimum spanning tree in all hypergraphs H_i and return of the at most s spanning trees those with minimum weight, this tree T will be, with probability at least $\frac{1}{2}$, a spanning tree in H such that $w(T) \leq (1 + \frac{\epsilon}{2})mst(H)$. The running time of this modified algorithm is, of course, still $\mathcal{O}(\log^2 n)$ while the number of processors increases by a factor of at most m to $\mathcal{O}(\frac{1}{\epsilon} \cdot m^2 \cdot n^{8.5})$. □

Corollary 11. *There exists a fully polynomial randomized (parallel and sequential) approximation scheme for finding a minimum spanning tree in 3-uniform hypergraphs.* □

Remark. It is not true that if Algorithm 8 outputs a spanning tree that this tree is then necessarily a *minimum* spanning tree. Consider the hypergraph $H = (V, F)$ with vertex set $V = \{1, \ldots, 7\}$ and six edges $\{1,2,7\}$, $\{3,4,7\}$, $\{5,6,7\}$, $\{1,2,3\}$, $\{3,4,5\}$, and $\{5,6,1\}$ the first three having weight 2 the remaining three having weight 1. One easily checks that H has exactly 3 minimum spanning trees (of weight 4) and that the algorithm therefore correctly determines $w_0 = 4$. However, by considering $H - f$ for the various edges f one also finds that the algorithm will return the three edges $\{1,2,7\}$, $\{3,4,7\}$, $\{5,6,7\}$. Which do form a spanning tree, but not a minimum one.

3 Approximation Algorithms for the Steiner Problem

Let $G = (V, E)$ be a graph and $K \subseteq V$ be a subset of the vertex set. A subgraph T of G is a *Steiner tree* for K, if T is a tree containing all vertices of K (*i.e.*, $K \subseteq V(T)$) such that all leaves of T are elements of K. A *Steiner minimum tree* for K in G is a Steiner tree T such that $|E(T)|$ (in unweighted graphs) resp. $w(T)$ (in weighted graphs) is minimum.

STEINER PROBLEM IN NETWORKS (SPN)
Input: A network (a weighted graph) $N = (V, E; w)$, and a set $K \subseteq V$.
Output: A Steiner minimum tree for K in N, that is a Steiner tree T such that $w(T) = \min\{w(T') \mid T' \text{ a Steiner tree for } K \text{ in } N\}$.

If the input is restricted to *unweighted* graphs $G = (V, E)$ and sets $K \subseteq V$ we speak of the STEINER PROBLEM IN GRAPHS (SPG). Given a graph $G = (V, E)$ or network $N = (V, E; w)$ and a terminal set K we denote by $smt(G, K)$ resp. $smt(N, K)$ the length of a Steiner minimum tree for K in G resp. N. Note that for $K = V$ the Steiner problem is exactly the minimum spanning tree problem. In the following we denote the length of a minimum spanning tree in a hypergraph H by $mst(H)$.

The computational complexity of the Steiner tree problem in graphs and networks varies considerably with the structure of the underlying graph and the cardinality of the terminal set K. While it can be easily solved whenever $|K| = 2$ or $K = V$ and for certain classes of graphs, it is rather difficult in general.

The Steiner tree problem in networks was among the first problems shown to be $\mathcal{NP}$-hard in the seminal paper of Karp [12]. Bern and Plassmann [3] then proved that even the special problem with edge weights restricted to the values 1 and 2 is $\mathcal{MAXSNP}$-hard. A consequence of the new characterization of the class $\mathcal{NP}$ by Arora, Lund, Motwani, Sudan and Szegedy [1] is thus that there exists no polynomial time approximation scheme for the Steiner tree problem, unless $\mathcal{P}=\mathcal{NP}$. Hence, unless $\mathcal{P}=\mathcal{NP}$ the best performance ratio attainable by a polynomial time algorithm is a constant larger than 1. It remains a challenging question how close to 1 this performance ratio can be.

Let $N = (V, E; w)$ be a network and $K \subseteq V$ a terminal set. With respect to this Steiner problem we define for all $r \geq 2$ a weighted hypergraph $H_r(N, K) = (K, F_r; w_r)$ on the vertex set K as follows. The edge set F_r consists of all subsets of K of cardinality at most r, and the weight $w_r(f)$ of an edge $f \in F_r$ is the length of a Steiner minimum tree for f in the network N.

Fact 12. *Let $N = (V, E, w)$ be a network with terminal set K and let $H_r(N, K)$ be as defined above. Then*

$$mst(H_r(N, K)) \geq smt(N, K). \tag{2}$$

Proof. Let T be a minimum spanning tree in $H_r(N, K)$. For every edge f in T choose a Steiner minimum tree T_f for f in N. Then the fact that T is a spanning tree in $H_r(N, K)$ implies that $S = \bigcup_{f \in T} T_f$ is a connected subgraph of (V, E) which contains all vertices in K. □

Reconsidering the proof of Fact 12, we observe that given a spanning tree T in $H_r(N, K)$ one can easily determine a Steiner tree for K in the original network N of at most the same length: We just take the union of Steiner minimum trees for all $f \in T$, and delete edges until the obtained graph is a tree in which all leaves are terminals.

Let ρ_r denote the least upper bound of the ratio $mst(H_r(N, K))/smt(N, K)$ for all networks $N = (V, E; w)$ and terminal sets $K \subseteq V$. One easily sees that $\rho_2 = 2$, cf. e.g. [22] or [14]. Obtaining ρ_3 is considerably more difficult. Zelikovsky [23] shows that $\rho_3 \leq 5/3$ and considering appropriate binary trees one deduces easily that in fact $\rho_3 = 5/3$. For general $r \in \mathbb{N}$, Du, Zhang, and Feng [9] proved that $\rho_{2^r} \leq 1 + \frac{1}{r}$ implying in particular that $\rho_r \to 1$ for $r \to \infty$ and, finally,

Borchers and Du [4] proved that $\rho_r = ((t+1)2^t + l)/(t2^t + l)$ for $r = 2^t + l$ and $0 \leq l < 2^t$.

Observe that for all constants $r \in \mathbb{N}$ the hypergraph $H_r(N, K)$ can be constructed in polynomial time: There are less than k^r subsets of K of cardinality at most r, where $k = |K|$. For each of these subsets a Steiner minimum tree can be found in $\mathcal{O}(n^2 \log n + nm)$ with the algorithm of Dreyfus and Wagner [8]. A plausible approach for designing an approximation algorithm is therefore to solve the minimum spanning tree problem in $H_r(N, K)$ and deduce from that a Steiner tree for K in N as outlined above. This would give an approximation algorithm with ratio ρ_r. As, however, finding minimum spanning trees in r-uniform hypergraphs is $\mathcal{NP}$-complete for all $r \geq 4$ this reduction to the spanning tree problem is not really helpful[4] – except for the case $k = 3$.

Zelikovsky [23] showed that in the special case of $H_3(N, K)$ one can use a greedy approach to find a spanning tree T in $H_3(N, K)$ of length at most $\frac{1}{2}[mst(H_2(N, K)) + mst(H_3(N, K))]$ and thus a Steiner tree of size at most $\frac{1}{2}(\rho_2 + \rho_3) = 11/6$ times the length of a Steiner minimum tree.

Berman and Ramaiyer [2] found a different procedure which makes also use of the hypergraphs $H_r(N, K)$ for $r > 3$. They obtained an algorithm with performance ratio

$$\rho_2 - \sum_{i=3}^{h} \frac{\rho_{i-1} - \rho_i}{i - 1} \quad \approx \; 1.734.$$

for all $h \geq 3$. Zelikovsky [24] invented a so-called relative greedy heuristic for approximating $mst(H_r(N, K))$ that yields an approximation algorithm for the length of a Steiner minimum tree with performance ratio $1 + \ln 2 \approx 1.693$. A slight further improvement, then, led to a ratio of $1.644 + \varepsilon$ for any positive $\varepsilon > 0$, see Karpinski and Zelikovsky [13].

In order to use the algorithm of the previous section for solving the spanning tree problem in $H_3(N, K)$ we have to reduce the spanning tree problem in hypergraphs with edges containing *at most* three vertices to a corresponding problem in a 3-uniform hypergraph, *i.e.*, a hypergraph where all edges consist of *exactly* three vertices. This, however, can easily be achieved:

Reducing the spanning tree problem to 3-uniform hypergraphs. Let $H = (V, F; w)$ be a weighted hypergraph on n vertices such that every edge contains at most three vertices. Construct a weighted 3-uniform hypergraph $\tilde{H} = (\tilde{V}, \tilde{F}, \tilde{w})$ as follows. The vertex set $\tilde{V}$ consists of all vertices of V plus $n - 1$ new vertices $z_1, \ldots, z_{n-1}$, and

$$\begin{aligned} \tilde{F} = F &\cup \{e \cup \{z_i\} \mid e \in F, |e| = 2, 1 \leq i \leq n-1\} \\ &\cup \{v \cup \{z_i, z_j\} \mid v \in V, 1 \leq i < j \leq n-1\}. \end{aligned}$$

New triples containing exactly one z-vertex will be called type I triples, while those containing two z-vertices are of type II. The weight function $\tilde{w}$ is defined

[4] In fact, this is not quite obvious at this point, as the reduction from the Steiner tree problem generates only special spanning tree problems. It is, however, easy to see that also these special spanning tree problems are $\mathcal{NP}$-hard to solve, cf. e.g. [20].

as follows:

$$\tilde{w}(f) = \begin{cases} w(f) & \text{if } f \in F, \\ w(e) + M & \text{if } f \text{ is a type I triple with } e \in F \text{ and } e \subseteq f, \\ 2M & \text{if } f \text{ is a type II triple.} \end{cases}$$

Fact 13. *The above reduction has the following properties:*

(i) Every spanning tree T in H gives rise to a spanning tree $\tilde{T}$ in $\tilde{H}$ of weight $\tilde{w}(\tilde{T}) = w(T) + (n-1)M$ by replacing every edge of cardinality 2 in T by a type I triple (by adding different z-vertices) and adding type II triples until every z-vertex is contained in exactly one triple.

(ii) If $\tilde{T}$ is a spanning tree in $\tilde{H}$ of weight $\tilde{w}(\tilde{T}) < nM$ then every z-vertex is covered by exactly one triple of $\tilde{T}$. Thus, $\tilde{T}$ corresponds to a spanning tree T of H of weight $w(T) = \tilde{w}(\tilde{T}) - (n-1)M$.

(iii) Let $M \geq n \cdot \max_{f \in F} w(f)$. Then a minimum spanning tree in $\tilde{H}$ of length less than nM corresponds to a minimum spanning tree of H, whereas the fact that the length of a minimum spanning tree in $\tilde{H}$ is at least nM indicates that H contains no spanning tree. Furthermore, if H is the complete hypergraph (e.g. the hypergraph $H_3(N,K)$ in the reduction from the Steiner tree problem) then choosing $M = \max_{f \in F} w(f)$ suffices to achieve these properties. □

Algorithm 14 (STEINER TREE PROBLEM IN GRAPHS)
Input: A connected network $N = (V, E; w)$ and a terminal set $K \subseteq V$.
Output: A Steiner tree for K or FAILURE.

1. Compute the hypergraph $H_3(N,K)$;
2. Transform the hypergraph $H_3(N,K)$ to the corresponding 3-uniform hypergraph $\tilde{H}$;
3. Use the algorithm of the previous section to find a spanning tree $\tilde{T}$ in $\tilde{H}$;
   If this algorithm returns FAILURE then stop;
4. Transform $\tilde{T}$ to a spanning tree T in $H_3(N,K)$ according to Fact 13;
5. Transform T into a Steiner tree S for K in N; return S.

For an implementation of step 1 observe first that a Steiner minimum tree for a two element set $\{x,y\}$ is just a shortest x-y path. Furthermore, a Steiner minimum tree for a three element set $\{x,y,z\}$ is the union of three shortest paths. Namely, a shortest x-w path plus a shortest y-w path plus a shortest z-w path, where w is an appropriate vertex in V. As the all-pairs shortest path problem can be solved in $\mathcal{O}(\log^2 n)$ time on $\mathcal{O}(n^3)$ processors, we conclude that step 1 can certainly be achieved in $\mathcal{O}(\log^2 n)$ time and $\mathcal{O}(n^4)$ processors. Step 2 is easily implemented within the same time bound. Note that the weight of an edge in the hypergraph $H_3(N,K)$ (and hence also in the hypergraph $\tilde{H}$) is less than $n \cdot \max_{e \in E} w(e)$. Theorem 9 thus implies that $\mathcal{O}(\log^2 n)$ time and $\mathcal{O}(n^{11.5} \cdot \max_{f \in F} w(f))$ processors suffice for step 3. As steps 4 and 5 are again easily implemented within this time bound, we obtain the following theorem.

Theorem 15. *Algorithm 14 is a randomized parallel algorithm which returns with probability at least $\frac{1}{2}$ a Steiner tree S for K such that $w(S) \leq \frac{5}{3} smt(N,K)$.*

The algorithms runs in $\mathcal{O}(\log^2 n)$ time, using $\mathcal{O}(n^{11.5} \cdot max_{f\in F}\, w(f))$ processors. In particular, if the weight function w is polynomially bounded in n or if the weights are given in unary, this yields an $\mathcal{RNC}$-approximation algorithm for the STEINER PROBLEM IN NETWORKS *with performance ratio* 5/3. □

Corollary 16. *There exists an $\mathcal{RNC}$-approximation algorithm with performance ratio* 5/3 *for the* STEINER PROBLEM IN GRAPHS. □

For networks we can also use the algorithm of Corollary 10 instead of those of Theorem 9 to obtain an $\mathcal{RNC}$-approximation scheme for performance ratio 5/3:

Corollary 17. *For every $\epsilon > 0$ there exists a randomized parallel algorithm that given a network $N = (V, E; \ell)$ on n vertices and a terminal set K returns in $\mathcal{O}(\log^2 n)$ steps, using $poly(n + \frac{1}{\epsilon})$ processors, with probability at least $\frac{1}{2}$ a Steiner tree T such that*

$$\ell(T) \leq \frac{5}{3}(1 + \epsilon)smt(N, K).$$ □

Of course, the algorithm can also be implemented as a sequential algorithm. Here one can also use the fact that the determinant of an $n \times n$ matrix containing l bit integers can be computed on a random access machine in $\mathcal{O}(n^{\alpha} l \log l)$ bit operations, where $\mathcal{O}(n^{\alpha})$ is the number of arithmetic operations required to multiply two $n \times n$ matrices. Currently, the best know value is $\alpha < 2.376$ [7].

Corollary 18. *There exists a randomized sequential $\mathcal{O}(n^{8+\alpha} \log n)$ approximation algorithm with performance ratio* 5/3 *for the* STEINER PROBLEM IN GRAPHS. □

Corollary 19. *For every $\epsilon > 0$ there exists a randomized sequential $\mathcal{O}(\frac{\log \frac{1}{\epsilon}}{\epsilon} \cdot n^{11+\alpha} \log n)$ approximation algorithm with performance ratio $5/3 + \epsilon$ for the* STEINER PROBLEM IN NETWORKS. □

In comparison, the best deterministic sequential approximation algorithm for the STEINER PROBLEM IN NETWORKS of Karpinski and Zelikovsky [13] has a performance ratio of ≈ 1.644 which is slightly better than 5/3 ≈ 1.667. However, this performance guarantee is only achieved in the limit. More precisely, building on the work of Zelikovsky [24] , Karpinski and Zelikovsky [13] define a class $(\mathcal{A}_k)$ of approximation algorithms such that the running time of $\mathcal{A}_k$ is bounded by $\mathcal{O}(n^k)$ and the performance ratio of $\mathcal{A}_k$ tends to 1.644 for k tending to infinity. For reasonable "small" k, say k up to 20, the performance ratio is, however, still larger than 5/3.

References

1. S. ARORA, C. LUND, R. MOTWANI, M. SUDAN, AND M. SZEGEDY, Proof verification and hardness of approximation problems, *Proc. 33rd Annual IEEE Symp. Foundations of Computer Science* (1992), 14–23.

2. P. Berman and V. Ramaiyer, Improved Approximations for the Steiner Tree Problem, *Journal of Algorithms* **17** (1994), 381–408.
3. M. Bern and P. Plassmann, The Steiner problem with edge lengths 1 and 2, *Information Processing Letters* **32** (1989), 171–176.
4. A. Borchers and D.-Z. Du, The k-Steiner ratio in graphs, *Proc. 27th Annual ACM Symp. on the Theory of Computing* (1995), 641–649.
5. P.M. Camerini, G. Galbiati, and F. Maffioli, Random pseudo-polynomial algorithms for exact matroid problems, *Journal of Algorithms* **13** (1992), 258–273.
6. Choukhmane El-Arbi, Une heuristique pour le problème de l'arbre de Steiner, *R.A.I.R.O. Recherche opérationnelle* **12** (1978), 207–212.
7. D. Coppersmith and S. Winograd, Matrix multiplication via arithmetic progressions, *Proc. 19th Annual ACM Symp. on Theory of Computing* (1987), 1–6.
8. S.E. Dreyfus and R.A. Wagner, The Steiner problem in graphs, *Networks* **1** (1972), 195–207.
9. D.-Z. Du, Y.-J. Zhang and Q. Feng, On better heuristic for Euclidean Steiner minimum trees, *Proc. 32nd Annual IEEE Symp. on Foundations of Computer Science* (1991), 431–439.
10. H.N. Gabow, Z. Galil, T.H. Spencer, Efficient implementation of graph algorithms using contraction, *Proc. 25th Annual IEEE Symp. on Foundations of Computer Science* (1984), 347–357.
11. H.N. Gabow and M. Stallmann, An augmenting path algorithm for linear matroid parity, *Combinatorica* **6** (1986), 123–150.
12. R. Karp, Reducibility among combinatorial problems, in: *Complexity of computer computations* (Miller, R.E., Thatcher, J.W., eds.), Plenum Press, 1972, 85-103.
13. M. Karpinski and A.Z. Zelikovsky, New approximation algorithms for the Steiner tree problem, *Electr. Colloq. Comput. Compl., TR95-030*, 1995.
14. L. Kou, G. Markowsky, and L. Berman, A fast algorithm for Steiner trees, *Acta Informatica* **15** (1981), 141–145.
15. S. Lang, Algebra, Addison-Wesley Publishing Company, 1993.
16. L. Lovász, The matroid matching problem, *Algebraic Methods in Graph Theory*, Colloquia Mathematica Societatis János Bolyai, Szeged (Hungary), 1978.
17. L. Lovász, On determinants, matchings and random algorithms, *Fund. Comput. Theory* **79** (1979), 565–574.
18. K. Mulmuley, U. Vazirani, and V. Vazirani, Matching is as easy as matrix inversion, *Combinatorica* **7** (1987), 105–113.
19. V. Pan, Fast and efficient algorithms for the exact inversion of integer matrices, *Fifth Annual Foundations of Software Technology and Theoretical Computer Science Conference* (1985), LNCS 206, 504–521.
20. H.J. Prömel and A. Steger, *The Steiner Tree Problem. A Tour through Graphs, Algorithms, and Complexity*, Vieweg Verlag, Wiesbaden, to appear 1997.
21. F. Suraweera and P. Bhattacharya, An $\mathcal{O}(\log m)$ parallel algorithm for the minimum spanning tree problem, *Inf. Proc. Lett.* **45** (1993), 159–163.
22. H. Takahashi and A. Matsuyama, An approximate solution for the Steiner problem in graphs, *Math. Japonica* **24** (1980), 573–577.
23. A.Z. Zelikovsky, An 11/6-approximation algorithm for the network Steiner problem, *Algorithmica* **9** (1993), 463–470.
24. A.Z. Zelikovsky, Better approximation algorithms for the network and Euclidean Steiner tree problems, Technical Report, Kishinev, 1995.

Pattern Matching in Trace Monoids

(Extended Abstract)

Jochen Messner*

Abt. Theoretische Informatik, Universität Ulm, 89069 Ulm, Germany
E-mail: messner@informatik.uni-ulm.de

Abstract. An algorithm is presented solving the factor problem in trace monoids. Given two traces represented by words, the algorithm determines in linear time whether the first trace is a factor of the second one. The space used for this task is linear in the length of the first word. Similar to the Knuth-Morris-Pratt Algorithm for the factor problem on words, the algorithm simulates a finite automaton determined by the first word on the second word. To develop the algorithm, we examine overlaps of two traces, and extensible trace pairs (which represent still extensible prefixes of a searched factor appearing in some other trace), and show that both structures are lattices.

1 Introduction

The pattern matching problem in free monoids is an extensively studied problem in computer science. For two words $v, x \in A^*$ it is asked whether there are words $u, w \in A^*$ such that $x = uvw$, i.e., it is asked whether v is a factor of x. There are several linear time algorithms solving the problem. The algorithm which was given by Knuth, Morris, and Pratt [14] has close connections to the theory of finite automata (see [1]): First from the first word v a so called failure function is computed as a table in time linear to the length of v; after this first stage the failure function is used to simulate on the second word in linear time a finite automaton accepting the language $A^* \cdot v \cdot A^* = \{uvw \mid u, w \in A^*\}$. Altogether the time used is linear in the input size, and even does not depend on the size of the alphabet, because the automaton is not constructed explicitly.

In this paper we use a quite similar approach to the factor problem in trace monoids. Trace monoids, also called free partially commutative monoids, have been studied in combinatorics in [3]. In [16] Mazurkiewicz considered them as a suitable mathematical model for concurrent systems. Given a finite set of actions (an alphabet), some of the actions are considered independent (e.g., they may use different resources). Because the order of independent actions is irrelevant, one identifies sequences of actions (i.e., words) which can be transformed to each other by exchanging adjacent independent actions. This yields an equivalence relation $\sim_I$ on words which is, in fact, a congruence; a congruence class is called trace, consequently the monoid of the congruence classes is a trace monoid. It

* The most part of this work was done while the author was a member of the section Theoretical Computer Science at the University of Stuttgart.

Reischuk, Morvan (Eds.): STACS'97 Proceedings, LNCS 1200

is determined uniquely by the generating alphabet and the relation I of independent letters. A free monoid is obtained as a special case, when all letters are pairwise dependent. Trace monoids have been studied in many publications. Good starting points are [8], [9], and [7].

The factor problem in a trace monoid M is to decide for two words, whether the trace l, represented by the first word, is a factor of the trace t, represented by the second word (where l is called factor of t, when $t = pls$ for some traces p, s). A linear-time algorithm for the problem using space linear in the input size was given in [15]. This space-complexity is not always desirable. So, for example, a control component of a concurrent system may need to recognize certain subsequences in a sequence of actions (modulo independence) without remembering all executed actions. Therefore we consider in this paper an approach more closely related to finite automata. From the first word v, in a first stage, several structures are computed in linear time. These structures are used in the second stage to simulate on the second word a finite automaton recognizing the language $\{x \in A^* \mid x \sim_I uvw \text{ for some words } u, w\}$ which is the set of words representing the traces in $M \cdot l \cdot M = \{pls \mid p, s \in M\}$. The time used for this simulation is linear in the length of the second word such that altogether the algorithm needs linear time. Because of this similarity we consider the presented algorithm as a generalization of the Knuth-Morris-Pratt Algorithm to trace monoids, although there is in general no correspondent to the failure function for traces (cf. [17] for related results). Already in [13] Hashiguchi and Yamada had this approach. However, the algorithm they proposed to solve the factor problem produces an incorrect answer in some cases, as we show by an example.

The organization of the paper is as follows: We first introduce basic notions. A representation of traces is obtained by the general embedding theorem using projections. In Section 3 we study the set of prefixes and suffixes of some trace. Both sets form lattices with the prefix (resp. suffix) orders. We observe that projections are morphisms for those lattices which allows us to deduce easily that the intersection of a set of prefixes with a set of suffixes (called overlaps of two traces) still forms a lattice. Using these results, we finally develop a finite automaton recognizing the language of traces containing a given trace l as a suffix. In Section 4 we present an algorithm computing the transition function of this automaton which allows us to simulate the automaton in linear time. This simulation solves the suffix problem (the algorithm was already given in [13], we obtain an improved time complexity). In Section 5 we obtain a finite automaton recognizing the traces containing a given trace l as a factor. A linear-time simulation of this automaton solves the factor problem. However, for some trace monoids the simulation presented needs time and space exponential in the alphabet-size. In Section 6 we investigate extensible trace pairs which have close relations to the states reached in the automaton. We show that they form a lattice—an observation which may lead to an improvement of the presented algorithm remaining efficient even when the alphabet is a part of the input.

Although the main purpose of this paper is the investigation of the factor problem, the results obtained for overlaps (cf. [17]) and extensible trace pairs (cf. [13]) are interesting on their own.

2 Preliminaries

In the following, some notions for trace monoids are given. In this extended abstract we refer to [8] for further notations and definitions related to traces. A suitable representation of traces is obtained by the embedding theorem, which allows us to represent a trace uniquely by a tuple of words. We also give some notations on finite automata and a very brief description of some properties of the pattern-matching algorithm of Knuth, Morris, and Pratt. As we use standard notions of [2], we give no introduction to lattice or poset theory, here.

2.1 Free and Free Partially Commutative Monoids

Throughout this extended abstract A denotes a finite alphabet, $I \subseteq A \times A$ denotes a irreflexive and symmetric independence relation, and $D = (A \times A) - I$ the corresponding dependence relation; the graph (A,D) is called dependence alphabet and $\mathrm{M}(A,D)$ denotes the corresponding trace monoid.

A trace $l \in \mathrm{M}(A,D)$ is called factor of $t \in \mathrm{M}(A,D)$, when $t = pls$ for some traces $p, s \in \mathrm{M}(A,D)$; l is called prefix (suffix) of t when p (resp. s) can be chosen to be λ. The set of prefixes (suffixes) of t is denoted by $\mathrm{Pre}(t)$ (resp. $\mathrm{Suf}(t)$). Levi's Lemma on factorizations of words has the following generalization for traces. For a proof of Proposition 1 see [7, Proposition 1.3.1] (see also [6]).

Proposition 1. *Let $p, s, p', s' \in M(A,D)$ with $ps = p's'$. Then there are uniquely determined traces $x, y, y', z \in M(A,D)$ with $alph(y) \times alph(y') \subseteq I$ and $p = xy$, $p' = xy'$, $s = y'z$ and $s' = yz$.*

For a subgraph $H = (A',D')$ of (A,D), π_H denotes the projection of $\mathrm{M}(A,D)$ onto $\mathrm{M}(A',D')$, i.e., $\pi_H : \mathrm{M}(A,D) \to \mathrm{M}(A',D')$ is an homomorphism such that for $a \in A$, $\pi_H(a) = a$ if $a \in A'$, and $\pi_H(a) = \lambda$ else. For $B \subseteq A$, we use π_B to denote $\pi_{(B,D_{|B})}$, where $D_{|B} = D \cap (B \times B)$. A family $\mathcal{G}$ of subgraphs of (A,D) is called covering, when $(A,D) = \bigcup_{G \in \mathcal{G}} G$. A subset S of $\mathcal{P}(A)$, where $\mathcal{P}(A)$ is the set of all subsets of $\mathcal{A}$, may denote the family $\{(B, D_{|B}) \mid B \in S\}$ and is called covering accordingly. A covering consisting only of cliques is called clique-covering. The set $\{\{a,b\} \mid (a,b) \in D\}$ is a trivial clique-covering of (A,D).

Proposition 2 is a generalization of [6, Proposition 1.1] given in [7], where it was called general embedding theorem. $\prod_i M_i$ denotes the direct product of the monoids M_i. An element of $\prod_i M_i$ is uniquely denoted by $(t_i)_i$ with $t_i \in M_i$.

Proposition 2. *Let $\mathcal{G}$ be family of subgraphs of (A,D). $\mathcal{G}$ is a covering if, and only if, the mapping $\pi : M(A,D) \to \prod_{G \in \mathcal{G}} M(G)$ defined by $\pi(t) = (\pi_G(t))_{G \in \mathcal{G}}$ is an embedding (i.e., an injective homomorphism).*

For a covering $\mathcal{G}$ of (A,D) we call the tuple $(\pi_G(t))_{G \in \mathcal{G}}$ the tuple-representative of $t \in \mathrm{M}(A,D)$. Extending terminology of [4], a tuple which is a tuple-representative of some trace is called reconstructible. Choosing $\mathcal{G}$ to be a clique-covering, we obtain a version of Proposition 2 already given in [10] which allows us to represent a trace uniquely by a tuple of words. Clearly, this tuple is computable in time linear to its size, given a word representing the trace (as model of computation we generally assume a RAM with a uniform cost criterion, see e.g. [18]).

2.2 Automata

A (nondeterministic) finite automaton, shortly called automaton, is a tuple $\mathcal{A} = (Q, A, \delta, q_0, F)$ consisting of the finite state-set Q, the alphabet A, the transition relation $\delta \subseteq Q \times A \times Q$, the initial state $q_0 \in Q$, and the set of final states $F \subseteq Q$. We write $p \xrightarrow{w}_{\mathcal{A}} q$, when there is a transition from p to q by $w \in A^*$. $\mathcal{A}$ is called M(A,D)-automaton, when $p \xrightarrow{w}_{\mathcal{A}} q$ implies $p \xrightarrow{v}_{\mathcal{A}} q$ for any $v \sim_I w$ (actually it suffices to consider only $w = ab$, $v = ba$ for $(a, b) \in I$). In this case we simply write $p \xrightarrow{t}_{\mathcal{A}} q$, where $t = [w]_{\sim_I}$ is the trace represented by w. $L = \bigcup_{q \in F} \{t \in \mathrm{M}(A,D) \mid q_0 \xrightarrow{t}_{\mathcal{A}} q\}$ is the trace language recognized by the M(A,D)-automaton $\mathcal{A}$. For further definitions on automata see [11].

2.3 The Algorithm of Knuth-Morris-Pratt

By the algorithm of Knuth-Morris-Pratt (see [14], [1]) it is decidable in linear time whether a word $v \in A^*$ is a factor of the word $w \in A^*$ using $\mathcal{O}(|v|)$ space. The basis of the algorithm is the so called failure function $\phi_v : \mathrm{Pre}(v) - \{\lambda\} \rightarrow \mathrm{Pre}(v)$, where $\phi_v(p)$ is the longest word $s \neq p$ which is a suffix and a prefix of p. The failure function ϕ_v can be calculated as a table in time linear to $|v|$. (Notice that a prefix p of v may be uniquely represented by its length $|p|$).

The complete deterministic automaton $\mathcal{A}_v = (\mathrm{Pre}(v), A, \varphi_v, \lambda, v)$, where for $p \in \mathrm{Pre}(v)$, $a \in A$, $\varphi_v(p, a)$ is the longest word in $\mathrm{Pre}(v) \cap \mathrm{Suf}(pa)$, is the minimal automaton recognizing the language $A^* \cdot v$. Using the failure function ϕ_v, the transition function φ_v is computable efficiently by the following relation

$$\varphi_v(p, a) = \begin{cases} pa & \text{if } pa \in \mathrm{Pre}(v), \\ \lambda & \text{if } p = \lambda \text{ and } a \notin \mathrm{Pre}(v), \\ \varphi_v(\phi_v(p), a) & \text{else.} \end{cases}$$

It is clear that the computation of $\varphi_v(p, a)$, for any prefix p of v, $a \in A$, needs time linear to the number of times the failure function is used in the computation. Because each use of ϕ_v shortens the input to φ_v, the time used is linear to $|pa| - |\varphi_v(p, a)|$.

3 Prefixes and Suffixes

We examine now the set of prefixes and the set of suffixes of some trace, both of which form a lattice. For $l, t \in \mathrm{M}(A,D)$, we write $l \leq_p t$ ($l \leq_s t$), when l is a prefix (resp. suffix) of t. It is clear that $\leq_s$ and $\leq_p$ are partial orders. See [12] for an investigation of the poset $(\mathrm{M}(A,D), \leq_p)$. Lemma 3 shows that the structures $(\mathrm{Pre}(l), \leq_p)$ and $(\mathrm{Suf}(l), \leq_s)$ are lattices for any trace l. A proof for the prefix-case can be found in [5], the result for suffixes is obtained by symmetry.

Lemma 3. *Let $t, t' \in \mathrm{M}(A,D)$ both be prefixes (suffixes) of the same trace. Then there are uniquely determined traces $x, y, y' \in \mathrm{M}(A,D)$ with $\mathrm{alph}(y) \times \mathrm{alph}(y') \subseteq I$, $t = xy$, and $t' = xy'$ (resp. $t = yx$, and $t' = y'x$). Further, x is the greatest lower bound for t and t' in the prefix (resp. suffix) order in M(A,D) and xyy' (resp. $yy'x$) is the least upper bound for t and t' in the prefix (resp. suffix) order.*

The least upper bound in the prefix (suffix) order will be denoted by $\sqcup_p$ (resp. $\sqcup_s$), the greatest lower bound by $\sqcap_p$ (resp. $\sqcap_s$). We say that $t \sqcup_p t'$ is not defined, when there is no trace s such that $t \leq_p s$ and $t' \leq_p s$ (similar for $\sqcup_s$).

Because projections are monoid homomorphisms, they are poset morphisms with respect to the prefix and the suffix order. Also independence of traces is preserved by projections. By Lemma 3 we obtain that projections are also morphisms for the lattices of prefixes (resp. suffixes):

Lemma 4. *Let $t, t' \in M(A,D)$ such that $t \sqcup_p t'$ (resp. $t \sqcup_s t'$) is defined. Let $G \subseteq (A,D)$. Then $\pi_G(t \sqcup_p t') = \pi_G(t) \sqcup_p \pi_G(t')$, and $\pi_G(t \sqcap_p t') = \pi_G(t) \sqcap_p \pi_G(t')$ (respectively, $\pi_G(t \sqcup_s t') = \pi_G(t) \sqcup_s \pi_G(t')$ and $\pi_G(t \sqcap_s t') = \pi_G(t) \sqcap_s \pi_G(t')$).*

3.1 The Lattice of Overlaps

The traces in $\mathrm{Pre}(l) \cap \mathrm{Suf}(t)$ are called overlaps of l and t. We show that the prefix and suffix order coincides for the overlaps of two traces and that there is a unique maximal overlap denoted by $\sqcup(\mathrm{Pre}(l) \cap \mathrm{Suf}(t))$. The notion of overlap used here is a slight generalization of the same notion in [17]. In [17] sets $\mathrm{Pre}(x) \cap \mathrm{Suf}(x)$ for some trace x were examined and it was shown that such a set forms a lattice. This lattice is equal to $(\mathrm{Pre}(l) \cap \mathrm{Suf}(t), \leq_p)$ when choosing $x = \sqcup(\mathrm{Pre}(l) \cap \mathrm{Suf}(t))$. Compared to [17] we obtain a very short proof, due to Lemma 4.

Observe for words $u, v \in A^*$ that $u \sqcup_p v$ (resp. $u \sqcup_s v$) is just the longest word of both (if defined). Thus, if $u \sqcup_p v$ and $u \sqcup_s v$ both exist, $u \sqcup_p v = u \sqcup_s v$, and $u \sqcap_p v = u \sqcap_s v$. We get the same result for traces.

Lemma 5. *Let $r, s \in \mathrm{Pre}(l) \cap \mathrm{Suf}(t)$. Then $r \sqcup_p s = r \sqcup_s s$ and $r \sqcap_p s = r \sqcap_s s$.*

Proof. Let C be a clique-covering of (A,D) and let $\pi : \mathrm{M}(A,D) \to \prod_{C \in \mathcal{C}} C^*$ be an embedding like in Proposition 2. We have by Lemma 4

$$\pi(r \sqcup_p s) = (\pi_C(r) \sqcup_p \pi_C(s))_{C \in \mathcal{C}} = (\pi_C(r) \sqcup_s \pi_C(s))_{C \in \mathcal{C}} = \pi(r \sqcup_s s).$$

As π is injective, the equality results. Replace $\sqcup$ by $\sqcap$ to show $r \sqcap_p s = r \sqcap_s s$.

Due to Lemma 5, we may just write $\sqcup$ for the suffix (resp. prefix) supremum in sets $\mathrm{Pre}(l) \cap \mathrm{Suf}(t)$. We write $x \leq_o y$ if $x \leq_p y$ and $x \leq_s y$. It is clear that $(\mathrm{M}(A,D), \leq_o)$ is a poset. As a corollary of Lemma 5 we obtain

Theorem 6. *Let $l, t \in M(A,D)$. Then the poset $(\mathrm{Pre}(l) \cap \mathrm{Suf}(t), \leq_p) = (\mathrm{Pre}(l) \cap \mathrm{Suf}(t), \leq_s) = (\mathrm{Pre}(l) \cap \mathrm{Suf}(t), \leq_o)$ is a lattice.*

3.2 An Automaton Recognizing $\mathbf{M}(A,D) \cdot l$

We present now a minimal deterministic automaton recognizing the language $\mathrm{M}(A,D) \cdot l = \{t \in \mathrm{M}(A,D) \mid l \leq_s t\}$. Clearly, l is a suffix of t if, and only if, $l = \sqcup(\mathrm{Pre}(l) \cap \mathrm{Suf}(t))$. We examine the extension of t by a letter a:

Lemma 7. *Let $a \in \Sigma$, $t, l \in M(A,D)$ and $p = \sqcup(\mathrm{Pre}(l) \cap \mathrm{Suf}(t))$. Then*

$$\sqcup(\mathrm{Pre}(l) \cap \mathrm{Suf}(ta)) = \sqcup(\mathrm{Pre}(l) \cap \mathrm{Suf}(pa)).$$

To prove Lemma 7 check that $\mathrm{Pre}(l) \cap \mathrm{Suf}(ta) = \mathrm{Pre}(l) \cap \mathrm{Suf}(pa)$. By induction we obtain

Theorem 8. *Let $l \in M(A,D)$. Let $\Phi_l : Pre(l) \times A \to Pre(l)$ be defined by*

$$\Phi_l(p,a) \quad = \quad \sqcup(Pre(l) \cap Suf(pa)).$$

Then $\mathcal{A}_l = (Pre(l), A, \Phi_l, \lambda, \{l\})$ is a minimal $M(A,D)$-automaton recognizing $M(A,D) \cdot l$. For $p,p' \in Pre(l)$ it holds $p \xrightarrow{t}_{\mathcal{A}_l} p'$ iff $p' = \sqcup(Pre(l) \cap Suf(pt))$.

4 An Algorithm for the Suffix Problem

Now we present an algorithm computing the transition function of $\mathcal{A}_l$. On input $p \in \mathrm{Pre}(l)$ and $a \in A$ the algorithm outputs $\Phi_l(p,a)$ in time linear to $|pa| - |\Phi_l(p,a)|$. This time complexity yields linear time complexity when simulating the automaton $\mathcal{A}_l$ on some input $x = a_1 \dots a_n$. Therefore one obtains a linear-time algorithm for the suffix problem in M(*A*,*D*) which is to decide on input of two words $v, x \in A^*$, whether the trace l represented by v is a suffix of the trace t represented by x. Implicitly, the computation of Φ_l was already given in [13, Algorithm 5.2]. We improve the time-complexity of this algorithm from $\mathcal{O}(|A|^3 \cdot |vx|)$ ([13, Theorem 5.1]) to $\mathcal{O}(|A| \cdot |vx|)$ in Theorem 21.

A state $p \in \mathrm{Pre}(l)$ will be represented by its tuple-representative. We give some preliminary results on this representation. Let $\mathcal{C}$ be a covering of (*A*,*D*) (not necessarily a clique-covering). Extending terminology of [4] a tuple $(u_C)_{C \in \mathcal{C}}$ is said to be quasi-reconstructible, when $|u_C|_a = |u_{C'}|_a$ for any $C, C' \in \mathcal{C}$, $a \in C \cap C'$. Proposition 9 is a slight generalization of [10, Proposition 1.6 (ii)].

Proposition 9. *Let $\mathcal{C}$ be a covering of (A,D), $t \in M(A,D)$. A tuple $(u_C)_{C \in \mathcal{C}} \in \prod_{C \in \mathcal{C}} M(C)$ represents a prefix (suffix) of t if, and only if,*
(i) it is quasi-reconstructible and
(ii) u_C is a prefix (resp. suffix) of $\pi_C(t)$ for all $C \in \mathcal{C}$.

As a direct consequence one obtains a useful tool to prove a prefix (resp. suffix) relation between two traces.

Corollary 10. *Let $\mathcal{C}$ be a covering of (A,D), $p,t \in M(A,D)$. Then p is a prefix (suffix) of t if and only if $\pi_C(p)$ is a prefix (resp. suffix) of $\pi_C(t)$ for all $C \in \mathcal{C}$.*

In the following $\mathcal{C}$ denotes a clique-covering of (*A*,*D*) (we consider first a trivial one which can be computed in time $\mathcal{O}(|D|)$). A state $p \in \mathrm{Pre}(l)$ is represented by its tuple-representative $(\pi_C(p))_{C \in \mathcal{C}}$ (to be more precise, a prefix of the word $\pi_C(l)$ is represented by its length). To obtain this representation we need to obtain $(\pi_C(l))_{C \in \mathcal{C}}$ in a first phase. Also the failure functions $\phi_{\pi_C(l)}$ for $C \in \mathcal{C}$ have to be calculated. These initializations can be done in time linear to $\sum_{C \in \mathcal{C}} |\pi_C(l)| \leq |A| \cdot |l|$ (notice, in a reasonable clique-covering a letter appears in at most $|A|$ cliques). The structure $(p_C)_{C \in \mathcal{C}}$ with $p_C \in \mathrm{Pre}(\pi_C(l))$ will be denoted $\mathcal{C}$-tuple. While computing Φ_l the $\mathcal{C}$-tuple p may not be reconstructible, but this will hold before and after any call to Φ_l. So we will in most cases identify a $\mathcal{C}$-tuple with the represented prefix. We give the algorithm here:

function $\Phi_l(p : \mathcal{C}\text{-tuple}, a : \text{letter}) : \mathcal{C}\text{-tuple}$ $\quad (* \; p \in \text{Pre}(l) \; *)$
 for each $C \in \mathcal{C}$ with $a \in C$ **do** $p_C := \varphi_{\pi_C(l)}(p_C, a)$
 $\mathcal{I} := \{C \in \mathcal{C} \mid \exists C' \in \mathcal{C}, \exists b \in C \cap C' \; |p_C|_b > |p_{C'}|_b\}$
 while $\mathcal{I} \neq \emptyset$ **do**
  for some $C \in \mathcal{I}$ **do** $p_C := \phi_{\pi_C(l)}(p_C)$
  $\mathcal{I} := \{C \in \mathcal{C} \mid \exists C' \in \mathcal{C}, \exists b \in C \cap C' \; |p_C|_b > |p_{C'}|_b\}$
 endwhile
 return p.

Lemma 11 reflects the basic considerations for the correctness of the algorithm.

Lemma 11. *Let $l, t \in M(A,D)$, $p = \sqcup(Pre(l) \cap Suf(t))$, and let $\mathcal{C}$ be a covering of (A,D). For each $C \in \mathcal{C}$ let p_C be a prefix of $\pi_C(l)$ and a suffix of $\pi_C(t)$ such that $\pi_C(p) \leq_o p_C$. Then $\pi_C(p) = p_C$ for all $C \in \mathcal{C}$ if, and only if, the tuple $(p_C)_{C \in \mathcal{C}}$ is quasi-reconstructible. Further, for all $C, C' \in \mathcal{C}$, if $|p_{C'}|_b < |p_C|_b$ for some $b \in C \cap C'$, then $\pi_C(p) <_o p_C$ (i.e., $\pi_C(p) \leq_o p_C$ and $p_C \neq \pi_C(p)$).*

Let $p' = \sqcup(\text{Pre}(l) \cap \text{Suf}(pa))$, the value which has to be computed. For the while-loop we obtain the invariant: $\pi_C(p') \leq_o p_C$ for all $C \in \mathcal{C}$. This implies the correctness of the algorithm. During the computation of $\Phi_l(p, a)$ all failure functions are used at most $\sum_{C \in \mathcal{C}} |\pi_C(pa)| - |\pi_C(p')| \leq |A| \cdot (|pa| - |p'|)$ times altogether, because each application of some failure function $\phi_{\pi_C(l)}$ reduces the value of $|p_C|$. This yields a time bound for the algorithm. (Further considerations allow us to maintain the set $\mathcal{I}$ within the given bound).

Theorem 12. *After preprocessing (A,D) in time $\mathcal{O}(|D|)$, and a word $v \in A^*$ representing the trace $l \in M(A,D)$ in time $\mathcal{O}(|A| \cdot |l|)$, the computation of Φ_l needs, on input of $p \in Pre(l)$, and $a \in A$, at most $c \cdot (|pa| - |\Phi_l(p, a)| + 1)$ time for some $c \in \mathcal{O}(|A|)$.*

Now it is easy to deduce an algorithm for the suffix problem. One just has to simulate the automaton $\mathcal{A}_l$, where l is determined by the first input word v, on the second word x (cf. Algorithm 5.2 in [13]): In a first phase the structures depending from v and (A,D) have to be computed (see above). Let $x = a_1 \ldots a_n$ with $a_i \in A$ for $1 \leq i \leq n$. Set $p_0 = \lambda$, and compute successively the values $p_i = \Phi_l(p_{i-1}, a_i)$ for $i = 1$ to n. Test finally whether $p_n = l$.

The time used for the second phase is bounded above by $\sum_{i=0}^{n-1} c \cdot (|p_i a_{i+1}| - |p_{i+1}| + 1)$ for some $c \in \mathcal{O}(|A|)$. This equals $c \cdot (2n - |p_n|)$ which is in $\mathcal{O}(|A| \cdot |x|)$.

In [13, Algorithm 5.2] the loop is only repeated when $p_i \neq l$. Thus the algorithm decides whether l is a suffix of a trace represented by some prefix $a_1 \ldots a_i$ of x. The comparison $p_i = l$ needs only constant time when a set $\{C \in \mathcal{C} \mid p_{C,i} \neq \pi_C(l)\}$ is maintained, which can be done within the given time bound. Thus we obtain the same complexity for this modified algorithm.

Theorem 13. *On input of (A,D), and $v, x \in A^*$ it is decidable in time linear to $|A| \cdot |vx| + |D|$ using space linear to $|A| \cdot |v| + |D|$, whether the trace $l \in M(A,D)$ represented by v is a suffix of the trace represented by x (resp. some prefix of x).*

It is sometimes better to compute a covering of (A,D) by maximal cliques, i.e., cliques which don't remain cliques when including some other letter, which can be done by an efficient greedy algorithm. In this case, after calculating the covering, one even obtains the time-complexity $\mathcal{O}(|vx|)$, when M(A,D) is a free monoid.

In a free monoid an automaton for the suffix-language can easily be transformed to an automaton for the factor-language (just stay in the final state, when it is reached once). However, in a free partially commutative monoid this is not so easy. Assume, for example, a monoid M(A,D) with $a, b, c \in A$ and $(a, b) \in D$, $(b, c) \in I$. Let $l = ac \in$ M(A,D) and $x = abc \in A^*$. Because $x \sim_I acb$, l is a factor of the trace represented by x. However, l is not a suffix of a trace represented by some prefix $\{\lambda,\ a,\ ab,\ abc\}$ of x.

5 Solving the Factor Problem

We first construct (for $l \in$ M(A,D)) a finite M(A,D)-automaton recognizing M(A,D)$\cdot l \cdot$M(A,D) using a known result about the concatenation of recognizable trace-languages. Then we give an algorithm which simulates this automaton in linear time for fixed (A,D).

5.1 Recognizing M(A,D) $\cdot\, l \,\cdot$ M(A,D)

The concatenation of recognizable trace languages is constructively recognizable by Theorem 14 (the construction was given in the proof of [7, Proposition 2.2.1]).

Theorem 14. *For $i \in \{1, 2\}$ let $\mathcal{A}_i = (Q_i, A, \delta_i, q_{0i}, F_i)$ be a finite M(A,D)-automaton recognizing $L_i \in$ M(A,D). Then the trace-language $L_1 \cdot L_2$ is recognized by the nondeterministic finite M(A,D)-automaton*

$$\mathcal{A} = (Q_1 \times \mathcal{P}(A) \times Q_2,\ A,\ \delta,\ (q_{01}, \emptyset, q_{02}),\ F_1 \times \mathcal{P}(A) \times F_2),$$

where $((p, B, q), a, (p', B', q')) \in \delta$ (for $B, B' \subseteq A$) if, and only if,

(i) $p' = p \in Q_1$, $B' = B \cup \{a\}$, $(q, a, q') \in \delta_2$, or

(ii) $q' = q \in Q_2$, $B' = B$, $a \notin D(B)$, $(p, a, p') \in \delta_1$.

The constructed automaton is a product automaton of $\mathcal{A}_1$ and $\mathcal{A}_2$. On each input letter it is nondeterministically chosen, whether $\mathcal{A}_1$ or $\mathcal{A}_2$ consumes it. In the alphabetic component the letters already read by $\mathcal{A}_2$ are remembered. $\mathcal{A}_1$ may only consume letters independent of this set. It holds

$$(p, B, q) \xrightarrow{t}_{\mathcal{A}} (p', B', q')$$

if, and only if, for some $r, s \in$ M(A,D) such that $t = rs$, and $\mathrm{alph}(r) \times B \subseteq I$:

$$p \xrightarrow{r}_{\mathcal{A}_1} p', \quad q \xrightarrow{s}_{\mathcal{A}_2} q', \quad \text{and } B' = B \cup \mathrm{alph}(s).$$

We already know the automaton $\mathcal{A}_l$ recognizing M(A,D) $\cdot\, l$. M(A,D) itself, is recognized by the trivial automaton $(\{q\}, A, \{(q, a, q) \mid a \in A\}, q, \{q\})$. Using the construction in Theorem 14 we obtain the state set $\mathrm{Pre}(l) \times \mathcal{P}(A) \times \{q\}$. The third component can be omitted, as it is unique. The set of final states is then $\{l\} \times \mathcal{P}(A)$. Notice now that a final state (l, B') is reachable from a state $(p, B) \in \mathrm{Pre}(l) \times \mathcal{P}(A)$ by a trace t only if $\pi_{D(B)}(p) = \pi_{D(B)}(l)$. This yields

Theorem 15. *The trace language $M(A,D) \cdot l \cdot M(A,D)$ is recognized by the non-deterministic finite $M(A,D)$-automaton $N_l = (S_l, A, \delta, (\lambda, \emptyset), \{l\} \times \mathcal{P}(A))$, where*

$$S_l = \{(p, B) \in Pre(l) \times \mathcal{P}(A) \mid \pi_{D(B)}(p) = \pi_{D(B)}(l)\},$$

and $((p, B), a, (p', B')) \in \delta$ if, and only if, $(p, B), (p', B') \in S_l$, and either

(i) $p' = p, \quad B' = B \cup \{a\}$, *or*

(ii) $p' = \Phi_l(p, a), B' = B, \ a \notin D(B)$.

In the following we denote by S_l and N_l the notions given above. For a trace t let $S_l(t) = \{q \in S_l \mid (\lambda, \emptyset) \xrightarrow{t}_{N_l} q\}$. From the construction of N_l we obtain

Lemma 16. *Let $l, t \in M(A,D)$. Then $(p, B) \in S_l(t)$ if, and only if, $(p, B) \in S_l$ and for some traces r, s: $t = rs$, $p = \sqcup(Pre(l) \cap Suf(r))$, and $B = alph(s)$.*

5.2 An Algorithm for the Factor Problem

We give some preliminary results.

Lemma 17. *Let $l, t \in M(A,D)$, $p = \sqcup(Pre(l) \cap Suf(t))$, and let $\{H, G\}$ be a covering of (A,D). If $\pi_H(p) = \pi_H(l)$, $\pi_G(p) = \sqcup(Pre(\pi_G(l)) \cap Suf(\pi_G(t)))$.*

Proof. Let $p_G = \sqcup(\mathrm{Pre}(\pi_G(l)) \cap \mathrm{Suf}(\pi_G(t)))$. Observe that $\pi_G(p) \leq_o p_G$ and that $(p_G, \pi_H(p))$ is quasi-reconstructible. By Lemma 11 $\pi_G(p) = p_G$.

The following lemma shows that if $(p, B) \in S_l(t)$ then p is uniquely determined by $D(B)$ and some suffix of $\pi_{D(A-D(B))}(t)$. As a consequence there is at most one p such that $(p, B) \in S_l(t)$ for a given $B \subseteq A$. Thus, if (A,D) is fixed, the set $S_l(t)$ of states reachable in N_l by t has constant size for any $l, t \in M(A,D)$. The proof of the lemma is omitted here. Figure 1 may give some intuition.

Lemma 18. *Let $l, t \in M(A,D)$, $(p, B) \in S_l(t)$, and $\Gamma = D(A - D(B))$. Then $\pi_\Gamma(p) = \sqcup(Pre(\pi_\Gamma(l)) \cap Suf(\pi_\Gamma(t)))$.*

We define a notation for the corresponding suffixes and give some other notations which will be useful for the construction of the algorithm:

Definition 19. For $l, t \in M(A,D)$, $B \subseteq A$, let

$z_B(l, t) = \sqcup(\mathrm{Pre}(\pi_B(l)) \cap \mathrm{Suf}(\pi_B(t)))$,

$\mathcal{B}(l, t) = \{B \subseteq A \mid (p, B) \in S_l(t)\}$,

$\Gamma(B) = D(A - B)$.

Let further, for a family $\mathcal{B} \subseteq \mathcal{P}(A)$, $D(\mathcal{B}) = \{D(B) \mid B \in \mathcal{B}\}$.

By Lemma 18, the tuple $(z_{\Gamma(D(B))}(l, t), \pi_{D(B)}(l))$ is a tuple-representative of p, when $(p, B) \in S_l(t)$. The result is visualized in Figure 1. By Lemma 16 there is a trace s with $ps \leq_s t$ and $\mathrm{alph}(s) = B$. In the picture, the alphabet ranges on the vertical axis, the dotted line is drawn where $D(B)$ and $\Gamma(D(B))$ intersect.

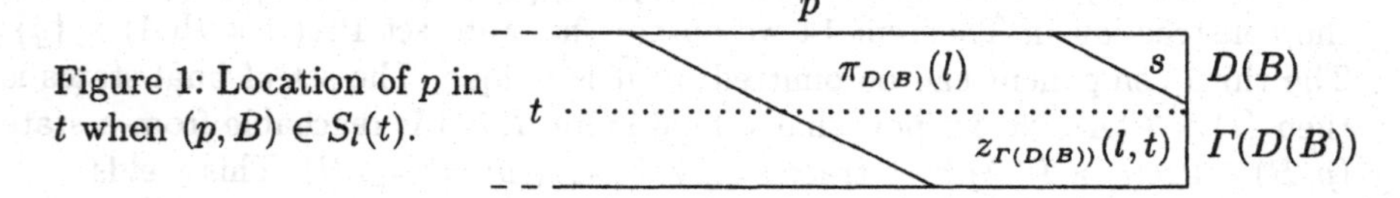

Figure 1: Location of p in t when $(p, B) \in S_l(t)$.

To decide whether a trace t is accepted by N_l, i.e., whether l is a factor of t, it suffices to examine if there is a $B \in \mathcal{B}(l,t)$ such that $z_{\Gamma(D(B))}(l,t) = \pi_{\Gamma(D(B))}(l)$. Lemma 20 shows how the set $D(\mathcal{B}(l,ta))$ can be computed from $D(\mathcal{B}(l,t))$, $a \in A$, and some $z_{\Gamma(D(B))}(l,t)$ for $B \subseteq A$.

Lemma 20. *Let $l,t \in M(A,D)$, $a \in A$, $\Delta' \in \mathcal{P}(A)$. Then $\Delta' \in D(\mathcal{B}(l,ta))$ if, and only if, there is a $\Delta \in D(\mathcal{B}(l,t))$ such that either*
(i) $\Delta' = \Delta \cup D(a)$ and $|z_{\Gamma(\Delta)}(l,t)|_b = |l|_b$ for all $b \in D(a) - \Delta$, or
(ii) $a \notin \Delta' = \Delta$, and the tuple $(\pi_\Delta(l), z_{\Gamma(\Delta)}(l,ta))$ is quasi-reconstructible.

Due to the limited space, the proof is omitted. The two cases of the lemma correspond to those of Theorem 15. The second case is obtained by use of Lemma 11.

Theorem 21. *On input $v,x \in A^*$ it is decidable in time linear to $|vx|$ using space linear in $|v|$, whether the trace $l \in M(A,D)$ represented by v is a factor of the trace $t \in M(A,D)$ represented by x.*

Proof. First preprocess v like in the proof of Theorem 12. Let $x = a_1 \ldots a_n$ with $a_i \in A$ for $1 \le i \le n$. Let $\mathcal{B}_0 = \{\emptyset\}$ $(= D(\mathcal{B}(l,\lambda)))$. Now proceed in n stages: for $1 \le i \le n$ let $\mathcal{B}_i$ be the union of the two sets

(i) $\{\Delta \cup D(a_i) \mid \Delta \in \mathcal{B}_{i-1}, \text{ and } |z_{\Gamma(\Delta),i-1}|_b = |l|_b \text{ for all } b \in D(a_i) - \Delta\}$
(ii) $\{\Delta \in \mathcal{B}_{i-1} \mid a_i \notin \Delta, \text{ and } (\pi_\Delta(l), z_{\Gamma(\Delta),i}) \text{ is quasi-reconstructible}\}$,

where the values $z_{\Gamma(\Delta),i}$ are for $\Delta \in D(\mathcal{P}(A))$ obtained by $z_{\Gamma(\Delta),0} = \lambda$, and

$$z_{\Gamma(\Delta),i} = \begin{cases} \Phi_{\pi_{\Gamma(\Delta)}(l)}(z_{\Gamma(\Delta),i-1}, a_i) & \text{if } a_i \in \Gamma(\Delta), \\ z_{\Gamma(\Delta),i-1} & \text{else.} \end{cases}$$

Finally test, whether there exists a $\Delta \in \mathcal{B}_n$ such that $z_{\Gamma(\Delta),n} = \pi_{\Gamma(\Delta)}(l)$.

Using Theorem 8 one deduces $z_{\Gamma(\Delta),i} = z_{\Gamma(\Delta)}(l, a_1 \ldots a_i)$. The correctness of the algorithm is thus due to Lemma 20 which implies by induction on i that $\mathcal{B}_i = D(\mathcal{B}(l, a_1 \ldots a_i))$ for $0 \le i \le n$.

Examine now complexity. Notice that a set $\mathcal{B}_i \subseteq D(\mathcal{P}(A))$ has constant size, as (A,D) is constant. Thus $\mathcal{B}_{i+1}$ can be computed from $\mathcal{B}_i$ in constant time using $z_{\Gamma(\Delta),i}$ and $z_{\Gamma(\Delta),i+1}$ for $\Delta \in \mathcal{B}_i$. The successive computation of $z_{\Gamma(\Delta),i}$, when done for $1 \le i \le n$, takes time linear to $|n|$ for each $\Delta \in D(\mathcal{P}(A))$, as can be seen by the considerations in the proof of Theorem 13. Together with the preprocessing of v this yields the time-complexity $\mathcal{O}(|vx|)$. For space-complexity notice that between stage i and $i+1$ only the values of $\mathcal{B}_i$, and $z_{\Gamma(\Delta),i}$ for $\Delta \in D(\mathcal{P}(A))$ have to be remembered. This needs additional constant space.

The set $D(\mathcal{B}(l,t))$ may equal $D(\mathcal{P}(A))$, so that the time-complexity of the given algorithm is, in general, exponential in $|A|$ when (A,D) is considered as a part of the input. However, if we have an upper bound k for the shortest path between any two vertices in (A,D) (consider only connected (A,D)), the algorithm remains efficient. Notice, in this case $|D(\mathcal{P}(A))| \le 2 + \sum_{i=1}^{k-1} \binom{|A|}{i} \le 2 + |A|^{k-1}$ which yields the time bound $\mathcal{O}(|A|^k \cdot |vx|)$ when using a trivial clique-covering. In the free monoid, for example, we have $D(\mathcal{P}(A)) = \{\emptyset, A\}$ and the time bound $\mathcal{O}(|A| \cdot |vx|)$.

If one adjusts [13, Algorithm 6.2] to our framework, one could roughly say that in that algorithm, between stage i and stage $i+1$, there is only one $B \subseteq A$ remembered (which should be maximal in $\mathcal{B}(l, a_1 \dots a_i)$). However, we are able to show that this information does not suffice to determine the next state correctly: Let $A = \{a, b, c, d, e\}$, and let the dependence relation D be the reflexive and symmetric closure of $\{(d,a),(a,b),(b,c),(c,e),(e,b)\}$. Let $l = adce$, $t = acebcec^{n-1}$, and $t' = acebec^n$ for some $n \geq 2$. Then $\mathcal{B}(l,t) = \{\emptyset, \{c\}, \{b,c,e\}\}$, and $\mathcal{B}(l,t') = \{\emptyset, \{b,c,e\}\}$ thus in both sets $B = \{b,c,e\}$ is maximal (notice also $z_B(l,t) = z_B(l,t')$ for $B \subseteq A$ which is due to the fact that $z_C(l,t) = z_C(l,t')$ for all trivial cliques $C \in D$). But we have $\mathcal{B}(l,ta) = \{\emptyset, \{c\}\}$, and $\{c\} \notin \mathcal{B}(l,t'a) = \{\emptyset\}$.

6 Extensible Trace Pairs

Extensible pairs were introduced in [13] to investigate the factor problem.

Definition 22. Let $l, t \in$ M(A,D). An extensible trace pair (short, extensible pair) of (l,t) is a pair $(p,s) \in \mathrm{Pre}(l) \times \mathrm{Suf}(t)$ with $ps \leq_s t$, and $\mathrm{alph}(p^{-1}l) \times \mathrm{alph}(s) \subseteq I$, where $p^{-1}l$ denotes the unique suffix of l with $p(p^{-1}l) = l$.

The extensible pairs of $(l,t) \in$ M(A,D) $\times$ M(A,D) form a lattice:

Theorem 23. *Let (p_1,s_1) and (p_2,s_2) be both extensible pairs of (l,t). Then $(p_1 \sqcup_p p_2, s_1 \sqcup_s s_2)$ and $(p_1 \sqcap_p p_2, s_1 \sqcap_s s_2)$ are extensible pairs of (l,t), too. Further, $p_1 s_1 \sqcup_s p_2 s_2 = (p_1 \sqcup_p p_2)(s_1 \sqcup_s s_2)$ and $p_1 s_1 \sqcap_s p_2 s_2 = (p_1 \sqcap_p p_2)(s_1 \sqcap_s s_2)$.*

Proof (sketch). Let $p = p_1 \sqcup_p p_2$ and $s = s_1 \sqcup_s s_2$. It is not hard to see that $\mathrm{alph}(p^{-1}l) \times \mathrm{alph}(s) \subseteq I$. Let $\mathcal{C}$ be a clique-covering of (A,D). Now show for each $C \in \mathcal{C}$ that the words $\pi_C(ps)$ and $\pi_C(p_1 s_1 \sqcup_s p_2 s_2)$ are equal which implies that $ps = p_1 s_1 \sqcup_s p_2 s_2$. Thus ps is a suffix of t. Similarly a proof for the infimum is obtained.

Theorem 23 was inspired by [13, Theorem 4.1 (1)] which erroneously states that for two extensible pairs (p_1,s_1), (p_2,s_2) of (l,t) there is an extensible pair (p,s) with $s_1, s_2 \leq_s s$ and $p_1, p_2 \leq_o p$. This fails in the free monoid $\{a,b\}^*$ with $t = aba$, $l = ab = p_1$, $s_1 = a$, $p_2 = a$, and $s_2 = \lambda$.

A pair $(p,B) \in \mathrm{Pre}(l) \times \mathcal{P}(A)$ is called extensible trace-alphabet-pair of (l,t) if (p,s) is an extensible pair of (l,t) for some $s \leq_s t$ with $B = \mathrm{alph}(s)$. It is easy to see how Theorem 23 is transfered to extensible trace-alphabet-pairs to show that they form a sublattice of the direct product of $(\mathrm{Pre}(l), \leq_p)$ and $(\mathcal{P}(A), \subseteq)$. Lemma 24 gives the relationship between extensible trace-alphabet-pairs of (l,t) and the states reachable in the automaton N_l by t. The proof is omitted.

Lemma 24. *Let $l, t \in M(A,D)$. Then (p,B) is an extensible trace-alphabet-pair of (l,t) if, and only if, $(p,B) \in S_l$ and, for a trace q, $(q,B) \in S_l(t)$ and $p \leq_o q$.*

We deduce that if (p_1,B_1), and (p_2,B_2) are both elements of $S_l(t)$ then $(p, B_1 \cup B_2)$, and $(q, B_1 \cap B_2)$ are in $S_l(t)$, too, for some prefixes p and q of l (in fact, one can show $q = p_1 \sqcap_p p_2$, but it is not clear whether $p = p_1 \sqcup_p p_2$). We obtain

Theorem 25. *Let $l, t \in M(A,D)$. Then $(\mathcal{B}(l,t), \cap, \cup)$ is a distributive lattice.*

Because $\mathcal{B}(l,t) \subseteq \mathcal{P}(A)$, this implies that the set $\mathcal{B}(l,t)$ can be fully represented by not more than $|A|$ of its elements (see [2]). Thus, in the algorithm of Theorem 21, between stage i and stage $i+1$ it suffices to remember only those elements of $\mathcal{B}(l, a_1 \dots a_i)$ (and some suffixes $z_B(l,t)$ for some $B \subseteq A$). However, it is not clear yet, how the elements representing $\mathcal{B}(l, a_1 \dots a_{i+1})$ can be computed efficiently from this information.

7 Acknowledgments

The author wishes to thank V. Diekert, A. Muscholl, and J. Torán for their advice when preparing the paper, and K. Hashiguchi for a discussion about the problem.

References

1. Aho, A. V., Hopcroft, J. E., and Ullman, J. D. *The Design and Analysis of Computer Algorithms*. Addison Wesley, 1974.
2. Birkhof, G. *Lattice Theory*. Amer. Math. Soc., Providence, RI, 1940.
3. Cartier, P., and Foata, D. *Problèmes combinatoires de commutation et réarrangements*. Lect. Notes in Math. 85. Springer, 1969.
4. Cori, R., and Métivier, Y. Recognizable subsets of some partially abelian monoids. *Theoretical Computer Science*, 35:179–189, 1985.
5. Cori, R. , Métivier, Y., and Zielonka, W. Asynchronous mappings and asynchronous cellular automata. *Information and Computation*, 106:159–202, 1993.
6. Cori, R., and Perrin, D. Automates et commutations partielles. *RAIRO Informatique Théorique et Applications*, 19:21–32, 1985.
7. Diekert, V. *Combinatorics on Traces*. LNCS 454. Springer, 1990.
8. Diekert, V., and Métivier, Y. Partial commutation and traces. In Rozenberg, G., and Salomaa, A., *Handbook on Formal Languages*, volume III. Springer, 1996.
9. Diekert, V., and Rozenberg G., editors. *The Book of Traces*. World Scientific, Singapore, 1995.
10. Duboc, C. On some equations in free partially commutative monoids. *Theoretical Computer Science*, 46:159–174, 1986.
11. Eilenberg, S. *Automata, Languages, and Machines*, volume A. Academic Press, New York and London, 1974.
12. Gastin, P., and Rozoy, B. The poset of infinitary traces. *Theoretical Computer Science*, 120:101–121, 1993.
13. Hashiguchi, K., and Yamada, K. String matching problems over free partially commutative monoids. *Information and Computation*, 101:131–149, 1992.
14. Knuth, D. E, Morris, J. H., and Pratt, V. R. Fast pattern matching in strings. *SIAM Journal on Computing*, 6:323–350, 1977.
15. Liu, H.-N., Wrathall, C., and Zeger, K. Efficient solution of some problems in free partially commutative monoids. *Information and Computation*, 89:180–198, 1990.
16. Mazurkiewicz, A. Concurrent program schemes and their interpretations. DAIMI Rep. PB 78, Aarhus University, 1977.
17. Otto, F., and Wrathall, C. Overlaps in free partially commutative monoids. *Journal of Computer and System Sciences*, 42:186–198, 1991.
18. Reischuk, K. R. *Einführung in die Komplexitätstheorie*. Teubner, Stuttgart, 1990.

Removing ε-Transitions in Timed Automata

Volker Diekert[1], Paul Gastin[2], and Antoine Petit[3]

[1] Inst. für Informatik, Universität Stuttgart, Breitwiesenstr. 20-22, D-70565 Stuttgart
[2] LITP, Université Paris 7, 2, place Jussieu, F-75251 Paris Cedex 05
[3] LSV, ENS de Cachan, 61, av. du Prés. Wilson, F-94235 Cachan Cedex

Abstract. Timed automata are among the most widely studied models for real-time systems. Silent transitions, i.e., ε-transitions, have already been proposed in the original paper on timed automata by Alur and Dill [2]. In [7] it is shown that ε-transitions can be removed, if they do not reset clocks; moreover ε-transitions strictly increase the power of timed automata, if there is a self-loop containing ε-transitions which reset some clocks. The authors of [7] left open the problem about the power of the ε-transitions which reset clocks, if they do not lie on any cycle.
The present paper settles this open question. Precisely, we prove that a timed automaton such that no ε-transition with nonempty reset set lies on any directed cycle can be effectively transformed into a timed automaton without ε-transitions. Interestingly, this main result holds under the assumption of non-Zenoness and it is false otherwise.
Besides, we develop a promising new technique based on a notion of *precise time* which allows to show that some timed languages are not recognizable by any ε-free timed automaton.

1 Introduction

A number of "real-life systems" demand time requirements which cannot easily be treated with the classical models based on transition systems. Therefore, new timed models have been introduced for the specification and verification of systems with quantitative properties. A natural way to define such a model is to consider some usual untimed model and to add a suitable notion of time.
We focus in this paper on the basic and natural model of so-called timed automata, proposed by Alur and Dill [2,3]. Since its introduction, this model has been intensively studied under several aspects: determinization [4], minimization [1], power of clocks [9], extensions of the model [5,6] and logical characterization [11] have been considered in particular. Moreover, this model has been used for verification and specification of real-time systems successfully, [8,10,12].
In the original paper [2] silent or internal actions (ε-transitions) of timed automata have been considered, but, somewhat surprisingly, they disappeared in most of the following papers on timed automata (even in the extended version [3]), until the recent work of [7]. It is shown there that ε-transitions strictly increase the power of timed automata, only if these ε-transitions are allowed to reset clocks (called ε-reset transitions in the following). The emptiness problem

Reischuk, Morvan (Eds.): STACS'97 Proceedings, LNCS 1200

of the class of timed automata with ε-transitions is still decidable, and its language class is more robust (e.g. closed under projection) than the class where ε-transitions are forbidden. Thus, the natural question to characterize the "useful" ε-transitions arises.
In [7], it is left as an open question to find when ε-reset transitions can be removed. The present paper settles this problem. Precisely, we prove that a timed automaton with ε-transitions can be effectively transformed into a timed automaton without ε-transitions, if no ε-reset transition lies on any directed cycle of the automaton. Moreover and surprisingly, this result holds only under the assumption of non-Zenoness, otherwise it becomes false. Our main construction is quite involved and leads to some huge state explosion. However, this is just a serious argument in favor of ε-transitions. We may use them in order to have a compact and concise specification for languages recognized by automata without ε-reset transitions, although no quantitative assertion about this statement can be given at this moment.
The example of [7], showing that ε-reset transitions increase the power, was based on the idea to have a self-loop of some ε-reset transition and the proof considered a path which uses several consecutive ε-reset transitions. We exhibit here a very simple timed automaton with some cycle containing an ε-reset transition and in which no path uses two consecutive ε-transitions and whose language cannot be accepted without ε-transition. To this purpose, we develop a new technique based on a notion of *precise time*, which appears to be very promising in its own right. This new notion yields a formal tool in order to show that some timed languages are not recognizable by any timed automaton (without ε-reset transitions).

2 Preliminaries

A *timed automaton* (over $\mathbb{R}$) is a tuple $\mathcal{A} = (Q, \Sigma, \delta, Q_0, F, R, X, C)$, where

Q is a finite set of states,
Σ is a finite alphabet and ε denotes the empty word of Σ^*,
δ is the transition relation explained below,
$Q_0 \subseteq Q$ is a subset of initial states,
$F \subseteq Q$ is a subset of final states,
$R \subseteq Q$ is a subset of repeated states,
X is a finite set of clocks, and
$C \subseteq \mathbb{R}$ is a finite set of constants.

A *constraint* is a propositional formula using the logical connectives $\{\vee, \wedge, \neg\}$ over atomic formulae of the form $x \# c$ or $x - y \# c$, for $x, y \in X$, $c \in C$, and $\# \in \{<, =, >\}$.
A transition of δ has the form $p \xrightarrow{A,a,\alpha} q$ where A is a constraint, $a \in \Sigma \cup \{\varepsilon\}$, $\alpha \subseteq X$, and $p, q \in Q$. If $a = \varepsilon$ and $\alpha = \emptyset$, then it is called an ε-transition without reset. If $a = \varepsilon$ and $\alpha \neq \emptyset$, then it is called an ε-reset transition. For the global time and the time values of clocks we shall use non-negative real numbers. For

a clock $x \in X$ and a time $t \in \mathbb{R}_+$, we denote by $x(t) \in \mathbb{R}_+$ the clock value of x. Initially, we are in some state $q_0 \in Q_0$ and the (global) time is $t_0 = 0$ with $x(0) = 0$ for all $x \in X$.

In the course of time all clocks run synchronously. However, executing a transition may reset some clocks to zero. Formally, assuming that the automaton has entered state p at time t with clock values $x(t)$, $x \in X$, then it may execute the transition $p \xrightarrow{A,a,\alpha} q$ at time $t' \geq t$, if the constraint A is satisfied with the clock value $x(t') = x(t) + (t' - t)$ for all clocks x. The execution switches to state q and enters this state at time t' with clock value $x(t') = x(t) + (t' - t)$ for all $x \in X \setminus \alpha$ and resets clock values $x(t') = 0$ for all $x \in \alpha$.

This leads to the notion of a run of the automaton:

$$p_0 \xrightarrow[t_1]{A_1, a_1, \alpha_1} p_1 \xrightarrow[t_2]{A_2, a_2, \alpha_2} p_2 \longrightarrow \cdots$$

The semantics is that $p_{i-1} \xrightarrow{A_i, a_i, \alpha_i} p_i$ has been executed at time t_i with $t_{i-1} \leq t_i$ for all $i \geq 1$ and $t_0 = 0$. In particular, for $i \geq 1$, the constraint A_i has been satisfied for the clock values $x(t_i)$ at time t_i of the execution of the i-th transition (before performing its reset operation).

The finite (infinite resp.) run is *accepted,* if both, $p_0 \in Q_0$ and it ends in a final state (and there are infinitely many states from R on the path resp.). Thus, we accept infinite runs by some Büchi condition. With every finite (infinite resp.) run we can associate in a natural way a finite (infinite resp.) timed ε-word

$$(a_1, t_1)(a_2, t_2) \cdots \in ((\Sigma \cup \{\varepsilon\}) \times \mathbb{R})^\infty.$$

Since an ε-transition is viewed as an invisible action we may cancel all pairs (a_i, t_i) where $a_i = \varepsilon$. In this way we obtain a timed word (which might be finite even if the underlying timed ε-word has been infinite):

$$(a_{i_1}, t_{i_1})(a_{i_2}, t_{i_2}) \cdots \in (\Sigma \times \mathbb{R})^\infty.$$

The timed language $L(\mathcal{A}) \subseteq (\Sigma \times \mathbb{R})^\infty$ accepted by the automaton $\mathcal{A}$ is the set of timed words associated with accepting runs.

A timed (ε-)word $(a_1, t_1)(a_2, t_2) \cdots$ is called a *Zeno word*, if it is infinite but the sequence $t_1, t_2, \ldots$ remains bounded. Let NZ be the set of non Zeno words. For some applications Zeno words are not wanted. It is possible to transform a timed automaton $\mathcal{A}$ into a timed automaton $\mathcal{A}'$ such that $L(\mathcal{A}') = L(\mathcal{A}) \cap NZ$. Note that this transformation is not as easy as it may appear because a finite timed word, which is indeed non Zeno, can be accepted by a run ending in a loop of ε-transitions where the underlying timed ε-word is Zeno.

Example 1. Consider the following automaton $\mathcal{A}_0$ where q_0 is both initial and repeated.

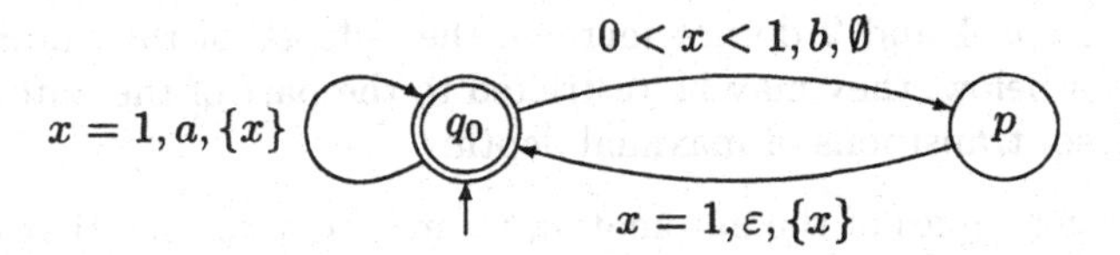

The accepted language can be described as follows: In each open time interval $(i, i+1)$, $i \geq 0$, there occurs at most one b. Moreover, there is an a at time $i+1$ if and only if there is no b in $(i, i+1)$. For the automaton $\mathcal{A}_0$, by construction, no infinite run yields any Zeno word.

In [7] it is shown that all ε-transitions without reset can be removed from the automaton. The technique is to shift the constraint of an ε-transition either to the previous or to the following (visible) transition. A priori, one could expect a similar technique to work in the example above. However we will see below in Corollary 9 that $L(\mathcal{A}_0)$ cannot be accepted by a timed automaton without ε-transition.

3 The main result

The main result of this paper is a construction how to remove all ε-transitions, if there is no directed cycle of the automaton including an ε-transition with reset.

Theorem 2. *Let $\mathcal{A}$ be a timed automaton such that no ε-reset transition lies on any directed cycle. Then we can effectively construct a timed automaton $\mathcal{A}'$ without any ε-transition such that*

$$L(\mathcal{A}) \subseteq L(\mathcal{A}')$$
$$L(\mathcal{A}) \cap NZ = L(\mathcal{A}') \cap NZ$$

The ε-*depth* of a timed automaton is defined as the maximal number of ε-reset transitions which can be found on some directed path through the automaton. An ε-reset transition is of ***maximal depth***, if it is the last one on such a directed path.
The proof of Theorem 2 will be done by induction on the ε-depth of the automaton. Our strategy is as follows. First (Steps 1 to 5), we transform $\mathcal{A}$ into some normal form without increasing the ε-depth. Then we explain how to remove all ε-reset transitions of maximal depth (this is the crucial part). We end up with an automaton $\mathcal{A}'$ (being not in normal form anymore) where the ε-depth is decreased by one. The result follows by induction.

Step 1: Remove all ε-transitions without reset by the procedure of [7]. Note that if the ε-depth of $\mathcal{A}$ is zero then the proof is done.

Step 2: Remove all constraints of the form $x - y \# c$ where x, y are clocks and c is some constant. Note that Step 1 introduces such constraints. By duplicating some transitions, we may also assume that all constraints are conjunction of atomic formulae of type $y \# c$ with $\# \in \{<, =, >\}$.

Note that Steps 1 and 2 do not increase the ε-depth of the automaton. Also, for the proof below, they may be restricted to the part of the automaton which follows ε-reset transitions of maximal depth.

Step 3: Using copies of the automaton, we may now assume that every ε-reset transition of maximal depth $p_1 \xrightarrow{A,\varepsilon,\alpha} p_2$ divides the automaton into two disjoint parts and the only bridge between these parts is this ε-reset transition (See the figure below). Moreover, we may assume that p_2 has no other in-going transition. Note that $\mathcal{A}_2$ contains no ε-transition and we may assume that it contains no initial state either.

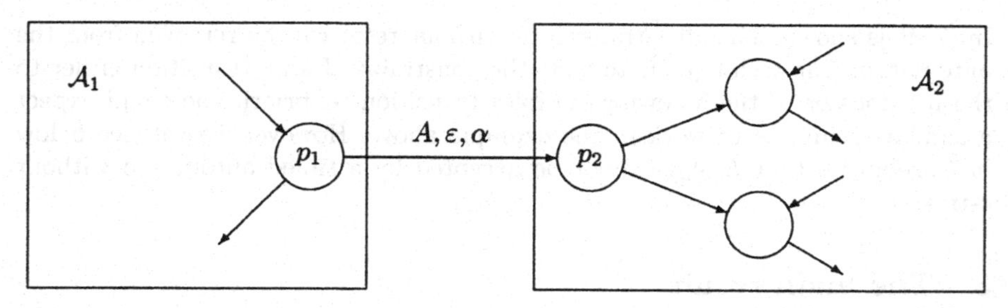

Step 4: Reduction of α to a single clock x which is never reset in $\mathcal{A}_2$. Again, this is an easy construction on timed automata which does not change $\mathcal{A}_1$ and hence does not change the ε-depth.

Step 5: Using copies of the clocks used in the constraint A which are reset with their originals inside $\mathcal{A}_1$ and are substituted to their originals in A, we may assume that the clocks used in A are different from x and are not reset in $\mathcal{A}_2$.

Important note: In order to make the following construction more readable we shall assume that the clock constraints of A and inside the subautomaton $\mathcal{A}_2$ are of the form $y < c$ or $y > c$, only. Thus, we do not consider the case $y = c$. The inclusion of such constraints would multiply the case distinction without giving any new insight. In fact, using a constraint $y = c$ would make life even easier, since then we know the exact value of clock y. In the same spirit we consider (and allow) ***strictly increasing*** time sequences, only.

Step 6: This is the main step of the construction. We will replace in $\mathcal{A}$ the automaton $\mathcal{A}_2$ by a new one $\mathcal{A}'_2$ which will not use the clock x and such that there will be a correspondence between the legal paths of the new automaton $\mathcal{A}'$ and the legal ones of $\mathcal{A}$. Then we will replace the ε-reset transition $p_1 \xrightarrow{A,\varepsilon,\alpha} p_2$ by the ε-transition without reset $p_1 \xrightarrow{A,\varepsilon,\emptyset} (p_2, L_0, U_0)$.

• The states of $\mathcal{A}'_2$ will be triples (p, L, U) where p is a state of $\mathcal{A}_2$ and L, U are its lower and upper attributes. Intuitively, these attributes will keep track of the possible interval for the clock x in the corresponding run of $\mathcal{A}$. More precisely, when we reach a state p of $\mathcal{A}_2$ with a legal run of $\mathcal{A}$, then the corresponding run of $\mathcal{A}'$ will reach the state (p, L, U) with the value $x(t)$ in the open interval $(L(t), U(t))$. Conversely, for each legal run of $\mathcal{A}'$ leading to the state (p, L, U)

and for each value in $(L(t), U(t))$, there exists a corresponding run in $\mathcal{A}$ leading to state p with this value for $x(t)$.
We will use two new clocks x_ℓ and x_u. The second one is reset on each transition which enters state p_1. Since we only consider *strictly increasing* time sequences, we can assume that the constraint A contains $x_u > 0$ and that each transition from p_2 contains $x > 0$ in its constraint.
We can write $A = \bigwedge_r m_r < x_r < m'_r$ with $m_r \in C \cup \{-\infty\}$ and $m'_r \in C \cup \{+\infty\}$. Let $L_0 = \max_r(x_r - m'_r)$ and $U_0 = \min_r(x_r - m_r)$ be the initial attributes given to the state p_2. Apart from the initial values L_0 and U_0, the possible values for L and U are $\{x_\ell + c \mid c \in C\}$ and $\{x_u + c \mid c \in C\}$ respectively. Hence, there are finitely many possible values and the automaton $\mathcal{A}'_2$ remains finite.
Moreover, a state (p, L, U) of $\mathcal{A}'_2$ is final (repeated resp.) if and only if the state p of $\mathcal{A}_2$ is final (repeated resp.).

• For each transition $p \xrightarrow{(a<x<b)\wedge B,\sigma,\beta} q$ of $\mathcal{A}_2$ where B does not contain the clock x and $a \in C \cup \{-\infty\}$ and $b \in C \cup \{+\infty\}$, we add the following transitions to the automaton $\mathcal{A}'_2$.

$$
\begin{aligned}
(p, L, U) &\xrightarrow{(a\le L<U\le b)\wedge B,\sigma,\beta} (p, L, U) \\
&\xrightarrow{(a\le L<b<U)\wedge B,\sigma,\beta\cup\{x_u\}} (p, L, x_u + b) \\
&\xrightarrow{(L<a<U\le b)\wedge B,\sigma,\beta\cup\{x_\ell\}} (p, x_\ell + a, U) \\
&\xrightarrow{(L<a<b<U)\wedge B,\sigma,\beta\cup\{x_\ell,x_u\}} (p, x_\ell + a, x_u + b)
\end{aligned}
\qquad (1)
$$

where $L_0 < a$ is an abbreviation for $\bigwedge_r (x_r - m'_r) < a$ and similarly for $a \le L_0$, $b < U_0$ and $U_0 \le b$.

Claim: All timed words (Zeno or non-Zeno) accepted by $\mathcal{A}$ are also accepted by $\mathcal{A}'$. Conversely, all non-Zeno timed words accepted by $\mathcal{A}'$ are also accepted by $\mathcal{A}$.

Note that, thanks to Step 3, we can do Steps 4 to 6 simultaneously on all ε-reset transitions of maximal depth. Hence we have reduced the ε-depth by one and Theorem 2 follows by induction. We will now prove the claim.

Proof. Let $\pi = q_0 \xrightarrow{A_1,\sigma_1,\alpha_1} q_1 \xrightarrow{A_2,\sigma_2,\alpha_2} q_2 \cdots$ be a path of $\mathcal{A}$ and let $t_1 t_2 \cdots$ be a timed sequence such that the timed ε-word $w = (\sigma_1, t_1)(\sigma_2, t_2)\cdots$ is accepted by π. Assume that the i-th transition of π is $p_1 \xrightarrow{A,\varepsilon,\{x\}} p_2$, hence the path from q_0 to $q_{i-1} = p_1$ runs in $\mathcal{A}_1$ while the path from $q_i = p_2$ runs in $\mathcal{A}_2$.
We will construct a path π' of $\mathcal{A}'$ which accepts precisely the same timed ε-word w. The path π' starts as the path π and its i-th transition enters the "initial" state of $\mathcal{A}'_2$:

$$q_0 \xrightarrow{A_1,\sigma_1,\alpha_1} q_1 \cdots \xrightarrow{A_{i-1},\sigma_{i-1},\alpha_{i-1}} q_{i-1} = p_1 \xrightarrow{A,\varepsilon,\emptyset} (p_2, L_0, U_0) = (q_i, L_i, U_i)$$

Since the constraint A has been satisfied at time t_i, we have $L_0(t_i) < 0 < U_0(t_i)$. Hence, we have the time invariant $L_0(t) < x(t) < U_0(t)$ for all $t \ge t_i$.

Assume that the path π' has been constructed up to state $(q_{j-1}, L_{j-1}, U_{j-1})$ and that the time invariant $L_{j-1}(t) < x(t) < U_{j-1}(t)$ for all $t \geq t_{j-1}$ holds. Assume also that the constraint $A_j = (a_j < x < b_j) \wedge B_j$ where B_j does not contain the clock x. Thanks to the time invariant, we can see that among the transitions described in (1), there is exactly one transition $(q_{j-1}, L_{j-1}, U_{j-1}) \xrightarrow{A'_j, \sigma_j, \alpha'_j} (q_j, L_j, U_j)$ whose constraint A'_j is true at time t_j. This transition is used to extend the path π'. One can easily verify that the time invariant $L_j(t) < x(t) < U_j(t)$ for all $t \geq t_j$ holds. We have thus obtain the desired path π' and we have proved the first part of the claim. Note that we do not need to assume that the time sequence $t_1 t_2 \cdots$ diverges for this part of the proof.

Conversely, let π' be a path of $\mathcal{A}'$ whose i-th transition is $p_1 \xrightarrow{A, \varepsilon, \emptyset} (p_2, L_0, U_0)$. Thus the path π' has the form

$$q_0 \xrightarrow{A_1, \sigma_1, \alpha_1} q_1 \cdots q_{i-1} \xrightarrow{A, \varepsilon, \emptyset} (q_i, L_i, U_i) \xrightarrow{A'_{i+1}, \sigma_{i+1}, \alpha'_{i+1}} (q_{i+1}, L_{i+1}, U_{i+1}) \cdots$$

Let $w = (\sigma_1, t_1)(\sigma_2, t_2) \cdots$ be a timed ε-word accepted by π'. For all $j > i$, let $q_{j-1} \xrightarrow{(a_j < x < b_j) \wedge B_j, \sigma_j, \alpha_j} q_j$ be the transition of $\mathcal{A}_2$ from which the corresponding transition of π' was obtained. We have thus constructed a path π of $\mathcal{A}$. The timed ε-word w is not necessarily accepted by the path π but we will see that we can find a time t'_i such that the corresponding timed ε-word w' is accepted by the path π. Since w and w' differs only by the time of the i-th action which is ε, this will prove the second part of the claim.

Let $I = \bigcap_{j \geq i} (t_j - U_j(t_j), t_j - L_j(t_j))$. By construction of $\mathcal{A}'_2$, since the constraint A'_j was satisfied at time t_j, we have $L_j(t_j) < U_j(t_j)$. Hence the open intervals considered in this intersection are nonempty. Moreover, we have $L_j(t_j) = \max(L_{j-1}(t_j), a_j)$. Hence, $L_j(t_j) \geq L_{j-1}(t_j) = L_{j-1}(t_{j-1}) + t_j - t_{j-1}$. Using the same argument for the upper attribute, we deduce that these intervals are decreasing.

We will see now that $I \neq \emptyset$. This is clear if the path π' is finite. Assume now that π is infinite and that the time sequence $t_1 t_2 \cdots$ diverges. Note that this is the only point where we need to use non-Zenoness.

By construction, $x > 0$ is part of the constraint A_{i+1}. Hence, $a_{i+1} \geq 0$ and $L_{i+1}(t_{i+1}) = \max(L_i(t_{i+1}), a_{i+1}) \geq 0$. Moreover, for all $j > i$ we have $L_j(t_j) \geq L_{i+1}(t_{i+1}) + t_j - t_{i+1} \geq t_j - t_{i+1}$. Let j_0 be such that $t_{j_0} - t_{i+1} > \max C$ where we recall that $C \subseteq \mathbb{R}$ is the set of (finite) constants which are used in the constraints of the automaton $\mathcal{A}$. We deduce that $L_j = L_{j_0}$ for all $j \geq j_0$. Now, for all $j \geq j_0$, we have $\max C < L_j(t_j) < U_j(t_j) = \min(U_{j-1}(t_j), b_j)$ and then $b_j = +\infty$. It follows that $U_j = U_{j_0}$ for all $j \geq j_0$. Finally, we have proved that $I = (t_{j_0} - U_{j_0}(t_{j_0}), t_{j_0} - L_{j_0}(t_{j_0}))$ which is non empty.

To complete the proof we will show that for all $t'_i \in I$, the timed ε-word w' obtained from w by replacing time t_i by t'_i is accepted by the path π.

First, $t'_i \in (t_i - U_i(t_i), t_i - L_i(t_i))$ therefore, $L_0(t'_i) = L_0(t_i) + t'_i - t_i < 0 < U_0(t_i) + t'_i - t_i = U_0(t'_i)$ and the constraint A is satisfied at time t'_i. Note that since by construction $x_u > 0$ is part of the constraint A and $x_u \in \alpha_{i-1}$, we deduce that $t_{i-1} < t'_i$.

Second, let $j > i$, we have $t'_i \in (t_j - U_j(t_j), t_j - L_j(t_j))$ therefore, $a_j \leq L_j(t_j) < t_j - t'_i < U_j(t_j) \leq b_j$ and the constraint $(a_j < x < b_j) \wedge B_j$ is satisfied at time t_j. Note that since by construction $x > 0$ is part of the constraint A_{i+1}, we deduce that $t'_i < t_{i+1}$. This concludes the proof of the claim.

We have seen that the non-Zenoness is needed in the proof above. One might wonder whether this hypothesis is really necessary. We will show that indeed, Theorem 2 becomes false if we allow Zeno-words. Consider the following automaton $\mathcal{A}$.

$$\to (p_1) \xrightarrow{true, \varepsilon, \{x\}} (p_2) \xrightarrow{0 < x, a, \{y\}} ((p_3)) \circlearrowleft x < 1 \wedge 0 < y, a, \{y\}$$

One can verify that the language accepted by this automaton is

$$L = \{(a, t_2)(a, t_3) \cdots \mid \exists \gamma > 0 \text{ such that } \forall i \geq 2, t_i < t_{i+1} < t_2 + 1 - \gamma\}$$

We will show that L is not recognizable by any timed automaton without ε-transitions, even if we allow for the automaton any set of constants which is a discrete subspace of $\mathbb{R}$.
Assume by contradiction that L is recognized by some ε-free automaton $\mathcal{A}'$. Let $C \subseteq \mathbb{R}$ be its set of constants assuming that C is a discrete subspace of $\mathbb{R}$ (e.g., C is finite). Let $t_2 > 0$ be an arbitrary positive real and choose $0 < \delta < 1$ such that $(t_2 + 1 - \delta, t_2 + 1) \cap (C \cup (t_2 + C)) = \emptyset$ and $\delta < \min\{c \in C \mid c > 0\}$.
Consider a timed word $w = (a, t_2)(a, t_3)(a, t_4) \cdots$ with $t_2 + 1 - \delta < t_i < t_{i+1}$ for all $i > 2$ and $\lim_{i \to \infty} t_i = t_2 + 1 - \gamma$ for some $0 < \gamma < \delta$.
Then we have $w \in L(\mathcal{A}')$ and there exists some accepting path π for the timed word w. Due to the choice of δ one can verify that the same path accepts also the timed word $w' = (a, t_2)(a, t_3 + \gamma)(a, t_4 + \gamma) \cdots$. However, since $\lim_{i \to \infty} t_i + \gamma = t_2 + 1$, we have $w' \notin L$ and therefore a contradiction.

4 A notion of precise time

Let $\mathcal{A}$ be a timed automaton which uses a set $C \subseteq \mathbb{Q}$ of *rational* constants. Let $C_{\max}$ be the maximum value of C. Let $\delta > 0$ be such that $C \subseteq \delta\mathbb{Z}$. Consider a finite or infinite path

$$\pi = p_0 \xrightarrow{A_1, a_1, \alpha_1} p_1 \xrightarrow{A_2, a_2, \alpha_2} p_2 \cdots$$

through $\mathcal{A}$ such that all constraints are conjunctions of atomic formulae $x \# c$ with $\# \in \{<, =, >\}$ for some clocks x and constants c. Note that one can always transform a timed automaton in such a way that this is true for all transitions of the automaton.
By $TS(\pi)$ we denote the set of possible time sequences $t_0 t_1 t_2 \cdots$ such that $0 = t_0 \leq t_1 \leq t_2 \leq \cdots$ and

$$(p_0) \xrightarrow[t_1]{A_1, a_1, \alpha_1} (p_1) \xrightarrow[t_2]{A_2, a_2, \alpha_2} (p_2) \to \cdots$$

defines a π-run of $\mathcal{A}$. Let us assume that $TS(\pi) \neq \emptyset$. By $TS_n(\pi)$ we denote the set of values $r_n \in \mathbb{R}$ such that there exists $t_0t_1t_2 \cdots t_{n-1}t_n \cdots \in TS(\pi)$ with $r_n = t_n$. Assuming that n is not greater than the length of π, we have $TS_n(\pi) \neq \emptyset$.

Definition 3. A time $t \in \mathbb{R}$ is called a *precise time* of π, if $TS_n(\pi) = \{t\}$ for some $n \geq 1$.

The aim of the present section is to show the following theorem.

Theorem 4. *Let π, δ and $C_{\max}$ be as above. Assume that $TS(\pi) \neq \emptyset$ and let $t \in TS_n(\pi)$. If t is a precise time of π then $t \in \delta\mathbb{N}$ and if $t > 0$ then the half-open interval $[t - C_{\max}, t)$ contains another precise time. Otherwise, $TS_n(\pi)$ is a non-empty open interval (r, s) with $r \in \delta\mathbb{N}$ and $s \in \delta\mathbb{N} \cup \{+\infty\}$.*

The main idea of the proof is to associate with the path π a directed graph G with edge weights such that the interval can be calculated from the graph. We first normalize the automaton in two simple steps as follows:

Step 1: We may assume that there is some clock x_0 such that $x_0 \geq 0$ is a constraint of A_i and $x_0 \in \alpha_i$ for all $i \geq 0$.

Step 2: We may assume that the path π is infinite. (If the path π is finite, enter a new state performing any action with the only constraint $x_0 > 0$. Loop in this state. This modification does not touch $TS_n(\pi)$, since $TS_n(\pi) \neq \emptyset$.)

We are now ready to define the graph G. For convenience we put $\alpha_0 = X$. The vertex set of the graph is $V = \{0, 1, 2, \ldots\} = \mathbb{N}$. There are two types of directed edges. Let $x \in X$ be a clock and $c \in C$ be a constant such that $x \in \alpha_i$ and $(x \# c) \in A_j$ for some $i < j$ and $x \notin (\alpha_{i+1} \cup \ldots \cup \alpha_{j-1})$.
If $(x = c) \in A_j$, then we define *strong arcs* with weights c and $-c$ respectively:

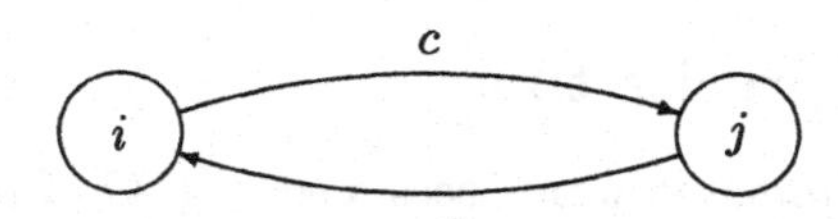

If $(x < c) \in A_j$, then we define a *soft arc* with weight c:

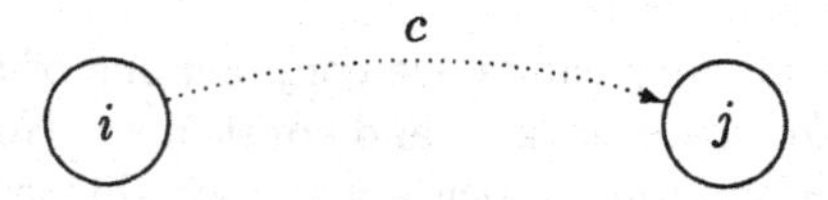

If, finally, $(x > c) \in A_j$ we define a soft arc with weight $-c$:

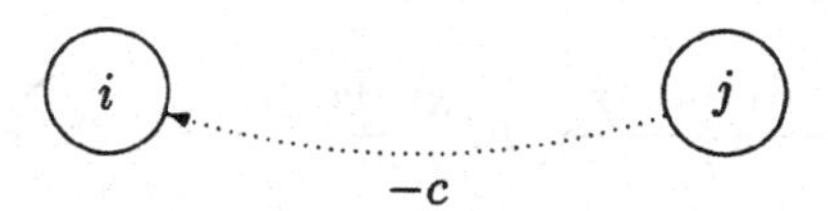

We define $i \sim j$, if i and j are connected by some path using strong edges, only. Clearly $\sim$ is an equivalence relation on V. An induction on the length of the path yields:

Lemma 5. *Let $t_0 t_1 t_2 \cdots \in TS(\pi)$ be the time sequence of some π-run; $i, j \in V$ and γ_{ij} the weight of some directed path of G from i to j.*

1. *If the path uses strong edges only, then we have $t_j - t_i = \gamma_{ij}$.*
2. *If the path uses at least one soft edge, then we have $t_j - t_i < \gamma_{ij}$.*

Let $i, j \in V$, assuming $\inf \emptyset = +\infty$, we define

$$c_{ij} = \inf\{\gamma_{ij} \mid \gamma_{ij} \text{ is the weight of a directed path from } i \text{ to } j\}$$

In order to prove the theorem, we need some few technical results:

Lemma 6. *Let $t_0 t_1 t_2 \cdots \in TS(\pi)$ be the time sequence of some π-run and let $i, j, m \in V$. Then,*

1. *$-c_{ji} \leq t_j - t_i \leq c_{ij}$.*
2. *$c_{ij} \in \delta\mathbb{Z} \cup \{+\infty\}$ and if $c_{ij} \neq +\infty$ then there exists a directed path in G from i to j of weight c_{ij}.*
3. *$i \sim j \iff t_j - t_i = c_{ij} \iff c_{ij} + c_{ji} = 0$.*
4. *$c_{ij} \leq c_{im} + c_{mj}$.*
5. *$c_{ij} = c_{im} + c_{mj}$ if $i \sim m$ or $m \sim j$.*
6. *$c_{ij} < c_{im} + c_{mj}$ if $i \sim j$ and $i \not\sim m$.*

Lemma 7. *We have*

$$TS(\pi) = \{t_0 t_1 t_2 \cdots \mid \forall i, j \in V : t_j - t_i = c_{ij} \text{ if } i \sim j \text{ and } t_j - t_i < c_{ij} \text{ otherwise}\}.$$

We can now state the decisive lemma.

Lemma 8. *We have*

$$TS_n(\pi) = \begin{cases} \{c_{0n}\} & \text{if } n \sim 0 \\ (-c_{n0}, c_{0n}) & \text{otherwise.} \end{cases}$$

Proof. If $n \sim 0$ then the result is a consequence of Lemma 6 (3) since $t_0 = 0$. Assume now that $n \not\sim 0$. For each $k \geq 0$ we will define inductively in stage k a subset $s(k) \subseteq V$ satisfying the following properties:

1. The $s(k)$ is a finite union of the equivalence classes.
2. For all $i \in s(k)$ a time t_i is defined.
3. For all $i, j \in s(k)$ we have

$$t_i - t_j = \begin{cases} c_{ij} & \text{if } i \sim j \\ t_j - t_i < c_{ij} & \text{otherwise.} \end{cases}$$

To begin with let $s(0) = [0] = \{i \in V \mid 0 \sim i\}$ the equivalence class of 0. For $i \in [0]$ define $t_i = c_{0i}$. Clearly 1), 2), and 3) are satisfied. To see 3) observe that for $i \sim j \sim 0$ we have by Lemma 6 $c_{ij} = c_{i0} + c_{0j}$ and $c_{0i} = -c_{i0}$.
Assume that $s(k)$ has been defined, $k \geq 0$. If $s(k) = V$ we stop the procedure. Otherwise, for each $m \in V \setminus s(k)$ define an open interval

$$I_m^{(k)} = (r_m^{(k)}, s_m^{(k)})$$

by $r_m^{(k)} = \max\{t_j - c_{mj} \mid t_j \in s(k)\}$ and $s_m^{(k)} = \min\{t_i + c_{im} \mid t_i \in s(k)\}$.
Note that for $i \sim j \in s(k)$ and $m \in V$, by Lemma 6, we have $t_i - c_{mi} = t_j - c_{mj}$ and $t_i + c_{im} = t_j + c_{jm}$. Hence the maximum and the minimum are taken over finite sets. The interval $I_m^{(k)}$ is not empty since $t_j - c_{mj} < t_i + c_{im}$ for all $i, j \in s(k), m \notin s(k)$: Indeed, for $i \sim j$ we have $t_j - t_i = c_{ij} < c_{im} + c_{mj}$; and for $i \not\sim j$ we have $t_j - t_i < c_{ij} \leq c_{im} + c_{mj}$. Hence the claim in both cases.
For $k = 0$ and $m \not\sim 0$ we have $I_m = (c_{m0}, c_{0m})$. Now consider $k+1$. If $k+1 = 1$ then choose $m = n$. For $k + 1 > 1$ choose any $m \notin s(k)$. Define $s(k+1) = s(k) \cup \{j \in V \mid j \sim m\}$ so that (1) holds for $s(k+1)$. Let t_m be any value of the open interval $I_m^{(k)}$. For all $j \sim m$ define $t_j = t_m + c_{mj}$. Now (2) holds for $s(k+1)$ as well. Finally, (3) is a direct consequence of $t_j \in I_j^{(k)}$ for all $j \sim m$ which can be easily checked using $t_m \in I_m^{(k)}$, the definition of t_j and Lemma 6.
Since $0 \not\sim n$, we can choose a sequence $m_0 = 0, m_1 = n, m_2, m_3, \ldots$ such that $s(k) = [0] \cup [n] \cup [m_2] \cup \cdots \cup [m_k]$ and $\bigcup_{k \geq 0} s(k) = V$. This defines values t_m for all $m \geq 0$ and thereby a sequence $t_0 t_1 t_2 \cdots$.
By property 3) above and Lemma 7 we see that $t_0 t_1 t_2 \cdots \in TS(\pi)$. The result follows since $m_1 = n$ and the only condition on t_n has been $t_n \in I_n^{(0)} = (-c_{n0}, c_{0n})$.

The proof of Theorem 4 is now easy. From Lemma 8 we deduce that the precise times of π are $\{c_{0n} \mid n \sim 0\} \subseteq \delta\mathbb{N}$. Moreover, assume that $t = c_{0n} > 0$ is a precise time. Since $n \sim 0$ there is a path in G composed of strong arcs from n to 0: $n = n_0 \longrightarrow n_1 \longrightarrow \cdots \longrightarrow n_k = 0$. Let $i = \inf\{j \mid c_{0n_j} < c_{0n}\}$. We have $c_{0n} - c_{0n_j} \leq c_{0n_{j-1}} - c_{0n_j} \leq C_{\max}$. Hence the precise time $c_{0n_j} \in [c_{0n} - C_{\max}, c_{0n})$. Now if $n \not\sim 0$, then by Lemma 8 we have $TS_n(\pi) = (-c_{n0}, c_{0n})$ and by Lemma 6 we obtain $-c_{n0} \in \delta\mathbb{N}$ and $c_{0n} \in \delta\mathbb{N} \cup \{+\infty\}$ which concludes the proof.

The theorem above yields a tool to prove that certain languages are not recognizable by timed automata (without ε-reset transitions resp.). The application we have in mind is the following simple consequence.

Corollary 9. *Every timed automaton recognizing the language $L(\mathcal{A}_0)$ from Ex. 1 has an ε-reset transition lying on some directed cycle.*

Proof. By contradiction. By Theorem 2 we may assume $L(\mathcal{A}_0)$ is recognized by some automaton without any ε-transition. Applying Theorem 4 we let $\delta > 0$ and $C_{\max}$ be the constants introduced at the beginning of Section 4. We find some $d \in \mathbb{N}$, $d \geq C_{\max}$ and an accepted word of the form

$$(b, \delta_1)(b, \delta_2) \cdots (b, \delta_{d-1})(a, d)(a, d+1) \cdots$$

such that $\delta_i \in (i-1, i) \setminus \delta\mathbb{N}$ for all $0 < i < d$. Let π be a path accepting this timed word. The time d must be precise contradicting Theorem 4.

Acknowledgment: We thank the anonymous referees for useful remarks which yield to a simplification of our original construction.

References

1. R. Alur, C. Courcoubetis, D.L. Dill, N. Halbwachs, and H. Wong-Toi. Minimization of timed transition systems. In *Proceedings of CONCUR'92*, number 630 in Lecture Notes in Computer Science. Springer Verlag, 1992.
2. R. Alur and D.L. Dill. Automata for modeling real-time systems. In *Proceedings of ICALP'90*, number 443 in Lecture Notes in Computer Science, pages 322–335. Springer Verlag, 1990.
3. R. Alur and D.L. Dill. A theory of timed automata. *Theoretical Computer Science*, 126:183–235, 1994.
4. R. Alur, L. Fix, and T.A. Henzinger. A determinizable class of timed automata. In *Proceedings of CAV'94*, number 818 in Lecture Notes in Computer Science, pages 1–13. Springer Verlag, 1994.
5. R. Alur and T.A. Henzinger. Back to the future: towards a theory of timed regular languages. In *Proceedings of FOCS'92*, Lecture Notes in Computer Science, pages 177–186. Springer Verlag, 1992.
6. B. Bérard. Untiming timed languages. *Information Processing Letters*, 55:129–135, 1995.
7. B. Bérard, P. Gastin, and A. Petit. On the power of non observable actions in timed automata. In *Proceedings of STACS'96*, number 1046 in Lecture Notes in Computer Science, pages 257–268. Springer Verlag, 1996.
8. C. Courcoubetis and M. Yannakakis. Minimum and maximum delay problems in real-time systems. In *Proceedings of CAV'91*, number 575 in Lecture Notes in Computer Science, pages 399–409. Springer Verlag, 1991.
9. T.A. Henzinger, P.W. Kopke, and H. Wong-Toi. The expressive power of clocks. In *Proceedings of ICALP'95*, number 944 in Lecture Notes in Computer Science, pages 335–346. Springer Verlag, 1995.
10. T.A. Henzinger, X. Nicollin, J. Sifakis, and S. Yovine. Symbolic model checking for real-time systems. *Information and Computation*, 111(2):193–244, 1994.
11. T. Wilke. Specifying timed state sequences in powerful decidable logics and timed automata. In H. Langmaack, W.-P. de Roever, and J. Vytopil, editors, *Formal Techniques in Real-Time and Fault-Tolerant Systems*, volume 863 of *Lecture Notes in Computer Science*. Springer Verlag, 1994.
12. H. Wong-Toi and G. Hoffmann. The control of dense real-time discrete event systems. In *Proceedings of the 30th IEEE Conf. on Decision and Control*, pages 1527–1528, 1991.

Probabilistic Proof Systems – A Survey*

Oded Goldreich
Department of Computer Science and Applied Mathematics
Weizmann Institute of Science, Rehovot, ISRAEL.
E-mail: oded@wisdom.weizmann.ac.il

Abstract. Various types of *probabilistic* proof systems have played a central role in the development of computer science in the last decade. In this exposition, we concentrate on three such proof systems — *interactive proofs*, *zero-knowledge proofs*, and *probabilistic checkable proofs* — stressing the essential role of randomness in each of them.

1 Introduction

The glory given to the creativity required to find proofs, makes us forget that it is the less glorified procedure of verification which gives proofs their value. Philosophically speaking, proofs are secondary to the verification procedure; whereas technically speaking, proof systems are defined in terms of their verification procedures.

The notion of a verification procedure assumes the notion of computation and furthermore the notion of efficient computation. This implicit assumption is made explicit in the definition of $\mathcal{NP}$, in which efficient computation is associated with (deterministic) polynomial-time algorithms.

Definition 1 (NP-proof systems): *Let $S \subseteq \{0,1\}^*$ and $\nu : \{0,1\}^* \times \{0,1\}^* \mapsto \{0,1\}$ be a function so that $x \in S$ if and only if there exists a $w \in \{0,1\}^*$ such that $\nu(x,w) = 1$. If ν is computable in time bounded by a polynomial in the length of its first argument then we say that S is an* NP-set *and that ν defines an* NP-proof system.

Traditionally, NP is defined as the class of NP-sets. Yet, each such NP-set can be viewed as a proof system. For example, consider the set of satisfiable Boolean formulae. Clearly, a satisfying assignment π for a formula ϕ constitutes an NP-proof for the assertion "ϕ is satisfiable" (the verification procedure consists of substituting the variables of ϕ by the values assigned by π and computing the value of the resulting Boolean expression).

The formulation of NP-proofs restricts the "effective" length of proofs to be polynomial in length of the corresponding assertions (since the running-time of

* Parts of the material presented in this survey have appeared in the *Proceedings of the International Congress of Mathematicians 1994*, Birkhäuser Verlag, Basel, 1995, pages 1395–1406.

Reischuk, Morvan (Eds.): STACS'97 Proceedings, LNCS 1200

the verification procedure is restricted to be polynomial in the length of the assertion). However, longer proofs may be allowed by padding the assertion with sufficiently many blank symbols. So it seems that NP gives a satisfactory formulation of proof systems (with efficient verification procedures). This is indeed the case if one associates efficient procedures with *deterministic* polynomial-time algorithms. However, we can gain a lot if we are willing to take a somewhat non-traditional step and allow *probabilistic* verification procedures. In particular,

- Randomized and interactive verification procedures, giving rize to *interactive proof systems*, seem much more powerful (i.e., "expressive") than their deterministic counterparts.
- Such randomized procedures allow the introduction of *zero-knowledge proofs* which are of great theoretical and practical interest.
- NP-proofs can be efficiently transformed into a (redundant) form which offers a trade-off between the number of locations examined in the NP-proof and the confidence in its validity (see *probabilistically checkable proofs*).

In all the abovementioned types of probabilistic proof systems, explicit bounds are imposed on the computational complexity of the verification procedure, which in turn is personified by the notion of a verifier. Furthermore, in all these proof systems, the verifier is allowed to toss coins and rule by statistical evidence. Thus, all these proof systems carry a probability of error; yet, this probability is explicitly bounded and, furthermore, can be reduced by successive application of the proof system.

Notational Conventions When presenting a proof system, we state all complexity bounds in terms of the length of the assertion to be proven (which is viewed as an input to the verifier). Namely, polynomial-time means time polynomial in the length of this assertion. Note that this convention is consistent with the definition of NP-proofs.

Denote by poly the set of all integer functions bounded by a polynomial and by log the set of all integer functions bounded by a logarithmic function (i.e., $f \in \texttt{log}$ iff $f(n) = O(\log n)$).

2 Interactive Proof Systems

In light of the growing acceptability of randomized and distributed computations, it is only natural to associate the notion of efficient computation with probabilistic and interactive polynomial-time computations. This leads naturally to the notion of an interactive proof system in which the verification procedure is interactive and randomized, rather than being non-interactive and deterministic. Thus, a "proof" in this context is not a fixed and static object but rather a randomized (dynamic) process in which the verifier interacts with the prover. Intuitively, one may think of this interaction as consisting of "tricky" questions asked by the verifier, to which the prover has to reply "convincingly". The above discussion, as well as the following definition, makes explicit reference to a prover, whereas a prover is only implicit in the traditional definitions of proof systems (e.g., NP-proofs).

2.1 Definition

Loosely speaking, an interactive proof is a game between a computationally bounded verifier and a computationally unbounded prover whose goal is to convince the verifier of the validity of some assertion. Specifically, the verifier is probabilistic polynomial-time. It is required that if the assertion holds then the verifier always accepts (i.e., when interacting with an appropriate prover strategy). On the other hand, if the assertion is false then the verifier must reject with probability at least $\frac{1}{2}$, no matter what strategy is being employed by the prover. A sketch of the formal definition is given in Item (1) below. Item (2) introduces additional complexity measures which can be ignored in first reading.

Definition 2 (Interactive Proofs – IP) [33]:

1. *An* interactive proof system for a set S *is a two-party game, between a* verifier *executing a probabilistic polynomial-time strategy* (denoted V) *and a* prover *which executes a computationally unbounded strategy* (denoted P), *satisfying*
   - Completeness*: For every $x \in S$ the verifier V always accepts after interacting with the prover P on common input x.*
   - Soundness*: For every $x \notin S$ and every potential strategy P^*, the verifier V rejects with probability at least $\frac{1}{2}$, after interacting with P^* on common input x.*
2. *For an integer function m, the complexity class $\mathcal{IP}(m(\cdot))$ consists of sets having an interactive proof system in which, on common input x, at most $m(|x|)$ messages are exchanged*[1] *between the parties. For a set of integer functions, M, we let $\mathcal{IP}(M)$ equal $\cup_{m \in M}\mathcal{IP}(m(\cdot))$. Finally, $\mathcal{IP} \stackrel{\text{def}}{=} \mathcal{IP}(\texttt{poly})$.*

In Item (1), we have followed the standard definition which specifies strategies for both the verifier and the prover. An alternative presentation only specifies the verifier's strategy while rephrasing the completeness condition as follows:

> there exists a prover strategy P so that, for every $x \in S$, the verifier V always accepts after interacting with P on common input x.

[1] We count the total number of messages exchanged regardless of the direction of communication. Thus, an interactive proof in which the verifier sends a single message answered by a single message of the prover corresponds to $\mathcal{IP}(2)$. Clearly, $\mathcal{NP} \subseteq \mathcal{IP}(1)$, yet the inclusion may be strict since the verifier may toss coins after receiving the prover's single message.

Arthur-Merlin games[2] introduced in [5] are a special case of interactive proofs; yet, as shown in [34], this restricted case has essentially[3] the same power as the general case previously introduced in [33]. Also, in some sources interactive proofs are defined so that two-sided error probability is allowed; yet, this does not increase their power [25].

2.2 The Role of Randomness

Randomness is essential to the formulation of interactive proofs; if randomness is not allowed (or if it is allowed but zero error is required in the soundness condition) then interactive proof systems collapse to NP-proof systems. The reason being that the prover can predict the verifier's part of the interaction and thus it suffices to let the prover send the full transcript of the interaction and let the verifier check that the interaction is indeed valid. (In case the verifier is not deterministic, the transcript sent by the prover may not match the outcome of the verifier coin tosses.) The moral is that there is no point to interact with predictable parties which are also computationally weaker[4].

2.3 The Power of Interactive Proofs

A simple example demonstrating the power of interactive proofs follows. Specifically, we present an interactive proof for proving that two graphs are not isomorphic[5]. It is not known whether such a statement can be proven via an NP-proof system.

Construction 1 (Interactive proof system for Graph Non-Isomorphism) [28]:

- Common Input: *A pair of two graphs, $G_1 = (V_1, E_1)$ and $G_2 = (V_2, E_2)$. Suppose, without loss of generality, that $V_1 = \{1, 2, ..., |V_1|\}$, and similarly for V_2.*
- Verifier's first step (V1): *The verifier selects at random one of the two input graphs, and sends to the prover a random isomorphic copy of this graph. Namely, the verifier selects uniformly $\sigma \in \{1,2\}$, and a random permutation π from the set of permutations over the vertex set V_σ. The verifier constructs a graph with vertex set V_σ and edge set*

$$E \stackrel{\text{def}}{=} \{\{\pi(u), \pi(v)\} : \{u,v\} \in E_\sigma\}$$

[2] In Arthur-Merlin games, the verifier must send the outcome of any coin it tosses (and thus need not send any other information).

[3] Here and in the next sentence, not only $\mathcal{IP}$ remains invariant under the various definitions, but also $\mathcal{IP}(m(\cdot))$, for every integer function satisfying $m(n) \geq 2$ for every n.

[4] This moral represents the prover's point of view. Certainly, from the verifier's point of view it is benefitial to interact with the prover, since it is computationally stronger.

[5] Two graphs, $G_1 = (V_1, E_1)$ and $G_2 = (V_2, E_2)$, are called *isomorphic* if there exists a 1-1 and onto mapping, ϕ, from the vertex set V_1 to the vertex set V_2 so that $\{u,v\} \in E_1$ if and only if $\{\phi(v), \phi(u)\} \in E_2$. The ("edge preserving") mapping ϕ, if existing, is called an *isomorphism* between the graphs.

and sends (V_σ, E) *to the prover.*

- Motivating Remark: *If the input graphs are non-isomorphic, as the prover claims, then the prover should be able to distinguish (not necessarily by an efficient algorithm) isomorphic copies of one graph from isomorphic copies of the other graph. However, if the input graphs are isomorphic then a random isomorphic copy of one graph is distributed identically to a random isomorphic copy of the other graph.*
- Prover's step: *Upon receiving a graph,* $G' = (V', E')$*, from the verifier, the prover finds a* $\tau \in \{1,2\}$ *so that the graph* G' *is isomorphic to the input graph* G_τ*. (If both* $\tau = 1, 2$ *satisfy the condition then* τ *is selected arbitrarily. In case no* $\tau \in \{1,2\}$ *satisfies the condition,* τ *is set to 0). The prover sends* τ *to the verifier.*
- Verifier's second step (V2): *If the message,* τ*, received from the prover equals* σ *(chosen in Step V1) then the verifier outputs 1 (i.e., accepts the common input). Otherwise the verifier outputs 0 (i.e., rejects the common input).*

The verifier's strategy presented above is easily implemented in probabilistic polynomial-time. We do not known of a probabilistic polynomial-time implementation of the prover's strategy, but this is not required. The motivating remark justifies the claim that Construction 1 constitutes an interactive proof system for the set of pairs of non-isomorphic graphs. Recall that the latter is a coNP-set (not known to be in $\mathcal{NP}$).

Interactive proofs are powerful enough to prove *any* coNP assertion (e.g., that a graph is not 3-colorable) [41]. Furthermore, the class of sets having interactive proof systems coincides with the class of sets that can be decided using a polynomial amount of work-space [49].

Theorem 1 [41, 49]: $\mathcal{IP} = \mathcal{PSPACE}$.

Recall that it is widely believed that $\mathcal{NP} \subset \mathcal{PSPACE}$. Thus, under this conjecture, interactive proofs are more powerful than NP-proofs.

Concerning the finer structure of the IP hierarchy it is known that this hierarchy has a "linear speed-up" property [8]. Namely, for every integer function, f, so that $f(n) \geq 2$ for all n, the class $\mathcal{IP}(O(f(\cdot)))$ collapses to the class $\mathcal{IP}(f(\cdot))$. In particular, $\mathcal{IP}(O(1))$ collapses to $\mathcal{IP}(2)$. It is conjectured that co$\mathcal{NP}$ is *not* contained in $\mathcal{IP}(2)$, and consequently that interactive proofs with unbounded number of message exchanges are more powerful than interactive proofs in which only a bounded (i.e., constant) number of messages are exchanged. Still, the class $\mathcal{IP}(2)$ contains sets not known to be in $\mathcal{NP}$; e.g., Graph Non-Isomorphism (as shown above).

2.4 How Powerful Should the Prover be?

Assume that a set S is in $\mathcal{IP}$. This means that there is a verifier V that can be convinced to accept any input in S but cannot be convinced to accept any input not in S (except with small probability). One may ask how powerful should a prover be so that it can convince the verifier V to accept any input in S. More interestingly, considering all possible verifiers which give rise to interactive proof systems for S, what is the minimum power required from a prover which satisfies the completeness requirement with respect to one of these verifiers? We stress that, unlike the case of computationally-sound proof systems (see Sec. 5), we do not restrict the power of the prover in the soundness condition but rather consider the minimum complexity of provers meeting the completeness condition. Specifically, we are interested in *relatively efficient* provers which meet the completeness condition. The term 'relatively efficient prover' has been given three different interpretations.

1. A prover is considered *relatively efficient* if, when given an auxiliary input (in addition to the common input in S), it works in (probabilistic) polynomial-time. Specifically, in case $S \in \mathcal{NP}$, the auxiliary input maybe an NP-proof that the common input is in the set[6]. This interpretation is adequate and in fact crucial for applications in which such an auxiliary input is available to the otherwise-polynomial-time parties. Typically, such auxiliary input is available in cryptographic applications in which parties wish to prove in (zero-knowledge) that they have conducted some computation correctly resulting in some string x. In these cases the NP-proof is just the transcript of the procedure by which x has been computed and thus the auxiliary input is available to the proving party. See [28].
2. A prover is considered *relatively efficient* if it can be implemented by a probabilistic polynomial-time oracle machine with oracle access to the set S itself. (Note that the prover in Construction 1 has this property.) This interpretation generalizes the notion of self-reducibility of NP-sets. (By self-reducibility of an NP-set we mean that the search problem of finding an NP-witness is polynomial-time reducible to deciding membership in the set.) See [12].
3. A prover is considered *relatively efficient* if it can be implemented by a probabilistic machine which runs in time which is polynomial in the deterministic complexity of the set. This interpretation relates the difficulty of convincing a "lazy verifier" to the complexity of finding the truth alone. Hence, in contrast to the first interpretation which is adequate in settings where assertions are generated along with their NP-proofs, the current interpretation is adequate in settings in which the prover is given only the assertion and has to find a proof to it by itself (before trying to convince a lazy verifier of its validity). See [43].

[6] Still, even in this case the interactive proof need not consist of the prover sending the auxiliary input to the verifier; e.g., an alternative procedure may allow the prover to be zero-knowledge (see Construction 2).

3 Zero-Knowledge Proof Systems

Zero-knowledge proofs, introduced in [33], are central to cryptography. Furthermore, zero-knowledge proofs are very intruiging from a conceptual point of view, since they exhibit an extreme contrast between being convinced of the validity of a statement and learning anything in addition while receiving such a convincing proof. Namely, zero-knowledge proofs have the remarkable property of being both convincing while yielding nothing to the verifier, beyond the fact that the statement is valid. Formally, the fact that "nothing is gained by the interaction" is captured by stating that whatever the verifier can efficiently compute after interacting with a zero-knowledge prover, can be efficiently computed from the assertion itself without interacting with anyone.

3.1 A Sample Definition

Zero-knowledge is a property of some interactive proof systems, or more acurately of some specified prover strategies. The formulation of the zero-knowledge condition considers two ensembles of probability distributions, each ensemble associates a probability distribution to each valid assertion. The first ensemble respresents the output distribution of the verifier after interacting with the specified prover strategy P, where the verifier is not necessarily employing the specified strategy (i.e., V) – but rather any efficient strategy. The second ensemble represents the output distribution of some probabilistic polynomial-time algorithm (which does not interact with anyone). The basic paradigm of zero-knowledge asserts that for every ensemble of the first type there exist a "similar" ensemble of the second type. The specific variants differ by the interpretation given to 'similarity'. The most strict interpretation, leading to *perfect zero-knowledge*, is that similarity means equality. Namely,

Definition 3 (perfect zero-knowledge) [33]: *A prover strategy, P, is said to be* perfect zero-knowledge *over a set S if for every probabilistic polynomial-time verifier strategy, V^*, there exists a probabilistic polynomial-time algorithm, M^*, such that*

$$(P,V^*)(x) = M^*(x) \ , \qquad \textit{for every } x \in S$$

where $(P,V^*)(x)$ is a random variable representing the output of verifier V^* after interacting with the prover P on common input x, and $M^*(x)$ is a random variable representing the output of machine M^* on input x.

A somewhat more relaxed interpretation, leading to *almost-perfect zero-knowledge*, is that similarity means statistical closeness (i.e., negligible difference between the ensembles). The most liberal interpretation, leading to the standard usage of the term zero-knowledge (and sometimes referred to as *computational zero-knowledge*), is that similarity means computational indistinguishability (i.e., failure of any efficient procedure to tell the two ensembles apart). Since the notion of computational indistinguishability is a fundamental one, it is indeed in place to present a definition of it.

Definition 4 (computational indistinguishability) [32, 50]: *An integer function, f, is called* negligible *if for every positive polynomial p and all sufficiently large n, it holds that $f(n) < \frac{1}{p(n)}$.* (Thus, multiplying a negligible function by any fixed polynomial yields a negiligible function.)
Two probability ensembles, $\{A_x\}_{x\in S}$ and $\{B_x\}_{x\in S}$, are indistinguishable *by an algorithm D if*

$$d(n) \stackrel{\text{def}}{=} \max_{x\in S\cap\{0,1\}^n} \{|prob(D(A_x)=1) - \text{Prob}(D(B_x)=1)|\}$$

is a negligible function. The ensembles $\{A_x\}_{x\in S}$ and $\{B_x\}_{x\in S}$ are computationally indistinguishable *if they are indistinguishable by every probabilistic polynomial-time algorithm.*

The definitions presented above are a simplified version of the actual definitions. For example, in order to guarantee that zero-knowledge is preserved under sequential composition it is necessary to slightly augment the definitions. For details see [30].

3.2 The Power of Zero-Knowledge

A simple example, demonstrating the power of zero-knowledge proofs, follows. Specifically, we will present a simple zero-knowledge proof for proving that a graph is 3-colorable[7]. The interactive proof will be described using "boxes" in which information can be hidden and later revealed. Such "boxes" can be implemented using one-way functions (see below).

Construction 2 (Zero-knowledge proof of 3-colorability) [28]:

- Common Input: *A simple graph $G=(V,E)$.*
- Prover's first step: *Let ψ be a 3-coloring of G. The prover selects a random permutation, π, over $\{1,2,3\}$, and sets $\phi(v) \stackrel{\text{def}}{=} \pi(\psi(v))$, for each $v \in V$. Hence, the prover forms a random relabelling of the 3-coloring ψ. The prover sends the verifier a sequence of $|V|$ locked and nontransparent boxes so that the vth box contains the value $\phi(v)$;*
- Verifier's first step: *The verifier uniformly selects an edge $\{u,v\} \in E$, and sends it to the prover;*
- Motivating Remark: *The verifier asks to inspect the colors of vertices u and v;*
- Prover's second step: *The prover sends to the verifier the keys to boxes u and v;*
- Verifier's second step: *The verifier opens boxes u and v, and accepts if and only if they contain two different elements in $\{1,2,3\}$;*

[7] A graph $G=(V,E)$ is said to be *3-colorable* if there exists a function $\pi: V \mapsto \{1,2,3\}$ so that $\pi(v) \neq \pi(u)$ for every $\{u,v\} \in E$. Such a function, π, is called a *3-coloring* of the graph.

The verifier strategy presented above is easily implemented in probabilistic polynomail-time. The same holds with respect to the prover's strategy, provided it is given a 3-coloring of G as auxiliary input. Clearly, if the input graph is 3-colorable then the prover can cause the verifier to accept always. On the other hand, if the input graph is not 3-colorable then any contents put in the boxes must be invalid on at least one edge, and consequently the verifier will reject with probability at least $\frac{1}{|E|}$. Hence, the above game exhibits a non-negligible gap in the accepting probabilities between the case of 3-colorable graphs and the case of non-3-colorable graphs. To increase the gap, the game may be repeated sufficiently many times (of course, using independent coin tosses in each repetition). The zero-knowledge property follows easily, in this abstract setting, since one can simulate the real interaction by placing a random pair of different colors in the boxes indicated by the verifier. This indeed demonstrates that the verifier learns nothing from the interaction (since it expects to see a random pair of different colors and indeed this is what it sees). We stress that this simple argument is not possible in the digital implementation since the boxes are not totally ineffected by their contents (but are rather effected, yet in an indistinguishable manner).

As stated above, the "boxes" need to be implemented digitally, and this is done using an adaquately defined "commitment scheme". Loosely speaking, such a scheme is a two phase game beteen a sender and a receiver so that after the first phase the sender is "committed" to a value and yet, at this stage, it is infeasible for the receiver to find out the committed value. The committed value will be revealed to the receiver in the second phase and it is guaranteed that the sender cannot reveal a value other than the one committed. Such commitment schemes can be implemented assuming the existence of one-way functions (i.e., loosely speaking, functions that are easy to compute but hard to invert, such as the multiplication of two large primes) [44, 37].

Using the fact that 3-colorability is NP-complete, one gets zero-knowledge proofs for any NP-set.

Theorem 2 [28]: *Assuming the existence of one-way functions, any NP-proof can be efficiently transformed into a* (computational) *zero-knowledge interactive proof.*

Theorem 2 has a dramatic effect on the design of cryptographic protocols (cf., [28, 29]). In a different vein and for the sake of elegancy, we mention that, using further ideas and under the same assumption, any interactive proof can be efficiently transformed into a zero-knowledge one [38, 13].

The above results may be contrasted with the results regarding the complexity of *almost-perfect* zero-knowledge proof systems; namely, that almost-perfect zero-knowledge proof systems exist only for sets in $\mathcal{IP}(2) \cap \text{co}\mathcal{IP}(2)$ [23, 2], and thus are unlikely to exist for all NP-sets. Also, a recent result seems to indicate that one-way functions are essential for the existence of zero-knowledge proofs for "hard" sets (i.e., sets which cannot be decided in average polynomial-time) [45].

3.3 The Role of Randomness

Again, randomness is essential to all the above mentioned (positive) results. Namely, if either verifier or prover is required to be deterministic then only BPP-sets can be proven in a zero-knowledge manner [30]. However, BPP-sets have trivial zero-knowledge proofs in which the prover sends nothing and the verifier just test the validity of the assertion by itself[8]. Thus, randomness is essential to the usefulness of zero-knowledge proofs.

4 Probabilistically Checkable Proof Systems

When viewed in terms of an interactive proof system, the probabilistically checkable proof setting consists of a prover which is memoryless. Namely, one can think of the prover as being an oracle and of the messages sent to it as being queries. A more appealing interpretation is to view the probabilistically checkable proof setting as an alternative way of generalizing $\mathcal{NP}$. Instead of receiving the entire proof and conducting a deterministic polynomial-time computation (as in the case of $\mathcal{NP}$), the verifier may toss coins and query the proof only at location of its choice. Potentially, this allows the verifier to utilize very long proofs (i.e., of super-polynomial length) or alternatively examine very few bits of an NP-proof.

4.1 Definition

Loosely speaking, a probabilistically checkable proof system consists of a probabilistic polynomial-time verifier having access to an oracle which represents a proof in redundent form. Typically, the verifier accesses only few of the oracle bits, and these bit positions are determined by the outcome of the verifier's coin tosses. Again, it is required that if the assertion holds then the verifier always accepts (i.e., when given access to an adaquate oracle); whereas, if the assertion is false then the verifier must reject with probability at least $\frac{1}{2}$, no matter which oracle is used. The basic definition of the PCP setting is given in Item (1) below. Yet, the complexity measures introduced in Item (2) are of key importance for the subsequent discussions, and should not be ignored.

Definition 5 (Probabilistic Checkable Proofs – PCP):

1. *A* probabilistic checkable proof system (pcp) for a set S *is a probabilistic polynomial-time oracle machine* (called verifier), *denoted* V, *satisfying*
   - Completeness: *For every $x \in S$ there exists an oracle set π_x so that V, on input x and access to oracle π_x, always accepts x.*

[8] Actually, this is slightly inaccurate since the resulting "interactive proof" may have two-sided error, whereas we have required interactive proofs to have only one-sided error. Yet, since the error can be made negligible by successive repetitions this issue is insignificant. Alternatively, one can use ideas in [25] to eliminate the error by letting the prover send some random-looking help.

– Soundness: *For every $x \notin S$ and every oracle set π, machine V, on input x and access to oracle π, rejects x with probability at least $\frac{1}{2}$.*

2. *Let r and q be integer functions. The complexity class $\mathcal{PCP}(r(\cdot), q(\cdot))$ consists of sets having a probabilistic checkable proof system in which the verifier, on any input of length n, makes at most $r(n)$ coin tosses and at most $q(n)$ oracle queries. We stress that here, as usual in complexity theory, the oracle answers are always binary (i.e., either 0 or 1). For sets of integer functions, R and Q, we let $\mathcal{PCP}(R, Q)$ equal $\cup_{r \in R, q \in Q} \mathcal{PCP}(r(\cdot), q(\cdot))$.*

The above model was suggested in [24] and shown related to a multi-prover model introduced previously in [14]. The fine complexity measures were introduced and motivated in [20], and further advocated in [4]. A related model was presented in [7], stressing the applicability to program checking.

We stress that the oracle π_x in a pcp system constitutes a proof in the standard mathematical sense[9]. Yet, this oracle has the extra property of enabling a lazy verifier, to toss coins, take its chances and "assess" the validity of the proof without reading all of it (but rather by reading a tiny portion of it).

4.2 The Power of Probabilistically Checkable Proofs

Clearly, $\mathcal{PCP}(\texttt{poly}, 0)$ equals co$\mathcal{RP}$, whereas $\mathcal{PCP}(0, \texttt{poly})$ equals $\mathcal{NP}$. It is easy to prove an upper bound on the non-deterministic time complexity of sets in the PCP hierarchy. In particular,

Proposition 1 : *$\mathcal{PCP}(\texttt{log}, \texttt{poly})$ is contained in $\mathcal{NP}$.*

These upper bounds turn out to be tight, but proving this is much more difficult (to say the least). The following result is a culmination of a sequence of great works [6, 7, 20, 4, 3].[10]

Theorem 3 : *$\mathcal{NP}$ is contained in $\mathcal{PCP}(\texttt{log}, O(1))$.*

Thus, probabilistically checkable proofs in which the verifier tosses only logarithmically many coins and makes only a constant number of queries exist for every set in the complexity class $\mathcal{NP}$. It follows that NP-proofs can be transformed into NP-proofs which offer a trade-off between the portion of the proof

---

[9] Jumping ahead, the oracles in pcp systems characterizing $\mathcal{NP}$ have the property of being NP proofs themselves.

[10] The sequence has started with the characterization of $\mathcal{PCP}(\texttt{poly}, \texttt{poly})$ as equal non-deterministic exponential-time [6], and continued with its scaled-down in [7, 20] which led to the $\mathcal{NP} \subseteq \mathcal{PCP}(\texttt{polylog}, \texttt{polylog})$ result of [20]. The first PCP-characterization of $\mathcal{NP}$, by which $\mathcal{NP} = \mathcal{PCP}(\texttt{log}, \texttt{log})$, has appeared in [4] and the cited result was obtained in [3]. This sequence of works, directly related to the stated theorem, was built on and inspired by works from various settings such as interactive proofs [41, 49, 22], program-checking [16, 26, 48], and private computation with oracles [9]. The constant (number of queries) in Theorem 3 has been subsequently improved and is currently 9; cf., [36].

being read and the confidence it offers. Specifically, if the verifier is willing to tolerate an error probability of ϵ then it suffices to let it examine $O(\log(1/\epsilon))$ bits of the (transformed) NP-proof. These bit locations need to be selected at random.

The characterization of $\mathcal{NP}$ in terms of probabilistically checkable proofs plays a central role in recent developments concerning the difficulty of approximation problems (cf., [20, 3, 42, 11] and [35, 36]). To demonstrate this relationship, we first note that Theorem 3 can be rephrased without mentioning the class $\mathcal{PCP}$ altogether. Instead, a new type of polynomial-time reductions, which we call *amplifying*, emerges.

Theorem 4 (Theorem 3 — Rephrased): *There exists a constant $\epsilon > 0$, and a polynomial-time computable function f, mapping the set of 3CNF formulae*[11] *to itself so that*

- *As usual, f maps satisfiable 3CNF formulae to satisfiable 3CNF formulae; and*
- *f maps non-satisfiable 3CNF formulae to* (non-satisfiable) *3CNF formulae for which every truth assignment satisfies at most a $1 - \epsilon$ fraction of the clauses.*

The function f is called an amplifying reduction.

proof sketch (Thm. 3 $\Rightarrow$ Thm. 4): We start by considering a pcp system for 3SAT, and use the fact that the pcp system given by the proof of Theorem 3 is non-adaptive (i.e., the queries are determined as a function of the input and the random-tape – and do not depend on answers to previous queries).[12] Next, we associate the bits of the oracle with Boolean variables and introduce a (constant size) Boolean formula for each possible outcome of the sequence of $O(\log n)$ coin tosses, describing whether the verifier would have accepted given this outcome. Finally, using auxiliary variables, we convert each of these formulae into a 3CNF formula and obtain (as the output of the reduction) the conjunction of all these polynomially many clauses. □

It is also easy to see that Theorem 4 implies Theorem 3: Given a reduction as in Thm. 4, we construct a pcp system for 3SAT by letting the verifier select a clause uniformly among the clauses of the reduced formula, and make three queries corresponding to the three variables in it. This yields a proof system with soundness error bounded by $1 - \epsilon$. Reducing the error by $O(1/\epsilon)$ successive applications of this proof system, we obtain Thm. 3.

As an immediate corollary to the formulation of Theorem 4 one concludes that it is NP-Hard to distinguish satisfiable 3CNF formulae from 3CNF formulae for

[11] A 3CNF formula is a Boolean formula consisting of a conjunction of clauses, where each clause is a disjunction of upto 3 literals. (A literal is variable or its negation.).

[12] Actually, it is not essential to use this fact, since one can easily convert any adaptive system into a non-adaptive one while incurring an exponential blowup in the query complexity (which in our case is a constant).

which no truth assignment satisfies at least a $1-\epsilon$ fraction of the clauses (as otherwise, using the reduction, one may decide membership in 3SAT). In general, probabilistic checkable proof systems for $\mathcal{NP}$ yield strong non-approximability results for various classical optimization problems. In particular, quite *tight* non-approximability results have been shown for MaxClique (cf., [35]), Chromatic Number (cf., [21]), Set Cover (cf., [19]), and Max-Exact-3SAT (cf., [36]).

4.3 The Role of Randomness

No trade-off between the number of bits examined and the confidence is possible if one requires the verifier to be deterministic. In particular, $\mathcal{PCP}(0, q(\cdot))$ contains only sets that are decidable by a deterministic algorithms of running time $2^{q(n)} \cdot \text{poly}(n)$. It follows that $\mathcal{PCP}(0, \log) = \mathcal{P}$. Furthermore, since it is unlikely that all NP-sets can be decided by (deterministic) algorithms of running time, say, $2^n \cdot \text{poly}(n)$, it follows that $\mathcal{PCP}(0, n)$ is unlikely to contain $\mathcal{NP}$.

5 Other Probabilistic Proof Systems

In this section, we shortly review some variants on the basic model of interactive proofs. These variants include models in which the prover is restricted in its choice of strategy, a model in which the prover-verifier interaction is restricted, and a model in which one proves "knowledge" of facts rather than their *validity*.

5.1 Restricting the Prover's Strategy

We stress that the restrictions discussed here refer to the strategies employed by the prover both in case it tries to prove valid assertions (i.e., the completeness condition) and in case it tries to fool the verifier to believe false statements (i.e., the soundness condition). Thus, the validity of the verifier decision (concerning false statements) depends on whether this restriction (concerning "cheating" prover strategies) really holds. The reason to consider these restricted models is that they enable to achieve results which are not possible in the general model of interactive proofs (cf., [14, 17, 39, 43]). We consider restrictions of two types: computational or physical.

We start with a physical restriction. In the so-called *multi-prover interactive proof* model, denoted MIP (cf., [14]), the prover is split into several (say, two) entities and the restriction (or assumption) is that these entities cannot interact with each other. Actually, the formulation allows them to coordinate their strategies prior to interacting with the verifier[13] but it is crucial that they don't exchange messages among themselves while interacting with the verifier. The multi-prover model is reminiscent of the common police procedure of isolating collaborating suspects and interrogating each of them separately. On the other hand, the multi-prover model is related to the PCP model [24]. Interestingly,

[13] This is implicit in the universal quantifier used in the soundness condition.

the multi-prover model allows to present (perfect) zero-knowledge proofs for all NP-sets, without relying on any comutational assumptions [14]. Furthermore, these proofs can be made very efficient in terms of communication complexity [18].

We now turn to computational restrictions. Since the effect of this restriction is more noticable in the soundness condition, we refer to these proof systems as being *computationally-sound.* Two variants have been suggested. In *argument* systems [17], the prover stategy is restricted to be probabilistic polynomial-time with auxiliary input (analogously to item (1) in Sec. 2.4). In *CS-proofs* [43], the prover stategy is restricted to be probabilistic and run in time polynomial in the time required to validate the assertion (analogously to item (3) in Sec. 2.4). Interestigly, computationally-sound interactive proofs can be much more communication-efficient than (regular) interactive proofs; cf. [39, 43, 27].

5.2 Non-Interactive Zero-Knowledge Proofs

Actualy the term "non-interactive" is somewhat misleading. The model, introduced in [15], consists of three entities: a prover, a verifier and a uniformly selected sequence of bits (which can be thought of as being selected by a trusted third party). Both verifier and prover can read the random sequence, and each can toss additional coins. The interaction consists of a single message sent from the prover to the verifier, who then is left with the decision (whether to accept or not). Based on some reasonable complexity assumptions, one may construct non-interactive zero-knowledge proof systems for every NP-set (cf., [15, 22, 40]).

5.3 Proofs of Knowledge

The concept of a proof of knowledge, introduced in [33], is very appealing; yet, its precise formulation is much more complex than one may expect (cf. [10]). Loosely speaking, a knowledge-verifier for a relation R guarantees the existence of a "knowledge extractor" that on input x and access to any interactive machine P^* outputs a y, so that $(x, y) \in R$, within complexity related to the probability that the verifier accepts x when interacting with P^*. By convincing such a knowledge-verifier, on common input x, one proves that he knows a y so that $(x, y) \in R$. It can be shown that the protocol which results by successively applying Construction 2 suffiently many time constitutes a "proof of knowledge" of a 3-coloring of the input graph.

5.4 Knowledge Complexity

Zero-knowledge is the lowest level of a knowledge-complexity hierarchy which quantifies the "knowledge revealed in an interaction" [33]. *Knowledge complexity* may be defined as the minimum number of oracle-queries required in order to (efficiently) simulate an interaction with the prover (cf. [31]). Results linking two different variants of this measure to other complexity measures are given in [1, 47], respectively.

Acknowledgement

I am grateful to Shafi Goldwasser for suggesting the essential role of randomness as the unifying theme for this exposition. Thanks also to Leonid Levin, Dana Ron, Madhu Sudan and Uri Zwick for commenting on earlier versions of this survey.

References

1. W. Aiello, M. Bellare and R. Venkatesan. Knowledge on the Average – Perfect, Statistical and Logarithmic. In *27th STOC*, pages 469–478, 1995.
2. W. Aiello and J. Hastad. Perfect Zero-Knowledge Languages can be Recognized in Two Rounds. In *28th FOCS*, pages 439–448, 1987.
3. S. Arora, C. Lund, R. Motwani, M. Sudan and M. Szegedy. Proof Verification and Intractability of Approximation Problems. In *33rd FOCS*, pages 14–23, 1992.
4. S. Arora and S. Safra. Probabilistic Checkable Proofs: A New Characterization of NP. In *33rd FOCS*, pages 1–13, 1992.
5. L. Babai. Trading Group Theory for Randomness. In *17th STOC*, pages 421–420, 1985.
6. L. Babai, L. Fortnow, and C. Lund. Non-Deterministic Exponential Time has Two-Prover Interactive Protocols. *Computational Complexity*, Vol. 1, No. 1, pages 3–40, 1991. Preliminary version in *31st FOCS*, 1990.
7. L. Babai, L. Fortnow, L. Levin, and M. Szegedy. Checking Computations in Polylogarithmic Time. In *23rd STOC*, pages 21–31, 1991.
8. L. Babai and S. Moran. Arthur-Merlin Games: A Randomized Proof System and a Hierarchy of Complexity Classes. *JCSS*, Vol. 36, pp. 254–276, 1988.
9. D. Beaver and J. Feigenbaum. Hiding Instances in Multioracle Queries. In *7th STACS*, Springer Verlag, LNCS Vol. 415, pages 37–48, 1990.
10. M. Bellare and O. Goldreich. On Defining Proofs of Knowledge. In *Crypto92*, Springer Verlag, LNCS Vol. 740, pages 390–420, 1992.
11. M. Bellare, O. Goldreich and M. Sudan. Free Bits, PCPs and Non-Approximability – Towards Tight Results. In *36th FOCS*, pages 422–431, 1995.
12. M. Bellare and S. Goldwasser. The Complexity of Decision versus Search. *SIAM Journal on Computing*, Vol. 23, pages 97–119, 1994.
13. M. Ben-Or, O. Goldreich, S. Goldwasser, J. Håstad, J. Kilian, S. Micali and P. Rogaway. Everything Provable is Probable in Zero-Knowledge. In *Crypto88*, Springer Verlag, LNCS Vol. 403, pages 37–56, 1990
14. M. Ben-Or, S. Goldwasser, J. Kilian and A. Wigderson. Multi-Prover Interactive Proofs: How to Remove Intractability Assumptions. In *20th STOC*, pages 113–131, 1988.
15. M. Blum, P. Feldman and S. Micali. Non-Interactive Zero-Knowledge and its Applications. In *20th STOC*, pages 103–112, 1988.
16. M. Blum, M. Luby and R. Rubinfeld. Self-Testing/Correcting with Applications to Numerical Problems. *JCSS*, Vol. 47, No. 3, pages 549–595, 1993.
17. G. Brassard, D. Chaum and C. Crépeau. Minimum Disclosure Proofs of Knowledge. *JCSS*, pages 156–189, 1988. Extended abstract, by Brassard and Crépeau, in *27th FOCS*, 1986.

18. C. Dwork, U. Feige, J. Kilian, M. Naor and S. Safra, Low Communication Perfect Zero Knowledge Two Provers Proof Systems. In *Crypto92*, Springer Verlag, LNCS Vol. 740, pages 215–227, 1992.
19. U. Feige. A Threshold of ln n for Approximating Set Cover. In *28th STOC*, pages 314–318, 1996.
20. U. Feige, S. Goldwasser, L. Lovász, S. Safra, and M. Szegedy. Approximating Clique is almost NP-complete. In *32nd FOCS*, pages 2–12, 1991.
21. U. Feige and J. Kilian. Zero knowledge and the chromatic number. In *11th IEEE Conference on Computational Complexity*, pages 278–287, 1996.
22. U. Feige, D. Lapidot, and A. Shamir. Multiple non-interactive zero knowledge proofs based on a single random string. In *31st FOCS*, pages 308–317, 1990.
23. L. Fortnow, The Complexity of Perfect Zero-Knowledge. In *19th STOC*, pages 204–209, 1987.
24. L. Fortnow, J. Rompel and M. Sipser. On the Power of Multi-Prover Interactive Protocols. In *Proc. 3rd IEEE Symp. on Structure in Complexity Theory*, pages 156–161, 1988.
25. M. Furer, O. Goldreich, Y. Mansour, M. Sipser, and S. Zachos, "On Completeness and Soundness in Interactive Proof Systems", *Advances in Computing Research: a research annual*, Vol. 5 (Randomness and Computation, S. Micali, ed.), pp. 429–442, 1989.
26. P. Gemmell, R. Lipton, R. Rubinfeld, M. Sudan, and A. Wigderson. Self-Testing/Correcting for Polynomials and for Approximate Functions. In *23th STOC*, pages 32–42, 1991.
27. O. Goldreich and J. Håstad. On the Message Complexity of Interactive Proof Systems. Available as TR96-018 of *ECCC*, `http://www.eccc.uni-trier.de/eccc/`, 1996.
28. O. Goldreich, S. Micali and A. Wigderson. Proofs that Yield Nothing but their Validity or All Languages in NP Have Zero-Knowledge Proof Systems. *JACM*, Vol. 38, No. 1, pages 691–729, 1991. Extended abstract in *27th FOCS*, 1986.
29. O. Goldreich, S. Micali and A. Wigderson. How to Play any Mental Game or a Completeness Theorem for Protocols with Honest Majority. In *19th STOC*, pages 218–229, 1987.
30. O. Goldreich and Y. Oren. Definitions and Properties of Zero-Knowledge Proof Systems. *Journal of Cryptology*, Vol. 7, No. 1, pages 1–32, 1994.
31. O. Goldreich and E. Petrank. Quantifying Knowledge Complexity. In *32nd FOCS*, pp. 59–68, 1991.
32. S. Goldwasser and S. Micali. Probabilistic Encryption. *JCSS*, Vol. 28, No. 2, pages 270–299, 1984. Extended abstract in *14th STOC*, 1982.
33. S. Goldwasser, S. Micali and C. Rackoff. The Knowledge Complexity of Interactive Proof Systems. *SIAM Journal on Computing*, Vol. 18, pages 186–208, 1989. Extended abstract in *17th STOC*, 1985.
34. S. Goldwasser and M. Sipser. Private Coins versus Public Coins in Interactive Proof Systems. In *18th STOC*, pages 59–68, 1986.
35. J. Håstad. Clique is hard to approximate within $n^{1-\epsilon}$. In *37th FOCS*, pages 627–636, 1996.
36. J. Håstad. Getting optimal in-approximability results. Unpublish manuscript, June 1996. (Revised October 1996.)
37. J. Håstad, R. Impagliazzo, L.A. Levin and M. Luby. Construction of Pseudorandom Generator from any One-Way Function. Manuscript, 1993. See preliminary versions by Impagliazzo et. al. in *21st STOC* and Håstad in *22nd STOC*.

38. R. Impagliazzo and M. Yung. Direct Zero-Knowledge Computations. In *Crypto87*, Springer Verlag, LNCS Vol. 293, pages 40–51, 1987.
39. J. Kilian. A Note on Efficient Zero-Knowledge Proofs and Arguments. In *24th STOC*, pages 723–732, 1992.
40. J. Kilian and E. Petrank. An Efficient Non-Interactive Zero-Knowledge Proof System for NP with General Assumptions. To appear in the *Journal of Cryptography*. Available as TR95-038 of *ECCC*, `http://www.eccc.uni-trier.de/eccc/`, 1995.
41. C. Lund, L. Fortnow, H. Karloff, and N. Nisan. Algebraic Methods for Interactive Proof Systems. *JACM*, Vol. 39, No. 4, pages 859–868, 1992. Preliminary version in *31st FOCS*, 1990.
42. C. Lund and M. Yannakakis. On the Hardness of Approximating Minimization Problems, In *25th STOC*, pages 286–293, 1993.
43. S. Micali. CS Proofs. In *35th FOCS*, pages 436–453, 1994.
44. M. Naor. Bit Commitment using Pseudorandom Generators. *Journal of Cryptology*, Vol. 4, pages 151–158, 1991.
45. R. Ostrovsky and A. Wigderson. One-Way Functions are essential for Non-Trivial Zero-Knowledge, In *Proc. 2nd Israel Symp. on Theory of Computing and Systems* (*ISTCS93*), IEEE Computer Society Press, pages 3–17, 1993.
46. C. H. Papadimitriou and M. Yannakakis. Optimization, Approximation, and Complexity Classes. In *20th STOC*, pages 229–234, 1988.
47. E. Petrank and G. Tardos. On the Knowledge Complexity of NP. In *37th FOCS*, pages 494–503, 1996.
48. R. Rubinfeld and M. Sudan. Robust Characterizations of Polynomials with Applications to Program Checking. *SIAM J. of Computing*, Vol. 25, No. 2, pages 252–271, 1996. Preliminary version in *3rd SODA*, 1992.
49. A. Shamir. IP=PSPACE. *JACM*, Vol. 39, No. 4, pages 869–877, 1992. Preliminary version in *31st FOCS*, 1990.
50. A.C. Yao. Theory and Application of Trapdoor Functions. In *23st FOCS*, pages 80–91, 1982.

List of Authors

Lecture Notes in Computer Science

For information about Vols. 1–1122

please contact your bookseller or Springer-Verlag

Vol. 1123: L. Bougé, P. Fraigniaud, A. Mignotte, Y. Robert (Eds.), Euro-Par'96. Parallel Processing. Proceedings, 1996, Vol. I. XXXIII, 842 pages. 1996.

Vol. 1124: L. Bougé, P. Fraigniaud, A. Mignotte, Y. Robert (Eds.), Euro-Par'96. Parallel Processing. Proceedings, 1996, Vol. II. XXXIII, 926 pages. 1996.

Vol. 1125: J. von Wright, J. Grundy, J. Harrison (Eds.), Theorem Proving in Higher Order Logics. Proceedings, 1996. VIII, 447 pages. 1996.

Vol. 1126: J.J. Alferes, L. Moniz Pereira, E. Orlowska (Eds.), Logics in Artificial Intelligence. Proceedings, 1996. IX, 417 pages. 1996. (Subseries LNAI).

Vol. 1127: L. Böszörményi (Ed.), Parallel Computation. Proceedings, 1996. XI, 235 pages. 1996.

Vol. 1128: J. Calmet, C. Limongelli (Eds.), Design and Implementation of Symbolic Computation Systems. Proceedings, 1996. IX, 356 pages. 1996.

Vol. 1129: J. Launchbury, E. Meijer, T. Sheard (Eds.), Advanced Functional Programming. Proceedings, 1996. VII, 238 pages. 1996.

Vol. 1130: M. Haveraaen, O. Owe, O.-J. Dahl (Eds.), Recent Trends in Data Type Specification. Proceedings, 1995. VIII, 551 pages. 1996.

Vol. 1131: K.H. Höhne, R. Kikinis (Eds.), Visualization in Biomedical Computing. Proceedings, 1996. XII, 610 pages. 1996.

Vol. 1132: G.-R. Perrin, A. Darte (Eds.), The Data Parallel Programming Model. XV, 284 pages. 1996.

Vol. 1133: J.-Y. Chouinard, P. Fortier, T.A. Gulliver (Eds.), Information Theory and Applications II. Proceedings, 1995. XII, 309 pages. 1996.

Vol. 1134: R. Wagner, H. Thoma (Eds.), Database and Expert Systems Applications. Proceedings, 1996. XV, 921 pages. 1996.

Vol. 1135: B. Jonsson, J. Parrow (Eds.), Formal Techniques in Real-Time and Fault-Tolerant Systems. Proceedings, 1996. X, 479 pages. 1996.

Vol. 1136: J. Diaz, M. Serna (Eds.), Algorithms – ESA '96. Proceedings, 1996. XII, 566 pages. 1996.

Vol. 1137: G. Görz, S. Hölldobler (Eds.), KI-96: Advances in Artificial Intelligence. Proceedings, 1996. XI, 387 pages. 1996. (Subseries LNAI).

Vol. 1138: J. Calmet, J.A. Campbell, J. Pfalzgraf (Eds.), Artificial Intelligence and Symbolic Mathematical Computation. Proceedings, 1996. VIII, 381 pages. 1996.

Vol. 1139: M. Hanus, M. Rogriguez-Artalejo (Eds.), Algebraic and Logic Programming. Proceedings, 1996. VIII, 345 pages. 1996.

Vol. 1140: H. Kuchen, S. Doaitse Swierstra (Eds.), Programming Languages: Implementations, Logics, and Programs. Proceedings, 1996. XI, 479 pages. 1996.

Vol. 1141: H.-M. Voigt, W. Ebeling, I. Rechenberg, H.-P. Schwefel (Eds.), Parallel Problem Solving from Nature – PPSN IV. Proceedings, 1996. XVII, 1.050 pages. 1996.

Vol. 1142: R.W. Hartenstein, M. Glesner (Eds.), Field-Programmable Logic. Proceedings, 1996. X, 432 pages. 1996.

Vol. 1143: T.C. Fogarty (Ed.), Evolutionary Computing. Proceedings, 1996. VIII, 305 pages. 1996.

Vol. 1144: J. Ponce, A. Zisserman, M. Hebert (Eds.), Object Representation in Computer Vision. Proceedings, 1996. VIII, 403 pages. 1996.

Vol. 1145: R. Cousot, D.A. Schmidt (Eds.), Static Analysis. Proceedings, 1996. IX, 389 pages. 1996.

Vol. 1146: E. Bertino, H. Kurth, G. Martella, E. Montolivo (Eds.), Computer Security – ESORICS 96. Proceedings, 1996. X, 365 pages. 1996.

Vol. 1147: L. Miclet, C. de la Higuera (Eds.), Grammatical Inference: Learning Syntax from Sentences. Proceedings, 1996. VIII, 327 pages. 1996. (Subseries LNAI).

Vol. 1148: M.C. Lin, D. Manocha (Eds.), Applied Computational Geometry. Proceedings, 1996. VIII, 223 pages. 1996.

Vol. 1149: C. Montangero (Ed.), Software Process Technology. Proceedings, 1996. IX, 291 pages. 1996.

Vol. 1150: A. Hlawiczka, J.G. Silva, L. Simoncini (Eds.), Dependable Computing – EDCC-2. Proceedings, 1996. XVI, 440 pages. 1996.

Vol. 1151: Ö. Babaoğlu, K. Marzullo (Eds.), Distributed Algorithms. Proceedings, 1996. VIII, 381 pages. 1996.

Vol. 1152: T. Furuhashi, Y. Uchikawa (Eds.), Fuzzy Logic, Neural Networks, and Evolutionary Computation. Proceedings, 1995. VIII, 243 pages. 1996. (Subseries LNAI).

Vol. 1153: E. Burke, P. Ross (Eds.), Practice and Theory of Automated Timetabling. Proceedings, 1995. XIII, 381 pages. 1996.

Vol. 1154: D. Pedreschi, C. Zaniolo (Eds.), Logic in Databases. Proceedings, 1996. X, 497 pages. 1996.

Vol. 1155: J. Roberts, U. Mocci, J. Virtamo (Eds.), Broadbank Network Teletraffic. XXII, 584 pages. 1996.

Vol. 1156: A. Bode, J. Dongarra, T. Ludwig, V. Sunderam (Eds.), Parallel Virtual Machine – EuroPVM '96. Proceedings, 1996. XIV, 362 pages. 1996.

Vol. 1157: B. Thalheim (Ed.), Conceptual Modeling – ER '96. Proceedings, 1996. XII, 489 pages. 1996.

Vol. 1158: S. Berardi, M. Coppo (Eds.), Types for Proofs and Programs. Proceedings, 1995. X, 296 pages. 1996.

Vol. 1159: D.L. Borges, C.A.A. Kaestner (Eds.), Advances in Artificial Intelligence. Proceedings, 1996. XI, 243 pages. (Subseries LNAI).

Vol. 1160: S. Arikawa, A.K. Sharma (Eds.), Algorithmic Learning Theory. Proceedings, 1996. XVII, 337 pages. 1996. (Subseries LNAI).

Vol. 1161: O. Spaniol, C. Linnhoff-Popien, B. Meyer (Eds.), Trends in Distributed Systems. Proceedings, 1996. VIII, 289 pages. 1996.

Vol. 1162: D.G. Feitelson, L. Rudolph (Eds.), Job Scheduling Strategies for Parallel Processing. Proceedings, 1996. VIII, 291 pages. 1996.

Vol. 1163: K. Kim, T. Matsumoto (Eds.), Advances in Cryptology – ASIACRYPT '96. Proceedings, 1996. XII, 395 pages. 1996.

Vol. 1164: K. Berquist, A. Berquist (Eds.), Managing Information Highways. XIV, 417 pages. 1996.

Vol. 1165: J.-R. Abrial, E. Börger, H. Langmaack (Eds.), Formal Methods for Industrial Applications. VIII, 511 pages. 1996.

Vol. 1166: M. Srivas, A. Camilleri (Eds.), Formal Methods in Computer-Aided Design. Proceedings, 1996. IX, 470 pages. 1996.

Vol. 1167: I. Sommerville (Ed.), Software Configuration Management. VII, 291 pages. 1996.

Vol. 1168: I. Smith, B. Faltings (Eds.), Advances in Case-Based Reasoning. Proceedings, 1996. IX, 531 pages. 1996. (Subseries LNAI).

Vol. 1169: M. Broy, S. Merz, K. Spies (Eds.), Formal Systems Specification. XXIII, 541 pages. 1996.

Vol. 1170: M. Nagl (Ed.), Building Tightly Integrated Software Development Environments: The IPSEN Approach. IX, 709 pages. 1996.

Vol. 1171: A. Franz, Automatic Ambiguity Resolution in Natural Language Processing. XIX, 155 pages. 1996. (Subseries LNAI).

Vol. 1172: J. Pieprzyk, J. Seberry (Eds.), Information Security and Privacy. Proceedings, 1996. IX, 333 pages. 1996.

Vol. 1173: W. Rucklidge, Efficient Visual Recognition Using the Hausdorff Distance. XIII, 178 pages. 1996.

Vol. 1174: R. Anderson (Ed.), Information Hiding. Proceedings, 1996. VIII, 351 pages. 1996.

Vol. 1175: K.G. Jeffery, J. Král, M. Bartošek (Eds.), SOFSEM'96: Theory and Practice of Informatics. Proceedings, 1996. XII, 491 pages. 1996.

Vol. 1176: S. Miguet, A. Montanvert, S. Ubéda (Eds.), Discrete Geometry for Computer Imagery. Proceedings, 1996. XI, 349 pages. 1996.

Vol. 1177: J.P. Müller, The Design of Intelligent Agents. XV, 227 pages. 1996. (Subseries LNAI).

Vol. 1178: T. Asano, Y. Igarashi, H. Nagamochi, S. Miyano, S. Suri (Eds.), Algorithms and Computation. Proceedings, 1996. X, 448 pages. 1996.

Vol. 1179: J. Jaffar, R.H.C. Yap (Eds.), Concurrency and Parallelism, Programming, Networking, and Security. Proceedings, 1996. XIII, 394 pages. 1996.

Vol. 1180: V. Chandru, V. Vinay (Eds.), Foundations of Software Technology and Theoretical Computer Science. Proceedings, 1996. XI, 387 pages. 1996.

Vol. 1181: D. Bjørner, M. Broy, I.V. Pottosin (Eds.), Perspectives of System Informatics. Proceedings, 1996. XVII, 447 pages. 1996.

Vol. 1182: W. Hasan, Optimization of SQL Queries for Parallel Machines. XVIII, 133 pages. 1996.

Vol. 1183: A. Wierse, G.G. Grinstein, U. Lang (Eds.), Database Issues for Data Visualization. Proceedings, 1995. XIV, 219 pages. 1996.

Vol. 1184: J. Waśniewski, J. Dongarra, K. Madsen, D. Olesen (Eds.), Applied Parallel Computing. Proceedings, 1996. XIII, 722 pages. 1996.

Vol. 1185: G. Ventre, J. Domingo-Pascual, A. Danthine (Eds.), Multimedia Telecommunications and Applications. Proceedings, 1996. XII, 267 pages. 1996.

Vol. 1186: F. Afrati, P. Kolaitis (Eds.), Database Theory - ICDT'97. Proceedings, 1997. XIII, 477 pages. 1997.

Vol. 1187: K. Schlechta, Nonmonotonic Logics. IX, 243 pages. 1997. (Subseries LNAI).

Vol. 1188: T. Martin, A.L. Ralescu (Eds.), Fuzzy Logic in Artificial Intelligence. Proceedings, 1995. VIII, 272 pages. 1997. (Subseries LNAI).

Vol. 1189: M. Lomas (Ed.), Security Protocols. Proceedings, 1996. VIII, 203 pages. 1997.

Vol. 1190: S. North (Ed.), Graph Drawing. Proceedings, 1996. XI, 409 pages. 1997.

Vol. 1191: V. Gaede, A. Brodsky, O. Günther, D. Srivastava, V. Vianu, M. Wallace (Eds.), Constraint Databases and Applications. Proceedings, 1996. X, 345 pages. 1996.

Vol. 1192: M. Dam (Ed.), Analysis and Verification of Multiple-Agent Languages. Proceedings, 1996. VIII, 435 pages. 1997.

Vol. 1193: J.P. Müller, M.J. Wooldridge, N.R. Jennings (Eds.), Intelligent Agents III. XV, 401 pages. 1997. (Subseries LNAI).

Vol. 1196: L. Vulkov, J. Waśniewski, P. Yalamov (Eds.), Numerical Analysis and Its Applications. Proceedings, 1996. XIII, 608 pages. 1997.

Vol. 1197: F. d'Amore, P.G. Franciosa, A. Marchetti-Spaccamela (Eds.), Graph-Theoretic Concepts in Computer Science. Proceedings, 1996. XI, 410 pages. 1997.

Vol. 1198: H.S. Nwana, N. Azarmi (Eds.), Software Agents and Soft Computing: Towards Enhancing Machine Intelligence. XIV, 298 pages. 1997. (Subseries LNAI).

Vol. 1199: D.K. Panda, C.B. Stunkel (Eds.), Communication and Architectural Support for Network-Based Parallel Computing. Proceedings, 1997. X, 269 pages. 1997.

Vol. 1200: R. Reischuk, M. Morvan (Eds.), STACS 97. Proceedings, 1997. XIII, 614 pages. 1997.

Vol. 1201: O. Maler (Ed.), Hybrid and Real-Time Systems. Proceedings, 1997. IX, 417 pages. 1997.

Vol. 1202: P. Kandzia, M. Klusch (Eds.), Cooperative Information Agents. Proceedings, 1997. IX, 287 pages. 1997. (Subseries LNAI).

Vol. 1203: G. Bongiovanni, D.P. Bovet, G. Di Battista (Eds.), Algorithms and Complexity. Proceedings, 1997. VIII, 311 pages. 1997.

Vol. 1204: H. Mössenböck (Ed.), Modular Programming Languages. Proceedings, 1997. X, 379 pages. 1997.

Lecture Notes in Computer Science

This series reports new developments in computer science research and teaching, quickly, informally, and at a high level. The timeliness of a manuscript is more important than its form, which may be unfinished or tentative. The type of material considered for publication includes

- drafts of original papers or monographs,
- technical reports of high quality and broad interest,
- advanced-level lectures,
- reports of meetings, provided they are of exceptional interest and focused on a single topic.

Publication of Lecture Notes is intended as a service to the computer science community in that the publisher Springer-Verlag offers global distribution of documents which would otherwise have a restricted readership. Once published and copyrighted they can be cited in the scientific literature.

Manuscripts

Lecture Notes are printed by photo-offset from the master copy delivered in camera-ready form. Manuscripts should be no less than 100 and preferably no more than 500 pages of text. Authors of monographs and editors of proceedings volumes receive 50 free copies of their book. Manuscripts should be printed with a laser or other high-resolution printer onto white paper of reasonable quality. To ensure that the final photo-reduced pages are easily readable, please use one of the following formats:

Font size	Printing area		Final size
(points)	(cm)	(inches)	(%)
10	12.2 x 19.3	4.8 x 7.6	100
12	15.3 x 24.2	6.0 x 9.5	80

On request the publisher will supply a leaflet with more detailed technical instructions or a TEX macro package for the preparation of manuscripts.

Manuscripts should be sent to one of the series editors or directly to:

Springer-Verlag, Computer Science Editorial I, Tiergartenstr. 17,
D-69121 Heidelberg, Germany

ISSN 0302-9743